World Headquarters

Jones and Bartlett Publishers	Jones and Bartlett	Jones and Bartlett
40 Tall Pine Drive	Publishers Canada	Publishers International
Sudbury, MA 01776	6339 Ormindale Way	Barb House, Barb Mews
978-443-5000	Mississauga, Ontario	London W6 7PA
info@jbpub.com	L5V 1J2	United Kingdom
www.jbpub.com	Canada	

Jones and Bartlett's books and products are available through most bookstores and online booksellers. To contact Jones and Bartlett Publishers directly, call 800-832-0034, fax 978-443-8000, or visit our website www.jbpub.com.

Substantial discounts on bulk quantities of Jones and Bartlett's publications are available to corporations, professional associations, and other qualified organizations. For details and specific discount information, contact the special sales department at Jones and Bartlett via the above contact information or send an email to specialsales@jbpub.com.

Production Credits
Acquisitions Editor: Timothy Anderson
Editorial Assistant: Melissa Potter
Production Director: Amy Rose
Production Assistant: Ashlee Hazeltine
Senior Marketing Manager: Andrea DeFronzo
V.P., Manufacturing and Inventory Control: Therese Connell
Composition: Northeast Compositors, Inc.
Cover and Title Page Design: Kristin E. Parker
Cover Image: © Marco Andras/Orange Stock/age fotostock
Printing and Binding: Malloy, Inc.
Cover Printing: Malloy, Inc.

Library of Congress Cataloging-in-Publication Data
Hein, James L.
Discrete structures, logic, and computability / James L. Hein. – 3rd ed.
p. cm.
Includes bibliographical references and index.
ISBN-13: 978-0-7637-7206-2 (hardcover)
ISBN-10: 0-7637-7206-2 (hardcover)
1. Computer science–Mathematics. 2. Logic programming. 3. Data structures (Computer science) 4. Logic, Symbolic and mathematical. 5. Computable functions. I. Title.
QA76.9.M35.H44 2010
004.01'51–dc22
2008055744

6048

Printed in the United States of America
13 12 11 10 09   10 9 8 7 6 5 4 3 2

THIRD EDITION

Discrete Structures, Logic, and Computability

James L. Hein

Professor Emeritus
Portland State University

JONES AND BARTLETT PUBLISHERS

Sudbury, Massachusetts

BOSTON   TORONTO   LONDON   SINGAPORE

Preface

*The last thing one discovers in writing a book
is what to put first.*

—Blaise Pascal (1623–1662)

Discrete Structures, Logic, and Computability is written for the prospective computer scientist, computer engineer, or applied mathematician who wants to learn the ideas that underlie computer science. The topics come from the fields of mathematics, logic, and computer science itself. I have attempted to give elementary introductions to those ideas and techniques that are necessary to understand and practice the art and science of computing. The book contains all the topics for discrete structures in the reports of the IEEE/ACM Joint Task Force on Computing Curricula for computer science programs and for computer engineering programs.

Although the book is intended for future computer scientists, computer engineers, or applied mathematicians, it may also be suitable for a wider audience. For example, it could be used in courses for students who intend to teach discrete mathematics or computer science in high school.

Structure and Method

The structure of the third edition continues to support the spiral (i.e., iterative or nonlinear) method for learning. The spiral method is a "just in time" approach. In other words, start by introducing just enough basic information about a topic so that students can do something with it. Then revisit the topic whenever new skills or knowledge about the topic are needed for students to solve problems in other topics that have been introduced in the same way. The process continues as much as possible for each topic.

Topics that are revisited with the spiral approach include logic, sets, relations, graphs, counting, number theory, cryptology, algorithm analysis, complexity, algebra, languages, and machines. Therefore, many traditional topics are dispersed throughout the text to places where they fit naturally with the techniques under discussion.

The logic coverage is much more extensive than in current books at this level. Logic is of fundamental importance in computer science not only for its use in problem solving, but also for its use in formal specification of programs, formal verification of programs, and for its growing use in many areas such as databases, artificial intelligence, robotics, automatic reasoning systems, and logic programming languages.

Logic is covered in a spiral manner. For example, informal proof techniques are introduced in the first section of Chapter 1. Then we use informal logic without much comment until Chapter 4, where inductive proof techniques are presented. After the informal use of logic is well in hand, we move to the formal aspects of logic in Chapters 6 and 7, where equivalence proofs and rule-based proofs are introduced. Formal logic is applied to proving correctness properties of programs in Chapter 8, where we also introduce higher forms of logic. We introduce automatic reasoning and logic programming in Chapter 9. Logic programming is used to construct simple machine interpreters in Chapters 11, 12, and 13.

The coverage of algebraic structures differs from that in other texts. In Chapter 10 we give elementary introductions to algebras and algebraic techniques that apply directly to computer science. In addition to the traditional topic of Boolean algebra, we introduce algebras for abstract data types, relational algebra for relational databases, functional algebra for reasoning about programs, congruences with applications to cryptology, and a few other algebraic ideas that are directly applicable to computing problems. In Chapter 11 we introduce the algebra of regular expressions for simplifying representations of regular languages.

In Chapters 11 through 14 the computing topics of languages, machines, and computation are presented at a new, more elementary, level. The last section of the book gives an elementary introduction to computational complexity.

Changes for the Third Edition

◆ Expanded coverage of discrete probability to a full section in Chapter 5. Applications of a posteriori probability include analyzing software errors. The section now includes an introduction to finite Markov chains with an application to software development.

◆ Expanded coverage of techniques for working with summations to find closed forms in Chapter 5. Coverage also includes an introduction to approximations and harmonic numbers.

◆ Expanded coverage of rates of growth in Chapter 5. Reworked the order of presentation by starting with Big Oh and Big Omega for upper and lower bound approximations. Then Big Theta is introduced as having both properties and as a notation for approximations. A large class of divide and conquer recurrences are shown to have easily approximated solutions that can be derived from a powerful master theorem on divide and conquer recurrences.

♦ Made major changes to formal reasoning in Chapter 6. A set of proof rules for "Natural Deduction" is given so that formal reasoning closely models the way we reason informally. Within the section, the material is presented "just in time" so that students can get to work quickly.

♦ Made several changes in Chapter 7 aimed at making quantifiers easier to understand and thus easier for students to apply to formal proofs.

♦ Simplified the discussion of context-free parsing in Chapter 12 by omitting the table-construction algorithms.

♦ Improved the discussion of context-sensitive grammars in Chapter 14.

♦ Over 100 new examples and exercises have been added. Over 20 new subtopic headings have been added to help clarify the presentation. Answers are provided for about half of the exercises, and they are identified with bold numbers.

I hope that this edition has no errors. But I do wish to apologize in advance for any errors found in the book. They were not intentional, and I would appreciate hearing about them. As always, we should read the printed word with some skepticism.

Note to the Student

Most problems in computer science involve one of the following five questions:

Can the problem be solved by a computer program?

If not, can you modify the problem so that it can be solved by a program?

If so, can you write a program to solve the problem?

Can you convince another person that your program is correct?

Can you convince another person that your program is efficient?

One goal of the book is that you obtain a better understanding of these questions together with a better ability to answer them. The book's ultimate goal is that you gain self-reliance and confidence in your own ability to solve problems, just like the self-reliance and confidence you have in your ability to ride a bike.

Supplements

Student and instructor supplements can be found on the text's catalog page at www.jbpub.com/catalog/9780763772062.

Using the Book

The book can be read by anyone with a good background in high school mathematics. Thus it could also be used at the college freshman level or at the advanced high school level. As with most books, there are some dependencies among the topics. But you should feel free to jump into the book at whatever topic suits your fancy and then refer back to unfamiliar definitions if necessary. We should note that some approximation methods in Chapter 5 rely on elementary calculus, but the details can be skipped over without losing the main points.

The topics in the book can be presented in a variety of ways, depending on the length of the course, the emphasis, and student background. The major portion of the text has been taught for several years to sophomores at Portland State University in the form of a two-term course (80 lecture hours) in discrete structures covering topics in Chapters 1 through 10 and a one-term course (40 lecture hours) in foundations of computing covering topics in Chapters 11 through 14.

Acknowledgments

Many people helped me create this book in the first place and many people have influenced the content and form of the third edition. Thanks go especially to the students and teachers who have kept my email quite interesting over the past several years with questions, suggestions, and criticisms. They have all influenced this book. At Jones and Bartlett, I thank Tim Anderson for maintaining the entrepreneurial spirit and Amy Rose for keeping me on track. I also thank my wife, Janice, for her constant help and support.

J.L.H.
Portland, Oregon

Contents

Elementary Notions and Notations

'Excellent!' I cried. 'Elementary,' said he.
—Watson in *The Crooked Man*
by Arthur Conan Doyle (1859–1930)

To communicate, we sometimes need to agree on the meaning of certain terms. If the same idea is mentioned several times in a discussion, we often replace it with some shorthand notation. The choice of notation can help us avoid wordiness and ambiguity, and it can help us achieve conciseness and clarity in our written and oral expression.

Since much of our communication involves reasoning about things, we'll begin the chapter with a short discussion about the notions of informal proof. The rest of the chapter is devoted to introducing the basic notions and notations for sets, tuples, graphs, and trees. The treatment here is introductory in nature, and we'll expand on these ideas in later chapters as the need arises.

chapter guide

1.1 (A Proof Primer) introduces some proof techniques that are used throughout the book. We'll practice each technique with a proof about numbers.

1.2 (Sets) introduces the basic ideas about sets. We'll see how to compare them and how to combine them, and we'll introduce some elementary ways to count them. We'll also introduce bags, which are like sets but which might contain repeated occurrences of elements, and we'll have a little discussion about why we should stick with uncomplicated sets.

1.3 (Ordered Structures) introduces some basic ideas about ordered structures and how to represent them. Tuples are introduced as a notation for ordered information. We'll introduce the notions and notations for lists, strings, and relations. We'll also see some elementary ways to count tuples.

1.4 (Graphs and Trees) introduces the basic ideas about graphs and trees. We'll discuss ways to represent them, ways to traverse them, and we'll see a famous algorithm for constructing a spanning tree for a graph.

1.1   A Proof Primer

For our purposes an *informal proof* is a demonstration that some statement is true. We normally communicate an informal proof in an English-like language that mixes everyday English with symbols that appear in the statement to be proved. In the next few paragraphs we'll discuss some basic techniques for doing informal proofs. These techniques will come in handy in trying to understand someone's proof or in trying to construct a proof of your own, so keep them in your mental tool kit.

We'll start off with a short refresher on logical statements followed by a short discussion about numbers. This will give us something to talk about when we look at examples of informal proof techniques.

1.1.1   Statements and Truth Tables

For this primer we'll consider only statements that are either true or false. To indicate that a statement is true (i.e., it is a truth) we assign it the *truth value* T (or True). Similarly, to indicate that a statement is false (i.e., it is a falsity) we assign it the *truth value* F (or False). We'll start by discussing some familiar ways to structure such statements.

Negation

If S represents some statement, then the *negation* of S is the statement "not S," whose truth value is opposite that of S. We can represent this relationship with a *truth table* in which each row gives a possible truth value for S and the corresponding truth value for not S:

S	not S
T	F
F	T

We often paraphrase the negation of a statement to make it more understandable. For example, to negate the statement "Earth is a star," we normally say, "Earth is not a star," or "It is not the case that Earth is a star," rather than "Not Earth is a star."

We should also observe that negation relates statements about *every* case with statements about *some* case. For example, the statement "Not every planet has a moon" has the same meaning as "Some planet does not have a moon." Similarly, the statement "It is not the case that some planet is a star" has the same meaning as "Every planet is not a star."

A	B	A and B	A or B
T	T	T	T
T	F	F	T
F	T	F	T
F	F	F	F

Figure 1.1   Truth tables.

Conjunction and Disjunction

The *conjunction* of A and B is the statement "A and B," which is true when both A and B are true. The *disjunction* of A and B is the statement "A or B," which is true if either or both of A and B are true. The truth tables for conjunction and disjunction are given in Figure 1.1.

Sometimes we paraphrase conjunctions and disjunctions. For example, instead of "Earth is a planet and Mars is a planet," we might write "Earth and Mars are planets." Instead of "x is positive or y is positive," we might write "Either x or y is positive."

Conditional Statements

Many statements are written in the general form "If A then B," where A and B are also statements. Such a statement is called a *conditional statement* in which A is the *antecedent* and B is the *consequent*. The statement can also be read as "A implies B." Sometimes, but not often, it is read as "A is a sufficient condition for B," or "B is a necessary condition for A." The truth table for a conditional statement is contained in Figure 1.2.

Let's make a few comments about this table. Notice that the conditional is false only when the antecedent is true and the consequent is false. It's true in the other three cases. The conditional truth table gives some people fits because they interpret "If A then B" to mean "B can be proved from A," which assumes that A and B are related in some way. But we've all heard statements like "If the moon is made of green cheese, then $1 = 2$." We nod our heads and agree that the statement is true, even though there is no relationship between the antecedent and consequent. Similarly, we shake our heads and don't agree with a statement like "If $1 = 1$, then the moon is made of green cheese."

A	B	if A then B
T	T	T
T	F	F
F	T	T
F	F	T

Figure 1.2   Truth table.

When the antecedent of a conditional is false, we say that the conditional is *vacuously* true. For example, the statement "If $1 = 2$, then $39 = 12$" is vacuously true because the antecedent is false. If the consequent is true, we say that the conditional is *trivially* true. For example, the statement "If $1 = 2$, then $2 + 2 = 4$" is trivially true because the consequent is true. We leave it to the reader to convince at least one person that the conditional truth table is defined properly.

The *converse* of "If A then B" is "If B then A." The converse does not always have the same truth value. For example, we know that the following statement about numbers is true.

$$\text{If } x > 0 \text{ and } y > 0, \text{ then } x + y > 0.$$

The converse of this statement is

$$\text{If } x + y > 0, \text{ then } x > 0 \text{ and } y > 0.$$

This converse is false. For example, let $x = 3$ and $y = -2$. Then the statement becomes "If $3 + (-2) > 0$, then $3 > 0$ and $-2 > 0$," which is false.

Equivalent Statements

Sometimes it's convenient to write a statement in a different form but with the same truth value. Two statements are said to be *equivalent* if they have the same truth value for any assignment of truth values to the variables that occur in the statements.

We can combine negation with either conjunction or disjunction to obtain the following pairs of equivalent statements.

$$\text{"not } (A \text{ and } B)\text{" is equivalent to "(not } A) \text{ or (not } B).\text{"}$$
$$\text{"not } (A \text{ or } B)\text{" is equivalent to "(not } A) \text{ and (not } B).\text{"}$$

For example, the statement "not $(x > 0$ and $y > 0)$" is equivalent to the statement "$x \leq 0$ or $y \leq 0$." The statement "not $(x > 0$ or $y > 0)$" is equivalent to the statement "$x \leq 0$ and $y \leq 0$."

Conjunctions and disjunctions distribute over each other in the following sense:

$$\text{"}A \text{ and } (B \text{ or } C)\text{" is equivalent to "}(A \text{ and } B) \text{ or } (A \text{ and } C).\text{"}$$
$$\text{"}A \text{ or } (B \text{ and } C)\text{" is equivalent to "}(A \text{ or } B) \text{ and } (A \text{ or } C).\text{"}$$

For example, the statement "$0 < x$ and $(x < 4$ or $x < 9)$" is equivalent to the statement "$0 < x < 4$ or $0 < x < 9$." The statement "$x > 0$ or $(x > 2$ and $x > 1)$" is equivalent to "$(x > 0$ or $x > 2)$ and $(x > 0$ or $x > 1)$."

The *contrapositive* of the conditional statement "If A then B" is the equivalent statement "If not B then not A." For example, the statement

$$\text{If } (x > 0 \text{ and } y > 0), \text{ then } x + y > 0$$

is equivalent to the statement

$$\text{If } x + y \leq 0, \text{ then } (x \leq 0 \text{ or } y \leq 0).$$

We can also express the conditional statement "If A then B" in terms of the equivalent statement "(not A) or B." For example, the statement

$$\text{If } x > 0 \text{ and } y > 0, \text{ then } x + y > 0$$

is equivalent to the statement

$$(x \leq 0 \text{ or } y \leq 0) \text{ or } x + y > 0.$$

Since we can express a conditional in terms of negation and disjunction, it follows that the statements "not (If A then B)" and "A and (not B)" are equivalent. For example, the statement

It is not the case that if Earth is a planet, then Earth is a star

is equivalent to the statement

Earth is a planet and Earth is not a star.

Let's summarize the equivalences that we have discussed. Each row of the following table contains two equivalent statements.

Table of Equivalent Statements	
not (A and B)	(not A) or (not B)
not (A or B)	(not A) and (not B)
A and (B or C)	(A and B) or (A and C)
A or (B and C)	(A or B) and (A or C)
if A then B	if not B then not A
if A then B	(not A) or B
not (if A then B)	A and (not B)

1.1.2 Something to Talk About

To discuss proof techniques, we need something to talk about when giving sample proofs. Since numbers are familiar to everyone, that's what we'll talk about. But to make sure that we all start on the same page, we'll review a little terminology.

The numbers that we'll be discussing are called *integers*, and we can list them as follows:

$$\dots, -4, -3, -2, -1, 0, 1, 2, 3, 4, \dots .$$

The integers in the following list are called *even* integers:

$$..., -4, -2, 0, 2, 4, ... \ .$$

The integers in the following list are called *odd* integers:

$$..., -3, -1, 1, 3, ... \ .$$

So every integer is either even or odd but not both. In fact, every even integer has the form $2n$ for some integer n. Similarly, every odd integer has the form $2n + 1$ for some integer n.

Divisibility and Prime Numbers

An integer d *divides* an integer n if $d \neq 0$ and there is an integer k such that $n = dk$. For example, 3 divides 18 because we can write $18 = (3)(6)$. But 5 does not divide 18 because there is no integer k such that $18 = 5k$. The following list shows all the divisors of 18.

$$-18, -9, -6, -3, -2, -1, 1, 2, 3, 6, 9, 18.$$

Some alternative words for *d divides n* are *d is a divisor of n* or *n is divisible by d*. We often denote the fact that d divides n with the following shorthand notation:

$$d|n.$$

For example, we have $-9|9$, $-3|9$, $-1|9$, $1|9$, $3|9$, and $9|9$. Here are two properties of divisibility that we'll record for future use.

Divisibility Properties   (1.1)

   a. If $d|a$ and $a|b$, then $d|b$.

   b. If $d|a$ and $d|b$, then $d|(ax + by)$ for any integers x and y.

An integer $p > 1$ is called a *prime* number if 1 and p are its only positive divisors. For example, the first eight prime numbers are

$$2, 3, 5, 7, 11, 13, 17, 19.$$

Prime numbers have many important properties and they have many applications in computer science. But for now all we need to know is the definition of a prime.

1.1.3   Proof Techniques

Now that we have something to talk about, we'll discuss some fundamental proof techniques and give some sample proofs for each technique.

Proof by Exhaustive Checking

When a statement asserts that each of a finite number of things has a certain property, then we might be able to prove the statement by checking that each thing has the stated property. For example, suppose someone says, "If n is an integer and $2 \leq n \leq 7$, then $n^2 + 2$ is not divisible by 4." We can prove the statement by *exhaustive checking*. For $2 \leq n \leq 7$, the corresponding values of $n^2 + 2$ are

$$6, 11, 18, 27, 38, 51.$$

We can check that these numbers are not divisible by 4. For another example, suppose someone says, "If n is an integer and $2 \leq n \leq 500$, then $n^2 + 2$ is not divisible by 4." Again, this statement can be proved by exhaustive checking, but perhaps by a computer rather than a person.

Exhaustive checking cannot be used to prove a statement that requires infinitely many things to check. For example, consider the statement, "If n is an integer, then $n^2 + 2$ is not divisible by 4." This statement is true, but there are infinitely many things to check. So another proof technique will be required. We'll get to it after a few more paragraphs.

An example that proves a statement false is often called a *counterexample*. Sometimes counterexamples can be found by exhaustive checking. For example, consider the statement, "Every odd number greater than 1 that is not prime has the form $2 + p$ for some prime p." We can observe that the statement is false because 27 is a counterexample.

Conditional Proof

Many statements that we wish to prove are in conditional form or can be rephrased in conditional form. The *direct approach* to proving a conditional of the form "if A then B," starts with the assumption that the antecedent A is true. The next step is to find a statement that is implied by the assumption or known facts. Each step proceeds in this fashion to find a statement that is implied by any of the previous statements or known facts. The *conditional proof* ends when the consequent B is reached.

example 1.1 A Proof About Sums

We'll prove the following general statement about integers:

The sum of any two odd integers is an even integer.

We can rephrase the statement in the conditional form

If x and y are odd integers, then $x + y$ is an even integer.

Proof: Assume that x and y are odd integers. It follows that x and y can be written in the the form $x = 2k + 1$ and $y = 2m + 1$ for some integers k and m. Now substitute for x and y in $x + y$ to obtain

$$x + y = (2k + 1) + (2m + 1) = 2k + 2m + 2 = 2(k + m + 1).$$

Since the expression on the right-hand side contains 2 as a factor, it represents an even integer. QED.

end example

---

example **1.2 A Divisibility Proof**

We'll prove that if 3 divides an even number, then 6 divides the number too. In other words, if $3|2n$ for some integer n, then $6|2n$.

Proof: Assume that $3|2n$. This means that $2n = 3k$ for some integer k. Notice that we can write $2n = 3k = 2k + k$. So $2n = 2k + k$. Solve for k to obtain $k = 2n - 2k = 2(n - k)$, which says that k is an even number. Now we can substitute for k to obtain

$$2n = 3k = 3(2(n - k)) = 6(n - k).$$

Therefore, $6|2n$. QED.
   Notice that the proof also implies that if $3|2n$, then $3|n$. From the preceding proof just notice that since $2n = 6(n - k)$, we can divide both sides by 2 to get $n = 3(n - k)$. So $3|n$.

end example

Proving the Contrapositive

Recall that a conditional statement "if A then B" and its contrapositive "if not B then not A" have the same truth table. So a proof of one is also a proof of the other. Sometimes the contrapositive is easier to prove. Here's an example.

example **1.3 An Odd Proof**

We'll prove the following statement about the integers:

If x^2 is odd, then x is odd.

To prove the statement, we'll prove its contrapositive:

If x is even, then x^2 is even.

Proof: Assume that x is even. It follows that $x = 2k$ for some integer k. Now square x and substitute for x to obtain

$$x^2 = (2k)^2 = 4k^2 = 2(2k^2).$$

The expression on the right side of the equation represents an even number. Therefore x^2 is even. QED.

end example

Proof by Contradiction

A *contradiction* is a false statement. A *proof by contradiction* starts out by assuming that the statement to be proved is false. Then an argument is made that reaches a contradiction. So we conclude that the original statement is true. This technique is known formally as *reductio ad absurdum*.

Proof by contradiction is often the method of choice because we can wander wherever the proof takes us to find a contradiction. We'll give two examples to show the wandering that can take place.

example **1.4 A Not-Divisible Proof**

We'll prove the following statement about divisibility:

If n is an integer, then $n^2 + 2$ is not divisible by 4.

Proof: Assume the statement is false. Then $4 \mid (n^2 + 2)$ for some integer n. This means that $n^2 + 2 = 4k$ for some integer k. We'll consider the two cases where n is even and where n is odd. If n is even, then $n = 2m$ for some integer m. Substituting for n we obtain

$$4k = n^2 + 2 = (2m)^2 + 2 = 4m^2 + 2.$$

We can divide both sides of the equation by 2 to obtain

$$2k = 2m^2 + 1.$$

This says that an even number ($2k$) is equal to an odd number ($2m^2 + 1$), which is a contradiction. Therefore, n cannot be even. Now assume n is odd. Then $n = 2m + 1$ for some integer m. Substituting for n we obtain

$$4k = n^2 + 2 = (2m + 1)^2 + 2 = 4m^2 + 4m + 3.$$

Isolate 3 on the right side of the equation to obtain

$$4k - 4m^2 - 4m = 3.$$

This is a contradiction because the left side is even and the right side is odd. Therefore, n cannot be odd. QED.

end example

 1.5   Prime Numbers

We'll prove the following statement about integers:

> Every integer greater than 1 is divisible by a prime.

Proof: Assume the statement is false. Then some integer $n > 1$ is not divisible by a prime. Since a prime divides itself, n cannot be a prime. So there is at least one integer d such that $d|n$ and $1 < d < n$. Assume that d is the smallest divisor of n between 1 and n. Now d is not prime, else it would be a prime divisor of n. So there is an integer a such that $a|d$ and $1 < a < d$. Since $a|d$ and $d|n$, it follows from (1.1a) that $a|n$. But now we have $a|n$ and $1 < a < d$, which contradicts the assumption that d is the smallest such divisor of n. QED.

end example

If and Only If Proofs

The statement "A if and only if B" is shorthand for the two statements "If A then B" and "If B then A." The abbreviation "A *iff* B" is often used for "A if and only if B." Instead of "A iff B," some people write "A is a necessary and sufficient condition for B" or "B is a necessary and sufficient condition for A." Remember that two proofs are required for an iff statement, one for each conditional statement.

example **1.6   An Iff Proof**

We'll prove the following iff statement about integers:

$$x \text{ is odd if and only if } 8 \mid (x^2 - 1).$$

To prove this iff statement, we must prove the following two statements:

a. If x is odd, then $8 \mid (x^2 - 1)$.

b. If $8 \mid (x^2 - 1)$, then x is odd.

Proof of (a): Assume x is odd. Then we can write x in the form $x = 2k + 1$ for some integer k. Substituting for x in $x^2 - 1$ gives

$$x^2 - 1 = 4k^2 + 4k = 4k(k + 1).$$

Since k and $k + 1$ are consecutive integers, one is odd and the other is even, so the product $k(k + 1)$ is even. So $k(k + 1) = 2m$ for some integer m. Substituting for $k(k + 1)$ gives

$$x^2 - 1 = 4k(k + 1) = 4(2m) = 8m.$$

Therefore, $8 \mid (x^2 - 1)$, so Part (a) is proven.

Proof of (b): Assume $8 \mid (x^2 - 1)$. Then $x^2 - 1 = 8k$ for some integer k. Therefore, we have $x^2 = 8k + 1 = 2(4k) + 1$, which has the form of an odd integer. So x^2 is odd, and it follows from Example 1.3 that x is odd, so Part (b) is proven. Therefore, the iff statement is proven. QED.

end example

Sometimes we encounter iff statements that can be proven by using statements that are related to each other by iff. Then a proof can be constructed as a sequence of iff statements. For example, to prove A iff B we might be able to find a statement C such that A iff C and C iff B are both true. Then we can conclude that A iff B is true. The proof could then be put in the form A iff C iff B.

example 1.7 Two Proofs in One

We'll prove the following statement about integers:

$$x \text{ is odd if and only if } x^2 + 2x + 1 \text{ is even.}$$

Proof: The following sequence of iff statements connects the left side to the right side. (The reason for each step is given in parentheses.)

$$
\begin{aligned}
x \text{ is odd iff } & \; x = 2k + 1 \text{ for some integer } k & \text{(definition)} \\
\text{iff } & \; x + 1 = 2k + 2 \text{ for some integer } k & \text{(algebra)} \\
\text{iff } & \; x + 1 = 2m \text{ for some integer } m & \text{(algebra)} \\
\text{iff } & \; x + 1 \text{ is even} & \text{(definition)} \\
\text{iff } & \; (x + 1)^2 \text{ is even} & \text{(Exercise 8a)} \\
\text{iff } & \; x^2 + 2x + 1 \text{ is even} & \text{(algebra) QED}
\end{aligned}
$$

end example

On Constructive Existence

If a statement asserts that some object exists, then we can try to prove the statement in either of two ways. One way is to use proof by contradiction, in which we assume that the object does not exist and then come up with some kind of contradiction. The second way is to construct an instance of the object. In either case we know that the object exists, but the second way also gives us an instance of the object. Computer science leans toward the construction of objects by algorithms. So the *constructive approach* is usually preferred, although it's not always possible.

Important Note

Always try to write out your proofs. Use complete sentences that describe your reasoning. If your proof seems to consist only of a bunch of equations or expressions, you still need to describe how they contribute to the proof. Try to write your proofs the same way you would write a letter to a friend who wants to understand what you have written.

◢ Exercises

1. See whether you can convince yourself, or a friend, that the conditional truth table is correct by making up English sentences of the form "If A then B."

2. Verify that the truth tables for each of the following pairs of statements are identical.
   a. "not (A and B)" and "(not A) or (not B)."
   b. "not (A or B)" and "(not A) and (not B)."
   c. "if A then B" and "if (not B) then (not A)."
   d. "if A then B" and "(not A) or B."
   e. "not (if A then B)" and "A and (not B)."

3. Prove or disprove each of the following statements by exhaustive checking.
   a. There is a prime number between 45 and 54.
   b. The product of any two of the four numbers 2, 3, 4, and 5 is even.
   c. Every odd integer between 2 and 26 is either prime or the product of two primes.
   d. If $d|ab$, then $d|a$ or $d|b$.
   e. If m and n are integers, then $(3m + 2)(3n + 2)$ has the form $(3k + 2)$ for some integer k.

4. Prove each of the following statements about the integers.
   a. If x and y are even, then $x + y$ is even.
   b. If x is even and y is odd, then $x + y$ is odd.
   c. If x and y are odd, then $x - y$ is even.
   d. If $3n$ is even, then n is even.

5. Write down the converse of the following statement about integers:

$$\text{If } x \text{ and } y \text{ are odd, then } x - y \text{ is even.}$$

   Is the statement that you wrote down true or false? Prove your answer.

6. Prove each of the following statements, where m and n are integers.

    a. If $x = 3m + 4$ and $y = 3n + 4$, then $xy = 3k + 4$ for some integer k.

    b. If $x = 5m + 6$ and $y = 5n + 6$, then $xy = 5k + 6$ for some integer k.

    c. If $x = 7m + 8$ and $y = 7n + 8$, then $xy = 7k + 8$ for some integer k.

7. Prove each of the following statements about divisibility of integers.

    a. If $d \mid (da + b)$, then $d \mid b$.

    b. If $d \mid (a + b)$ and $d \mid a$, then $d \mid b$.

    c. (1.1a) If $d \mid a$ and $a \mid b$ then $d \mid b$.

    d. (1.1b) If $d \mid a$ and $d \mid b$, then $d \mid (ax + by)$ for any integers x and y.

    e. If $5 \mid 2n$, then $5 \mid n$.

8. Prove each of the following iff statements about integers.

    a. x is even if and only if x^2 is even.

    b. xy is odd if and only if x is odd and y is odd.

    c. x is odd if and only if $x^2 + 6x + 9$ is even.

    d. $m \mid n$ and $n \mid m$ if and only if $n = m$ or $n = -m$.

9. Prove that any positive integer has the form $2^k n$, where k is a nonnegative integer and n is an odd integer.

10. Prove that if x is an integer that is a multiple of 3 and also a multiple of 5, then x is a multiple of 15. In other words, show that if $3 \mid x$ and $5 \mid x$, then $15 \mid x$. *Hint:* Look at Example 1.2.

1.2 Sets

In our everyday discourse we sometimes run into the problem of trying to define a word in terms of other words whose definitions may include the word we are trying to define. That's the problem we have in trying to define the word *set*. To illustrate the point, we often think of some (perhaps all) of the words

set, collection, bunch, group, class

as synonyms for each other. We pick up the meaning for such a word intuitively by seeing how it is used.

1.2.1 Definition of a Set

We'll simply say that a *set* is a collection of things called its *elements, members,* or *objects*. Sometimes the word *collection* is used in place of *set* to clarify a sentence. For example, "a collection of sets" seems clearer than "a set of sets." We say that a set contains its elements, or that the elements belong to the set, or that the elements are in the set. If S is a set and x is an element in S, then we write

$$x \in S.$$

If x is not an element of S, then we write $x \notin S$. If $x \in S$ and $y \in S$, we often denote this fact by the shorthand notation

$$x, \, y \in S.$$

Describing Sets

To describe a set we need to describe its elements in some way. One way to define a set is to explicitly name its elements. A set defined in this way is denoted by listing its elements, separated by commas, and surrounding the listing with braces. For example, the set S consisting of the letters x, y, and z is denoted by

$$S = \{x, \, y, \, z\}.$$

Sets can have other sets as elements. For example, the set $A = \{x, \, \{x, \, y\}\}$ has two elements. One element is x, and the other element is $\{x, \, y\}$. So we can write $x \in A$ and $\{x, \, y\} \in A$.

We often use the three-dot ellipsis, $\ldots$, to informally denote a sequence of elements that we do not wish to write down. For example, the set

$$\{1, \, 2, \, 3, \, 4, \, 5, \, 6, \, 7, \, 8, \, 9, \, 10, \, 11, \, 12\}$$

can be denoted in several different ways with ellipses, two of which are

$$\{1, \, 2, \, \ldots, \, 12\} \text{ and } \{1, \, 2, \, 3, \, \ldots, 11, \, 12\}.$$

The set with no elements is called the *empty set*—some people refer to it as the *null set*. The empty set is denoted by { } or more often by the symbol

$$\varnothing.$$

A set with one element is called a *singleton*. For example, $\{a\}$ and $\{b\}$ are singletons.

Equality of Sets

Two sets are *equal* if they have the same elements. We denote the fact that two sets A and B are equal by writing

$$A = B.$$

It follows from the definition of equality that there is no particular order or arrangement of the elements in a set. For example, since $\{u, g, h\}$ and $\{h, u, g\}$ have the same elements, we have

$$\{u, g, h\} = \{h, u, g\}.$$

It also follows that there are no repeated occurrences of elements in a set. For example, since $\{h, u, g, h\}$ and $\{h, u, g\}$ have the same elements, we have

$$\{h, u, g, h\} = \{h, u, g\}.$$

So $\{h, u, g, h\}$, $\{u, g, h\}$, and $\{h, u, g\}$ are different representations of the same set.

If the sets A and B are not equal, we write

$$A \neq B.$$

For example, $\{a, b, c\} \neq \{a, b\}$ because c is an element of only one of the sets. We also have $\{a\} \neq \varnothing$ because the empty set doesn't have any elements.

Before we go any further let's record the two important characteristics of sets that we have discussed.

Two Characteristics of Sets

1. There are no repeated occurrences of elements.

2. There is no particular order or arrangement of the elements.

Finite and Infinite Sets

Suppose we start counting the elements of a set S one element per second of time with a stop watch. If $S = \varnothing$, then we don't need to start, because there are no elements to count. But if $S \neq \varnothing$, we agree to start the counting after we have started the timer. If a point in time is reached when all the elements of S have been counted, then we stop the timer, or in some cases we might need to have one of our descendants stop the timer. In this case we say that S is a *finite* set. If the counting never stops, then S is an *infinite* set. All the examples that we have discussed to this point are finite sets. We will discuss counting finite and infinite sets in other parts of the book as the need arises.

Natural Numbers and Integers

Familiar infinite sets are sometimes denoted by listing a few of the elements followed by an ellipsis. We reserve some letters to denote specific sets that we'll refer to throughout the book. For example, the set of *natural numbers* will be denoted by $\mathbb{N}$[1] and the set of *integers* by $\mathbb{Z}$. So we can write

$$\mathbb{N} = \{0, 1, 2, 3, \ldots\} \quad \text{and} \quad \mathbb{Z} = \{\ldots, -3, -2, -1, 0, 1, 2, 3, \ldots\}.$$

---

[1]Some people consider the natural numbers to be the set $\{1, 2, 3, \ldots\}$. If you are one of these people, then think of $\mathbb{N}$ as the nonnegative integers.

Describing Sets by Properties

Many sets are hard to describe by listing elements. Examples that come to mind are the *rational numbers*, which we denote by $\mathbb{Q}$, and the *real numbers*, which we denote by $\mathbb{R}$. Instead of listing the elements, we can often describe a property that the elements of the set satisfy. For example, the set of odd integers consists of integers having the form $2k + 1$ for some integer k.

If P is a property, then the set S whose elements have property P is denoted by writing

$$S = \{x \,|\, x \text{ has property } P\}.$$

We read this as "S is the set of all x such that x has property P." For example, if we let *Odd* be the set of odd integers, then we can describe *Odd* in several ways.

$$
\begin{aligned}
Odd &= \{\ldots, -5, -3, -1, 1, 3, 5, \ldots\} \\
&= \{x \,|\, x \text{ is an odd integer}\} \\
&= \{x \,|\, x = 2k + 1 \text{ for some integer } k\} \\
&= \{x \,|\, x = 2k + 1 \text{ for some } k \in \mathbb{Z}\}.
\end{aligned}
$$

Of course, we can also describe finite sets by finding properties that they possess. For example,

$$\{1, 2, \ldots, 12\} = \{x \,|\, x \in \mathbb{N} \text{ and } 1 \leq x \leq 12\}.$$

We can also describe a set by writing expressions for the elements. For example, the set *Odd* has the following additional descriptions.

$$
\begin{aligned}
Odd &= \{2k + 1 \,|\, k \text{ is an integer}\}. \\
&= \{2k + 1 \,|\, k \in \mathbb{Z}\}.
\end{aligned}
$$

Subsets

If A and B are sets and every element of A is also an element of B, then we say that A is a *subset* of B and write

$$A \subseteq B.$$

For example, we have $\{a, b\} \subseteq \{a, b, c\}$, $\{0, 1, 2\} \subseteq \mathbb{N}$, and $\mathbb{N} \subseteq \mathbb{Z}$. It follows from the definition that every set A is a subset of itself. Thus we have $A \subseteq A$. It also follows from the definition that the empty set is a subset of any set A. So we have $\varnothing \subseteq A$. Can you see why? We'll leave this as an exercise.

If $A \subseteq B$ and there is some element in B that does not occur in A, then A is called a *proper* subset of B and we can write

$$A \subset B.$$

For example, $\{a, b\} \subseteq \{a, b, c\}$ and we can also write $\{a, b\} \subset \{a, b, c\}$. We can also write $\mathbb{N} \subset \mathbb{Z}$.

Remember that the idea of subset is different from the idea of membership. For example, if $A = \{a, b, c\}$, then $\{a\} \subseteq A$ and $a \in A$. But $\{a\} \notin A$ and a is not a subset of A.

The Power Set

The collection of all subsets of a set S is called the *power set* of S, which we denote by power(S). For example, if $S = \{a, b, c\}$, then the power set of S can be written as follows:

$$\text{power}(S) = \{\varnothing, \{a\}, \{b\}, \{c\}, \{a, b\}, \{a, c\}, \{b, c\}, S\}.$$

An interesting programming problem is to construct the power set of a finite set. We'll discuss this problem later, once we've developed some tools to help build an easy solution.

Venn Diagrams

In dealing with sets, it's often useful to draw a picture in order to visualize the situation. A *Venn diagram*—named after the logician John Venn (1834–1923)—consists of one or more closed curves in which the interior of each curve represents a set. For example, the Venn diagram in Figure 1.3 represents the fact that A is a proper subset of B and x is an element of B that does not occur in A.

Proof Strategies with Subsets and Equality

Subsets allow us to give a precise definition of set equality: Two sets are equal if and only if they are subsets of each other. In more concise form we can write

---

Equality of Sets   (1.2)

$$A = B \text{ if and only if } A \subseteq B \text{ and } B \subseteq A$$

Let's record three useful proof strategies for comparing two sets.

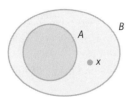

Figure 1.3   Venn diagram of proper subset $A \subset B$.

Statement to Prove	Proof Strategy
$A \subseteq B$	For arbitrary $x \in A$, show that $x \in B$.
A is not a subset of B	Find an element $x \in A$ such that $x \notin B$.
$A = B$	Show that $A \subseteq B$ and show that $B \subseteq A$.

example   **1.8   A Subset Proof**

We'll show that $A \subseteq B$, where A and B are defined as follows:

$$A = \{x \mid x \text{ is a prime number and } 42 \le x \le 51\},$$
$$B = \{x \mid x = 4k + 3 \text{ and } k \in \mathbb{N}\}.$$

We start the proof by letting $x \in A$. Then either $x = 43$ or $x = 47$. We can write $43 = 4(10) + 3$ and $47 = 4(11) + 3$. So in either case, x has the form of an element of B. Thus $x \in B$. Therefore, $A \subseteq B$.

end example

example   **1.9   A Not-Subset Proof**

We'll show that A and B are not subsets of each other, where A and B are defined by

$$A = \{3k + 1 \mid k \in \mathbb{N}\} \text{ and } B = \{4k + 1 \mid k \in \mathbb{N}\}.$$

By listing a few elements from each set we can write A and B as follows:

$$A = \{1, 4, 7, ...\} \text{ and } B = \{1, 5, 9, ...\}.$$

Now it's easy to prove that A is not a subset of B because $4 \in A$ and $4 \notin B$. Similarly, B is not a subset of A because $5 \in B$ and $5 \notin A$.

end example

example   **1.10   An Equal Sets Proof**

We'll show that $A = B$, where A and B are defined as follows:

$$A = \{x \mid x \text{ is prime and } 12 \le x \le 18\},$$
$$B = \{x \mid x = 4k + 1 \text{ and } k \in \{3, 4\}\}.$$

First we'll show that $A \subseteq B$. Let $x \in A$. Then either $x = 13$ or $x = 17$. We can write $13 = 4(3) + 1$ and $17 = 4(4) + 1$. It follows that $x \in B$. Therefore, $A \subseteq B$. Next we'll show that $B \subseteq A$. Let $x \in B$. It follows that either $x = 4(3) + 1$ or $x = 4(4) + 1$. In either case, x is a prime number between 12 and 18. Therefore, $B \subseteq A$. So $A = B$.

end example

1.2.2 Operations on Sets

We'll discuss the operations of union, intersection, and complement, all of which combine sets to form new sets.

Union of Sets

The *union* of two sets A and B is the set of all elements that are either in A or in B or in both A and B. The union is denoted by $A \cup B$ and we can give the following formal definition.

Union of Sets   (1.3)

$$A \cup B = \{x \mid x \in A \text{ or } x \in B\}.$$

The use of the word "or" in the definition is taken to mean "either or both." For example, if $A = \{a, b, c\}$ and $B = \{c, d\}$, then $A \cup B = \{a, b, c, d\}$. The union of two sets A and B is represented by the shaded regions of the Venn diagram in Figure 1.4.

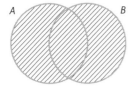

Figure 1.4   Venn diagram of $A \cup B$.

The following properties give some basic facts about the union operation. We'll prove one of the facts in the next example and leave the rest as exercises.

Properties of Union   (1.4)
   a. $A \cup \varnothing = A$.
   b. $A \cup B = B \cup A$.   ($\cup$ is commutative.)
   c. $A \cup (B \cup C) = (A \cup B) \cup C$.   ($\cup$ is associative.)
   d. $A \cup A = A$.
   e. $A \subseteq B$ if and only if $A \cup B = B$.

---

example **1.11 A Subset Condition**

We'll prove property (1.4e):

$$A \subseteq B \text{ if and only if } A \cup B = B.$$

Proof: Since this is an if and only if statement we have two statements to prove. First we'll prove that $A \subseteq B$ implies $A \cup B = B$. Assume that $A \subseteq B$. With this assumption we must show that $A \cup B = B$. Let $x \in A \cup B$. It follows that $x \in A$ or $x \in B$. Since we have assumed that $A \subseteq B$, it follows that $x \in B$. Thus $A \cup B \subseteq B$. But since we always have $B \subseteq A \cup B$, it follows from (1.2) that $A \cup B = B$. So the first part is proven. Next we'll prove that $A \cup B = B$ implies $A \subseteq B$. Assume that $A \cup B = B$. If $x \in A$, then $x \in A \cup B$. Since we are assuming that $A \cup B = B$, it follows that $x \in B$. Therefore $A \subseteq B$. So the second part is proven. QED.

end example

Intersection of Sets

The *intersection* of two sets A and B is the set of all elements that are in both A and B. The intersection is denoted by $A \cap B$ and we can give the following formal definition.

Intersection of Sets   (1.5)

$$A \cap B = \{x \mid x \in A \text{ and } x \in B\}.$$

For example, if $A = \{a, b, c\}$ and $B = \{c, d\}$, then $A \cap B = \{c\}$. If $A \cap B = \varnothing$, then A and B are said to be *disjoint*. The nonempty intersection of two sets A and B is represented by the shaded region of the Venn diagram in Figure 1.5.

The following properties give some basic facts about the intersection operation. We'll leave the proofs as exercises.

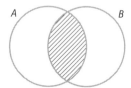

Figure 1.5   Venn diagram of $A \cap B$.

Properties of Intersection (1.6)

a. $A \cap \varnothing = \varnothing$.

b. $A \cap B = B \cap A$.   ($\cap$ is commutative.)

c. $A \cap (B \cap C) = (A \cap B) \cap C$.   ($\cap$ is associative.)

d. $A \cap A = A$.

e. $A \subseteq B$ if and only if $A \cap B = A$.

Extending Union and Intersection

The union operation extends to larger collections of sets as the set of elements that occur in at least one of the underlying sets. The intersection operation also extends to larger collections of sets as the set of elements that occur in all of the underlying sets.

Such unions and intersections can be represented in several ways. For example, if $A_1, \ldots, A_n$ are sets, then the union is denoted by $A_1 \cup \ldots \cup A_n$ and the intersection is denoted by $A_1 \cap \ldots \cap A_n$. The following symbols can also be used to denote these sets.

$$\bigcup_{k=1}^{n} A_k \quad \text{and} \quad \bigcap_{k=1}^{n} A_k.$$

If we have an infinite collection of sets $A_1, A_2, \ldots, A_n, \ldots$, then the union and intersection are denoted by $A_1 \cup \ldots \cup A_n \cup \ldots$ and $A_1 \cap \ldots \cap A_n \cap \ldots$. The following symbols can also be used to denote these sets.

$$\bigcup_{k=1}^{\infty} A_k \quad \text{and} \quad \bigcap_{k=1}^{\infty} A_k.$$

If I is a set of indices such that A_k is a set for each $k \in I$, then the union and intersection of the sets in the collection can be denoted by the following symbols.

$$\bigcup_{k \in I} A_k \quad \text{and} \quad \bigcap_{k \in I} A_k.$$

example **1.12 Unions and Intersections**

1. In a local book club with 10 members let A_k be the set of book titles read by member k. Then each title in $A_1 \cup \ldots \cup A_{10}$ has been read by at least one member of the club and $A_1 \cap \ldots \cap A_{10}$ is the set of titles read by every member.

2. Let A_k be the set of error messages that occur on the kth test of a piece of software. Then each error message in $A_1 \cup \ldots \cup A_{100}$ has occurred in at least one of the 100 tests and $A_1 \cap \ldots \cap A_{100}$ is the set of error messages that occur in every test.

3. A car company manufactures 15 different models and keeps track of the names of the parts used in each model in the sets $C_1, \ldots, C_{15}$. Then $C_1 \cup \ldots \cup C_{15}$ is the set (hopefully a small set) of different part names used by the company and $C_1 \cap \ldots \cap C_{15}$ is the set (hopefully a large set) of part names common to each model.

4. Any natural number has a binary representation. For example, $0 = 0$, $1 = 1$, $2 = 10$, $3 = 11$, $4 = 100$, $5 = 101$, and so on. Suppose we let $B_1 = \{0, 1\}$ and for $k \geq 2$ we let

$$B_k = \{x \,|\, x \in \mathbb{N} \quad \text{and} \quad 2^{k-1} \leq x < 2^k\}.$$

For example, $B_2 = \{2, 3\}$ and $B_3 = \{4, 5, 6, 7\}$. Notice that each number in B_k has a binary representation that uses no fewer than k digits. Notice also that $\mathbb{N}$ is the infinite union $B_1 \cup \ldots \cup B_k \ldots$. We can also write

$$\mathbb{N} = \bigcup_{k=1}^{\infty} B_k.$$

end example

Difference of Sets

If A and B are sets, then the *difference* $A - B$ (also called the *relative complement* of B in A) is the set of elements in A that are not in B, which we can describe with the following definition.

Difference of Sets	(1.7)
$$A - B = \{x \mid x \in A \text{ and } x \notin B\}.$$	

For example, if $A = \{a, b, c\}$ and $B = \{c, d\}$, then $A - B = \{a, b\}$. We can picture the difference $A - B$ of two general sets A and B by the shaded region of the Venn diagram in Figure 1.6.

A natural extension of the difference $A - B$ is the *symmetric difference* of sets A and B, which is the union of $A - B$ with $B - A$ and is denoted by $A \oplus B$. For example, if $A = \{a, b, c\}$ and $B = \{c, d\}$, then $A \oplus B = \{a, b, d\}$. The set $A \oplus B$ is represented by the shaded regions of the Venn diagram in Figure 1.7.

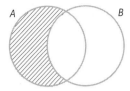

Figure 1.6   Venn diagram of $A - B$.

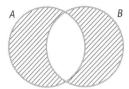

Figure 1.7   Venn diagram of $A \oplus B$.

We can define the symmetric difference by using the "exclusive" form of "or" as given in the following formal definition:

Symmetric Difference of Sets   (1.8)

$$A \oplus B = \{x \mid x \in A \text{ or } x \in B \text{ but not both}\}.$$

As is usually the case, there are many relationships to discover. For example, it's easy to see that

$$A \oplus B = (A \cup B) - (A \cap B).$$

Can you verify that $(A \oplus B) \oplus C = A \oplus (B \oplus C)$? For example, try to draw two Venn diagrams, one for each side of the equation.

Complement of a Set

If the discussion always refers to sets that are subsets of a particular set U, then U is called the *universe of discourse*, and the difference $U - A$ is called the *complement* of A, which we denote by A'. For example, if the universe is the set of integers and A is the set of even integers, then A' is the set of odd integers. For another example, the Venn diagram in Figure 1.8 pictures the universe U as a rectangle, with two subsets A and B, where the shaded region represents the complement $(A \cup B)'$.

Combining Set Operations

There are many useful properties that combine different set operations. Venn diagrams are often quite useful in trying to visualize sets that are constructed

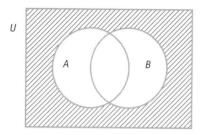

Figure 1.8   Venn diagram of $(A \cup B)'$.

with different operations. For example, the set $A \cap (B \cup C)$ is represented by the shaded regions of the Venn diagram in Figure 1.9.

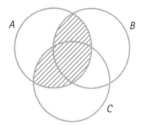

Figure 1.9   Venn diagram of $A \cap (B \cup C)$.

Here are two distributive properties and two absorption properties that combine the operations of union and intersection. We'll prove one of the facts in the next example and leave the rest as exercises.

Combining Properties of Union and Intersection   (1.9)
   a. $A \cap (B \cup C) = (A \cap B) \cup (A \cap C)$.   ($\cap$ distributes over $\cup$.)
   b. $A \cup (B \cap C) = (A \cup B) \cap (A \cup C)$.   ($\cup$ distributes over $\cap$.)
   c. $A \cap (A \cup B) = A$.   (absorption law)
   d. $A \cup (A \cap B) = A$.   (absorption law)

 1.13 A Distributive Proof

We'll prove the following statement (1.9a):

$$A \cap (B \cup C) = (A \cap B) \cup (A \cap C).$$

Proof: We'll show that $x \in A \cap (B \cup C)$ if and only if $x \in (A \cap B) \cup (A \cap C)$.

$$x \in A \cap (B \cup C) \text{ iff } x \in A \text{ and } x \in B \cup C$$
$$\text{iff } x \in A, \text{ and either } x \in B \text{ or } x \in C$$
$$\text{iff either } (x \in A \text{ and } x \in B) \text{ or } (x \in A \text{ and } x \in C)$$
$$\text{iff } x \in (A \cap B) \cup (A \cap C). \quad \text{QED.}$$

end example

Here are some basic properties that combine the complement operation with union, intersection, and subset. We'll prove one of the facts in the next example and leave the rest as exercises.

Properties of Complement (1.10)

a. $(A')' = A$.
b. $\varnothing' = U$ and $U' = \varnothing$.
c. $A \cap A' = \varnothing$ and $A \cup A' = U$.
d. $A \subseteq B$ if and only if $B' \subseteq A'$.
e. $(A \cup B)' = A' \cap B'$. (De Morgan's law)
f. $(A \cap B)' = A' \cup B'$. (De Morgan's law)
g. $A \cap (A' \cup B) = A \cap B$. (absorption law)
h. $A \cup (A' \cap B) = A \cup B$. (absorption law)

---

example **1.14 Subset Conditions**

We'll prove the following statement (1.10d):

$$A \subseteq B \text{ if and only if } B' \subseteq A'.$$

Proof: In this case we're able to connect the two sides of the iff statement with a sequence of iff statements. Be sure that you know the reason for each step.

$$A \subseteq B \text{ iff } x \in A \text{ implies } x \in B$$
$$\text{iff } x \notin B \text{ implies } x \notin A$$
$$\text{iff } x \in B' \text{ implies } x \in A'$$
$$\text{iff } B' \subseteq A'. \quad \text{QED.}$$

end example

1.2.3 Counting Finite Sets

Let's apply some of our knowledge about sets to counting finite sets. The size of a set S is called its *cardinality*, which we'll denote by

$$|S|.$$

For example, if $S = \{a, b, c\}$, then $|S| = |\{a, b, c\}| = 3$. We can say "the cardinality of S is 3," or "3 is the cardinal number of S," or simply "S has three elements."

Counting by Inclusion and Exclusion

Suppose we want to count the union of two finite sets A and B. Since A and B might have some elements in common, it follows that $|A \cup B| \leq |A| + |B|$. For example, if $A = \{1, 2, 3, 4, 5\}$ and $B = \{3, 4, 5, 6\}$, then $A \cup B = \{1, 2, 3, 4, 5, 6\}$. Therefore, $|A| = 5$, $|B| = 4$, and $|A \cup B| = 6$. In this case, we have $|A \cup B| < |A| + |B|$. The reason for the inequality is that the expression $|A| + |B|$ counts the intersection $A \cap B = \{2, 3, 4\}$ twice. So to find the formula for $|A \cup B|$ we need to subtract one copy of $|A \cap B|$. This example is the idea for the following counting rule for the union of two finite sets.

Union Rule (1.11)

$$|A \cup B| = |A| + |B| - |A \cap B|.$$

A popular name for the union rule is the *principle of inclusion and exclusion*. The name is appropriate because the rule says to add (include) the count of each individual set. But in so doing we've counted the intersection twice. So we must subtract (exclude) the count of the intersection.

example **1.15 Keeping Track of Tools**

1. A set of 20 tools is available for two people working on a project. One person uses 15 tools and the other person uses 12 tools. What is the minimum number of tools that they share? The union of the two sets of tools must be less than or equal to 20. So it follows that they shared at least 7 tools. To see this, let A and B be the two sets of tools used by the two workers, where $|A| = 15$ and $|B| = 12$. Then $A \cup B$ is a subset of the set of 20 tools. So we have $|A \cup B| \leq 20$. But now we can use the union rule (1.11).

   $$20 \geq |A \cup B| = |A| + |B| - |A \cap B| = 15 + 12 - |A \cap B| = 27 - |A \cap B|.$$

   So $|A \cap B| \geq 7$. In other words, the two workers shared at least 7 tools.

2. Now suppose a third worker joins the project and this worker uses a set C of 18 tools. What is the least number of tools shared by all three workers? In other words, we want to find the least value for the quantity $|A \cap B \cap C|$. We might try to use the fact that $|A \cap B| \geq 7$. Notice that we can apply (1.11) to the union $(A \cap B) \cup C$ to obtain our result from the equation

$$|(A \cap B) \cup C| = |A \cap B| + |C| - |A \cap C \cap B|.$$

Since $(A \cap B) \cup C$ is a subset of the set of 20 available tools, it must be the case that $|(A \cap B) \cup C| \leq 20$. So we can make the following calculation.

$$20 \geq |(A \cap B) \cup C| = |A \cap B| + |C| - |A \cap B \cap C|$$
$$\geq 7 + 18 - |A \cap B \cap C|.$$

This tells us that $20 \geq 7 + 18 - |A \cap B \cap C|$. So $|A \cap B \cap C| \geq 5$. In other words, the three workers share at least 5 tools.

3. Let's do one more step. Suppose a fourth worker joins the project and uses a set D of 16 tools. What is the least number of tools shared by all four workers? In other words, we want to find the least value for the quantity $|A \cap B \cap C \cap D|$. If we want, we can use the previous information by considering the union $(A \cap B \cap C) \cup D$. This union is a subset of the set of 20 available tools. So $|(A \cap B \cap C) \cup D| \leq 20$. Now we can use (1.11) as before.

$$20 \geq |(A \cap B \cap C) \cup D| = |A \cap B \cap C| + |D| - |A \cap B \cap C \cap D|$$
$$\geq 5 + 16 - |A \cap B \cap C \cap D|.$$

So $|A \cap B \cap C \cap D| \geq 1$. In other words, the four workers share at least one tool. We'll see other techniques for solving this problem in Examples 1.16 and 1.19.

end example

example 1.16 Using Divide and Conquer

There are many ways to solve a problem. For example, suppose in Example 1.15 that we start with the four workers and ask right off the bat the minimum number of tools that they share. Then we might "divide and conquer" the problem by first working with A and B, as above, to obtain $|A \cap B| \geq 7$. Then work with C and D to obtain $|C \cap D| \geq 14$. Now we'll use the union of the two intersections $(A \cap B) \cup (C \cap D)$ and apply (1.11) to find the solution.

$$20 \geq |(A \cap B) \cup (C \cap D)| = |A \cap B| + |C \cap D| - |A \cap B \cap C \cap D|$$
$$\geq 7 + 14 - |A \cap B \cap C \cap D|.$$

Therefore, we have $|A \cap B \cap C \cap D| \geq 1$, as before.

end example

 1.17 Extending the Union Rule

The union rule (1.11) extends to three or more sets. For example, the following calculation gives the union rule for three finite sets. (Notice how the calculation uses properties of union and intersection together with the union rule itself.)

$$|A \cup B \cup C| \qquad\qquad (1.12)$$
$$= |A \cup (B \cup C)|$$
$$= |A| + |B \cup C| - |A \cap (B \cup C)|$$
$$= |A| + |B| + |C| - |B \cap C| - |A \cap (B \cup C)|$$
$$= |A| + |B| + |C| - |B \cap C| - |(A \cap B) \cup (A \cap C)|$$
$$= |A| + |B| + |C| - |B \cap C| - |A \cap B| - |A \cap C| + |A \cap B \cap C|.$$

end example

Counting the Difference of Two Sets

Suppose we need to count the difference of two finite sets A and B. Since $A - B$ might have fewer elements than A, it follows that $|A - B| \leq |A|$. For example, if $A = \{1, 2, 3, 4, 5, 6\}$ and $B = \{5, 6, 7\}$, then $A - B = \{1, 2, 3, 4\}$. So $|A - B| = 4$ and $|A| = 6$. In this case we have $|A - B| < |A|$. The reason for the inequality is that we did not subtract from $|A|$ the number of elements in A that are removed because they are also in B. In other words, we did not subtract $|A \cap B| = 2$ from $|A|$. This example is the idea for the following rule to count the difference of two finite sets.

Difference Rule	(1.13)
$$|A - B| = |A| - |A \cap B|.$$	

It's easy to discover this rule by drawing a Venn diagram. There are also two special cases of (1.13). The first case involves subsets and the second case involves disjoint sets.

$$\text{If } B \subseteq A, \text{ then } |A - B| = |A| - |B|. \qquad\qquad (1.14)$$
$$\text{If } A \cap B = \varnothing, \text{ then } |A - B| = |A|. \qquad\qquad (1.15)$$

example 1.18 Eating Desserts

A tour group visited a buffet for lunch that had three kinds of desserts: cheese-cake, chocolate cake, and apple pie. A survey after lunch asked the following questions in the order listed, along with the responses. Who ate cheesecake? 30 people raised their hands. Who ate cheesecake and chocolate cake? 12 people raised their hands. Who ate cheesecake and apple pie? 15 people raised their hands. Who ate all three desserts? Eight people raised their hands. Let A, B, and C be the sets of people who ate the three desserts, respectively. Then the data from the survey tells us that

$$|A| = 30, |A \cap B| = 12, |A \cap C| = 15, \text{ and } |A \cap B \cap C| = 8.$$

Can we find out how many people ate cheesecake but not the other two desserts? The question asks for the size of the set $A - (B \cup C)$. We can compute this value using both the difference rule (1.13) and the union rule (1.11) as follows.

$$\begin{aligned}|A - (B \cup C)| &= |A| - |A \cap (B \cup C)| \\ &= |A| - |(A \cap B) \cup (A \cap C)| \\ &= |A| - (|A \cap B| + |(A \cap C)| - |A \cap B \cap C|) \\ &= 30 - (12 + 15 - 8) \\ &= 11 \text{ people ate only cheese cake.}\end{aligned}$$

end example

example 1.19 Counting with Complements

Complements can be useful in solving problems like those of Example 1.15. Let's revisit the problem of finding the least value for $|A \cap B \cap C \cap D|$ where A, B, C, and D are subsets of the given set of 20 tools with $|A| = 15$, $|B| = 12$, $|C| = 18$, and $|D| = 16$. We'll use complements taken over a universe together with De Morgan's laws. In our problem the universe is the set of 20 tools. We want to find the least value for $|A \cap B \cap C \cap D|$. By one of De Morgan's laws, we have

$$(A \cap B \cap C \cap D)' = A' \cup B' \cup C' \cup D'.$$

Using the given sets A, B, C, and D, it follows that $|A'| = 5$, $|B'| = 8$, $|C'| = 2$, and $|D'| = 4$. We know from (1.11) that the cardinality of a union of sets is less than or equal to the sum of the cardinalities of the sets. So we have

$$|A' \cup B' \cup C' \cup D'| \le |A'| + |B'| + |C'| + |D'| = 5 + 8 + 2 + 4 = 19.$$

By using De Morgan's law we can restate this inequality as $|(A \cap B \cap C \cap D)'| \le 19$. Therefore, we obtain $|A \cap B \cap C \cap D| \ge 1$, as before.

end example

1.2.4 Bags (Multisets)

A *bag* (or *multiset*) is a collection of objects that may contain repeated occurrences of elements. Here are the important characteristics.

Two Characteristics of Bags

1. There may be repeated occurrences of elements.

2. There is no particular order or arrangement of the elements.

To differentiate bags from sets, we'll use brackets to enclose the elements. For example, $[h, u, g, h]$ is a bag with four elements. Two bags A and B are *equal* if the number of occurrences of each element in A or B is the same in either bag. If A and B are equal bags, we write $A = B$. For example, $[h, u, g, h] = [h, h, g, u]$, but $[h, u, g, h] \neq [h, u, g]$.

We can also define the subbag notion. Define A to be a *subbag* of B, and write $A \subseteq B$, if the number of occurrences of each element x in A is less than or equal to the number of occurrences of x in B. For example, $[a, b] \subseteq [a, b, a]$, but $[a, b, a]$ is not a subbag of $[a, b]$. It follows from the definition of a subbag that two bags A and B are equal if and only if A is a subbag of B and B is a subbag of A.

If A and B are bags, we define the *sum* of A and B, denoted by $A + B$, as follows: If x occurs m times in A and n times in B, then x occurs $m + n$ times in $A + B$. For example,

$$[2, 2, 3] + [2, 3, 3, 4] = [2, 2, 2, 3, 3, 3, 4].$$

We can define union and intersection for bags also (we will use the same symbols as for sets). Let A and B be bags, and let m and n be the number of times x occurs in A and B, respectively. Put the larger of m and n occurrences of x in $A \cup B$. Put the smaller of m and n occurrences of x in $A \cap B$. For example, we have

$$[2, 2, 3] \cup [2, 3, 3, 4] = [2, 2, 3, 3, 4]$$

and

$$[2, 2, 3] \cap [2, 3, 3, 4] = [2, 3].$$

example 1.20 Least and Greatest

Let $p(x)$ denote the bag of prime numbers that occur in the prime factorization of the natural number x. For example, we have

$$p(54) = [2, 3, 3, 3] \text{ and } p(12) = [2, 2, 3].$$

Let's compute the union and intersection of these two bags. The union gives $p(54) \cup p(12) = [2, 2, 3, 3, 3] = p(108)$, and 108 is the least common multiple of 54 and 12 (i.e., the smallest positive integer that they both divide). Similarly, we get $p(54) \cap p(12) = [2, 3] = p(6)$, and 6 is the greatest common divisor of 54 and 12 (i.e., the largest positive integer that divides them both). Can we discover anything here? It appears that $p(x) \cup p(y)$ and $p(x) \cap p(y)$ compute the least common multiple and the greatest common divisor of x and y. Can you convince yourself?

end example

1.2.5 Sets Should Not Be Too Complicated

Set theory was created by the mathematician Georg Cantor (1845–1918) during the period 1874 to 1895. Later some contradictions were found in the theory. Everything works fine as long as we don't allow sets to be too complicated. Basically, we never allow a set to be defined by a test that checks whether a set is a member of itself. If we allowed such a thing, then we could not decide some questions of set membership. For example, suppose we define the set T as follows:

$$T = \{A \mid A \text{ is a set and } A \notin A\}.$$

In other words, T is the set of all sets that are not members of themselves. Now ask the question "Is $T \in T$?" If so, then the condition for membership in T must hold. But this says that $T \notin T$. On the other hand, if we assume that $T \notin T$, then we must conclude that $T \in T$. In either case we get a contradiction. This example is known as *Russell's paradox*—after the philosopher and mathematician Bertrand Russell (1872–1970).

This kind of paradox led to a more careful study of the foundations of set theory. For example, Whitehead and Russell [1910] developed a theory of sets based on a hierarchy of levels that they called *types*. The lowest type contains individual elements. Any other type contains only sets whose elements are from the next lower type in the hierarchy. We can list the hierarchy of types as T_0, $T_1, \ldots, T_k, \ldots$, where T_0 is the lowest type containing individual elements and in general T_{k+1} is the type consisting of sets whose elements are from T_k. So any set in this theory belongs to exactly one type T_k for some $k \geq 1$.

As a consequence of the definition, we can say that $A \notin A$ for all sets A in the theory. To see this, suppose A is a set of type T_{k+1}. This means that the elements of A are of type T_k. If we assume that $A \in A$, we would have to conclude that A is also a set of type T_k. This says that A belongs to the two types T_k and T_{k+1}, contrary to the fact that A must belong to exactly one type.

Let's examine why Russell's paradox can't happen in this new theory of sets. Since $A \notin A$ for all sets A in the theory, the original definition of T can be simplified to $T = \{A \mid A \text{ is a set}\}$. This says that T contains all sets. But T itself isn't even a set in the theory because it contains sets of different types. In order for T to be a set in the theory, each A in T must belong to the same type. For example, we could pick some type T_k and define $T = \{A \mid A \text{ has type } T_k\}$. This says that T is a set of type T_{k+1}. Now since T is a set in the theory, we know that $T \notin T$. But this fact doesn't lead us to any kind of contradictory statement.

 Exercises

Describing Sets

1. The set $\{x \mid x$ is a vowel$\}$ can also be described by $\{a, e, i, o, u\}$. Describe each of the following sets by listing its elements.

    a. $\{x \mid x \in \mathbb{N}$ and $0 < x < 8\}$.
    b. $\{2k + 1 \mid k$ is an even integer between 1 and 10$\}$.
    c. $\{x \mid x$ is an odd prime less than 20$\}$.
    d. $\{x \mid x$ is a month ending with the letter "y"$\}$.
    e. $\{x \mid x$ is a letter in the words MISSISSIPPI RIVER$\}$.
    f. $\{x \mid x \in \mathbb{N}$ and x divides 24$\}$.

2. The set $\{a, e, i, o, u\}$ can also be described by $\{x \mid x$ is a vowel$\}$. Describe each of the following sets in terms of a property of its elements.

    a. The set of dates in the month of January.
    b. $\{1, 3, 5, 7, 9, 11, 13, 15\}$.
    c. $\{1, 4, 9, 16, 25, 36, 49, 64\}$.
    d. The set of even integers $\{..., -4, -2, 0, 2, 4,...\}$.

Subsets

3. Let $A = \{a, \varnothing\}$. Answer true or false for each of the following statements.

    a. $a \in A$.   **b.** $\{a\} \in A$.   **c.** $a \subseteq A$.   **d.** $\{a\} \subseteq A$.
    e. $\varnothing \subseteq A$.   **f.** $\varnothing \in A$.   **g.** $\{\varnothing\} \subseteq A$.   **h.** $\{\varnothing\} \in A$.

4. Explain why $\varnothing \subseteq A$ for every set A.

5. Find two finite sets A and B such that $A \in B$ and $A \subseteq B$.

6. Write down the power set for each of the following sets.

    a. $\{x, y, z, w\}$.   **b.** $\{a, \{a, b\}\}$.   **c.** $\varnothing$.
    d. $\{\varnothing\}$.   **e.** $\{\{a\}, \varnothing\}$.

7. For each collection of sets, find the smallest set A such that the collection is a subset of power(A).

    a. $\{\{a\}, \{b, c\}\}$.   **b.** $\{\{a\}, \{\varnothing\}\}$.   **c.** $\{\{a\}, \{\{a\}\}\}$.
    d. $\{\{a\}, \{\{b\}\}, \{a, b\}\}$.

8. Let $A = \{6k + 5 \mid k \in \mathbb{N}\}$ and $B = \{3k + 2 \mid k \in \mathbb{N}\}$.

    a. Show that $A \subseteq B$.
    b. Show that $A \neq B$.

Set Operations

9. For each case find the value of x that makes the equation hold.

    a. $\{1, 2, 3, x\} \cup \{2, 3, 5, 6\} = \{1, 2, 3, 4, 5, 6\}$.

    b. $\{1, 2, 3, x\} \cap \{3, 5, 6\} = \{3, 6\}$.

    c. $\{1, 2, 3, x\} - \{1, 2, 6\} = \{3, 5\}$.

    d. $\{1, 2, 3, 5, 7\} - \{1, 2, x\} = \{3, 5\}$.

10. Find the set A that satisfies the following two properties:

$$A \cup \{3, 4, 5, 6\} = \{1, 2, 3, 4, 5, 6\}$$

$$A \cap \{3, 4, 5, 6\} = \{3, 4\}.$$

11. Find the set A that satisfies the following two properties:

$$A - \{1, 2, 4, 5\} = \{3, 6, 7\}$$

$$\{1, 2, 4, 5\} - A = \{1, 4\}.$$

12. Find the set A that satisfies the following four properties:

$$A \cap \{2, 3, 5, 6\} = \{2, 5\}$$

$$A \cup \{4, 5, 6, 7\} = \{1, 2, 4, 5, 6, 7\}$$

$$\{2, 4\} \subseteq A$$

$$A \subseteq \{1, 2, 4, 5, 6, 8\}.$$

13. For each of the following expressions, use a Venn diagram representing a universe U and two subsets A and B. Shade the part of the diagram that corresponds to the given set.

    a. A'.

    b. B'.

    c. $(A \cup B)'$.

    d. $A' \cap B'$.

    e. $A' \cup B'$.

    f. $(A \cap B)'$.

14. Each Venn diagram in the following figure represents a set whose regions are indicated by the letter x. Find an expression for each of the three sets in terms of set operations. Try to simplify your answers.

a.

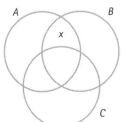

b.

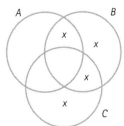

c.

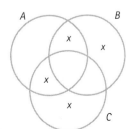

15. (Sets of Divisors) For each natural number n, let

$D_n = \{d \mid d \in \mathbb{N} \text{ and } d \text{ divides } n \text{ with no remainder}\}.$

    a. Find D_0, D_6, D_{12}, and D_{18}.

    b. Verify that $D_{12} \cap D_{18} = D_6$.

    c. Show that if $m \mid n$, then $D_m \subseteq D_n$.

    d. Convince yourself that $\displaystyle\bigcup_{k=1}^{\infty} D_{2k} = \mathbb{N} - \{0\}$.

16. (Sets of multiples) For each natural number n, let $M_n = \{kn \mid k \in \mathbb{N}\}$.

    a. Find M_0, M_1, M_3, M_4, and M_{12}.

    b. Verify that $M_3 \cap M_4 = M_{12}$.

    c. Show that if $m \mid n$, then $M_n \subseteq M_m$.

    d. Convince yourself that $\displaystyle\bigcup_{k=1}^{\infty} M_{2k+1} = \mathbb{N} - \{2^n \mid n \in \mathbb{N}\}$.

Counting Finite Sets

17. Discover an inclusion exclusion formula for the number of elements in the union of four sets A, B, C, and D.

18. Given three sets A, B, and C. Suppose the union of the three sets has cardinality 280. Suppose also that $|A| = 100$, $|B| = 200$, and $|C| = 150$. And suppose we also know $|A \cap B| = 50$, $|A \cap C| = 80$, and $|B \cap C| = 90$. Find the cardinality of the intersection of the three given sets.

19. Suppose A, B, and C represent three bus routes through a suburb of your favorite city. Let A, B, and C also be sets whose elements are the bus stops for the corresponding bus route. Suppose A has 25 stops, B has 30 stops, and C has 40 stops. Suppose further that A and B share (have in common) 6 stops, A and C share 5 stops, and B and C share 4 stops. Lastly, suppose that A, B, and C share 2 stops. Answer each of the following questions.

    a. How many distinct stops are on the three bus routes?

    b. How many stops for A are not stops for B?

    c. How many stops for A are not stops for both B and C?

    d. How many stops for A are not stops for any other bus?

20. Suppose a highway survey crew noticed the following information about 500 vehicles: In 100 vehicles the driver was smoking, in 200 vehicles the driver was talking to a passenger, and in 300 vehicles the driver was tuning the radio. Further, in 50 vehicles the driver was smoking and talking, in 40 vehicles the driver was smoking and tuning the radio, and in 30 vehicles the driver was talking and tuning the radio. What can you say about the number of drivers who were smoking, talking, and tuning the radio?

21. Suppose a survey revealed that 70 percent of the population visited an amusement park and 80 percent visited a national park. At least what percentage of the population visited both?

22. Suppose that 100 senators voted on three separate senate bills as follows: 70 percent of the senators voted for the first bill, 65 percent voted for the second bill, and 60 percent voted for the third bill. At least what percentage of the senators voted for all three bills?

23. Suppose that 25 people attended a conference with three sessions, where 15 people attended the first session, 18 the second session, and 12 the third session. At least how many people attended all three sessions?

Bags

24. Find the union and intersection of each of the following pairs of bags.

   a. $[x, y]$ and $[x, y, z]$.
   b. $[x, y, x]$ and $[y, x, y, x]$.
   c. $[a, a, a, b]$ and $[a, a, b, b, c]$.
   d. $[1, 2, 2, 3, 3, 4, 4]$ and $[2, 3, 3, 4, 5]$.
   e. $[x, x, [a, a], [a, a]]$ and $[a, a, x, x]$.
   f. $[a, a, [b, b], [a, [b]]]$ and $[a, a, [b], [b]]$.

25. Find a bag B that solves the following two simultaneous bag equations:

$$B \cup [2, 2, 3, 4] = [2, 2, 3, 3, 4, 4, 5],$$
$$B \cap [2, 2, 3, 4, 5] = [2, 3, 4, 5].$$

26. How would you define the difference operation for bags? Try to make your definition agree with the difference operation for sets whenever the bags are like sets (without repeated occurrences of elements).

Proofs and Challenges

27. Prove each of the following facts about the union operation (1.4). Use subset arguments that are written in complete sentences.

   a. $A \cup \varnothing = A$.
   b. $A \cup B = B \cup A$.
   c. $A \cup A = A$.
   d. $A \cup (B \cup C) = (A \cup B) \cup C$.

28. Prove each of the following facts about the intersection operation (1.6). Use subset arguments that are written in complete sentences.

   a. $A \cap \varnothing = \varnothing$.

   b. $A \cap B = B \cap A$.

   c. $A \cap (B \cap C) = (A \cap B) \cap C$.

   d. $A \cap A = A$.

   e. $A \subseteq B$ if and only if $A \cap B = A$.

29. Prove that $\operatorname{power}(A \cap B) = \operatorname{power}(A) \cap \operatorname{power}(B)$.

30. Prove that $A \cup (B \cap C) = (A \cup B) \cap (A \cup C)$.

31. Prove each of the following absorption laws (1.9) twice. The first proof should use subset arguments. The second proof should use an already known result.

   a. $A \cap (B \cup A) = A$.

   b. $A \cup (B \cap A) = A$.

32. Show that $(A \cap B) \cup C = A \cap (B \cup C)$ if and only if $C \subseteq A$.

33. Give a proof or a counterexample for each of the following statements.

   a. $A \cap (B \cup A) = A \cap B$.

   b. $A - (B \cap A) = A - B$.

   c. $A \cap (B \cup C) = (A \cup B) \cap (A \cup C)$.

   d. $A \oplus A = A$.

34. Prove each of the following properties of the complement (1.10).

   a. $(A')' = A$.

   b. $\varnothing' = U$ and $U' = \varnothing$.

   c. $A \cap A' = \varnothing$ and $A \cup A' = U$.

   d. $(A \cup B)' = A' \cap B'$.   (De Morgan's law)

   e. $(A \cap B)' = A' \cup B'$.   (De Morgan's law)

   f. $A \cap (A' \cup B) = A \cap B$.   (absorption law)

   g. $A \cup (A' \cap B) = A \cup B$.   (absorption law)

35. Try to find a description of a set A satisfying the equation $A = \{a, A, b\}$. Notice in this case that $A \in A$.

1.3 Ordered Structures

In the previous section we saw that sets and bags are used to represent unordered information. In this section we'll introduce some notions and notations for structures that have some kind of ordering to them.

1.3.1 Tuples

When we write down a sentence, it always has a sequential nature. For example, in the previous sentence the word "When" is the first word, the word "we" is the second word, and so on. Informally, a *tuple* is a collection of things, called its *elements*, where there is a first element, a second element, and so on. The elements of a tuple are also called *members*, *objects*, or *components*. We'll denote a tuple by writing down its elements, separated by commas, and surrounding everything with the two symbols "(" and ")". For example, the tuple (12, R, 9) has three elements. The first element is 12, the second element is the letter R, and the third element is 9. The beginning sentence of this paragraph can be represented by the following tuple:

$$(\text{When, we, write, down, } \ldots, \text{ sequential, nature}).$$

If a tuple has n elements, we say that its *length* is n, and we call it an *n-tuple*. So the tuple (8, k, hello) is a 3-tuple, and $(x_1, \ldots, x_8)$ is an 8-tuple. The 0-tuple is denoted by ( ), and we call it the *empty* tuple. A 2-tuple is often called an *ordered pair*, and a 3-tuple might be called an *ordered triple*. Other words used in place of the word *tuple* are *vector* and *sequence*, possibly modified by the word *ordered*.

Two n-tuples $(x_1, \ldots, x_n)$ and $(y_1, \ldots, y_n)$ are said to be *equal* if $x_i = y_i$ for $1 \leq i \leq n$, and we denote this by $(x_1, \ldots, x_n) = (y_1, \ldots, y_n)$. Thus the ordered pairs (3, 7) and (7, 3) are not equal, and we write $(3, 7) \neq (7, 3)$. Since tuples convey the idea of order, they are different from sets and bags. Here are some examples:

Sets: $\{b, a, t\} = \{t, a, b\}$.
Bags: $[t, o, o, t] = [o, t, t, o]$.
Tuples: $(t, o, o, t) \neq (o, t, t, o)$ and $(b, a, t) \neq (t, a, b)$.

Here are the two important characteristics of tuples.

Two Characteristics of Tuples

1. There may be repeated occurrences of elements.

2. There is an order or arrangement of the elements.

The rest of this section introduces structures that are represented as tuples. We'll also see in the next section that graphs and trees are often represented using tuples.

Cartesian Product of Sets

We often need to represent information as a set of tuples, where the elements in each tuple come from known sets. Such a set is called a Cartesian product, in honor of René Descartes (1596–1650), who introduced the idea of graphing ordered pairs. The Cartesian product is also referred to as the *cross product*. Here's the formal definition.

Definition of Cartesian Product

If A and B are sets, then the *Cartesian product* of A and B, which is denoted by $A \times B$, is the set of all ordered pairs (a, b) such that $a \in A$ and $b \in B$. In other words, we have

$$A \times B = \{(a, b) \mid a \in A \text{ and } b \in B\}.$$

For example, if $A = \{x, y\}$ and $B = \{0, 1\}$, then

$$A \times B = \{(x, 0), (x, 1), (y, 0), (y, 1)\}.$$

Suppose we let $A = \varnothing$ and $B = \{0, 1\}$ and then ask the question: "What is $A \times B$?" If we apply the definition of Cartesian product, we must conclude that there are no ordered pairs with first elements from the empty set. Therefore, $A \times B = \varnothing$. So it's easy to generalize and say that $A \times B$ is nonempty if and only if both A and B are nonempty sets. The Cartesian product of two sets is easily extended to any number of sets $A_1, \ldots, A_n$ by writing

$$A_1 \times \cdots \times A_n = \{(x_1, \ldots, x_n) \mid x_i \in A_i\}.$$

If all the sets A_i in a Cartesian product are the same set A, then we use the abbreviated notation $A^n = A \times \cdots \times A$. With this notation we have the following definitions for the sets A^1 and A^0:

$$A^1 = \{(a) \mid a \in A\} \text{ and } A^0 = \{(\ )\}.$$

So we must conclude that $A^1 \neq A$ and $A^0 \neq \varnothing$.

example **1.21 Some Products**

Let $A = \{a, b, c\}$. Then we have the following Cartesian products:

$$A^0 = \{()\},$$
$$A^1 = \{(a), (b), (c)\},$$
$$A^2 = \{(a, a), (a, b), (a, c), (b, a), (b, b), (b, c), (c, a), (c, b), (c, c)\},$$

A^3 is bigger yet, with twenty-seven 3-tuples.

end example

When working with tuples, we need the ability to randomly access any component. The components of an n-tuple can be indexed in several different ways depending on the problem at hand. For example, if $t \in A \times B \times C$, then we might represent t in any of the following ways:

$$(t_1, t_2, t_3),$$
$$(t(1), t(2), t(3)),$$
$$(t[1], t[2], t[3]),$$
$$(t(A), t(B), t(C)),$$
$$(A(t), B(t), C(t)).$$

Let's look at an example that shows how Cartesian products and tuples are related to some familiar objects of programming.

example 1.22   Arrays, Matrices, and Records

In computer science, a 1-dimensional array of size n with elements in the set A is an n-tuple in the Cartesian product A^n. So we can think of the Cartesian product A^n as the set of all 1-dimensional arrays of size n over A. If $x = (x_1, \ldots, x_n)$, then the component x_i is usually denoted—in programming languages—by $x[i]$.

A 2-dimensional array—also called a *matrix*—can be thought of as a table of objects that are indexed by rows and columns. If x is a matrix with m rows and n columns, we say that x is an m by n matrix. For example, if x is a 3 by 4 matrix, then x can be represented by the following diagram:

$$x = \begin{bmatrix} x_{11} & x_{12} & x_{13} & x_{14} \\ x_{21} & x_{22} & x_{23} & x_{24} \\ x_{31} & x_{32} & x_{33} & x_{34} \end{bmatrix}.$$

We can also represent x as a 3-tuple whose components are 4-tuples as follows:

$$x = ((x_{11}, x_{12}, x_{13}, x_{14}), (x_{21}, x_{22}, x_{23}, x_{24}), (x_{31}, x_{32}, x_{33}, x_{34})).$$

In programming, the component x_{ij} is usually denoted by $x[i, j]$. We can think of the Cartesian product $(A^4)^3$ as the set of all 2-dimensional arrays over A with 3 rows and 4 columns. Of course, this idea extends to higher dimensions. For example, $((A^5)^7)^4$ represents the set of all 3-dimensional arrays over A consisting of 4-tuples whose components are 7-tuples whose components are 5-tuples of elements of A.

For another example, we can think of $A \times B$ as the set of all records, or structures, with two fields A and B. For a record $r = (a, b) \in A \times B$ the components a and b are normally denoted by $r.A$ and $r.B$.

end example

There are at least three nice things about tuples: They are easy to understand; they are basic building blocks for the representation of information; and they are easily implemented by a computer, which we'll discuss in the next example.

example **1.23** **Computer Repesentation of Tuples**

Computers represent tuples in contiguous cells of memory so that each component can be accessed quickly. For example, suppose that each component of the tuple $x = (x_1, \ldots, x_n)$ needs M memory cells to store it. If B is the beginning address of memory allocated for the tuple x, then x_1 is at location B, x_2 is at location $B + M$, and in general, x_k is at location

$$B + M(k - 1).$$

So each component x_k of x can be accessed in the amount of time that it takes to evaluate $B + M(k - 1)$.

For multidimensional arrays, the access time is also fast. For example, suppose x is a 3 by 4 matrix represented in the following "row-major" form as a 3-tuple of rows, where each row is a 4-tuple.

$$x = ((x_{11}, x_{12}, x_{13}, x_{14}), (x_{21}, x_{22}, x_{23}, x_{24}), (x_{31}, x_{32}, x_{33}, x_{34})).$$

Suppose that each component of x needs M memory cells. If B is the beginning address of memory allocated for x, then x_{11} is at location B, x_{21} is at location $B + 4M$, and x_{31} is at location $B + 8M$. The location of an arbitrary element x_{jk} is given by the expression

$$B + 4M(j - 1) + M(k - 1).$$

Expressions such as this are called *address polynomials*. Each component x_{jk} can be accessed in the amount of time that it takes to evaluate the address polynomial, which is close to a constant for any j and k.

end example

1.3.2   Lists

A *list* is a finite ordered sequence of zero or more elements that can be repeated. At this point a list seems just like a tuple. So what's the difference between tuples and lists? The difference—a big one in computer science—is in what parts can be randomly accessed. In the case of tuples we can randomly access any component in a constant amount of time. In the case of lists we can randomly access only two things in a constant amount of time: the first component of a list, which is called its *head*, and the list made up of everything except the first component, which is called its *tail*.

So we'll use a different notation for lists. We'll denote a list by writing down its elements, separated by commas, and surrounding everything with the two symbols "⟨" and "⟩". The *empty list* is denoted by

$$\langle \, \rangle.$$

The number of elements in a list is called its *length*. For example, the list

$$\langle w,\ x,\ y,\ z\rangle,$$

has length 4, its head is w, and its tail is the list $\langle x,\ y,\ z\rangle$. If L is a list, we'll use the notation

$$\text{head}(L)\ \text{and}\ \text{tail}(L)$$

to denote the head of L and the tail of L. For example,

$$\text{head}(\langle w,\ x,\ y,\ z\rangle) = w,$$
$$\text{tail}(\langle w,\ x,\ y,\ z\rangle) = \langle x,\ y,\ z\rangle.$$

Notice that the empty list $\langle\rangle$ does not have a head or tail.

An important computational property of lists is the ability to easily construct a new list by adding a new element at the head of an existing list. The name *cons* will be used to denote this construction operation. If h is an element of some kind and L is a list, then

$$\text{cons}(h,\ L)$$

denotes the list whose head is h and whose tail is L. Here are a few examples:

$$\text{cons}(w,\ \langle x,\ y,\ z\rangle) = \langle w,\ x,\ y,\ z\rangle,$$
$$\text{cons}(a,\ \langle\ \rangle) = \langle a\rangle.$$
$$\text{cons}(\text{this},\ \langle \text{is},\ \text{helpful}\rangle) = \langle \text{this},\ \text{is},\ \text{helpful}\rangle.$$

The operations head, tail, and cons can be done efficiently and dynamically during the execution of a program. The three operations are related by the following equation for any nonempty list L.

$$\text{cons}(\text{head}(L),\ \text{tail}(L)) = L.$$

There is no restriction on the kind of object that a list can contain. In fact, it is often quite useful to represent information in the form of lists whose elements may be lists, and the elements of those lists may be lists, and so on. Here are a few examples of such lists, together with their heads and tails.

L	$\text{head}(L)$	$\text{tail}(L)$
$\langle a, \langle b\rangle\rangle$	a	$\langle\langle b\rangle\rangle$
$\langle\langle a\rangle, \langle b, a\rangle\rangle$	$\langle a\rangle$	$\langle\langle b, a\rangle\rangle$
$\langle\langle\langle\ \rangle, a, \langle\ \rangle\rangle, b, \langle\ \rangle\rangle$	$\langle\langle\ \rangle, a, \langle\ \rangle\rangle$	$\langle b, \langle\ \rangle\rangle$

If all the elements of a list L are from a particular set A, then L is said to be a *list over A*. For example, each of the following lists is a list over $\{a,\ b,\ c\}$.

$$\langle\ \rangle,\ \langle a\rangle,\ \langle a,\ b\rangle,\ \langle b,\ a\rangle,\ \langle b,\ c,\ a,\ b,\ c\rangle.$$

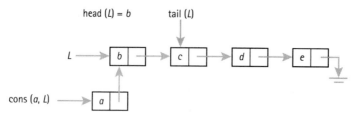

Figure 1.10   Memory representation of a list.

We'll denote the collection of all lists over A by

$$\text{lists}(A).$$

There are at least four nice things about lists: They are easy to understand; they are basic building blocks for the representation of information; they are easily implemented by a computer; and they are easily manipulated by a computer program. We'll discuss this in the next example.

example 1.24 Computer Representation of Lists

A simple way to represent a list in a computer is to allocate a block of memory for each element of the list. The block of memory contains the element together with an address (called a pointer or link) to the next block of memory for the next element of the list. In this way there is no need to have list elements next to each other in memory, so that the creation and deletion of list elements can occur dynamically during the execution of a program.

For example, let's consider the list $L = \langle b, c, d, e \rangle$. Figure 1.10 shows the memory representation of L in which each arrow represents an address (i.e., a pointer or link) and each box represents a block of memory containing an element of the list and an arrow pointing to the address of the next box. The last arrow in the box for e points to the "ground" symbol to signify the end of the list. Empty lists point to the ground symbol, too. The figure also shows head$(L) = b$ and tail$(L) = \langle c, d, e \rangle$. So head and tail are easily calculated from L. The figure also shows how the cons operation constructs a new list cons$(a, L) = \langle a, b, c, d, e \rangle$ by allocating a new block of memory to contain a and a pointer to L.

end example

1.3.3 Strings and Languages

A *string* is a finite ordered sequence of zero or more elements that are placed next to each other in juxtaposition. The individual elements that make up a string are taken from a finite set called an *alphabet*. If A is an alphabet, then a string of elements from A is said to be a string over A. For example, here are a few strings over $\{a, b, c\}$:

$$a, \ ba, \ bba, \ aacabb.$$

The string with no elements is called the *empty string*, and we denote it by the Greek capital letter lambda:

$$\Lambda.$$

The number of elements that occur in a string s is called the *length* of s, which we sometimes denote by

$$|s|.$$

For example, $|\Lambda| = 0$ and $|aacabb| = 6$ over the alphabet $\{a, b, c\}$.

Concatenation of Strings

The operation of placing two strings s and t next to each other to form a new string st is called *concatenation*. For example, if aab and ba are two strings over the alphabet $\{a, b\}$, then the concatenation of aab and ba is the string

$$aabba.$$

We should note that if the empty string occurs as part of another string, then it does not contribute anything new to the string. In other words, if s is a string, then

$$s\Lambda = \Lambda s = s.$$

Strings are used in the world of written communication to represent information: computer programs; written text in all the languages of the world; and formal notation for logic, mathematics, and the sciences.

There is a strong association between strings and lists because both are defined as finite sequences of elements. This association is important in computer science because computer programs must be able to recognize certain kinds of strings. This means that a string must be decomposed into its individual elements, which can then be represented by a list. For example, the string $aacabb$ can be represented by the list $\langle a, a, c, a, b, b \rangle$. Similarly, the empty string Λ can be represented by the empty list $\langle \ \rangle$.

Languages (Sets of Strings)

A *language* is a set of strings. If A is an alphabet, then a *language* over A is a set of strings over A. The set of all strings over A is denoted by A^*. So any language over A is a subset of A^*. Four simple examples of languages over an alphabet A are the sets $\varnothing$, $\{\Lambda\}$, A, and A^*.

For example, if $A = \{a\}$, then these four simple languages over A become

$$\varnothing, \{\Lambda\}, \{a\}, \text{ and } \{\Lambda, a, aa, aaa, \dots\}.$$

For any natural number n the concatenation of a string s with itself n times is denoted by s^n. For example,

$$s^0 = \Lambda, \; s^1 = s, \; s^2 = ss, \text{ and } s^3 = sss.$$

example 1.25 Language Representations

The exponent notation allows us to represent some languages in a nice concise manner. Here are a few examples of languages.

$$\{a^n \mid n \in \mathbb{N}\} = \{\Lambda, a, aa, aaa, \dots\}.$$
$$\{ab^n \mid n \in \mathbb{N}\} = \{a, ab, abb, abbb, \dots\}.$$
$$\{a^n b^n \mid n \in \mathbb{N}\} = \{\Lambda, ab, aabb, aaabbb, \dots\}$$
$$\{(ab)^n \mid n \in \mathbb{N}\} = \{\Lambda, ab, abab, ababab, \dots\}.$$

end example

example 1.26 Numerals

A *numeral* is a written number. In terms of strings, we can say that a numeral is a nonempty string of symbols that represents a number. Most of us are familiar with the following three numeral systems. The *Roman numerals* represent the set of positive integers by using the alphabet

$$\{I, V, X, L, C, D, M\}.$$

The *decimal numerals* represent the set of natural numbers by using the alphabet

$$\{0, 1, 2, 3, 4, 5, 6, 7, 8, 9\}.$$

The *binary numerals* represent the natural numbers by using the alphabet

$$\{0, 1\}.$$

For example, the Roman numeral MDCLXVI, the decimal numeral 1666, and the binary numeral 11010000010 all represent the same number.

end example

Products of Languages

Since languages are sets of strings, they can be combined by the usual set operations of union, intersection, difference, and complement. But there is another important way to combine languages.

We can combine two languages L and M to obtain the set of all concatenations of strings in L with strings in M. This new language is called the *product* of L and M and is denoted by LM. For example, if $L = \{ab, ac\}$ and $M = \{a, bc, abc\}$, then the product LM is the language

$$LM = \{aba, abbc, ababc, aca, acbc, acabc\}.$$

We can give the following formal definition for the product of two languages.

Product of Languages

The product of languages L and M is the language

$$LM = \{st \mid s \in L \text{ and } t \in M\}.$$

It's easy to see, from the definition of product, that the following simple properties hold for any language L.

$$L\{\Lambda\} = \{\Lambda\}L = L.$$
$$L\varnothing = \varnothing L = \varnothing.$$

It's also easy to see that the product is associative. In other words, if L, M, and N are languages, then $L(MN) = (LM)N$. Thus we can write down products without using parentheses. On the other hand, it's easy to see that the product is not commutative. In other words, we can find two languages L and M such that $LM \neq ML$.

For any natural number n, the product of a language L with itself n times is denoted by L^n. In other words, we have

$$L^n = \{s_1 s_2 \dots s_n \mid s_k \in L \text{ for each } k\}.$$

The special case when $n = 0$ has the following definition.

$$L^0 = \{\Lambda\}.$$

For example, if $L = \{a, bb\}$, then we have the following four products.

$$L^0 = \{\Lambda\},$$
$$L^1 = L = \{a, bb\},$$
$$L^2 = LL = \{aa, abb, bba, bbbb\},$$
$$L^3 = LL^2 = \{aaa, aabb, abba, abbbb, bbaa, bbabb, bbbba, bbbbbb\}.$$

Closure of a Language

If L is a language, then the *closure* of L, denoted by L^*, is the set of all possible concatenations of strings from L. In other words, we have

$$L^* = L^0 \cup L^1 \cup L^2 \cup \cdots \cup L^n \cup \cdots.$$

So $x \in L^*$ if and only if $x \in L^n$ for some n. Therefore, we have

$$x \in L^* \text{ if and only if either } x = \Lambda \text{ or } x = l_1 l_2 \ldots l_n$$

for some $n \geq 1$, where $l_k \in L$ for $1 \leq k \leq n$.

If L is a language, then the *positive closure* of L, which is denoted by L^+, is defined by

$$L^+ = L^1 \cup L^2 \cup L^3 \cup \cdots.$$

It follows from the definition that $L^* = L^+ \cup \{\Lambda\}$. But it's not necessarily true that $L^+ = L^* - \{\Lambda\}$. For example, if $L = \{\Lambda, a\}$, then $L^+ = L^*$.

We should observe that any alphabet A is itself a language and its closure coincides with our original definition of A^* as the set of all strings over A. The following properties give some basic facts about the closure of languages. We'll prove one of the facts in the next example and leave the rest as exercises.

Properties of Closure (1.16)

  a. $\{\Lambda\}^* = \varnothing^* = \{\Lambda\}$.

  b. $\Lambda \in L$ if and only if $L^+ = L^*$.

  c. $L^* = L^* L^* = (L^*)^*$.

  d. $(L^* M^*)^* = (L^* \cup M^*)^* = (L \cup M)^*$.

  e. $L(ML)^* = (LM)^* L$.

---

example 1.27 Products and Closure

We'll prove Part (1.16e). We'll start by examining the structure of an arbitrary string $x \in L(ML)^*$. Since $L(ML)^*$ is the product of L and $(ML)^*$, we can write $x = ly$, where $l \in L$ and $y \in (ML)^*$. Since $y \in (ML)^*$, it follows that $y \in (ML)^n$ for some n. If $n = 0$, then $y = \Lambda$ and we have $x = ly = l\Lambda = l \in L$.

If $n > 0$, then $y = w_1 \ldots w_n$, where $w_k \in ML$ for $1 \leq k \leq n$. So we can write each w_k in the form $w_k = m_k l_k$, where $m_k \in M$ and $l_k \in L$. Now we can collect our facts and write x as a concatenation of strings from L and M.

$$x = ly \qquad \text{where } l \in L \text{ and } y \in (ML)^*$$
$$= l(w_1 \ldots w_n) \qquad \text{where } l \in L \text{ and each } w_k \in (ML)$$
$$= l(m_1 l_1 \ldots m_n l_n) \qquad \text{where } l \in L \text{ and each } l_k \in L, \text{ and } m_k \in M.$$

Since we can group strings with parentheses any way we want, we can put things back together in the following order.

$$= (lm_1l_1 \ldots m_n)l_n \qquad \text{where } l \in L \text{ and each } l_k \in L, \text{ and } m_k \in M$$
$$= (z_1 \ldots z_n)l_n \qquad \text{where each } z_k \in LM \text{ and } l_n \in L$$
$$= ul_n \qquad \text{where } u \in (LM)^* \text{ and } l_n \in L.$$

So $x \in (LM)^*L$. Therefore, $L(ML)^* \subseteq (LM)^*L$. The argument is reversible. So we have $L(ML)^* = (LM)^*L$. QED.

end example

example **1.28** **Decimal Numerals**

The product is a useful tool for describing languages in terms of simpler languages. For example, suppose we need to describe the language L of all strings of the form $a.b$, where a and b are decimal numerals. For example, the strings 0.45, 1.569, 000.34000 are elements of L. If we let

$$D = \{.\} \text{ and } N = \{0, 1, 2, 3, 4, 5, 6, 7, 8, 9\},$$

we can then describe L as the following product in terms of D and N.

$$L = N(N)^* \, DN(N)^*.$$

end example

1.3.4 Relations

Ideas such as kinship, connection, and association of objects are keys to the concept of a relation. Informally, a *relation* is a set of n-tuples, where the elements in each tuple are related in some way.

For example, the parent–child relation can be described as the following set of ordered pairs:

$$\text{isParentOf} = \{(x, y) \mid x \text{ is a parent of } y\}.$$

For another example, recall from geometry that if the sides of a right triangle have lengths x, y, and z, where z is the hypotenuse, then $x^2 + y^2 = z^2$. Any 3-tuple of positive real numbers (x, y, z) with this property is called a *Pythagorean triple*. For example, $(1, \sqrt{3}, 2)$ and $(3, 4, 5)$ are Pythagorean triples. The Pythagorean triple relation can be described as the following set of ordered triples:

$$PT = \{(x, y, z) \mid x^2 + y^2 = z^2\}.$$

When we discuss relations in terms of where the tuples come from, there is some terminology that can be helpful.

Definition of Relation

If R is a subset of $A_1 \times \cdots \times A_n$, then R is said to be an *n-ary relation on* (or *over*) $A_1 \times \cdots \times A_n$. If R is a subset of A^n, then we say R is an *n-ary relation on* A. Instead of 1-ary, 2-ary, and 3-ary, we say *unary*, *binary*, and *ternary*.

For example, the isParentOf relation is a binary relation on the set of people and the Pythagorean triple relation is a ternary relation on the set of positive real numbers. In formal terms, if P is the set of all people who are living or who have ever lived, then

$$\text{isParentOf} \subseteq P \times P.$$

Similarly, if we let $\mathbb{R}^+$ denote the set of positive real numbers, then

$$PT \subseteq \mathbb{R}^+ \times \mathbb{R}^+ \times \mathbb{R}^+.$$

Since there are many subsets of a set, there can be many relations. The smallest relation is the empty set $\varnothing$, which is called the *empty relation*. The largest relation is $A_1 \times \cdots \times A_n$ itself, which is called the *universal relation*.

If R is a relation and $(x_1, \ldots, x_n) \in R$, this fact is often denoted by the *prefix expression*

$$R(x_1, \ldots, x_n).$$

For example, $PT(1, \sqrt{3}, 2)$ can be written as $(1, \sqrt{3}, 2) \in PT$.

If R is a binary relation, then the statement $(x, y) \in R$ can be denoted by $R(x, y)$, but it is often denoted by the *infix expression*

$$x \ R \ y.$$

For example, "John isParentOf Mary" can be written as (John, Mary) $\in$ isParentOf.

We use many binary relations without even thinking about it. For example, we use the less-than relation on numbers m and n by writing $m < n$ instead of $(m, n) \in <$ or $< (m, n)$. We also use equality without thinking about it as a binary relation. The *equality relation* on a set A is the set

$$\{(x, x) \mid x \in A\}.$$

For example, if $A = \{a, b, c\}$, then the equality relation on A is the set $\{(a, a), (b, b), (c, c)\}$. We normally denote equality by the symbol $=$ and we write $a = a$ instead of $(a, a) \in =$ or $= (a, a)$.

Since unary relations are sets of 1-tuples, we usually dispense with the tuple notation and simply write a unary relation as a set of elements. For example, instead of writing $R = \{(2), (3), (5), (7)\}$ we write $R = \{2, 3, 5, 7\}$. So $R(2)$ and $2 \in R$ mean the same thing.

Relational Databases

A *relational database* is a collection of facts that are represented by tuples in such a way that the tuples can be accessed in various ways to answer queries about the facts. To accomplish these tasks each component of a tuple must have an associated name, called an *attribute*.

For example, suppose we have a database called Borders that describes the foreign countries and large bodies of water that border each state of the United States. The following table represents a sample of the information in the database with attribute names State, Foreign, and Water.

Borders

State	Foreign	Water
Washington	Canada	Pacific Ocean
Minnesota	Canada	Lake Superior
Wisconsin	None	Lake Michigan
Oregon	None	Pacific Ocean
Maine	Canada	Atlantic Ocean
Michigan	Canada	Lake Superior
Michigan	Canada	Lake Huron
Michigan	Canada	Lake Michigan
California	Mexico	Pacific Ocean
Arizona	Mexico	None

There is no special order to the rows of a relational database. So the table can be represented as a set of tuples.

$$\text{Borders} \quad = \quad \{ (\text{Washington, Canada, Pacific Ocean}),$$
$$(\text{Minnesota, Canada, Lake Superior}),$$
$$(\text{Wisconsin, None, Lake Michigan}),\dots \}.$$

 1.29 Questions about Borders

Let's look at a few questions or queries that can be asked about the Borders database. Each question can be answered by describing a set or a relation. For example, suppose we ask the question

What states border Mexico?

The answer is the set

$$\{x \mid (x, \text{Mexico}, z) \in \text{Borders, for some } z\}.$$

Here are a few more questions that we'll leave as exercises.

> What bodies of water border Michigan?
>
> What states border the Pacific Ocean?
>
> What states are landlocked?
>
> What relation represents the state–water pairs?
>
> What state–water pairs contain a state bordering Canada?

end example

Queries can be answered not only by describing a set or a relation as in the example, but also by describing an expression in terms of basic operations that construct new relations by selecting certain tuples, by eliminating certain attributes, or by combining attributes of two relations. We'll discuss these basic operations on relational databases in Section 10.4.

1.3.5   Counting Tuples

How can we count a set of tuples, lists, or strings? Since tuples, lists, and strings represent finite ordered sequences of objects, the only difference is how we represent them, not whether there are more of one kind than another. For example, over the set $\{a, b\}$ there are eight 3-tuples, eight lists of length 3, and eight strings of length 3. So without any loss of generality we'll discuss counting sets of tuples. The main tools that we'll use are the rules for counting Cartesian products of finite sets.

The Product Rule

Suppose we need to know the cardinality of $A \times B$ for finite sets A and B. In other words, we want to know how many 2-tuples are in $A \times B$. For example, suppose that $A = \{a, b, c\}$ and $B = \{0, 1, 2, 3\}$. The sets A and B are small enough so that we can write down all 12 of the tuples. The exercise might also help us notice that each element of A can be paired with any one of the four elements in B. Since there are three elements in A, it follows that

$$|A \times B| = (3)(4) = 12.$$

This is an example of a general counting technique called the product rule, which we'll state as follows for any two finite sets A and B.

Product Rule   **(1.17)**

$$|A \times B| = |A||B|.$$

It's easy to see that (1.17) generalizes to a Cartesian product of three or more finite sets. For example, $A \times B \times C$ and $A \times (B \times C)$ are not actually equal because an arbitrary element in $A \times B \times C$ is a 3-tuple (a, b, c), while an arbitrary element in $A \times (B \times C)$ is a 2-tuple $(a, (b, c))$. Still the two sets have the same cardinality. Can you convince yourself of this fact? Now proceed as follows:

$$
\begin{aligned}
|A \times B \times C| &= |A \times (B \times C)| \\
&= |A|\,|B \times C| \\
&= |A|\,|B|\,|C| .
\end{aligned}
$$

The extension of (1.17) to any number of sets allows us to obtain other useful formulas for counting tuples of things. For example, for any finite set A and any natural number n we have the following product rule.

$$
|A^n| = |A|^n . \tag{1.18}
$$

Counting Strings as Tuples

We can use product rules to count strings as well as tuples because a string can be represented as a tuple. In each of the following examples, the problem to be solved is expressed in terms of strings.

example 1.30 **Counting All Strings**

Suppose we need to count the number of strings of length 5 over the alphabet $A = \{a, b, c\}$. Any string of length 5 can be considered as a 5-tuple. For example, the string $abcbc$ can be represented by the tuple (a, b, c, b, c). So the number of strings of length 5 over A equals the number of 5-tuples over A, which by product rule (1.18) is

$$
|A^5| = |A|^5 = 3^5 = 243.
$$

end example

example 1.31 **Strings with Restrictions**

Suppose we need to count the number of strings of length 6 over the alphabet $A = \{a, b, c, d\}$ that begin with either a or c and contain at least one occurrence of b.

Since strings can be represented by tuples, we'll count the number of 6-tuples over A that begin with either a or c and contain at least one occurrence of b. We'll break up the problem into two simpler problems. First, let U be the set of 6-tuples over A that begin with a or c. In other words, $U = \{a, c\} \times A^5$. Next, let S be the subset of U consisting of those 6-tuples that do not contain any occurrences of b. In other words, $S = \{a, c\} \times \{a, c, d\}^5$. Then the set

$U - S$ is the desired set of 6-tuples over A that begin with either a or c and contain at least one occurrence of b. So we have

$$
\begin{aligned}
|U - S| &= |U| - |S| && \text{(by 1.14)} \\
&= |\{a, c\} \times A^5| - |\{a, c\} \times \{a, c, d\}^5| \\
&= |\{a, c\}\,||A|^5 - |\{a, c\}\,||\{a, c, d\}|^5 && \text{(by 1.17 and 1.18)} \\
&= 2(4^5) - 2(3^5) \\
&= 1,562.
\end{aligned}
$$

end example

example **1.32 Strings with More Restrictions**

We'll count the number of strings of length 6 over $A = \{a,\ b,\ c,\ d\}$ that start with a or c and contain at least one occurrence of either b or d.

As in the previous example, let U be the set of 6-tuples over A that start with a or c. Then $U = \{a,\ c\} \times A^5$ and $|U| = 2(4^5)$. Next, let S be the subset of U whose 6-tuples do not contain any occurrences of b and do not contain any occurrences of d. So $S = \{a,\ c\}^6$ and $|S| = 2^6$. Then the set $U - S$ is the desired set of 6-tuples over A that begin with either a or c and contain at least one occurrence of either b or d. So by (1.14) we have

$$
\begin{aligned}
|U - S| &= |U| - |S| \\
&= 2(4^5) - 2^6 \\
&= 1,984.
\end{aligned}
$$

end example

example **1.33 Strings with More Restrictions**

Suppose we need to count the number of strings of length 6 over $A = \{a,\ b,\ c,\ d\}$ that start with a or c and contain at least one occurrence of b and at least one occurrence of d.

In this case we'll break up the problem into three simpler problems. First, let U be the set of 6-tuples that start with a or c. So $U = \{a,\ c\} \times A^5$ and $|U| = 2(4^5)$. Next, let S be the subset of U whose 6-tuples don't contain b. So $S = \{a,\ c\} \times \{a,\ c,\ d\}^5$ and $|S| = 2(3^5)$. Similarly, let T be the subset of U whose 6-tuples don't contain d. So $T = \{a,\ c\} \times \{a,\ b,\ c\}^5$ and $|T| = 2(3^5)$. Then the set $U - (S \cup T)$ is the desired set of 6-tuples that start with a or c and contain at least one occurrence of b and at least one occurrence of d. The cardinality of this set has the form

$$
\begin{aligned}
|U - (S \cup T)| &= |U| - |S \cup T| && \text{(by 1.14)} \\
&= |U| - (|S| + |T| - |S \cap T|). && \text{(by 1.11)}
\end{aligned}
$$

We'll be done if we can calculate the cardinality of $S \cap T$. Notice that

$$
\begin{aligned}
S \cap T &= \{a, c\} \times \{a, c, d\}^5 \cap \{a, c\} \times \{a, b, c\}^5 \\
&= \{a, c\} \times \{a, c\}^5 \\
&= \{a, c\}^6 .
\end{aligned}
$$

So $|S \cap T| = 2^6$. Now we can complete the calculation of $|U - (S \cup T)|$.

$$
\begin{aligned}
|U - (S \cup T)| &= |U| - (|S| + |T| - |S \cap T|) \\
&= 2\left(4^5\right) - \left[2\left(3^5\right) + 2\left(3^5\right) - 2^6\right] \\
&= 1,140.
\end{aligned}
$$

end example

 Exercises

Tuples

1. Write down all possible 3-tuples over the set $\{x, y\}$.

2. Let $A = \{a, b, c\}$ and $B = \{a, b\}$. Compute each of the following sets.

   a. $A \times B$.   **b.** $B \times A$.   **c.** A^0.

   d. A^1.   **e.** A^2.   **f.** $A^2 \cap (A \times B)$.

Lists

3. Write down all possible lists of length 2 or less over the set $A = \{a, b\}$.

4. Find the head and tail of each list.

   a. $\langle a \rangle$.   **b.** $\langle a, b, c \rangle$.   **c.** $\langle \langle a, b \rangle, c \rangle$.

   d. $\langle \langle a, b \rangle, \langle a, c \rangle \rangle$.

5. For positive integers m and n let D be the list of integers greater than 1 that divide both m and n, where D is ordered from smallest to largest. For example, if $m = 12$ and $n = 18$, then $D = \langle 2, 3, 6 \rangle$. We'll combine this information into a list $\langle m, n, D \rangle = \langle 12, 18, \langle 2, 3, 6 \rangle \rangle$. Construct $\langle m, n, D \rangle$ for each of the following cases.

   a. $m = 24$ and $n = 60$.

   b. $m = 36$ and $n = 36$.

   c. $m = 14$ and $n = 15$.

6. Write down all possible lists over $\{a, b\}$ that can be represented with five symbols, where the symbols that we count are a or b or $\langle$ or $\rangle$. For example, $\langle a, \langle \rangle \rangle$ uses five of the symbols. Can you do the same for lists over A that have six symbols? There are quite a few of them.

Strings and Languages

7. Write down all possible strings of length 2 over the set $A = \{a, b, c\}$.

8. Let $L = \{\Lambda, abb, b\}$ and $M = \{bba, ab, a\}$. Evaluate each of the following language expressions.

    a. LM.   b. ML.   **c.** L^0.   d. L^1.   **e.** L^2.

9. Use your wits to solve each of the following language equations for the unknown language.

    a. $\{\Lambda, a, ab\} L = \{b, ab, ba, aba, abb, abba\}$.
    b. $L \{a, b\} = \{a, baa, b, bab\}$.
    c. $\{a, aa, ab\} L = \{ab, aab, abb, aa, aaa, aba\}$.
    d. $L \{\Lambda, a\} = \{\Lambda, a, b, ab, ba, aba\}$.
    e. $\{a, b\} L = \{a, b, aba, bba\}$.
    f. $L \{b, \Lambda, ab\} = \{abb, ab, abab, bab, b, bb\}$.

10. Let L and M be two languages. For each of the following languages describe the general structure of a string x by writing it as a concatenation of strings that are in either L or M.

    a. LML.   b. LM^*.   **c.** $(L \cup M)^*$.
    d. $(L \cap M)^*$.   **e.** L^*M^*.   f. $(LM)^*$.

11. Try to describe each of the following languages in some way.

    a. $\{a, b\}^* \cap \{b, c\}^*$.   b. $\{a, b\}^* - \{b\}^*$.   **c.** $\{a, b, c\}^* - \{a\}^*$.

Relations

12. Represent each relation as a set by listing each individual tuple.

    a. $\{(d, 12) \mid d > 0 \text{ and } d \text{ divides } 12\}$.
    b. $\{(d, n) \mid d, n \in \{2, 3, 4, 5, 6\} \text{ and } d \text{ divides } n \}$.
    c. $\{(x, y, z) \mid x = y + z, \text{ where } x, y, z \in \{1, 2, 3\}\}$.
    d. Let $(x, y) \in S$ if and only if $x \leq y$ and $x, y \in \{1, 2, 3\}$.
    e. Let $(x, y) \in U$ if and only if $x \in \{a, b\}$ and $y \in \{1, 2\}$.

13. Each of the following database queries refers to the Borders relational database given prior to Example 1.29. Express each answer by defining a set or relation in the same manner as the answer given in Example 1.29.

    a. What bodies of water border Michigan?

    b. What states border the Pacific Ocean?

    c. What states are landlocked?

    d. What relation represents the state–water pairs?

    e. What state–water pairs contain a state bordering Canada?

Counting Tuples

14. For each of the following cases, find the number of strings over the alphabet $\{a, b, c, d, e\}$ that satisfy the given conditions.

    a. Length 4, begins with a or b, contains at least one c.

    b. Length 5, begins with a, ends with b, contains at least one c or d.

    c. Length 6, begins with d, ends with b or d, contains no c's.

    d. Length 6, contains at least one a and at least one b.

15. Find a formula for the number of strings of length n over an alphabet A such that each string contains at least one occurrence of a letter from a subset B of A. Express the answer in terms of $|A|$ and $|B|$.

Proofs and Challenges

16. Prove each of the following statements about combining set operations with Cartesian product.

    a. $(A \cup B) \times C = (A \times C) \cup (B \times C)$.

    b. $(A - B) \times C = (A \times C) - (B \times C)$.

    c. Find and prove a similar equality using the intersection operation.

17. Let L, M, and N be languages. Prove each of the following properties of the product operation on languages.

    a. $L\{\Lambda\} = \{\Lambda\}L = L$.

    b. $L\varnothing = \varnothing L = \varnothing$.

    c. $L(M \cup N) = LM \cup LN$ and $(M \cup N)L = ML \cup NL$.

    d. $L(M \cap N) \subseteq LM \cap LN$ and $(M \cap N)L \subseteq ML \cap NL$.

18. Let L and M be languages. Prove each of the following statements about the closure of languages (1.16).

    a. $\{\Lambda\}^* = \varnothing^* = \{\Lambda\}$.

    b. $\Lambda \in L$ if and only if $L^+ = L^*$.

    c. $L^* = L^*L^* = (L^*)^*$.

    d. $(L^*M^*)^* = (L^* \cup M^*)^* = (L \cup M)^*$.

19. (*Tuples Are Special Sets*). We can define the concept of tuples in terms of sets. For example, we'll define

$$() = \varnothing, \ (x) = \{x\}, \quad \text{and} \quad (x, \ y) = \{\{x\}, \ \{x, \ y\}\}.$$

Use this definition to verify each of the following statements.

a. Show that $(3, \ 7) \neq (7, \ 3)$.

b. Show that $(x, \ y) = (u, \ v)$ if and only if $x = u$ and $y = v$.

c. Find an example to show that the definition $(x, \ y) = \{x, \ \{y\}\}$ will not distinguish between distinct 2-tuples.

20. (*Tuples Are Special Sets*). Continuing with Exercise 19, we can define a 3-tuple in terms of sets by letting S be the set representing the ordered pair $(x, \ y)$ from Exercise 19. Then define

$$(x, \ y, \ z) = \{\{S\}, \ \{S, \ z\}\}.$$

a. Write down the complete set to represent $(x, \ y, \ z)$.

b. Show that $(a, \ b, \ c) = (d, \ e, \ f)$ if and only if $a = d, \ b = e$, and $c = f$.

c. Find an example to show that the definition

$$(x, \ y, \ z) = \{\{x\}, \ \{x, \ y\}, \ \{x, \ y, \ z\}\}$$

will not distinguish between distinct 3-tuples.

Note: We could continue in this manner and define n-tuples as sets for any natural number n. Although defining a tuple as a set is not at all intuitive, it does illustrate how sets can be used as a foundation from which to build objects and ideas. It also shows why good notation is so important for communicating ideas.

21. Use Example 1.23 as a guide to find the address polynomial for an arbitrary element in each of the following cases. Assume that all indexes start with 1, that the beginning address is B, and each component needs M memory cells.

a. A matrix of size 3 by 4 stored in column-major form as a 4-tuple of columns, each of which is a 3-tuple.

b. A matrix of size m by n stored in row-major form.

c. A three-dimensional array of size l by m by n stored as an l-tuple where each component is an m by n matrix stored in row-major form.

1.4 Graphs and Trees

When we think about graphs we might think about pictures of some kind that are used to represent information. The graphs that we'll discuss can be thought

Figure 1.11   Graphs.

about in the same way. But we need to describe them in a little more detail if they are to be much use to us. We'll also see that trees are special kinds of graphs.

1.4.1   Definition of a Graph

Informally, a *graph* is a set of objects in which some of the objects are connected to each other in some way. The objects are called *vertices* or *nodes*, and the connections are called *edges*. For example, the United States can be represented by a graph where the vertices are states and the edges are the common borders between adjacent states. In this case, Hawaii and Alaska would be vertices without any edges connected to them. We say that two vertices are *adjacent* if there is an edge connecting them.

Picturing a Graph

We can picture a graph in several ways. For example, Figure 1.11 shows two ways to represent the graph with vertices 1, 2, and 3 and edges connecting 1 to 2 and 1 to 3.

example 1.34   States and Provinces

Figure 1.12 represents a graph, where the vertices are those states in the United States and those provinces in Canada that border the Pacific Ocean or that border states and provinces that touch the Pacific Ocean, and where an edge denotes a common border.

end example

Coloring a Graph

An interesting problem dealing with maps is to try to color a map with the fewest number of colors subject to the restriction that any two adjacent areas must have distinct colors. From a graph point of view, this means that any two distinct adjacent vertices must have different colors. Before reading any further, try to color the graph in Figure 1.12 with the fewest colors. It's usually easier to represent the colors by numbers like 1, 2, ...

A graph is *n-colorable* if there is an assignment of n colors to its vertices such that any two distinct adjacent vertices have distinct colors. The *chromatic number* of a graph is the smallest n for which it is n-colorable. For example, the chromatic number of the graph in Figure 1.12 is 3. A graph whose edges are the

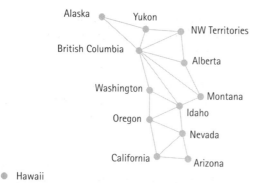

Figure 1.12   Graph of states and provinces.

connections between all pairs of distinct vertices is called a *complete graph*. It's easy to see that the chromatic number of a complete graph with n vertices is n.

A graph is *planar* if it can be drawn on a plane such that no edges intersect. For example, the graph in Figure 1.12 is planar. A complete graph with four vertices is planar, but a complete graph with five vertices is not planar. See whether you can convince yourself of these facts. A fundamental result on graph coloring—which remained an unproven conjecture for over 100 years—states that *every planar graph is 4-colorable*. The result was proven in 1976 by Kenneth Appel and Wolfgang Haken [1976, 1977]. They used a computer to test over 1900 special cases.

More Terminology

A *directed graph* (*digraph* for short) is a graph where each edge points in one direction. For example, the vertices could be cities and the edges could be the one-way air routes between them. For digraphs we use arrows to denote the edges. For example, Figure 1.13 shows two ways to represent the digraph with three vertices a, b, and c and edges from a to b, c to a, and c to b.

The *degree* of a vertex is the number of edges that it touches. However, we add two to the degree of a vertex if it has a *loop*, which is an edge that starts and ends at the same vertex. For directed graphs the *indegree* of a vertex is the number of edges pointing at the vertex, whereas the *outdegree* of a vertex is the number of edges pointing away from the vertex. In a digraph a vertex is called

Figure 1.13   Directed graphs.

a *source* if its indegree is zero and a *sink* if its outdegree is zero. For example, in the digraph of Figure 1.13, c is a source and b is a sink.

If a graph has more than one edge between some pair of vertices, the graph is called a *multigraph*, or a *directed multigraph* in case the edges point in the same direction. For example, there are usually two or more road routes between most cities. So a graph representing road routes between a set of cites is most likely a multigraph.

Representaions of Graphs

From a computational point of view, we need to represent graphs as data. This is easy to do because we can define a graph in terms of tuples, sets, and bags. For example, we can define a graph G as an ordered pair (V, E), where V is a set of vertices and E is a set or bag of edges. If G is a digraph, then the edges in E can be represented by ordered pairs, where (a, b) represents the edge with an arrow from a to b. In this case the set E of edges is a subset of $V \times V$. In other words, E is a binary relation on V. For example, the digraph in Figure 1.13 has vertex set $\{a, b, c\}$ and edge set

$$\{(a, b), (c, b), (c, a)\}.$$

If G is a directed multigraph, then we can represent the edges as a bag (or multiset) of ordered pairs. For example, the bag $[(a, b), (a, b), (b, a)]$ represents three edges: two from a to b and one from b to a.

If a graph is not directed, we have more ways to represent the edges. We could still represent an edge as an ordered pair (a, b) and agree that it represents an undirected line between a and b. But we can also represent an edge between vertices a and b by a set $\{a, b\}$. For example, the graph in Figure 1.11 has vertex set $\{1, 2, 3\}$ and edge set $\{\{1, 2\}, \{1, 3\}\}$.

Weighted Graphs

We often encounter graphs that have information attached to each edge. For example, a good road map places distances along the roads between major intersections. A graph is called *weighted* if each edge is assigned a number, called a *weight*. We can represent an edge (a, b) that has weight w by the 3-tuple

$$(a, b, w).$$

In some cases we might want to represent an unweighted graph as a weighted graph. For example, if we have a multigraph in which we wish to distinguish between multiple edges that occur between two vertices, then we can assign a different weight to each edge, thereby creating a weighted multigraph.

Graphs and Binary Relations

We can observe from our discussion of graphs that any binary relation R on a set A can be thought of as a digraph $G = (A, R)$ with vertices A and edges R. For example, let $A = \{1, 2, 3\}$ and

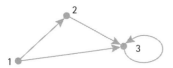

Figure 1.14   Digraph of binary relation.

$$R = \{(1, 2), (1, 3), (2, 3), (3, 3)\}.$$

Figure 1.14 shows the digraph corresponding to this binary relation. Representing a binary relation as a graph is often quite useful in trying to establish properties of the relation.

Subgraphs

Sometimes we need to discuss graphs that are part of other graphs. A graph (V', E') is a *subgraph* of a graph (V, E) if $V' \subseteq V$ and $E' \subseteq E$. For example, the four graphs in Figure 1.15 are subgraphs of the graph in Figure 1.14.

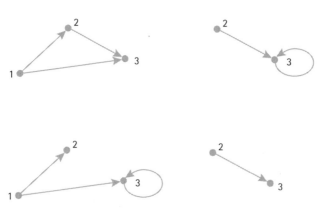

Figure 1.15   Subgraphs.

1.4.2   Paths in Graphs

Problems that use graphs often involve moving from one vertex to another along a sequence of edges, where each edge shares a vertex with the next edge in the sequence. In formal terms, a *path* from x_0 to x_n is a sequence of edges that we denote by a sequence of vertices $x_0, x_1, \ldots, x_n$ such that there is an edge from x_{i-1} to x_i for $1 \leq i \leq n$. A path allows the possibility that some edge or some vertex occurs more than once. A *cycle* is a path whose beginning and ending vertices are equal and in which no edge occurs more than once. A graph with no cycles is called *acyclic*. The *length* of the path $x_0, \ldots, x_n$ is the number n of edges.

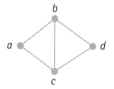

Figure 1.16   Sample graph.

Let's emphasize here that the definitions for path and cycle apply to both graphs and directed graphs.

example **1.35 Paths in a Graph**

We'll examine a few paths in the graph pictured in Figure 1.16.

1. The path b, c, d, b, a visits b twice. The length of the path is 4.

2. The path a, b, c, b, d visits b twice and uses the edge between b and c twice. The length of the path is 4.

3. The path a, b, c, a is a cycle of length 3.

4. The path a, b, a is not a cycle because the edge from a to b occurs twice. The path has length 2.

end example

Connected Graphs

Now we come to an idea that needs a separate definition for each type of graph. A graph is *connected* if there is a path between every pair of vertices. For directed graphs there are two kinds of connectedness. A digraph is *weakly connected* if, when direction is ignored, the resulting undirected graph is connected. A digraph is *strongly connected* if there is a directed path between every pair of vertices. For example, the digraphs in Figure 1.13 and Figure 1.14 are weakly connected but not strongly connected.

Two Graph Problems

Let's look at two graph problems that you may have seen. For the first problem you are to trace the first diagram in Figure 1.17 without taking your pencil off the paper and without retracing any line.

After some fiddling, it's easy to see that it can be done by starting at one of the bottom two corners and finishing at the other bottom corner. The second

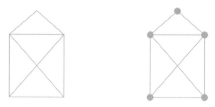

Figure 1.17   Tracing a graph.

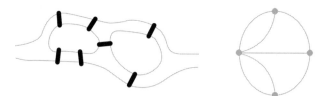

Figure 1.18   Seven bridges of Königsberg.

diagram in Figure 1.17 emphasizes the graphical nature of the problem. From this point of view we can say that there is a path that travels each edge exactly once.

The second famous problem is named after the seven bridges of Königsberg that, in the early 1800s, connected two islands in the river Pregel to the rest of the town. The problem is to tour the town by walking across each of the seven bridges exactly once. In Figure 1.18 we've pictured the two islands and seven bridges of Königsberg together with a multigraph representing the situation. The vertices of the multigraph represent the four land areas and the edges represent the seven bridges.

The mathematician Leonhard Euler (1707–1783) proved that there aren't any such paths by finding a general condition for such paths to exist. In his honor, any path that contains each edge of a graph exactly once is called an *Euler path*. For example, the graph for the tracing problem has an Euler path, but the seven bridges problem does not. We can go one step further and define an *Euler circuit* to be any path that begins and ends at the same vertex and contains each edge of a graph exactly once. There are no Euler circuits in the graphs of Figure 1.17 and Figure 1.18. We'll discuss conditions for the existence of Euler paths and Euler circuits in the exercises.

1.4.3 Graph Traversals

A *graph traversal* starts at some vertex v and visits all vertices x that can be reached from v by traveling along some path from v to x. If a vertex has already been visited, it is not visited again. It follows that any traversal of a connected graph will visit all its vertices. Similarly, any traversal of a strongly connected digraph will visit all its vertices.

Let's look at two popular traversal algorithms: *breadth-first traversal* and *depth-first traversal.*

Breadth–First Traversal

To describe *breadth-first traversal*, we'll let visit(v, k) denote any procedure that visits every vertex x not yet visited for which there is a length k path from v to x. If the graph has n vertices, a breadth-first traversal starting at v can be described as follows:

$$\textbf{for } k := 0 \textbf{ to } n - 1 \textbf{ do } \text{visit}(v, k) \textbf{ od}.$$

Since we haven't specified how visit(v, k) does it's job, there are usually several different traversals from any given starting vertex.

example **1.36 Breadth-First Traversals**

We'll do some breadth-first traversals of the graph in Figure 1.19. If we start at vertex a, then there are four possible breadth-first traversals, which we've represented by the following strings:

$a\ b\ c\ d\ e\ f\ g$ $a\ b\ c\ d\ f\ e\ g$ $a\ c\ b\ d\ e\ f\ g$ $a\ c\ b\ d\ f\ e\ g$.

Of course, we can start a breadth-first traversal at any vertex. For example, one of several breadth-first traversals that start with d is represented by the string

$$d\ b\ e\ f\ a\ g\ c.$$

end example

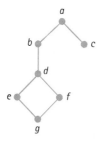

Figure 1.19   Sample graph.

Depth–First Traversal

We can describe *depth-first traversal* with a recursive procedure—one that calls itself. Let $DF(v)$ denote the depth-first procedure that traverses the graph starting at vertex v. Then $DF(v)$ has the following definition.

$$DF(v): \quad \textbf{if } v \text{ has not been visited } \textbf{then}$$
$$\text{visit } v;$$
$$\textbf{for } \text{each edge from } v \text{ to } x \textbf{ do } DF(x) \textbf{ od}$$
$$\textbf{fi}$$

Since we haven't specified how each edge from v to x is picked in the **for** loop, there are usually several different traversals from any given starting vertex.

example | **1.37 Depth-First Traversals**

We'll do some depth-first traversals of the graph represented in Figure 1.19. For example, starting from vertex a in the graph, there are four possible depth-first traversals, which are represented by the following strings:

$$a\ b\ d\ e\ g\ f\ c \qquad a\ b\ d\ f\ g\ e\ c \qquad a\ c\ b\ d\ e\ g\ f \qquad a\ c\ b\ d\ f\ g\ e.$$

end example

1.4.4 Trees

From an informal point of view, a *tree* is a structure that looks like a real tree. For example, a family tree and an organizational chart for a business are both trees. From a formal point of view we can say that a *tree* is a graph that is connected and has no cycles. Such a graph can be drawn to look like a real tree. The vertices and edges of a tree are called *nodes* and *branches*, respectively.

In computer science, and some other areas, trees are usually pictured as upside down versions of real trees, as in Figure 1.20. For trees represented this way, the node at the top is called the *root*. The nodes that hang immediately below a given node are its *children*, and the node immediately above a given node is its *parent*. If a node is childless, then it is a *leaf*. The *height* or *depth* of a tree is the length of the longest path from the root to the leaves. The *level* of a node is the length of the path from the root to the node. The path from a node to the root contains all the *ancestors* of the node. Any path from a node to a leaf contains *descendants* of the node.

A tree with a designated root is often called a *rooted tree*. Otherwise, it is called a *free tree* or an *unrooted tree*. We'll use the term *tree* and let the context indicate the type of tree.

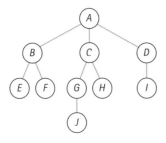

Figure 1.20   Sample tree.

Figure 1.21   A subtree.

example 1.38 Parts of a Tree

We'll make some observations about the tree in Figure 1.20. The root is A. The children of A are B, C, and D. The parent of F is B. The leaves of the tree are E, F, J, H, and I. The height or depth of the tree is 3. The level of H is 2. The ancestors of J are A, C, and G. The descendants of C are G, J, and H.

end example

Subtrees

If x is a node in a tree T, then x together with all its descendants forms a tree S with x as its root. S is called a *subtree* of T. If y is the parent of x, then S is sometimes called a subtree of y.

example 1.39 A Subtree

The tree pictured in Figure 1.21 is a subtree of the tree in Figure 1.20. Since A is the parent of B, we can also say that this tree is a subtree of node A.

end example

Ordered and Unordered Trees

If we don't care about the ordering of the children of a tree, then the tree is called an *unordered tree*. A tree is *ordered* if there is a unique ordering of the children of each node. For example, any algebraic expression can be represented as an ordered tree.

example 1.40 Representing Algebraic Expressions

The expression $x - y$ can be represented by a tree whose root is the minus sign and with two subtrees, one for x on the left and one for y on the right. Ordering is important here because the subtraction operation is not commutative. For

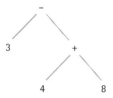
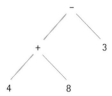

Figure 1.22   Tree for $3 - (4 + 8)$.   **Figure 1.23**   Tree for $(4 + 8) - 3$.

example, Figure 1.22 represents the expression $3 - (4 + 8)$ and Figure 1.23 represents the expression $(4 + 8) - 3$.

end example

Representations of Trees

How can we represent a tree as a data object? The key to any representation is that we should be able to recover the tree from its representation. One method is to let the tree be a list whose first element is the root and whose next elements are lists that represent the subtrees of the root.

For example, the tree with a single node r is represented by $\langle r \rangle$, and the list representation of the tree for the algebraic expression $a - b$ is

$$\langle -, \langle a \rangle, \langle b \rangle \rangle.$$

For another example, the list representation of the tree for the arithmetic expression $3 - (4 + 8)$ is

$$\langle -, \langle 3 \rangle, \langle +, \langle 4 \rangle, \langle 8 \rangle \rangle \rangle.$$

For a more complicated example, let's consider the tree represented by the following list.

$$T = \langle r, \langle b, \langle c \rangle, \langle d \rangle \rangle, \langle x, \langle y, \langle z \rangle \rangle, \langle w \rangle \rangle, \langle e, \langle u \rangle \rangle \rangle.$$

Notice that T has root r, which has the following three subtrees:

$$\langle b, \langle c \rangle, \langle d \rangle \rangle$$
$$\langle x, \langle y, \langle z \rangle \rangle, \langle w \rangle \rangle$$
$$\langle e, \langle u \rangle \rangle.$$

Similarly, the subtree $\langle b, \langle c \rangle, \langle d \rangle \rangle$ has root b, which has two children c and d. We can continue in this way to recover the picture of T in Figure 1.24.

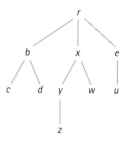

Figure 1.24   Sample tree.

1.41 Computer Representation of Trees

Let's represent a tree as a list and then see what it looks like in computer memory. For example, let T be the following tree.

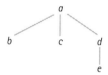

We'll represent the tree as the list $T = \langle a, \langle b \rangle, \langle c \rangle, \langle d, \langle e \rangle \rangle \rangle$. Figure 1.25 shows the representation of T in computer memory, where we have used the same notation as Example 1.24.

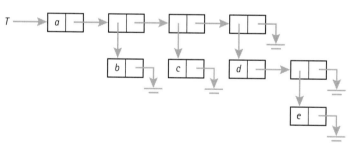

Figure 1.25   Computer representation of a tree.

Binary Trees

A *binary tree* is an ordered tree that may be empty or else has the property that each node has two subtrees, called the *left* and *right* subtrees of the node, which are binary trees. We can represent the empty binary tree by the empty

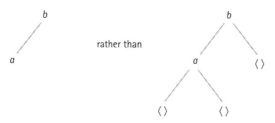

Figure 1.26   Simplified binary tree.

list $\langle \ \rangle$. Since each node has two subtrees, we represent nonempty binary trees as 3-element lists of the form

$$\langle L, \ x, \ R \rangle,$$

where x is the root, L is the left subtree, and R is the right subtree. For example, the tree with one node x is represented by the list $\langle \langle \ \rangle, \ x, \ \langle \ \rangle \rangle$.

When we draw a picture of a binary tree, it is common practice to omit the empty subtrees. For example, the binary tree represented by the list

$$\langle \langle \langle \ \rangle, \ a, \ \langle \ \rangle \rangle, \ b, \ \langle \ \rangle \rangle$$

is usually, but not always, pictured as the simpler tree in Figure 1.26.

example 1.42 Computer Representation of Binary Trees

Let's see what the representation of a binary tree as a list looks like in computer memory. For example, let T be the following tree.

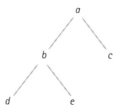

Figure 1.27 shows the representation of T in computer memory, where each block of memory contains a node and pointers to the left and right subtrees.

end example

Binary Search Trees

Binary trees can be used to represent sets whose elements have some ordering. Such a tree is called a *binary search tree* and has the property that for each node

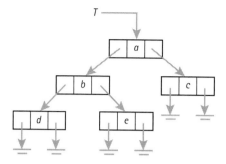

Figure 1.27   Computer representation of a binary tree.

Figure 1.28   Binary search tree.

of the tree, each element in its left subtree precedes the node element and each element in its right subtree succeeds the node element.

example **1.43 A Binary Search Tree**

The binary search tree in Figure 1.28 holds three-letter abbreviations for six of the months, where we are using the dictionary ordering of the words. So the correct order is FEB, JAN, JUL, NOV, OCT, SEP.

This binary search tree has depth 2. There are many other binary search trees to hold these six months. Find another one that has depth 2. Then find one that has depth 3.

end example

1.4.5 Spanning Trees

A *spanning tree* for a connected graph is a subgraph that is a tree and contains all the vertices of the graph. For example, Figure 1.29 shows a graph followed by two of its spanning trees. This example shows that a graph can have many spanning trees. A *minimal spanning tree* for a connected weighted graph is a spanning tree such that the sum of the edge weights is minimum among all spanning trees.

Figure 1.29   Two spanning trees.

Prim's Algorithm

A famous algorithm, due to Prim [1957], constructs a minimal spanning tree for any undirected connected weighted graph. Starting with any vertex, the algorithm searches for an edge of minimum weight connected to the vertex. It adds the edge to the tree and then continues by trying to find new edges of minimum weight such that one vertex is in the tree and the other vertex is not. Here's an informal description of the algorithm.

Prim's Algorithm

Construct a minimal spanning tree for an undirected connected weighted graph. The variables: V is the vertex set of the graph; W is the vertex set and S is the edge set of the spanning tree.

1. Initialize $S := \varnothing$.

2. Pick any vertex $v \in V$ and set $W := \{v\}$.

3. **while** $W \neq V$ **do**

   Find a minimum weight edge $\{x, y\}$, where $x \in W$ and
   $y \in V - W$;

   $S := S \cup \{\{x, y\}\}$;

   $W := W \cup \{y\}$

   od

Of course, Prim's algorithm can also be used to find a spanning tree for an unweighted graph. Just assign a weight of 1 to each edge of the graph. Or modify the first statement in the **while** loop to read "Find an edge $\{x, y\}$ such that $x \in W$ and $y \in V - W$."

example   **1.44   A Minimal Spanning Tree**

We'll construct a minimal spanning tree for the following weighted graph.

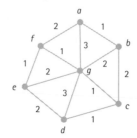

To see how the algorithm works we'll do a trace of each step showing the values of the variables S and W. The algorithm gives us several choices since it is not implemented as a computer program. So we'll start with the letter a since it's the first letter of the alphabet.

S	W
$\{\}$	$\{a\}$
$\{\{a,b\}\}$	$\{a,b\}$
$\{\{a,b\},\{b,c\}\}$	$\{a,b,c\}$
$\{\{a,b\},\{b,c\},\{c,d\}\}$	$\{a,b,c,d\}$
$\{\{a,b\},\{b,c\},\{c,d\},\{c,g\}\}$	$\{a,b,c,d,g\}$
$\{\{a,b\},\{b,c\},\{c,d\},\{c,g\},\{g,f\}\}$	$\{a,b,c,d,g,f\}$
$\{\{a,b\},\{b,c\},\{c,d\},\{c,g\},\{g,f\},\{f,e\}\}$	$\{a,b,c,d,g,f,e\}$

The algorithm stops because $W = V$. So S is a spanning tree.

Exercises

Graphs

1. Draw a picture of a graph that represents those states of the United States and those provinces of Canada that touch the Atlantic Ocean or border states or provinces that do.

2. Find planar graphs with the smallest possible number of vertices that have chromatic numbers of 1, 2, 3, and 4.

3. What is the chromatic number of the graph representing the map of the United States? Explain your answer.

4. Draw a picture of the directed graph that corresponds to each of the following binary relations.

   a. $\{(a,\ a),\ (b,\ b),\ (c,\ c)\}$.
   b. $\{(a,\ b),\ (b,\ b),\ (b,\ c),\ (c,\ a)\}$.
   c. The relation $\leq$ on the set $\{1, 2, 3\}$.

5. Given the following graph:

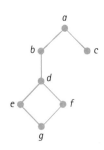

    a. Write down all breadth-first traversals that start at vertex *f*.

    b. Write down all depth-first traversals that start at vertex *f*.

6. Given the following graph.

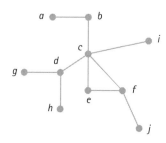

    a. Write down one breadth-first traversal that starts at vertex *f*.

    b. Write down one depth-first traversal that starts at vertex *f*.

7. Given the following graph:

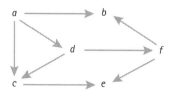

    a. Write down one breadth-first traversal that starts at vertex *a*.

    b. Write down one depth-first traversal that starts at vertex *a*.

8. Convince yourself that in any set of six people there are three people who are either mutual friends or mutual strangers. *Hint:* Consider a complete graph with six vertices, where the vertices are people and an edge is colored red or blue to denote two friends or two strangers. Show that the graph must contain either a red-edged triangle or a blue-edged triangle.

Trees

9. Given the algebraic expression $a \times (b + c) - (d \mathbin{/} e)$. Draw a picture of the tree representation of this expression. Then convert the tree into a list representation of the expression.

10. Draw a picture of the ordered tree that is represented by the list

$$\langle a, \langle b, \langle c \rangle, \langle d, \langle e \rangle \rangle \rangle, \langle r, \langle s \rangle, \langle t \rangle \rangle, \langle x \rangle \rangle.$$

11. Draw a picture of a binary search tree containing the three-letter abbreviations for the 12 months of the year in dictionary order. Make sure that your tree has the least possible depth.

12. Find two distinct minimal spanning trees for the following weighted graph.

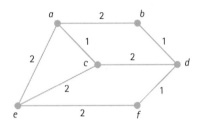

13. For the weighted graph in Example 1.44, find two distinct minimal spanning trees that are different from the spanning tree given.

14. For each case find a binary tree of height 2 with 5 nodes that satisfies the given condition with respect to traversals starting at the root.

    a. Some depth-first traversal is the same as some breadth-first traversal.

    b. Every depth-first traversal is different from every breadth-first traversal.

15. For each case find a binary tree of height 3 with 7 nodes that satisfies the given condition with respect to traversals starting at the root.

    a. Some depth-first traversal is the same as some breadth-first traversal.

    b. Every depth-first traversal is different from every breadth-first traversal.

16. Use examples to convince yourself of the following general facts.

    a. The maximum number of leaves in a binary tree of height n is 2^n.

    b. The maximum number of nodes in a binary tree of height n is $2^{n+1} - 1$.

17. A ternary tree is a tree with at most three children for each node. Use examples to convince yourself of the following general facts.

    a. The maximum number of leaves in a ternary tree of height n is 3^n.

    b. The maximum number of nodes in a ternary tree of height n is $(3^{n+1} - 1)/2$.

Path Problems

18. Try to find a necessary and sufficient condition for an undirected graph to have an Euler path. *Hint:* Look at the degrees of the vertices.

19. Try to find a necessary and sufficient condition for an undirected graph to have an Euler circuit. *Hint:* Look at the degrees of the vertices.

1.5 Chapter Summary

We normally prove things informally, and we use a variety of proof techniques: proof by exhaustive checking, conditional proof, proving the contrapositive, and proof by contradiction.

Sets are characterized by lack of order and no repeated occurrences of elements. There are easy techniques for comparing sets by subset and by equality. Sets can be combined by the operations of union, intersection, difference, and complement. Venn diagrams are useful for representing these operations. Two useful rules for counting sets are the union rule—also called the inclusion-exclusion principle—and the difference rule.

Bags—also called multisets—are characterized by lack of order, and they may contain repeated occurrences of elements.

Tuples are characterized by order, and they may contain repeated occurrences of elements. Many useful structures are related to tuples. Cartesian products of sets are collections of tuples. Lists are similar to tuples except that lists can be accessed only by head and tail. Strings are like lists except that elements from an alphabet are placed next to each other in juxtaposition. Languages are sets of strings and they can be combined by concatenating strings as well as by set operations. Relations are sets of tuples that are related in some way. A useful rule for counting tuples is the product rule.

Graphs are characterized by a set of vertices and a set of edges, where the edges may be undirected or directed. Graphs can be colored, and they can be traversed. Trees are special graphs that look like real trees. Prim's algorithm constructs a minimal spanning tree for an undirected, connected, weighted graph.

chapter 2

Facts about Functions

Leibniz himself attributed all of his mathematical discoveries to improvements in notation.

—Gottfried Wilhelm von Leibniz (1646–1716)[1]
From *The Nature of Mathematics* by Philip E. B. Jourdain

Functions can often make life simpler. In this chapter we'll start with the basic notions and notations for functions. Then we'll study some functions that are especially important in computer science. Since programs can be functions and functions can be programs, we'll spend some time discussing techniques for constructing new functions from simpler ones. We'll discuss properties of functions that are useful in problem solving and we'll see how functions are used to describe the notion of countability.

chapter guide

2.1 (Definitions and Examples) introduces the basic ideas of functions—what they are, and how to represent them. We'll concentrate on studying functions that are especially useful to computer scientists: floor, ceiling, gcd, mod, and log.

2.2 (Constructing Functions) introduces the important idea of composition as a way to combine functions to construct new functions. We'll see that the map function is a useful tool for constructing functions that calculate lists.

2.3 (Properties of Functions) introduces three important properties of functions—injective, surjective, and bijective. We'll see how these properties are used when we discuss the pigeonhole principle, cryptology, and hash functions.

2.4 (Countability) gives a brief introduction to techniques for comparing infinite sets. We'll discuss the ideas of countable and uncountable sets. We'll introduce the diagonalization technique, and we'll discuss whether we can compute everything.

---

[1]Leibniz introduced the word "function" around 1692. He is responsible for such diverse ideas as binary arithmetic, symbolic logic, combinatorics, and calculus. Around 1694 he built a calculating machine that could add and multiply.

2.1 Definitions and Examples

In this section we'll give the definition of a function along with various ways to describe functions. Then we'll get to the main task of studying functions that are very useful in computer science: floor, ceiling, gcd, mod, and log.

2.1.1 Definition of a Function

Suppose A and B are sets and for each element in A we associate exactly one element in B. Such an association is called a *function* from A to B. The main idea is that each element of A is associated with *exactly one* element of B. In other words, if $x \in A$ is associated with $y \in B$, then x is not associated with any other element of B.

Functions are normally denoted by letters like f, g, and h or other descriptive names or symbols. If f is a function from A to B and f associates the element $x \in A$ with the element $y \in B$, then we write $f(x) = y$ or $y = f(x)$. The expression $f(x)$ is read, "f of x," or "f at x," or "f applied to x." When $f(x) = y$, we often say, "f maps x to y." Some other words for "function" are *mapping*, *transformation*, and *operator*.

Describing Functions

Functions can be described in many ways. Sometimes a formula will do the job. For example, the function f from $\mathbb{N}$ to $\mathbb{N}$ that maps every natural number x to its square can be described by the following formula:

$$f(x) = x^2.$$

Other times, we'll have to write down all possible associations. For example, the following associations define a function g from $A = \{a, b, c\}$ to $B = \{1, 2, 3\}$:

$$g(a) = 1, \ g(b) = 1, \text{ and } g(c) = 2.$$

We can also describe a function by drawing a figure. For example, Figure 2.1 shows three ways to represent the function g. The top figure uses Venn diagrams together with a digraph. The lower-left figure is a digraph. The lower-right figure is the familiar Cartesian graph, in which each ordered pair $(x, g(x))$ is plotted as a point.

Figure 2.2 shows two associations that are not functions. Be sure to explain why these associations do not represent functions from A to B.

Terminology

To communicate with each other about functions, we need to introduce some more terminology. If f is a function from A to B, we denote this by writing

$$f : A \to B.$$

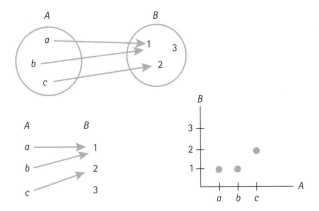

Figure 2.1   Three ways to describe the same function.

Figure 2.2   Two associations that are *not* functions.

The set A is the *domain* of f and the set B is the *codomain* of f. We also say that f has *type* $A \to B$. The expression $A \to B$ denotes the set of all functions from A to B.

If $f(x) = y$, then x is called an *argument* of f, and y is called a *value* of f. If the domain of f is the Cartesian product $A_1 \times \cdots \times A_n$, we say f has *arity* n or f has n *arguments*. In this case, if $(x_1, \ldots, x_n) \in A_1 \times \cdots \times A_n$, then

$$f(x_1, \ldots, x_n)$$

denotes the value of f at $(x_1, \ldots, x_n)$. A function f with two arguments is called a *binary* function and we have the option of writing $f(x, y)$ in the popular *infix* form $x\ f\ y$. For example, $4 + 5$ is usually preferable to $+(4, 5)$.

Ranges, Images, and Pre-Images

At times it is necessary to discuss certain subsets of the domain and codomain of a function $f : A \to B$. The *range* of f, denoted by range(f), is the set of elements in the codomain B that are associated with some element of A. In other words, we have

$$\text{range}(f) = \{f(a) \mid a \in A\} .$$

For any subset $S \subseteq A$, the *image* of S under f, denoted by $f(S)$, is the set of elements in B that are associated with some element of S. In other words, we have

$$f(S) = \{f(x) \mid x \in S\}\,.$$

Notice that we always have the special case $f(A) = \text{range}(f)$. Notice also that images allow us to think of f not only as a function from A to B, but also as a function from power(A) to power(B).

For any subset $T \subseteq B$ the *pre-image* of T under f, denoted by $f^{-1}(T)$, is the set of elements in A that associate with elements of T. In other words, we have

$$f^{-1}(T) = \{a \in A \mid f(a) \in T\}\,.$$

Notice that we always have the special case $f^{-1}(B) = A$. Notice also that pre-images allow us to think of f^{-1} as a function from power(B) to power(A).

---

example **2.1 Sample Notations**

Consider the function $f : \{a,\, b,\, c\} \to \{1,\, 2,\, 3\}$ defined by $f(a) = 1$, $f(b) = 1$, and $f(c) = 2$. We can make the following observations.

> f has type $\{a,\, b,\, c\} \to \{1,\, 2,\, 3\}$.
> The domain of f is $\{a,\, b,\, c\}$.
> The codomain of f is $\{1,\, 2,\, 3\}$.
> The range of f is $\{1,\, 2\}$.

Some sample images are

> $f(\{a\}) = \{1\}$,
> $f(\{a,\, b\}) = \{1\}$,
> $f(A) = f(\{a,\, b,\, c\}) = \{1,\, 2\} = \text{range}(f)$.

Some sample pre-images are

> $f^{-1}(\{1,\, 3\}) = \{a,\, b\}$,
> $f^{-1}(\{3\}) = \varnothing$,
> $f^{-1}(B) = f^{-1}(\{1,\, 2,\, 3\}) = \{a,\, b,\, c\} = A$.

end example

---

example **2.2 Functions and Not Functions**

Let P be the set of all people, alive or dead. We'll make some associations and discuss whether each is a function of type $P \to P$.

1. $f(x)$ is a parent of x.

   In this case f is *not* a function of type $P \rightarrow P$ because people have two parents. For example, if q has mother m and father p, then $f(q) = m$ and $f(q) = p$, which is contrary to the requirement that each domain element be associated with exactly one codomain element.

2. $f(x)$ is the mother of x.

   In this case f *is* a function of type $P \rightarrow P$ because each person has exactly one mother. In other words, each $x \in P$ maps to exactly one person, the mother of x. If m is a mother, what is the pre-image of the set $\{m\}$ under f?

3. $f(x)$ is the oldest child of x.

   In this case f is *not* a function of type $P \rightarrow P$ because some person has no children. Therefore, $f(x)$ is not defined for some $x \in P$.

4. $f(x)$ is the set of all children of x.

   In this case f is *not* a function of type $P \rightarrow P$ because each person is associated with a set of people rather than a person. However, f *is* a function of type $P \rightarrow \text{power}(P)$. Can you see why?

   end example

example 2.3 Tuples Are Functions

Any ordered sequence of objects can be thought of as a function. For example, the tuple (22, 14, 55, 1, 700, 67) can be thought of as a listing of the values of the function

$$f : \{0, 1, 2, 3, 4, 5\} \rightarrow \mathbb{N}$$

where f is defined by the equality

$$(f(0), f(1), f(2), f(3), f(4), f(5)) = (22, 14, 55, 1, 700, 67).$$

Similarly, any infinite sequence of objects can also be thought of as a function. For example, suppose that $(b_0, b_1, \ldots, b_n, \ldots)$ is an infinite sequence of objects from a set S. Then the sequence can be thought of as a listing of values in the range of the function $f : \mathbb{N} \rightarrow S$ defined by $f(n) = b_n$.

end example

example **2.4 Functions and Binary Relations**

Any function can be defined as a special kind of binary relation. A function $f\colon A \to B$ is a binary relation $f \subseteq A \times B$ with the property that for each $a \in A$ there is a unique $b \in B$ such that $(a, b) \in f$. The uniqueness condition can be described as: If $(a, b), (a, c) \in f$, then $b = c$. We normally write $f(a) = b$ instead of $(a, b) \in f$ or $f(a, b)$.

end example

Equality of Functions

Two functions are equal if they have the same type and the same values for each domain element. In other words, if f and g are functions of type $A \to B$, then f and g are said to be *equal* if $f(x) = g(x)$ for all $x \in A$. If f and g are equal, we write

$$f = g.$$

For example, suppose f and g are functions of type $\mathbb{N} \to \mathbb{N}$ defined by the formulas $f(x) = x + x$ and $g(x) = 2x$. It's easy to see that $f = g$.

Defining a Function by Cases

Functions can often be defined by cases. For example, the absolute value function "abs" has type $\mathbb{R} \to \mathbb{R}$, and can be defined by the following rule:

$$\mathrm{abs}(x) = \begin{cases} x & \text{if } x \geq 0 \\ -x & \text{if } x < 0. \end{cases}$$

A definition by cases can also be written in terms of the *if-then-else* rule. For example, we can write the preceding definition in the following form:

$$\mathrm{abs}(x) = \text{if } x \geq 0 \text{ then } x \text{ else } -x.$$

The if-then-else rule can be used more than once if there are several cases to define. For example, suppose we want to classify the roots of a quadratic equation having the following form:

$$ax^2 + bx + c = 0.$$

We can define the function "classifyRoots" to give the appropriate statements as follows:

classifyRoots$(a, b, c) = $ if $b^2 - 4ac > 0$ then
    "The roots are real and distinct."
  else if $b^2 - 4ac < 0$ then
    "The roots are complex conjugates."
  else
    "The roots are real and repeated."

Partial Functions

A *partial function* from A to B is like a function except that it might not be defined for some elements of A. In other words, some elements of A might not be associated with any element of B. But we still have the requirement that if $x \in A$ is associated with $y \in B$, then x can't be associated with any other element of B. For example, we know that division by zero is not allowed. Therefore, $\div$ is a partial function of type $\mathbb{R} \times \mathbb{R} \to \mathbb{R}$ because $\div$ is undefined for all pairs of the form $(x, 0)$.

To avoid confusion when discussing partial functions, we use the term *total function* to mean a function that is defined on all its domain. Any partial function can be transformed into a total function. One simple technique is to shrink the domain to the set of elements for which the partial function is defined. For example, $\div$ is a total function of type $\mathbb{R} \times (\mathbb{R} - \{0\}) \to \mathbb{R}$.

A second technique keeps the domain the same but increases the size of the codomain. For example, suppose $f : A \to B$ is a partial function. Pick some symbol that is not in B, say $\# \notin B$, and assign $f(x) = \#$ whenever $f(x)$ is not defined. Then we can think of f as a total function of type $A \to B \cup \{\#\}$. In programming, the analogy would be to pick an error message to indicate that an incorrect input string has been received.

2.1.2   Floor and Ceiling Functions

Let's discuss two important functions that "integerize" real numbers by going down or up to the nearest integer. The *floor* function has type $\mathbb{R} \to \mathbb{Z}$ and is defined by setting $\text{floor}(x)$ to the closest integer less than or equal to x. For example, $\text{floor}(8) = 8$, $\text{floor}(8.9) = 8$, and $\text{floor}(-3.5) = -4$. A useful shorthand notation for $\text{floor}(x)$ is

$$\lfloor x \rfloor.$$

The *ceiling* function also has type $\mathbb{R} \to \mathbb{Z}$ and is defined by setting $\text{ceiling}(x)$ to the closest integer greater than or equal to x. For example, $\text{ceiling}(8) = 8$, $\text{ceiling}(8.9) = 9$, and $\text{ceiling}(-3.5) = -3$. The shorthand notation for $\text{ceiling}(x)$ is

$$\lceil x \rceil.$$

Figure 2.3 gives a few sample values for the floor and ceiling functions.

x	-2.0	-1.7	-1.3	-1.0	-0.7	-0.3	0.0	0.3	0.7	1.0	1.3	1.7	2.0
$\lfloor x \rfloor$	-2	-2	-2	-1	-1	-1	0	0	0	1	1	1	2
$\lceil x \rceil$	-2	-1	-1	-1	0	0	0	1	1	1	2	2	2

Figure 2.3   Some floor and ceiling values.

One consequence of the definitions for floor and ceiling is that if $x < y$, then $\lfloor x \rfloor \leq \lfloor y \rfloor$ and $\lceil x \rceil \leq \lceil y \rceil$. The following characterizations are also direct consequences of the definitions, where x is a real number and n is an integer.

$$\lfloor x \rfloor = n \quad \text{iff} \quad n \leq x < n+1 \quad \text{iff} \quad x - 1 < n \leq x.$$
$$\lceil x \rceil = n \quad \text{iff} \quad n - 1 < x \leq n \quad \text{iff} \quad x \leq n < x + 1.$$

Properties of Floor and Ceiling

There are many interesting and useful properties of floor and ceiling, some of which are relationships between floor and ceiling. For example,

$$\lfloor x \rfloor = \lceil x \rceil \text{ if and only if } x \text{ is an integer.}$$

If x is not an integer, then there is some integer n such that $n < x < n+1$. Therefore, $\lfloor x \rfloor = n$ and $\lceil x \rceil = n + 1 = \lfloor x \rfloor + 1$. So we can say that

$$\lceil x \rceil = \lfloor x \rfloor + 1 \text{ if and only if } x \notin \mathbb{Z}.$$

Floor and ceiling can be defined in terms of each other by observing the following equality, where x is any real number.

$$\lceil -x \rceil = -\lfloor x \rfloor.$$

So for each floor property there is a similar property for ceiling.

There are an enormous number of interesting and useful properties that involve floor and ceiling. The basic properties in the following list can be quite useful in determining other properties. We'll leave them as exercises and we'll use some of them in the next two examples.

Floor and Ceiling Properties   (2.1)
The following properties hold when x is a real number and n is an integer.

Floor Properties

a. $\lfloor x + n \rfloor = \lfloor x \rfloor + n$.

b. $\lfloor x \rfloor < n$ iff $x < n$.

c. $n \leq \lfloor x \rfloor$ iff $n \leq x$.

Ceiling Properties

d. $\lceil x + n \rceil = \lceil x \rceil + n$.

e. $n < \lceil x \rceil$ iff $n < x$.

f. $\lceil x \rceil \leq n$ iff $x \leq n$.

---

example 2.5 A Divide and Conquer Equality

We can often solve a problem by dividing it up into two smaller problems. For example, we might try to find the largest number in a list of numbers by splitting the list into two sublists of about equal size. Then we can have two people (or algorithms) work on the lists separately to find the largest number in each sublist after which we can compare the two numbers to find the largest. For example, if the size of the list is 15, then the sublists should have sizes 7 and 8. It's nice to know that we can obtain 7 and 8 from 15 by calculating the floor and ceiling of $15/2$. This is an application of the following equality, where n is any integer.

$$n = \lfloor n/2 \rfloor + \lceil n/2 \rceil.$$

To see that this equality holds, we'll consider two cases. If n is even, then $n = 2k$ for some integer k. So we have $\lfloor n/2 \rfloor = \lfloor 2k/2 \rfloor = \lfloor k \rfloor = k$. Similarly, $\lceil n/2 \rceil = k$. So the equation holds. If n is odd, then $n = 2k+1$ for some integer k. In this case, we have $\lfloor n/2 \rfloor = \lfloor (2k+1)/2 \rfloor = \lfloor k + 1/2 \rfloor$. Now use property (2.1a) to obtain $\lfloor k + 1/2 \rfloor = k + \lfloor 1/2 \rfloor = k + 0 = k$. Similarly, we have $\lceil n/2 \rceil = \lceil (2k+1)/2 \rceil = \lceil k + 1/2 \rceil$. Now use property (2.1d) to obtain $\lceil k + 1/2 \rceil = k + \lceil 1/2 \rceil = k + 1$. So the equation holds in this case, too. QED.

end example

example 2.6 A Property of Fractions

The following two properties hold for floor and ceiling, where x is a real number and p is a positive integer.

$$\lfloor \lfloor x \rfloor /p \rfloor = \lfloor x/p \rfloor \text{ and } \lceil \lceil x \rceil /p \rceil = \lceil x/p \rceil.$$

We'll prove the floor property. If x is an integer, the equation holds. So assume x is not an integer. Then $\lfloor x \rfloor < x$. Since p is positive, we have $\lfloor x \rfloor /p < x/p$ and so it follows that $\lfloor \lfloor x \rfloor /p \rfloor \le \lfloor x/p \rfloor$. Now assume, by way of contradiction, that $\lfloor \lfloor x \rfloor /p \rfloor < \lfloor x/p \rfloor$. Then we can use (2.1b) to conclude that $\lfloor x \rfloor /p < \lfloor x/p \rfloor$. So we have $\lfloor x \rfloor /p < \lfloor x/p \rfloor \le x/p$. Now multiply the inequality by p to obtain the inequality $\lfloor x \rfloor < p\lfloor x/p \rfloor \le x$. But $p\lfloor x/p \rfloor$ is an integer. So the inequality tells us that we have an integer larger than $\lfloor x \rfloor$ and still less than or equal to x, contrary to the definition of $\lfloor x \rfloor$. QED.

end example

2.1.3 Greatest Common Divisor

Let's recall from Section 1.1 that an integer d *divides* an integer n if $d \ne 0$ and there is an integer k such that $n = dk$, and we denote this fact with $d \mid n$. Our focus here will be on the largest of all common divisors for two integers.

Definiton of Greatest Common Divisor

The *greatest common divisor* of two integers, not both zero, is the largest integer that divides them both. We denote the greatest common divisor of a and b by

$$\gcd(a,\, b).$$

For example, the common divisors of 12 and 18 are ± 1, ± 2, ± 3, ± 6. So the greatest common divisor of 12 and 18 is 6, and we write $\gcd(12,\, 18) = 6$. Other examples are $\gcd(-44,\, -12) = 4$ and $\gcd(5,\, 0) = 5$. If $a \neq 0$, then $\gcd(a,\, 0) = |a|$. An important and useful special case occurs when $\gcd(a,\, b) = 1$. In this case a and b are said to be *relatively prime*. For example, 9 and 4 are relatively prime.

Properties of GCD

The following properties give some basic facts about the greatest common divisor function. We'll discuss them below and also in the exercises.

Greatest Common Divisor Properties   (2.2)

a. $\gcd(a,\, b) = \gcd(b,\, a) = \gcd(a,\, -b)$.

b. $\gcd(a,\, b) = \gcd(b,\, a - bq)$ for any integer q.

c. If $g = \gcd(a,\, b)$, then there are integers x and y such that $g = ax + by$.

d. If $d \mid ab$ and $\gcd(d,\, a) = 1$, then $d \mid b$.

e. If c is a common divisor of a and b, then c divides $\gcd(a,\, b)$.

Property (2.2a) confirms that the ordering of the arguments doesn't matter and that negative numbers have positive greatest common divisors. For example, $\gcd(-4,\, -6) = \gcd(-4,\, 6) = \gcd(6,\, -4) = \gcd(6,\, 4) = 2$. We'll see shortly how property (2.2b) can help us compute greatest common divisors. Property (2.2c) says that we can write $\gcd(a,\, b)$ as a linear combination of a and b. For example, $\gcd(15,\, 9) = 3$, and we can write 3 in terms of 15 and 9 as

$$3 = \gcd(15,\, 9) = 15(2) + 9(-3).$$

We'll see how to do this in Example 2.7. Properties (2.2d) and (2.2e) are divisibility properties that we'll be using later.

Computing GCD

Now let's get down to brass tacks and describe an algorithm to compute the greatest common divisor. Most of us recall from elementary school that we can divide an integer a by a nonzero integer b to obtain two other integers, a quotient q and a remainder r, which satisfy an equation like the following:

$$a = bq + r.$$

For example, if $a = -16$ and $b = 3$, then we can write many equations, each with different values for q and r. For example, the following four equations all have the form $a = bq + r$:

$$-16 = 3 \cdot (-4) + (-4)$$
$$-16 = 3 \cdot (-5) + (-1)$$
$$-16 = 3 \cdot (-6) + 2$$
$$-16 = 3 \cdot (-7) + 5.$$

In mathematics and computer science the third equation is by far the most useful. In fact it's a result of a theorem called the *division algorithm*, which we'll state for the record.

Division Algorithm
If a and b are integers and $b \neq 0$, then there are unique integers q and r such that $a = bq + r$, where $0 \leq r < |b|$.

The division algorithm together with property (2.2b) gives us the seeds of an algorithm to compute greatest common divisors. Suppose a and b are integers and $b \neq 0$. The division algorithm gives us the equation $a = bq + r$, where $0 \leq r < |b|$. Solving the equation for r gives $r = a - bq$. This fits the form of (2.2b). So we have the nice equation

$$\gcd(a, b) = \gcd(b, a - bq) = \gcd(b, r).$$

The important point about this equation is that the numbers in $\gcd(b, r)$ are getting closer to zero. Let's see how we can use this equation to compute the greatest common divisor. For example, to compute $\gcd(315, 54)$, we apply the division algorithm to obtain the equation $315 = 54 \cdot 5 + 45$. Thus we know that

$$\gcd(315, 54) = \gcd(54, 45).$$

Now apply the division algorithm again to obtain $54 = 45 \cdot 1 + 9$. So we have

$$\gcd(315, 54) = \gcd(54, 45) = \gcd(45, 9).$$

Continuing, we have $45 = 9 \cdot 5 + 0$, which extends our computation to

$$\gcd(315, 54) = \gcd(54, 45) = \gcd(45, 9) = \gcd(9, 0) = 9.$$

The algorithm that we have been demonstrating is called *Euclid's algorithm*. Since greatest common divisors are always positive, we'll describe the algorithm to calculate $\gcd(a, b)$ for the case in which a and b are natural numbers that are not both zero.

Euclid's Algorithm (2.3)

Input natural numbers a and b, not both zero, and output $\gcd(a, b)$.

while $b > 0$ **do**

    Construct $a = bq + r$, where $0 \leq r < b$;   (by the division algorithm)

    $a := b$;

    $b := r$

od;

Output a.

We can use Euclid's algorithm to show how property (2.2c) is satisfied. The idea is to keep track of the equations $a = bq + r$ from each execution of the loop. Then work backward through the equations to solve for $\gcd(a, b)$ in terms of a and b. Here's an example.

example 2.7 Writing gcd(*a*, *b*) as a Linear Combination

We'll write $\gcd(315, 54)$ as a linear combination of 315 and 54 to demonstrate property (2.2c). In our preceding calculation of $\gcd(315, 54)$ we obtained the three equations

$$315 = 54 \cdot 5 + 45$$
$$54 = 45 \cdot 1 + 9$$
$$45 = 9 \cdot 5 + 0.$$

Starting with the second equation, we can solve for 9. Then we can use the first equation to replace 45. The result is an expression for $9 = \gcd(315, 54)$ written in terms of 315 and 54 as $9 = 315 \cdot (-1) + 54 \cdot 6$.

end example

2.1.4 The Mod Function

If a and b are integers, where $b > 0$, then the division algorithm states that there are two unique integers q and r such that

$$a = bq + r \qquad \text{where} \qquad 0 \leq r < b.$$

We say that q is the quotient and r is the remainder upon division of a by b. The remainder $r = a - bq$ is the topic of interest.

Definition of the Mod Function

If a and b are integers with $b > 0$, then the remainder upon the division of a by b is denoted

$$a \bmod b$$

---

x	0	1	2	3	4	5	6	7	8	9	10	11	12	13	14	15
x mod 1	0	0	0	0	0	0	0	0	0	0	0	0	0	0	0	0
x mod 2	0	1	0	1	0	1	0	1	0	1	0	1	0	1	0	1
x mod 3	0	1	2	0	1	2	0	1	2	0	1	2	0	1	2	0
x mod 4	0	1	2	3	0	1	2	3	0	1	2	3	0	1	2	3
x mod 5	0	1	2	3	4	0	1	2	3	4	0	1	2	3	4	0

Figure 2.4   Sample values of the mod function.

If we agree to fix n as a positive integer, then x mod n takes values in the set $\{0, 1, \ldots, n-1\}$, which is the set of possible remainders obtained upon division of any integer x by n. For example, each row of the table in Figure 2.4 gives some sample values for x mod n.

We sometimes let $\mathbb{N}_n$ denote the set

$$\mathbb{N}_n = \{0, 1, 2, \ldots, n-1\}.$$

For example, $\mathbb{N}_0 = \varnothing$, $\mathbb{N}_1 = \{0\}$, and $\mathbb{N}_2 = \{0, 1\}$. So for fixed n, the function f defined by $f(x) = x$ mod n has type $\mathbb{Z} \to \mathbb{N}_n$.

A Formula for Mod

Can we find a formula for a mod b in terms of a and b? Sure. We have the following formula

$$a \text{ mod } b = r = a - bq, \qquad \text{where} \qquad 0 \le r < b.$$

So we'll have a formula for a mod b if we can find a formula for the quotient q in terms of a and b. Starting with the inequality $0 \le r < b$ we have the following sequence of inequalities.

$$0 \le r < b,$$
$$-b < -r \le 0,$$
$$a - b < a - r \le a,$$
$$\frac{a-b}{b} < \frac{a-r}{b} \le \frac{a}{b},$$
$$\frac{a}{b} - 1 < \frac{a-r}{b} \le \frac{a}{b},$$
$$\frac{a}{b} - 1 < q \le \frac{a}{b}, \qquad \text{since } q = \frac{a-r}{b}.$$

Since q is an integer, the last inequality implies that q can be written as the floor expression

$$q = \lfloor a/b \rfloor.$$

Since $r = a - bq$, we have the following representation of r when $b > 0$.

$$r = a - b\lfloor a/b \rfloor.$$

This gives us a formula for the mod function.

Formula for the Mod Function

$$a \bmod b = a - b\lfloor a/b \rfloor.$$

Properties of Mod

The mod function has many properties. For example, the definition of mod tells us that $0 < x \bmod n < n$ for any integer x. So for any integer x we have

$$(x \bmod n) \bmod n = x \bmod n$$

and

$$x \bmod n = x \text{ iff } 0 \le x < n.$$

The following list contains some of the most useful properties of the mod function.

Mod Function Properties   (2.4)

   a. $x \bmod n = y \bmod n$ iff n divides $x - y$ iff $(x - y) \bmod n = 0$.

   b. $(x + y) \bmod n = ((x \bmod n) + (y \bmod n)) \bmod n$.

   c. $(xy) \bmod n = ((x \bmod n)(y \bmod n)) \bmod n$.

   d. If $ax \bmod n = ay \bmod n$ and $\gcd(a, n) = 1$, then $x \bmod n = y \bmod n$.

   e. If $\gcd(a, n) = 1$, then $1 \bmod n = ax \bmod n$ for some integer x.

We'll prove Parts (a) and (d) and discuss the other properties in the exercises. Using the definition of mod we can write

$$x \bmod n = x - nq_1 \text{ and } y \bmod n = y - nq_2$$

for some integers q_1 and q_2. Now we have a string of iff statements.

$$
\begin{aligned}
x \bmod n = y \bmod n \quad &\text{iff} \quad x - nq_1 = y - nq_2 \\
&\text{iff} \quad x - y = n(q_1 - q_2) \\
&\text{iff} \quad n \text{ divides } (x - y) \\
&\text{iff} \quad (x - y) \bmod n = 0.
\end{aligned}
$$

So Part (a) is true.

For Part (d), assume that $ax \bmod n = ay \bmod n$ and $\gcd(a, n) = 1$. By Part (a) we can say that n divides $(ax - ay)$. So n divides the product $a(x - y)$.

Since $\gcd(a, n) = 1$, it follows from (2.2d) that n divides $(x - y)$. So again by Part (a) we have $x \bmod n = y \bmod n$. QED.

example 2.8 Converting Decimal to Binary

How can we convert a decimal number to binary? For example, the decimal number 53 has the binary representation 110101. The rightmost bit (binary digit) in this representation of 53 is 1 because 53 is an odd number. In general, we can find the rightmost bit (binary digit) of the binary representation of a natural decimal number x by evaluating the expression $x \bmod 2$. In our example, 53 mod $2 = 1$, which is the rightmost bit.

So we can apply the division algorithm, dividing 53 by 2, to obtain the rightmost bit as the remainder. This gives us the equation

$$53 = 2 \cdot 26 + 1.$$

Now do the same thing for the quotient 26 and the succeeding quotients.

$$
\begin{aligned}
53 &= 2 \cdot 26 + 1 \\
26 &= 2 \cdot 13 + 0 \\
13 &= 2 \cdot 6 + 1 \\
6 &= 2 \cdot 3 + 0 \\
3 &= 2 \cdot 1 + 1 \\
1 &= 2 \cdot 0 + 1 \\
0. & \quad \text{(done)}
\end{aligned}
$$

We can read off the remainders in the above equations from bottom to top to obtain the binary representation 110101 for 53. An important point to notice is that we can represent any natural number x in the form

$$x = 2\lfloor x/2 \rfloor + x \bmod 2.$$

So an algorithm to convert x to binary can be implemented with the floor and mod functions.

end example

2.1.5 The Log Function

The "log" function—which is shorthand for logarithm—measures the size of exponents. We start with a positive real number $b \neq 1$. If x is a positive real number, then

$$\log_b x = y \text{ means } b^y = x,$$

and we say, "log base b of x is y."

Figure 2.5   Sample binary tree.

x	1	2	4	8	16	32	64	128	256	512	1024
$\log_2 x$	0	1	2	3	4	5	6	7	8	9	10

Figure 2.6   Sample log values.

The base-2 log function $\log_2$ occurs frequently in computer science because many algorithms make binary decisions (two choices) and binary trees are useful data structures. For example, suppose we have a binary search tree with 16 nodes having the structure shown in Figure 2.5. The depth of the tree is 4, so a maximum of 5 comparisons are needed to find any element in the tree. Notice in this case that $16 = 2^4$, so we can write the depth in terms of the number of nodes: $4 = \log_2 16$.

Figure 2.6 gives a few choice values for the $\log_2$ function. Of course, $\log_2$ takes real values also. For example, if $8 < x < 16$, then

$$3 < \log_2 x < 4.$$

For any real number $b > 1$, the function $\log_b$ is an increasing function with the positive real numbers as its domain and the real numbers as its range. In this case the graph of $\log_b$ has the general form shown in Figure 2.7. What does the graph look like if $0 < b < 1$?

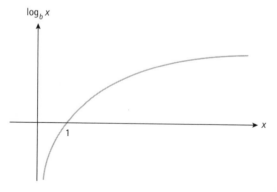

Figure 2.7   Graph of a log function.

Properties of Log

The log function has many properties. For example, it's easy to see that

$$\log_b 1 = 0 \text{ and } \log_b b = 1.$$

The following list contains some of the most useful properties of the log function. We'll leave the proofs as exercises in applying the definition of log.

Log Function Properties   (2.5)

a. $\log_b (b^x) = x$.

b. $\log_b (x \, y) = \log_b x + \log_b y$.

c. $\log_b (x^y) = y \log_b x$.

d. $\log_b (x/y) = \log_b x - \log_b y$.

e. $\log_a x = (\log_a b)(\log_b x)$.   (change of base)

f. $b^{\log_b x} = x$.

g. $a^{\log_b c} = c^{\log_b a}$.

These properties have many simple consequences. For example, if we let $x = 1$ in (2.5d), we obtain $\log_b(1/y) = -\log_b y$. And if we let $x = a$ in (2.5e), we obtain $1 = (\log_a b)(\log_b a)$. The next example shows how to use the properties to find approximate values for log expressions.

> example **2.9 Estimating a Log**

Suppose we need to evaluate the expression $\log_2(2^7 3^4)$. Make sure you can justify each step in the following evaluation:

$$\log_2(2^7 3^4) = \log_2(2^7) + \log_2(3^4) = 7 \log_2(2) + 4 \log_2(3) = 7 + 4 \log_2(3).$$

Now we can easily estimate the value of the expression. We know that $1 = \log_2(2) < \log_2(3) < \log_2(4) = 2$. Therefore, $1 < \log_2(3) < 2$. Thus we have the following estimate of the answer:

$$11 < \log_2(2^7 3^4) < 15.$$

> end example

The Natural Logarithm

From calculus the number $e = 2.71828\ldots$ is the irrational number with the property that e^x is its own derivative and $\log_e x$ has derivative $1/x$. The base-e logarithm is called the *natural logarithm* and is most often denoted by

$$\ln x.$$

The number e can be approximated in many ways. For example, it is the limit of a sequence of numbers $2, (3/2)^2, (4/3)^3, (5/4)^4, (6/5)^5, \ldots, (1 + 1/n)^n, \ldots,$ which is expressed as

$$e = \lim_{n \to \infty} \left(1 + \frac{1}{n}\right)^n.$$

 Exercises

Definitions and Examples

1. Describe all possible functions for each of the following types.

   a. $\{a, b\} \to \{1\}$.

   b. $\{a\} \to \{1, 2, 3\}$.

   c. $\{a, b\} \to \{1, 2\}$.

   d. $\{a, b, c\} \to \{1, 2\}$.

2. Suppose we have a function $f : \mathbb{N} \to \mathbb{N}$ defined by $f(x) = 2x + 1$. Describe each of the following sets, where E and O denote the sets of even and odd natural numbers, respectively.

   a. range(f).   **b.** $f(E)$.   **c.** $f(O)$.

   d. $f(\varnothing)$.   **e.** $f^{-1}(E)$.   **f.** $f^{-1}(O)$.

Floor, Ceiling, GCD, Mod, and Log

3. Evaluate each of the following expressions.

   a. $\lfloor -4.1 \rfloor$.   **b.** $\lceil -4.1 \rceil$.   **c.** $\lfloor 4.1 \rfloor$.   **d.** $\lceil 4.1 \rceil$.

4. Find an integer n and a number x such that $n\lfloor x \rfloor \neq \lfloor nx \rfloor$ and $n\lceil x \rceil \neq \lceil nx \rceil$.

5. Find numbers x and y such that $\lfloor x+y \rfloor \neq \lfloor x \rfloor + \lfloor y \rfloor$ and $\lceil x+y \rceil \neq \lceil x \rceil + \lceil y \rceil$.

6. Evaluate each of the following expressions.

   a. gcd(-12, 15).   **b.** gcd(98, 35).   **c.** gcd(872, 45).

7. Find gcd(296, 872) and write the answer in the form $296x + 872y$.

8. Evaluate each of the following expressions.

   a. 15 mod 12.   **b.** -15 mod 12.

   c. -12 mod 15.   **d.** -21 mod 15.

9. Let $f : \mathbb{N}_6 \to \mathbb{N}_6$ be defined by $f(x) = 2x$ mod 6. Find the image under f of each of the following sets:

   a. $\varnothing$.   **b.** $\{0, 3\}$.   **c.** $\{2, 5\}$.

   d. $\{3, 5\}$.   **e.** $\{1, 2, 3\}$.   **f.** $\mathbb{N}_6$.

10. For a real number x, let $\text{trunc}(x)$ denote the truncation of x, which is the integer obtained from x by deleting the part of x to the right of the decimal point.

   a. Write the floor function in terms of trunc.

   b. Write the ceiling function in terms of trunc.

11. For integers x and $y \neq 0$, let $f(x, y) = x - y\,\text{trunc}(x/y)$, where trunc is from Exercise 10. How does f compare to the mod function?

12. Does it make sense to extend the definition of the mod function to real numbers? What would be the range of the function f defined by $f(x) = x \bmod 2.5$?

13. Use properties of log to evaluate each of the following expressions.

   a. $\log_5 625$.
   b. $\log_2 8192$.
   c. $\log_3 (1/27)$.
   d. $4^{\log_2 5}$.
   e. $2(1/5)^{\log_5 2}$.
   f. $\log_{\sqrt{2}} 2$.

14. Use properties of functions to verify each of the following equations.

   a. $(a/b)^{\log_b c} = (1/c)^{a^{\log_b c}}$.
   b. $a^k (n/b^k)^{\log_b a} = n^{\log_b a}$.
   c. $a^k (n/b^{2k})^{\log_b a} = (n/b^k)^{\log_b a}$.
   d. $\lfloor \log_b x \rfloor + 1 = \lfloor \log_b bx \rfloor$.

Proofs and Challenges

15. For a subset S of a set U, the *characteristic* function $\chi_S : U \rightarrow \{0, 1\}$ is a test for membership in S and is defined by

$$\chi_S(x) = \text{if } x \in S \text{ then } 1 \text{ else } 0.$$

   a. Verify that the following equation is correct for subsets A and B of U.

$$\chi_{A \cup B}(x) = \chi_A(x) + \chi_B(x) - \chi_A(x)\chi_B(x)$$

   b. Find a formula for $\chi_{A \cap B}(x)$ in terms of $\chi_A(x)$ and $\chi_B(x)$.

   c. Find a formula for $\chi_{A - B}(x)$ in terms of $\chi_A(x)$ and $\chi_B(x)$.

16. Given a function $f : A \rightarrow A$. An element $a \in A$ is called a *fixed point* of f if $f(a) = a$. Find the set of fixed points for each of the following functions.

   a. $f : A \rightarrow A$ where $f(x) = x$.
   b. $f : \mathbb{N} \rightarrow \mathbb{N}$ where $f(x) = x + 1$.
   c. $f : \mathbb{N}_6 \rightarrow \mathbb{N}_6$ where $f(x) = 2x \bmod 6$.
   d. $f : \mathbb{N}_6 \rightarrow \mathbb{N}_6$ where $f(x) = 3x \bmod 6$.

17. Prove that $\lceil -x \rceil = -\lfloor x \rfloor$.

18. Prove each of the following statements, where x is a real number and n is an integer.

    a. $\lfloor x + n \rfloor = \lfloor x \rfloor + n$.

    b. $\lceil x + n \rceil = \lceil x \rceil + n$.

19. Prove that $\lceil n/2 \rceil = \lfloor (n+1)/2 \rfloor$ for n a natural number.

20. Prove each of the following statements about inequalities with the floor and ceiling, where x is a real number and n is an integer.

    a. $\lfloor x \rfloor < n$ iff $x < n$.

    b. $n < \lceil x \rceil$ iff $n < x$.

    c. $n \leq \lfloor x \rfloor$ iff $n \leq x$.

    d. $\lceil x \rceil \leq n$ iff $x \leq n$.

21. Prove each of the following properties for any numbers x and y

    a. $\lfloor x \rfloor + \lfloor y \rfloor \leq \lfloor x + y \rfloor$.

    b. $\lceil x \rceil + \lceil y \rceil \geq \lceil x + y \rceil$.

22. Find the values of x that satisfy each of the following equations.

    a. $\lceil (x-1)/2 \rceil = \lfloor x/2 \rfloor$.

    b. $\lceil (x-1)/3 \rceil = \lfloor x/3 \rfloor$.

23. Given a polynomial $p(x) = a_0 + a_1 x + a_2 x^2 + \ldots + a_n x^n$, where each coefficient a_i is a nonnegative integer. If we know the value $p(b)$ for some b that is larger than the sum of the coefficients, then we can find all the coefficients by using the division algorithm. Find $p(x)$ for each of the following cases where $p(b)$ is given. *Hint:* Look at Example 2.8, but use b in place of 2.

    a. $p(7) = 162$.

    b. $p(4) = 273$.

    c. $p(6) = 180$.

    d. $p(20) = 32000502$.

24. Use the definition of the logarithm function to prove each of the following facts.

    a. $\log_b(b^x) = x$.

    b. $\log_b(xy) = \log_b x + \log_b y$.

    c. $\log_b(x^y) = y \log_b x$.

    d. $\log_b(x/y) = \log_b x - \log_b y$.

    e. $\log_a x = (\log_a b)(\log_b x)$. (change of base)

    f. $b^{\log_b x} = x$.

    g. $a^{\log_b c} = c^{\log_b a}$.

25. Prove each of the following facts about greatest common divisors.

    a. $\gcd(a, b) = \gcd(b, a) = \gcd(a, -b)$.

    b. $\gcd(a, b) = \gcd(b, a - bq)$ for any integer q.

    c. If $d \,|\, ab$ and $\gcd(d, a) = 1$, then $d \,|\, b$.   *Hint:* Use (2.2c).

    d. If c is a common divisor of a and b, then c divides $\gcd(a, b)$. *Hint:* Use (2.2c).

26. Given the result of the division algorithm $a = bq + r$, where $0 \le r < |b|$, prove the following statement:

$$\text{If } b < 0 \text{ then } r = a - b\lceil a/b\rceil.$$

27. Let $f : A \to B$ be a function, and let E and F be subsets of A. Prove each of the following facts about images.

    a. $f(E \cup F) = f(E) \cup f(F)$.

    b. $f(E \cap F) \subseteq f(E) \cap f(F)$.

    c. Find an example to show that Part (b) can be a proper subset.

28. Let $f : A \to B$ be a function, and let G and H be subsets of B. Prove each of the following facts about pre-images.

    a. $f^{-1}(G \cup H) = f^{-1}(G) \cup f^{-1}(H)$.

    b. $f^{-1}(G \cap H) = f^{-1}(G) \cap f^{-1}(H)$.

    c. $E \subseteq f^{-1}(f(E))$.

    d. $f(f^{-1}(G)) \subseteq G$.

    e. Find examples to show that Parts (c) and (d) can be a proper subsets.

29. Prove each of the following properties of the mod function. *Hint:* Use (2.4a) for Parts (a) and (b), and use (2.2c) and Parts (a) and (b) for Part (c).

    a. $(x + y) \bmod n = ((x \bmod n) + (y \bmod n)) \bmod n$.

    b. $(xy) \bmod n = ((x \bmod n)(y \bmod n)) \bmod n$.

    c. If $n > 0$ and $\gcd(a, n) = 1$, then there is an integer x such that $1 = ax \bmod n$.

30. We'll start a proof of (2.2c): If $g = \gcd(a, b)$, then there are integers x and y such that $g = ax + by$. Proof: Let $S = \{ax + by \mid x, y \in \mathbb{Z} \text{ and } ax + by > 0\}$ and let d be the smallest number in S. Then there are integers x and y such that $d = ax + by$. The idea is to show that $g = d$. Since $g \,|\, a$ and $g \,|\, b$, it follows from (1.1b) that $g \,|\, d$. So $g \le d$. If we can show that $d \,|\, a$ and $d \,|\, b$, then we must conclude that $d = g$ because g is the greatest common divisor of a and b. Finish the proof by showing that $d \,|\, a$ and $d \,|\, b$. *Hint:* To show $d \,|\, a$, write $a = dq + r$, where $0 \le r < d$. Argue that r must be 0.

31. For each natural number n, let $D_n = \{d \mid d \in \mathbb{N} \text{ and } d \text{ divides } n \text{ with no remainder}\}$. Show that if $m, n \in \mathbb{N}$ and $d = \gcd(m, n)$, then $D_m \cap D_n = D_d$.

32. Show that any common multiple of two integers is a multiple of the least common multiple of the two integers. In other words, in terms of divisibility, if $x \mid n$ and $y \mid n$ and m is the smallest number such that $x \mid m$ and $y \mid m$, then $m \mid n$.

33. For each natural number n, let $M_n = \{kn \mid k \in \mathbb{N}\}$. Show that if $m, n \in \mathbb{N}$ and k is the least common multiple of m and n, then there is $k \in \mathbb{N}$ such that $M_m \cap M_n = M_k$. *Hint:* You might want to use the result of Exercise 25.

34. The following equalities hold when x is a real number and p is a positive integer.

$$\lfloor \lfloor x \rfloor /p \rfloor = \lfloor x/p \rfloor \text{ and } \lceil \lceil x \rceil /p \rceil = \lceil x/p \rceil.$$

We proved the first equality in Example 2.6 by way of contradiction. An alternative direct proof can be written using the division algorithm. Fill in the missing parts of the following argument for Part (a): If x is an integer, the equation holds. So assume x is not an integer. Then let $n = \lfloor x \rfloor$ and it follows that $n < x < n + 1$. So $0 < x - n < 1$. Use the division algorithm to write $n = pq + r$, where $0 \le r < p$, which we can write as $0 \le r \le p - 1$. Divide the equation by p to obtain $n/p = q + r/p$, where $0 \le r/p \le (p-1)/p$. Then we can write $x = n + (x - n)$ and divide this equation by p to obtain the equation $x/p = n/p + (x - n)/p$. Now substitute for n/p to obtain

$$x/p = q + r/p + x - n)/p.$$

Take the floor to get $\lfloor x/p \rfloor = \lfloor q + r/p + (x-n)/p \rfloor = q + \lfloor r/p + (x-n)/p \rfloor$. Finish the proof by showing that $\lfloor r/p + (x-n)/p \rfloor = 0$ and $\lfloor x/p \rfloor = q = \lfloor n/p \rfloor = \lfloor \lfloor x \rfloor /p \rfloor$. For practice, write a proof of the ceiling property.

35. Prove each of the following properties of floor, where x is a real number.

   a. $\lfloor x \rfloor + \lfloor x + 1/2 \rfloor = \lfloor 2x \rfloor$.
   b. $\lfloor x \rfloor + \lfloor x + 1/3 \rfloor + \lfloor x + 2/3 \rfloor = \lfloor 3x \rfloor$.
   c. State and prove a generalization of Parts (a) and (b).

2.2 Constructing Functions

We often construct a new function by combining other simpler functions in some way. The combining method that we'll discuss in this section is called composition. We'll see that it is a powerful tool to create new functions. We'll also introduce the map function as a useful tool for displaying a list of values of a function. Many programming systems and languages rely on the ideas of this section.

2.2.1 Composition of Functions

Composition of functions is a natural process that we often use without even thinking. For example, the expression floor($\log_2(5)$) involves the composition of the two functions floor and $\log_2$. To evaluate the expression, we first evaluate $\log_2(5)$, which is a number between 2 and 3. Then we apply the floor function to this number, obtaining the value 2.

Definition of Composition

The *composition* of two functions f and g is the function denoted by $f \circ g$ and defined by

$$(f \circ g)(x) = f(g(x)).$$

Notice that composition makes sense only for values of x in the domain of g such that $g(x)$ is in the domain of f. So if $g : A \to B$ and $f : C \to D$ and $B \subseteq C$, then the composition $f \circ g$ makes sense. In other words, for every $x \in A$ it follows that $g(x) \in B$, and since $B \subseteq C$ it follows that $f(g(x)) \in D$. It also follows that $f \circ g : A \to D$.

For example, we'll consider the floor and $\log_2$ functions. These functions have types

$$\log_2 : \mathbb{R}^+ \to \mathbb{R} \text{ and floor} : \mathbb{R} \to \mathbb{Z},$$

where $\mathbb{R}^+$ denotes the set of positive real numbers. So for any positive real number x, the expression $\log_2(x)$ is a real number and thus floor($\log_2(x)$) is an integer. So the composition floor $\circ \log_2$ is defined and

$$\text{floor} \circ \log_2 : \mathbb{R}^+ \to \mathbb{Z}.$$

Composition of functions is associative. In other words, if f, g, and h are functions of the right type such that $(f \circ g) \circ h$ and $f \circ (g \circ h)$ make sense, then

$$(f \circ g) \circ h = f \circ (g \circ h).$$

This is easy to establish by noticing that the two expressions $((f \circ g) \circ h)(x)$ and $(f \circ (g \circ h))(x)$ are equal:

$$((f \circ g) \circ h)(x) = (f \circ g)(h(x)) = f(g(h(x))).$$
$$(f \circ (g \circ h))(x) = f((g \circ h)(x)) = f(g(h(x))).$$

So we can feel free to write the composition of three or more functions without the use of parentheses.

But we should observe that composition is usually not a commutative operation. For example, suppose that f and g are defined by $f(x) = x + 1$ and

$g(x) = x^2$. To show that $f \circ g \neq g \circ f$, we only need to find one number x such that $(f \circ g)(x) \neq (g \circ f)(x)$. We'll try $x = 5$ and observe that

$$(f \circ g)(5) = f(g(5)) = f(5^2) = 5^2 + 1 = 26.$$
$$(g \circ f)(5) = g(f(5)) = g(5 + 1) = (5 + 1)^2 = 36.$$

A function that always returns its argument is called an *identity* function. For a set A we sometimes write "id_A" to denote the identity function defined by $\mathrm{id}_A(a) = a$ for all $a \in A$. If $f : A \rightarrow B$, then we always have the following equation.

$$f \circ \mathrm{id}_A = f = \mathrm{id}_B \circ f.$$

The Sequence, Distribute, and Pairs Functions

We'll describe here three functions that are quite useful as basic tools to construct more complicated functions that involve lists.

The *sequence* function "seq" has type $\mathbb{N} \rightarrow \mathrm{lists}(\mathbb{N})$ and is defined as follows for any natural number n:

$$\mathrm{seq}(n) = \langle 0, 1, \ldots, n \rangle.$$

For example, $\mathrm{seq}(0) = \langle 0 \rangle$ and $\mathrm{seq}(4) = \langle 0, 1, 2, 3, 4 \rangle$.

The *distribute* function "dist" has type $A \times \mathrm{lists}(B) \rightarrow \mathrm{lists}(A \times B)$. It takes an element x from A and a list y from $\mathrm{lists}(B)$ and returns the list of pairs made up by pairing x with each element of y. For example,

$$\mathrm{dist}(x, \langle r, s, t \rangle) = \langle (x, r), (x, s), (x, t) \rangle.$$

The *pairs* function takes two lists of equal length and returns the list of pairs of corresponding elements. For example,

$$\mathrm{pairs}(\langle a, b, c \rangle, \langle d, e, f \rangle) = \langle (a, d), (b, e), (c, f) \rangle.$$

Since the domain of pairs is a proper subset of $\mathrm{lists}(A) \times \mathrm{lists}(B)$, it is a partial function of type $\mathrm{lists}(A) \times \mathrm{lists}(B) \rightarrow \mathrm{lists}(A \times B)$.

Composing Functions with Different Arities

Composition can also occur between functions with different arities. For example, suppose we define the following function.

$$f(x, y) = \mathrm{dist}(x, \mathrm{seq}(y)).$$

In this case dist has two arguments and seq has one argument. For example, we'll evaluate the expression $f(5, 3)$.

$$f(5, 3) = \mathrm{dist}(5, \mathrm{seq}(3))$$
$$= \mathrm{dist}(5, \langle 0, 1, 2, 3 \rangle)$$
$$= \langle (5, 0), (5, 1), (5, 2), (5, 3) \rangle.$$

In the next example we'll show that the definition $f(x, y) = \text{dist}(x, \text{seq}(y))$ is a special case of the following more general form of *composition*, where X can be replaced by any number of arguments.

$$f(X) = h(g_1(X), \ldots, g_n(X)).$$

example **2.10 Distribute a Sequence**

We'll show that the definition $f(x, y) = \text{dist}(x, \text{seq}(y))$ fits the general form of composition. To make it fit the form, we'll define the functions $\text{one}(x, y) = x$ and $\text{two}(x, y) = y$. Then we have the following representation of f.

$$\begin{aligned} f(x, y) &= \text{dist}(x, \text{seq}(y)) \\ &= \text{dist}(\text{one}(x, y), \text{seq}(\text{two}(x, y))) \\ &= \text{dist}(\text{one}(x, y), (\text{seq} \circ \text{two})(x, y)). \end{aligned}$$

The last expression has the general form of composition

$$f(X) = h(g_1(X), g_2(X)),$$

where $X = (x, y)$, $h = \text{dist}$, $g_1 = \text{one}$, and $g_2 = \text{seq} \circ \text{two}$.

end example

example **2.11 The Max Function**

Suppose we define the function "max," to return the maximum of two numbers as follows.

$$\text{max}(x, y) = \text{if } x < y \text{ then } y \text{ else } x.$$

Then we can use max to define the function "max3," which returns the maximum of three numbers, by the following composition.

$$\text{max3}(x, y, z) = \text{max}(\text{max}(x, y), z).$$

end example

We can often construct a function by first writing down an informal definition and then proceeding by stages to transform the definition into a formal one that suits our needs. For example, we might start with an informal definition of some function f such as

$$f(x) = \text{expression involving } x.$$

Now we try to transform the right side of the equality into an expression that has the degree of formality that we need. For example, we might try to reach a composition of known functions as follows:

$$f(x) = \text{expression involving } x$$
$$= \text{another expression involving } x$$
$$= \ldots$$
$$= g(h(x)).$$

From a programming point of view, our goal would be to find an expression that involves known functions that already exist in the programming language being used. Let's do some examples to demonstrate how composition can be useful in solving problems.

example **2.12 Minimum Depth of a Binary Tree**

Suppose we want to find the minimum depth of a binary tree in terms of the numbers of nodes. Figure 2.8 lists a few sample cases in which the trees are as compact as possible, which means that they have the least depth for the number of nodes. Let n denote the number of nodes. Notice that when $4 \le n < 8$, the depth is 2. Similarly, the depth is 3 whenever $8 \le n < 16$.

At the same time we know that $\log_2(4) = 2$, $\log_2(8) = 3$, and for $4 \le n < 8$ we have $2 \le \log_2(n) < 3$. So $\log_2(n)$ almost works as the depth function. The problem is that the depth must be exactly 2 whenever $4 \le n < 8$. Can we make this happen? Sure—just apply the floor function to $\log_2(n)$ to get $\text{floor}(\log_2(n)) = 2$ if $4 \le n < 8$. This idea extends to the other intervals that make up $\mathbb{N}$. For example, if $8 \le n < 16$, then $\text{floor}(\log_2(n)) = 3$.

So it makes sense to define our minimum depth function as the composition of the floor function and the $\log_2$ function:

$$\text{minDepth}(n) = \text{floor}(\log_2(n)).$$

end example

example **2.13 A List of Pairs**

Suppose we want to construct a definition for the following function in terms of known functions.

$$f(n) = \langle (0, 0), (1, 1), \ldots, (n, n) \rangle \qquad \text{for any } n \in \mathbb{N}.$$

Binary tree	Nodes	Depth
	1	0
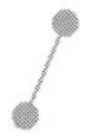	2	1
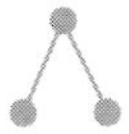	3	1
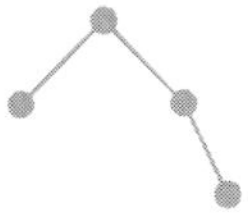	4	2
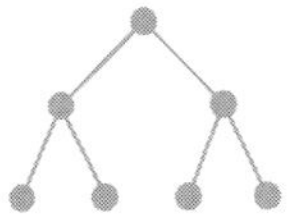	7	2
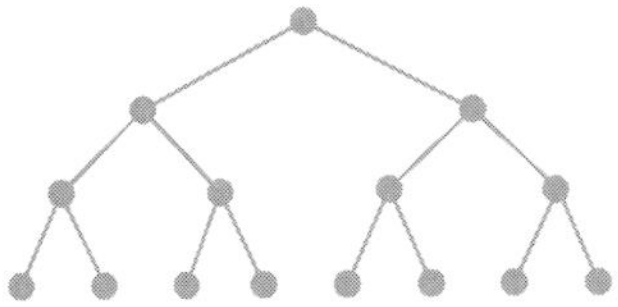	15	3

Figure 2.8 Compact binary trees.

Starting with this informal definition, we'll transform it into a composition of known functions.

$$\begin{aligned} f(n) &= \langle (0,0), (1,1), \dots, (n,n) \rangle \\ &= \text{pairs}(\langle 0,1,\dots,n\rangle, \langle 0,1,\dots,n\rangle) \\ &= \text{pairs}(\text{seq}(n), \text{seq}(n)). \end{aligned}$$

Can you figure out the type of f?

end example

example 2.14 Another List of Pairs

Suppose we want to construct a definition for the following function in terms of known functions.

$$g(k) = \langle (k, 0), (k, 1), \dots, (k, k) \rangle \qquad \text{for any } k \in \mathbb{N}.$$

Starting with this informal definition, we'll transform it into a composition of known functions.

$$g(k) = \langle (k, 0), (k, 1), \ldots, (k, k) \rangle$$
$$= \text{dist}(k, \langle 0, 1, \ldots, k \rangle)$$
$$= \text{dist}(k, \text{seq}(k)).$$

Can you figure out the type of g?

2.2.2   The Map Function

We sometimes need to compute a list of values of a function. A useful tool to accomplish this is the *map* function. It takes a function $f : A \to B$ and a list of elements from A and it returns the list of elements from B constructed by applying f to each element of the given list from A. Here is the definition.

Definition of the Map Function

Let f be a function with domain A and let $\langle x_1, \ldots, x_n \rangle$ be a list of elements from A. Then

$$\text{map}(f, \langle x_1, \ldots, x_n \rangle) = \langle f(x_1), \ldots, f(x_n) \rangle.$$

So the type of the map function can be written as

$$(A \to B) \times \text{lists}(A) \to \text{lists}(B).$$

Here are some example calculations.

$$\text{map}(\text{floor}, \langle -1.5, -0.5, 0.5, 1.5, 2.5 \rangle)$$
$$= \langle \text{floor}(-1.5), \text{floor}(-0.5), \text{floor}(0.5), \text{floor}(1.5), \text{floor}(2.5) \rangle$$
$$= \langle -2, -1, 0, 1, 2 \rangle.$$
$$\text{map}(\text{floor} \circ \log_2, \langle 2, 3, 4, 5 \rangle)$$
$$= \langle \text{floor}(\log_2(2)), \text{floor}(\log_2(3)), \text{floor}(\log_2(4)), \text{floor}(\log_2(5)) \rangle$$
$$= \langle 1, 1, 2, 2 \rangle.$$
$$\text{map}(+, \langle (1, 2), (3, 4), (5, 6), (7, 8), (9, 10) \rangle)$$
$$= \langle +(1, 2), +(3, 4), +(5, 6), +(7, 8), +(9, 10) \rangle$$
$$= \langle 3, 7, 11, 15, 19 \rangle.$$

The map function is an example of a *higher-order* function, which is any function that either has a function as an argument or has a function as a value.

The composition operation is also higher-order because it has functions as arguments and returns a function as a result.

example **2.15 A List of Squares**

Suppose we want to compute sequences of squares of natural numbers, such as 0, 1, 4, 9, 16. In other words, we want to compute $f : \mathbb{N} \to \text{lists}(\mathbb{N})$ defined by $f(n) = \langle 0, 1, 4, \ldots, n^2 \rangle$. We'll present two solutions. For the first solution we'll define $s(x) = x \cdot x$ and then construct a definition for f in terms of map, s, and seq as follows.

$$
\begin{aligned}
f(n) &= \langle 0, 1, 4, \ldots, n^2 \rangle \\
&= \langle s(0), s(1), s(2), \ldots, s(n) \rangle \\
&= \text{map}(s, \langle 0, 1, 2, \ldots, n \rangle) \\
&= \text{map}(s, \text{seq}(n)).
\end{aligned}
$$

For the second solution we'll construct a definition for f without using the function s that we defined for the first solution.

$$
\begin{aligned}
f(n) &= \langle 0, 1, 4, \ldots, n^2 \rangle \\
&= \langle 0 \cdot 0, 1 \cdot 1, 2 \cdot 2, \ldots, n \cdot n \rangle \\
&= \text{map}(\cdot, \langle (0, 0), (1, 1), (2, 2), \ldots, (n, n) \rangle) \\
&= \text{map}(\cdot, \text{pairs}(\langle 0, 1, 2, \ldots, n \rangle, \langle 0, 1, 2, \ldots, n \rangle)) \\
&= \text{map}(\cdot, \text{pairs}(\text{seq}(n), \text{seq}(n))).
\end{aligned}
$$

end example

example **2.16 Graphing with Map**

Suppose we have a function f defined on the closed interval $[a, b]$ and we have a list of numbers $\langle x_0, \ldots, x_n \rangle$ that form a regular partition of $[a, b]$. We want to find the following sequence of $n + 1$ points:

$$
\langle (x_0, f(x_0)), \ldots, (x_n, f(x_n)) \rangle.
$$

The partition is defined by $x_i = a + dk$ for $0 \le k \le n$, where $d = (b - a)/n$. So the sequence is a function of a, d, and n. If we can somehow create the two lists $\langle x_0, \ldots, x_n \rangle$ and $\langle f(x_0), \ldots, f(x_n) \rangle$, then the desired sequence of points can be obtained by applying the pairs function to these two sequences.

Let "makeSeq" be the function that returns the list $\langle x_0, \ldots, x_n \rangle$. We'll start by trying to define makeSeq in terms of functions that are already at hand. First we write down the desired value of the expression, makeSeq(a, d, n) and then try to gradually transform the value into an expression involving known functions and the arguments a, d, and n.

$$\text{makeSeq}(a, d, n)$$
$$= \langle x_0, x_1, \ldots, x_n \rangle$$
$$= \langle a, a + d, a + 2d, \ldots, a + nd \rangle$$
$$= \text{map}(+, \langle (a, 0), (a, d), (a, 2d), \ldots, (a, nd) \rangle)$$
$$= \text{map}(+, \text{dist}(a, \langle 0, d, 2d, \ldots, nd \rangle))$$
$$= \text{map}(+, \text{dist}(a, \text{map}(\cdot, \langle (d, 0), (d, 1), (d, 2), \ldots, (d, n) \rangle))))$$
$$= \text{map}(+, \text{dist}(a, \text{map}(\cdot, \text{dist}(d, \langle 0, 1, 2, \ldots, n \rangle)))))$$
$$= \text{map}(+, \text{dist}(a, \text{map}(\cdot, \text{dist}(d, \text{seq}(n)))))).$$

The last expression contains only known functions and the arguments a, d, and n. So we have a definition for makeSeq. Now it's an easy matter to build the second list. Just notice that

$$\langle f(x_0), \ldots, f(x_n) \rangle = \text{map}(f, \langle x_0, x_1, \ldots, x_n \rangle)$$
$$= \text{map}(f, \text{makeSeq}(a, d, n)).$$

Now let "makeGraph" be the name of the function that returns the desired sequence of points. Then makeGraph can be written as follows:

$$\text{makeGraph}(f, a, d, n) = \langle (x_0, f(x_0)), \ldots, (x_n, f(x_n)) \rangle$$
$$= \text{pairs}(\text{makeSeq}(a, d, n), \text{map}(f, \text{makeSeq}(a, d, n))).$$

This gives us a definition of makeGraph in terms of known functions and the variables f, a, d, and n.

end example

Exercises

Composing Functions

1. Evaluate each of the following expressions.

    a. floor($\log_2 17$).

    b. ceiling($\log_2 25$).

    c. gcd(14 mod 6, 18 mod 7).

    d. gcd(12, 18) mod 5.

    e. dist(4, seq(3)).

    f. pairs(seq(3), seq(3)).

    g. dist(+, pairs(seq(2), seq(2))).

2. In each case find the compositions $f \circ g$ and $g \circ f$, and find an integer x such that $f(g(x)) \neq g(f(x))$.

    a. $f(x) = \text{ceiling}(x/2)$ and $g(x) = 2x$.
    b. $f(x) = \text{floor}(x/2)$ and $g(x) = 2x + 1$.
    c. $f(x) = \gcd(x, 10)$ and $g(x) = x \bmod 5$.

3. Let $f(x) = x^2$ and $g(x, y) = x + y$. Find compositions that use the functions f and g for each of the following expressions.

    a. $(x + y)^2$.   **b.** $x^2 + y^2$.   **c.** $(x + y + z)^2$.   **d.** $x^2 + y^2 + z^2$.

4. Describe the set of natural numbers x satisfying each equation.

    a. $\text{floor}(\log_2(x)) = 7$.
    b. $\text{ceiling}(\log_2(x)) = 7$.

5. Find a definition for the function max4 that calculates the maximum value of four numbers. Use only composition and the function max that returns the maximum value of two numbers.

6. Find a formula for the number of binary digits in the binary representation of a nonzero natural number x. *Hint:* Notice, for example, that the numbers from 4 through 7 require three binary digits, while the numbers 8 through 15 require four binary digits, and so on.

Composing with the Map Function

7. Evaluate each expression:

    a. $\text{map}(\text{floor} \circ \log_2, \langle 1, 2, 3, \ldots, 16 \rangle)$.
    b. $\text{map}(\text{ceiling} \circ \log_2, \langle 1, 2, 3, \ldots, 16 \rangle)$.

8. Suppose that $f : \mathbb{N} \to \text{lists}(\mathbb{N})$ is defined by $f(n) = \langle 0, 2, 4, 6, \ldots, 2n \rangle$. For example, $f(5) = \langle 0, 2, 4, 6, 8, 10 \rangle$. In each case find a definition for f as a composition of the listed functions.

    a. map, +, pairs, seq.
    b. map, ·, dist, seq.

9. For each of the following functions, construct a definition of the function as a composition of known functions. Assume that all of the variables are natural numbers.

    a. $f(n, k) = \langle n, n + 1, n + 2, \ldots, n + k \rangle$.
    b. $f(n, k) = \langle 0, k, 2k, 3k, \ldots, nk \rangle$.
    c. $f(n, m) = \langle n, n + 1, n + 2, \ldots, m - 1, m \rangle$,   where $n \leq m$.

    d. $f(n) = \langle n, n - 1, n - 2, \ldots, 1, 0 \rangle$.
    e. $f(n) = \langle (0, n), (1, n - 1), \ldots, (n - 1, 1), (n, 0) \rangle$.
    f. $f(n) = \langle 1, 3, 5, \ldots, 2n + 1 \rangle$.
    g. $f(g, n) = \langle (0, g(0)), (1, g(1)), \ldots, (n, g(n)) \rangle$.
    h. $f(g, \langle x_1, x_2, \ldots, x_n \rangle) = \langle (x_1, g(x_1)), (x_2, g(x_2)), \ldots, (x_n, g(x_n)) \rangle$.
    i. $f(g, h, \langle x_1, \ldots, x_n \rangle) = \langle (g(x_1), h(x_1)), \ldots, (g(x_n), h(x_n)) \rangle$.

10. We defined $\text{seq}(n) = \langle 0, 1, 2, 3, \ldots, n \rangle$. Suppose we want the sequence to start with the number 1. In other words, for $n > 0$, we want to define a function $f(n) = \langle 1, 2, 3, \ldots, n \rangle$. Find a definition for f as a composition of the functions map, +, dist, and seq.

Proofs

11. Prove that $\text{tail}(\text{dist}(x, \text{seq}(n))) = \text{dist}(x, \text{tail}(\text{seq}(n)))$, where n is a positive integer.

12. Prove that $\lfloor \log_2(x) \rfloor = \lfloor \log_2(\lfloor x \rfloor) \rfloor$ for $x \geq 1$.

2.3 Properties of Functions

Functions that satisfy one or both of two special properties can be very useful in solving a variety of computational problems. One property is that distinct elements map to distinct elements. The other property is that the range is equal to the codomain. We'll discuss these properties in more detail and give some examples where they are useful.

2.3.1 Injections and Surjections

Injective Property

A function $f : A \to B$ is called *injective* (also *one-to-one*, or an *embedding*) if it maps distinct elements of A to distinct elements of B. Another way to say this is that f is injective if $x \neq y$ implies $f(x) \neq f(y)$. Yet another way to say this is that f is injective if $f(x) = f(y)$ implies $x = y$. An injective function is called an *injection*.

    For example, Figure 2.9 illustrates an injection from a set A to a set B.

Surjective Property

A function $f : A \to B$ is called *surjective* (also *onto*) if the range of f is the codomain B. Another way to say this is that f is surjective if each element

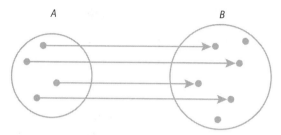

Figure 2.9   An injection.

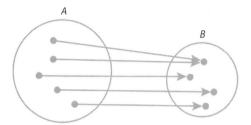

Figure 2.10   A surjection.

$b \in B$ can be written as $b = f(x)$ for some element $x \in A$. A surjective function is called a *surjection*.

For example, Figure 2.10 pictures a surjection from A to B.

example 2.17 Injective or Surjective

We'll give a few examples of functions that have one or the other of the injective and surjective properties.

1. The function $f : \mathbb{R} \to \mathbb{Z}$ defined by $f(x) = \lceil x + 1 \rceil$ is surjective because for any $y \in \mathbb{Z}$ there is a number in $\mathbb{R}$, namely $y - 1$, such that $f(y - 1) = y$. But f is not injective because, for example, $f(3.5) = f(3.6)$.

2. The function $f : \mathbb{N}_8 \to \mathbb{N}_8$ defined by $f(x) = 2x \bmod 8$ is not injective because, for example, $f(0) = f(4)$. f is not surjective because the range of f is only the set $\{0, 2, 4, 6\}$.

3. Let $g : \mathbb{N} \to \mathbb{N} \times \mathbb{N}$ be defined by $g(x) = (x, x)$. Then g is injective because if $x, y \in \mathbb{N}$ and $x \neq y$, then $g(x) = (x, x) \neq (y, y) = g(y)$. But g is not surjective because, for example, nothing maps to $(0, 1)$.

4. The function $f : \mathbb{N} \times \mathbb{N} \to \mathbb{N}$ defined by $f(x, y) = 2x + y$ is surjective. To see this, notice that any $z \in \mathbb{N}$ is either even or odd. If z is even, then $z = 2k$ for some $k \in \mathbb{N}$, so $f(k, 0) = z$. If z is odd, then $z = 2k + 1$ for some

$k \in \mathbb{N}$, so $f(k, 1) = z$. Thus f is surjective. But f is not injective because, for example, $f(0, 2) = f(1, 0)$.

end example

2.3.2   Bijections and Inverses

A function is called *bijective* if it is both injective and surjective. Another term for bijective is "one-to-one and onto." A bijective function is called a *bijection* or a "one-to-one correspondence."

For example, Figure 2.11 pictures a bijection from A to B.

example **2.18   A Bijection**

Let $(0, 1) = \{x \in \mathbb{R} \mid 0 < x < 1\}$ and let $\mathbb{R}^+$ denote the set of positive real numbers. We'll show that the function $f : (0, 1) \to \mathbb{R}^+$ defined by

$$f(x) = \frac{x}{1 - x}$$

is a bijection. To show that f is an injection, let $f(x) = f(y)$. Then

$$\frac{x}{1 - x} = \frac{y}{1 - y},$$

which can be cross multiplied to get $x - xy = y - xy$. Subtract $-xy$ from both sides to get $x = y$. Thus f is injective. To show that f is surjective, let $y > 0$ and try to find $x \in (0, 1)$ such that $f(x) = y$. We'll solve the equation

$$\frac{x}{1 - x} = y.$$

Cross multiply and solve for x to obtain

$$x = \frac{y}{y + 1}.$$

It follows that $f(y/(y+1)) = y$, and since $y > 0$, it follows that $0 < y/(y+1) < 1$. Thus f is surjective. Therefore, f is a bijection.

end example

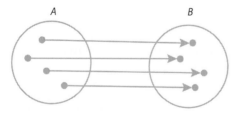

Figure 2.11   A bijection.

Inverse Functions

Bijections always come in pairs. If $f : A \to B$ is a bijection, then there is a function $g : B \to A$, called the *inverse* of f, defined by $g(b) = a$ if $f(a) = b$. Of course, the inverse of f is also a bijection and we have $g(f(a)) = a$ for all $a \in A$ and $f(g(b)) = b$ for all $b \in B$. In other words, $g \circ f = \mathrm{id}_A$ and $f \circ g = \mathrm{id}_B$.

We should note that there is exactly one inverse of any bijection f. Suppose that g and h are two inverses of f. Then for any $x \in B$ we have

$$
\begin{aligned}
g(x) &= g(\mathrm{id}_B(x)) \\
&= g(f(h(x))) \qquad &\text{(since } f \circ h = \mathrm{id}_B) \\
&= \mathrm{id}_A\,(h(x)) \qquad &\text{(since } g \circ f = \mathrm{id}_A) \\
&= h(x).
\end{aligned}
$$

This tells us that $g = h$. The inverse of f is often denoted by the symbol f^{-1}. So if f is a bijection and $f(a) = b$, then $f^{-1}(b) = a$. Notice the close relationship between the equation $f^{-1}(b) = a$ and the pre-image equation $f^{-1}(\{b\}) = \{a\}$.

example 2.19 **Inverses**

1. For any real number $b > 1$ the log function $\log_b x$ and the exponential function b^x are inverses of each other because $\log_b b^x = x$ and $b^{\log_b x} = x$.

2. Let *Odd* and *Even* be the sets of odd and even natural numbers, respectively. The function $f : Odd \to Even$ defined by $f(x) = x - 1$ is a bijection. Check it out. The inverse of f can be defined by $f^{-1}(x) = x + 1$. Notice that $f^{-1}(f(x)) = f^{-1}(x - 1) = (x - 1) + 1 = x$.

3. The function $f : \mathbb{N}_5 \to \mathbb{N}_5$ defined by $f(x) = 2x \bmod 5$ is bijective because, $f(0) = 0$, $f(1) = 2$, $f(2) = 4$, $f(3) = 1$, and $f(4) = 3$. The inverse of f can be defined by $f^{-1}(x) = 3x \bmod 5$. For example, $f^{-1}(f(4)) = 3f(4) \bmod 5 = 9 \bmod 5 = 4$. Check the other values too.

end example

The fact that the function $f : \mathbb{N}_5 \to \mathbb{N}_5$ defined by $f(x) = 2x \bmod 5$ (in the preceding example) is a bijection is an instance of an interesting and useful fact about the mod function and inverses. Here is the result.

The Mod Function and Inverses   (2.6)

Let $n \geq 1$ and let $f : \mathbb{N}_n \to \mathbb{N}_n$ be defined as follows, where a and b are integers.

$$
f(x) = (ax + b) \bmod n.
$$

Continued ➡

➡ ➡

> Then f is a bijection if and only if $\gcd(a, n) = 1$. When this is the case, the inverse function f^{-1} is defined by
>
> $$f^{-1}(x) = (kx + c) \bmod n,$$
>
> where c is an integer such that $f(c) = 0$, and k is an integer such that $1 = ak + nm$ for some integer m.

Proof: We'll prove the iff part of the statement and leave the form of the inverse as an exercise. If $n = 1$, then $f : \{0\} \rightarrow \{0\}$ is a bijection and $\gcd(a, 1) = 1$. So we'll let $n > 1$.

Assume that f is a bijection and show that $\gcd(a, n) = 1$. Then f is surjective, so there are numbers s, $c \in \mathbb{N}_n$ such that $f(s) = 1$ and $f(c) = 0$. Using the definition of f, these equations become

$$(as + b) \bmod n = 1 \text{ and } (ac + b) \bmod n = 0.$$

Therefore, there are integers q_1 and q_2 such that the two equations become

$$as + b + nq_1 = 1 \text{ and } ac + b + nq_2 = 0.$$

Solve the second equation for b to get $b = -ac - nq_2$, and substitute for b in the first equation to get

$$1 = a(s - c) + n(q_1 - q_2).$$

Since $\gcd(a, n)$ divides both a and n, it divides the right side of the equation and so must also divide 1. Therefore, $\gcd(a, n) = 1$.

Now assume that $\gcd(a, n) = 1$ and show that f is a bijection. Since $\mathbb{N}_n$ is finite, we need only show that f is an injection to conclude that it is a bijection. So let $x, y \in \mathbb{N}_n$ and let $f(x) = f(y)$. Then

$$(ax + b) \bmod n = (ay + b) \bmod n,$$

which by (2.4a) implies that n divides $(ax + b) - (ay + b)$. Therefore, n divides $a(x - y)$, and since $\gcd(a, n) = 1$, we conclude from (2.2d) that n divides $x - y$. But the only way for n to divide $x - y$ is for $x - y = 0$ because both $x, y \in \mathbb{N}_n$. Thus $x = y$, and it follows that f is injective, hence also surjective, and therefore bijective. QED.

Relationships

An interesting property that relates injections and surjections is that if there is an injection from A to B, then there is a surjection from B to A, and conversely. A less surprising fact is that if the composition $f \circ g$ makes sense and if both f and g have one of the properties injective, surjective, or bijective, then $f \circ g$ also has the property. We'll list these facts for the record.

Injective and Surjective Relationships (2.7)

a. If f and g are injective, then $g \circ f$ is injective.

b. If f and g are surjective, then $g \circ f$ is surjective.

c. If f and g are bijective, then $g \circ f$ is bijective.

d. There is an injection from A to B if and only if there is a surjection from B to A.

Proof: We'll prove (2.7d) and leave the others as exercises. Suppose that f is an injection from A to B. We'll define a function g from B to A. Since f is an injection, it follows that for each $b \in \text{range}(f)$ there is exactly one $a \in A$ such that $b = f(a)$. In this case, we define $g(b) = a$. For each $b \in B - \text{range}(f)$ we have the freedom to let g map b to any element of A that we like. So g is a function from B to A and we defined g so that $\text{range}(g) = A$. Thus g is surjective.

For the other direction, assume that g is a surjection from B to A. We'll define a function f from A to B. Since g is a surjection, it follows that for each $a \in A$, the pre-image $g^{-1}(\{a\}) \neq \varnothing$. So we can pick an element $b \in g^{-1}(\{a\})$ and define $f(a) = b$. Thus f is a function from A to B. Now if x, $y \in A$ and $x \neq y$, then $g^{-1}(\{x\}) \cap g^{-1}(\{y\}) = \varnothing$. Since $f(x) \in g^{-1}(\{x\})$ and $f(y) \in g^{-1}(\{y\})$, it follows that $f(x) \neq f(y)$. Thus f is injective. QED.

2.3.3 The Pigeonhole Principle

We're going to describe a useful rule that we often use without thinking. For example, suppose 21 pieces of mail are placed in 20 mailboxes. Then one mailbox receives at least two pieces of mail. This is an example of the *pigeonhole principle*, where we think of the pieces of mail as pigeons and the mailboxes as pigeonholes.

Pigeonhole Principle

If m pigeons fly into n pigeonholes where $m > n$, then one pigeonhole will have two or more pigeons.

We can describe the pigeonhole principle in more formal terms as follows: If A and B are finite sets with $|A| > |B|$, then every function from A to B maps at least two elements of A to a single element of B. This is the same as saying that no function from A to B is an injection.

This simple idea is used often in many different settings. We'll be using it at several places in the book.

example **2.20 Pigeonhole Examples**

Here are a few sample statements that can be justified by the pigeonhole principle.

1. The "musical chairs" game is played with n people and $n - 1$ chairs for them to sit on when the music stops.

2. In a group of eight people, two were born on the same day of the week.

3. If a six-sided die is tossed seven times, one side will come up twice.

4. If a directed graph with n vertices has a path of length n or longer, then the path must pass through some vertex at least twice. This implies that the graph contains a cycle.

5. In any set of $n + 1$ integers, there are two numbers that have the same remainder on division by n. This follows because there are only n remainders possible on division by n.

6. The decimal expansion of any rational number contains a repeating sequence of digits (they might be all zeros). For example, $359/495 = 0.7252525\ldots$, $7/3 = 2.333\ldots$, and $2/5 = 0.4000\ldots$. To see this, let m/n be a rational number. Divide m by n until all the digits of m are used up. This gets us to the decimal point. Now continue the division by n for $n + 1$ more steps. This gives us $n + 1$ remainders. Since there are only n remainders possible on division by n, the pigeonhole principle tells us that one of the remainders will be repeated. So the sequence of remainders between the repeated remainders will be repeated forever. This causes the corresponding sequence of digits in the decimal expansion to be repeated forever.

end example

2.3.4 Simple Ciphers

Bijections and inverse functions play an important role when working with systems (called ciphers) to encipher and decipher information. We'll give a few examples to illustrate the connections. For ease of discussion we'll denote the 26 letters of the lowercase alphabet by the set $\mathbb{N}_{26} = \{0, 1, 2, \ldots, 25\}$, where we identify a with 0, b with 1, and so on.

To get things started we'll describe a cipher that enciphers a message by adding 5 to each letter. For example, the message 'abc' becomes 'fgh'. The cipher is easy to write once we figure out how to wrap around the end of the alphabet. For example, to shift the letter z (i.e., 25) by 5 letters we need to come up with the letter e (i.e., 4). All we need to do is add the two numbers mod 26:

$$(25 + 5) \bmod 26 = 4.$$

So we can define the cipher as a function f as follows:

$$f(x) = (x + 5) \bmod 26.$$

It follows from (2.6) that f is a bijection of type $\mathbb{N}_{26} \to \mathbb{N}_{26}$. So we have a cipher for transforming letters. For example, the message 'hello' transforms to 'mjqqt'. To decipher the message we need to reverse the process, which is easy in this case because the inverse of f is easy to guess.

$$f^{-1}(x) = (x - 5) \bmod 26.$$

For example, to see that e maps back to z, we can observe that 4 maps to 25.

$$f^{-1}(4) = (4 - 5) \bmod 26 = (-1) \bmod 26 = 25.$$

A cipher *key* is the information needed to encipher or decipher a message. The cipher we've been talking about is called an *additive* cipher and its key is 5. A cryptanalyst who intercepts the message 'mjqqt' can easily check whether it was created by an additive cipher. An additive cipher is an example of a *monoalphabetic* cipher, which is a cipher that replaces repeated occurrences of a character in a message by the same character from the cipher alphabet.

A *multiplicative* cipher is a monoalphabetic cipher that enciphers each letter by using a multiplier as a key. For example, suppose we define the cipher

$$g(x) = 3x \bmod 26.$$

In this case the key is 3. For example, this cipher maps a to a, c to g, and m to k. Is g a bijection? You can convince yourself that it is by exhaustive checking. But it's easier to use (2.6). Since $\gcd(3, 26) = 1$ it follows that g is a bijection. What about deciphering? Again, (2.6) comes to the rescue to tell us the form of g^{-1}. Since we can write $\gcd(3, 26) = 1 = 3(9) + 26(-1)$, and since $g(0) = 0$, it follows that we can define g^{-1} as

$$g^{-1}(x) = 9x \bmod 26.$$

There are some questions to ask about multiplicative ciphers. Which keys act as an identity (not changing the message)? Is there always one letter that never changes no matter the key? Do fractions work as keys? What about decoding (i.e., deciphering) a message? Do you need a new deciphering algorithm?

An *affine* cipher is a monoalphabetic cipher that combines additive and multiplicative ciphers. For example, with the pair of keys (3, 5) we can define f as

$$f(x) = (3x + 5) \bmod 26.$$

We can use (2.6) to conclude that f is a bijection because $\gcd(3, 26) = 1$. So we can also decipher messages with f^{-1}, which we can construct using (2.6) as

$$f^{-1}(x) = (9x + 7) \bmod 26.$$

Some ciphers leave one or more letters fixed. For example, an additive cipher that shifts by a multiple of 26 will leave all letters fixed. A multiplicative cipher always sends 0 to 0, so one letter is fixed. But what about an affine cipher of the form $f(x) = (ax + b) \bmod 26$? When can we be sure that no letters are fixed? In other words, when can we be sure that $f(x) \neq x$ for all $x \in \mathbb{N}_{26}$? The answer is, when $\gcd(a - 1, 26)$ does not divide b. Here is the general result that we've been discussing.

The Mod Function and Fixed Points   (2.8)

Let $n \geq 1$ and let $f : \mathbb{N}_n \to \mathbb{N}_n$ be defined as follows, where a and b are integers.

$$f(x) = (ax + b) \bmod n.$$

Then f has no fixed points (i.e., f changes every letter of an alphabet) if and only if $\gcd(a - 1, n)$ does not divide b.

This result follows from an old and easy result from number theory, and we'll discuss it in the exercises. Let's see how the result helps our cipher problem.

example **2.21 Simple Ciphers**

The function $f(x) = (3x + 5) \bmod 26$ does not have any fixed points because $\gcd(3 - 1, 26) = \gcd(2, 26) = 2$, and 2 does not divide 5. It's nice to know that we don't have to check all 26 values of f.

On the other hand, the function $f(x) = (3x + 4) \bmod 26$ has fixed points because $\gcd(3 - 1, 26) = 2$ and 2 divides 4. For this example, we can observe that $f(11) = 11$ and $f(24) = 24$. So in terms of our association of letters with numbers we would have $f(l) = l$ and $f(y) = y$.

end example

Whatever cipher we use, we always have some questions: Is it a bijection? What is the range of values for the keys? Is it hard to decipher an intercepted message?

2.3.5 Hash Functions

Suppose we wish to retrieve some information stored in a table of size n with indexes 0, 1, ..., $n - 1$. The items in the table can be very general things. For example, the items might be strings of letters, or they might be large records with many fields of information. To look up a table item we need a *key* to the information we desire.

For example, if the table contains records of information for the 12 months of the year, the keys might be the three-letter abbreviations for the 12 months. To look up the record for January, we would present the key Jan to a lookup program. The program uses the key to find the table entry for the January record of information. Then the information would be available to us.

An easy way to look up the January record is to search the table until the key Jan is found. This might be OK for a small table with 12 entries. But it may be impossibly slow for large tables with thousands of entries. Here is the general problem that we want to solve:

Given a key, find the table entry containing the key without searching.

This may seem impossible at first glance. But let's consider a way to use a function to map each key directly to its table location.

Definition of a Hash Function

A *hash function* is a function that maps a set S of keys to a finite set of table indexes, which we'll assume are 0, 1, ..., $n - 1$. A table whose information is found by a hash function is called a *hash table*.

---

example 2.22 A Hash Function

Let S be the set of three-letter abbreviations for the months of the year. We might define a hash function $f : S \rightarrow \{0, 1, \ldots, 11\}$ in the following way.

$$f(XYZ) = (\text{ord}(X) + \text{ord}(Y) + \text{ord}(Z)) \bmod 12,$$

where $\text{ord}(X)$ denotes the integer value of the ASCII code for X. (The ASCII values for A to Z and a to z are 65 to 90 and 97 to 122, respectively.) For example, we'll compute the value for the key Jan.

$$f(\text{Jan}) = (\text{ord}(J) + \text{ord}(a) + \text{ord}(n)) \mod 12$$
$$= (74 + 97 + 110) \mod 12$$
$$= 5.$$

Most programming languages have efficient implementations of the ord and mod functions, so hash functions constructed from them are quite fast. Here is the listing of all the values of f.

Jan	Feb	Mar	Apr	May	Jun	Jul	Aug	Sep	Oct	Nov	Dec
5	5	0	3	7	1	11	9	8	6	7	4

end example

We should notice the function f in Example 2.22 is not injective. For example, $f(\text{Jan}) = f(\text{Feb}) = 5$. So if we use f to construct a hash table, we can't put the information for January and February at the same address. Let's discuss this problem.

Collisions

If a hash function is injective, then it maps every key to the index of the hash table where the information is stored and no searching is involved. Often this is not possible. When two keys map to the same table index, the result is called a *collision*. So if a hash function is not injective, it has collisions. In Example 2.22 the hash function has collisions $f(\text{Jan}) = f(\text{Feb})$ and $f(\text{May}) = f(\text{Nov})$.

When collisions occur, we store the information for one of the keys in the common table location and must find some other location for the other keys. There are many ways to find the location for a key that has collided with another key. One technique is called *linear probing*. With this technique the program searches the remaining locations in a "linear" manner. For example, if location k is the collision index, then the following sequence of table locations is searched.

$$(k + 1) \bmod n, \ (k + 2) \bmod n, \ (k + 3) \bmod n, \ldots.$$

example **2.23 A Hash Table**

We'll use the hash function f from Example 2.22 to construct a hash table for the months of the year by placing the three-letter abreviations in the table one by one, starting with Jan and continuing to Dec. We'll use linear probing to resolve collisions that occur in the process. For example, since $f(\text{Jan}) = 5$, we place Jan in position 5 of the table. Next, since $f(\text{Feb}) = 5$ and since position 5 is full, we look for the next available position and place Feb in position 6. Continuing in this way, we eventually construct the following hash table, where entries in parentheses need some searching to be found.

0	1	2	3	4	5	6	7	8	9	10	11
Mar	Jun	(Nov)	Apr	Dec	Jan	(Feb)	May	Sep	Aug	(Oct)	Jul

There are many questions. Can we find an injection so there are no collisions? If we increased the size of the table, would it give us a better chance of finding an injection? If the table size is increased, can we scatter the elements so that collisions can be searched for in less time?

end example

Searching successive locations may not be the best way to resolve collisions if it takes too many comparisons to find an open location. A simple alternative is to use a "gap" between table locations in order to "scatter" or "hash" the information to other parts of the table. Let g be a gap, where $1 \leq g < n$. Then the following sequence of table locations is searched in case a collision occurs at location k:

$$(k + g) \bmod n, \ (k + 2g) \bmod n, \ (k + 3g) \bmod n, \ldots.$$

Some problems can occur if we're not careful with our choice of g. For example, suppose $n = 12$ and $g = 4$. Then the search sequence can skip some table entries. For example, if $k = 7$, then the first 12 indexes in the previous sequence are

$$11, \ 3, \ 7, \ 11, \ 3, \ 7, \ 11, \ 3, \ 7, \ 11, \ 3, \ 7.$$

So we would miss table entries 0, 1, 2, 4, 5, 6, 8, 9, and 10. Let's try another value for g. Suppose we try $g = 5$. If $k = 7$, then the first 12 indexes in the previous sequence are

$$0, \ 5, \ 10, \ 3, \ 8, \ 1, \ 6, \ 11, \ 4, \ 9, \ 2, \ 7.$$

In this case we cover the entire set $\{0, 1, \ldots, 11\}$.

Can we always find a search sequence that hits all the elements of $\{0, 1, \ldots, n - 1\}$? Happily, the answer is yes. Just pick g and n so that $\gcd(g, n) = 1$. Then (2.6) tells us that the function f defined by $f(x) = (gx + k) \bmod n$ is a bijection. For example, if we pick n to be a prime number, then $\gcd(g, n) = 1$ for any g in the interval $1 \leq g < n$. That's why table sizes are often prime numbers, even though the data set may have fewer entries than the table size.

There are many ways to define hash functions and to resolve collisions. The paper by Cichelli [1980] examines some bijective hash functions.

 Exercises

Injections, Surjections, and Bijections

1. The fatherOf function from *People* to *People* is neither injective nor surjective. Why?

2. For each of the following cases, construct a function satisfying the given condition, where the domain and codomain are chosen from the sets

$$A = \{a, \ b, \ c\}, \ B = \{x, \ y, \ z\}, \ C = \{1, \ 2\}.$$

   a. Injective but not surjective.
   b. Surjective but not injective.
   c. Bijective.
   d. Neither injective nor surjective.

3. For each of the following types, compile some statistics: the number of functions of that type; the number that are injective; the number that are surjective; the number that are bijective; the number that are neither injective, surjective, nor bijective.

    a. $\{a, b, c\} \rightarrow \{1, 2\}$.
    b. $\{a, b\} \rightarrow \{1, 2, 3\}$.
    c. $\{a, b, c\} \rightarrow \{1, 2, 3\}$.

4. Show that each function $f : \mathbb{N} \rightarrow \mathbb{N}$ has the listed properties.

    a. $f(x) = 2x$.   (injective and not surjective)
    b. $f(x) = x + 1$.   (injective and not surjective)
    c. $f(x) = \text{floor}(x/2)$.   (surjective and not injective)
    d. $f(x) = \text{ceiling}(\log_2 (x + 1))$.   (surjective and not injective)
    e. $f(x) =$ if x is odd then $x - 1$ else $x + 1$.   (bijective)

5. Determine whether each function is injective or surjective.

    a. $f : \mathbb{R} \rightarrow \mathbb{Z}$, where $f(x) = \text{floor}(x)$.
    b. $f : \mathbb{N} \rightarrow \mathbb{N}$, where $f(x) = x \bmod 10$.
    c. $f : \mathbb{Z} \rightarrow \mathbb{N}$ defined by $f(x) = |x + 1|$.
    d. $\text{seq} : \mathbb{N} \rightarrow \text{lists}(\mathbb{N})$.
    e. $\text{dist} : A \times \text{lists}(B) \rightarrow \text{lists}(A \times B)$.
    f. $f : A \rightarrow \text{power}(A)$, A is any set, and $f(x) = \{x\}$.
    g. $f : \text{lists}(A) \rightarrow \text{power}(A)$, A is finite, and $f((x_1, \ldots, x_n)) = \{x_1, \ldots, x_n\}$.
    h. $f : \text{lists}(A) \rightarrow \text{bags}(A)$, A is finite, and $f((x_1, \ldots, x_n)) = [x_1, \ldots, x_n]$.
    i. $f : \text{bags}(A) \rightarrow \text{power}(A)$, A is finite, and $f([x_1, \ldots, x_n]) = \{x_1, \ldots, x_n\}$.

6. Let $\mathbb{R}^+$ and $\mathbb{R}^-$ denote the sets of positive and negative real numbers, respectively. If $a, b \in \mathbb{R}$ and $a < b$, let $(a, b) = \{x \in \mathbb{R} \mid a < x < b\}$. Show that each of the following functions is a bijection.

    a. $f : (0, 1) \rightarrow (a, b)$ defined by $f(x) = (b - a)x + a$.
    b. $f : \mathbb{R}^+ \rightarrow (0, 1)$ defined by $f(x) = 1/(x + 1)$.
    c. $f : (1/2, 1) \rightarrow \mathbb{R}^+$ defined by $f(x) = 1/(2x - 1) - 1$.
    d. $f : (0, 1/2) \rightarrow \mathbb{R}^-$ defined by $f(x) = 1/(2x - 1) + 1$.
    e. $f : (0, 1) \rightarrow \mathbb{R}$ defined by $f(x) = \begin{cases} 1/(2x - 1) - 1 & \text{if } 1/2 < x < 1 \\ 0 & \text{if } x = 1/2 \\ 1/(2x - 1) + 1 & \text{if } 0 < x < 1/2 \end{cases}$.

The Pigeonhole Principle

7. Use the pigeonhole principle for each of the following statements.

    a. How many people are needed in a group to say that three were born on the same day of the week?

    b. How many people are needed in a group to say that four were born in the same month?

    c. Why does any set of 10 nonempty strings over $\{a, b, c\}$ contain two different strings whose starting letters agree and whose ending letters agree?

    d. Find the size needed for a set of nonempty strings over $\{a, b, c, d\}$ to contain two strings whose starting letters agree and whose ending letters agree.

8. Use the pigeonhole principle to verify each of the following statements.

    a. In any set of 11 natural numbers, there are two numbers whose decimal representations contain a common digit.

    b. In any set of four numbers picked from the set $\{1, 2, 3, 4, 5, 6\}$, there are two numbers whose sum is seven.

    c. If five distinct numbers are chosen from the set $\{1, 2, 3, 4, 5, 6, 7, 8\}$, then two of the numbers chosen are consecutive (i.e., of the form n and $n + 1$). *Hint:* List the five numbers chosen as, x_1, x_2, x_3, x_4, x_5 and list the successors as $x_1 + 1, x_2 + 1, x_3 + 1, x_4 + 1, x_5 + 1$. Are there more than eight numbers listed?

Simple Ciphers and the Mod Function

9. Each of the following functions has the form $f(x) = (ax + b) \bmod n$. Assume that each function has type $\mathbb{N}_n \to \mathbb{N}_n$, so that we can think of f as a cipher for an alphabet represented by the numbers $0, 1, \ldots, n - 1$. Use (2.6) to determine whether each function is a bijection, and, if so, construct its inverse. Then use (2.8) to determine whether the function has fixed points (i.e., letters that don't change), and, if so, find them.

    a. $f(x) = 2x \bmod 6$.

    b. $f(x) = 2x \bmod 5$.

    c. $f(x) = 5x \bmod 6$.

    d. $f(x) = (3x + 2) \bmod 6$.

    e. $f(x) = (2x + 3) \bmod 7$.

    f. $f(x) = (5x + 3) \bmod 12$.

    g. $f(x) = (25x + 7) \bmod 16$.

10. Think of the letters A to Z as the numbers 0 to 25 and let f be a cipher of the form $f(x) = (ax + b) \bmod 26$.

   a. Use (2.6) to find all values of a $(0 \leq a < 26)$ that will make f bijective.

   b. For the values of a in Part (a) that make f bijective, use (2.8) to find a general statement about the values of b $(0 \leq b < 26)$ that will ensure that f maps each letter to a different letter.

Hash Functions

11. Let $S = \{$one, two, three, four, five, six, seven, eight, nine$\}$ and let $f : S \rightarrow \mathbb{N}_9$ be defined by $f(x) = (3|x|) \bmod 9$, where $|x|$ means the number of letters in x. For each of the following gaps, construct a hash table that contains the strings of S by choosing a string for entry in the table by the order that it is listed in S. Resolve collisions by linear probing with the given gap and observe whether all strings can be placed in the table.

   a. Gap $= 1$.   **b.** Gap $= 2$.   **c.** Gap $= 3$.

12. Repeat Exercise 11 for the set $S = \{$Monday, Tuesday, Wednesday, Thursday, Friday, Saturday, Sunday$\}$ and the function $f: S \rightarrow \mathbb{N}_7$ defined by $f(x) = (2|x| + 3) \bmod 7$.

13. Repeat Exercise 11 for the set $S = \{$January, February, March, April, May, June, July, August$\}$ and $f: S \rightarrow \mathbb{N}_8$ defined by $f(x) = (|x| + 3) \bmod 8$.

Proofs and Challenges

14. Find integers a and b such that the function $f : \mathbb{N}_{12} \rightarrow \mathbb{N}_{12}$ defined by $f(x) = (ax + b) \bmod 12$ is bijective and $f^{-1} = f$.

15. Let $f : A \rightarrow B$ and $g : B \rightarrow C$. Prove each of the following statements.

   a. If f and g are injective, then $g \circ f$ is injective.

   b. If f and g are surjective, then $g \circ f$ is surjective.

   c. If f and g are bijective, then $g \circ f$ is bijective.

16. Let f and g be bijections of type $A \rightarrow A$ such that $g(f(x)) = x$ for all $x \in A$. Prove that $f(g(x)) = x$ for all $x \in A$.

17. Assume that the functions f and g can be formed into a composition $g \circ f$.

   a. If $g \circ f$ is surjective, what can you say about f or g ?

   b. If $g \circ f$ is injective, what can you say about f or g ?

18. Let $g : A \rightarrow B$ and $h : A \rightarrow C$ and let f be defined by $f(x) = (g(x), h(x))$. Show that each of the following statements holds.

   a. If f is surjective, then g and h are surjective. Find an example to show that the converse is false.

   b. If g or h is injective, then f is injective. Find an example to show that the converse is false.

19. Prove that the equation $ax \bmod n = b \bmod n$ has a solution x if and only if $\gcd(a, n)$ divides b.

20. Use the result of Exercise 19 to prove (2.8): Let $n \geq 1$ and $f : \mathbb{N}_n \to \mathbb{N}_n$ be defined by $f(x) = (ax + b) \bmod n$. Then f has no fixed points if and only if $\gcd(a - 1, n)$ does not divide b.

21. Prove the second part of (2.6). In other words, assume the following facts.

> $f : \mathbb{N}_n \to \mathbb{N}_n$ is defined by $f(x) = (ax + b) \bmod n$.
> f is a bijection, which also means that $\gcd(a, n) = 1$.
> c is an integer such that $f(c) = 0$.
> k is an integer such that $1 = ak + nm$ for some integer m.
> $g : \mathbb{N}_n \to \mathbb{N}_n$ is defined by $g(x) = (kx + c) \bmod n$.

Prove that $g = f^{-1}$.

2.4 Countability

Let's have a short discussion about counting sets that may not be finite. We'll have to examine what it means to count an infinite set and what it means to compare the size of infinite sets. In so doing we'll find some useful techniques that can be applied to questions in computer science. For example, we'll see as a consequence of our discussions that there are some limits on what can be computed. We'll start with some simplifying notation.

2.4.1 Comparing the Size of Sets

Let A and B be sets. If there is a bijection between A and B, we'll denote the fact by writing

$$|A| = |B|.$$

In this case we'll say that A and B have the *same size* or have the *same cardinality*, or are *equipotent*.

example **2.24 Cardinality of a Finite Set**

Let $A = \{(x + 1)^3 \mid x \in \mathbb{N} \text{ and } 1 \leq (x + 1)^3 \leq 3000\}$. Let's find the cardinality of A. After a few calculations we can observe that

$$(0 + 1)^3 = 1, (1 + 1)^3 = 8, \ldots, (13 + 1)^3 = 2744, \text{ and } (14 + 1)^3 = 3375.$$

So we have a bijection $f : \{0, 1, \ldots, 13\} \to A$, where $f(x) = (x + 1)^3$. Therefore, $|A| = |\{0, 1, \ldots, 13\}| = 14$.

end example

2.25   Cardinality of an Infinite Set

Let *Odd* denote the set of odd natural numbers. Then the function $f : \mathbb{N} \to Odd$ defined by $f(x) = 2x + 1$ is a bijection. So *Odd* and $\mathbb{N}$ have the same size and we write $|Odd| = |\mathbb{N}|$.

If there is an injection from A to B, we'll denote the fact by writing

$$|A| \leq |B|.$$

In this case we'll say that the size, or cardinality, of A is *less than or the same as* that of B. Recall that there is an injection from A to B if and only if there is a surjection from B to A. So $|A| \leq |B|$ also means that there is a surjection from B to A.

If there is an injection from A to B but *no* bijection between them, we'll denote the fact by writing

$$|A| < |B|.$$

In this case we'll say that the size, or cardinality, of A *is less than* that of B. So $|A| < |B|$ means that $|A| \leq |B|$ and $|A| \neq |B|$.

2.4.2   Sets That Are Countable

Informally, a set is countable if its elements can be counted in a step-by-step fashion (e.g., count one element each second), even if it takes as many seconds as there are natural numbers. Let's clarify the idea by relating sets that can be counted to subsets of the natural numbers.

If A is a finite set with n elements, then we can represent the elements of A by listing them in the following sequence:

$$x_0, x_1, x_2, \ldots, x_{n-1}.$$

If we associate each x_k with the subscript k, we get a bijection between A and the set $\{0, 1, \ldots, n - 1\}$.

Suppose A is an infinite set such that we can represent all the elements of A by listing them in the following infinite sequence:

$$x_0, x_1, x_2, \ldots, x_n, \ldots.$$

If we associate each x_k with the subscript k, we get a bijection between A and the set $\mathbb{N}$ of natural numbers.

Definition of Countable

The preceding descriptions give us the seeds for a definition of countable. A set is *countable* if it is finite or if there is a bijection between it and $\mathbb{N}$. In the latter case, the set is said to be *countably infinite*. In terms of size we can say that a set S is countable if $|S| = |\{0, 1, \ldots, n-1\}|$ for some natural number n or $|S| = |\mathbb{N}|$. If a set is not countable, it is said to be *uncountable*.

Countable Properties (2.9)

a. Every subset of $\mathbb{N}$ is countable.

b. S is countable if and only if $|S| \leq |\mathbb{N}|$.

c. Any subset of a countable set is countable.

d. Any image of a countable set is countable.

Proof. We'll prove (a) and (b) and leave (c) and (d) as exercises. Let S be a subset of $\mathbb{N}$. If S is finite, then it is countable by definition. So assume that S is infinite. Now since S is a set of natural numbers, it has a smallest element that we'll represent by x_0. Next, we'll let x_1 be the smallest element of the set $S - \{x_0\}$. We'll continue in this manner, letting x_n be the smallest element of $S - \{x_0, \ldots, x_{n-1}\}$. In this way we obtain an infinite listing of the elements of S:

$$x_0, x_1, x_2, \ldots, x_n, \ldots.$$

Notice that each element $m \in S$ is in the listing because there are at most m elements of S that are less than m. So m must be represented as one of the elements $x_0, x_1, x_2, \ldots, x_m$ in the sequence. The association x_k to k gives a bijection between S and $\mathbb{N}$. So $|S| = |\mathbb{N}|$ and thus S is countable.

(b) If S is countable then $|S| \leq |\mathbb{N}|$ by definition. So assume that $|S| \leq |\mathbb{N}|$. Then there is an injection $f : S \to \mathbb{N}$. So $|S| = |f(S)|$. Since $f(S)$ is a subset of $\mathbb{N}$, it is countable by (a). Therefore $f(S)$ is either finite or $|f(S)|=|\mathbb{N}|$. So S is either finite or $|S| = |f(S)|=|\mathbb{N}|$. QED

Techniques to Show Countability

An interesting and useful fact about countablity is that the set $\mathbb{N} \times \mathbb{N}$ is countable. We'll state it for the record.

Theorem (2.10)

$\mathbb{N} \times \mathbb{N}$ is a countable set.

Proof: We need to describe a bijection between $\mathbb{N} \times \mathbb{N}$ and $\mathbb{N}$. We'll do this by arranging the elements of $\mathbb{N} \times \mathbb{N}$ in such a way that they can be easily counted. One way to do this is shown in the following listing, where each row lists a sequence of tuples in $\mathbb{N} \times \mathbb{N}$ followed by a corresponding sequence of natural numbers.

$$
\begin{array}{lcl}
(0,0), & \longleftrightarrow & 0, \\
(0,1),(1,0), & \longleftrightarrow & 1, 2 \\
(0,2),(1,1),(2,0), & \longleftrightarrow & 3, 4, 5, \\
\quad \vdots & \quad \vdots & \vdots \\
(0,n),\cdots & \longleftrightarrow & (n^2 + n)/2, \cdots \\
\quad \vdots & \quad \vdots & \vdots
\end{array}
$$

Notice that each row of the listing contains all the tuples whose components add up to the same number. For example, the sequence $(0, 2)$, $(1, 1)$, $(2, 0)$ consists of all tuples whose components add up to 2. So we have a bijection between $\mathbb{N} \times \mathbb{N}$ and $\mathbb{N}$. Therefore, $\mathbb{N} \times \mathbb{N}$ is countable. QED.

We should note that the bijection described in (2.10) is called *Cantor's pairing function*. It maps each pair of natural numbers (x, y) to the natural number

$$
\frac{(x + y)^2 + 3x + y}{2}.
$$

We can use (2.10) to prove the following result that the union of a countable collection of countable sets is countable.

Counting Unions of Countable Sets   (2.11)

If S_0, S_1, ..., S_n, ... is a sequence of countable sets, then the union

$$
S_0 \cup S_1 \cup \cdots \cup S_n \cup \cdots
$$

is a countable set.

Proof: Since each set is countable, its elements can be indexed by natural numbers. So for each set S_n we'll list its elements as x_{n0}, x_{n1}, x_{n2}, ... If S_n is a finite set then we'll list one of its elements repeatedly to make the listing infinite. In the same way, if there are only finitely many sets, then we'll list one of the sets repeatedly to make the sequence infinite. In this way we can associate each tuple (m, n) in $\mathbb{N} \times \mathbb{N}$ with an element x_{mn} in the union of the given sets. The mapping may not be a bijection since some elements of the union might be repeated in the listings. But the mapping is a surjection from $\mathbb{N} \times \mathbb{N}$ to the union of the sets. So, since $\mathbb{N} \times \mathbb{N}$ is countable, it follows from (2.9d) that the union is countable. QED.

2.26 Countability of the Rationals

We'll show that the set $\mathbb{Q}$ of rational numbers is countable by showing that $|\mathbb{Q}| = |\mathbb{N}|$. Let $\mathbb{Q}^+$ denote the set of positive rational numbers. So we can represent $\mathbb{Q}^+$ as the following set of fractions, where repetitions are included (e.g., $1/1$ and $2/2$ are both elements of the set).

$$\mathbb{Q}^+ = \{m/n \mid m,\ n \in \mathbb{N} \text{ and } n \neq 0\}.$$

Now we'll associate each fraction m/n with the tuple (m, n) in $\mathbb{N} \times \mathbb{N}$. This association is an injection, so we have $|\mathbb{Q}^+| \leq |\mathbb{N} \times \mathbb{N}|$. Since $\mathbb{N} \times \mathbb{N}$ is countable, it follows that $\mathbb{Q}^+$ is countable. In the same way, the set $\mathbb{Q}^-$ of negative rational numbers is countable. Now we can write $\mathbb{Q}$ as the union of three countable sets:

$$\mathbb{Q} = \mathbb{Q}^+ \cup \{0\} \cup \mathbb{Q}^-.$$

Since each set in the union is countable, it follows from (2.11) that the union of the sets is countable.

Counting Strings

An important consequence of (2.11) is the following fact about the countability of the set of all strings over a finite alphabet.

Theorem   **(2.12)**

The set A^* of all strings over a finite alphabet A is countably infinite.

Proof: For each $n \in \mathbb{N}$, let A_n be the set of strings over A having length n. It follows that A^* is the union of the sets $A_0, A_1, \ldots, A_n, \ldots$. Since each set A_n is finite, we can apply (2.11) to conclude that A^* is countable. QED.

2.4.3 Diagonalization

Let's discuss a classic construction technique, called *diagonalization*, which is quite useful in several different settings that deal with counting. The technique was introduced by Cantor when he showed that the real numbers are uncountable. Here is a general description of diagonalization.

Diagonalization   **(2.13)**

Let A be an alphabet with two or more symbols and let $S_0, S_1, \ldots, S_n, \ldots$, be a countable listing of sequences of the form $S_n = (a_{n0}, a_{n1}, \ldots, a_{nn}, \ldots)$, where $a_{ni} \in A$. The sequences are listed as the rows of the following infinite matrix.

Continued ➤

➡ ➡

	0	1	2	$\cdots$	n	$\cdots$
S_0	a_{00}	a_{01}	a_{02}	$\cdots$	a_{0n}	$\cdots$
S_1	a_{10}	a_{11}	a_{12}	$\cdots$	a_{1n}	$\cdots$
S_2	a_{20}	a_{21}	a_{22}	$\cdots$	a_{2n}	$\cdots$
$\vdots$	$\vdots$	$\vdots$	$\vdots$	$\ddots$	$\vdots$	$\vdots$
S_n	a_{n0}	a_{n1}	a_{n2}	$\cdots$	a_{nn}	$\cdots$
$\vdots$	$\vdots$	$\vdots$	$\vdots$	$\vdots$	$\vdots$	$\ddots$

Then there is a sequence $S = (a_0, a_1, a_2, \ldots, a_n, \ldots)$ over A that is *not* in the original list. We can construct S from the list of diagonal elements (a_{00}, a_{11}, a_{22}, ..., a_{nn}, ...) by changing each element so that $a_n \neq a_{nn}$ for each n. Therefore, S differs from each S_n at the nth element. For example, pick two elements $x, y \in A$ and define

$$a_n = \begin{cases} x & \text{if } a_{nn} = y \\ y & \text{if } a_{nn} \neq y. \end{cases}$$

Uncountable Sets

Now we're in position to give some examples of uncountable sets. We'll demonstrate the method of Cantor, which uses proof by contradiction together with the diagonalization technique.

example   2.27 Uncountability of the Reals

We'll show that the set $\mathbb{R}$ of real numbers is uncountable. It was shown in Exercise 6e of Section 2.3 that there is a bijection between R and the set U of real numbers between 0 and 1. So $|\mathbb{R}| = |U|$ and we need only show that U is uncountable. Assume, by way of contradiction, that U is countable. Then we can list all the numbers between 0 and 1 as a countable sequence

$$r_0, \ r_1, \ r_2, \ \ldots, \ r_n, \ \ldots.$$

Each real number between 0 and 1 can be represented as an infinite decimal. So for each n there is a representation $r_n = 0.d_{n0}d_{n1}\ldots d_{nn}\ldots$, where each d_{ni} is a decimal digit. Since we can also represent r_n by the sequence of decimal digits $(d_{n0}, d_{n1}, \ldots, d_{nn}, \ldots)$, it follows by diagonalization (2.13) that there is an infinite decimal that is not in the list. For example, we'll choose the digits 1 and 2 to construct the number $s = 0.s_0s_1s_2\ldots$ where

$$s_k = \begin{cases} 1 & \text{if } d_{kk} = 2 \\ 2 & \text{if } d_{kk} \neq 2. \end{cases}$$

So $0 < s < 1$ and s differs from each r_n at the nth decimal place. Thus s is not in the list, contrary to our assumption that we have listed all numbers in U. So the set U is uncountable, and hence $\mathbb{R}$ is also uncountable.

end example

example 2.28 Natural Number Functions

How many different functions are there from $\mathbb{N}$ to $\mathbb{N}$? We'll show that the set of all such functions is uncountable. Assume, by way of contradiction, that the set is countable. Then we can list all the functions type $\mathbb{N} \to \mathbb{N}$ as $f_0, f_1, \ldots f_n, \ldots$. We'll represent each function f_n as the sequence of its values $(f_n(0), f_n(1), \ldots, f_n(n), \ldots)$. Now (2.13) tells us there is a function missing from the list, which contradicts our assumption that all functions are in the list. So the set of all functions of type $\mathbb{N} \to \mathbb{N}$ is uncountable.

For example, we might choose the numbers 1, 2 $\in \mathbb{N}$ and define a function $g : \mathbb{N} \to \mathbb{N}$ by

$$g(n) = \begin{cases} 1 & \text{if } f_n(n) = 2 \\ 2 & \text{if } f_n(n) \neq 2. \end{cases}$$

Then the sequence of values $(g(0), g(1), \ldots, g(n), \ldots)$ is different from each of the sequences for the listed functions because $g(n) \neq f_n(n)$ for each n. There are many different ways to find a function that it is not in the list. For example, we could define g by

$$g(n) = f_n(n) + 1.$$

It follows that $g(n) \neq f_n(n)$ for each n. So g cannot be in the list $f_0, f_1, \ldots f_n, \ldots$.

end example

2.4.4 Limits on Computability

Let's have a short discussion about whether there are limits on what can be computed. As another application of (2.12) we can answer the question: How many programs can be written in your favorite programming language? The answer is countably infinite. Here is the result.

Theorem	(2.14)
The set of programs for a programming language is countably infinite.	

Proof: One way to see this is to consider each program as a finite string of symbols over a fixed finite alphabet A. For example, A might consist of all characters that can be typed from a keyboard. Now we can proceed as in the proof of (2.12). For each natural number n, let P_n denote the set of all programs that are strings of length n over A. For example, the program

$$\{\text{print}(\text{'help'})\}$$

is in P_{15} because it's a string of length 15. So the set of all programs is the union of the sets $P_0, P_1, \ldots, P_n, \ldots$. Since each P_n is finite, hence countable, it follows from (2.11) that the union is countable. QED

Not Everything Is Computable

Since there are "only" a countable number of computer programs, it follows that there are limits on what can be computed. For example, there are an uncountable number of functions of type $\mathbb{N} \to \mathbb{N}$. So there are programs to calculate only a countable subset of these functions.

Can any real number be computed to any given number of decimal places? The answer is no. The reason is that there are "only" a countable number of computer programs (2.14) but the set of real numbers is uncountable. Therefore, there are only a countable number of computable numbers in $\mathbb{R}$ because each computable number needs a program to compute it. If we remove the computable numbers from $\mathbb{R}$, the resulting set is still uncountable. Can you see why? So most real numbers cannot be computed.

The rational numbers can be computed, and there are also many irrational numbers that can be computed. Pi is the most famous example of a computable irrational number. In fact, there are countably infinitely many computable irrational numbers.

Higher Cardinalities

It's easy to find infinite sets having many different cardinalities because Cantor proved that there are more subsets of a set than there are elements of the set. In other words, for any set A, we have the following result.

Theorem	(2.15)
$$|A| < |\text{power}(A)|.$$	

We know this is true for finite sets. But it's also true for infinite sets. We'll discuss the proof in an exercise. Notice that if A is countably infinite, then we can conclude from (2.15) that power(A) is uncountable. So, for example, we can conclude that power(N) is uncountable.

For another example, we might wonder how many different languages there are over a finite alphabet such as $\{a, b\}$. Since a language over $\{a, b\}$ is a set of strings over $\{a, b\}$, it follows that such a language is a subset of $\{a, b\}^*$, the set of all strings over $\{a, b\}$. So the set of all languages over $\{a, b\}$ is power($\{a, b\}^*$). From (2.12) we can conclude that $\{a, b\}^*$ is countably infinite. In other words, we have $|\{a, b\}^*| = |\mathbb{N}|$. So we can use (2.15) to obtain

$$|\mathbb{N}| = |\{a, b\}^*| < |\text{power}(\{a, b\}^*)|.$$

Therefore, power($\{a, b\}^*$) is uncountable, which is the same as saying that there are uncountably many languages over the alphabet $\{a, b\}$. Of course, this generalizes to any finite alphabet. So we have the following statement.

Theorem   (2.16)

There are uncountably many languages over a finite alphabet.

We can use (2.15) to find infinite sequences of sets of higher and higher cardinality. For example, we have

$$|\mathbb{N}| < |\text{power}(\mathbb{N})| < |\text{power}(\text{power}(\mathbb{N}))| < \cdots.$$

Can we associate these sets with more familiar sets? Sure, it can be shown that $|\mathbb{R}| = |\text{power}(\mathbb{N})|$, which we'll discuss in an exercise. So we have

$$|\mathbb{N}| < |\mathbb{R}| < |\text{power}(\text{power}(\mathbb{N}))| < \cdots.$$

Is there any "well-known" set S such that $|S| = |\text{power}(\text{power}(\mathbb{N}))|$? Since the real numbers are hard enough to imagine, how can we comprehend all the elements in power(power(N))? Luckily, in computer science we will seldom, if ever, have occasion to worry about sets having higher cardinality than the set of real numbers.

The Continuum Hypothesis

We'll close the discussion with a question: Is there a set S whose cardinality is between that of $\mathbb{N}$ and that of the real numbers $\mathbb{R}$? In other words, does there exist a set S such that $|\mathbb{N}| < |S| < |\mathbb{R}|$? The answer is that no one knows. Interestingly, it has been shown that people who assume that the answer is yes won't run into any contradictions by using the assumption in their reasoning. Similarly, it has been shown that people who assume that the answer is no won't run into any contradictions by using the assumption in their arguments! The assumption that the answer is no is called the *continuum hypothesis*.

If we accept the continuum hypothesis, then we can use it as part of a proof technique. For example, suppose that for some set S we can show that $|\mathbb{N}| \leq |S| < |\mathbb{R}|$. Then we can conclude that $|\mathbb{N}| = |S|$ by the continuum hypothesis.

 Exercises

Finite Sets

1. Find the cardinality of each set by establishing a bijection between it and a set of the form $\{0, 1, \ldots, n\}$.

   a. $\{2x + 5 \mid x \in \mathbb{N} \text{ and } 1 \leq 2x + 5 \leq 98\}$.
   b. $\{x^2 \mid x \in \mathbb{N} \text{ and } 0 \leq x^2 \leq 500\}$.
   c. $\{2, 5, 8, 11, 14, 17, \ldots, 44, 47\}$.

Countable Infinite Sets

2. Show that each of the following sets is countable by establishing a bijection between the set and $\mathbb{N}$.

   a. The set of even natural numbers.
   b. The set of negative integers.
   c. The set of strings over $\{a\}$.
   d. The set of lists over $\{a\}$ that have even length.
   e. The set $\mathbb{Z}$ of integers.
   f. The set of odd integers.
   g. The set of even integers.

3. Use (2.11) to show that each of the following sets is countable by describing the set as a union of countable sets.

   a. The set of strings over $\{a, b\}$ that have odd length.
   b. The set of all lists over $\{a, b\}$.
   c. The set of all binary trees over $\{a, b\}$.
   d. $\mathbb{N} \times \mathbb{N} \times \mathbb{N}$.

Diagonalization

4. For each countable set of infinite sequences, use diagonalization (2.13) to construct an infinite sequence of the same type that is not in the set.

   a. $\{(f_n(0), f_n(1), \ldots, f_n(n), \ldots) \mid f_n(k) \in \{\text{hello, world}\} \text{ for } n, k \in \mathbb{N}\}$.
   b. $\{(f(n, 0), f(n, 1), \ldots, f(n, n), \ldots) \mid f(n, k) \in \{a, b, c\} \text{ for } n, k \in \mathbb{N}\}$.
   c. $\{\{a_{n0}, a_{n1}, \ldots, a_{nn}, \ldots \} \mid a_{nk} \in \{2, 4, 6, 8\} \text{ for } n, k \in \mathbb{N}\}$.

5. To show that power($\mathbb{N}$) is uncountable, we can proceed by way of contradiction. Assume that it is countable, so that all the subsets of $\mathbb{N}$ can be listed S_0, S_1, S_2, ..., S_n, ... Complete the proof by finding a way to represent each subset of $\mathbb{N}$ as an infinite sequence of 1's and 0's, where 1 means true and 0 means false. Then a contradiction arises by using diagonalization (2.13) to construct an infinite sequence of the same type that represents a subset of $\mathbb{N}$ that is not listed.

Proofs and Challenges

6. Show that if A is uncountable and B is a countable subset of A, then the set $A - B$ is uncountable.

7. Prove each statement about countable sets:

    a. Any subset of a countable set is countable.

    b. Any image of a countable set is countable.

8. Let A be a countably infinite alphabet $A = \{a_0, a_1, a_2, ... \}$. Let A^* denote the set of all strings over A. For each $n \in \mathbb{N}$, let A_n denote the set of all strings in A^* having length n.

    a. Show that A_n is countable for $n \in \mathbb{N}$. *Hint:* Use (2.11).

    b. Show that A^* is countable. *Hint:* Use (2.11) and Part (a).

9. Let finite($\mathbb{N}$) denote the set of all finite subsets of $\mathbb{N}$. Use (2.11) to show that finite($\mathbb{N}$) is countable.

10. We'll start a proof that $|A| < |\text{power}(A)|$ for any set A. Proof: Since each element $x \in A$ can be associated with $\{x\} \in \text{power}(A)$, it follows that $|A| \leq |\text{power}(A)|$. To show that $|A| < |\text{power}(A)|$ we'll assume, by way of contradiction, that there is a bijection $A \to \text{power}(A)$. So each $x \in A$ is associated with a subset S_x of A. Now, define the following subset of A.

$$S = \{x \in A \mid x \notin S_x\}.$$

Since S is a subset of A, our assumed bijection tells us that there must be an element y in A that is associated with S. In other words, $S_y = S$. Find a contradiction by observing where y is and where it is not.

2.5 Chapter Summary

Functions allow us to associate different sets of objects. They are characterized by associating each domain element with a unique codomain element. For any function $f : A \to B$, subsets of the domain A have images in the codomain B

and subsets of B have pre-images in A. The image of A is the range of f. Partial functions need not be defined for all domain elements.

Some functions that are particularly useful in computer science are floor, ceiling, greatest common divisor (with the associated division algorithm), mod, and log.

Composition is a powerful tool for constructing new functions from known functions. Three functions that are useful in programming with lists are sequence, distribute, and pairs. The map function is a useful tool for computing lists of values of a function.

Three important properties of functions that allow us to compare sets are injective, surjective, and bijective. These properties are useful in describing the pigeonhole principle and in working with ciphers and hash functions. These properties are also useful in comparing the cardinality of sets.

A set is countable if it is finite or has the same cardinality as the set of natural numbers. Countable unions of countable sets are countable. The set of all computer programs is countable. The diagonalization technique can be used to show that some countable listings are not exhaustive. It can also be used to show that some sets, such as the real numbers, are uncountable. So we can't compute all the real numbers. Any set has smaller cardinality than its power set, even when the set is infinite.

chapter 3

Construction Techniques

When we build, let us think that we build forever.

　　　　　—John Ruskin (1819–1900)

To construct an object, we need some kind of description. If we're lucky, the description might include a construction technique. Otherwise, we may need to use our wits and our experience to construct the object. This chapter focuses on gaining some construction experience.

　　The only way to learn a technique is to use it on a wide variety of problems. We'll present each technique in the framework of objects that occur in computer science, and as we go along, we'll extend our knowledge of these objects. We'll begin by introducing the technique of inductive definition for sets. Then we'll discuss techniques for describing recursively defined functions and procedures. Last but not least, we'll introduce grammars for describing sets of strings.

　　There are usually two parts to solving a problem. The first is to guess at a solution and the second is to verify that the guess is correct. The focus of this chapter is to introduce techniques to help us make good guesses. We'll usually check a few cases to satisfy ourselves that our guesses are correct. In the next chapter we'll study inductive proof techniques that can be used to actually prove correctness of claims about objects constructed by the techniques of this chapter.

chapter guide

3.1 (Inductively Defined Sets) introduces the inductive definition technique. We'll apply the technique by defining various sets of numbers, strings, lists, binary trees, and Cartesian products.

3.2 (Recursive Functions and Procedures) introduces the technique of recursive definition for functions and procedures. We'll apply the technique to functions and procedures that process numbers, strings, lists, and binary trees.

We'll solve the repeated element problem and the power set problem, and we'll construct some functions for infinite sequences.

3.3 (Grammars) introduces the idea of a grammar as a way to describe a language. We'll see that grammars describe the strings of a language in an inductive fashion, and we'll see that they provide recursive rules for testing whether a string belongs to a language.

3.1 Inductively Defined Sets

When we write down an informal statement such as $A = \{3, 5, 7, 9, \dots\}$, most of us will agree that we mean the set $A = \{2k + 3 \mid k \in \mathbb{N}\}$. Another way to describe A is to observe that $3 \in A$, that $x \in A$ implies $x + 2 \in A$, and that the only way an element gets in A is by the previous two steps. This description has three ingredients, which we'll state informally as follows:

1. There is a starting element (3 in this case).

2. There is a construction operation to build new elements from existing elements (addition by 2 in this case).

3. There is a statement that no other elements are in the set.

Definition of Inductive Definition

This process is an example of an *inductive definition* of a set. The set of objects defined is called an *inductive set*. An inductive set consists of objects that are constructed, in some way, from objects that are already in the set. So nothing can be constructed unless there is at least one object in the set to start the process. Inductive sets are important in computer science because the objects can be used to represent information and the construction rules can often be programmed. We give the following formal definition.

An inductive definition of a set S consists of three steps: (3.1)

Basis: Specify one or more elements of S.

Induction: Give one or more rules to construct new elements of S from existing elements of S.

Closure: State that S consists exactly of the elements obtained by the basis and induction steps. This step is usually assumed rather than stated explicitly.

The closure step is a very important part of the definition. Without it, there could be lots of sets satisfying the first two steps of an inductive definition. For example, the two sets $\mathbb{N}$ and $\{3, 5, 7, \dots\}$ both contain the number 3, and if x

is in either set, then so is $x + 2$. It's the closure statement that tells us that the only set defined by the basis and induction steps is $\{3, 5, 7, \ldots\}$. So the closure statement tells us that we're defining exactly one set, namely, the smallest set satisfying the basis and induction steps. We'll always omit the specific mention of closure in our inductive definitions.

The *constructors* of an inductive set are the basis elements and the rules for constructing new elements. For example, the inductive set $\{3, 5, 7, 9, \ldots\}$ has two constructors, the number 3 and the operation of adding 2 to a number.

For the rest of this section we'll use the technique of inductive definition to construct sets of objects that are often used in computer science.

3.1.1 Numbers

The set of natural numbers $\mathbb{N} = \{0, 1, 2, \ldots\}$ is an inductive set. Its basis element is 0, and we can construct a new element from an existing one by adding the number 1. So we can write an inductive definition for $\mathbb{N}$ in the following way.

Basis: $0 \in \mathbb{N}$.

Induction: If $n \in \mathbb{N}$, then $n + 1 \in \mathbb{N}$.

The constructors of $\mathbb{N}$ are the integer 0 and the operation that adds 1 to an element of $\mathbb{N}$. The operation of adding 1 to n is called the *successor* function, which we write as

$$\text{succ}(n) = n + 1.$$

Using the successor function, we can rewrite the induction step in the above definition of $\mathbb{N}$ in the alternative form

$$\text{If } n \in \mathbb{N}, \text{ then succ}(n) \in \mathbb{N}.$$

So we can say that $\mathbb{N}$ is an inductive set with two constructors, 0 and succ.

example 3.1 Some Familiar Odd Numbers

We'll give an inductive definition of $A = \{1, 3, 7, 15, 31, \ldots\}$. Of course, the basis case should place 1 in A. If $x \in A$, then we can construct another element of A with the expression $2x + 1$. So the constructors of A are the number 1 and the operation of multiplying by 2 and adding 1. An inductive definition of A can be written as follows:

Basis: $1 \in A$.

Induction: If $x \in A$, then $2x + 1 \in A$.

end example

example 3.2 Some Even and Odd Numbers

Is the following set inductive?

$$A = \{2, 3, 4, 7, 8, 11, 15, 16, \ldots\}.$$

It might be easier if we think of A as the union of the two sets

$$B = \{2, 4, 8, 16, \ldots\} \text{ and } C = \{3, 7, 11, 15, \ldots\}.$$

Both of these sets are inductive. The constructors of B are the number 2 and the operation of multiplying by 2. The constructors of C are the number 3 and the operation of adding by 4. We can combine these definitions to give an inductive definition of A.

Basis: $2, 3 \in A$.

Induction: If $x \in A$ and x is odd, then $x + 4 \in A$.

If $x \in A$ and x is even, then $2x \in A$.

This example shows that there can be more than one basis element, more than one induction rule, and tests can be included.

end example

example 3.3 Communicating with a Robot

Suppose we want to communicate the idea of the natural numbers to a robot that knows about functions, has a loose notion of sets, and can follow an inductive definition. Symbols like 0, 1, $\ldots$, and $+$ make no sense to the robot. How can we convey the idea of $\mathbb{N}$? We'll tell the robot that N is the name of the set we want to construct.

Suppose we start by telling the robot to put the symbol 0 in N. For the induction case we need to tell the robot about the successor function. We tell the robot that $s : N \rightarrow N$ is a function, and whenever an element $x \in N$, then put the element $s(x) \in N$. After a pause, the robot says, "$N = \{0\}$ because I'm letting s be the function defined by $s(0) = 0$."

Since we don't want $s(0) = 0$, we have to tell the robot that $s(0) \neq 0$. Then the robot says, "$N = \{0, s(0)\}$ because $s(s(0)) = 0$." So we tell the robot that $s(s(0)) \neq 0$. Since this could go on forever, let's tell the robot that $s(x)$ does not equal any previously defined element. Do we have it? Yes. The robot responds with "$N = \{0, s(0), s(s(0)), s(s(s(0))), \ldots\}$." So we can give the robot the following definition:

Basis: $0 \in N$.

Induction: If $x \in N$, then put $s(x) \in N$, where $s(x) \neq 0$ and $s(x)$ is not equal to any previously defined element of N.

This definition of the natural numbers—along with a closure statement—is due to the mathematician and logician Giuseppe Peano (1858–1932).

example 3.4 Communicating with Another Robot

Suppose we want to define the natural numbers for a robot that knows about sets and can follow an inductive definition. How can we convey the idea of $\mathbb{N}$ to the robot? Since we can use only the notation of sets, let's use $\varnothing$ to stand for the number 0.

What about the number 1? Can we somehow convey the idea of 1 using the empty set? Let's let $\{\varnothing\}$ stand for 1. What about 2? We can't use $\{\varnothing, \varnothing\}$, because $\{\varnothing, \varnothing\} = \{\varnothing\}$. Let's let $\{\varnothing, \{\varnothing\}\}$ stand for 2 because it has two distinct elements. Notice the little pattern we have going: If s is the set standing for a number, then $s \cup \{s\}$ stands for the successor of the number.

Starting with $\varnothing$ as the basis element, we have an inductive definition. Letting *Nat* be the set that we are defining for the robot, we have the following inductive definition.

Basis: $\varnothing \in Nat$.

Induction: If $s \in Nat$, then $s \cup \{s\} \in Nat$.

For example, since 2 is represented by the set $\{\varnothing, \{\varnothing\}\}$, the number 3 is represented by the set

$$\{\varnothing, \{\varnothing\}\} \cup \{\{\varnothing, \{\varnothing\}\}\} = \{\varnothing, \{\varnothing\}, \{\varnothing, \{\varnothing\}\}\}.$$

This is not fun. After a while we might try to introduce some of our own notation to the robot. For example, we'll introduce the decimal numerals in the following way.

$$0 = \varnothing,$$
$$1 = 0 \cup \{0\},$$
$$2 = 1 \cup \{1\},$$
$$\vdots$$

Now we can think about the natural numbers in the following way.

$$0 = \varnothing,$$
$$1 = 0 \cup \{0\} = \varnothing \cup \{0\} = \{0\},$$
$$2 = 1 \cup \{1\} = \{0\} \cup \{1\} = \{0, 1\},$$
$$\vdots$$

Therefore, each number is the set of numbers that precedes it.

3.1.2 Strings

We often define strings of things inductively without even thinking about it. For example, in high school algebra we might say that an algebraic expression is either a number or a variable, and if A and B are algebraic expressions, then so are (A), $A + B$, $A - B$, AB, and $A \div B$. So the set of algebraic expressions is a set of strings. For example, if x and y are variables, then the following strings are algebraic expressions.

$$x, \quad y, \quad 25, \quad 25x, \quad x + y, \quad (4x + 5y), \quad (x + y)(2yx), \quad 3x \div 4.$$

If we like, we can make our definition more formal by specifying the basis and induction parts. For example, if we let E denote the set of algebraic expressions as we have described them, then we have the following inductive definition for E.

Basis: If x is a variable or a number, then $x \in E$.

Induction: If A, $B \in E$, then (A), $A + B$, $A - B$, AB, $A \div B \in E$.

Let's recall that for an alphabet A, the set of all strings over A is denoted by A^*. This set has the following inductive definition.

All Strings over A   (3.2)

Basis: $\Lambda \in A^*$.

Induction: If $s \in A^*$ and $a \in A$, then $as \in A^*$.

We should note that when we place two strings next to each other in juxtaposition to form a new string, we are concatenating the two strings. So, from a computational point of view, concatenation is the operation we are using to construct new strings.

Recall that any set of strings is called a *language*. If A is an alphabet, then any language over A is one of the subsets of A^*. Many languages can be defined inductively. Here are some examples.

example **3.5 Three Languages**

We'll give an inductive definition for each of three languages.

1. $S = \{a, ab, abb, abbb, \dots\} = \{ab^n \mid n \in \mathbb{N}\}$.

Informally, we can say that the strings of S consist of the letter a followed by zero or more b's. But we can also say that the letter a is in S, and if x is a string in S, then so is xb. This gives us an inductive definition for S.

Basis: $a \in S$.

Induction: If $x \in S$, then $xb \in S$.

2. $S = \{\Lambda, ab, aabb, aaabbb, \ldots\} = \{a^n b^n \mid n \in \mathbb{N}\}$.

Informally, we can say that the strings of S consist of any number of a's followed by the same number of b's. But we can also say that the empty string Λ is in S, and if x is a string in S, then so is axb. This gives us an inductive definition for S.

Basis: $\Lambda \in S$.

Induction: If $x \in S$, then $axb \in S$.

3. $S = \{\Lambda, ab, abab, ababab, \ldots\} = \{(ab)^n \mid n \in \mathbb{N}\}$.

Informally, we can say that the strings of S consist of any number of ab pairs. But we can also say that the empty string Λ is in S, and if x is a string in S, then so is abx. This gives us an inductive definition for S.

Basis: $\Lambda \in S$.

Induction: If $x \in S$, then $abx \in S$.

end example

example **3.6 Decimal Numerals**

Let's give an inductive definition for the set of decimal numerals. Recall that a decimal numeral is a nonempty string of decimal digits. For example, 2340 and 002965 are decimal numerals. If we let D denote the set of decimal numerals, we can describe D by saying that any decimal digit is in D, and if x is in D and d is a decimal digit, then dx is in D. This gives us the following inductive definition for D:

Basis: $\{0, 1, 2, 3, 4, 5, 6, 7, 8, 9\} \subseteq D$.

Induction: If $x \in D$ and d is a decimal digit, then $dx \in D$.

end example

3.1.3 Lists

Recall that a list is an ordered sequence of elements. Let's try to find an inductive definition for the set of lists with elements from a set A. In Chapter 1 we denoted the set of all lists over A by lists(A), and we'll continue to do so. We also mentioned that from a computational point of view the only parts of a nonempty list that can be accessed randomly are its *head* and its *tail*. Head and tail are sometimes called *destructors*, since they are used to destroy a list (take it apart). For example, the list $\langle x, y, z \rangle$ has x as its head and $\langle y, z \rangle$ as its tail, which we write as

$$\text{head}(\langle x, y, z \rangle) = x \quad \text{and} \quad \text{tail}(\langle x, y, z \rangle) = \langle y, z \rangle.$$

We also introduced the operation "cons" to construct lists, where if h is an element and t is a list, the new list whose head is h and whose tail is t is represented by the expression

$$\text{cons}(h, t).$$

So cons is a constructor of lists. For example, we have

$$\text{cons}(x, \langle y, z \rangle) = \langle x, y, z \rangle$$
$$\text{cons}(x, \langle \, \rangle) = \langle x \rangle.$$

The operations cons, head, and tail work nicely together. For example, we can write

$$\langle x, y, z \rangle = \text{cons}(x, \langle y, z \rangle) = \text{cons}(\text{head}(\langle x, y, z \rangle), \text{tail}(\langle x, y, z \rangle)).$$

So if L is any nonempty list, then we have the equation

$$L = \text{cons}(\text{head}(L), \text{tail}(L)).$$

Now we have the proper tools, so let's get down to business and write an inductive definition for lists(A). Informally, we can say that lists(A) is the set of all ordered sequences of elements taken from the set A. But we can also say that $\langle \, \rangle$ is in lists(A), and if L is in lists(A), then so is cons(a, L) for any a in A. This gives us an inductive definition for lists(A), which we can state formally as follows.

All Lists over A   (3.3)

Basis: $\langle \, \rangle \in \text{lists}(A)$.

Induction: If $x \in A$ and $L \in \text{lists}(A)$, then $\text{cons}(x, L) \in \text{lists}(A)$.

3.7 List Membership

Let $A = \{a, b\}$. We'll use (3.3) to show how some lists become members of lists(A). The basis case puts $\langle\ \rangle \in$ lists(A). Since $a \in A$ and $\langle\ \rangle \in$ lists(A), the induction step gives

$$\langle a \rangle = \text{cons}(a, \langle\ \rangle) \in \text{lists}(A).$$

In the same way we get $\langle b \rangle \in$ lists(A). Now since $a \in A$ and $\langle a \rangle \in$ lists(A), the induction step puts $\langle a, a \rangle \in$ lists(A). Similarly, we get $\langle b, a \rangle$, $\langle a, b \rangle$, and $\langle b, b \rangle$ as elements of lists(A), and so on.

end example

A Notational Convenience

It's convenient when working with lists to use an infix notation for cons to simplify the notation for list expressions. We'll use the double colon symbol ::, so that the infix form of cons(x, L) is

$$x :: L.$$

For example, the list $\langle a, b, c \rangle$ can be constructed using cons as

$$\text{cons}(a, \text{cons}(b, \text{cons}(c, \langle\ \rangle))) = \text{cons}(a, \text{cons}(b, \langle c \rangle))$$
$$= \text{cons}(a, \langle b, c \rangle)$$
$$= \langle a, b, c \rangle.$$

Using the infix form, we construct $\langle a, b, c \rangle$ as follows:

$$a :: (b :: (c :: \langle\ \rangle)) = a :: (b :: \langle c \rangle) = a :: \langle b, c \rangle = \langle a, b, c \rangle.$$

The infix form of cons allows us to omit parentheses by agreeing that :: is right associative. In other words, $a :: b :: L = a :: (b :: L)$. Thus we can represent the list $\langle a, b, c \rangle$ by writing

$$a :: b :: c :: \langle\ \rangle \text{ instead of } a :: (b :: (c :: \langle\ \rangle)).$$

Many programming problems involve processing data represented by lists. The operations cons, head, and tail provide basic tools for writing programs to create and manipulate lists. So they are necessary for programmers. Now let's look at a few examples.

3.8 Lists of Binary Digits

Suppose we need to define the set S of all nonempty lists over the set $\{0, 1\}$ with the property that adjacent elements in each list are distinct. We can get an idea about S by listing a few elements:

$$S = \{\langle 0 \rangle, \langle 1 \rangle, \langle 1, 0 \rangle, \langle 0, 1 \rangle, \langle 0, 1, 0 \rangle, \langle 1, 0, 1 \rangle, \ldots\}.$$

Let's try $\langle 0 \rangle$ and $\langle 1 \rangle$ as basis elements of S. Then we can construct a new list from a list $L \in S$ by testing whether head(L) is 0 or 1. If head(L) = 0, then we place 1 at the left of L. Otherwise, we place 0 at the left of L. So we can write the following inductive definition for S.

Basis: $\langle 0 \rangle, \langle 1 \rangle \in S$.

Induction: If $L \in S$ and head(L) = 0, then cons(1, L) $\in S$.
If $L \in S$ and head(L) = 1, then cons(0, L) $\in S$.

The infix form of these induction rules looks like

If $L \in S$ and head(L) = 0, then $1 :: L \in S$.
If $L \in S$ and head(L) = 1, then $0 :: L \in S$.

end example

example **3.9 Lists of Letters**

Suppose we need to define the set S of all lists over $\{a, b\}$ that begin with the single letter a followed by zero or more occurrences of b. We can describe S informally by writing a few of its elements:

$$S = \{\langle a \rangle, \langle a, b \rangle, \langle a, b, b \rangle, \langle a, b, b, b \rangle, \ldots \}.$$

It seems appropriate to make $\langle a \rangle$ the basis element of S. Then we can construct a new list from any list $L \in S$ by attaching the letter b on the right end of L. But cons places new elements at the left end of a list. We can overcome the problem in the following way:

$$\text{If } L \in S, \text{ then cons}(a, \text{cons}(b, \text{tail}(L))) \in S.$$

In infix form the statement reads as follows:

$$\text{If } L \in S, \text{ then } a :: b :: \text{tail}(L) \in S.$$

For example, if $L = \langle a \rangle$, then we construct the list

$$a :: b :: \text{tail}(\langle a \rangle) = a :: b :: \langle \, \rangle = a :: \langle b \rangle = \langle a, b \rangle.$$

So we have the following inductive definition of S:

Basis: $\langle a \rangle \in S$.

Induction: If $L \in S$, then $a :: b :: \text{tail}(L) \in S$.

end example

example 3.10 **All Possible Lists**

Can we find an inductive definition for the set of all possible lists over $\{a, b\}$, including lists that can contain other lists? Suppose we start with lists having a small number of symbols, including the symbols $\langle$ and $\rangle$. Then, for each $n \geq 2$, we can write down the lists made up of n symbols (not including commas). Figure 3.1 shows these listings for the first few values of n.

If we start with the empty list $\langle \; \rangle$, then with a and b we can construct three more lists as follows:

$$a :: \langle \; \rangle = \langle a \rangle \, ,$$
$$b :: \langle \; \rangle = \langle b \rangle \, ,$$
$$\langle \; \rangle :: \langle \; \rangle = \langle \langle \; \rangle \rangle \, .$$

Now if we take these three lists together with $\langle \; \rangle$, then with a and b we can construct many more lists. For example,

$$a :: \langle a \rangle = \langle a, a \rangle \, ,$$
$$\langle a \rangle :: \langle \; \rangle = \langle \langle a \rangle \rangle \, ,$$
$$\langle \langle \; \rangle \rangle :: \langle b \rangle = \langle \langle \langle \; \rangle \rangle , b \rangle \, ,$$
$$\langle b \rangle :: \langle \langle \; \rangle \rangle = \langle \langle b \rangle , \langle \; \rangle \rangle \, .$$

Using this idea, we'll make an inductive definition for the set S of all possible lists over $\{a, b\}$.

Basis: $\langle \; \rangle \in S.$ $\hspace{4cm}$ (3.4)

Induction: If $x \in \{a, b\} \cup S$ and $L \in S$, then $x :: L \in S$.

end example

2	3	4	5	6
$\langle \; \rangle$	$\langle a \rangle$	$\langle \langle \; \rangle \rangle$	$\langle \langle a \rangle \rangle$	$\langle \langle \langle \; \rangle \rangle \rangle$
	$\langle b \rangle$	$\langle a, a \rangle$	$\langle \langle b \rangle \rangle$	$\langle \langle \; \rangle , \langle \; \rangle \rangle$
		$\langle a, b \rangle$	$\langle \langle \; \rangle , a \rangle$	$\langle a, a, \langle \; \rangle \rangle$
		$\langle b, a \rangle$	$\langle \langle \; \rangle , b \rangle$	$\langle a, \langle \; \rangle , a \rangle$
		$\langle b, b \rangle$	$\langle a, \langle \; \rangle \rangle$	$\langle \langle \; \rangle , a, a \rangle$
			$\langle b, \langle \; \rangle \rangle$	$\langle a, b, \langle \; \rangle \rangle$
			$\langle a, a, a \rangle$	$\langle a, b, a, b \rangle$
			$\vdots$	$\vdots$

Figure 3.1 A listing of lists by size.

3.1.4 Binary Trees

Recall that a binary tree is either empty or it has a root with a left and right subtree, each of which is a binary tree. This is an informal inductive description of the set of binary trees. To give a formal definition and to work with binary trees, we need some operations to pick off parts of a tree and to construct new trees.

In Chapter 1 we represented binary trees by lists, where the empty binary tree is denoted by $\langle \ \rangle$ and a nonempty binary tree is denoted by the list $\langle L, x, R \rangle$, where x is the root, L is the left subtree, and R is the right subtree. This gives us the ingredients for a more formal inductive definition of the set of all binary trees.

For convenience we'll let tree(L, x, R) denote the binary tree with root x, left subtree L, and right subtree R. If we still want to represent binary trees as lists, then of course we can write

$$\text{tree}(L, x, R) = \langle L, x, R \rangle.$$

Now suppose A is any set. Then we can describe the set B of all binary trees whose nodes come from A by saying that $\langle \ \rangle$ is in B, and if L and R are in B, then so is tree(L, a, R) for any a in A. This gives us an inductive definition, which we can state formally as follows.

All Binary Trees over A   (3.5)

    Basis: $\langle \ \rangle \in B$.

Induction: If $x \in A$ and $L, R \in B$, then tree(L, x, R) $\in B$.

We also have destructor operations for binary trees. We'll let *left*, *root*, and *right* denote the operations that return the left subtree, the root, and the right subtree, respectively, of a nonempty tree. For example, if

$$T = \text{tree}(L, x, R), \text{ then left}(T) = L, \text{root}(T) = x, \text{ and right}(T) = R.$$

So for any nonempty binary tree T we have

$$T = \text{tree}(\text{left}(T), \text{root}(T), \text{right}(T)).$$

example **3.11** **Binary Trees of Twins**

Let $A = \{0, 1\}$. Suppose we need to work with the set *Twins* of all binary trees T over A that have the following property: The left and right subtrees of each node in T are identical in structure and node content. For example, *Twins* contains the empty tree and any single-node tree. *Twins* also contains the two trees shown in Figure 3.2.

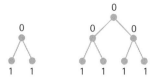

Figure 3.2   Twins as subtrees.

We can give an inductive definition of *Twins* by simply making sure that each new tree has the same left and right subtrees. Here's the definition:

Basis: $\langle \, \rangle \in$ *Twins*.

Induction: If $x \in A$ and $T \in$ *Twins*, then tree(T, x, T) $\in$ *Twins*.

end example

example **3.12 Binary Trees of Opposites**

Let $A = \{0, 1\}$, and suppose that *Opps* is the set of all nonempty binary trees T over A with the following property: The left and right subtrees of each node of T have identical structures, but the 0's and 1's are interchanged. For example, each single node tree is in *Opps*, as well as the two trees shown in Figure 3.3.

Since our set does not include the empty tree, the two singleton trees with nodes 1 and 0 should be the basis trees in *Opps*. The inductive definition of *Opps* can be given as follows:

Basis: tree($\langle \, \rangle$, 0, $\langle \, \rangle$), tree($\langle \, \rangle$, 1, $\langle \, \rangle$) $\in$ *Opps*.

Induction: Let $x \in A$ and $T \in$ *Opps*.
 If root(T) = 0, then
  tree(T, x, tree(right(T), 1, left(T))) $\in$ *Opps*.
 Otherwise,
  tree(T, x, tree(right(T), 0, left(T))) $\in$ *Opps*.

Does this definition work? Try out some examples. See whether the definition builds the four possible three-node trees.

end example

Figure 3.3   Opposites as subtrees.

3.1.5   Cartesian Products of Sets

Let's consider the problem of finding inductive definitions for subsets of the Cartesian product of two sets. For example, the set $\mathbb{N} \times \mathbb{N}$ can be defined inductively by starting with the pair $(0, 0)$ as the basis element. Then, for any pair (x, y) in the set, we can construct the following two pairs.

$$(x, y + 1) \text{ and } (x + 1, y).$$

It seems clear that this definition will define all elements of $\mathbb{N} \times \mathbb{N}$, although some pairs will be defined more than once. For example, the graph in Figure 3.4 shows four points. Notice that the point for the pair $(x + 1, y + 1)$ is constructed from $(x, y + 1)$, but it is also constructed from $(x + 1, y)$.

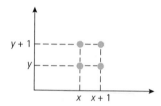

Figure 3.4   Four integer points.

---

example   **3.13   Cartesian Product**

A Cartesian product can be defined inductively if at least one set of the product can be defined inductively. For example, if A is any set, then we have the following inductive definition of $\mathbb{N} \times A$:

Basis: $(0, a) \in \mathbb{N} \times A$ for all $a \in A$.

Induction: If $(x, y) \in \mathbb{N} \times A$, then $(x + 1, y) \in \mathbb{N} \times A$.

end example

---

example   **3.14   Part of a Plane**

Let $S = \{(x, y)\mid x, y \in \mathbb{N} \text{ and } x \leq y\}$. From the point of view of a plane, S is the set of points in the first quadrant with integer coordinates on or above the main diagonal. We can define S inductively as follows:

Basis: $(0, 0) \in S$.

Induction: If $(x, y) \in S$, then $(x, y + 1)$, $(x + 1, y + 1) \in S$.

For example, we can use $(0, 0)$ to construct $(0, 1)$ and $(1, 1)$. From $(0, 1)$ we construct $(0, 2)$ and $(1, 2)$. From $(1, 1)$ we construct $(1, 2)$ and $(2, 2)$. So some pairs get defined more than once.

end example

example **3.15 Describing an Area**

Suppose we need to describe some area as a set of points. From a computational point of view, the area will be represented by discrete points, like pixels on a computer screen. So we can think of the area as a set of ordered pairs (x, y) forming a subset of $\mathbb{N} \times \mathbb{N}$.

To keep things simple we'll describe the area A under the curve of a function f between two points a and b on the x-axis. Figure 3.5 shows a general picture of the area A.

So the area A can be described as the following set of points, where $a, b \in \mathbb{N}$, and $f : \mathbb{N} \to \mathbb{N}$.

$$A = \{(x, y) \mid x, y \in \mathbb{N}, \ a \leq x \leq b, \text{ and } 0 \leq y \leq f(x)\}.$$

There are several ways we might proceed to give an inductive definition of A. For example, we can start with the point $(a, 0)$ on the x-axis. From $(a, 0)$ we can construct the column of points above it and the point $(a + 1, 0)$, from which the next column of points can be constructed. Here's the definition.

Basis: $(a, 0) \in A$.

Induction: If $(x, 0) \in A$ and $x < b$, then $(x + 1, 0) \in A$.
If $(x, y) \in A$ and $y < f(x)$, then $(x, y + 1) \in A$.

For example, the column of points $(a, 0), (a, 1), (a, 2), \ldots, (a, f(a))$ is constructed by starting with the basis point $(a, 0)$ and by repeatedly using the second if-then statement. The first if-then statement constructs the points on the x-axis that are then used to construct the other columns of points. Notice with this definition that each pair is constructed exactly once.

end example

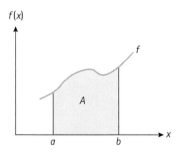

Figure 3.5   Area under a curve.

■ Exercises

Numbers

1. For each of the following inductive definitions, start with the basis element and construct ten elements in the set.

   a.   *Basis:* $3 \in S$.

   Induction: If $x \in S$, then $2x - 1 \in S$.

   b.   *Basis:* $1 \in S$.

   Induction: If $x \in S$, then $2x, 2x + 1 \in S$.

2. Find an inductive definition for each set S.

   a. $\{1, 3, 5, 7, \ldots\}$.
   b. $\{0, 2, 4, 6, 8, \ldots\}$.
   c. $\{-3, -1, 1, 3, 5, \ldots\}$.
   d. $\{\ldots, -7, -4, -1, 2, 5, 8, \ldots\}$.
   e. $\{1, 4, 9, 16, 25, \ldots\}$.
   f. $\{1, 3, 7, 15, 31, 63, \ldots\}$.

3. Find an inductive definition for each set S.

   a. $\{4, 7, 10, 13, \ldots\} \cup \{3, 6, 9, 12, \ldots\}$.
   b. $\{3, 4, 5, 8, 9, 12, 16, 17, \ldots\}$. *Hint:* Write the set as a union.

4. Find an inductive definition for each set S.

   a. $\{x \in \mathbb{N} \mid \text{floor}(x/2) \text{ is even}\}$.
   b. $\{x \in \mathbb{N} \mid \text{floor}(x/2) \text{ is odd}\}$.
   c. $\{x \in \mathbb{N} \mid x \bmod 5 = 2\}$.
   d. $\{x \in \mathbb{N} \mid 2x \bmod 7 = 3\}$.

5. The following inductive definition was given in Example 3.4, the second robot example.

   Basis: $\varnothing \in Nat$.

   Induction: If $s \in Nat$, then $s \cup \{s\} \in Nat$.

   In Example 3.4 we identified natural numbers with the elements of *Nat* by setting $0 = \varnothing$ and $n = n \cup \{n\}$ for $n \neq 0$. Show that $4 = \{0, 1, 2, 3\}$.

Strings

6. Find an inductive definition for each set S of strings.

a. $\{a^n bc^n \mid n \in \mathbb{N}\}$.

b. $\{a^{2n} \mid n \in \mathbb{N}\}$.

c. $\{a^{2n+1} \mid n \in \mathbb{N}\}$.

d. $\{a^m b^n \mid m, n \in \mathbb{N}\}$.

e. $\{a^m bc^n \mid m, n \in \mathbb{N}\}$.

f. $\{a^m b^n \mid m, n \in \mathbb{N}, \text{ where } m > 0\}$.

g. $\{a^m b^n \mid m, n \in \mathbb{N}, \text{ where } n > 0\}$.

h. $\{a^m b^n \mid m, n \in \mathbb{N}, \text{ where } m > 0 \text{ and } n > 0\}$.

i. $\{a^m b^n \mid m, n \in \mathbb{N}, \text{ where } m > 0 \text{ or } n > 0\}$.

j. $\{a^{2n} \mid n \in \mathbb{N}\} \cup \{b^{2n+1} \mid n \in \mathbb{N}\}$.

k. $\{s \in \{a, b\}^* \mid s \text{ has the same number of } a\text{'s and } b\text{'s}\}$.

7. Find an inductive definition for each set S of strings. Note that a *palindrome* is a string that is the same when written in reverse order.

    a. Even palindromes over the set $\{a, b\}$.

    b. Odd palindromes over the set $\{a, b\}$.

    c. All palindromes over the set $\{a, b\}$.

    d. The binary numerals.

8. Let the letters a, b, and c be constants; let the letters x, y, and z be variables; and let the letters f and g be functions of arity 1. We can define the set of terms over these symbols by saying that any constant or variable is a term and if t is a term, then so are $f(t)$ and $g(t)$. Find an inductive definition for the set T of terms.

Lists

9. For each of the following inductive definitions, start with the basis element and construct five elements in the set.

    a.   *Basis:* $\langle a \rangle \in S$.

        Induction: If $x \in S$, then $b :: x \in S$.

    b.   *Basis:* $\langle 1 \rangle \in S$.

        Induction: If $x \in S$, then $2 \cdot \text{head}(x) :: x \in S$.

10. Find an inductive definition for each set S of lists. Use the cons constructor.

    a. $\{\langle a \rangle, \langle a, a \rangle, \langle a, a, a \rangle, \dots\}$.

    b. $\{\langle 1 \rangle, \langle 2, 1 \rangle, \langle 3, 2, 1 \rangle, \dots\}$.

    c. $\{\langle a, b \rangle, \langle b, a \rangle, \langle a, a, b \rangle, \langle b, b, a \rangle, \langle a, a, a, b \rangle, \langle b, b, b, a \rangle, \dots\}$.

    d. $\{L \mid L \text{ has even length over } \{a\}\}$.

    e. $\{L \mid L \text{ has even length over } \{0, 1, 2\}\}$.

    f. $\{L \mid L \text{ has even length over a set } A\}$.

    g. $\{L \mid L \text{ has odd length over } \{a\}\}$.

    h. $\{L \mid L \text{ has odd length over } \{0, 1, 2\}\}$.

    i. $\{L \mid L \text{ has odd length over a set } A\}$.

11. Find an inductive definition for each set S of lists. You may use the "consR" operation, where $\text{consR}(L, x)$ is the list constructed from the list L by adding a new element x on the right end. Similarly, you may use the "headR" and "tailR" operations, which are like head and tail but look at things from the right side of a list.

    a. $\{\langle a \rangle, \langle a, b \rangle, \langle a, b, b \rangle, \dots \}$.

    b. $\{\langle 1 \rangle, \langle 1, 2 \rangle, \langle 1, 2, 3 \rangle, \dots \}$.

    c. $\{ L \in \text{lists}(\{a, b\}) \mid L \text{ has the same number of } a\text{'s and } b\text{'s}\}$.

12. Find an inductive definition for the set S of all lists over $A = \{a, b\}$ that alternate a's and b's. For example, the lists $\langle \ \rangle$, $\langle a \rangle$, $\langle b \rangle$, $\langle a, b, a \rangle$, and $\langle b, a \rangle$ are in S. But $\langle a, a \rangle$ is not in S.

Binary Trees

13. Given the following inductive definition for a set S of binary trees. Start with the basis element and draw pictures of four binary trees in the set. Don't draw the empty subtrees.

    Basis: $\text{tree}(\langle \ \rangle, a, \langle \ \rangle) \in S$.

  Induction: If $T \in S$, then $\text{tree}(\text{tree}(\langle \ \rangle, a, \langle \ \rangle), a, T) \in S$.

14. Find an inductive definition for the set B of binary trees that represent arithmetic expressions that are either numbers in $\mathbb{N}$ or expressions that use operations $+$ or $-$.

15. Find an inductive definition for the set B of nonempty binary trees over $\{a\}$ in which each non-leaf node has two subtrees, one of which is a leaf and the other of which is either a leaf or a member of B.

Cartesian Products

16. Given the following inductive definition for a subset B of $\mathbb{N} \times \mathbb{N}$.

    Basis: $(0, 0) \in B$.

  Induction: If $(x, y) \in B$, then $(x + 1, y)$, $(x + 1, y + 1) \in B$.

    a. Describe the set B as a set of the form $\{(x, y) \mid \text{some property holds}\}$.

    b. Describe those elements in B that get defined in more than one way.

17. Find an inductive definition for each subset S of $\mathbb{N} \times \mathbb{N}$.

    a. $S = \{(x, y) \mid y = x \text{ or } y = x + 1\}$.

    b. $S = \{(x, y) \mid x \text{ is even and } y \leq x/2\}$.

18. Find an inductive definition for each product set S.

    a. $S = \text{lists}(A) \times \text{lists}(A)$ for some set A.

    b. $S = A \times \text{lists}(A)$.

    c. $S = \mathbb{N} \times \text{lists}(\mathbb{N})$.

    d. $S = \mathbb{N} \times \mathbb{N} \times \mathbb{N}$.

Proofs and Challenges

19. Let A be a set. Suppose O is the set of binary trees over A that contains an odd number of nodes. Similarly, let E be the set of binary trees over A that contains an even number of nodes. Find inductive definitions for O and E. *Hint:* You can use O when defining E, and you can use E when defining O.

20. Use Example 3.15 as a guide to construct an inductive definition for the set of points in $\mathbb{N} \times \mathbb{N}$ that describe the area A between two curves f and g defined as follows for two natural numbers a and b:

$$A = \{(x,\, y)\, |x,\, y \in \mathbb{N},\, a \leq x \leq b,\, \text{and}\, g(x) \leq y \leq f(x)\}.$$

21. Prove that a set defined by (3.1) is countable if the basis elements in Step 1 are countable, the outside elements, if any, used in Step 2 are countable, and the rules specified in Step 2 are finite.

3.2 Recursive Functions and Procedures

Since we're going to be constructing functions and procedures in this section, we'd better agree on the idea of a procedure. From a computer science point of view a *procedure* is a program that performs one or more actions. So there is no requirement to return a specific value. For example, the execution of a statement like $\text{print}(x,\, y)$ will cause the values of x and y to be printed. In this case, two actions are performed, and no values are returned. A procedure may also return one or more values through its argument list. For example, a statement like $\text{allocate}(m,\, a,\, s)$ might perform the action of allocating a block of m memory cells and return the values a and s, where a is the beginning address of the block and the s tells whether the allocation was successful.

Definition of Recursively Defined

A function or a procedure is said to be *recursively defined* if it is defined in terms of itself. In other words, a function f is recursively defined if at least one value $f(x)$ is defined in terms of another value $f(y)$, where $x \neq y$. Similarly, a procedure P is recursively defined if the actions of P for some argument x are defined in terms of the actions of P for another argument y, where $x \neq y$.

Many useful recursively defined functions have domains that are inductively defined sets. Similarly, many recursively defined procedures process elements from inductively defined sets. For these cases there are very useful construction techniques. Let's describe the two techniques.

Constructing a Recursively Defined Function   (3.6)

If S is an inductively defined set, then we can construct a function f with domain S as follows:

1. For each basis element $x \in S$, specify a value for $f(x)$.

2. Give rules that, for any inductively defined element $x \in S$, will define $f(x)$ in terms of previously defined values of f.

Any function constructed by (3.6) is recursively defined because it is defined in terms of itself by the induction part of the definition. In a similar way we can construct a recursively defined procedure to process the elements of an inductively defined set.

Constructing a Recursively Defined Procedure   (3.7)

If S is an inductively defined set, we can construct a procedure P to process the elements of S as follows:

1. For each basis element $x \in S$, specify a set of actions for $P(x)$.

2. Give rules that, for any inductively defined element $x \in S$, will define the actions of $P(x)$ in terms of previously defined actions of P.

In the following paragraphs we'll see how (3.6) and (3.7) can be used to construct recursively defined functions and procedures over a variety of inductively defined sets. Most of our examples will be functions. But we'll define a few procedures too.

3.2.1 Numbers

Let's see how some number functions can be defined recursively. To illustrate the idea, suppose we want to calculate the sum of the first n natural numbers for any $n \in \mathbb{N}$. Letting $f(n)$ denote the desired sum, we can write the informal definition

$$f(n) = 0 + 1 + 2 + \cdots + n.$$

We can observe, for example, that $f(0) = 0$, $f(1) = 1$, $f(2) = 3$, and so on. After a while we might notice that $f(3) = f(2) + 3 = 6$ and $f(4) = f(3) + 4 = 10$.

In other words, when $n > 0$, the definition can be transformed in the following way:

$$
\begin{aligned}
f(n) &= 0 + 1 + 2 + \cdots + n \\
&= (0 + 1 + 2 + \cdots + (n-1)) + n \\
&= f(n-1) + n.
\end{aligned}
$$

This gives us the recursive part of a definition of f for any $n > 0$. For the basis case we have $f(0) = 0$. So we can write the following recursive definition for f:

$$
\begin{aligned}
f(0) &= 0, \\
f(n) &= (n-1) + n \quad \text{for } n > 0.
\end{aligned}
$$

There are two alternative forms that can be used to write a recursive definition. One form expresses the definition as an *if-then-else* equation. For example, f can be described in the following way.

$$
f(n) = \text{if } n = 0 \text{ then } 0 \text{ else } f(n-1) + n.
$$

Another form expresses the definition as equations whose left sides determine which equation to use in the evaluation of an expression rather than a conditional like $n > 0$. Such a form is called a *pattern-matching* definition because the equation chosen to evaluate $f(x)$ is determined uniquely by which left side $f(x)$ matches. For example, f can be described in the following way.

$$
\begin{aligned}
f(0) &= 0, \\
f(n+1) &= f(n) + n + 1.
\end{aligned}
$$

For example, $f(3)$ matches $f(n+1)$ with $n = 2$, so we would choose the second equation to evaluate $f(3) = f(2) + 3$, and so on.

A recursively defined function can be evaluated by a technique called *unfolding* the definition. For example, we'll evaluate the expression $f(4)$.

$$
\begin{aligned}
f(4) &= f(3) + 4 \\
&= f(2) + 3 + 4 \\
&= f(1) + 2 + 3 + 4 \\
&= f(0) + 1 + 2 + 3 + 4 \\
&= 0 + 1 + 2 + 3 + 4 \\
&= 10.
\end{aligned}
$$

example 3.16 Using the Floor Function

Let $f : \mathbb{N} \to \mathbb{N}$ be defined in terms of the floor function as follows:

$$f(0) = 0,$$
$$f(n) = f\left(\text{floor}\left(n/2\right)\right) + n \quad \text{for } n > 0.$$

Notice in this case that $f(n)$ is not defined in terms of $f(n-1)$ but rather in terms of $f(\text{floor}(n/2))$. For example, $f(16) = f(8) + 16$. The first few values are $f(0) = 0$, $f(1) = 1$, $f(2) = 3$, $f(3) = 4$, and $f(4) = 7$. We'll calculate $f(25)$.

$$\begin{aligned} f(25) &= f(12) + 25 \\ &= f(6) + 12 + 25 \\ &= f(3) + 6 + 12 + 25 \\ &= f(1) + 3 + 6 + 12 + 25 \\ &= f(0) + 1 + 3 + 6 + 12 + 25 \\ &= 0 + 1 + 3 + 6 + 12 + 25 \\ &= 47. \end{aligned}$$

end example

example 3.17 Adding Odd Numbers

Let $f : \mathbb{N} \to \mathbb{N}$ denote the function to add up the first n odd natural numbers. So f has the following informal definition.

$$f(n) = 1 + 3 + \cdots + (2n + 1).$$

For example, the definition tells us that $f(0) = 1$. For $n > 0$ we can make the following transformation of $f(n)$ into an expression in terms of $f(n-1)$:

$$\begin{aligned} f(n) &= 1 + 3 + \cdots + (2n + 1) \\ &= (1 + 3 + \cdots + (2(n-1) + 1)) + (2n + 1) \\ &= f(n-1) + 2n + 1. \end{aligned}$$

So we can make the following recursive definition of f:

$$f(0) = 1,$$
$$f(n) = f(n-1) + 2n + 1 \quad \text{for } n > 0.$$

Alternatively, we can write the recursive part of the definition as

$$f(n+1) = f(n) + 2n + 3.$$

We can also write the definition in the following if-then-else form.

$$f(n) = \text{if } n = 0 \text{ then } 1 \text{ else } f(n-1) + 2n + 1.$$

Here is the evaluation of $f(3)$ using the if-then-else definition:

$$
\begin{aligned}
f(3) &= f(2) + 2(3) + 1 \\
&= f(1) + 2(2) + 1 + 2(3) + 1 \\
&= f(0) + 2(1) + 1 + 2(2) + 1 + 2(3) + 1 \\
&= 1 + 2(1) + 1 + 2(2) + 1 + 2(3) + 1 \\
&= 1 + 3 + 5 + 7 \\
&= 16.
\end{aligned}
$$

end example

 3.18 The Rabbit Problem

The *Fibonacci numbers* are the numbers in the sequence

$$0, 1, 1, 2, 3, 5, 8, 13, \ldots$$

where each number after the first two is computed by adding the preceding two numbers. These numbers are named after the mathematician Leonardo Fibonacci, who in 1202 introduced them in his book *Liber Abaci*, in which he proposed and solved the following problem: Starting with a pair of rabbits, how many pairs of rabbits can be produced from that pair in a year if it is assumed that every month each pair produces a new pair that becomes productive after one month?

For example, if we don't count the original pair and assume that the original pair needs one month to mature and that no rabbits die, then the number of new pairs produced each month for 12 consecutive months is given by the sequence

$$0, 1, 1, 2, 3, 5, 8, 13, 21, 34, 55, 89.$$

The sum of these numbers, which is 232, is the number of pairs of rabbits produced in one year from the original pair.

Fibonacci numbers seem to occur naturally in many unrelated problems. Of course, they can also be defined recursively. For example, letting fib(n) be the nth Fibonacci number, we can define fib recursively as follows:

$$
\begin{aligned}
\text{fib}(0) &= 0, \\
\text{fib}(1) &= 1, \\
\text{fib}(n) &= \text{fib}(n-1) + \text{fib}(n-2) \quad \text{for } n \geq 2.
\end{aligned}
$$

The third line could be written in pattern-matching form as

$$\text{fib}(n+2) = \text{fib}(n+1) + \text{fib}(n).$$

The definition of fib in if-then-else form looks like

$$\text{fib}\,(n) = \text{if } n = 0 \text{ then } 0$$
$$\text{else if } n = 1 \text{ then } 1$$
$$\text{else fib}\,(n-1) + \text{fib}\,(n-2)\,.$$

end example

3.2.2   Strings

Let's see how some string functions can be defined recursively. To illustrate the idea, suppose we want to calculate the complement of any string over the alphabet $\{a, b\}$. For example, the complement of the string *bbab* is *aaba*.

Let $f(x)$ be the complement of x. To find a recursive definition for f we'll start by observing that an arbitrary string over $\{a, b\}$ is either Λ or has the form ay or by for some string y. So we'll define the result of f applied to each of these forms as follows:

$$f(\Lambda) = \Lambda,$$
$$f(ax) = bf(x)\,,$$
$$f(bx) = af(x)\,.$$

For example, we'll evaluate $f(bbab)$:

$$f(bbab) = af(bab)$$
$$= aaf(ab)$$
$$= aabf(b)$$
$$= aabaf(\Lambda)$$
$$= abba\Lambda$$
$$= aaba.$$

Here are some more examples.

example **3.19** **Prefixes of Strings**

Consider the problem of finding the longest common prefix of two strings. A string p is a *prefix* of the string x if x can be written in the form $x = ps$ for some string s. For example, *aab* is the longest common prefix of the two strings *aabbab* and *aababb*.

For two strings x and y over $\{a, b\}$, let $f(x, y)$ be the longest common prefix of x and y. To find a recursive definition for f we can start by observing that an arbitrary string over $\{a, b\}$ is either the empty string Λ or has the form as or *bs* for some string *s*. In other words, the strings over $\{a, b\}$ are an inductively defined set. Here is a definition of f in pattern-matching form:

$$f(\Lambda, x) = \Lambda,$$
$$f(x, \Lambda) = \Lambda,$$
$$f(as, bt) = \Lambda,$$
$$f(bs, at) = \Lambda,$$
$$f(as, at) = af(s, t),$$
$$f(bs, bt) = bf(s, t).$$

We can put the definition in if-then-else form as follows:

$$f(x, y) = \text{if } x = \Lambda \text{ or } y = \Lambda \text{ then } \Lambda$$
$$\text{else if } x = as \text{ and } y = at \text{ then } af(s, t)$$
$$\text{else if } x = bs \text{ and } y = bt \text{ then } bf(s, t)$$
$$\text{else } \Lambda.$$

We'll demonstrate the definition of f by calculating $f(aabbab, aababb)$:

$$f(aabbab, aababb) = af(abbab, ababb)$$
$$= aaf(bbab, babb)$$
$$= aabf(bab, abb)$$
$$= aab\Lambda$$
$$= aab.$$

end example

example 3.20 **Converting Natural Numbers to Binary**

Recall from Section 2.1 that we can represent a natural number x as

$$x = 2(\text{floor}(x/2)) + x \bmod 2.$$

This formula can be used to create a binary representation of x because $x \bmod 2$ is the rightmost bit of the representation. The next bit is found by computing $\text{floor}(x/2) \bmod 2$. The next bit is $\text{floor}(\text{floor}(x/2)/2) \bmod 2$, and so on. For example, we'll compute the binary representation of 13.

$$
\begin{array}{rclcl}
13 & = & 2\,\text{floor}(13/2) + 13 \bmod 2 & = & 2\,(6) + 1 \\
6 & = & 2\,\text{floor}(6/2) + 6 \bmod 2 & = & 2\,(3) + 0 \\
3 & = & 2\,\text{floor}(3/2) + 3 \bmod 2 & = & 2\,(1) + 1 \\
1 & = & 2\,\text{floor}(1/2) + 1 \bmod 2 & = & 2\,(0) + 1
\end{array}
$$

We can read off the remainders in reverse order to obtain 1101, which is the binary representation of 13.

Let's try to use this idea to write a recursive definition for the function "binary" to compute the binary representation for a natural number. If $x = 0$ or $x = 1$, then x is its own binary representation. If $x > 1$, then the binary representation of x is that of $\text{floor}(x/2)$ with the bit $x \bmod 2$ attached on the right end. So our recursive definition of binary can be written as follows, where "cat" is the string concatenation function.

$$\text{binary}(0) = 0, \tag{3.8}$$
$$\text{binary}(1) = 1,$$
$$\text{binary}(x) = \text{cat}(\text{binary}(\text{floor}(x/2)), x \bmod 2) \quad \text{for } x > 1.$$

The definition can be written in if-then-else form as

$$\text{binary}(x) = \text{if } x = 0 \text{ or } x = 1 \text{ then } x$$
$$\text{else cat}(\text{binary}(\text{floor}(x/2)), x \bmod 2).$$

For example, we'll unfold the definition to calculate binary(13):

$$\begin{aligned}
\text{binary}(13) &= \text{cat}(\text{binary}(6), 1) \\
&= \text{cat}(\text{cat}(\text{binary}(3), 0), 1) \\
&= \text{cat}(\text{cat}(\text{cat}(\text{binary}(1), 1), 0), 1) \\
&= \text{cat}(\text{cat}(\text{cat}(1, 1), 0), 1) \\
&= \text{cat}(\text{cat}(11, 0), 1) \\
&= \text{cat}(110, 1) \\
&= 1101.
\end{aligned}$$

end example

3.2.3 Lists

Let's see how some functions that use lists can be defined recursively. To illustrate the idea, suppose we need to define the function $f : \mathbb{N} \to \text{lists}(\mathbb{N})$ that computes the following backward sequence:

$$f(n) = \langle n, n - 1, \ldots, 1, 0 \rangle.$$

With a little help from the cons function for lists, we can transform the informal definition of $f(n)$ into a computable expression in terms of $f(n - 1)$:

$$\begin{aligned}
f(n) &= \langle n, n - 1, \ldots, 1, 0 \rangle \\
&= \text{cons}(n, \langle n - 1, \ldots, 1, 0 \rangle) \\
&= \text{cons}(n, f(n - 1)).
\end{aligned}$$

Therefore, f can be defined recursively by

$$f(0) = \langle 0 \rangle,$$
$$f(n) = \text{cons}(n, f(n-1)) \text{ for } n > 0.$$

This definition can be written in if-then-else form as

$$f(n) = \text{if } n = 0 \text{ then } \langle 0 \rangle \text{ else cons}(n, f(n-1)).$$

To see how the evaluation works, look at the unfolding that results when we evaluate $f(3)$:

$$
\begin{aligned}
f(3) &= \text{cons}(3, f(2)) \\
&= \text{cons}(3, \text{cons}(2, f(1))) \\
&= \text{cons}(3, \text{cons}(2, \text{cons}(1, f(0)))) \\
&= \text{cons}(3, \text{cons}(2, \text{cons}(1, \langle 0 \rangle))) \\
&= \text{cons}(3, \text{cons}(2, \langle 1, 0 \rangle)) \\
&= \text{cons}(3, \langle 2, 1, 0 \rangle) \\
&= \langle 3, 2, 1, 0 \rangle.
\end{aligned}
$$

We haven't given a recursively defined procedure yet. So let's give one for the problem we've been discussing. For example, suppose that $P(n)$ prints out the numbers in the list $\langle n, n-1, \ldots, 0 \rangle$. A recursive definition of P can be written as follows.

$$
\begin{aligned}
P(n): \quad &\textbf{if } n = 0 \textbf{ then } \text{print}(0) \\
&\textbf{else} \\
&\quad \text{print}(n); \\
&\quad P(n-1) \\
&\textbf{fi.}
\end{aligned}
$$

example 3.21 Length of a List

Let S be a set and let "length" be the function of type lists$(S) \to \mathbb{N}$, which returns the number of elements in a list. We can define length recursively by noticing that the length of an empty list is zero and the length of a nonempty list is one plus the length of its tail. A definition follows.

$$\text{length}(\langle \ \rangle) = 0,$$
$$\text{length}(\text{cons}(x, t)) = 1 + \text{length}(t).$$

Recall that the infix form of cons(x, t) is $x :: t$. So we could just as well write the second equation as

$$\text{length}(x :: t) = 1 + \text{length}(t).$$

Also, we could write the recursive part of the definition with a condition as follows:

$$\text{length}(L) = 1 + \text{length}(\text{tail}(L)) \quad \text{for } L \neq \langle \, \rangle.$$

In if-then-else form the definition can be written as follows:

$$\text{length}(L) = \text{if } L = \langle \, \rangle \text{ then } 0 \text{ else } 1 + \text{length}(\text{tail}(L)).$$

The length function can be evaluated by unfolding its definition. For example, we'll evaluate length($\langle a, b, c \rangle$).

$$\begin{aligned}
\text{length}(\langle a, b, c \rangle) &= 1 + \text{length}(\langle b, c \rangle) \\
&= 1 + 1 + \text{length}(\langle c \rangle) \\
&= 1 + 1 + 1 + \text{length}(\langle \, \rangle) \\
&= 1 + 1 + 1 + 0 \\
&= 3.
\end{aligned}$$

end example

example **3.22   The Distribute Function**

Suppose we want to write a recursive definition for the distribute function, which we'll denote by "dist." Recall, for example, that

$$\text{dist}(a, \langle b, c, d, e \rangle) = \langle (a, b), (a, c), (a, d), (a, e) \rangle.$$

To discover the recursive part of the definition, we'll rewrite the example equation by splitting up the lists into head and tail components as follows:

$$\begin{aligned}
\text{dist}(a, \langle b, c, d, e \rangle) &= \langle (a, b), (a, c), (a, d), (a, e) \rangle \\
&= (a, b) :: \langle (a, c), (a, d), (a, e) \rangle \\
&= (a, b) :: \text{dist}(a, \langle c, d, e \rangle).
\end{aligned}$$

That's the key to the recursive part of the definition. Since we are working with lists, the basis case is dist($a, \langle \, \rangle$), which we define as $\langle \, \rangle$. So the recursive definition can be written as follows:

$$\text{dist}(x, \langle \, \rangle) = \langle \, \rangle,$$
$$\text{dist}(x, h :: T) = (x, h) :: \text{dist}(x, T).$$

For example, we'll evaluate the expression dist($3, \langle 10, 20 \rangle$):

$$\begin{aligned}
\text{dist}(3, \langle 10, 20 \rangle) &= (3, 10) :: \text{dist}(3, \langle 20 \rangle) \\
&= (3, 10) :: (3, 20) :: \text{dist}(3, \langle \, \rangle) \\
&= (3, 10) :: (3, 20) :: \langle \, \rangle \\
&= (3, 10) :: \langle (3, 20) \rangle \\
&= \langle (3, 10), (3, 20) \rangle.
\end{aligned}$$

An if-then-else definition of dist takes the following form:

$$\text{dist}(x, L) = \text{if } L = \langle\,\rangle \text{ then } \langle\,\rangle$$
$$\text{else } (x, \text{head}(L)) :: \text{dist}(x, \text{tail}(L)).$$

end example

example **3.23 The Pairs Function**

Recall that the "pairs" function creates a list of pairs of corresponding elements from two lists. For example,

$$\text{pairs}(\langle a,\ b,\ c\rangle,\ \langle 1,\ 2,\ 3\rangle) = \langle (a,\ 1),\ (b,\ 2),\ (c,\ 3)\rangle.$$

To discover the recursive part of the definition, we'll rewrite the example equation by splitting up the lists into head and tail components as follows:

$$\text{pairs}(\langle a, b, c\rangle, \langle 1, 2, 3\rangle) = \langle (a, 1), (b, 2), (c, 3)\rangle$$
$$= (a, 1) :: \langle (b, 2), (c, 3)\rangle$$
$$= (a, 1) :: \text{pairs}(\langle b, c\rangle, \langle 2, 3\rangle).$$

Now the pairs function can be defined recursively by the following equations:

$$\text{pairs}\,(\langle\,\rangle, \langle\,\rangle) = \langle\,\rangle,$$
$$\text{pairs}\,(x :: T, y :: U) = (x, y) :: \text{pairs}(T, U).$$

For example, we'll evaluate the expression pairs($\langle a,\ b\rangle$, $\langle 1,\ 2\rangle$):

$$\text{pairs}\,(\langle a, b\rangle, \langle 1, 2\rangle) = (a, 1) :: \text{pairs}\,(\langle b\rangle, \langle 2\rangle)$$
$$= (a, 1) :: (b, 2) :: \text{pairs}\,(\langle\,\rangle, \langle\,\rangle)$$
$$= (a, 1) :: (b, 2) :: \langle\,\rangle$$
$$= (a, 1) :: \langle (b, 2)\rangle$$
$$= \langle (a, 1), (b, 2)\rangle.$$

end example

example **3.24 The ConsRight Function**

Suppose we need to give a recursive definition for the sequence function. Recall, for example, that seq(4) = $\langle 0, 1, 2, 3, 4\rangle$. Good old "cons" doesn't seem up to the task. For example, if we somehow have computed seq(3), then cons(4, seq(3)) = $\langle 4, 0, 1, 2, 3\rangle$. It would be nice if we had a constructor to place an element on the right of a list, just as cons places an element on the left of a list. We'll write a definition for the function "consR" to do just that. For example, we want

$$\text{consR}(\langle a,\ b,\ c \rangle,\ d) = \langle a,\ b,\ c,\ d \rangle.$$

We can get an idea of how to proceed by rewriting the previous equation as follows in terms of the infix form of cons:

$$\begin{aligned}
\text{consR}(\langle a,b,c \rangle, d) &= \langle a,b,c,d \rangle \\
&= a :: \langle b,c,d \rangle \\
&= a :: \text{consR}(\langle b,c \rangle, d).
\end{aligned}$$

So the clue is to split the list $\langle a,\ b,\ c \rangle$ into its head and tail. We can write the recursive definition of consR using the if-then-else form as follows:

$$\begin{aligned}
\text{consR}(L, a) = \ &\text{if } L = \langle\ \rangle \text{ then } \langle a \rangle \\
&\text{else head}(L) :: \text{consR}(\text{tail}(L), a).
\end{aligned}$$

This definition can be written in pattern-matching form as follows:

$$\begin{aligned}
\text{consR}(\langle\ \rangle, a) &= a :: \langle\ \rangle, \\
\text{consR}(b :: T, a) &= b :: \text{consR}(T, a).
\end{aligned}$$

For example, we can construct the list $\langle x,\ y \rangle$ with consR as follows:

$$\begin{aligned}
\text{consR}(\text{consR}(\langle\ \rangle, x), y) &= \text{consR}(x :: \langle\ \rangle, y) \\
&= x :: \text{consR}(\langle\ \rangle, y) \\
&= x :: y :: \langle\ \rangle \\
&= x :: \langle y \rangle \\
&= \langle x, y \rangle.
\end{aligned}$$

end example

example 3.25 Concatenation of Lists

An important operation on lists is the concatenation of two lists into a single list. Let "cat" denote the concatenation function. Its type is lists$(A) \times$ lists(A) $\to$ lists(A). For example,

$$\text{cat}(\langle a,\ b \rangle,\ \langle c,\ d \rangle) = \langle a,\ b,\ c,\ d \rangle.$$

Since both arguments are lists, we have some choices to make. Notice, for example, that we can rewrite the equation as follows:

$$\begin{aligned}
\text{cat}(\langle a,b \rangle, \langle c,d \rangle) &= \langle a,b,c,d \rangle \\
&= a :: \langle b,c,d \rangle \\
&= a :: \text{cat}(\langle b \rangle, \langle c,d \rangle).
\end{aligned}$$

So the recursive part can be written in terms of the head and tail of the first argument list. Here's an if-then-else definition for cat.

$$\text{cat}(L, M) = \text{if } L = \langle \, \rangle \text{ then } M$$
$$\text{else head}(L) :: \text{cat}(\text{tail}(L), M).$$

We'll unfold the definition to evaluate the expression $\text{cat}(\langle a,\ b\rangle, \langle c,\ d\rangle)$:

$$\begin{aligned}
\text{cat} (\langle a, b\rangle, \langle c, d\rangle) &= a :: \text{cat} (\langle b\rangle, \langle c, d\rangle) \\
&= a :: b :: \text{cat} (\langle \, \rangle, \langle c, d\rangle) \\
&= a :: b :: \langle c, d\rangle \\
&= a :: \langle b, c, d\rangle \\
&= \langle a, b, c, d\rangle.
\end{aligned}$$

end example

example **3.26 Sorting a List by Insertion**

Let's define a function to sort a list of numbers by repeatedly inserting a new number into an already sorted list of numbers. Suppose "insert" is a function that does this job. Then the sort function itself is easy. For a basis case, notice that the empty list is already sorted. For the recursive case we sort the list $x :: L$ by inserting x into the list obtained by sorting L. The definition can be written as follows:

$$\text{sort} (\langle \, \rangle) = \langle \, \rangle,$$
$$\text{sort} (x :: L) = \text{insert} (x, \text{sort} (L)).$$

Everything seems to make sense as long as insert does its job. We'll assume that whenever the number to be inserted is already in the list, then a new copy will be placed to the left of the one already there. Now let's define insert. Again, the basis case is easy. The empty list is sorted, and to insert x into $\langle \, \rangle$, we simply create the singleton list $\langle x\rangle$. Otherwise—if the sorted list is not empty—either x belongs on the left of the list, or it should actually be inserted somewhere else in the list. An if-then-else definition can be written as follows:

$$\text{insert} (x, S) = \text{if } S = \langle \, \rangle \text{ then } \langle x\rangle$$
$$\text{else if } x \leq \text{ head} (S) \text{ then } x :: S$$
$$\text{else head} (S) :: \text{insert} (x, \text{tail} (S)).$$

Notice that insert works only when S is already sorted. For example, we'll unfold the definition of insert$(3, \langle 1, 2, 6, 8 \rangle)$:

$$
\begin{aligned}
\text{insert } (3, \langle 1, 2, 6, 8 \rangle) &= 1 :: \text{insert } (3, \langle 2, 6, 8 \rangle) \\
&= 1 :: 2 :: \text{insert } (3, \langle 6, 8 \rangle) \\
&= 1 :: 2 :: 3 :: \langle 6, 8 \rangle \\
&= \langle 1, 2, 3, 6, 8 \rangle .
\end{aligned}
$$

end example

example **3.27 The Map Function**

We'll construct a recursive definition for the map function. Recall, for example that

$$
\text{map}(f, \langle a, b, c \rangle) = \langle f(a), f(b), f(c) \rangle .
$$

Since the second argument is a list, it makes sense to define the basis case as $\text{map}(f, \langle \, \rangle) = \langle \, \rangle$. To discover the recursive part of the definition, we'll rewrite the example equation as follows:

$$
\begin{aligned}
\text{map}(f, \langle a, b, c \rangle) &= \langle f(a), f(b), f(c) \rangle \\
&= f(a) :: \langle f(b), f(c) \rangle \\
&= f(a) :: \text{map}(f, \langle b, c \rangle).
\end{aligned}
$$

So the recursive part can be written in terms of the head and tail of the input list. Here's an if-then-else definition for map.

$$
\begin{aligned}
\text{map}(f, L) = \ &\text{if } L = \langle \, \rangle \text{ then } \langle \, \rangle \\
&\text{else } f(\text{head}(L)) :: \text{map}(f, \text{tail}(L)).
\end{aligned}
$$

For example, we'll evaluate the expression map$(f, \langle a, b, c \rangle)$.

$$
\begin{aligned}
\text{map } (f, \langle a, b, c \rangle) &= f(a) :: \text{map } (f, \langle b, c \rangle) \\
&= f(a) :: f(b) :: \text{map } (f, \langle c \rangle) \\
&= f(a) :: f(b) :: f(c) :: \text{map } (f, \langle \, \rangle) \\
&= f(a) :: f(b) :: f(c) :: \langle \, \rangle \\
&= \langle f(a), f(b), f(c) \rangle .
\end{aligned}
$$

end example

3.2.4 Binary Trees

Let's look at some functions that compute properties of binary trees. To start, suppose we need to know the number of nodes in a binary tree. Since the set of binary trees over a particular set can be defined inductively, we should be able to come up with a recursively defined function that suits our needs. Let "nodes" be the function that returns the number of nodes in a binary tree. Since the empty tree has no nodes, we have nodes($\langle \, \rangle$) = 0. If the tree is not empty, then the number of nodes can be computed by adding 1 to the number of nodes in the left and right subtrees. The definition of nodes can be written as follows:

$$\text{nodes}\left(\langle \, \rangle\right) = 0,$$
$$\text{nodes}\left(\text{tree}\left(L, a, R\right)\right) = 1 + \text{nodes}\left(L\right) + \text{nodes}\left(R\right).$$

If we want the corresponding if-then-else form of the definition, it looks like

$$\text{nodes}\left(T\right) = \text{if } T = \langle \, \rangle \text{ then } 0$$
$$\text{else } 1 + \text{nodes}\left(\text{left}\left(T\right)\right) + \text{nodes}\left(\text{right}\left(T\right)\right).$$

For example, we'll evaluate nodes(T) for $T = \langle\langle\langle \, \rangle, a, \langle \, \rangle \rangle, b, \langle \, \rangle \rangle$:

$$\text{nodes}\left(T\right) = 1 + \text{nodes}\left(\langle\langle \, \rangle, a, \langle \, \rangle\rangle\right) + \text{nodes}\left(\langle \, \rangle\right)$$
$$= 1 + 1 + \text{nodes}\left(\langle \, \rangle\right) + \text{nodes}\left(\langle \, \rangle\right) + \text{nodes}\left(\langle \, \rangle\right)$$
$$= 1 + 1 + 0 + 0 + 0$$
$$= 2.$$

example 3.28 A Binary Search Tree

Suppose we have a binary search tree whose nodes are numbers, and we want to add a new number to the tree, under the assumption that the new tree is still a binary search tree. A function to do the job needs two arguments, a number x and a binary search tree T. Let the name of the function be "insert."

The basis case is easy. If $T = \langle \, \rangle$, then return tree($\langle \, \rangle$, x, $\langle \, \rangle$). The recursive part is straightforward. If $x < \text{root}(T)$, then we need to replace the subtree left(T) by insert(x, left(T)). Otherwise, we replace right(T) by insert(x, right(T)). Notice that repeated elements are entered to the right. If we didn't want to add repeated elements, then we could simply return T whenever $x = \text{root}(T)$. The if-then-else form of the definition is

$$\text{insert}\left(x, T\right) = \text{if } T = \langle \, \rangle \text{ then tree}\left(\langle \, \rangle, x, \langle \, \rangle\right)$$
$$\text{else if } x < \text{root}\left(T\right) \text{ then}$$
$$\text{tree}\left(\text{insert}\left(x, \text{left}\left(T\right)\right), \text{root}\left(T\right), \text{right}\left(T\right)\right)$$
$$\text{else}$$
$$\text{tree}\left(\text{left}\left(T\right), \text{root}\left(T\right), \text{insert}\left(x, \text{right}\left(T\right)\right)\right).$$

Now suppose we want to build a binary search tree from a given list of numbers in which the numbers are in no particular order. We can use the insert function as the main ingredient in a recursive definition. Let "makeTree" be the name of the function. We'll use two variables to describe the function, a binary search tree T and a list of numbers L.

$$\text{makeTree}\,(T, L) = \text{if } L = \langle\;\rangle \text{ then } T \tag{3.9}$$
$$\text{else makeTree}\,(\text{insert}\,(\text{head}\,(L)\,, T)\,, \text{tail}\,(L))\,.$$

To construct a binary search tree with this function, we apply makeTree to the pair of arguments $(\langle\;\rangle, L)$. As an example, the reader should unfold the definition for makeTree($\langle\;\rangle$, $\langle 3, 2, 4\rangle$).

The function makeTree can be defined another way. Suppose we consider the following definition for constructing a binary search tree:

$$\text{makeTree}\,(T, L) = \text{if } L = \langle\;\rangle \text{ then } T \tag{3.10}$$
$$\text{else insert}\,(\text{head}\,(L)\,, \text{makeTree}\,(T, \text{tail}\,(L)))\,.$$

You should evaluate the expression makeTree($\langle\;\rangle$, $\langle 3, 2, 4\rangle$) by unfolding this alternative definition. It should help explain the difference between the two definitions.

end example

Traversing Binary Trees

There are several useful ways to list the nodes of a binary tree. The three most popular methods of traversing a binary tree are called *preorder, inorder,* and *postorder.* We'll start with the definition of a preorder traversal.

Preorder Traversal

The *preorder* traversal of a binary tree starts by visiting the root. Then there is a preorder traversal of the left subtree followed by a preorder traversal of the right subtree.

For example, the preorder listing of the nodes of the binary tree in Figure 3.6 is $\langle a, b, c, d, e\rangle$. It's common practice to write the listing without any punctuation symbols as

$$a\ b\ c\ d\ e.$$

Figure 3.6   A binary tree.

example **3.29 A Preorder Procedure**

Since binary trees are inductively defined, we can easily write a recursively defined procedure to output the preorder listing of a binary tree. For example, the following recursively defined procedure prints the preorder listing of its argument T.

$$\text{Preorder}(T): \quad \textbf{if } T \neq \langle \rangle \textbf{ then}$$
$$\text{print}(\text{root}(T));$$
$$\text{Preorder}(\text{left}(T));$$
$$\text{Preorder}(\text{right}(T))$$
$$\textbf{fi.}$$

end example

example **3.30 A Preorder Function**

Now let's write a function to compute the preorder listing of a binary tree. Letting "preOrd" be the name of the preorder function, a definition can be written as follows:

$$\text{preOrd}\,(\langle\,\rangle) = \langle\,\rangle,$$
$$\text{preOrd}\,(\text{tree}\,(L, x, R)) = x :: \text{cat}\,(\text{preOrd}\,(L), \text{preOrd}\,(R))\,.$$

The if-then-else form of preOrd can be written as follows:

$$\text{preOrd}\,(T) = \text{if } T = \langle\,\rangle \text{ then } \langle\,\rangle$$
$$\text{else root}\,(T) :: \text{cat}\,(\text{preOrd}\,(\text{left}\,(T)), \text{preOrd}\,(\text{right}\,(T)))\,.$$

We'll evaluate the expression preOrd(T) for the tree $T = \langle\langle\langle\,\rangle,\, a,\, \langle\,\rangle\rangle,\, b,\, \langle\,\rangle\rangle$:

$$\text{preOrd}\,(T) = b :: \text{cat}\,(\text{preOrd}\,(\langle\langle\,\rangle, a\,\langle\,\rangle\rangle), \text{preOrd}\,(\langle\,\rangle))$$
$$= b :: \text{cat}\,(a :: \text{cat}\,(\text{preOrd}\,(\langle\,\rangle), \text{preOrd}\,(\langle\,\rangle)), \text{preOrd}\,(\langle\,\rangle))$$
$$= b :: \text{cat}\,(a :: \langle\,\rangle, \langle\,\rangle)$$
$$= b :: \text{cat}\,(\langle a\rangle, \langle\,\rangle)$$
$$= b :: \langle a\rangle$$
$$= \langle b, a\rangle\,.$$

end example

The definitions for the inorder and postorder traversals of a binary tree are similar to the preorder traversal. The only difference is when the root is visited during the traversal.

Inorder Traversal

The *inorder* traversal of a binary tree starts with an inorder traversal of the left subtree. Then the root is visited. Lastly, there is an inorder traversal of the right subtree.

For example, the inorder listing of the tree in Figure 3.6 is

$$b\ a\ d\ c\ e.$$

Postorder Traversal

The *postorder* traversal of a binary tree starts with a postorder traversal of the left subtree and is followed by a postorder traversal of the right subtree. Lastly, the root is visited.

The postorder listing of the tree in Figure 3.6 is

$$b\ d\ e\ c\ a.$$

We'll leave the construction of the inorder and postorder procedures and functions as exercises.

3.2.5 Two More Problems

We'll look at two more problems, each of which requires a little extra thinking on the way to a solution.

The Repeated Element Problem

Suppose we want to remove repeated elements from a list. Depending on how we proceed, there might be different solutions. For example, we can remove the repeated elements from the list $\langle u,\ g,\ u,\ h,\ u \rangle$ in three ways, depending on which occurrence of u we keep: $\langle u,\ g,\ h \rangle$, $\langle g,\ u,\ h \rangle$, or $\langle g,\ h,\ u \rangle$. We'll solve the problem by always keeping the leftmost occurrence of each element. Let "remove" be the function that takes a list L and returns the list remove(L), which has no repeated elements and contains the leftmost occurrence of each element of L.

To start things off, we can say remove($\langle\ \rangle$) = $\langle\ \rangle$. Now if $L \neq \langle\ \rangle$, then L has the form $L = b :: M$ for some list M. In this case, the head of remove(L) should be b. The tail of remove(L) can be obtained by removing all occurrences of b from M and then removing all repeated elements from the resulting list. So we need a new function to remove all occurrences of an element from a list.

Let removeAll(b, M) denote the list obtained from M by removing all occurrences of b. Now we can write a definition for the remove function as follows:

$$\text{remove}(\langle \, \rangle) = \langle \, \rangle,$$
$$\text{remove}(b :: M) = b :: \text{remove}(\text{removeAll}(b, M)).$$

We can rewrite the solution in if-then-else form as follows:

$$\text{remove}(L) = \text{if } L = \langle \, \rangle \text{ then } \langle \, \rangle$$
$$\text{else head}(L) :: \text{remove}(\text{removeAll}(\text{head}(L), \text{tail}(L))).$$

To complete the task, we need to define the "removeAll" function. The basis case is removeAll(b, $\langle \, \rangle$) = $\langle \, \rangle$. If $M \neq \langle \, \rangle$, then the value of removeAll(b, M) depends on head(M). If head(M) = b, then throw it away and return the value of removeAll(b, tail(M)). But if head(M) $\neq$ b, then it's a keeper. So we should return the value head(M) :: removeAll(b, tail(M)). We can write the definition in if-then-else form as follows:

$$\text{removeAll}(b, M) = \text{if } M = \langle \, \rangle \text{ then } \langle \, \rangle$$
$$\text{else if head}(M) = b \text{ then}$$
$$\text{removeAll}(b, \text{tail}(M))$$
$$\text{else}$$
$$\text{head}(M) :: \text{removeAll}(b, \text{tail}(M)).$$

We'll evaluate the expression removeAll(b, $\langle a, b, c, b \rangle$):

$$\text{removeAll}(b, \langle a, b, c, b \rangle) = a :: \text{removeAll}(b, \langle b, c, b \rangle)$$
$$= a :: \text{removeAll}(b, \langle c, b \rangle)$$
$$= a :: c :: \text{removeAll}(b, \langle b \rangle)$$
$$= a :: c :: \text{removeAll}(b, \langle \, \rangle)$$
$$= a :: c :: \langle \, \rangle$$
$$= a :: \langle c \rangle$$
$$= \langle a, c \rangle.$$

Try to write out each unfolding step in the evaluation of the expression remove($\langle b, a, b \rangle$). Be sure to start writing at the left-hand edge of your paper.

The Power Set Problem

Suppose we want to construct the power set of a finite set. One solution uses the fact that power($\{x\} \cup T$) is the union of power(T) and the set obtained from power(T) by adding x to each of its elements. Let's see whether we can discover a solution technique by considering a small example. Let $S = \{a, b, c\}$. Then we can write power(S) as follows:

$$\text{power}(S) = \{\{\}, \{a\}, \{b\}, \{c\}, \{a, b\}, \{a, c\}, \{b, c\}, \{a, b, c\}\}$$
$$= \{\{\}, \{b\}, \{c\}, \{b, c\}\} \cup \{\{a\}, \{a, b\}, \{a, c\}, \{a, b, c\}\}.$$

We've written power$(S) = A \cup B$, where B is obtained from A by adding the element a to each set in A. If we represent S as the list $\langle a, b, c \rangle$, then we can restate the definition for power(S) as the concatenation of the following two lists:

$$\langle \langle \, \rangle, \langle b \rangle, \langle c \rangle, \langle b, c \rangle \rangle \quad \text{and} \quad \langle \langle a \rangle, \langle a, b \rangle, \langle a, c \rangle, \langle a, b, c \rangle \rangle.$$

The first of these lists is power$(\langle b, c \rangle)$. The second list can be obtained from power$(\langle b, c \rangle)$ by working backward to the answer as follows:

$$\langle \langle a \rangle, \langle a, b \rangle, \langle a, c \rangle, \langle a, b, c \rangle \rangle = \langle a :: \langle \, \rangle, a :: \langle b \rangle, a :: \langle c \rangle, a :: \langle b, c \rangle \rangle$$
$$= \text{map}(::, \langle \langle a, \langle \, \rangle \rangle, \langle a, \langle b \rangle \rangle, \langle a, \langle c \rangle \rangle, \langle a, \langle b, c \rangle \rangle \rangle)$$
$$= \text{map}(::, \text{dist}(a, \text{power}(\langle b, c \rangle))).$$

This example is the key to the recursive part of the definition. Using the fact that power$(\langle \, \rangle) = \langle \langle \, \rangle \rangle$ as the basis case, we can write down the following definition for power:

$$\text{power}(\langle \, \rangle) = \langle \langle \, \rangle \rangle,$$
$$\text{power}(a :: T) = \text{cat}(\text{power}(T), \text{map}(::, \text{dist}(a, \text{power}(T)))).$$

The if-then-else form of the definition can be written as follows:

$$\text{power}(L) = \text{if } L = \langle \, \rangle \text{ then } \langle \langle \, \rangle \rangle \text{ else}$$
$$\text{cat}(\text{power}(\text{tail}(L)), \text{map}(::, \text{dist}(\text{head}(L), \text{power}(\text{tail}(L))))).$$

We'll evaluate the expression power$(\langle a, b \rangle)$. The first step yields the equation

$$\text{power}(\langle a, b \rangle) = \text{cat}(\text{power}(\langle b \rangle), \text{map}(::, \text{dist}(a, \text{power}(\langle b \rangle)))).$$

Now we'll evaluate power$(\langle b \rangle)$ and substitute it in the preceding equation:

$$\text{power}(\langle b \rangle) = \text{cat}(\text{power}(\langle \, \rangle), \text{map}(::, \text{dist}(b, \text{power}(\langle \, \rangle))))$$
$$= \text{cat}(\langle \langle \, \rangle \rangle, \text{map}(::, \text{dist}(b, \langle \langle \, \rangle \rangle)))$$
$$= \text{cat}(\langle \langle \, \rangle \rangle, \text{map}(::, \langle \langle b, \langle \, \rangle \rangle \rangle))$$
$$= \text{cat}(\langle \langle \, \rangle \rangle, \langle b :: \langle \, \rangle \rangle)$$
$$= \text{cat}(\langle \langle \, \rangle \rangle, \langle \langle b \rangle \rangle)$$
$$= \langle \langle \, \rangle, \langle b \rangle \rangle.$$

Now we can continue with the evaluation of power($\langle a,\, b \rangle$):

$$
\begin{aligned}
\text{power} \left(\langle a, b \rangle \right) &= \text{cat} \left(\text{power} \left(\langle b \rangle \right), \text{map} \left(::, \text{dist} \left(a, \text{power} \left(\langle b \rangle \right) \right) \right) \right) \\
&= \text{cat} \left(\langle \langle \ \rangle, \langle b \rangle \rangle, \text{map} \left(::, \text{dist} \left(a, \langle \langle \ \rangle, \langle b \rangle \rangle \right) \right) \right) \\
&= \text{cat} \left(\langle \langle \ \rangle, \langle b \rangle \rangle, \text{map} \left(::, \langle \langle a, \langle \ \rangle \rangle, \langle a, \langle b \rangle \rangle \rangle \right) \right) \\
&= \text{cat} \left(\langle \langle \ \rangle, \langle b \rangle \rangle, \langle a :: \langle \ \rangle, a :: \langle b \rangle \rangle \right) \\
&= \text{cat} \left(\langle \langle \ \rangle, \langle b \rangle \rangle, \langle \langle a \rangle, \langle a, b \rangle \rangle \right) \\
&= \langle \langle \ \rangle, \langle b \rangle, \langle a \rangle, \langle a, b \rangle \rangle \, .
\end{aligned}
$$

3.2.6 Infinite Sequences

Let's see how some infinite sequences can be defined recursively. To illustrate the idea, suppose the function "ints" returns the following infinite sequence for any integer x:

$$\text{ints}(x) = \langle x, x + 1, x + 2, \ldots \rangle.$$

We'll assume that the list operations of cons, head, and tail work for infinite sequences. For example, the following relationships hold.

$$
\begin{aligned}
\text{ints} \left(x \right) &= x :: \text{ints} \left(x + 1 \right), \\
\text{head} \left(\text{ints} \left(x \right) \right) &= x, \\
\text{tail} \left(\text{ints} \left(x \right) \right) &= \text{ints} \left(x + 1 \right).
\end{aligned}
$$

Even though the definition of ints does not conform to (3.6), it is still recursively defined because it is defined in terms of itself. If we executed the definition, an infinite loop would construct the infinite sequence. For example, ints(0) would construct the infinite sequence of natural numbers as follows:

$$
\begin{aligned}
\text{ints} \left(0 \right) &= 0 :: \text{ints} \left(1 \right) \\
&= 0 :: 1 :: \text{ints} \left(2 \right) \\
&= 0 :: 1 :: 2 :: \text{ints} \left(3 \right) \\
&= \ldots .
\end{aligned}
$$

In practice, an infinite sequence is used as an argument and it is evaluated only when some of its values are needed. Once the needed values are computed, the evaluation stops. This is an example of a technique called *lazy evaluation*. For example, the following function returns the nth element of an infinite sequence s.

$$\text{get}(n,\, s) = \text{if } n = 1 \text{ then head}(s) \text{ else get}(n - 1, \text{tail}(s)).$$

example **3.31 Picking Elements**

We'll get the third element from the infinite sequence ints(6) by unfolding the expression get(3, ints(6)).

$$
\begin{aligned}
\operatorname{get}(3, \operatorname{ints}(6)) &= \operatorname{get}(2, \operatorname{tail}(\operatorname{ints}(6))) \\
&= \operatorname{get}(1, \operatorname{tail}(\operatorname{tail}(\operatorname{ints}(6)))) \\
&= \operatorname{head}(\operatorname{tail}(\operatorname{tail}(\operatorname{ints}(6)))) \\
&= \operatorname{head}(\operatorname{tail}(\operatorname{tail}(6 :: \operatorname{ints}(7)))) \\
&= \operatorname{head}(\operatorname{tail}(\operatorname{ints}(7))) \\
&= \operatorname{head}(\operatorname{tail}(7 :: \operatorname{ints}(8))) \\
&= \operatorname{head}(\operatorname{ints}(8)) \\
&= \operatorname{head}(8 :: \operatorname{ints}(9)) \\
&= 8.
\end{aligned}
$$

end example

example **3.32 Summing**

Suppose we need a function to sum the first n elements in an infinite sequence s of integers. The following definition does the trick:

$$\operatorname{sum}(n, s) = \text{if } n = 0 \text{ then } 0 \text{ else } \operatorname{head}(s) + \operatorname{sum}(n - 1, \operatorname{tail}(s)).$$

We'll compute the sum of the first three numbers in ints(4):

$$
\begin{aligned}
\operatorname{sum}(3, \operatorname{ints}(4)) &= 4 + \operatorname{sum}(2, \operatorname{ints}(5)) \\
&= 4 + 5 + \operatorname{sum}(1, \operatorname{ints}(6)) \\
&= 4 + 5 + 6 + \operatorname{sum}(0, \operatorname{ints}(7)) \\
&= 4 + 5 + 6 + 0 \\
&= 15.
\end{aligned}
$$

end example

example **3.33 The Sieve of Eratosthenes**

Suppose we want to study prime numbers. For example, we might want to find the 500th prime, we might want to find the difference between the 500th and 501st primes, and so on. One way to proceed might be to define functions to extract information from the following infinite sequence of all prime numbers.

$$Primes = \langle 2, 3, 5, 7, 11, 13, 17, \ldots \rangle.$$

We'll construct this infinite sequence by the method of Eratosthenes (called *the sieve of Eratosthenes*). The method starts with the infinite sequence ints(2):

$$\text{ints}(2) = \langle 2,\ 3,\ 4,\ 5,\ 6,\ 7,\ 8,\ 9,\ 10,\ \dots \rangle.$$

The next step removes all multiples of 2 (except 2) to obtain the infinite sequence

$$\langle 2,\ 3,\ 5,\ 7,\ 9,\ 11,\ 13,\ 15,\ \dots \rangle.$$

The next step removes all multiples of 3 (except 3) to obtain the infinite sequence

$$\langle 2,\ 3,\ 5,\ 7,\ 11,\ 13,\ 17,\ \dots \rangle.$$

The process continues in this way.

We can construct the desired infinite sequence of primes once we have the function to remove multiples of a number from an infinite sequence. If we let remove(n, s) denote the infinite sequence obtained from s by removing all multiples of n, then we can define the sieve process as follows for an infinite sequence s of numbers:

$$\text{sieve}(s) = \text{head}(s) :: \text{sieve}(\text{remove}(\text{head}(s),\ \text{tail}(s))).$$

But we need to define the remove function. Notice that for natural numbers m and n with $n > 0$ that we have the following equivalences:

$$m \text{ is a multiple of } n \text{ iff } n \text{ divides } m \text{ iff } m \bmod n = 0.$$

This allows us to write the following definition for the remove function:

$$\text{remove}\,(n, s) = \text{if head}\,(s)\ \bmod\ n = 0 \text{ then remove}\,(n, \text{tail}\,(s))$$
$$\text{else head}\,(s) :: \text{remove}\,(n, \text{tail}\,(s)).$$

Then our desired sequence of primes is represented by the expression

$$\textit{Primes} = \text{sieve}(\text{ints}(2)).$$

In the exercises we'll evaluate some functions dealing with primes.

Exercises

Evaluating Recursively Defined Functions

1. Given the following definition for the nth Fibonacci number:

$$\text{fib}\,(0) = 0,$$
$$\text{fib}\,(1) = 1,$$
$$\text{fib}\,(n) = \text{fib}\,(n - 1) + \text{fib}\,(n - 2) \quad \text{if } n > 1.$$

Write down each step in the evaluation of fib(4).

2. Given the following definition for the length of a list:

$$\text{length}(L) = \text{if } L = \langle \, \rangle \text{ then } 0 \text{ else } 1 + \text{length}(\text{tail}(L)).$$

Write down each step in the evaluation of $\text{length}(\langle r, s, t, u \rangle)$.

3. For each of the two definitions of "makeTree" given by (3.9) and (3.10), write down all steps to evaluate $\text{makeTree}(\langle \, \rangle, \langle 3, 2, 4 \rangle)$.

Numbers

4. Construct a recursive definition for each of the following functions, where all variables represent natural numbers.

    a. $f(n) = 0 + 2 + 4 + \cdots + 2n.$
    b. $f(n) = \text{floor}(0/2) + \text{floor}(1/2) + \cdots + \text{floor}(n/2).$
    c. $f(k, n) = \gcd(1, n) + \gcd(2, n) + \cdots + \gcd(k, n)$ for $k > 0.$
    d. $f(n) = (0 \bmod 2) + (1 \bmod 3) + \cdots + (n \bmod (n + 2)).$
    e. $f(n, k) = 0 + k + 2k + \cdots + nk.$
    f. $f(n, k) = k + (k + 1) + (k + 2) + \cdots + (k + n).$

Strings

5. Construct a recursive definition for each of the following string functions for strings over the alphabet $\{a, b\}$.

    a. $f(x)$ returns the reverse of $x.$
    b. $f(x) = xy$, where y is the reverse of $x.$
    c. $f(x, y)$ tests whether x is a prefix of $y.$
    d. $f(x, y)$ tests whether $x = y.$
    e. $f(x)$ tests whether x is a palindrome.

Lists

6. Construct a recursive definition for each of the following functions that involve lists. Use the infix form of cons in the recursive part of each definition. In other words, write $h :: t$ in place of $\text{cons}(h, t)$.

    a. $f(n) = \langle 2n, 2(n - 1), \ldots, 2, 0 \rangle$, over natural numbers.
    b. $f(z) = \text{if } z \geq 0 \text{ then } \langle z, z - 1, \ldots, 0 \, \rangle \text{ else } \langle z, z + 1, \ldots, 0 \, \rangle$, over integers.
    c. $\max(L)$ is the maximum value in the nonempty list L of numbers.
    d. $f(x, \langle a_0, \ldots, a_n \rangle) = a_0 + a_1 x + a_2 x^2 + \cdots + a_n x^n.$
    e. $f(L)$ is the list of elements x in list L that have property $P.$

f. $f(a, \langle x_1, \ldots, x_n \rangle) = \langle x_1 + a, \ldots, x_n + a \rangle$.

g. $f(a, \langle (x_1, y_1), \ldots, (x_n, y_n) \rangle) = \langle (x_1 + a, y_1), \ldots, (x_n + a, y_n) \rangle$.

h. $f(g, \langle x_1, \ldots, x_n \rangle) = \langle (x_1, g(x_1)), \ldots, (x_n, g(x_n)) \rangle$.

i. $f(n) = \langle (0, n), (1, n-1), \ldots, (n-1, 1), (n, 0) \rangle$. *Hint:* Use Part (g).

j. $f(g, h, \langle x_1, \ldots, x_n \rangle) = \langle (g(x_1), h(x_1)), \ldots, (g(x_n), h(x_n)) \rangle$.

Using Cat or ConsR

7. Construct a recursive definition for each of the following functions that involve lists. Use the cat operation or consR operation in the recursive part of each definition. (Notice that for any list L and element x we have cat(L, $\langle x \rangle$) = consR(L, x).)

    a. $f(n) = \langle 0, 1, \ldots, n \rangle$.

    b. $f(n) = \langle 0, 2, 4, \ldots, 2n \rangle$.

    c. $f(n) = \langle 1, 3, 5, \ldots, 2n + 1 \rangle$.

    d. $f(n, k) = \langle n, n + 1, n + 2, \ldots, n + k \rangle$.

    e. $f(n, k) = \langle 0, k, 2k, 3k, \ldots, nk \rangle$.

    f. $f(g, n) = \langle (0, g(0)), (1, g(1)), \ldots, (n, g(n)) \rangle$.

    g. $f(n, m) = \langle n, n + 1, n + 2, \ldots, m - 1, m \rangle$,   where $n \leq m$.

8. Let *insert* be a function that extends any binary function so that it evaluates a list of two or more arguments. For example,

$$\text{insert}(+, \langle 1, 4, 2, 9 \rangle) = 1 + (4 + (2 + 9)) = 16.$$

Write a recursive definition for insert(f, L), where f is any binary function and L is a list of two or more arguments.

9. Write a recursive definition for the function "eq" to check two lists for equality.

10. Write recursive definitions for the following list functions.

    a. The function "last" that returns the last element of a nonempty list. For example, last($\langle a, b, c \rangle$) = c.

    b. The function "front" that returns the list obtained by removing the last element of a nonempty list. For example, front($\langle a, b, c \rangle$) = $\langle a, b \rangle$.

11. Write down a recursive definition for the function "pal" that tests a list of letters to see whether their concatenations form a palindrome. For example, pal($\langle r, a, d, a, r \rangle$) = true since *radar* is a palindrome. *Hint:* Use the functions of Exercise 10.

12. Solve the repeated element problem with the restriction that we want to keep the rightmost occurrence of each repeated element. *Hint:* Use the functions of Exercise 10.

Binary Trees

13. Given the algebraic expression $a + (b \cdot (d + e))$, draw a picture of the binary tree representation of the expression. Then write down the preorder, inorder, and postorder listings of the tree. Are any of the listings familiar to you?

14. Write down recursive definitions for each of the following procedures to print the nodes of a binary tree.

    a. In: Prints the nodes of a binary tree from an inorder traversal.

    b. Post: Prints the nodes of a binary tree from a postorder traversal.

15. Write down recursive definitions for each of the following functions. Include both the pattern-matching and if-then-else forms for each definition.

    a. leaves: Returns the number of leaf nodes in a binary tree.

    b. inOrd: Returns the inorder listing of nodes in a binary tree.

    c. postOrd: Returns the postorder listing of nodes in a binary tree.

16. Construct a recursive definition for each of the following functions that involve binary trees. Represent binary trees as lists where $\langle\ \rangle$ is the empty tree and any nonempty binary tree has the form $\langle L,\ r,\ R \rangle$, where r is the root and L and R are its left and right subtrees. Assume that nodes are numbers.

    a. $f(T)$ is the sum of the nodes of T.

    b. $f(T)$ is the depth of T, where the empty tree has depth -1.

    c. $f(T)$ is the list of nodes in T that have property p.

    d. $f(T)$ is the maximum value of the nodes in the nonempty binary tree T.

Trees and Algebraic Expressions

17. Recall from Section 1.4 that any algebraic expression can be represented as a tree and the tree can be represented as a list whose head is the root and whose tail is the list of operands in the form of trees. For example, the algebraic expression $a \times b + f(c, d, e)$, can be represented by the list

$$\langle +, \langle \times, \langle a \rangle, \langle b \rangle \rangle, \langle f, \langle c \rangle, \langle d \rangle, \langle e \rangle \rangle \rangle.$$

    a. Draw the picture of the tree for the given algebraic expression.

    b. Construct a recursive definition for the function "post" that takes an algebraic expression written in the form of a list, as above, and returns the list of nodes in the algebraic expression tree in postfix notation. For example,

$$\text{post}(\langle +, \langle \times, \langle a \rangle, \langle b \rangle \rangle, \langle f, \langle c \rangle, \langle d \rangle, \langle e \rangle \rangle \rangle) = \langle a, b, \times, c, d, e, f, + \rangle.$$

Relations as Lists of Tuples

18. Construct a recursive definition for each of the following functions that involve lists of tuples. If x is an n-tuple, then x_k represents the kth component of x.

   a. $f(k,\ L)$ is the list of kth components x_k of tuples x in the list L.

   b. $\mathrm{sel}(k,\ a,\ L)$ is the list of tuples x in the list L such that $x_k = a$.

Sets Represented as Lists

19. Write a recursive definition for each of the following functions, in which the input arguments are sets represented as lists. Use the primitive operations of cons, head, and tail to build your functions (along with functions already defined):

   a. isMember. For example, isMember($a, \langle b, a, c \rangle$) is true.

   b. isSubset. For example, isSubset($\langle a, b \rangle, \langle b, c, a \rangle$) is true.

   c. areEqual. For example, areEqual($\langle a, b \rangle, \langle b, a \rangle$) is true.

   d. union. For example, union($\langle a, b \rangle, \langle c, a \rangle$) = $\langle a, b, c \rangle$.

   e. intersect. For example, intersect($\langle a, b \rangle, \langle c, a \rangle$) = $\langle a \rangle$.

   f. difference. For example, difference($\langle a, b, c \rangle, \langle b, d \rangle$) = $\langle a, c \rangle$.

Challenges

20. Conway's challenge sequence is defined recursively as follows:

Basis: $f(1) = f(2) = 1$.
Recursion: $f(n) = f(f(n-1)) + f(n - f(n-1))$ for $n > 2$.

Calculate the first 17 elements $f(1), f(2), \ldots, f(17)$. The article by Mallows [1991] contains an account of this sequence.

21. Let $\mathrm{fib}(k)$ denote the kth Fibonacci number, and let

$$\mathrm{sum}(k) = 1 + 2 + \cdots + k.$$

Write a recursive definition for the function $f : \mathbb{N} \to \mathbb{N}$ defined by $f(n) = \mathrm{sum}(\mathrm{fib}(n))$. *Hint:* Write down several examples, such as $f(0), f(1), f(2)$, $f(3), f(4), \ldots$. Then try to find a way to write $f(4)$ in terms of $f(3)$. This might help you discover a pattern.

22. Write a function in if-then-else form to produce the Cartesian product set of two finite sets. You may assume that the sets are represented as lists.

23. We can approximate the square root of a number by using the Newton–Raphson method, which gives an infinite sequence of approximations to the square root of x by starting with an initial guess g. We can define the sequence with the following function:

$$\mathrm{sqrt}(x,\, g) = g :: \mathrm{sqrt}(x,\, (0.5)(g + (x/g))).$$

Find the first three numbers in each of the following infinite sequences, and compare the values with the square root obtained with a calculator.

    a. $\mathrm{sqrt}(4, 1)$.   **b.** $\mathrm{sqrt}(4, 2)$.   **c.** $\mathrm{sqrt}(4, 3)$.

    d. $\mathrm{sqrt}(2, 1)$.   **e.** $\mathrm{sqrt}(9, 1)$.   f. $\mathrm{sqrt}(9, 5)$.

24. Find a definition for each of the following infinite sequence functions.

    a. Square: Squares each element in a sequence of numbers.

    b. Diff: Finds the difference of the nth and mth numbers of a sequence.

    c. Prod: Finds the product of the first n numbers of a sequence.

    d. Add: Adds corresponding elements of two numeric sequences.

    e. $\mathrm{Skip}(x, k) = \langle x, x + k, x + 2k, x + 3k, \ldots \rangle$.

    f. Map: Applies a function to each element of a sequence.

    g. ListOf: Finds the list of the first n elements of a sequence.

25. Evaluate each of the following expressions by unfolding the definitions for *Primes* and remove from Example 3.33.

    a. head(*Primes*).

    b. tail(*Primes*) until reaching the value sieve(remove $(2, \mathrm{ints}\,(3)))$.

    c. remove$(2, \mathrm{ints}\,(0))$ until reaching the value $1 :: 2 :: \mathrm{remove}(2, \mathrm{ints}\,(4))$.

26. Suppose we define the function $f : \mathbb{N} \to \mathbb{N}$ by

$$f(x) = \text{if } x > 10 \text{ then } x - 10 \text{ else } f(f(x + 11)).$$

This function is recursively defined even though it is not defined by (3.6). Give a simple definition of the function.

3.3   Grammars

Informally, a grammar is a set of rules used to define the structure of the strings in a language. Grammars are important in computer science not only for defining programming languages, but also for defining data sets for programs. Typical applications try to build algorithms that test whether or not an arbitrary string belongs to some language. In this section we'll see that grammars provide a convenient and useful way to describe languages in a fashion similar to an inductive definition, which we discussed in Section 3.1. We'll also see that grammars provide a technique to test whether a string belongs to a language in a fashion similar to the calculation of a recursively defined function, which we described in Section 3.2. So let's get to it.

3.3.1 Recalling English Grammar

We can think of an English sentence as a string of characters if we agree to let the alphabet consist of the usual letters together with the blank character, period, comma, and so on. To *parse* a sentence means to break it up into parts that conform to a given grammar.

For example, if an English sentence consists of a subject followed by a predicate, then the sentence

"The big dog chased the cat."

would be broken up into two parts, a subject and a predicate, as follows:

$$\text{subject} = \text{The big dog},$$
$$\text{predicate} = \text{chased the cat}.$$

To denote the fact that a sentence consists of a subject followed by a predicate we'll write the following *grammar rule*:

$$\text{sentence} \rightarrow \text{subject predicate}.$$

If we agree that a subject can be an article followed by either a noun or an adjective followed by a noun, then we can break up "The big dog" into smaller parts. The corresponding grammar rule can be written as follows:

$$\text{subject} \rightarrow \text{article adjective noun}.$$

Similarly, if we agree that a predicate is a verb followed by an object, then we can break up "chased the cat" into smaller parts. The corresponding grammar rule can be written as follows:

$$\text{predicate} \rightarrow \text{verb object}.$$

This is the kind of activity that can be used to detect whether or not a sentence is grammatically correct.

A parsed sentence is often represented as a tree, called the *parse tree* or *derivation tree*. The parse tree for "The big dog chased the cat" is pictured in Figure 3.7.

3.3.2 Structure of Grammars

Now that we've recalled a bit of English grammar, let's describe the general structure of grammars for arbitrary languages. If L is a language over an alphabet A, then a grammar for L consists of a set of *grammar rules* of the form

$$\alpha \rightarrow \beta,$$

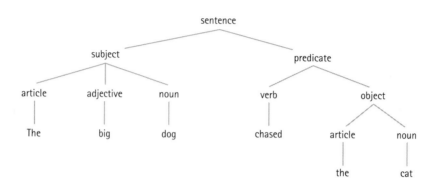

Figure 3.7   Parse tree.

where α and β denote strings of symbols taken from A and from a set of grammar symbols disjoint from A.

The grammar rule $\alpha \to \beta$ is often called a *production*, and it can be read in several different ways as

replace α by β,

α produces β,

α rewrites to β,

α reduces to β.

Every grammar has a special grammar symbol called the *start symbol*, and there must be at least one production with the left side consisting of only the start symbol. For example, if S is the start symbol for a grammar, then there must be at least one production of the form

$$S \to \beta.$$

A Beginning Example

Let's give an example of a grammar for a language and then discuss the process of deriving strings from the productions. Let $A = \{a,\ b,\ c\}$. Then a grammar for the language A^* can be described by the following four productions:

$$
\begin{aligned}
S &\to \Lambda \\
S &\to aS \\
S &\to bS \\
S &\to cS.
\end{aligned}
\tag{3.11}
$$

How do we know that this grammar describes the language A^*? We must be able to describe each string of the language in terms of the grammar rules. For example, let's see how we can use the productions (3.11) to show that the string $aacb$ is in A^*. We'll begin with the start symbol S. Next we'll replace S

by the right side of production $S \rightarrow aS$. We chose production $S \rightarrow aS$ because *aacb* matches the right-hand side of $S \rightarrow aS$ by letting $S = acb$. The process of replacing S by aS is called a *derivation step*, which we denote by writing

$$S \Rightarrow aS.$$

A *derivation* is a sequence of derivation steps. The right-hand side of this derivation contains the symbol S. So we again replace S by aS using the production $S \rightarrow aS$ a second time. This results in the derivation

$$S \Rightarrow aS \Rightarrow aaS.$$

The right-hand side of this derivation contains S. In this case we'll replace S by the right side of $S \rightarrow cS$. This gives the derivation

$$S \Rightarrow aS \Rightarrow aaS \Rightarrow aacS.$$

Continuing, we replace S by the right side of $S \rightarrow bS$. This gives the derivation

$$S \Rightarrow aS \Rightarrow aaS \Rightarrow aacS \Rightarrow aacbS.$$

Since we want this derivation to produce the string *aacb*, we now replace S by the right side of $S \rightarrow \Lambda$. This gives the desired derivation of the string *aacb*:

$$S \Rightarrow aS \Rightarrow aaS \Rightarrow aacS \Rightarrow aacbS \Rightarrow aacb\Lambda = aacb.$$

Each step in a derivation corresponds to attaching a new subtree to the parse tree whose root is the start symbol. For example, the parse trees corresponding to the first three steps of our example are shown in Figure 3.8. The completed derivation and parse tree are shown in Figure 3.9.

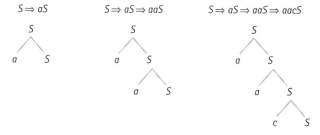

Figure 3.8   Partial derivations and parse trees.

$S \Rightarrow aS \Rightarrow aaS \Rightarrow aacS \Rightarrow aacbS \Rightarrow aacb\Lambda = aacb$

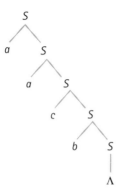

Figure 3.9   Derivation and parse tree.

Definition of a Grammar

Now that we've introduced the idea of a grammar, let's take a minute to describe the four main ingredients of any grammar.

The Four Parts of a Grammar   (3.12)

1. An alphabet N of grammar symbols called *nonterminals*.

2. An alphabet T of symbols called *terminals*. The terminals are distinct from the nonterminals.

3. A specific nonterminal S, called the *start* symbol.

4. A finite set of productions of the form $\alpha \rightarrow \beta$, where α and β are strings over the alphabet $N \cup T$ with the restriction that α is not the empty string. There is at least one production with only the start symbol S on its left side. Each nonterminal must appear on the left side of some production.

Assumption: In this chapter, all grammar productions will have a single nonterminal on the left side. In Chapter 14 we'll see examples of grammars that allow productions to have strings of more than one symbol on the left side.

When two or more productions have the same left side, we can simplify the notation by writing one production with alternate right sides separated by the vertical line |. For example, the four productions (3.11) can be written in the following shorthand form:

$$S \rightarrow \Lambda \mid aS \mid bS \mid cS,$$

and we say, "S can be replaced by either Λ , or aS, or bS, or cS."

We can represent a grammar G as a 4-tuple $G = (N, T, S, P)$, where P is the set of productions. For example, if P is the set of productions (3.11), then the grammar can be represented by the 4-tuple

$$(\{S\}, \{a, b, c\}, S, P).$$

The 4-tuple notation is useful for discussing general properties of grammars. But for a particular grammar it's common practice to write down only the productions of the grammar, where the nonterminals are uppercase letters and the first production listed contains the start symbol on its left side. For example, suppose we're given the following grammar:

$$S \rightarrow AB \tag{3.13}$$
$$A \rightarrow \Lambda \,|\, aA$$
$$B \rightarrow \Lambda \,|\, bB.$$

We can deduce that the nonterminals are S, A, and B, the start symbol is S, and the terminals are a and b.

3.3.3 Derivations

To discuss grammars further, we need to formalize things a bit. Suppose we're given some grammar. A string made up of terminals and/or nonterminals is called a *sentential form*. Now we can formalize the idea of a derivation.

Definition of Derivation

If x and y are sentential forms and $\alpha \rightarrow \beta$ is a production, then the replacement of α by β in $x\alpha y$ is called a *derivation step*, which we denote by writing

$$x\alpha y \Rightarrow x\beta y.$$

A *derivation* is a sequence of derivation steps.

The following three symbols with their associated meanings are used quite often in discussing derivations:

$$\Rightarrow \qquad \text{derives in one step,}$$
$$\Rightarrow^+ \qquad \text{derives in one or more steps,}$$
$$\Rightarrow^* \qquad \text{derives in zero or more steps.}$$

For example, let's consider the previous grammar (3.13) and the string *aab*.

The statement $S \Rightarrow^+ aab$ means that there exists a derivation of aab that takes one or more steps. For example, we have the derivation

$$S \Rightarrow AB \Rightarrow aAB \Rightarrow aaAB \Rightarrow aaB \Rightarrow aabB \Rightarrow aab.$$

In some grammars it may be possible to find several different derivations of the same string. Two kinds of derivations are worthy of note. A derivation is called a *leftmost derivation* if at each step the leftmost nonterminal of the sentential form is reduced by some production. Similarly, a derivation is called a *rightmost derivation* if at each step the rightmost nonterminal of the sentential form is reduced by some production. For example, the preceding derivation of *aab* is a leftmost derivation. Here's a rightmost derivation of *aab*:

$$S \Rightarrow AB \Rightarrow AbB \Rightarrow Ab \Rightarrow aAb \Rightarrow aaAb \Rightarrow aab.$$

The Language of a Grammar

Sometimes it can be quite difficult, or impossible, to write down a grammar for a given language. So we had better nail down the idea of the language that is associated with a grammar. If G is a grammar, then the *language of G* is the set of terminal strings derived from the start symbol of G. The language of G is denoted by

$$L(G).$$

We can also describe $L(G)$ more formally.

The Language of a Grammar (3.14)

If G is a grammar with start symbol S and set of terminals T, then the language of G is the set

$$L(G) = \{s \mid s \in T^* \text{ and } S \Rightarrow^+ s\}.$$

When we're trying to write a grammar for a language, we should at least check to see whether the language is finite or infinite. If the language is finite, then a grammar can consist of all productions of the form $S \to w$ for each string w in the language. For example, the language $\{a, ab\}$ can be described by the grammar $S \to a \mid ab$.

If the language is infinite, then some production or sequence of productions must be used repeatedly to construct the derivations. To see this, notice that there is no bound on the length of strings in an infinite language. Therefore, there is no bound on the number of derivation steps used to derive the strings. If the grammar has n productions, then any derivation consisting of $n + 1$ steps must use some production twice (by the pigeonhole principle).

For example, the infinite language $\{a^n b \mid n \geq 0\}$ can be described by the grammar

$$S \rightarrow b \mid aS.$$

To derive the string $a^n b$, we would use the production $S \rightarrow aS$ repeatedly—n times to be exact—and then stop the derivation by using the production $S \rightarrow b$. The situation is similar to the way we make inductive definitions for sets. For example, the production $S \rightarrow aS$ allows us to make the informal statement "If S derives w, then it also derives aw."

Recursive Productions

A production is called *recursive* if its left side occurs on its right side. For example, the production $S \rightarrow aS$ is recursive. A production $A \rightarrow \alpha$ is *indirectly recursive* if A derives a sentential form that contains A. For example, suppose we have the following grammar:

$$S \rightarrow b \mid aA$$
$$A \rightarrow c \mid bS.$$

The productions $S \rightarrow aA$ and $A \rightarrow bS$ are both indirectly recursive because of the following derivations:

$$S \Rightarrow aA \Rightarrow abS,$$
$$A \Rightarrow bS \Rightarrow baA.$$

A grammar is *recursive* if it contains either a recursive production or an indirectly recursive production. So we can make the following more precise statement about grammars for infinite languages:

A grammar for an infinite language must be recursive.

Now let's look at the opposite problem of describing the language of a grammar. We know—by definition—that the language of a grammar is the set of all strings derived from the grammar. But we can also make another interesting observation about any language defined by a grammar:

Any language defined by a grammar is an inductively defined set.

Let's see why this is the case for any grammar G. The following inductive definition does the job, where S denotes the start symbol of G. To simplify the description, we'll say that a derivation is *recursive* if some nonterminal occurs twice due to a recursive production or due to a series of indirectly recursive productions.

Constructing an Inductive Definition for $L(G)$ (3.15)

Basis: If $S \Rightarrow^+ w$ without using a recursive derivation, then put w in $L(G)$.

Induction: Let $w \in L(G)$ with derivation $S \Rightarrow^+ w$. If the derivation contains a nonterminal A from a recursive production $A \to \alpha A\beta$ or from an indirectly recursive production corresponding to $A \Rightarrow^+ \alpha A\beta$, then modify the derivation by using either $A \to \alpha A\beta$ or $A \Rightarrow^+ \alpha A\beta$ to obtain a new derivation $S \Rightarrow^+ x$ and put x in $L(G)$.

Proof: Let G be a grammar and let M be the inductive set defined by (3.15). We need to show that $M = L(G)$. It's clear that $M \subseteq L(G)$ because all strings in M are derived from the start symbol of G. Assume, by way of contradiction, that $M \neq L(G)$. In other words, we have $L(G) - M \neq \varnothing$. Since S derives all the elements of $L(G) - M$, there must be some string $w \in L(G) - M$ that has the shortest leftmost derivation among elements of $L(G) - M$. We can assume that this derivation is recursive. Otherwise, the basis case of (3.15) would force us to put $w \in M$, contrary to our assumption that $w \in L(G) - M$. So the leftmost derivation of w must have the following form, where s and t are terminal strings and α, β, and γ are sentential forms that don't include B:

$$S \Rightarrow^+ sB\gamma \Rightarrow^+ stB\beta\gamma \Rightarrow st\alpha\beta\gamma \Rightarrow^* w.$$

We can replace $sB\gamma \Rightarrow^+ stB\beta\gamma$ in this derivation with $sB\gamma \Rightarrow s\alpha\gamma$ to obtain the following derivation of a string u of terminals:

$$S \Rightarrow^+ sB\gamma \Rightarrow s\alpha\gamma \Rightarrow^* u.$$

This derivation is shorter than the derivation of w. So we must conclude that $u \in M$. Now we can apply the induction part of (3.15) to this latter derivation of u to obtain the derivation of w. This tells us that $w \in M$, contrary to our assumption that $w \notin M$. The only thing left for us to conclude is that our assumption that $M \neq L(G)$ was wrong. Therefore $M = L(G)$. QED.

example 3.34 From Grammar to Inductive Definition

Let's do a simple example to illustrate the use of (3.15). Suppose we're given the following grammar G:

$$S \to \Lambda \mid aB$$
$$B \to b \mid bB.$$

We'll give an inductive definition for $L(G)$. There are two derivations that don't contain recursive productions: $S \Rightarrow \Lambda$ and $S \Rightarrow aB \Rightarrow ab$. This gives us the basis part of the definition for $L(G)$.

Basis: Λ, $ab \in L(G)$.

Now let's find the induction part of the definition. The only recursive production of G is $B \rightarrow bB$. So any element of $L(G)$ whose derivation contains an occurrence of B must have the general form $S \Rightarrow aB \Rightarrow^+ ay$ for some string y. So we can use the production $B \rightarrow bB$ to add one more step to the derivation as follows:

$$S \Rightarrow aB \Rightarrow abB \Rightarrow^+ aby.$$

This gives us the induction step in the definition of $L(G)$.

Induction: If $ay \in L(G)$, then put aby in $L(G)$.

For example, the basis case tells us that $ab \in L(G)$ and the derivation $S \Rightarrow aB \Rightarrow ab$ contains an occurrence of B. So we add one more step to the derivation using the production $B \rightarrow bB$ to obtain the derivation

$$S \Rightarrow aB \Rightarrow abB \Rightarrow abb.$$

So $ab \in L(G)$ implies that $abb \in L(G)$, which in turn implies $ab^3 \in L(G)$, and so on. Thus we can conjecture with some confidence that $L(G)$ is the language $\{\Lambda\} \cup \{ab^n \mid n \in \mathbb{N}\}$.

> end example

3.3.4 Constructing Grammars

Now let's get down to business and construct some grammars. We'll start with a few simple examples, and then we'll give some techniques for combining simple grammars. We should note that a language might have more than one grammar. So we shouldn't be surprised when two people come up with two different grammars for the same language.

> example 3.35 **Three Simple Grammars**

We'll write a grammar for each of three simple languages. In each case we'll include a sample derivation of a string in the language. Test each grammar by constructing a few more derivations for strings.

1. $\{\Lambda, a, aa, \ldots, a^n, \ldots\} = \{a^n \mid n \in \mathbb{N}\}$.

   Notice that any string in this language is either Λ or of the form ax for some string x in the language. The following grammar will derive any of these strings:

   $$S \rightarrow \Lambda \mid aS.$$

For example, we'll derive the string aaa:

$$S \Rightarrow aS \Rightarrow aaS \Rightarrow aaaS \Rightarrow aaa.$$

2. $\{\Lambda, ab, aabb, \ldots, a^n b^n, \ldots\} = \{a^n b^n \mid n \in \mathbb{N}\}$.

Notice that any string in this language is either Λ or of the form axb for some string x in the language. The following grammar will derive any of these strings:

$$S \rightarrow \Lambda \mid aSb.$$

For example, we'll derive the string $aaabbb$:

$$S \Rightarrow aSb \Rightarrow aaSbb \Rightarrow aaaSbbb \Rightarrow aaabbb.$$

3. $\{\Lambda, ab, abab, \ldots, (ab)^n, \ldots\} = \{(ab)^n \mid n \in \mathbb{N}\}$.

Notice that any string in this language is either Λ or of the form abx for some string x in the language. The following grammar will derive any of these strings:

$$S \rightarrow \Lambda \mid abS.$$

For example, we'll derive the string $ababab$:

$$S \Rightarrow abS \Rightarrow ababS \Rightarrow abababS \Rightarrow ababab.$$

end example

Combining Grammars

Sometimes a language can be written in terms of simpler languages, and a grammar can be constructed for the language in terms of the grammars for the simpler languages. We'll concentrate here on the operations of union, product, and closure.

Combining Grammars (3.16)

Suppose M and N are languages whose grammars have disjoint sets of non-terminals. (Rename them if necessary.) Suppose also that the start symbols for the grammars of M and N are A and B, respectively. Then we have the following new languages and grammars:

Union Rule: The language $M \cup N$ starts with the two productions

$$S \rightarrow A \mid B.$$

Continued ➡

➡ ➡

Product Rule: The language MN starts with the production

$$S \to AB.$$

Closure Rule: The language M^* starts with the two productions

$$S \to AS \mid \Lambda.$$

---

example 3.36 Using the Union Rule

Let's write a grammar for the following language:

$$L = \{\Lambda, a, b, aa, bb, \ldots, a^n, b^n, \ldots\}.$$

After some thinking we notice that L can be written as a union $L = M \cup N$, where $M = \{a^n \mid n \in \mathbb{N}\}$ and $N = \{b^n \mid n \in \mathbb{N}\}$. Thus we can write the following grammar for L.

$$
\begin{aligned}
S &\to A \mid B & &\text{union rule,}\\
A &\to \Lambda \mid aA & &\text{grammar for } M,\\
B &\to \Lambda \mid bB & &\text{grammar for } N.
\end{aligned}
$$

end example

example 3.37 Using the Product Rule

We'll write a grammar for the following language:

$$L = \{a^m b^n \mid m, n \in \mathbb{N}\}.$$

After a little thinking we notice that L can be written as a product $L = MN$, where $M = \{a^m \mid m \in \mathbb{N}\}$ and $N = \{b^n \mid n \in \mathbb{N}\}$. Thus we can write the following grammar for L:

$$
\begin{aligned}
S &\to AB & &\text{product rule,}\\
A &\to \Lambda \mid aA & &\text{grammar for } M,\\
B &\to \Lambda \mid bB & &\text{grammar for } N.
\end{aligned}
$$

end example

example 3.38 Using the Closure Rule

We'll construct a grammar for the language $L = \{aa, bb\}^*$. If we let $M = \{aa, bb\}$, then $L = M^*$. Thus we can write the following grammar for L.

$$S \rightarrow AS \mid \Lambda \qquad \text{closure rule,}$$
$$A \rightarrow aa \mid bb \qquad \text{grammar for } M.$$

We can simplify this grammar by substituting for A to obtain the following grammar:

$$S \rightarrow aaS \mid bbS \mid \Lambda.$$

end example

example 3.39 Decimal Numerals

We can find a grammar for the language of decimal numerals by observing that a decimal numeral is either a digit or a digit followed by a decimal numeral. The following grammar rules reflect this idea:

$$S \rightarrow D \mid DS$$
$$D \rightarrow 0 \mid 1 \mid 2 \mid 3 \mid 4 \mid 5 \mid 6 \mid 7 \mid 8 \mid 9.$$

We can say that S is replaced by either D or DS, and D can be replaced by any decimal digit. A derivation of the numeral 7801 can be written as follows:

$$S \Rightarrow DS \Rightarrow 7S \Rightarrow 7DS \Rightarrow 7DDS \Rightarrow 78DS \Rightarrow 780S \Rightarrow 780D \Rightarrow 7801.$$

This derivation is not unique. For example, another derivation of 7801 can be written as follows:

$$S \Rightarrow DS \Rightarrow DDS \Rightarrow D8S \Rightarrow D8DS \Rightarrow D80S \Rightarrow D80D \Rightarrow D801 \Rightarrow 7801.$$

end example

example 3.40 Even Decimal Numerals

We can find a grammar for the language of decimal numerals for the even natural numbers by observing that each numeral must have an even digit on its right side. In other words, either it's an even digit or it's a decimal numeral followed by an even digit. The following grammar will do the job:

$$S \rightarrow E \mid NE$$
$$N \rightarrow D \mid DN$$
$$E \rightarrow 0 \mid 2 \mid 4 \mid 6 \mid 8$$
$$D \rightarrow 0 \mid 1 \mid 2 \mid 3 \mid 4 \mid 5 \mid 6 \mid 7 \mid 8 \mid 9.$$

For example, the even numeral 136 has the derivation

$$S \Rightarrow NE \Rightarrow N6 \Rightarrow DN6 \Rightarrow DD6 \Rightarrow D36 \Rightarrow 136.$$

end example

example 3.41 Identifiers

Most programming languages have identifiers for names of things. Suppose we want to describe a grammar for the set of identifiers that start with a letter of the alphabet followed by zero or more letters or digits. Let S be the start symbol. Then the grammar can be described by the following productions:

$$S \rightarrow L \mid LA$$
$$A \rightarrow LA \mid DA \mid \Lambda$$
$$L \rightarrow a \mid b \mid \ldots \mid z$$
$$D \rightarrow 0 \mid 1 \mid \ldots \mid 9.$$

We'll give a derivation of the string $a2b$ to show that it is an identifier.

$$S \Rightarrow LA \Rightarrow aA \Rightarrow aDA \Rightarrow a2A \Rightarrow a2LA \Rightarrow a2bA \Rightarrow a2b.$$

end example

example 3.42 Some Rational Numerals

Let's find a grammar for those rational numbers that have a finite decimal representation. In other words, we want to describe a grammar for the language of strings having the form $m.n$ or $-m.n$, where m and n are decimal numerals. For example, 0.0 represents the number 0. Let S be the start symbol. We can start the grammar with the two productions

$$S \rightarrow N.N \mid -N.N.$$

To finish the job, we need to write some productions that allow N to derive a decimal numeral. We can use the following renamed productions from Example 3.39:

$$N \rightarrow D \mid DN$$
$$D \rightarrow 0 \mid 1 \mid 2 \mid 3 \mid 4 \mid 5 \mid 6 \mid 7 \mid 8 \mid 9.$$

end example

example 3.43   **Palindromes**

We can write a grammar for the set of all palindromes over an alphabet A. Recall that a palindrome is a string that is the same when written in reverse order. For example, let $A = \{a, b, c\}$. Let P be the start symbol. Then the language of palindromes over the alphabet A has the grammar

$$P \rightarrow aPa \mid bPb \mid cPc \mid a \mid b \mid c \mid \Lambda.$$

For example, the palindrome $abcba$ can be derived as follows:

$$P \Rightarrow aPa \Rightarrow abPba \Rightarrow abcba.$$

end example

3.3.5   Meaning and Ambiguity

Most of the time we attach meanings to the strings in our lives. For example, the string 3+4 means 7 to most people. The string 3–4–2 may have two distinct meanings to two different people. One person may think that

$$3\text{–}4\text{–}2 = (3\text{–}4)\text{–}2 = \text{–}3,$$

while another person might think that

$$3\text{–}4\text{–}2 = 3\text{–}(4\text{–}2) = 1.$$

If we have a grammar, then we can define the *meaning* of any string in the grammar's language to be the parse tree produced by a derivation. We can often write a grammar so that each string in the grammar's language has exactly one meaning (i.e., one parse tree). When this is not the case, we have an ambiguous grammar. Here's the formal definition.

> **Definition of Ambiguous Grammar**
>
> A grammar is said to be *ambiguous* if its language contains some string that has two different parse trees. This is equivalent to saying that some string has two distinct leftmost derivations or that some string has two distinct rightmost derivations.

To illustrate the ideas, we'll look at some grammars for simple arithmetic expressions. For example, suppose we define a set of arithmetic expressions by the grammar

$$E \rightarrow a \mid b \mid E\text{–}E.$$

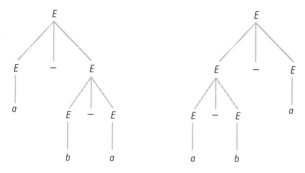

Figure 3.10   Parse trees for an ambiguous string.

The language of this grammar contains strings like a, b, $b-a$, $a-b-a$, and $b-b-a-b$. This grammar is ambiguous because it has a string, namely, $a-b-a$, that has two distinct parse trees as shown in Figure 3.10.

Since having two distinct parse trees means the same thing as having two distinct leftmost derivations, it's no problem to find the following two distinct leftmost derivations of $a-b-a$.

$$E \Rightarrow E - E \Rightarrow a - E \Rightarrow a - E - E \Rightarrow a - b - E \Rightarrow a - b - a.$$
$$E \Rightarrow E - E \Rightarrow E - E - E \Rightarrow a - E - E \Rightarrow a - b - E \Rightarrow a - b - a.$$

The two trees in Figure 3.10 reflect the two ways we could choose to evaluate $a-b-a$. The first tree indicates the meaning

$$a\text{-}b\text{-}a = a\text{-}(b\text{-}a),$$

while the second tree indicates

$$a\text{-}b\text{-}a = (a\text{-}b)\text{-}a.$$

How can we make sure there is only one parse tree for every string in the language? We can try to find a different grammar for the same set of strings. For example, suppose we want $a-b-a$ to mean $(a-b)-a$. In other words, we want the first minus sign to be evaluated before the second minus sign. We can give the first minus sign higher precedence than the second by introducing a new nonterminal as shown in the following grammar:

$$E \rightarrow E - T \,|\, T$$
$$T \rightarrow a \,|\, b.$$

Notice that T can be replaced in a derivation only by either a or b. Therefore, every derivation of $a-b-a$ produces a unique parse tree.

■ Exercises

Derivations

1. Given the following grammar:

$$S \to D \mid DS$$
$$D \to 0 \mid 1 \mid 2 \mid 3 \mid 4 \mid 5 \mid 6 \mid 7 \mid 8 \mid 9.$$

    a. Find the production used in each step of the following derivation.

$$S \Rightarrow DS \Rightarrow 7S \Rightarrow 7DS \Rightarrow 7DDS \Rightarrow 78DS \Rightarrow 780S \Rightarrow 780D \Rightarrow 7801.$$

    b. Find a leftmost derivation of the string 7801.
    c. Find a rightmost derivation of the string 7801.

2. Given the following grammar:

$$S \to S[S] \mid \Lambda.$$

For each of the following strings, construct a leftmost derivation, a rightmost derivation, and a parse tree.

    a. [ ].   **b.** [[ ]].   **c.** [ ] [ ].   **d.** [[ ] [[ ]]].

Constructing Grammars

3. Find a grammar for each of the following languages.

    a. $\{bb, bbbb, bbbbbb, \dots\} = \left\{(bb)^{n+1} \mid n \in \mathbb{N}\right\}.$
    b. $\{a, ba, bba, bbba, \dots\} = \{b^n a \mid n \in \mathbb{N}\}.$
    c. $\{\Lambda, ab, abab, ababab, \dots\} = \{(ab)^n \mid n \in \mathbb{N}\}.$
    d. $\{bb, bab, baab, baaab, \dots\} = \{ba^n b \mid n \in \mathbb{N}\}.$
    e. $\left\{ab, abab, \dots, (ab)^{n+1}, \dots\right\} = \left\{(ab)^{n+1} \mid n \in \mathbb{N}\right\}.$
    f. $\{ab, aabb, \dots, a^n b^n, \dots\} = \{a^{n+1} b^{n+1} \mid n \in \mathbb{N}\}.$
    g. $\left\{b, bbb, \dots, b^{2n+1}, \dots\right\} = \{b^{2n+1} \mid n \in \mathbb{N}\}.$
    h. $\{b, abc, aabcc, \dots, a^n bc^n, \dots\} = \{a^n bc^n \mid n \in \mathbb{N}\}.$
    i. $\{ac, abc, abbc, \dots, ab^n c, \dots\} = \{ab^n c \mid n \in \mathbb{N}\}.$
    j. $\{\Lambda, aa, aaaa, \dots, a^{2n}, \dots\} = \{a^{2n} \mid n \in \mathbb{N}\}.$

4. Find a grammar for each language.

    a. $\{a^m b^n \mid m,\ n \in \mathbb{N}\}$.

    b. $\{a^m b c^n \mid m,\ n \in \mathbb{N}\}$.

    c. $\{a^m b^n \mid m,\ n \in \mathbb{N},\ \text{where } m > 0\}$.

    d. $\{a^m b^n \mid m,\ n \in \mathbb{N},\ \text{where } n > 0\}$.

    e. $\{a^m b^n \mid m,\ n \in \mathbb{N},\ \text{where } m > 0 \text{ and } n > 0\}$.

5. Find a grammar for each language.

    a. The even palindromes over $\{a,\ b,\ c\}$.

    b. The odd palindromes over $\{a,\ b,\ c\}$.

    c. $\{a^{2n} \mid n \in \mathbb{N}\} \cup \{b^{2n+1} \mid n \in \mathbb{N}\}$.

    d. $\{a^n b c^n \mid n \in \mathbb{N}\} \cup \{b^m a^n \mid m,\ n \in \mathbb{N}\}$

    e. $\{a^m b^n \mid m,\ n \in \mathbb{N},\ \text{where } m > 0 \text{ or } n > 0\}$.

Mathematical Expressions

6. Find a grammar for each of the following languages.

    a. The set of binary numerals that represent odd natural numbers.

    b. The set of binary numerals that represent even natural numbers.

    c. The set of decimal numerals that represent odd natural numbers.

7. Find a grammar for each of the following languages.

    a. The set of arithmetic expressions that are constructed from decimal numerals, +, and parentheses. Examples: 17, 2+3, (3+(4+5)), and 5+9+20.

    b. The set of arithmetic expressions that are constructed from decimal numerals, − (subtraction), and parentheses, with the property that each expression has only one meaning. For example, 9−34−10 is not allowed.

8. Let the letters a, b, and c be constants; let the letters x, y, and z be variables; and let the letters f and g be functions of arity 1. We can define the set of terms over these symbols by saying that any constant or variable is a term and if t is a term, then so are $f(t)$ and $g(t)$.

    a. Find a grammar for the set of terms.

    b. Find a derivation for the expression $f(g(f(x)))$.

9. Let the letters a, b, and c be constants; let the letters x, y, and z be variables; and let the letters f and g be functions of arity 1 and 2, respectively. We can define the set of terms over these symbols by saying that any constant or variable is a term and if s and t are terms, then so are $f(t)$ and $g(s,\ t)$.

    a. Find a grammar for the set of terms.

    b. Find a derivation for the expression $f(g(x,\ f(b)))$.

10. Find a grammar to capture the precedence $\cdot$ over $+$ in the absence of paren-theses. For example, the meaning of $a + b \cdot c$ should be $a + (b \cdot c)$.

Ambiguity

11. Show that each of the following grammars is ambiguous. In other words, find a string that has two different parse trees (equivalently, two different leftmost derivations or two different rightmost derivations).

    a. $S \rightarrow a \mid SbS$.

    b. $S \rightarrow abB \mid AB$ and $A \rightarrow \Lambda \mid Aa$ and $B \rightarrow \Lambda \mid bB$.

    c. $S \rightarrow aS \mid Sa \mid a$.

    d. $S \rightarrow aS \mid Sa \mid b$.

    e. $S \rightarrow S[S]S \mid \Lambda$.

    f. $S \rightarrow Ab \mid A$ and $A \rightarrow b \mid bA$.

Challenges

12. Find a grammar for the language of all strings over $\{a, b\}$ that have the same number of a's and b's.

13. For each grammar, try to find an equivalent grammar that is not ambiguous.

    a. $S \rightarrow a \mid SbS$.

    b. $S \rightarrow abB \mid AB$ and $A \rightarrow \Lambda \mid Aa$ and $B \rightarrow \Lambda \mid bB$.

    c. $S \rightarrow a \mid aS \mid Sa$.

    d. $S \rightarrow b \mid aS \mid Sa$.

    e. $S \rightarrow S[S]S \mid \Lambda$.

    f. $S \rightarrow Ab \mid A$ and $A \rightarrow b \mid bA$.

14. For each grammar, find an equivalent grammar that has no occurrence of Λ on the right side of any rule.

    a. $S \rightarrow AB$ b. $S \rightarrow AcAB$

        $A \rightarrow Aa \mid a$ $A \rightarrow aA \mid \Lambda$

        $B \rightarrow Bb \mid \Lambda$. $B \rightarrow bB \mid b$.

15. For each grammar G, use (3.15) to find an inductive definition for $L(G)$.

    a. $S \rightarrow \Lambda \mid aaS$.

    b. $S \rightarrow a \mid aBc$ and $B \rightarrow b \mid bB$.

3.4   Chapter Summary

This chapter covered some basic construction techniques that apply to many objects of importance to computer science.

Inductively defined sets are characterized by a basis step, an induction step, and a closure clause that is always assumed without comment. The constructors of an inductively defined set are the elements listed in the basis step and the rules specified in the induction step. Many sets of objects used in computer science can be defined inductively—numbers, strings, lists, binary trees, and Cartesian products of sets.

A recursively defined function is defined in terms of itself. Most recursively defined functions have domains that are inductively defined sets. These functions are normally defined by a basis case and a recursive case. The situation is similar for recursively defined procedures. Some infinite sequence functions can be defined recursively. Recursively defined functions and procedures yield powerful programs that are simply stated.

Grammars provide useful ways to describe languages. Grammar productions are used to derive the strings of a language. Any grammar for an infinite language must contain at least one production that is recursive or indirectly recursive. Grammars for different languages can be combined to form new grammars for unions, products, and closures of the languages. Some grammars are ambiguous.

chapter 4

Equivalence, Order, and Inductive Proof

Good order is the foundation of all things.
                    —Edmund Burke (1729–1797)

Classifying things and ordering things are activities in which we all engage from time to time. Whenever we classify or order a set of things, we usually compare them in some way. That's how binary relations enter the picture.

In this chapter we'll discuss some special properties of binary relations that are useful for solving comparison problems. We'll introduce techniques to construct binary relations with the properties that we need. We'll discuss the idea of equivalence by considering properties of the equality relation. We'll also study the properties of binary relations that characterize our intuitive ideas about ordering. We'll also see that ordering is the fundamental ingredient needed to discuss inductive proof techniques.

chapter guide

4.1 (Properties of Binary Relations) introduces some of the desired properties of binary relations and shows how to construct new relations by composition and closure. We'll see how the results apply to solving path problems in graphs.

4.2 (Equivalence Relations) concentrates on the idea of equivalence. We'll see that equivalence is closely related to partitioning of sets. We'll show how to generate equivalence relations, we'll solve a typical equivalence problem, and we'll see an application for finding a spanning tree for a graph.

4.3 (Order Relations) introduces the idea of order. We'll discuss partial orders and how to sort them. We'll introduce well-founded orders and show some techniques for constructing them. Ordinal numbers are also introduced.

4.4 (Inductive Proof) introduces inductive proof techniques. We'll discuss the technique of mathematical induction for proving statements indexed by the natural numbers. Then we'll extend the discussion to inductive proof techniques for any well-founded set.

4.1 Properties of Binary Relations

Recall that the statement "R is a binary relation on the set A" means that R relates certain pairs of elements of A. Thus R can be represented as a set of ordered pairs (x, y), where $x, y \in A$. In other words, R is a subset of the Cartesian product $A \times A$. When $(x, y) \in R$, we also write $x \ R \ y$.

Binary relations that satisfy certain special properties can be very useful in solving computational problems. So let's discuss these properties.

Three Special Properties

For a binary relation R on a set A, we have the following definitions.

   a. R is *reflexive* if $x \ R \ x$ for all $x \in A$.

   b. R is *symmetric* if $x \ R \ y$ implies $y \ R \ x$ for all $x, y \in A$.

   c. R is *transitive* if $x \ R \ y$ and $y \ R \ z$ implies $x \ R \ z$ for all $x, y, z \in A$.

Since a binary relation can be represented by a directed graph, we can describe the three properties in terms of edges: R is reflexive if there is an edge from x to x for each $x \in A$; R is symmetric if for each edge from x to y, there is also an edge from y to x. R is transitive if whenever there are edges from x to y and from y to z, there must also be an edge from x to z.

There are two more properties that we will also find to be useful.

Two More Properties

For a binary relation R on a set A, we have the following definitions.

   a. R is *irreflexive* if $(x, x) \notin R$ for all $x \in A$.

   b. R is *antisymmetric* if $x \ R \ y$ and $y \ R \ x$ implies $x = y$ for all $x, y \in A$.

From a graphical point of view we can say that R is irreflexive if there are no loop edges from x to x for all $x \in A$; and R is antisymmetric if whenever there is an edge from x to y with $x \neq y$, then there is no edge from y to x.

Many well-known relations satisfy one or more of the properties that we've been discussing. So we better look at a few examples.

example 4.1 **Five Binary Relations**

Here are some sample binary relations with the properties that they satisfy.

a. The equality relation on any set is reflexive, symmetric, transitive, and antisymmetric.

b. The $<$ relation on real numbers is transitive, irreflexive, and antisymmetric.

c. The $\leq$ relation on real numbers is reflexive, transitive, and antisymmetric.

d. The "is parent of" relation is irreflexive and antisymmetric.

e. The "has the same birthday as" relation is reflexive, symmetric, and transitive.

end example

4.1.1 Composition of Relations

Relations can often be defined in terms of other relations. For example, we can describe the "is grandparent of" relation in terms of the "is parent of" relation by saying that "a is grandparent of c" if and only if there is some b such that "a is parent of b" and "b is parent of c". This example demonstrates the fundamental idea of composing binary relations.

Definition of Composition

If R and S are binary relations, then the *composition* of R and S, which we denote by $R \circ S$, is the following relation:

$$R \circ S = \{(a, c) \mid (a, b) \in R \text{ and } (b, c) \in S \text{ for some element } b\}.$$

From a directed graph point of view, if we find an edge from a to b in the graph of R and we find an edge from b to c in the graph of S, then we must have an edge from a to c in the graph of $R \circ S$.

example 4.2 **Grandparents**

To construct the "isGrandparentOf" relation we can compose "isParentOf" with itself.

$$\text{isGrandparentOf} = \text{isParentOf} \circ \text{isParentOf}.$$

Similarly, we can construct the "isGreatGrandparentOf" relation by the following composition:

$$\text{isGreatGrandparentOf} = \text{isGrandparentOf} \circ \text{isParentOf}.$$

end example

example **4.3 Numeric Relations**

Suppose we consider the relations "less," "greater," "equal," and "notEqual" over the set $\mathbb{R}$ of real numbers. We want to compose some of these relations to see what we get. For example, let's verify the following equality.

$$\text{greater} \circ \text{less} = \mathbb{R} \times \mathbb{R}.$$

For any pair (x, y), the definition of composition says that x (greater $\circ$ less) y if and only if there is some number z such that x greater z and z less y. We can write this statement more concisely as follows:

$$x \ (> \circ <) \ y \text{ iff there is some number } z \text{ such that } x > z \text{ and } z < y.$$

We know that for any two real numbers x and y there is always another number z that is less than both. So the composition must be the whole universe $\mathbb{R} \times \mathbb{R}$. Many combinations are possible. For example, it's easy to verify the following two equalities:

$$\text{equal} \circ \text{notEqual} = \text{notEqual},$$
$$\text{notEqual} \circ \text{notEqual} = \mathbb{R} \times \mathbb{R}.$$

end example

Other Combining Methods

Since relations are just sets (of ordered pairs), they can also be combined by the usual set operations of union, intersection, difference, and complement.

example **4.4 Combining Relations**

The following samples show how we can combine some familiar numeric relations. Check out each one with a few example pairs of numbers.

$$\text{equal} \cap \text{less} = \varnothing,$$
$$\text{equal} \cap \text{lessOrEqual} = \text{equal},$$
$$(\text{lessOrEqual})' = \text{greater},$$
$$\text{greaterOrEqual} - \text{equal} = \text{greater},$$
$$\text{equal} \cup \text{greater} = \text{greaterOrEqual},$$
$$\text{less} \cup \text{greater} = \text{notEqual}.$$

end example

Let's list some fundamental properties of combining relations. We'll leave the proofs of these properties as exercises.

Properties of Combining Relations (4.1)

a. $R \circ (S \circ T) = (R \circ S) \circ T.$ (associativity)

b. $R \circ (S \cup T) = R \circ S \cup R \circ T.$

c. $R \circ (S \cap T) \subseteq R \circ S \cap R \circ T.$

Notice that Part (c) is stated as a set containment rather than an equality. For example, let R, S, and T be the following relations:

$$R = \{(a, b), (a, c)\}, \ S = \{(b, b)\}, \ T = \{(b, c), (c, b)\}.$$

Then $S \cap T = \varnothing$, $R \circ S = \{(a, b)\}$, and $R \circ T = \{(a, c), (a, b)\}$. Therefore

$$R \circ (S \cap T) = \varnothing \text{ and } R \circ S \cap R \circ T = \{(a, b)\}.$$

So (4.1c) isn't always an equality. But there are cases in which equality holds. For example, if $R = \varnothing$ or if $S \subseteq T$, then (4.1c) is an equality.

Representations

If R is a binary relation on A, then we'll denote the composition of R with itself n times by writing

$$R^n.$$

For example, if we compose isParentOf with itself, we get some familiar names as follows:

$$\text{isParentOf}^2 = \text{isGrandparentOf},$$
$$\text{isParentOf}^3 = \text{isGreatGrandparentOf}.$$

We mentioned in Chapter 1 that binary relations can be thought of as digraphs and, conversely, that digraphs can be thought of as binary relations. In other words, we can think of (x, y) as an edge from x to y in a digraph and as a member of a binary relation. So we can talk about the digraph of a binary relation.

An important and useful representation of R^n is as the digraph consisting of all edges (x, y) such that there is a path of length n from x to y. For example, if $(x, y) \in R^2$, then $(x, z), (z, y) \in R$ for some element z. This says that there is a path of length 2 from x to y in the digraph of R.

example **4.5 Compositions**

Let $R = \{(a, b), (b, c), (c, d)\}$. The digraphs shown in Figure 4.1 are the digraphs for the three relations R, R^2, and R^3.

end example

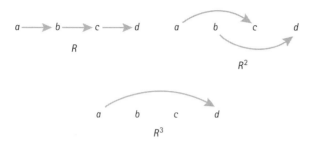

Figure 4.1   Composing a relation.

Let's give a more precise definition of R^n using induction. (Notice the interesting choice for R^0.)

$$R^0 = \{(a, a) \mid a \in A\} \quad \text{(basic equality)}$$
$$R^{n+1} = R^n \circ R.$$

We defined R^0 as the basic equality relation because we want to infer the equality $R^1 = R$ from the definition. To see this, observe the following evaluation of R^1:

$$R^1 = R^{0+1} = R^0 \circ R = \{(a, a) \mid a \in A\} \circ R = R.$$

We also could have defined $R^{n+1} = R \circ R^n$ instead of $R^{n+1} = R^n \circ R$ because composition of binary operations is associative by (4.1a).

Let's note a few other interesting relationships between R and R^n.

Inheritance Properties   (4.2)

   a. If R is reflexive, then R^n is reflexive.

   b. If R is symmetric, then R^n is symmetric.

   c. If R is transitive, then R^n is transitive.

On the other hand, if R is irreflexive, then it may not be the case that R^n is irreflexive. Similarly, if R is antisymmetric, it may not be the case that R^n is antisymmetric. We'll examine these statements in the exercises.

example **4.6 Integer Relations**

Let $R = \{(x, y) \in \mathbb{Z} \times \mathbb{Z} \mid x + y \text{ is odd}\}$. We'll calculate R^2 and R^3. To calculate R^2, we'll examine an arbitrary element $(x, y) \in R^2$. This means there is an element z such that $(x, z) \in R$ and $(z, y) \in R$. So $x + z$ is odd and $z + y$ is odd. We know that a sum is odd if and only if one number is even and the other number is odd. If x is even, then since $x + z$ is odd, it follows that z is odd. So, since $z + y$ is odd, it follows that y is even. Similarly, if x is odd, the same kind of reasoning shows that y is odd. So we have

$$R^2 = \{(x,\, y) \in \mathbb{Z} \times \mathbb{Z} \mid x \text{ and } y \text{ are both even or both odd}\}.$$

To calculate R^3, we'll examine an arbitrary element $(x,\, y) \in R^3$. This means there is an element z such that $(x,\, z) \in R$ and $(z,\, y) \in R^2$. In other words, $x + z$ is odd and z and y are both even or both odd. If x is even, then since $x + z$ is odd, it follows that z is odd. So y must be odd. Similarly, if x is odd, the same kind of reasoning shows that y is even. So if $(x,\, y) \in R^3$, then one of x and y is even and the other is odd. In other words, $x + y$ is odd. Therefore

$$R^3 = R.$$

We don't have to go to higher powers now because, for example,

$$R^4 = R^3 \circ R = R \circ R = R^2.$$

end example

4.1.2 Closures

We've seen how to construct a new binary relation by composing two existing binary relations. Now we'll see how to construct a new binary relation by adding pairs of elements to an existing binary relation so that the new relation satisfies some particular property. For example, from the "isParentOf" relation for a family, we could construct the "isAncestorOf" relation for the family by adding the isGrandParentOf pairs, the isGreatGrandParentOf pairs, and so on. But for some properties this process cannot always be done. For example, if $R = \{(a, b), (b, a)\}$, then R is symmetric, so it cannot be a subset of any antisymmetric relation. To discuss this further we need to introduce the idea of closure.

Definition of Closure

If R is a binary relation and p is a property that can be satisfied by adding pairs of elements to R, then the p *closure* of R is the smallest binary relation containing R that has property p.

Three properties of interest for which closures always exist are reflexive, symmetric, and transitive. To introduce each of the three closures we'll use the following relation on the set $A = \{a, b, c\}$:

$$R = \{(a,\, a),\, (a,\, b),\, (b,\, a),\, (b,\, c)\}.$$

Notice that R is not reflexive, not symmetric, and not transitive. So the closures of R that we construct will all contain R as a proper subset.

Reflexive Closure

If R is a binary relation on A, then the *reflexive closure* of R, which we'll denote by $r(R)$, can be constructed by including all pairs (x, x) that are not already in R. Recall that the relation $\{(x, x) \mid x \in A\}$ is called the equality relation on A and it is also denoted by R^0. So we can say that

$$r(R) = R \cup R^0.$$

In our example, the pairs (b, b) and (c, c) are missing from R. So $r(R)$ is R together with these two pairs.

$$r(R) = \{(a, a), (a, b), (b, a), (b, c), (b, b), (c, c)\}.$$

Symmetric Closure

If R is a binary relation, then the *symmetric closure* of R, which we'll denote by $s(R)$, must include all pairs (x, y) for which $(y, x) \in R$. The set $\{(x, y) \mid (y, x) \in R\}$ is called the *converse* of R, which we'll denote by R^c. So we can say that

$$s(R) = R \cup R^c.$$

Notice that R is symmetric if and only if $R = R^c$.

In our example, the only problem is with the pair $(b, c) \in R$. Once we include the pair (c, b) we'll have $s(R)$.

$$s(R) = \{(a, a), (a, b), (b, a), (b, c), (c, b)\}.$$

Transitive Closure

If R is a binary relation, then the *transitive closure* of R is denoted by $t(R)$. We'll use our example to show how to construct $t(R)$. Notice that R contains the pairs (a, b) and (b, c), but (a, c) is not in R. Similarly, R contains the pairs (b, a) and (a, b), but (b, b) is not in R. So $t(R)$ must contain the pairs (a, c) and (b, b). Is there some relation that we can union with R that will add the two needed pairs? The answer is yes, it's R^2. Notice that

$$R^2 = \{(a, a), (a, b), (b, a), (b, b), (a, c)\}.$$

It contains the two missing pairs along with three other pairs that are already in R. Thus we have

$$t(R) = R \cup R^2 = \{(a, a), (a, b), (b, a), (b, c), (a, c), (b, b)\}.$$

To get some further insight into constructing the transitive closure, we need to look at another example. Let $A = \{a, b, c, d\}$, and let R be the following relation.

$$R = \{(a, b), (b, c), (c, d)\}.$$

To compute $t(R)$, we need to add the three pairs (a, c), (b, d), and (a, d). In this case, $R^2 = \{(a, c), (b, d)\}$. So the union of R with R^2 is missing (a, d).

Can we find another relation to union with R and R^2 that will add this missing pair? Notice that $R^3 = \{(a, d)\}$. So for this example, $t(R)$ is the union

$$t(R) = R \cup R^2 \cup R^3$$
$$= \{(a, b), (b, c), (c, d), (a, c), (b, d), (a, d)\}.$$

As the examples show, $t(R)$ is a bit more difficult to construct than the other two closures.

Constructing the Three Closures

The three closures can be calculated by using composition and union. Here are the construction techniques.

Constructing Closures   (4.3)

If R is a binary relation over a set A, then:

a. $r(R) = R \cup R^0$ (R^0 is the equality relation.)

b. $s(R) = R \cup R^c$ (R^c is the converse relation.)

c. $t(R) = R \cup R^2 \cup R^3 \cup \cdots$.

d. If A is finite with n elements, then $t(R) = R \cup R^2 \cup \cdots \cup R^n$.

Let's discuss (4.3d). If $(x, y) \in R^m$, then there is a path of length m from x to y in the digraph representation of R. If $m > n$, then some element of A occurs at least twice on the path from x to y. So there is a shorter path from x to y. Thus $(x, y) \in R^k$ for some $k < m$. This argument can be repeated, if necessary, until we find that $(x, y) \in R^k$ for some $k \leq n$. So nothing new gets added to $t(R)$ by adding powers of R that are higher than n.

Sometimes we don't have to compute all the powers of R. For example, let $A = \{a, b, c, d, e\}$ and $R = \{(a, b), (b, c), (b, d), (d, e)\}$. The digraphs of R and $t(R)$ are drawn in Figure 4.2. Convince yourself that $t(R) = R \cup R^2 \cup R^3$. In other words, the relations R^4 and R^5 don't add anything new. In fact, you should verify that $R^4 = R^5 = \varnothing$.

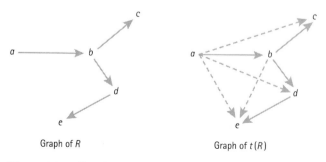

Graph of R Graph of $t(R)$

Figure 4.2 R and its transitive closure.

example **4.7 A Big Transitive Closure**

Let $A = \{a, b, c\}$ and $R = \{(a, b), (b, c), (c, a)\}$. Then we have

$$R^2 = \{(a, c), (c, b), (b, a)\} \text{ and } R^3 = \{(a, a), (b, b), (c, c)\}.$$

So the transitive closure of R is the union

$$t(R) = R \cup R^2 \cup R^3 = A \times A.$$

end example

example **4.8 A Small Transitive Closure**

Let $A = \{a, b, c\}$ and $R = \{(a, b), (b, c), (c, b)\}$. Then we have

$$R^2 = \{(a, c), (b, b), (c, c)\} \text{ and } R^3 = \{(a, b), (b, c), (c, b)\} = R.$$

So the transitive closure of R is the union of the sets, which gives

$$t(R) = \{(a, b), (b, c), (c, b), (a, c), (b, b), (c, c)\}.$$

end example

example **4.9 Generating Less-Than**

Suppose $R = \{(x, x + 1) \mid x \in \mathbb{N}\}$. Then $R^2 = \{(x, x + 2) \mid x \in \mathbb{N}\}$. In general, for any natural number $k > 0$ we have

$$R^k = \{(x, x + k) \mid x \in \mathbb{N}\}.$$

Since $t(R)$ is the union of all these sets, it follows that $t(R)$ is the familiar "less" relation over $\mathbb{N}$. Just notice that if $x < y$, then $y = x + k$ for some k, so the pair (x, y) is in R^k.

end example

example **4.10 Closures of Numeric Relations**

We'll list some closures for the numeric relations "less" and "notEqual" over the set $\mathbb{N}$ of natural numbers.

$$r\,(\text{less}) = \text{lessOrEqual},$$
$$s\,(\text{less}) = \text{notEqual},$$
$$t\,(\text{less}) = \text{less},$$
$$r\,(\text{notEqual}) = \mathbb{N} \times \mathbb{N},$$
$$s\,(\text{notEqual}) = \text{notEqual},$$
$$t\,(\text{notEqual}) = \mathbb{N} \times \mathbb{N}.$$

end example

Properties of Closures

Some properties are retained by closures. For example, we have the following results, which we'll leave as exercises:

Inheritance Properties (4.4)

a. If R is reflexive, then $s(R)$ and $t(R)$ are reflexive.

b. If R is symmetric, then $r(R)$ and $t(R)$ are symmetric.

c. If R is transitive, then $r(R)$ is transitive.

Notice that (4.4c) doesn't include the statement "$s(R)$ is transitive" in its conclusion. To see why, we can let $R = \{(a, b), (b, c), (a, c)\}$. It follows that R is transitive. But $s(R)$ is not transitive because, for example, we have (a, b), $(b, a) \in s(R)$ and $(a, a) \notin s(R)$.

Sometimes, it's possible to take two closures of a relation and not worry about the order. Other times, we have to worry. For example, we might be interested in the double closure $r(s(R))$, which we'll denote by $rs(R)$. Do we get the same relation if we interchange r and s and compute $sr(R)$? The inheritance properties (4.4) should help us see that the answer is yes. Here are the facts:

Double Closure Properties (4.5)

a. $rt(R) = tr(R)$.

b. $rs(R) = sr(R)$.

c. $st(R) \subseteq ts(R)$.

Notice that (4.5c) is not an equality. To see why, let $A = \{a, b, c\}$, and consider the relation $R = \{(a, b), (b, c)\}$. Then $st(R)$ and $ts(R)$ are

$$st(R) = \{(a,b), (b,a), (b,c), (c,b), (a,c), (c,a)\}.$$
$$ts(R) = A \times A.$$

Therefore, $st(R)$ is a proper subset of $ts(R)$. Of course, there are also situations in which $st(R) = ts(R)$. For example, if $R = \{(a,a), (b,b), (c,c), (a,b), (b,c)\}$, then you can verify that $st(R) = ts(R)$.

Before we finish this discussion of closures, we should remark that the symbols R^+ and R^* are often used to denote the closures $t(R)$ and $rt(R)$.

4.1.3   Path Problems

Suppose we need to write a program that inputs two points in a city and outputs a bus route between the two points. A solution to the problem depends on the definition of "point." For example, if a point is any street intersection, then the solution may be harder than in the case in which a point is a bus stop.

This problem is an instance of a path problem. Let's consider some typical path problems in terms of a digraph.

Some Path Problems   **(4.6)**

Given a digraph and two of its vertices i and j.

   a. Find out whether there is a path from i to j. For example, find out whether there is a bus route from i to j.

   b. Find a path from i to j. For example, find a bus route from i to j.

   c. Find a path from i to j with the minimum number of edges. For example, find a bus route from i to j with the minimum number of stops.

   d. Find a shortest path from i to j, where each edge has a nonnegative weight. For example, find the shortest bus route from i to j, where shortest might refer to distance or time.

   e. Find the length of a shortest path from i to j. For example, find the number of stops (or the time or miles) on the shortest bus route from i to j.

Each problem listed in (4.6) can be phrased as a question and the same question is often asked over and over again (e.g., different people asking about the same bus route). So it makes sense to get the answers in advance if possible. We'll see how to solve each of the problems in (4.6).

Adjacency Matrix

A useful way to represent a binary relation R over a finite set A (equivalently, a digraph with vertices A and edges R) is as a special kind of matrix called an *adjacency matrix* (or incidence matrix). For ease of notation we'll assume that $A = \{1,..., n\}$ for some n. The adjacency matrix for R is an n by n matrix M with entries defined as follows:

$$M_{ij} = \text{if } (i, j) \in R \text{ then } 1 \text{ else } 0.$$

example   **4.11   An Adjacency Matrix**

Consider the relation $R = \{(1, 2), (2, 3), (3, 4), (4, 3)\}$ over $A = \{1, 2, 3, 4\}$. We can represent R as a directed graph or as an adjacency matrix M. Figure 4.3 shows the two representations.

end example

$$
1 \longrightarrow 2 \longrightarrow 3 \longrightarrow 4 \qquad M = \begin{bmatrix} 0 & 1 & 0 & 0 \\ 0 & 0 & 1 & 0 \\ 0 & 0 & 0 & 1 \\ 0 & 0 & 1 & 0 \end{bmatrix}
$$

Figure 4.3   Directed graph and adjacency matrix.

If we look at the digraph in Figure 4.3, it's easy to see that R is neither reflexive, symmetric, nor transitive. We can see from the matrix M in Figure 4.3 that R is not reflexive because there is at least one zero on the main diagonal formed by the elements M_{ii}. Similarly, R is not symmetric because a reflection on the main diagonal is not the same as the original matrix. In other words, there are indices i and j such that $M_{ij} \neq M_{ji}$. R is not transitive, but there isn't any visual pattern in M that corresponds to transitivity.

It's an easy task to construct the adjacency matrix for $r(R)$: Just place 1's on the main diagonal of the adjacency matrix. It's also an easy task to construct the adjacency matrix for $s(R)$. We'll leave this one as an exercise.

Warshall's Algorithm for Transitive Closure

Let's look at an interesting algorithm to construct the adjacency matrix for $t(R)$. The idea, of course, is to repeat the following process until no new edges can be added to the adjacency matrix: If (i, k) and (k, j) are edges, then construct a new edge (i, j). The following algorithm to accomplish this feat with three **for**-loops is due to Warshall [1962].

Warshall's Algorithm for Transitive Closure (4.7)

Let M be the adjacency matrix for a relation R over $\{1,..., n\}$. The algorithm replaces M with the adjacency matrix for $t(R)$.

> **for** $k := 1$ **to** n **do**
> > **for** $i := 1$ **to** n **do**
> > > **for** $j := 1$ **to** n **do**
> > > > **if** $(M_{ik} = M_{kj} = 1)$ **then** $M_{ij} := 1$
>
> **od od od**

---

example **4.12 Applying Warshall's Algorithm**

We'll apply Warshall's algorithm to find the transitive closure of the relation R given in Example 4.11. So the input to the algorithm will be the adjacency matrix M for R shown in Figure 4.3. The four matrices in Figure 4.4 show how

$$
M \rightarrow
\begin{matrix} k=1 \\ \begin{bmatrix} 0 & 1 & 0 & 0 \\ 0 & 0 & 1 & 0 \\ 0 & 0 & 0 & 1 \\ 0 & 0 & 1 & 0 \end{bmatrix} \end{matrix}
\rightarrow
\begin{matrix} k=2 \\ \begin{bmatrix} 0 & 1 & 1 & 0 \\ 0 & 0 & 1 & 0 \\ 0 & 0 & 0 & 1 \\ 0 & 0 & 1 & 0 \end{bmatrix} \end{matrix}
\rightarrow
\begin{matrix} k=3 \\ \begin{bmatrix} 0 & 1 & 1 & 1 \\ 0 & 0 & 1 & 1 \\ 0 & 0 & 0 & 1 \\ 0 & 0 & 1 & 1 \end{bmatrix} \end{matrix}
\rightarrow
\begin{matrix} k=4 \\ \begin{bmatrix} 0 & 1 & 1 & 1 \\ 0 & 0 & 1 & 1 \\ 0 & 0 & 1 & 1 \\ 0 & 0 & 1 & 1 \end{bmatrix} \end{matrix}
$$

Figure 4.4   Matrix transformations via Warshall's algorithm.

Warshall's algorithm transforms M into the adjacency matrix for $t(R)$. Each matrix represents the value of M for the given value of k after the inner i and j loops have executed. To get some insight into how Warshall's algorithm works, draw the four digraphs for the adjacency matrices in Figure 4.4.

end example

Now we have an easy way find out whether there is a path from i to j in a digraph. Let R be the set of edges in the digraph. First we represent R as an adjacency matrix. Then we apply Warshall's algorithm to construct the adjacency matrix for $t(R)$. Now we can check to see whether there is a path from i to j in the original digraph by checking M_{ij} in the adjacency matrix M for $t(R)$. So we have all the solutions to problem (4.6a).

Floyd's Algorithm for Length of Shortest Path

Let's look at problem (4.6e). Can we compute the length of a shortest path in a weighted digraph? Sure. Let R denote the set of edges in the digraph. We'll represent the digraph as a *weighted adjacency matrix* M as follows: First of all, we set $M_{ii} = 0$ for $1 \le i \le n$ because we're not interested in the shortest path from i to itself. Next, for each edge $(i, j) \in R$ with $i \ne j$, we set M_{ij} to be the nonnegative weight for that edge. Lastly, if $(i, j) \notin R$ with $i \ne j$, then we set $M_{ij} = \infty$, where ∞ represents some number that is larger than the sum of all the weights on all the edges of the digraph.

example   4.13 A Weighted Adjacency Matrix

The diagram in Figure 4.5 represents the weighted adjacency matrix M for a weighted digraph over the vertex set $\{1, 2, 3, 4, 5, 6\}$.

end example

Now we can present an algorithm to compute the shortest distances between vertices in a weighted digraph. The algorithm, due to Floyd [1962], modifies

	1	2	3	4	5	6
1	0	10	10	∞	20	10
2	∞	0	∞	30	∞	∞
3	∞	∞	0	30	∞	∞
4	∞	∞	∞	0	∞	∞
5	∞	∞	∞	40	0	∞
6	∞	∞	∞	∞	5	0

Figure 4.5   Sample weighted adjacency matrix.

the weighted adjacency matrix M so that M_{ij} is the shortest distance between distinct vertices i and j. For example, if there are two paths from i to j, then the entry M_{ij} denotes the smaller of the two path weights. So again, transitive closure comes into play. Here's the algorithm.

Floyd's Algorithm for Shortest Distances   (4.8)

Let M be the weighted adjacency matrix for a weighted digraph over the set $\{1,..., n\}$. The algorithm replaces M with a weighted adjacency matrix that represents the shortest distances between distinct vertices.

$$\textbf{for } k := 1 \textbf{ to } n \textbf{ do}$$
$$\textbf{for } i := 1 \textbf{ to } n \textbf{ do}$$
$$\textbf{for } j := 1 \textbf{ to } n \textbf{ do}$$
$$M_{ij} := \min\{M_{ij},\, M_{ik} + M_{kj}\}$$
$$\textbf{od od od}$$

---

example **4.14 Applying Floyd's Algorithm**

We'll apply Floyd's algorithm to the weighted adjacency matrix in Figure 4.5. The result is given in Figure 4.6. The entries M_{ij} that are not zero and not ∞ represent the minimum distances (weights) required to travel from i to j in the original digraph.

end example

Let's summarize our results so far. Algorithm (4.8) creates a matrix M that allows us to easily answer two questions: Is there a path from i to j for distinct vertices i and j? Yes, if $M_{ij} \neq \infty$. What is the distance of a shortest path from i to j? It's M_{ij} if $M_{ij} \neq \infty$.

	1	2	3	4	5	6
1	0	10	10	40	15	10
2	∞	0	∞	30	∞	∞
3	∞	∞	0	30	∞	∞
4	∞	∞	∞	0	∞	∞
5	∞	∞	∞	40	0	∞
6	∞	∞	∞	45	5	0

Figure 4.6   The result of Floyd's algorithm.

Floyd's Algorithm for Finding the Shortest Path

Now let's try to find a shortest path. We can make a slight modification to (4.8) to compute a "path" matrix P, which will hold the key to finding a shortest path. We'll initialize P to be all zeros. The algorithm will modify P so that $P_{ij} = 0$ means that the shortest path from i to j is the edge from i to j and $P_{ij} = k$ means that a shortest path from i to j goes through k. The modified algorithm, which computes M and P, is stated as follows:

Shortest Distances and Shortest Paths Algorithm   (4.9)

Let M be the weighted adjacency matrix for a weighted digraph over the set $\{1,..., n\}$. Let P be the n by n matrix of zeros. The algorithm replaces M by a matrix of shortest distances and it replaces P by a path matrix.

> **for** $k := 1$ **to** n **do**
>   **for** $i := 1$ **to** n **do**
>   **for** $j := 1$ **to** n **do**
>   **if** $M_{ik} + M_{kj} < M_{ij}$ **then**
> $M_{ij} := M_{ik} + M_{kj}$;
> $P_{ij} := k$
> **od od od fi**

---

example **4.15 The Path Matrix**

We'll apply (4.9) to the weighted adjacency matrix in Figure 4.5. The algorithm produces the matrix M in Figure 4.6, and it produces the path matrix P given in Figure 4.7.

For example, the shortest path between 1 and 4 passes through 2 because $P_{14} = 2$. Since $P_{12} = 0$ and $P_{24} = 0$, the shortest path between 1 and 4 consists of the sequence 1, 2, 4. Similarly, the shortest path between 1 and 5 is the sequence 1, 6, 5, and the shortest path between 6 and 4 is the sequence 6, 5, 4. So once

	1	2	3	4	5	6
1	0	0	0	2	6	0
2	0	0	0	0	0	0
3	0	0	0	0	0	0
4	0	0	0	0	0	0
5	0	0	0	0	0	0
6	0	0	0	5	0	0

Figure 4.7   A path matrix.

we have matrix P from (4.9), it's an easy matter to compute a shortest path between two points. We'll leave this as an exercise.

end example

Let's make a few observations about Example 4.15. We should note that there is another shortest path from 1 to 4, namely, 1, 3, 4. The algorithm picked 2 as the intermediate point of the shortest path because the outer index k increments from 1 to n. When the computation got to $k = 3$, the value M_{14} had already been set to the minimal value, and P_{24} had been set to 2. So the condition of the if-then statement was false, and no changes were made. Therefore, P_{ij} gets the value of k closest to 1 whenever there are two or more values of k that give the same value to the expression $M_{ik} + M_{kj}$, and that value is less than M_{ij}.

Application Notes

Before we finish with this topic, let's make a couple of comments. If we have a digraph that is not weighted, then we can still find shortest distances and shortest paths with (4.8) and (4.9). Just let each edge have weight 1. Then the matrix M produced by either (4.8) or (4.9) will give us the length of a shortest path, and the matrix P produced by (4.9) will allow us to find a path of shortest length.

If we have a weighted graph that is not directed, then we can still use (4.8) and (4.9) to find shortest distances and shortest paths. Just modify the weighted adjacency matrix M as follows: For each edge between i and j having weight d, set $M_{ij} = M_{ji} = d$.

 Exercises

Properties

1. Write down all of the properties that each of the following binary relations satisfies from among the five properties reflexive, symmetric, transitive, irreflexive, and antisymmetric.

    a. The similarity relation on the set of triangles.

    b. The congruence relation on the set of triangles.

    c. The relation on people that relates people with the same parents.

    d. The subset relation on sets.

    e. The if and only if relation on the set of statements that may be true or false.

    f. The relation on people that relates people with bachelor's degrees in computer science.

    g. The "is brother of" relation on the set of people.

    h. The "has a common national language with" relation on countries.

    i. The "speaks the primary language of" relation on the set of people.

    j. The "is father of" relation on the set of people.

2. Write down all of the properties that each of the following relations satisfies from among the properties reflexive, symmetric, transitive, irreflexive, and antisymmetric.

    a. $R = \{(a, b) \mid a^2 + b^2 = 1\}$ over the real numbers.

    b. $R = \{(a, b) \mid a^2 = b^2\}$ over the real numbers.

    c. $R = \{(x, y) \mid x \bmod y = 0 \text{ and } x, y \in \{1, 2, 3, 4\}\}$.

    d. $R = \{(x, y) \mid x \text{ divides } y\}$ over the positive integers.

    e. $R = \{(x, y) \mid \gcd(x, y) = 1\}$ over the positive integers.

3. Explain why each of the following relations has the properties listed.

    a. The empty relation $\varnothing$ over any set is irreflexive, symmetric, antisymmetric, and transitive.

    b. For any set A, the universal relation $A \times A$ is reflexive, symmetric, and transitive. If $|A| = 1$, then $A \times A$ is also antisymmetric.

4. For each of the following conditions, find the smallest relation over the set $A = \{a, b, c\}$ that satisfies the stated properties.

    a. Reflexive but not symmetric and not transitive.

    b. Symmetric but not reflexive and not transitive.

    c. Transitive but not reflexive and not symmetric.

    d. Reflexive and symmetric but not transitive.

    e. Reflexive and transitive but not symmetric.

    f. Symmetric and transitive but not reflexive.

    g. Reflexive, symmetric, and transitive.

Composition

5. Write down suitable names for each of the following compositions.

    a. isChildOf ∘ isChildOf.

    b. isSisterOf ∘ isParentOf.

    c. isSonOf ∘ isSiblingOf.

    d. isChildOf ∘ isSiblingOf ∘ isParentOf.

6. Suppose we define $x \, R \, y$ to mean "x is the father of y and y has a brother." Write R as the composition of two well-known relations.

7. For each of the following properties, find a binary relation R such that R has the property but R^2 does not.

    a. Irreflexive.

    b. Antisymmetric.

8. Given the relation "less" over the natural numbers $\mathbb{N}$, describe each of the following compositions as a set of the form $\{(x, y) \mid \text{property}\}$.

    a. less ∘ less.

    b. less ∘ less ∘ less.

9. Given the three relations "less," "greater," and "notEqual" over the natural numbers $\mathbb{N}$, find each of the following compositions.

    a. less ∘ greater.

    b. greater ∘ less.

    c. notEqual ∘ less.

    d. greater ∘ notEqual.

10. Let $R = \{(x, y) \in \mathbb{Z} \times \mathbb{Z} \mid x + y \text{ is even}\}$. Find R^2.

Closure

11. Describe the reflexive closure of the empty relation $\varnothing$ over a set A.

12. Find the symmetric closure of each of the following relations over the set $\{a, b, c\}$.

    a. $\varnothing$.

    b. $\{(a, b), (b, a)\}$.

    c. $\{(a, b), (b, c)\}$.

    d. $\{(a, a), (a, b), (c, b), (c, a)\}$.

13. Find the transitive closure of each of the following relations over the set $\{a, b, c, d\}$.

    a. $\varnothing$.

    b. $\{(a, b), (a, c), (b, c)\}$.

    c. $\{(a, b), (b, a)\}$.

    d. $\{(a, b), (b, c), (c, d), (d, a)\}$.

14. Let $R = \{(x, y) \in \mathbb{Z} \times \mathbb{Z} \mid x + y$ is odd$\}$. Use the results of Example 4.6 to calculate $t(R)$.

15. Find an appropriate name for the transitive closure of each of the following relations.

   a. isParentOf.

   b. isChildOf.

   c. $\{(x + 1, x) \mid x \in \mathbb{N}\}$.

Path Problems

16. Suppose G is the following weighted digraph, where the triple (i, j, d) represents edge (i, j) with distance d:

$$\{(1, 2, 20), (1, 4, 5), (2, 3, 10), (3, 4, 10), (4, 3, 5), (4, 2, 10)\}.$$

   a. Draw the weighted adjacency matrix for G.

   b. Use (4.9) to compute the two matrices representing the shortest distances and the shortest paths in G.

17. Write an algorithm to compute the shortest path between two points of a weighted digraph from the matrix P produced by (4.9).

18. How many distinct path matrices can describe the shortest paths in the following graph, where it is assumed that all edges have weight $= 1$?

19. Write algorithms to perform each of the following actions for a binary relation R represented as an adjacency matrix.

   a. Check R for reflexivity.

   b. Check R for symmetry.

   c. Check R for transitivity.

   d. Compute $r(R)$.

   e. Compute $s(R)$.

Proofs and Challenges

20. For each of the following properties, show that if R has the property, then so does R^2.

   a. Reflexive.

   b. Symmetric.

   c. Transitive.

21. For the "less" relation over $\mathbb{N}$, show that $st(\text{less}) \neq ts(\text{less})$.

22. Prove each of the following statements about binary relations.

    a. $R \circ (S \circ T) = (R \circ S) \circ T.$ (associativity)

    b. $R \circ (S \cup T) = R \circ S \cup R \circ T.$

    c. $R \circ (S \cap T) \subseteq R \circ S \cap R \circ T.$

23. Let A be a set, R be any binary relation on A, and E be the equality relation on A. Show that $E \circ R = R \circ E = R$.

24. Prove each of the following statements about a binary relation R over a set A.

    a. If R is reflexive, then $s(R)$ and $t(R)$ are reflexive.

    b. If R is symmetric, then $r(R)$ and $t(R)$ are symmetric.

    c. If R is transitive, then $r(R)$ is transitive.

25. Prove each of the following statements about a binary relation R over a set A.

    a. $rt(R) = tr(R).$

    b. $rs(R) = sr(R).$

    c. $st(R) \subseteq ts(R).$

26. A binary relation R over a set A is *asymmetric* if $(x \, R \, y$ and $y \, R \, x)$ is false for all $x, y \in A$. Prove the following statements.

    a. If R is asymmetric, then R is irreflexive.

    b. If R is asymmetric, then R is antisymmetric.

    c. R is asymmetric if and only if R is irreflexive and antisymmetric.

4.2 Equivalence Relations

The word "equivalent" is used in many ways. For example, we've all seen statements like "Two triangles are equivalent if their corresponding angles are equal." We want to find some general properties that describe the idea of "equivalence."

The Equality Problem

We'll start by discussing the idea of "equality" because, to most people, "equal" things are examples of "equivalent" things, whatever meaning is attached to the word "equivalent." Let's consider the following problem.

The Equality Problem

Write a computer program to check whether two objects are equal.

    What is equality? Does it depend on the elements of the set? Why is equality important? What are some properties of equality? We all have an intuitive notion of what equality is because we use it all the time. Equality is important in computer science because programs use equality tests on data. If a programming language doesn't provide an equality test for certain data, then the programmer may need to implement such a test.

The simplest equality on a set A is basic equality: $\{(x, x) \mid x \in A\}$. But most of the time we use the word "equality" in a much broader context. For example, suppose A is the set of arithmetic expressions made from natural numbers and the symbol $+$. Thus A contains expressions like $3 + 7$, 8, and $9 + 3 + 78$. Most of us already have a pretty good idea of what equality means for these expressions. For example, we probably agree that $3 + 2$ and $2 + 1 + 2$ are equal. In other words, two expressions (*syntactic* objects) are equal if they have the same value (meaning or *semantics*), which is obtained by evaluating all $+$ operations.

Are there some fundamental properties that hold for any definition of equality on a set A? Certainly we want to have $x = x$ for each element x in A (the basic equality on A). Also, whenever $x = y$, it ought to follow that $y = x$. Lastly, if $x = y$ and $y = z$, then $x = z$ should hold. Of course, these are the three properties reflexive, symmetric, and transitive.

Most equalities are more than just basic equality. That is, they equate different syntactic objects that have the same meaning. In these cases the symmetric and transitive properties are needed to convey our intuitive notion of equality. For example, the following statements are true if we let "=" mean "has the same value as":

$$\text{If } 2 + 3 = 1 + 4, \text{ then } 1 + 4 = 2 + 3.$$

$$\text{If } 2 + 5 = 1 + 6 \text{ and } 1 + 6 = 3 + 4, \text{ then } 2 + 5 = 3 + 4.$$

4.2.1   Definition and Examples

Now we're ready to define equivalence. Any binary relation that is reflexive, symmetric, and transitive is called an *equivalence* relation. Sometimes people refer to an equivalence relation as an RST relation in order to remember the three properties.

Equivalence relations are all around us. Of course, the basic equality relation on any set is an equivalence relation. Similarly, the notion of equivalent triangles is an equivalence relation.

For another example, suppose we relate two books in the Library of Congress if their call numbers start with the same letter. (This is an instance in which it seems to be official policy to have a number start with a letter.) This relation is clearly an equivalence relation. Each book is related to itself (reflexive). If book A and book B have call numbers that begin with the same letter, then so do books B and A (symmetric). If books A and B have call numbers beginning with the same letter and books B and C have call numbers beginning with the same letter, then so do books A and C (transitive).

example   **4.16   Sample Equivalence Relations**

Here are a few more samples of equivalence relations, where the symbol $\sim$ denotes each relation.

a. For the set of integers, let $x \sim y$ mean $x + y$ is even.

b. For the set of nonzero rational numbers, let $x \sim y$ mean $xy > 0$.

c. For the set of rational numbers, let $x \sim y$ mean $x - y$ is an integer.

d. For the set of triangles, let $x \sim y$ mean x and y are similar.

e. For the set of integers, let $x \sim y$ mean $x \bmod 4 = y \bmod 4$.

f. For the set of binary trees, let $x \sim y$ mean x and y have the same depth.

g. For the set of binary trees, let $x \sim y$ mean x and y have the same number of nodes.

h. For the set of real numbers, let $x \sim y$ mean $x^2 = y^2$.

i. For the set of people, let $x \sim y$ mean x and y have the same mother.

j. For the set of TV programs, let $x \sim y$ mean x and y start at the same time and day.

end example

Intersections and Equivalence Relations

We can always verify that a binary relation is an equivalence relation by checking that the relation is reflexive, symmetric, and transitive. But in some cases we can determine equivalence by other means. For example, we have the following intersection result, which we'll leave as an exercise.

Intersection Property of Equivalence (4.10)

If E and F are equivalence relations on the set A, then $E \cap F$ is an equivalence relation on A.

The practical use of (4.10) comes about when we notice that a relation $\sim$ on a set A is defined in the following form, where E and F are relations on A.

$$x \sim y \text{ iff } x \, E \, y \text{ and } x \, F \, y.$$

This is just another way of saying that $x \sim y$ iff $(x, y) \in E \cap F$. So if we can show that E and F are equivalence relations, then (4.10) tells us that $\sim$ is an equivalence relation.

 4.17 Equivalent Binary Trees

Suppose we define the relation $\sim$ on the set of binary trees by

$$x \sim y \text{ iff } x \text{ and } y \text{ have the same depth and the same number of nodes.}$$

From Example 4.16 we know that "has the same depth as" and "has the same number of nodes as" are both equivalence relations. Therefore, $\sim$ is an equivalence relation.

end example

Equivalence Relations from Functions (Kernel Relations)

A very powerful technique for obtaining equivalence relations comes from the fact that any function defines a natural equivalence relation on its domain by relating elements that map to the same value. In other words, for any function $f : A \rightarrow B$, we obtain an equivalence relation $\sim$ on A by

$$x \sim y \quad \text{iff} \quad f(x) = f(y).$$

It's easy to see that $\sim$ is an equivalence relation. The reflexive property follows because $f(x) = f(x)$ for all $x \in A$. The symmetric property follows because $f(x) = f(y)$ implies $f(y) = f(x)$. The transitive property follows because $f(x) = f(y)$ and $f(y) = f(z)$ implies $f(x) = f(z)$.

An equivalence relation defined in this way is called the *kernel relation* for *f*. Let's state the result for reference.

Kernel Relations   (4.11)

If f is a function with domain A, then the relation $\sim$ defined by

$$x \sim y \text{ iff } f(x) = f(y)$$

is an equivalence relation on A, and it is called the *kernel relation* of *f*.

For example, notice that the relation given in Part (e) of Example 4.16 is the kernel relation for the function $f(x) = x \bmod 4$. Thus Part (e) of Example 4.16 is an equivalence relation by (4.11). Several other parts of Example 4.16 are also kernel relations. The nice thing about kernel relations is that they are always equivalence relations. So there is nothing to check. For example, we can use (4.11) to generalize Part (e) of Example 4.16 to the following important result.

Mod Function Equivalence   (4.12)

If S is any set of integers and n is a positive integer, then the relation $\sim$ defined by

$$x \sim y \text{ iff } x \bmod n = y \bmod n$$

is an equivalence relation over S.

In many cases it's possible to show that a relation is an equivalence relation by rewriting its definition so that it is the kernel relation of some function.

example **4.18   A Numeric Equivalence Relation**

Suppose we're given the relation $\sim$ defined on integers by

$$x \sim y \text{ if and only if } x - y \text{ is an even integer.}$$

We'll show that $\sim$ is an equivalence relation by writing it as the kernel relation of a function. Notice that $x - y$ is even if and only if x and y are both even or both odd. We can test whether an integer x is even or odd by checking whether $x \bmod 2 = 0$ or 1. So we can write our original definition of $\sim$ in terms of the mod function:

$$
\begin{aligned}
x \sim y \quad &\text{iff} \quad x - y \text{ is an even integer} \\
&\text{iff} \quad x \text{ and } y \text{ are both even or both odd} \\
&\text{iff} \quad x \bmod 2 = y \bmod 2.
\end{aligned}
$$

We can now conclude that $\sim$ is an equivalence relation because it's the kernel relation of the function f defined by $f(x) = x \bmod 2$.

end example

The Equivalence Problem

We can generalize the equality problem to the following more realistic problem of equivalence.

> **The Equivalence Problem**
> Write a computer program to check whether two objects are equivalent.

example **4.19   Binary Trees with the Same Structure**

Suppose we need two binary trees to be equivalent whenever they have the same structure regardless of the values of the nodes. For binary trees S and T, let equiv(S, T) be true if S and T are equivalent and false otherwise. Here is a program to compute equiv.

$$
\begin{aligned}
\text{equiv}\,(S, T) = \ &\text{if } S = \langle\,\rangle \text{ and } T = \langle\,\rangle \text{ then True} \\
&\text{else if } S = \langle\,\rangle \text{ or } T = \langle\,\rangle \text{ then False} \\
&\text{else equiv}\,(\text{left}\,(S), \text{left}\,(T)) \text{ and equiv}\,(\text{right}\,(S), \text{right}\,(T)).
\end{aligned}
$$

end example

4.2.2 Equivalence Classes

The nice thing about an equivalence relation over a set is that it defines a natural way to group elements of the set into disjoint subsets. These subsets are called equivalence classes, and here's the definition.

Equivalence Class

Let R be an equivalence relation on a set S. If $a \in S$, then the *equivalence class* of a, denoted by $[a]$, is the subset of S consisting of all elements that are equivalent to a. In other words, we have

$$[a] = \{x \in S \mid x\ R\ a\}.$$

For example, we always have $a \in [a]$ because of the property $a\ R\ a$.

example **4.20 Equivalent Strings**

Consider the relation $\sim$ defined on strings over the alphabet $\{a, b\}$ by

$$x \sim y \text{ iff } x \text{ and } y \text{ have the same length.}$$

Notice that $\sim$ is an equivalence relation because it is the kernel relation of the length function. Some sample equivalences are $abb \sim bab$ and $ba \sim aa$. Let's look at a few equivalence classes.

$$[\Lambda] = \{\Lambda\}\,,$$
$$[a] = \{a, b\}\,,$$
$$[ab] = \{ab, aa, ba, bb\}\,,$$
$$[aaa] = \{aaa, aab, aba, baa, abb, bab, bba, bbb\}\,.$$

Notice that any member of an equivalence class can define the class. For example, we have

$$[a] = [b] = \{a, b\}\,,$$
$$[ab] = [aa] = [ba] = [bb] = \{ab, aa, ba, bb\}\,.$$

end example

Equivalence classes enjoy a very nice property, namely that any two such classes are either equal or disjoint. Here is the result in more formal terms.

Property of Equivalences   (4.13)

Let S be a set with an equivalence relation R. If $a, b \in S$, then either $[a] = [b]$ or $[a] \cap [b] = \varnothing$.

Proof: It suffices to show that $[a] \cap [b] \neq \varnothing$ implies $[a] = [b]$. If $[a] \cap [b] \neq \varnothing$, then there is a common element $c \in [a] \cap [b]$. It follows that cRa and cRb. From the symmetric and transitive properties of R, we conclude that aRb. To show that $[a] = [b]$, we'll show that $[a] \subseteq [b]$ and $[b] \subseteq [a]$. Let $x \in [a]$. Then xRa. Since aRb, the transitive property tells us that xRb, which implies that $x \in [b]$. Therefore, $[a] \subseteq [b]$. In an entirely similar manner we obtain $[b] \subseteq [a]$. Therefore, we have the desired result $[a] = [b]$. QED.

4.2.3 Partitions

By a *partition* of a set we mean a collection of nonempty subsets that are disjoint from each other and whose union is the whole set. For example, the set $S = \{0, 1, 2, 3, 4, 5, 6, 7, 8, 9\}$ can be partitioned in many ways, one of which consists of the following three subsets of S:

$$\{0, 1, 4, 9\}, \{2, 5, 8\}, \{3, 6, 7\}.$$

Notice that, if we wanted to, we could define an equivalence relation on S by saying that $x \sim y$ iff x and y are in the same set of the partition. In other words, we would have

$$[0] = \{0, 1, 4, 9\},$$
$$[2] = \{2, 5, 8\},$$
$$[3] = \{3, 6, 7\}.$$

We can do this for any partition of any set.

But something more interesting happens when we start with an equivalence relation on S. For example, let $\sim$ be the following relation on S:

$$x \sim y \text{ iff } x \bmod 4 = y \bmod 4.$$

This relation is an equivalence relation because it is the kernel relation of the function $f(x) = x \bmod 4$. Now let's look at some of the equivalence classes.

$$[0] = \{0, 4, 8\}.$$
$$[1] = \{1, 5, 9\}.$$
$$[2] = \{2, 6\}.$$
$$[3] = \{3, 7\}.$$

Notice that these equivalence classes form a partition of S. This is no fluke. It always happens for any equivalence relation on any set S. To see this, notice that if $s \in S$, then $s \in [s]$, which says that S is the union of the equivalence classes. We also know from (4.13) that distinct equivalence classes are disjoint.

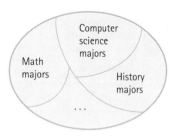

Figure 4.8   A partition of students.

Therefore, the set of equivalence classes forms a partition of S. Here's a summary of our discussion.

Equivalence Relations and Partitions   (4.14)

If R is an equivalence relation on the set S, then the equivalence classes form a partition of S. Conversely, if P is a partition of a set S, then there is an equivalence relation on S whose equivalence classes are sets of P.

For example, let S denote the set of all students at some university, and let M be the relation on S that relates two students if they have the same major. (Assume here that every student has exactly one major.) It's easy to see that M is an equivalence relation on S and each equivalence class is the set of all the students majoring in the same subject. For example, one equivalence class is the set of computer science majors. The partition of S is pictured by the Venn diagram in Figure 4.8.

example   **4.21   Partitioning a Set of Strings**

The relation from Example 4.20 is defined on the set $S = \{a, b\}^*$ of all strings over the alphabet $\{a, b\}$ by

$$x \sim y \text{ iff } x \text{ and } y \text{ have the same length.}$$

For each natural number n, the equivalence class $[a^n]$ contains all strings over $\{a, b\}$ that have length n. So we can write

$$S = \{a, b\}^* = [\Lambda] \cup [a] \cup [aa] \cup \cdots \cup [a^n] \cup \cdots.$$

end example

4.22 A Partition of the Natural Numbers

Let $\sim$ be the relation on the natural numbers defined by

$$x \sim y \text{ iff } \lfloor x/10 \rfloor = \lfloor y/10 \rfloor.$$

This is an equivalence relation because it is the kernel relation of the function $f(x) = \lfloor x/10 \rfloor$. After checking a few values we see that each equivalence class is a decade of numbers. For example,

$$[0] = \{0, 1, 2, 3, 4, 5, 6, 7, 8, 9\},$$
$$[10] = \{10, 11, 12, 13, 14, 15, 16, 17, 18, 19\},$$

and in general, for any natural number n,

$$[10n] = \{10n,\ 10n + 1,\ \dots,\ 10n + 9\}.$$

So we have $\mathbb{N} = [0] \cup [10] \cup \cdots \cup [10n] \cup \cdots$.

end example

 4.23 Partitioning with Mod 5

Let R be the equivalence relation on the integers $\mathbb{Z}$ defined by

$$a \ R \ b \text{ iff } a \bmod 5 = b \bmod 5.$$

After some checking we see that the partition of $\mathbb{Z}$ consists of the following five equivalence classes

$$[0] = \{\dots - 10, -5, 0, 5, 10, \dots\},$$
$$[1] = \{\dots - 9, -4, 1, 6, 11, \dots\},$$
$$[2] = \{\dots - 8, -3, 2, 7, 12, \dots\},$$
$$[3] = \{\dots - 7, -2, 3, 8, 13, \dots\},$$
$$[4] = \{\dots - 6, -1, 4, 9, 14, \dots\}.$$

Remember, it doesn't matter which element of a class is used to represent it. For example, $[0] = [5] = [-15]$. It is clear that the five classes are disjoint from each other and that $\mathbb{Z}$ is the union of the five classes.

end example

 4.24 Software Testing

If the input data set for a program is infinite, then the program can't be tested on every input. However, every program has a finite number of instructions. So we should be able to find a finite data set to cause all instructions of the program to be executed. For example, suppose p is the following program, where x is an integer and q, r, and s represent other parts of the program:

$$p(x): \quad \textbf{if } x > 0 \textbf{ then } q(x)$$
$$\textbf{else if } x \textbf{ is even then } r(x)$$
$$\textbf{else } s(x)$$
$$\textbf{fi}$$
$$\textbf{fi}$$

The condition "$x > 0$" causes a natural partition of the integers into the positives and the nonpositives. The condition "x is even" causes a natural partition of the nonpositives into the even nonpositives and the odd nonpositives. So we have a partition of the integers into the following three subsets:

$$\{1, 2, 3, \ldots\}, \{0, -2, -4, \ldots\}, \{-1, -3, -5, \ldots\}.$$

Now we can test the instructions in q, r, and s by picking three numbers, one from each set of the partition. For example, $p(1)$, $p(0)$, and $p(-1)$ will do the job. Of course, further partitioning may be necessary if q, r, or s contains further conditional statements. The equivalence relation induced by the partition relates two integers x and y if and only if $p(x)$ and $p(y)$ execute the same set of instructions.

end example

Refinement of a Partition

Suppose that P and Q are two partitions of a set S. If each set of P is a subset of a set in Q, then P is a *refinement* of Q. We also say P is *finer* than Q or Q is *coarser* than P. The finest of all partitions on S is the collection of singleton sets. The coarsest of all partitions of S is $\{S\}$.

For example, here is a listing of four partitions of $\{a, b, c, d\}$ that are successive refinements from the coarsest to finest:

$$\{\{a, b, c, d\}\} \qquad \text{(coarsest)}$$
$$\{\{a, b\}, \{c, d\}\}$$
$$\{\{a, b\}, \{c\}, \{d\}\}$$
$$\{\{a\}, \{b\}, \{c\}, \{d\}\} \quad \text{(finest)}.$$

example 4.25 **Partitioning with Mod 2**

Let R be the relation over $\mathbb{N}$ defined by

$$a \; R \; b \text{ iff } a \bmod 2 = b \bmod 2.$$

Then R is an equivalence relation because it is the kernel relation of the function f defined by $f(x) = x \bmod 2$. The corresponding partition of $\mathbb{N}$ consists of the two subsets

$$[0] = \{0, 2, 4, 6, \ldots\},$$
$$[1] = \{1, 3, 5, 7, \ldots\}.$$

Can we find a refinement of this partition? Sure. Let T be defined by

$$a \ T \ b \ \text{iff} \ a \bmod 4 = b \bmod 4.$$

T induces the following partition of $\mathbb{N}$ that is a refinement of the partition induced by R because we get the following four equivalence classes:

$$[0] = \{0, 4, 8, 12, \ldots\},$$
$$[1] = \{1, 5, 9, 13, \ldots\},$$
$$[2] = \{2, 6, 10, 14, \ldots\},$$
$$[3] = \{3, 7, 11, 15, \ldots\}.$$

This partition is indeed a refinement of the preceding partition. Can we find a refinement of this partition? Yes, because we can continue the process forever. Just let k be a power of 2 and define T_k by

$$a \ T_k \ b \ \text{iff} \ a \bmod k = b \bmod k.$$

So the partition for each T_{2k} is a refinement of the partition for T_k.

end example

An Intersection Property

We noted in (4.10) that the intersection of equivalence relations over a set A is also an equivalence relation over A. It also turns out that the equivalence classes for the intersection are intersections of equivalence classes for the given relations. Here is the statement and we'll leave the proof as an exercise.

Intersection Property of Equivalence   **(4.15)**

Let E and F be equivalence relations on a set A. Then the equivalence classes for the relation $E \cap F$ are of the form $[x] = [x]_E \cap [x]_F$, where $[x]_E$ and $[x]_F$ denote the equivalence classes of x for E and F, respectively.

---

example 4.26 **Intersecting Equivalence Relations**

Let $\sim$ be the relation on the natural numbers defined by

$$x \sim y \quad \text{iff} \quad \lfloor x/10 \rfloor = \lfloor y/10 \rfloor \text{ and } x + y \text{ is even}.$$

Notice that $\sim$ is the intersection of two relations E and F, where $x\,E\,y$ means $\lfloor x/10 \rfloor = \lfloor y/10 \rfloor$ and $x\,F\,y$ means $x + y$ is even. We can observe that $x + y$ is even if and only if $x \bmod 2 = y \bmod 2$. So both E and F are kernel relations of functions and thus are equivalence relations. Therefore, $\sim$ is an equivalence relation by (4.10). We computed the equivalence classes for E and F in Examples 4.22 and 4.25. The equivalence classes for E are of the following form for each natural number n.

$$[10n] = \{10n,\ 10n + 1,\ \ldots,\ 10n + 9\}.$$

The equivalence classes for F are

$$[0] = \{0, 2, 4, 6, \ldots\},$$
$$[1] = \{1, 3, 5, 7, \ldots\}.$$

By (4.15) the equivalence classes for $\sim$ have the following form for each n:

$$[10n] \cap [0] = \{10n,\ 10n + 2,\ 10n + 4,\ 10n + 6,\ 10n + 8\},$$
$$[10n] \cap [1] = \{10n + 1,\ 10n + 3,\ 10n + 5,\ 10n + 7,\ 10n + 9\}.$$

end example

example **4.27 Solving the Equality Problem**

If we want to define an equality relation on a set S of objects that do not have any established meaning, then we can use the basic equality relation $\{(x, x) \mid x \in S\}$. On the other hand, suppose a meaning has been assigned to each element of S. We can represent the meaning by a mapping m from S to a set of values V. In other words, we have a function $m : S \to V$. It's natural to define two elements of S to be equal if they have the same meaning. That is, we define $x = y$ if and only if $m(x) = m(y)$. This equality relation is just the kernel relation of m.

For example, let S denote the set of arithmetic expressions made from nonempty unary strings and the symbol $+$. For example, some typical expressions in S are 1, 11, 111, 1+1, 11+111+1. Now let's assign a meaning to each expression in S. Let $m(1^n) = n$ for each positive natural number n. If $e + e'$ is an expression of S, we define $m(e + e') = m(e) + m(e')$. We'll assume that $+$ is applied left to right. For example, the value of the expression $1 + 111 + 11$ can be calculated as follows:

$$
\begin{aligned}
m\left(1 + 111 + 11\right) &= m\left((1 + 111) + 11\right) \\
&= m\left(1 + 111\right) + m\left(11\right) \\
&= m\left(1\right) + m\left(111\right) + 2 \\
&= 1 + 3 + 2 \\
&= 6.
\end{aligned}
$$

If we define two expressions of S to be equal when they have the same meaning, then the desired equality relation on S is the kernel relation of m. So the partition of S induced by the kernel relation of m consists of the sets of expressions with equal values. For example, the equivalence class [1111] contains the eight expressions

$$1+1+1+1, \; 1+1+11, \; 1+11+1, \; 11+1+1, \; 11+11, \; 1+111, \; 111+1, \; 1111.$$

end example

4.2.4 Generating Equivalence Relations

Any binary relation can be considered as the *generator* of an equivalence relation obtained by adding just enough pairs to make the result reflexive, symmetric, and transitive. In other words, we can take the reflexive, symmetric, and transitive closures of the binary relation.

Does the order that we take closures make a difference? For example, what about $str(R)$? An example will suffice to show that $str(R)$ need not be an equivalence relation. Let $A = \{a, b, c\}$ and $R = \{(a, b), (a, c), (b, b)\}$. Then

$$str(R) = \{(a, a), (b, b), (c, c), (a, b), (b, a), (a, c), (c, a)\}.$$

This relation is reflexive and symmetric, but it's not transitive. On the other hand, we have $tsr(R) = A \times A$, which is an equivalence relation. As the next result shows, $tsr(R)$ is always an equivalence relation.

The Smallest Equivalence Relation (4.16)

If R is a binary relation on A, then $tsr(R)$ is the smallest equivalence relation that contains R.

Proof: The inheritance properties of (4.4) tell us that $tsr(R)$ is an equivalence relation. To see that it's the smallest equivalence relation containing R, we'll let T be an arbitrary equivalence relation containing R. Since $R \subseteq T$ and T is reflexive, it follows that $r(R) \subseteq T$. Since $r(R) \subseteq T$ and T is symmetric, it follows that $sr(R) \subseteq T$. Since $sr(R) \subseteq T$ and T is transitive, it follows that $tsr(R) \subseteq T$. So $tsr(R)$ is contained in every equivalence relation that contains R. Thus it's the smallest equivalence relation containing R. QED.

example **4.28 Family Trees**

Let R be the "is parent of" relation for a set of people. In other words, $(x, y) \in R$ iff x is a parent of y. Suppose we want to answer questions like the following:

Is x a descendant of y?
Is x an ancestor of y?

Are x and y related in some way?

What is the relationship between x and y?

Each of these questions can be answered from the given information by finding whether an appropriate path exists between x and y. But if we construct $t(R)$, then things get better because $(x, y) \in t(R)$ iff x is an ancestor of y. So we can find out whether x is an ancestor of y or x is a descendant of y by looking to see whether $(x, y) \in t(R)$ or $(y, x) \in t(R)$.

If we want to know whether x and y are related in some way, then we would have to look for paths in $t(R)$ taking each of x and y to a common ancestor. But if we construct $ts(R)$, then things get better because $(x, y) \in ts(R)$ iff x and y have a common ancestor. So we can find out whether x and y are related in some way by looking to see whether $(x, y) \in ts(R)$.

If x and y are related, then we might want to know the relationship. This question is asking for paths from x and y to a common ancestor, which can be done by searching $t(R)$ for the common ancestor and keeping track of each person along the way.

Notice also that the set of people can be partitioned into family trees by the equivalence relation $tsr(R)$. So the simple "is parent of" relation is the generator of an equivalence relation that constructs family trees.

end example

An Equivalence Problem

Suppose we have an equivalence relation over a set S that is generated by a given set of pairs. For example, the equivalence relation might be the family relationship "is related to" and the generators might be a set of parent-child pairs.

Can we represent the generators in such a way that we can find out whether two arbitrary elements of S are equivalent? If two elements are equivalent, can we find a sequence of generators to confirm the fact? The answer to both questions is yes. We'll present a solution due to Galler and Fischer [1964], which uses a special kind of tree structure to represent the equivalence classes.

The idea is to use the generating pairs to build the partition of S induced by the equivalence relation. For example, let $S = \{1, 2, \ldots, 10\}$, let $\sim$ denote the equivalence relation on S, and let the generators be the following pairs:

$$1 \sim 8, \, 4 \sim 5, \, 9 \sim 2, \, 4 \sim 10, \, 3 \sim 7, \, 6 \sim 3, \, 4 \sim 9.$$

To have something concrete in mind, let the numbers 1, 2, ..., 10 be people, let $\sim$ be "is related to," and let the generators be "parent $\sim$ child" pairs.

The construction process starts by building the following ten singleton equivalence classes to represent the partition of S caused by the reflexive property $x \sim x$.

$$\{1\}, \{2\}, \{3\}, \{4\}, \{5\}, \{6\}, \{7\}, \{8\}, \{9\}, \{10\}.$$

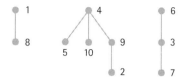

Figure 4.9   Equivalence classes as trees.

i	1	2	3	4	5	6	7	8	9	10
$p[i]$	0	9	6	0	4	0	3	1	4	4

Figure 4.10   Equivalence classes as an array.

Now we process the generators, one at a time. The generator $1 \sim 8$ is processed by forming the union of the equivalence classes that contain 1 and 8. In other words, the partition becomes

$$\{1, 8\}, \{2\}, \{3\}, \{4\}, \{5\}, \{6\}, \{7\}, \{9\}, \{10\}.$$

Continuing in this manner to process the other generators, we eventually obtain the partition of S consisting of the following three equivalence classes.

$$\{1, 8\}, \{2, 4, 5, 9, 10\}, \{3, 6, 7\}.$$

Representing Equivalence Classes

To answer questions about an equivalence relation, we need to consider its representation. We can represent each equivalence class in the partition as a tree, where the generator $a \sim b$ will be processed by creating the branch "a is the parent of b." For our example, if we process the generators in the order in which they are written, then we obtain the three trees in Figure 4.9.

A simple way to represent these trees is with a 10-tuple (a 1-dimensional array of size 10) named p, where $p[i]$ denotes the parent of i. We'll let $p[i] = 0$ mean that i is a root. Figure 4.10 shows the three equivalence classes represented by p.

Now it's easy to answer the question "Is $a \sim b$?" Just find the roots of the trees to which a and b belong. If the roots are the same, the answer is yes. If the answer is yes, then there is another question, "Can you find a sequence of equivalences to show that $a \sim b$?" One way to do this is to locate one of the numbers, say b, and rearrange the tree to which b belongs so that b becomes the root. This can be done easily by reversing the links from b to the root. Once we have b at the root, it's an easy matter to read off the equivalences from a to b. We'll leave it as an exercise to construct an algorithm to do the reversing.

For example, if we ask whether $5 \sim 2$, we find that 5 and 2 belong to the same tree. So the answer is yes. To find a set of equivalences to prove that $5 \sim 2$,

Figure 4.11   Proof that $5 \sim 2$.

we can, for example, reverse the links from 2 to the root of the tree. The before and after pictures are given in Figure 4.11.

Now it's an easy computation to traverse the tree from 5 to the root 2 and read off the equivalences $5 \sim 4$, $4 \sim 9$, and $9 \sim 2$.

Kruskal's Algorithm for Minimal Spanning Trees

In Chapter 1 we discussed Prim's algorithm to find a minimal spanning tree for a connected weighted undirected graph. Let's look an another such algorithm, due to Kruskal [1956], which uses equivalence classes.

The algorithm constructs a minimal spanning tree as follows: Starting with an empty tree, an edge $\{a, b\}$ of smallest weight is chosen from the graph. If there is no path in the tree from a to b, then the edge $\{a, b\}$ is added to the tree. This process is repeated with the remaining edges of the graph until the tree contains all vertices of the graph.

At any point in the algorithm, the edges in the spanning tree define an equivalence relation on the set of vertices of the graph. Two vertices a and b are equivalent iff there is a path between a and b in the tree. Whenever an edge $\{a, b\}$ is added to the spanning tree, the equivalence relation is modified by creating the equivalence class $[a] \cup [b]$. The algorithm ends when there is exactly one equivalence class consisting of all the vertices of the graph. Here are the steps of the algorithm.

Kruskal's Algorithm

1. Sort the edges of the graph by weight, and let L be the sorted list.

2. Let T be the minimal spanning tree and initialize $T := \varnothing$.

3. For each vertex v of the graph, create the equivalence class $[v] = \{v\}$.

4. while there are 2 or more equivalence classes **do**
   Let $\{a, b\}$ be the edge at the head of L;
   $L := \mathrm{tail}(L)$;
   if $[a] \neq [b]$ **then**
       $T := T \cup \{\{a, b\}\}$;
       Replace the equivalence classes $[a]$ and $[b]$ by $[a] \cup [b]$
   fi
   od

To implement the algorithm, we must find a representation for the equivalence classes. For example, we might use a parent array like the one we've been discussing.

 4.29 Minimal Spanning Trees

We'll use Kruskal's algorithm to construct a minimal spanning tree for the following weighted graph:

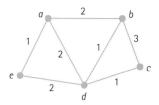

To see how the algorithm works, we'll do a trace of each step. We'll assume that the edges have been sorted by weight in the following order:

$$\{a, e\}, \{b, d\}, \{c, d\}, \{a, b\}, \{a, d\}, \{e, d\}, \{b, c\}.$$

The following table shows the value of the spanning tree T and the equivalence classes at each step, starting with the initialization values.

Spanning Tree T	Equivalence Classes
$\{\}$	$\{a\}, \{b\}, \{c\}, \{d\}, \{e\}$
$\{\{a, e\}\}$	$\{a, e\}, \{b\}, \{c\}, \{d\}$
$\{\{a, e\}, \{b, d\}\}$	$\{a, e\}, \{b, d\}, \{c\}$
$\{\{a, e\}, \{b, d\}, \{c, d\}\}$	$\{a, e\}, \{b, c, d\}$
$\{\{a, e\}, \{b, d\}, \{c, d\}, \{a, b\}\}$	$\{a, b, c, d, e\}$

The algorithm stops because there is only one equivalence class. So T is a spanning tree for the graph.

end example

Exercises

Properties

1. Verify that each of the following relations is an equivalence relation.

    a. $x \sim y$ iff x and y are points in a plane equidistant from a fixed point.

    b. $s \sim t$ iff s and t are strings with the same occurrences of each letter.

    c. $x \sim y$ iff $x + y$ is even, over the set of natural numbers.

    d. $x \sim y$ iff $x - y$ is an integer, over the set of rational numbers.

    e. $x \sim y$ iff $xy > 0$, over the set of nonzero rational numbers.

2. Each of the following relations is not an equivalence relation. In each case, find the properties that are not satisfied.

    a. $a \; R \; b$ iff $a + b$ is odd, over the set of integers.

    b. $a \; R \; b$ iff a/b is an integer, over the set of nonzero rational numbers.

    c. $a \; R \; b$ iff $|a - b| \le 5$, over the set of natural numbers.

    d. $a \; R \; b$ iff either $a \bmod 4 = b \bmod 4$ *or* $a \bmod 6 = b \bmod 6$, over $\mathbb{N}$.

    e. $a \; R \; b$ iff $x < a/10 < x + 1$ and $x \le b/10 < x + 1$ for some integer x.

Equivalence Classes

3. For each of the following functions f with domain $\mathbb{N}$, describe the equivalence classes of the kernel relation of f.

    a. $f(x) = 7$.

    b. $f(x) = x$.

    c. $f(x) = \text{floor}(x/2)$.

    d. $f(x) = \text{floor}(x/3)$.

    e. $f(x) = \text{floor}(x/4)$.

    f. $f(x) = \text{floor}(x/k)$ for a fixed positive integer k.

    g. $f(x) = $ if $0 \le x \le 10$ then 10 else $x - 1$.

4. For each of the following functions f, describe the equivalence classes of the kernel relation of f that partition the domain of f.

    a. $f : \mathbb{Z} \to \mathbb{N}$ is defined by $f(x) = |x|$.

    b. $f : \mathbb{R} \to \mathbb{Z}$ is defined by $f(x) = \text{floor}(x)$.

5. Describe the equivalence classes for each of the following relations on $\mathbb{N}$.

    a. $x \sim y$ iff $x \bmod 2 = y \bmod 2$ *and* $x \bmod 3 = y \bmod 3$.

    b. $x \sim y$ iff $x \bmod 2 = y \bmod 2$ *and* $x \bmod 4 = y \bmod 4$.

    c. $x \sim y$ iff $x \bmod 4 = y \bmod 4$ *and* $x \bmod 6 = y \bmod 6$.

6. Given the following set of words.

$$\{rot, \; tot, \; root, \; toot, \; roto, \; toto, \; too, \; to, \; otto\}.$$

    a. Let f be the function that maps a word to its set of letters. For the kernel relation of f, describe the equivalence classes.

    b. Let f be the function that maps a word to its bag of letters. For the kernel relation of f, describe the equivalence classes.

Spanning Trees

7. Use Kruskal's algorithm to find a minimal spanning tree for each of the following weighted graphs.

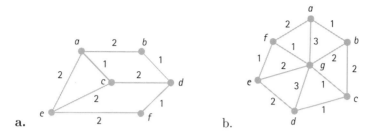

a.   **b.**

Proofs and Challenges

8. Let R be a relation on a set S such that R is symmetric and transitive and for each $x \in S$ there is an element $y \in S$ such that $x \, R \, y$. Prove that R is an equivalence relation (i.e., prove that R is reflexive).

9. Let E and F be equivalence relations on the set A, Show that $E \cap F$ is an equivalence relation on A.

10. Let E and F be equivalence relations on a set A and for each $x \in A$ let $[x]_E$ and $[x]_F$ denote the equivalence classes of x for E and F, respectively. Show that the equivalence classes for the relation $E \cap F$ are of the form $[x] = [x]_E \cap [x]_F$ for all $x \in A$.

11. Which relations among the following list are equal to $tsr(R)$, the smallest equivalence relation generated by R?

$$trs(R), \ str(R), \ srt(R), \ rst(R), \ rts(R).$$

12. In the equivalence problem we represented equivalence classes as a set of trees, where the nodes of the trees are the numbers 1, 2, ..., n. Suppose the trees are represented by an array $p[1]$, ..., $p[n]$, where $p[i]$ is the parent of i. Suppose also that $p[i] = 0$ when i is a root. Write a procedure that takes a node i and rearranges the tree that i belongs to so that i is the root, by reversing the links from the root to i.

13. (*Factoring a Function*). An interesting consequence of equivalence relations and partitions is that any function f can be factored into a composition of two functions, one an injection and one a surjection. For a function $f : A \to B$, let P be the partition of A by the kernel relation of f. Then define the function $s : A \to P$ by $s(a) = [a]$ and define $i : P \to B$ by $i([a]) = f(a)$. Prove that s is a surjection, i is an injection, and $f = i \circ s$.

4.3   Order Relations

Each day we see the idea of "order" used in many different ways. For example, we might encounter the expression $1 < 2$. We might notice that someone is older than someone else. We might be interested in the third component of the tuple (x, d, c, m). We might try to follow a recipe. Or we might see that the word "aardvark" resides at a certain place in the dictionary. The concept of order occurs in many different forms, but they all have the common idea of some object preceding another object.

Two Essential Properties of Order

Let's try to formally describe the concept of order. To have an ordering, we need a set of elements together with a binary relation having certain properties. What are these properties?

Well, our intuition tells us that if a, b, and c are objects that are ordered so that a precedes b and b precedes c, then we certainly want a to precede c. In other words, an ordering should be transitive. For example, if a, b, and c are natural numbers and $a < b$ and $b < c$, then we have $a < c$.

Our intuition also tells us that we don't want distinct objects preceding each other. In other words, if a and b are distinct objects and a precedes b, then b can't precede a. In still other words, if a precedes b and b precedes a then we better have $a = b$. For example, if a, b, and c are natural numbers and $a \leq b$ and $b \leq a$, we certainly want $a = b$. In other words, an ordering should be antisymmetric.

For example, over the natural numbers we recognize that the relation $<$ is an ordering and we notice that it is transitive and antisymmetric. Similarly, the relation $\leq$ is an ordering and we notice that it is transitive and antisymmetric. So the two essential properties of any kind of order are antisymmetric and transitive.

Let's look at how different orderings can occur in trying to perform the tasks of a recipe.

example **4.30   A Pancake Recipe**

Suppose we have the following recipe for making pancakes.

1. Mix the dry ingredients (flour, sugar, baking powder) in a bowl.

2. Mix the wet ingredients (milk, eggs) in a bowl.

3. Mix the wet and dry ingredients together.

4. Oil the pan. (It's an old pan.)

5. Heat the pan.

6. Make a test pancake and throw it away.

7. Make pancakes.

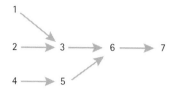

Figure 4.12   A pancake recipe.

Steps 1 through 7 indicate an ordering for the steps of the recipe. But the steps could also be done in some other order. To help us discover some other orders, let's define a relation R on the seven steps of the pancake recipe as follows:

$i\ R\ j$ means that step i must be done before step j.

Notice that R is antisymmetric and transitive. We can picture R as the digraph (without the transitive arrows) in Figure 4.12.

The graph helps us pick out different orders for the steps of the recipe. For example, the following ordering of steps will produce pancakes just as well.

4, 5, 2, 1, 3, 6, 7.

So there are several ways to perform the recipe. For example, three people could work in parallel doing tasks 1, 2, and 4 at the same time.

end example

This example demonstrates that different orderings for time-oriented tasks are possible whenever some tasks can be done at different times without changing the outcome. The orderings can be discovered by modeling the tasks by a binary relation R defined by

$i\ R\ j$ means that step i must be done before step j.

Notice that R is irreflexive because time-oriented tasks can't be done before themselves. If there are at least two tasks that are not related by R, as in Example 4.30, then there will be at least two different orderings of the tasks.

4.3.1 Partial Orders

Now let's get down to business and discuss the basic ideas and techniques of ordering. The two essential properties of order suffice to define the notion of partial order.

Definition of a Partial Order

A binary relation is called a *partial order* if it is antisymmetric, transitive, and either reflexive or irreflexive.

We should note that this definition is a bit more general than most definitions found in the literature. Some definitions require the reflexive property, while others require the irreflexive property. The reasons for requiring one of these properties are mostly historical and neither property is required to describe the idea of ordering. So if a partial order is reflexive and we wish to emphasize it, we'll call it a *reflexive partial order*. For example, $\leq$ is a reflexive partial order on the integers. If a partial order is irreflexive and we wish to emphasize it, we'll call it an *irreflexive partial order*. For example, $<$ is an irreflexive partial order on the integers.

Definition of a Partially Ordered Set

The set over which a partial order is defined is called a *partially ordered set*—or *poset* for short. If we want to emphasize the fact that R is the partial order that makes S a poset, we'll write $\langle S, R \rangle$ and call it a poset.

For example, in our pancake example we defined a partial order R on the set of recipe steps $\{1, 2, 3, 4, 5, 6, 7\}$. So we can say that $\langle \{1, 2, 3, 4, 5, 6, 7\}, R \rangle$ is a poset. There are many more examples of partial orders. For example, $\langle \mathbb{N}, < \rangle$ and $\langle \mathbb{N}, \leq \rangle$ are posets because the relations $<$ and $\leq$ are both antisymmetric and transitive.

Partial and Total

The word "partial" is used in the definition because we include the possibility that some elements may not be related to each other, as in the pancake recipe example. For another example, consider the subset relation on power($\{a, b, c\}$). Certainly the subset relation is antisymmetric and transitive. So we can say that $\langle \text{power}(\{a, b, c\}), \subseteq \rangle$ is a poset. Notice that there are some subsets that are not related. For example, $\{a, b\}$ and $\{a, c\}$ are not related by the relation $\subseteq$.

Suppose R is a binary relation on a set S and $x, y \in S$. We say that x and y are *comparable* if either $x\ R\ y$ or $y\ R\ x$. In other words, elements that are related are comparable. If every pair of distinct elements in a partial order are comparable, then the order is called a *total* order (also called a *linear* order). If R is a total order on the set S, then we also say that S is a *totally ordered set* or a *linearly ordered set*. For example, the natural numbers are totally ordered by both "less" and "lessOrEqual." In other words, $\langle \mathbb{N}, < \rangle$ and $\langle \mathbb{N}, \leq \rangle$ are totally ordered sets.

example **4.31   The Divides Relation**

Let's look at some interesting posets that can be defined by the divides relation, $|$. First we'll consider the set $\mathbb{N}$. If $a|b$ and $b|c$, then $a|c$. Thus $|$ is transitive. Also, if $a|b$ and $b|a$, then it must be the case that $a = b$. So $|$ is antisymmetric. Therefore,

$$\langle \mathbb{N}, | \rangle \text{ is a poset.}$$

But $\langle \mathbb{N}, | \rangle$ is not totally ordered because, for example, 2 and 3 are not comparable. To obtain a total order, we need to consider subsets of $\mathbb{N}$. For example, it's easy to see that for any m and n, either $2^m | 2^n$ or $2^n | 2^m$. Therefore,

$$\langle \{2^n \mid n \in \mathbb{N}\}, | \rangle \text{ is a totally ordered set.}$$

Let's consider some finite subsets of $\mathbb{N}$. For example, it's easy to see that

$$\langle \{1, 3, 9, 45\}, | \rangle \text{ is a totally ordered set.}$$

It's also easy to see that

$$\langle \{1, 2, 3, 4\}, | \rangle \text{ is a poset that is not totally ordered}$$

because 3 can't be compared to either 2 or 4.

end example

Notation for Partial Orders

When talking about partial orders, we'll often use the symbols

$$\prec \text{ and } \preceq$$

to stand for an irreflexive partial order and a reflexive partial order, respectively. We can read $a \prec b$ as "a is less than b," and we can read $a \preceq b$ as "a is less than or equal to b." The two symbols can be defined in terms of each other. For example, if $\langle A, \prec \rangle$ is a poset, then we can define the relation $\preceq$ in terms of $\prec$ by writing

$$\preceq \, = \, \prec \cup \{(x, x) \,|\, x \in A\}.$$

In other words, $\preceq$ is the reflexive closure of $\prec$. So $x \preceq y$ always means $x \prec y$ or $x = y$. Similarly, if $\langle B, \preceq \rangle$ is a poset, then we can define the relation $\prec$ in terms of $\preceq$ by writing

$$\prec \, = \, \preceq \, - \, \{(x, x) \,|\, x \in B\}.$$

Therefore, $x \prec y$ always means $x \preceq y$ and $x \neq y$. We also write the expression $y \succ x$ to mean the same thing as $x \prec y$.

Chains

A set of elements in a poset is called a *chain* if all the elements are comparable—linked—to each other. For example, any totally ordered set is itself a chain. A sequence of elements $x_1, x_2, x_3, \ldots$ in a poset is said to be a *descending chain* if $x_i \succ x_{i+1}$ for each $i \geq 1$. We can write the descending chain in the following familiar form:

$$x_1 \succ x_2 \succ x_3 \succ \cdots .$$

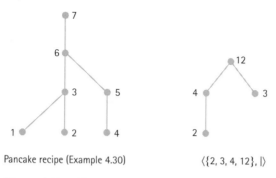

Pancake recipe (Example 4.30) $\langle\{2, 3, 4, 12\}, |\rangle$

Figure 4.13   Two poset diagrams.

For example, $4 > 2 > 0 > -2 > -4 > -6 > \ldots$ is a descending chain in $\langle \mathbb{Z}, < \rangle$. For another example, $\{a, b, c\} \supset \{a, b\} \supset \{a\} \supset \varnothing$ is a finite descending chain in $\langle \text{power}(\{a, b, c\}), \subseteq \rangle$. We can define an *ascending chain* of elements in a similar way. For example, $1 \mid 2 \mid 4 \mid \ldots \mid 2^n \mid \ldots$ is an ascending chain in the poset $\langle \mathbb{N}, | \rangle$.

Predecessors and Successors

If $x \prec y$, then we say that x is a *predecessor* of y, or y is a *successor* of x. Suppose that $x \prec y$ and there are no elements between x and y. In other words, suppose we have the following situation:

$$\{z \in A \mid x \prec z \prec y\} = \varnothing.$$

When this is the case, we say that x is an *immediate predecessor* of y, or y is an *immediate successor* of x. In a finite poset an element with a successor has an immediate successor. Some infinite posets also have this property. For example, every natural number x has an immediate successor $x + 1$ with respect to the "less" relation. But no rational number has an immediate successor with respect to the "less" relation.

Poset Diagrams

A poset can be represented by a special graph called a *poset diagram* or a *Hasse diagram*—after the mathematician Helmut Hasse (1898–1979). Whenever $x \prec y$ and x is an immediate predecessor of y, then place an edge (x, y) in the poset diagram with x at a lower level than y. A poset diagram can often help us observe certain properties of a poset. For example, the two poset diagrams in Figure 4.13 represent the pancake recipe poset from Example 4.30 and the poset $\langle\{2, 3, 4, 12\}, |\rangle$.

The three poset diagrams shown in Figure 4.14 are for the natural numbers and the integers with their usual orderings and for $\text{power}(\{a, b\})$ with the subset relation.

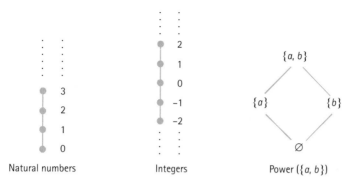

Figure 4.14   Three poset diagrams.

Maxima, Minima, and Bounds

When we have a partially ordered set, it's natural to use words like "minimal," "least," "maximal," and "greatest." Let's give these words some formal definitions.

Suppose that S is any nonempty subset of a poset P. An element $x \in S$ is called a *minimal element* of S if x has no predecessors in S. An element $x \in S$ is called the *least element* of S if $x \preceq y$ for all $y \in S$. For example, let's consider the poset $\langle \mathbb{N}, | \rangle$.

The subset $\{2, 4, 5, 10\}$ has two minimal elements, 2 and 5.
The subset $\{2, 4, 12\}$ has least element 2.
The set $\mathbb{N}$ has least element 1 because $1|x$ for all $x \in \mathbb{N}$.

For another example, let's consider the poset $\langle \text{power}(\{a, b, c\}), \subseteq \rangle$. The subset $\{\{a, b\}, \{a\}, \{b\}\}$ has two minimal elements, $\{a\}$ and $\{b\}$. The power set itself has least element $\varnothing$.

In a similar way we can define *maximal elements* and the *greatest element* of a subset of a poset. For example, let's consider the poset $\langle \mathbb{N}, | \rangle$.

The subset $\{2, 4, 5, 10\}$ has two maximal elements, 4 and 10.
The subset $\{2, 4, 12\}$ has greatest element 12.
The set $\mathbb{N}$ itself has greatest element 0 because $x|0$ for all $x \in \mathbb{N}$.

For another example, let's consider the poset $\langle \text{power}(\{a, b, c\}), \subseteq \rangle$. The subset $\{\varnothing, \{a\}, \{b\}\}$ has two maximal elements, $\{a\}$ and $\{b\}$. The power set itself has greatest element $\{a, b, c\}$.

Some sets may not have any minimal elements, yet still be bounded below by some element. For example, the set of positive rational numbers has no least element yet is bounded below by the number 0. Let's introduce some standard terminology that can be used to discuss ideas like this.

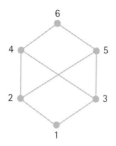

Figure 4.15   A poset diagram.

If S is a nonempty subset of a poset P, an element $x \in P$ is called a *lower bound* of S if $x \preceq y$ for all $y \in S$. An element $x \in P$ is called the *greatest lower bound* (or *glb*) of S if x is a lower bound and $z \preceq x$ for all lower bounds z of S. The expression glb(S) denotes the greatest lower bound of S, if it exists. For example, if we let $\mathbb{Q}^+$ denote the set of positive rational numbers, then over the poset $\langle \mathbb{Q}, \leq \rangle$ we have glb($\mathbb{Q}^+$) = 0.

In a similar way we define upper bounds for a subset S of the poset P. An element $x \in P$ is called an *upper bound* of S if $y \preceq x$ for all $y \in S$. An element $x \in P$ is called the *least upper bound* (or *lub*) of S if x is an upper bound and $x \preceq z$ for all upper bounds z of S. The expression lub(S) denotes the least upper bound of S, if it exists. For example, lub($\mathbb{Q}^+$) does not exist in $\langle \mathbb{Q}, \leq \rangle$.

For another example, in the poset $\langle \mathbb{N}, \leq \rangle$, every finite subset has a glb—the least element—and a lub—the greatest element. Every infinite subset has a glb but no upper bound.

Can subsets have upper bounds without having a least upper bound? Sure. Here's an example.

example **4.32 Upper Bounds**

Suppose the set $\{1, 2, 3, 4, 5, 6\}$ represents six time-oriented tasks. You can think of the numbers as chapters in a book, as processes to be executed on a computer, or as the steps in a recipe for making ice cream. In any case, suppose the tasks are partially ordered according to the poset diagram in Figure 4.15.

The subset $\{2, 3\}$ is bounded above, but it has no least upper bound. Notice that 4, 5, and 6 are all upper bounds of $\{2, 3\}$, but none of them is a least upper bound.

end example

Lattices

A *lattice* is a poset with the property that every pair of elements has a glb and a lub. So the poset of Example 4.32 is not a lattice. For example, $\langle \mathbb{N}, \leq \rangle$ is a

 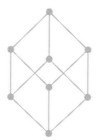

Figure 4.16   Two lattices.

lattice in which the glb of two elements is their minimum and the lub is their maximum. For another example, if A is any set, then $\langle \text{power}(A), \subseteq \rangle$ is a lattice, where $\text{glb}(X, Y) = X \cap Y$ and $\text{lub}(X, Y) = X \cup Y$. The word "lattice" is used because lattices that aren't totally ordered often have poset diagrams that look like "latticeworks" or "trellisworks."

For example, the two poset diagrams in Figure 4.16 represent lattices. These two poset diagrams can represent many different lattices. For example, the poset diagram on the left represents the lattice whose elements are the positive divisors of 36, ordered by the divides relation. In other words, it represents the lattice $\langle \{1, 2, 3, 4, 6, 9, 12, 18, 36\}, | \rangle$. See whether you can label the poset diagram with these numbers. The diagram on the right represents the lattice $\langle \text{power}(\{a, b, c\}), \subseteq \rangle$. It also represents the lattice whose elements are the positive divisors of 70, ordered by the divides relation. See whether you can label the poset diagram with both of these lattices. We'll give some more examples in the exercises.

4.3.2 Topological Sorting

A typical computing task is to sort a list of elements taken from a totally ordered set. Here's the problem statement.

The Sorting Problem

Find an algorithm to sort a list of elements from a totally ordered set.

For example, suppose we're given the list $\langle x_1, x_2, \ldots, x_n \rangle$, where the elements of the list are related by a total order relation R. We might sort the list by a program "sort," which we could call as follows:

$$\text{sort}(R, \langle x_1, x_2, \ldots, x_n \rangle).$$

For example, we should be able to obtain the following results with sort:

$$\text{sort}\,(<, \langle 8, 3, 10, 5 \rangle) = \langle 3, 5, 8, 10 \rangle,$$
$$\text{sort}\,(>, \langle 8, 3, 10, 5 \rangle) = \langle 10, 8, 5, 3 \rangle.$$

Programming languages normally come equipped with several totally ordered sets. If a total order R is not part of the language, then R must be implemented as a relational test, which can then be called into action whenever a comparison is required in the sorting algorithm.

Topological Sorting

Can a partially ordered set be sorted? The answer is yes if we broaden our idea of what sorting means. Here's the problem statement.

The Topological Sorting Problem

Find an algorithm to sort a list of elements from a partially ordered set.

How can we "sort" a list when some elements may not be comparable? Well, we try to find a listing that maintains the partial ordering, as in the pancake recipe from Example 4.30. Given a partial order R on a set, a list of elements from the set is *topologically sorted* if whenever two elements in the list satisfy a R b, then a is to the left of b in the list.

The ordering of a set of tasks is a topological sorting problem. For example, the list $\langle 4, 5, 2, 1, 3, 6, 7 \rangle$ is a topological sort of the steps in the pancake recipe from Example 4.30. Another example of a topological sort is the ordering of the chapters in a textbook in which the partial order is defined to be the dependence of one chapter upon another. In other words, we hope that we don't have to read some chapter further on in the book to understand what we're reading now.

Is there a technique to do topological sorting? Yes. Suppose R is a partial order on a finite set A. For each element $y \in A$, let $P(y)$ be the number of immediate predecessors of y, and let $S(y)$ be the set of immediate successors of y. Let *Sources* be the set of sources—minimal elements—in A. Therefore, y is a source if and only if $P(y) = 0$. A topological sort algorithm goes something like the following:

Topological Sort Algorithm   (4.17)

While the set of sources is not empty, do the following steps:

1. Output a source y.

2. For all z in $S(y)$, decrement $P(z)$; if $P(z) = 0$, then add z to *Sources*.

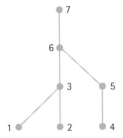

Figure 4.17   Poset of a pancake recipe.

A more detailed description of the algorithm can be given as follows:

Detailed Topological Sort

> **while** *Sources* $\neq \varnothing$ **do**
>> Pick a source y from *Sources;*
>> Output y;
>> **for** each z in $S(y)$ **do**
>>> $P(z) := P(z) - 1$;
>>> **if** $P(z) = 0$ **then** *Sources* := *Sources* $\cup \{z\}$
>>
>> **od**;
>> *Sources* := *Sources* $- \{y\}$;
>
> **od**

Let's do an example that includes some details on how the data for the algorithm might be represented.

example 4.33 A Topological Sort

We'll consider the steps of the pancake recipe from Example 4.30. Figure 4.17 shows the poset diagram for the steps of the recipe.

The initial set of sources is $\{1, 2, 4\}$. Letting P be an array of integers, we get the following initial table of predecessor counts:

i	1	2	3	4	5	6	7
$P(i)$	0	0	2	0	1	2	1

The following table is the initial table of successor sets S:

i	1	2	3	4	5	6	7
$S(i)$	$\{3\}$	$\{3\}$	$\{6\}$	$\{5\}$	$\{6\}$	$\{7\}$	$\varnothing$

You should trace the algorithm for these data representations.

end example

There is a very interesting and efficient implementation of algorithm (4.17) in Knuth [1968]. It involves the construction of a novel data structure to represent the set of sources, the sets $S(y)$ for each y, and the numbers $P(z)$ for each z.

4.3.3   Well-Founded Orders

Let's look at a special property of the natural numbers. Suppose we're given a descending chain of natural numbers that begins as follows:

$$29 > 27 > 25 > \cdots.$$

Can this descending chain continue forever? Of course not. We know that 0 is the least natural number, so the given chain must stop after only a finite number of terms. This is not an earthshaking discovery, but it is an example of the property of well-foundedness that we're about to discuss.

Definition of a Well-Founded Order

We're going to consider posets with the property that every descending chain of elements is finite. So we'll give these posets a name.

> **Well-Founded**
> A poset is said to be *well-founded* if every descending chain of elements is finite. In this case, the partial order is called a *well-founded order*.

For example, we've seen that $\mathbb{N}$ is a well-founded set with respect to the less relation $<$. In fact, any set of integers with a least element is well-founded by $<$. For example, the following three sets of integers are well-founded.

$$\{1, 2, 3, 4, \ldots\}, \{m \mid m \geq -3\}, \text{ and } \{5, 9, 13, 17, \ldots\}.$$

For another example, any collection of finite sets is well-founded by $\subseteq$. This is easy to see because any descending chain must start with a finite set. If the set has n elements, it can start a descending chain of at most $n + 1$ subsets. For example, the following expression displays a longest descending chain starting with the set $\{a, b, c\}$.

$$\{a, b, c\} \supset \{b, c\} \supset \{c\} \supset \varnothing.$$

So the power set of a finite set is well-founded with respect to $\subseteq$.

But many posets are not well-founded. For example, the integers and the positive rationals are not well-founded with respect to the less relation because they have infinite descending chains as the following examples show.

$$2 > 0 > -2 > -4 > \cdots$$

$$\frac{1}{2} > \frac{1}{3} > \frac{1}{4} > \frac{1}{5} > \cdots .$$

The power set of an infinite set is not well-founded by $\subseteq$. For example, if we let $S_k = \mathbb{N} - \{0, 1, \ldots, k\}$, then we obtain the following infinite descending chain in power($\mathbb{N}$):

$$S_0 \supset S_1 \supset S_2 \supset \cdots \supset S_k \supset \cdots .$$

Are well-founded sets good for anything? The answer is yes. We'll see in the next section that they are basic tools for inductive proofs. So we should get familiar with them. We'll do this by looking at another property that well-founded sets possess.

The Minimal Element Property

Does every subset of $\mathbb{N}$ have a least element? A quick-witted person might say, "Yes," and then think a minute and say, "except that the empty set doesn't have any elements, so it can't have a least element." Suppose the question is modified to "Does every nonempty subset of $\mathbb{N}$ have a least element?" Then a bit of thought will convince most of us that the answer is yes.

We might reason as follows: Suppose S is some nonempty subset of $\mathbb{N}$ and x_1 is some element of S. If x_1 is the least element of S, then we are done. So assume that x_1 is not the least element of S. Then x_1 must have a predecessor x_2 in S—otherwise, x_1 would be the least element of S. If x_2 is the least element of S, then we are done. If x_2 is not the least element of S, then it has a predecessor x_3 in S, and so on. If we continue in this manner, we will obtain a descending chain of distinct elements in S:

$$x_1 > x_2 > x_3 > \cdots .$$

This looks familiar. We already know that this chain of natural numbers can't be infinite. So it stops at some value, which must be the least element of S. So every nonempty subset of the natural numbers has a least element.

This property is not true for all posets. For example, the set of integers has no least element. The open interval of real numbers $(0, 1)$ has no least element. Also the power set of a finite set can have collections of subsets that have no least element.

Notice however that every collection of subsets of a finite set does contain a minimal element. For example, the collection $\{\{a\}, \{b\}, \{a, b\}\}$ has two minimal elements $\{a\}$ and $\{b\}$. Remember, the property that we are looking for must be true for all well-founded sets. So the existence of least elements is out; it's too restrictive.

But what about the existence of minimal elements for nonempty subsets of a well-founded set? This property is true for the natural numbers. (Least elements are certainly minimal.) It's also true for power sets of finite sets. In fact, this property is true for all well-founded sets, and we can state the result as follows:

Descending Chains and Minimality (4.18)

If A is a well-founded set, then every nonempty subset of A has a minimal element. Conversely, if every nonempty subset of A has a minimal element, then A is well-founded.

It follows from (4.18) that the property of finite descending chains is equivalent to the property of nonempty subsets having minimal elements. In other words, if a poset has one of the properties, then it also has the other property. Thus it is also correct to define a well-founded set to be a poset with the property that every nonempty subset has a minimal element. We will call this latter property the *minimum condition* on a poset.[1]

Whenever a well-founded set is totally ordered, then each nonempty subset has a single minimal element, the least element. Such a set is called a *well-ordered set*. So a well-ordered set is a totally ordered set such that every nonempty subset has a least element. For example, $\mathbb{N}$ is well-ordered by the "less" relation. Let's examine a few more total orderings to see whether they are well-ordered.

Lexicographic Ordering of Tuples

The linear ordering $<$ on $\mathbb{N}$ can be used to create the *lexicographic* order on $\mathbb{N}^k$, which is defined as follows.

$$(x_1, \ldots, x_k) \prec (y_1, \ldots, y_k)$$

if and only if there is an index $j \geq 1$ such that $x_j < y_j$ and for each $i < j$, $x_i = y_i$. This ordering is a total ordering on $\mathbb{N}^k$. It's also a well-ordering.

For example, the lexicographic order on $\mathbb{N} \times \mathbb{N}$ has least element $(0, 0)$. Every nonempty subset of $\mathbb{N} \times \mathbb{N}$ has a least element, namely, the pair (x, y) with the smallest value of x, where y is the smallest value among second components of pairs with x as the first component. For example, $(0, 10)$ is the least element in the set $\{(0, 10), (0, 11), (1, 0)\}$. Notice that $(1, 0)$ has infinitely many predecessors of the form $(0, y)$, but $(1, 0)$ has no immediate predecessor.

---

[1] Other names for a well-founded set are *poset with minimum condition, poset with descending chain condition,* and *Artinian poset,* after Emil Artin, who studied algebraic structures with the descending chain condition. Some people use the term *Noetherian,* after Emmy Noether, who studied algebraic structures with the ascending chain condition.

Lexicographic Ordering of Strings

Another type of lexicographic ordering involves strings. To describe it we need to define the prefix of a string. If a string x can be written as $x = uv$ for some strings u and v, then u is called a *prefix* of x. If $v \neq \Lambda$, then u is a *proper prefix* of x. For example, the prefixes of the string aba over the alphabet $\{a, b\}$ are Λ, a, ab, and aba. The proper prefixes of aba are Λ, a, and ab.

Definition of Lexicographic Ordering

Let A be a finite alphabet with some agreed-upon linear ordering. Then the *lexicographic* ordering on A^* is defined as follows: $x \prec y$ iff either x is a proper prefix of y or x and y have a longest common proper prefix u such that $x = uv$, $y = uw$, and head(v) precedes head(w) in A.

The lexicographic ordering on A^* is often called the *dictionary ordering* because it corresponds to the ordering of words that occur in a dictionary. The definition tells us that if $x \neq y$, then either $x \prec y$ or $y \prec x$. So the lexicographic ordering on A^* is a total (i.e., linear) ordering. It also follows that every string x has an immediate successor xa, where a is the first letter of A.

If A has at least two elements, then the lexicographic ordering on A^* is *not* well-ordered. For example, let $A = \{a, b\}$ and suppose that a precedes b. Then the elements in the set $\{a^n b \mid n \in \mathbb{N}\}$ form an infinite descending chain:

$$b \succ ab \succ aab \succ aaab \succ \cdots \succ a^n b \succ \cdots .$$

Notice also that b has no immediate predecessor because if $x \prec b$, then we have $x \prec xa \prec b$.

Standard Ordering of Strings

Now let's look at an ordering that is well-ordered. The *standard ordering on strings* uses a combination of length and the lexicographic ordering.

Definition of Standard Ordering

Let A be a finite alphabet with some agreed-upon linear ordering. The standard ordering on A^* is defined as follows, where $\prec_L$ denotes the lexicographic ordering on A^*:

$x \prec y$ iff either length(x) < length(y), or length(x) = length(y) and $x \prec_L y$.

It's easy to see that $\prec$ is a total order and every string has an immediate successor and an immediate predecessor. The standard ordering on A^* is also well-ordered because each string has a finite number of predecessors. For example, let $A = \{a, b\}$ and suppose that a precedes b. Then the first few elements in the standard order of A^* are given as follows:

$$\Lambda, a, b, aa, ab, ba, bb, aaa, aab, aba, abb, baa, bab, bba, bbb, \ldots .$$

Constructing Well-Founded Orderings

Collections of strings, lists, trees, graphs, or other structures that programs process can usually be made into well-founded sets by defining an appropriate order relation. For example, any finite set can be made into a well-founded set—actually a well-ordered set—by simply listing its elements in any order we wish, letting the leftmost element be the least element.

Let's look at some ways to build well-founded orderings for infinite sets. Suppose we want to define a well-founded order on some infinite set S. A simple and useful technique is to associate each element of S with some element in an existing well-founded set. For example, the natural numbers are well-founded by $<$. So we'll use them as a building block for well-founded constructions.

Constructing a Well-Founded Order   **(4.19)**

Given any function $f : S \to \mathbb{N}$, there is a well-founded order $\prec$ defined on S in the following way, where $x, y \in S$:

$$x \prec y \quad \text{means} \quad f(x) < f(y).$$

Does the new relation $\prec$ make S into a well-founded set? Sure. Suppose we have a descending chain of elements in S as follows:

$$x_1 \succ x_2 \succ x_3 \succ \cdots.$$

The chain must stop because $x \succ y$ is defined to mean $f(x) > f(y)$, and we know that any descending chain of natural numbers must stop. Let's look at a few more examples.

example **4.34 Some Well-Founded Orderings**

    a. Any set of lists, where $L \prec M$ means length(L) < length(M).

    b. Any set of strings, where $s \prec t$ means length(s) < length(t).

    c. Any set of trees, where $B \prec C$ means the number of nodes in B is less than the number of nodes in C.

    d. Any set of trees, where $B \prec C$ means the number of leaves in B is less than the number of leaves in C.

    e. Any set of nonempty trees, where $B \prec C$ means depth(B) < depth(C).

    f. Any set of people can be well-founded by the age at the last birthday of each person. What are the minimal elements?

    g. The set $\{\ldots, -3, -2, -1\}$ of negative integers, where $x \prec y$ means $x > y$.

end example

As the examples show, we can use known properties of structures to find useful well-founded orderings for sets of structures. The next example constructs a finite, hence well-founded, lexicographic order.

example 4.35 A Finite Lexicographic Order

Let $S = \{0, 1, 2, \ldots, m\}$. Then we can define a lexicographic ordering on the set S^k in a natural way. Since S is finite, it follows that the lexicographic ordering on S^k is well-founded. The least element is $(0, \ldots, 0)$, and the greatest element is $(m, \ldots, m)$. For example, if $k = 3$, then the immediate successor of any element can be defined as

$$\text{succ}\,((x, y, z)) = \text{if } z < m \text{ then } (x, y, z + 1)$$
$$\text{else if } y < m \text{ then } (x, y + 1, 0)$$
$$\text{else if } x < m \text{ then } (x + 1, 0, 0)$$
$$\text{else error (no successor).}$$

end example

Inductively Defined Sets are Well-Founded

It's easy to make an inductively defined set W into a well-founded set if no two elements are defined in terms of each other. We'll give two methods. Both methods let the basis elements of W be the minimal elements of the well-founded order.

Method 1: (4.20)

Define a function $f : W \to \mathbb{N}$ as follows:

1. $f(c) = 0$ for all basis elements c of W.

2. If $x \in W$ and x is constructed from elements $y_1, y_2, \ldots, y_n$ in W, then define $f(x) = 1 + \max\{f(y_1), f(y_2), \ldots, f(y_n)\}$.

Let $x \prec y$ mean $f(x) < f(y)$.

Since 0 is the least element of $\mathbb{N}$ and $f(c) = 0$ for all basis elements c of W, it follows that the basis elements of W are minimal elements under the ordering defined by (4.20). For example, if c is a basis element of W and if $x \prec c$, then $f(x) < f(c) = 0$, which can't happen with natural numbers. Therefore c is a minimal element of W.

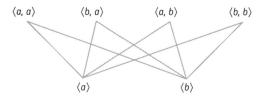

Figure 4.18   Part of a poset diagram.

example **4.36 A Well-Founded Ordering**

Let W be the set of all nonempty lists over $\{a, b\}$. First we'll give an inductive definition of W. The lists $\langle a \rangle$ and $\langle b \rangle$ are the basis elements of W. For the induction case, if $L \in W$, then the lists $\text{cons}(a, L)$ and $\text{cons}(b, L)$ are in W. Now we'll use (4.20) to make W into a well-founded set. The function f of (4.20) turns out to be $f(L) = \text{length}(L) - 1$. So for any lists L and M in W, we define $L \prec M$ to mean $f(L) < f(M)$, which means $\text{length}(L) - 1 < \text{length}(M) - 1$, which also means $\text{length}(L) < \text{length}(M)$. The diagram in Figure 4.18 shows the bottom two layers of a poset diagram for W with its two minimal lists $\langle a \rangle$ and $\langle b \rangle$.

end example

Method 1 can relate many elements. For example, to add one more level to the diagram in Figure 4.18, we have to include eight 3-element lists and draw 32 lines from the two element lists up to the three element lists. Sometimes it isn't necessary to have an ordering that relates so many elements. This brings us to the second method for defining a well-founded ordering on an inductively defined set W.

Method 2:   (4.21)

The ordering $\prec$ is defined as follows:

1. Let the basis elements of W be minimal elements.

2. If $x \in W$ and x is constructed from elements $y_1, y_2, \ldots, y_n$ in W, then define $y_i \prec x$ for each $i = 1, \ldots, n$.

The actual ordering is the transitive closure of $\prec$.

The ordering of (4.21) is well-founded because any x can be constructed from basis elements with finitely many constructions. Therefore, there can be no

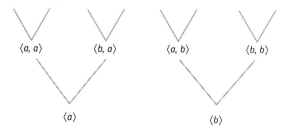

$\langle a, a \rangle$ $\langle b, a \rangle$ $\langle a, b \rangle$ $\langle b, b \rangle$

$\langle a \rangle$ $\langle b \rangle$

Figure 4.19   Part of a poset diagram.

infinite descending chain starting at x. With this ordering, there can be many pairs that are not related.

For example, we'll use the preceding example of nonempty lists over the set $\{a, b\}$. The picture in Figure 4.19 shows the bottom two levels of the poset diagram for the well-founded ordering constructed by (4.21).

Notice that each list has only two immediate successors. For example, the two successors of $\langle a \rangle$ are $\text{cons}(a, \langle a \rangle) = \langle a, a \rangle$ and $\text{cons}(b, \langle a \rangle) = \langle b, a \rangle$. The two successors of $\langle b, a \rangle$ are $\langle a, b, a \rangle$ and $\langle b, b, a \rangle$. This is much simpler than the ordering we got using (4.20).

Let's look at some examples of inductively defined sets that are well-founded sets by the method of (4.21).

example **4.37 Using One Part of a Product**

We'll define the set $\mathbb{N} \times \mathbb{N}$ inductively by using the first copy of $\mathbb{N}$. For the basis case we put $(0, n) \in \mathbb{N} \times \mathbb{N}$ for all $n \in \mathbb{N}$. For the induction case, whenever the pair $(m, n) \in \mathbb{N} \times \mathbb{N}$, we put $(m+1, n) \in \mathbb{N} \times \mathbb{N}$. The relation on $\mathbb{N} \times \mathbb{N}$ induced by this inductive definition and (4.21) is not linearly ordered.

For example, $(0, 0)$ and $(0, 1)$ are not related because they are both basis elements. Notice that any pair (m, n) is the beginning of a descending chain containing at most $m + 1$ pairs. For example, the following chain is the longest descending chain that starts with $(3, 17)$.

$$(3, 17), (2, 17), (1, 17), (0, 17).$$

end example

example **4.38 Using Both Parts of a Product**

Let's define the set $\mathbb{N} \times \mathbb{N}$ inductively by using both copies of $\mathbb{N}$. The single basis element is $(0, 0)$. For the induction case, if $(m, n) \in \mathbb{N} \times \mathbb{N}$, then put $(m + 1, n), (m, n + 1) \in \mathbb{N} \times \mathbb{N}$. Notice that each pair with both components nonzero is defined twice by this definition. The relation induced by this definition and (4.21) is nonlinear.

For example, the two pairs $(2, 1)$ and $(1, 2)$ are not related. Any pair (m, n) is the beginning of a descending chain of at most $m + n + 1$ pairs. For example, the following descending chain has maximum length among the descending chains that start at the pair $(2, 3)$.

$$(2,\ 3),\ (2,\ 2),\ (1,\ 2),\ (1,\ 1),\ (0,\ 1),\ (0,\ 0).$$

Can you find a different chain of the same length starting at $(2, 3)$?

4.3.4   Ordinal Numbers

We'll finish our discussion of order by introducing the *ordinal numbers*. These numbers are ordered, and they can be used to count things. An ordinal number may be represented as a set with certain properties. For example, any ordinal number x has an immediate successor defined by $\text{succ}(x) = x \cup \{x\}$. The expression $x + 1$ is also used to denote $\text{succ}(x)$. The natural numbers denote ordinal numbers when we define $0 = \varnothing$ and interpret $+$ as addition, in which case it's easy to see that

$$x + 1 = \{0, \ldots, x\}.$$

For example, $1 = \{0\}$, $2 = \{0,\ 1\}$, and $5 = \{0,\ 1,\ 2,\ 3,\ 4\}$. In this way, each natural number is an ordinal number, called a *finite ordinal*.

Now let's define some *infinite ordinals*. The first infinite ordinal is

$$\omega = \{0,\ 1,\ 2, \ldots\},$$

the set of natural numbers. The next infinite ordinal is

$$\omega + 1 = \text{succ}(\omega) = \omega \cup \{\omega\} = \{\omega,\ 0,\ 1, \ldots\}.$$

If α is an ordinal number, we'll write $\alpha + n$ in place of $\text{succ}^n(\alpha)$. So the first four infinite ordinals are ω, $\omega + 1$, $\omega + 2$, and $\omega + 3$. The infinite ordinals continue in this fashion. To get beyond this sequence of ordinals, we need to make a definition similar to the one for ω. The main idea is that any ordinal number is the union of all its predecessors. For example, we define $\omega 2 = \omega \cup \{\omega,\ \omega + 1,\ \ldots\}$. The ordinals continue with $\omega 2 + 1$, $\omega 2 + 2$, and so on. Of course, we can continue and define $\omega 3 = \omega 2 \cup \{\omega 2,\ \omega 2 + 1, \ldots\}$. After ω, $\omega 2$, $\omega 3, \ldots$ comes the ordinal ω^2. Then we get $\omega^2 + 1$, $\omega^2 + 2, \ldots$, and we eventually get $\omega^2 + \omega$. Of course, the process goes on forever.

We can order the ordinal numbers by defining $\alpha < \beta$ iff $\alpha \in \beta$. For example, we have $x < x + 1$ for any ordinal x because $x \in \text{succ}(x) = x + 1$. So we get the familiar ordering $0 < 1 < 2 < \ldots$ for the finite ordinals. For any finite ordinal n we have $n < \omega$ because $n \in \omega$. Similarly, we have $\omega < \omega + 1$, and for any finite ordinal n we have $\omega + n < \omega 2$. So it goes. There are also uncountable ordinals, the least of which is denoted by Ω. And the ordinals continue on after this too.

Although every ordinal number has an immediate successor, there are some ordinals that don't have any immediate predecessors. These ordinals are called *limit ordinals* because they are defined as "limits" or unions of all their predecessors. The limit ordinals that we've seen are

$$0, \omega, \omega2, \omega3, \ldots, \omega^2, \ldots, \Omega, \ldots.$$

An interesting fact about ordinal numbers states that for any set S there is a bijection between S and some ordinal number. For example, there is a bijection between the set $\{a, b, c\}$ and the ordinal number $3 = \{0, 1, 2\}$. For another example there are bijections between the set $\mathbb{N}$ of natural numbers and each of the ordinals $\omega, \omega + 1, \omega + 2, \ldots$. Some people define the cardinality of a set to be the least ordinal number that is bijective to the set. So we have $|\{a, b, c\}| = 3$ and $|\mathbb{N}| = \omega$.

More information about ordinal numbers—including ordinal arithmetic—can be found in the excellent book by Halmos [1960].

 Exercises

Partial Orders

1. Sometimes our intuition about a symbol can be challenged. For example, suppose we define the relation $\prec$ on the integers by saying that $x \prec y$ means $|x| < |y|$. Assign the value True or False to each of the following statements.

   a. $-7 \prec 7$.   b. $-7 \prec -6$.   **c.** $6 \prec -7$.   d. $-6 \prec 2$.

2. State whether each of the following relations is a partial order.

   a. isFatherOf.   b. isAncestorOf.   **c.** isOlderThan.
   d. isSisterOf.   **e.** $\{(a, b), (a, a), (b, a)\}$.   f. $\{(2, 1), (1, 3), (2, 3)\}$.

3. Draw a poset diagram for each of the following partially ordered relations.

   a. $\{(a, a), (a, b), (b, c), (a, c), (a, d) \}$.
   b. power($\{a, b, c\}$), with the subset relation.
   c. lists($\{a, b\}$), where $L \prec M$ if length(L) < length(M).
   d. The set of all binary trees over the set $\{a, b\}$ that contain either one or two nodes. Let $s \prec t$ mean that s is either the left or right subtree of t.

4. Suppose we wish to evaluate the following expression as a set of time-oriented tasks:

$$(f(x) + g(x))(f(x)g(x)).$$

We'll order the subexpressions by data dependency. In other words, an expression can't be evaluated until its data are available. So the subexpressions that occur in the evaluation process are

$$x, f(x), g(x), f(x) + g(x), f(x)g(x), \text{ and } (f(x) + g(x))(f(x)g(x)).$$

Draw the poset diagram for the set of subexpressions. Is the poset a lattice?

5. For any positive integer n, let D_n be the set of positive divisors of n. The poset $\langle D_n, | \rangle$ is a lattice. Describe the glb and lub for any pair of elements.

Well-Founded Property

6. Why is it true that every partially ordered relation over a finite set is well-founded?

7. For each set S, show that the given partial order on S is well-founded.

   a. Let S be a set of trees. Let $s \prec t$ mean that s has fewer nodes than t.

   b. Let S be a set of trees. Let $s \prec t$ mean that s has fewer leaves than t.

   c. Let S be a set of lists. Let $L \prec M$ mean that $\text{length}(L) < \text{length}(M)$.

8. Example 4.38 discussed a well-founded ordering for the set $\mathbb{N} \times \mathbb{N}$. Use this ordering to construct two distinct descending chains that start at the pair $(4, 3)$, both of which have maximum length.

9. Suppose we define the relation $\prec$ on $\mathbb{N} \times \mathbb{N}$ as follows:

$$(a, b) \prec (c, d) \quad \text{if and only if} \quad \max\{a, b\} < \max\{c, d\}.$$

Is $\mathbb{N} \times \mathbb{N}$ well-founded with respect to $\prec$?

Topological Sorting

10. Trace the topological sort algorithm (4.17) for the pancake recipe in Example 4.30 by starting with the source 1. There are several possible answers because any source can be output by the algorithm.

11. Describe a way to perform a topological sort that uses an adjacency matrix to represent the partial order.

Proofs and Challenges

12. Show that the two properties irreflexive and transitive imply the antisymmetric property. So an irreflexive partial order can be defined by just the two properties irreflexive and transitive.

13. Prove the two statements of (4.18).

14. For a poset P, a function $f : P \to P$ is said to be *monotonic* if $x \preceq y$ implies $f(x) \preceq f(y)$ for all $x, y \in P$. For each poset and function definition, determine whether the function is monotonic.

  a. $\langle \mathbb{N}, < \rangle, f(x) = 2x + 3$.   **b.** $\langle \mathbb{N}, < \rangle, f(x) = x^2$.

  c. $\langle \mathbb{Z}, < \rangle, f(x) = x^2$.   **d.** $\langle \mathbb{N}, | \rangle, f(x) = 2x + 3$.

  e. $\langle \mathbb{N}, | \rangle, f(x) = x^2$.   **f.** $\langle \mathbb{N}, | \rangle, f(x) = x \bmod 5$.

  g. $\langle \text{power}(A), \subseteq \rangle$ for some set A, $f(X) = A - X$.

  h. $\langle \text{power}(\mathbb{N}), \subseteq \rangle, f(X) = \{n \in \mathbb{N} \mid n \text{ divides } x \text{ for some } x \in X\}$.

4.4 Inductive Proof

When discussing properties of things we deal not only with numbers, but also with structures such as strings, lists, trees, graphs, programs, and more complicated structures constructed from them. Do the objects that we construct have the properties that we expect? Does a program halt when it's supposed to halt and give the proper answer?

To answer these questions, we must find ways to reason about the objects that we construct. This section concentrates on a powerful proof technique called inductive proof. We'll see that the technique springs from the idea of a well-founded set that we discussed in Section 4.3.

4.4.1 Proof by Mathematical Induction

Suppose we want to find the value of the sum $2 + 4 + \cdots + 2n$ for any natural number n. Consider the following two programs written by two different students to calculate this sum:

$$f(n) = \text{if } n = 0 \text{ then } 0 \text{ else } f(n-1) + 2n$$
$$g(n) = n(n+1).$$

Are these programs correct? That is, do they both compute the correct value of the sum $2 + 4 + \cdots + 2n$? We can test a few cases such as $n = 0$, $n = 1$, $n = 2$ until we feel confident that the programs are correct. Or maybe we just can't get any feeling of confidence in these programs. Is there a way to prove, once and for all, that these programs are correct for all natural numbers n? Let's look at the second program. If it's correct, then the following equation must be true for all natural numbers n:

$$2 + 4 + \cdots + 2n = n(n+1).$$

Certainly we don't have the time to check it for the infinity of natural numbers. Is there some other way to prove it? Happily, we will be able to prove the infinitely many cases in just two steps with a technique called *proof by induction*, which we discuss next. If you don't want to see why it works, you can skip ahead to (4.23).

A Basis for Mathematical Induction

Interestingly, the technique that we present is based on the fact that any nonempty subset of the natural numbers has a least element. Recall that this is the same as saying that any descending chain of natural numbers is finite. In fact, this is just a statement that $\mathbb{N}$ is a well-founded set. In fact we can generalize a bit. Let m be an integer, and let W be the following set.

$$W = \{m,\ m + 1,\ m + 2, \dots \}.$$

Every nonempty subset of W has a least element. Let's see whether this property can help us find a tool to prove infinitely many things in just two steps. First, we state the following result, which forms a basis for the inductive proof technique.

A Basis for Mathematical Induction (4.22)

Let $m \in \mathbb{Z}$ and $W = \{m,\ m + 1,\ m + 2, \dots \}$. Let S be a subset of W such that the following two conditions hold.

1. $m \in S$.

2. Whenever $k \in S$, then $k + 1 \in S$.

Then $S = W$.

Proof: We'll prove $S = W$ by contradiction. Suppose $S \neq W$. Then $W - S$ has a least element x because every nonempty subset of W has a least element. The first condition of (4.22) tells us that $m \in S$. So it follows that $x > m$. Thus $x - 1 \geq m$, and it follows that $x - 1 \in S$. Thus we can apply the second condition to obtain $(x - 1) + 1 \in S$. In other words, we are forced to conclude that $x \in S$. This is a contradiction, since we can't have both $x \in S$ and $x \in W - S$ at the same time. Therefore $S = W$. QED.

We should note that there is an alternative way to think about (4.22). First, notice that W is an inductively defined set. The basis case is $m \in W$. The inductive step states that whenever $k \in W$, then $k + 1 \in W$. Now we can appeal to the closure part of an inductive definition, which can be stated as follows: If S is a subset of W and S satisfies the basis and inductive steps for W, then $S = W$. From this point of view, (4.22) is just a restatement of the closure part of the inductive definition of W.

The Principle of Mathematical Induction

Let's put (4.22) into a practical form that can be used as a proof technique for proving that infinitely many cases of a statement are true. The technique is called the *principle of mathematical induction,* which we state as follows.

The Principle of Mathematical Induction   (4.23)

Let $m \in \mathbb{Z}$. To prove that $P(n)$ is true for all integers $n \geq m$, perform the following two steps:

1. Prove that $P(m)$ is true.

2. Assume that $P(k)$ is true for an arbitrary $k \geq m$. Prove that $P(k + 1)$ is true.

Proof: Let $W = \{n \mid n \geq m\}$, and let $S = \{n \mid n \geq m \text{ and } P(n) \text{ is true}\}$. Assume that we have performed the two steps of (4.23). Then S satisfies the hypothesis of (4.22). Therefore $S = W$. So $P(n)$ is true for all $n \geq m$. QED.

The principle of mathematical induction contains a technique to prove that infinitely many statements are true in just two steps. This proof technique is just what we need to prove our opening example about computing a sum of even natural numbers.

example **4.39 A Correct Formula**

Let's prove that the following equation is true for all natural numbers $n \geq 1$:

$$2 + 4 + \cdots + 2n = n(n + 1).$$

Proof: To see how to use (4.23), we can let $P(n)$ denote the above equation. Now we need to perform two steps. First, we have to show that $P(1)$ is true. Second, we have to assume that $P(k)$ is true and then prove that $P(k + 1)$ is true. When $n = 1$, the equation becomes the true statement

$$2 = 1(1 + 1).$$

Therefore, $P(1)$ is true. Now assume that $P(k)$ is true. This means that we assume that the following equation is true:

$$2 + 4 + \cdots + 2k = k(k + 1).$$

To prove that $P(k + 1)$ is true, start on the left side of the equation for the expression $P(k + 1)$:

$$
\begin{aligned}
2 + 4 + \cdots + 2k + 2\,(k+1) &= (2 + 4 + \cdots + 2k) + 2\,(k+1) \quad \text{(associate)} \\
&= k\,(k+1) + 2\,(k+1) \qquad\qquad \text{(assumption)} \\
&= (k+1)\,(k+2) \\
&= (k+1)\,[(k+1)+1]\,.
\end{aligned}
$$

The last term is the right-hand side of $P(k + 1)$. Thus $P(k + 1)$ is true. So we have performed both steps of (4.23). Therefore, $P(n)$ is true for all natural numbers $n \geq 1$. QED.

end example

example   4.40   A Correct Recursively Defined Function

We'll show that the following function computes $2 + 4 + \cdots + 2n$ for any natural number n:

$$
f(n) = \text{if } n = 0 \text{ then } 0 \text{ else } f(n-1) + 2n.
$$

Proof: For each $n \in \mathbb{N}$, let $P(n)$ be the statement "$f(n) = 2 + 4 + \cdots + 2n$." We want to show that $P(n)$ is true for all $n \in \mathbb{N}$. To start, notice that $f(0) = 0$. Thus $P(0)$ is true. Now assume that $P(k)$ is true for an arbitrary $k \geq 0$. To prove that $P(k + 1)$ is true, we'll start on the left side of $P(k + 1)$ to obtain

$$
\begin{aligned}
f\,(k+1) &= f\,(k+1-1) + 2\,(k+1) \qquad \text{(definition of } f) \\
&= f\,(k) + 2\,(k+1) \\
&= (2 + 4 + \cdots + 2k) + 2\,(k+1) \qquad \text{(assumption)} \\
&= 2 + 4 + \cdots + 2\,(k+1)\,.
\end{aligned}
$$

The last term is the right-hand side of $P(k + 1)$. Therefore, $P(k + 1)$ is true. So we have performed both steps of (4.23). It follows that $P(n)$ is true for all $n \in \mathbb{N}$. In other words, $f(n) = 2 + 4 + \cdots + 2n$ for all $n \in \mathbb{N}$. QED.

end example

A Classic Example: Arithmetic Progressions

When Gauss—mathematician Karl Friedrich Gauss (1777–1855)—was a 10-year-old boy, his schoolmaster, Buttner, gave the class an arithmetic progression of numbers to add up to keep them busy. We should recall that an *arithmetic progression* is a sequence of numbers where each number differs from its successor by the same constant. Gauss wrote down the answer just after Buttner finished writing the problem. Although the formula was known to Buttner, no child of 10 had ever discovered it.

For example, suppose we want to add up the seven numbers in the following arithmetic progression:

$$3, 7, 11, 15, 19, 23, 27.$$

The trick is to notice that the sum of the first and last numbers, which is 30, is the same as the sum of the second and next to last numbers, and so on. In other words, if we list the numbers in reverse order under the original list, each column totals to 30.

$$
\begin{array}{ccccccc}
3 & 7 & 11 & 15 & 19 & 23 & 27 \\
27 & 23 & 19 & 15 & 11 & 7 & 3 \\
\hline
30 & 30 & 30 & 30 & 30 & 30 & 30
\end{array}
$$

If S is the sum of the progression, then $2S = 7(30) = 210$. So $S = 105$.

The Sum of an Arithmetic Progression

The example illustrates a use of the following formula for the sum of an arithmetic progression of n numbers $a_1, a_2, \ldots, a_n$.

Sum of an Arithmetic Progression (4.24)

$$a_1 + a_2 + \cdots + a_n = \frac{n\,(a_1 + a_n)}{2}.$$

Proof: We'll prove it by induction. Let $P(n)$ denote Equation (4.24). We'll show that $P(n)$ is true for all natural numbers $n \geq 1$. Starting with $P(1)$, we obtain

$$a_1 = \frac{(a_1 + a_1)}{2}.$$

Since this equation is true, $P(1)$ is true. Next we'll assume that $P(n)$ is true, as stated in (4.24), and try to prove the statement $P(n + 1)$, which is

$$a_1 + a_2 + \cdots + a_n + a_{n+1} = \frac{(n + 1)\,(a_1 + a_{n+1})}{2}.$$

Since the progression $a_1, a_2, \ldots, a_{n+1}$ is arithmetic, there is a constant d such that it can be written in the following form, where $a = a_1$.

$$a, a + d, a + 2d, \ldots, a + nd.$$

In other words, $a_k = a + (k-1)d$ for $1 \le k \le n+1$. Starting with the left-hand side of the equation, we obtain

$$a_1 + a_2 + \cdots + a_n + a_{n+1} = (a_1 + a_2 + \cdots + a_n) + a_{n+1}$$

$$= \frac{n(a_1 + a_n)}{2} + a_{n+1} \qquad \text{(induction)}$$

$$= \frac{n(a + a + (n-1)d)}{2} + (a + nd) \quad \text{(write in terms of } d\text{)}$$

$$= \frac{2na + n(n-1)d + 2a + 2nd}{2}$$

$$= \frac{2a(n+1) + (n+1)nd}{2}$$

$$= \frac{(n+1)(2a + nd)}{2}$$

$$= \frac{(n+1)(a_1 + a_{n+1})}{2}.$$

Therefore, $P(n+1)$ is true. So by (4.23), Equation (4.24) is correct for all arithmetic progressions of n numbers. QED.

We should observe that (4.24) can be used to calculate the sum of the arithmetic progression $2, 4, \ldots, 2n$ in Example 4.38. The best known arithmetic progression is $1, 2, \ldots, n$ and we can use (4.24) to calculate its sum.

A Well-Known Sum (4.25)

$$1 + 2 + \cdots + n = \frac{n(n+1)}{2}.$$

A Classic Example: Geometric Progressions

A *geometric progression* is a sequence of numbers where the ratio of each number and its successor is the same constant, called the common ratio. For example, the following sequence is a geometric progression:

$$3, 6, 12, 24, 48, 96.$$

The common ratio for this progression is 2 and we can write it in the following form.

$$3, 3 \cdot 2, 3 \cdot 2^2, 2 \cdot 2^3, 2 \cdot 2^4, 2 \cdot 2^5.$$

Any geometric progression beginning with the number a can be written in the following form, where r is the common ratio.

$$a, ar, ar^2, ar^3, \ldots, ar^n.$$

The common ratio can be negative, too. For example, the following progression is geometric with $a = 3$ and $r = -2$:

$$3, -6, 12, -24, 48, -96.$$

The Sum of a Geometric Progression

We're interested in a formula for the sum of a geometric progression a, ar, ar^2, ..., ar^n, where $r \neq 1$. In other words, we want a formula for the sum

$$a + ar + ar^2 + \cdots + ar^n.$$

We can find a formula for the sum by multiplying the given expression by the term $r - 1$ to obtain the equation

$$(r - 1)(a + ar + ar^2 + \cdots + ar^n) = a(r^{n+1} - 1).$$

Now divide both sides by $r - 1$ to obtain the following formula for the sum of a geometric progression, where $r \neq 1$.

The Sum of a Geometric Progression (4.26)

$$a + ar + ar^2 + \cdots + ar^n = \frac{a(r^{n+1} - 1)}{r - 1}.$$

We'll give an induction proof of the formula.

Proof: If $n = 0$, then both sides are a. So assume that the formula is true for n, and prove that it is true for $n + 1$. Starting with the left-hand side, we have

$$
\begin{aligned}
a + ar + ar^2 + \cdots + ar^n + ar^{n+1} &= (a + ar + ar^2 + \cdots + ar^n) + ar^{n+1} \\
&= \frac{a(r^{n+1} - 1)}{r - 1} + ar^{n+1} \\
&= \frac{a(r^{n+1} - 1) + (r - 1)ar^{n+1}}{r - 1} \\
&= \frac{a(r^{(n+1)+1} - 1)}{r - 1}.
\end{aligned}
$$

Thus, by (4.23), the formula is true for all natural numbers n. QED.

The most popular geometric progression starts with $a = 1$. In this case we obtain the following sum.

A Well-Known Sum (4.27)

$$1 + r + r^2 + \cdots + r^n = \frac{r^{n+1} - 1}{r - 1}.$$

Sometimes, (4.23) does not have enough horsepower to do the job. For example, we might need to assume more than is allowed by (4.23), or we might be dealing with structures that are not numbers, such as lists, strings, or binary trees, and there may be no easy way to apply (4.23). The solution to many of these problems is a stronger version of induction based on well-founded sets. That's next.

4.4.2 Proof by Well-Founded Induction

Let's extend the idea of inductive proof to well-founded sets. Recall that a well-founded set is a poset whose nonempty subsets have minimal elements or, equivalently, every descending chain of elements is finite. We'll start by noticing an easy extension of (4.22) to the case of well-founded sets. If you aren't interested in why the method works, you can skip ahead to (4.29).

The Basis of Well-Founded Induction   **(4.28)**

Let W be a well-founded set, and let S be a subset of W satisfying the following two conditions.

1. S contains all the minimal elements of W.

2. Whenever an element $x \in W$ has the property that all its predecessors are elements of S, then $x \in S$.

Then $S = W$.

Proof: The proof is by contradiction. Suppose $S \neq W$. Then $W - S$ has a minimal element x. Since x is a minimal element of $W - S$, each predecessor of x cannot be in $W - S$. In other words, each predecessor of x must be in S. The second condition now forces us to conclude that $x \in S$. This is a contradiction, since we can't have both $x \in S$ and $x \in W - S$ at the same time. Therefore, $S = W$. QED.

You might notice that Condition 1 of (4.28) was not used in the proof. This is because it's a consequence of Condition 2 of (4.28). We'll leave this as an exercise (something about an element that doesn't have any predecessors). Condition 1 is stated explicitly because it helps to understand the ideas, and students are advised to begin an inductive proof by establishing it separately.

The Technique of Well-Founded Induction

Let's find a more practical form of (4.28) that gives us a technique for proving a collection of statements of the form $P(x)$ for each x in a well-founded set W. The technique is called *well-founded induction*.

Well-Founded Induction (4.29)

Let $P(x)$ be a statement for each x in the well-founded set W. To prove $P(x)$ is true for all $x \in W$, perform the following two steps:

1. Prove that $P(m)$ is true for all minimal elements $m \in W$.

2. Let x be an arbitrary element of W, and assume that $P(y)$ is true for all elements y that are predecessors of x. Prove that $P(x)$ is true.

Proof: Let $S = \{x \mid x \in W \text{ and } P(x) \text{ is true}\}$. Assume that we have performed the two steps of (4.29). Then S satisfies the hypothesis of (4.28). Therefore $S = W$. In other words, $P(x)$ is true for all $x \in W$. QED.

Second Principle of Mathematical Induction

Now we can state a corollary of (4.29), which lets us make a bigger assumption than we were allowed in (4.23). The principle is sometimes called "course-of-values induction."

Second Principle of Mathematical Induction (4.30)

Let $m \in \mathbb{Z}$. To prove that $P(n)$ is true for all integers $n \geq m$, perform the following two steps:

1. Prove that $P(m)$ is true.

2. Assume that n is an arbitrary integer $n > m$, and assume that $P(k)$ is true for all k in the interval $m \leq k < n$. Prove that $P(n)$ is true.

Proof: Let $W = \{n \mid n \geq m\}$. Notice that W is a well-founded set (actually well-ordered) whose least element is m. Let $S = \{n \mid n \in W \text{ and } P(n) \text{ is true}\}$. Assume that Steps 1 and 2 have been performed. Then $m \in S$, and if $n > m$ and all predecessors of n are in S, then $n \in S$. Therefore, $S = W$, by (4.29). QED.

example **4.41 Products of Primes**

We'll prove the following well-known result about prime numbers.

Every natural number $n \geq 2$ is prime or a product of prime numbers.

Proof: For $n \geq 2$, let $P(n)$ be the statement "n is prime or a product of prime numbers." We need to show that $P(n)$ is true for all $n \geq 2$. Since 2 is prime, it follows that $P(2)$ is true. So Step 1 of (4.30) is finished. For Step 2 we'll assume that $n > 2$ and $P(k)$ is true for $2 \leq k < n$. With this assumption we must show that $P(n)$ is true. If n is prime, then $P(n)$ is true. So assume that n is not prime.

Then $n = xy$, where $2 \leq x < n$ and $2 \leq y < n$. By our assumption, $P(x)$ and $P(y)$ are both true, which means that x and y are products of primes. Therefore, n is a product of primes. So $P(n)$ is true. Now (4.30) implies that $P(n)$ is true for all $n \geq 2$. QED.

Notice that we can't use (4.23) for the proof because its induction assumption is the single statement that $P(n-1)$ is true. We need the stronger assumption that $P(k)$ is true for $2 \leq k < n$ to allow us to say that $P(x)$ and $P(y)$ are true.

end example

Things You Must Do

Let's pause and make a few comments about inductive proof. Remember, when you are going to prove something with an inductive proof technique, there are always two distinct steps to be performed. First prove the basis case, showing that the statement is true for each minimal element. Now comes the second step. The most important part about this step is making an assumption. Let's write it down for emphasis.

You are required to make an assumption in the inductive step of a proof.

Some people find it hard to make assumptions. But inductive proof techniques require it. So if you find yourself wondering about what to do in an inductive proof, here are two questions to ask yourself: "Have I made an induction assumption?" If the answer is yes, ask the question, "Have I used the induction assumption in my proof?" Let's write it down for emphasis:

In the inductive step, MAKE AN ASSUMPTION and then USE IT.

Look at the previous examples and find the places where the basis case was proved, where the assumption was made, and where the assumption was used. Do the same thing as you read through the remaining examples.

4.4.3   A Variety of Examples

Now let's do some examples that do not involve numbers. Thus we'll be using well-founded induction (4.29). We should note that some people refer to well-founded induction as "structural induction" because well-founded sets can contain structures other than numbers, such as lists, strings, binary trees, and Cartesian products of sets. Whatever it's called, let's see how to use it.

example   **4.42   Correctness of MakeSet**

The following function is supposed to take any list K as input and return the list obtained by removing all repeated occurrences of elements from K:

$$\text{makeSet}\,(\langle\,\rangle) = \langle\,\rangle,$$
$$\text{makeSet}\,(a :: L) = \text{if isMember}\,(a, L) \text{ then makeSet}\,(L)$$
$$\text{else } a :: \text{makeSet}\,(L).$$

We'll assume that isMember correctly checks whether an element is a member of a list. Let $P(K)$ be the statement "makeSet(K) is a list obtained from K by removing its repeated elements." We'll prove that $P(K)$ is true for any list K.

Proof: We'll define a well-founded ordering on lists by letting $K \prec M$ mean length$(K) <$ length(M). So the basis element is $\langle\ \rangle$. The definition of makeSet tells us that makeSet$(\langle\ \rangle) = \langle\ \rangle$. Thus $P(\langle\ \rangle)$ is true. Next, we'll let K be an arbitrary nonempty list and assume that $P(L)$ is true for all lists $L \prec K$. In other words, we're assuming that makeSet(L) has no repeated elements for all lists $L \prec K$. We need to show that $P(K)$ is true. In other words, we need to show that makeSet(K) has no repeated elements. Since K is nonempty, we can write $K = a :: L$. There are two cases to consider. If isMember(a, L) is true, then the definition of makeSet gives

$$\text{makeSet}(K) = \text{makeSet}(a :: L) = \text{makeSet}(L).$$

Since $L \prec K$, it follows that $P(L)$ is true. Therefore $P(K)$ is true. If isMember(a, L) is false, then the definition of makeSet gives

$$\text{makeSet}(K) = \text{makeSet}(a :: L) = a :: \text{makeSet}(L).$$

Since $L \prec K$, it follows that $P(L)$ is true. Since isMember(a, L) is false, it follows that the list $a :: \text{makeSet}(L)$ has no repeated elements. Thus $P(K)$ is true. Therefore, (4.29) implies that $P(K)$ is true for all lists K. QED.

end example

example 4.43 Using a Lexicographic Ordering

We'll prove that the following function computes the number $|x - y|$ for any natural numbers x and y:

$$f(x,\ y) = \text{if } x = 0 \text{ then } y \text{ else if } y = 0 \text{ then } x \text{ else } f(x - 1,\ y - 1).$$

In other words, we'll prove that $f(x,\ y) = |x - y|$ for all $(x,\ y)$ in $\mathbb{N} \times \mathbb{N}$.

Proof: We'll use the well-founded set $\mathbb{N} \times \mathbb{N}$ with the lexicographic ordering. For the basis case, we'll check the formula for the least element $(0,\ 0)$ to get $f(0,\ 0) = 0 = |0 - 0|$. For the induction case, let $(x,\ y) \in \mathbb{N} \times \mathbb{N}$ and assume that $f(u,\ v) = |u - v|$ for all $(u,\ v) \prec (x,\ y)$. We must show $f(x,\ y) = |x - y|$. The case where $x = 0$ is taken care of by observing that $f(0,\ y) = y = |0 - y|$. Similarly, if $y = 0$, then $f(x,\ 0) = x = |x - 0|$. The only case remaining is $x \neq 0$ and $y \neq 0$. In this case the definition of f gives $f(x,\ y) = f(x - 1,\ y - 1)$. The lexicographic ordering gives $(x - 1,\ y - 1) \prec (x,\ y)$. So it follows by induction that $f(x - 1, y - 1) = |(x - 1) - (y - 1)|$. Putting the two equations together we obtain the following result.

$$
\begin{aligned}
f(x, y) &= f(x - 1, y - 1) & \text{(definition of } f) \\
&= |(x - 1) - (y - 1)| & \text{(induction assumption)} \\
&= |x - y|.
\end{aligned}
$$

The result now follows from (4.29). QED.

end example

Inductive Proof with One of Several Variables

Sometimes the claims that we wish to prove involve two or more variables, but we only need one of the variables in the proof. For example, suppose we need to show that $P(x, y)$ is true for all $(x, y) \in A \times B$ where the set A is inductively defined. We can perform the following steps, where y denotes an arbitrary element in B:

1. Show that $P(m, y)$ is true for minimal elements $m \in A$.

2. Assume that $P(a, y)$ is true for all predecessors a of x. Show that $P(x, y)$ is true.

The form of the statement $P(x, y)$ often gives us a clue as to whether we can use induction with a single variable. Here are some examples.

example 4.44 **Induction with a Single Variable**

Suppose we want to prove that the following function computes the number y^{x+1} for any natural numbers x and y:

$$f(x, y) = \text{if } x = 0 \text{ then } y \text{ else } f(x - 1, y) \cdot y.$$

In other words, we want to prove that $f(x, y) = y^{x+1}$ for all (x, y) in $\mathbb{N} \times \mathbb{N}$. We'll use induction with the variable x because it's changing in the definition of f.

Proof: For the basis case the definition of f gives $f(0, y) = y = y^{0+1}$. So the basis case is proved. For the induction case, assume that $x > 0$ and $f(n, y) = y^{n+1}$ for $n < x$. We must show that $f(x, y) = y^{x+1}$. The definition of f and the induction assumption give us the following equations:

$$
\begin{aligned}
f(x, y) &= f(x - 1, y) \cdot y & \text{(definition of } f) \\
&= y^{x-1+1} \cdot y & \text{(induction assumption)} \\
&= y^{x+1}.
\end{aligned}
$$

The result now follows from (4.29). QED.

end example

4.45 Inserting an Element in a Binary Search Tree

Let's prove that the following insert function does its job. Given a number x and a binary search tree T, the function returns a binary search tree obtained by inserting x in T.

$$\text{insert}\,(x, T) = \text{if } T = \langle\,\rangle \text{ then tree}\,(\langle\,\rangle, x, \langle\,\rangle)$$
$$\text{else if } x < \text{root}\,(T) \text{ then}$$
$$\text{tree}\,(\text{insert}\,(x, \text{left}\,(T)), \text{root}\,(T), \text{right}\,(T))$$
$$\text{else}$$
$$\text{tree}\,(\text{left}\,(T), \text{root}\,(T), \text{insert}\,(x, \text{right}\,(T))).$$

The claim that we wish to prove is,

insert(x, T) is a binary search tree for all binary search trees T.

Proof: We'll use induction on the binary tree variable T. Our ordering of binary search trees will be based on the number of nodes in a tree. The empty binary tree $\langle\,\rangle$ is a binary search tree with no nodes. In this case, we have insert(x, $\langle\,\rangle$) = tree($\langle\,\rangle$, x, $\langle\,\rangle$), which is a single node binary tree and thus also a binary search tree. So the basis case is true. Now let T be a nonempty binary search tree of the form $T = \text{tree}(L, y, R)$ and, since L and R each have fewer nodes than T, we can assume that insert(x, L) and insert(x, R) are binary search trees. With these assumptions we must show that insert(x, T) is a binary search tree. There are two cases to consider, depending on whether $x < y$. First, suppose $x < y$. Then we have

insert(x, T) = tree(insert(x, L), y, R).

By the induction assumption it follows that insert(x, L) is a binary search tree. Thus insert(x, T) is a binary search tree. We obtain a similar result if $x \geq y$. It follows from (4.29) that insert(x, T) is a binary search tree for all binary search trees T. QED.

On Using "Well-Founded"

We often see induction proofs that don't mention the word "well-founded." For example, we might see a statement such as: "We will use induction on the depth of the trees." In such a case the induction assumption might be stated something like "Assume that $P(T)$ is true for all trees T with depth less than n." Then a proof is given that uses the assumption to prove that $P(T)$ is true for an arbitrary tree of depth n. Even though the term "well-founded" may not be mentioned in a proof, there is always a well-founded ordering lurking underneath the surface.

Before we leave the subject of inductive proof, let's discuss how we can use inductive proof to help us tell whether inductive definitions of sets are correct.

Proofs about Inductively Defined Sets

Recall that a set S is inductively defined by a basis case, an inductive case, and a closure case (which we never state explicitly). The closure case says that S is the smallest set satisfying the basis and inductive cases. The closure case can also be stated in practical terms as follows.

Closure Property of Inductive Definitions   (4.31)

If S is an inductively defined set and T is a set that also satisfies the basis and inductive cases for the definition of S, and if $T \subseteq S$, then it must be the case that $T = S$.

We can use this closure property to see whether an inductive definition correctly characterizes a given set. For example, suppose we have an inductive definition for a set named S, we have some other description of a set named T, and we wish to prove that T and S are the same set. Then we must prove three things:

1. Prove that T satisfies the basis case of the inductive definition.

2. Prove that T satisfies the inductive case of the inductive definition.

3. Prove that $T \subseteq S$. This can often be accomplished with an induction proof.

example 4.46 Describing an Inductive Set

Suppose we have the following inductive definition for a set S:

Basis: $1 \in S$.

Induction: If $x \in S$, then $x + 2 \in S$.

This gives us a pretty good description of S. For example, suppose someone tells us that $S = \{2k + 1 \mid k \in \mathbb{N}\}$. It seems reasonable. Can we prove it? Let's give it a try. To clarify the situation, we'll let $T = \{2k + 1 \mid k \in \mathbb{N}\}$ and prove that $T = S$. We'll be done if we can show that T satisfies the basis and induction cases for S and that $T \subseteq S$. Then the closure property of inductive definitions will tell us that $T = S$.

Proof: Observe that $1 = 2 \cdot 0 + 1 \in T$ and if $x \in T$, then $x = 2k + 1$ and it follows that $x + 2 = 2(k + 1) + 1 \in T$. So T satisfies the inductive definition. Next we need to show that $T \subseteq S$. In other words, show that $2k + 1 \in S$ for all $k \in \mathbb{N}$. This calls for an induction proof. If $k = 0$, we have $2 \cdot 0 + 1 = 1 \in S$. Now assume that $2k + 1 \in S$ and show that $2(k + 1) + 1 \in S$. Since $2k + 1 \in S$, the inductive definition tells us that $(2k + 1) + 2 \in S$

so that we have $2(k + 1) + 1 = (2k + 1) + 2 \in S$. Therefore, $2k + 1 \in S$ for all $k \in \mathbb{N}$, which proves that $T \subseteq S$. So we've proven the three things that allow us to conclude—by the closure property of inductive definitions—that $T = S$. QED.

end example

example 4.47 A Correct Grammar

Suppose we're asked to find a grammar for the language $\{ab^n \mid n \in \mathbb{N}\}$, and we write the following grammar G.

$$S \to a \mid Sb.$$

This grammar seems to do the job. But how do we know for sure? One way is to use (3.15) to create an inductive definition for $L(G)$, the language of G. Then we can try to prove that $L(G) = \{ab^n \mid n \in \mathbb{N}\}$. Using (3.15) we see that the basis case is $a \in L(G)$ because of the derivation $S \Rightarrow a$.

For the induction case, if $x \in L(G)$ with derivation $S \Rightarrow^+ x$, then we can add one step to the derivation by using the recursive production $S \to Sb$ to obtain the derivation $S \Rightarrow Sb \Rightarrow^+ xb$. So we obtain the following inductive definition for $L(G)$.

Basis: $a \in L(G)$.

Induction: If $x \in L(G)$, then $xb \in L(G)$.

Now we'll prove that $\{ab^n \mid n \in \mathbb{N}\} = L(G)$. For ease of notation we'll let $M = \{ab^n \mid n \in \mathbb{N}\}$. So we must prove that $M = L(G)$. By (4.31) we must show that M satisfies the basis and induction cases and that $M \subseteq L(G)$. Then by the closure property of inductive definitions we will infer that $M = L(G)$.

Proof: Since $a = ab^0 \in M$, it follows that the basis case of the inductive definition holds. For the induction case, let $x \in M$. Then $x = ab^n$ for some number $n \in \mathbb{N}$. Thus $xb = ab^{n+1} \in M$. Therefore, M satisfies the inductive definition. Now we'll show that $M \subseteq L(G)$ with an induction proof. For $n = 0$, we have $ab^0 = a \in L(G)$. Now assume that $ab^n \in L(G)$. Then the definition of $L(G)$ tells us that $ab^n b \in L(G)$. But $ab^{n+1} = ab^n b$. So $ab^{n+1} \in L(G)$. Therefore, $M \subseteq L(G)$. Now the closure property of inductive definitions gives us our conclusion $M = L(G)$. QED.

end example

■ Exercises

Numbers

1. Find the sum of the arithmetic progression $12, 26, 40, 54, 68, \ldots, 278$.

2. Use induction to prove each of the following equations.

   a. $1 + 3 + 5 + \cdots + (2n - 1) = n^2$.

   b. $5 + 9 + 11 + \cdots + (2n + 3) = n^2 + 4n$.

   c. $3 + 7 + 11 + \cdots + (4n - 1) = n(2n + 1)$.

   d. $2 + 6 + 10 + \cdots + (4n - 2) = 2n^2$.

   e. $1 + 5 + 9 + \cdots + (4n + 1) = (n + 1)(2n + 1)$.

   f. $2 + 8 + 24 + \cdots + n2^n = (n - 1)2^{n+1} + 2$.

   g. $1^2 + 2^2 + \cdots + n^2 = \dfrac{n(n + 1)(2n + 1)}{6}$.

   h. $2 + 6 + 12 + \cdots + n(n + 1) = \dfrac{n(n + 1)(n + 2)}{3}$.

   i. $2 + 6 + 12 + \cdots + (n^2 - n) = \dfrac{n(n^2 - 1)}{3}$.

   j. $(1 + 2 + \cdots + n)^2 = 1^3 + 2^3 + \cdots + n^3$.

3. The *Fibonacci numbers* are defined by $F_0 = 0$, $F_1 = 1$, and $F_n = F_{n-1} + F_{n-2}$ for $n \geq 2$. Use induction to prove each of the following statements.

   a. $F_0 + F_1 + \cdots + F_n = F_{n+2} - 1$.

   b. $F_{n-1}F_{n+1} - F_n^2 = (-1)^n$.

   c. $F_{m+n} = F_{m-1}F_n + F_m F_{n+1}$. *Hint:* Use the lexicographic ordering of $\mathbb{N} \times \mathbb{N}$.

   d. If $m \mid n$ then $F_m \mid F_n$. *Hint:* Use the fact that $n = mk$ for some k and show the result by induction on k with the help of Part (c).

4. The *Lucas numbers* are defined by $L_0 = 2$, $L_1 = 1$, and $L_n = L_{n-1} + L_{n-2}$ for $n \geq 2$. The sequence begins as $2, 1, 3, 4, 7, 11, 18, \ldots$. These numbers are named after the mathematician Édouard Lucas (1842–1891). Use induction to prove each of the following statements.

   a. $L_0 + L_1 + \cdots + L_n = L_{n+2} - 1$.

   b. $L_n = F_{n-1} + F_{n+1}$ for $n \geq 1$, where F_n is the nth Fibonacci number.

5. Let $\text{sum}(n) = 1 + 2 + \cdots + n$ for all natural numbers n. Give an induction proof to show that the following equation is true for all natural numbers m and n: $\text{sum}(m + n) = \text{sum}(m) + \text{sum}(n) + mn$.

6. We know that $1 + 2 = 3$, $4 + 5 + 6 = 7 + 8$, and $9 + 10 + 11 + 12 = 13 + 14 + 15$. Show that we can continue these equations forever. *Hint:* The left side of each equation starts with a number of the form n^2. Formulate a general summation for each side, and then prove that the two sums are equal.

Structures

7. Let $R = \{(x,\ x + 1) \mid x \in \mathbb{N}\}$ and let L be the "less than" relation on $\mathbb{N}$. Prove that $t(R) = L$.

8. Use induction to prove that a finite set with n elements has 2^n subsets.

9. Use induction to prove that the function f computes the length of a list:

$$f(L) = \text{if } L = \langle\ \rangle \text{ then } 0 \text{ else } 1 + f(\text{tail}(L)).$$

10. Use induction to prove that each function performs its stated task.

    a. The function g computes the number of nodes in a binary tree:

$$g(T) = \text{if } T = \langle\ \rangle \text{ then } 0$$
$$\text{else } 1 + g(\text{left}(T)) + g(\text{right}(T)).$$

    b. The function h computes the number of leaves in a binary tree:

$$h(T) = \text{if } T = \langle\ \rangle \text{ then } 0$$
$$\text{else if } T = \text{tree}(\langle\ \rangle, x, \langle\ \rangle) \text{ then } 1$$
$$\text{else } h(\text{left}(T)) + h(\text{right}(T)).$$

11. Suppose we have the following two procedures to write out the elements of a list. One claims to write the elements in the order listed, and one writes out the elements in reverse order. Prove that each is correct.

    a. forward(L): if $L \neq \langle\ \rangle$ then {print(head(L)); forward(tail(L))}.

    b. back(L): if $L \neq \langle\ \rangle$ then {back(tail(L)); print(head(L))}.

12. The following function "sort" takes a list of numbers and returns a sorted version of the list (from lowest to highest), where "insert" places an element correctly into a sorted list:

$$\text{sort}(\langle\ \rangle) = \langle\ \rangle,$$
$$\text{sort}(x :: L) = \text{insert}(x, \text{sort}(L)).$$

    a. Assume that the function insert is correct. That is, if S is sorted, then insert(x, S) is also sorted. Prove that sort is correct.

    b. Prove that the following definition for insert is correct. That is, prove that insert(x, S) is sorted for all sorted lists S.

$$\text{insert}(x, S) = \text{if } S = \langle\ \rangle \text{ then } \langle x \rangle$$
$$\text{else if } x \leq \text{head}(S) \text{ then } x :: S$$
$$\text{else head}(S) :: \text{insert}(x, \text{tail}(S)).$$

13. Show that the following function g correctly computes the greatest common divisor for each pair of positive integers x and y: *Hint:* (2.2a and 2.2b) might be useful.

$$g(x, y) = \text{if } x = y \text{ then } x$$
$$\text{else if } x > y \text{ then } g(x - y, y)$$
$$\text{else } g(x, y - x).$$

14. The following program is supposed to input a list of numbers L and output a binary search tree containing the numbers in L:

$$f(L) = \text{if } L = \langle\, \rangle \text{ then } \langle\, \rangle$$
$$\text{else insert}(\text{head}(L), f(\text{tail}(L))).$$

Assume that insert$(x,\ T)$ correctly returns the binary search tree obtained by inserting the number x in the binary search tree T. Prove the following claim: $f(M)$ is a binary search tree for all lists M.

15. The following program is supposed to return the list obtained by removing the first occurrence of x from the list L.

$$\text{delete}(x, L) = \text{if } L = \langle\, \rangle \text{ then } \langle\, \rangle$$
$$\text{else if } x = \text{head}(L) \text{ then tail}(L)$$
$$\text{else head}(L) :: \text{delete}(x, \text{tail}(L)).$$

Prove that delete performs as expected.

16. The following function claims to remove all occurrences of an element from a list:

$$\text{removeAll}(a, L) = \text{if } L = \langle\, \rangle \text{ then } L$$
$$\text{else if } a = \text{head}(L) \text{ then removeAll}(a, \text{tail}(L))$$
$$\text{else head}(L) :: \text{removeAll}(a, \text{tail}(L)).$$

Prove that removeAll satisfies the claim.

17. Let r stand for the removeAll function from Exercise 16. Prove the following property of r for all elements a, b, and all lists L:

$$r(a, r(b,\ L)) = r(b, r(a,\ L)).$$

18. The following program computes a well-known function called *Ackermann's function. Note:* If you try out this function, don't let x and y get too large.

$$f(x, y) = \text{if } x = 0 \text{ then } y + 1$$
$$\text{else if } y = 0 \text{ then } f(x - 1, 1)$$
$$\text{else } f(x - 1, f(x, y - 1)).$$

Prove that f is defined for all pairs (x, y) in $\mathbb{N} \times \mathbb{N}$. *Hint:* Use the lexicographic ordering on $\mathbb{N} \times \mathbb{N}$. This gives the single basis element $(0, 0)$. For the induction assumption, assume that $f(x', y')$ is defined for all (x', y') such that $(x', y') \prec (x, y)$. Then show that $f(x, y)$ is defined.

19. Let the function "isMember" be defined as follows for any list L:

$$\text{isMember}\,(a, L) = \text{if } L = \langle\,\rangle \text{ then False}$$
$$\text{else if } a = \text{head}\,(L) \text{ then True}$$
$$\text{else isMember}\,(a, \text{tail}\,(L))\,.$$

    a. Prove that isMember is correct. That is, show that isMember(a, L) is true if and only if a occurs as an element of L.

    b. Prove that the following equation is true for all lists L when $a \neq b$, where removeAll is the function from Exercise 16:

$$\text{isMember}(a, \text{removeAll}(b, L)) = \text{isMember}(a, L).$$

20. Use induction to prove that the following concatenation function is associative.

$$\text{cat}\,(x, y) = \text{if } x = \langle\,\rangle \text{ then } y$$
$$\text{else head}\,(x) :: \text{cat}\,(\text{tail}\,(x)\,, y)\,.$$

In other words, show that cat(x, cat(y, z)) = cat(cat(x, y), z) for all lists x, y, and z.

21. Two students came up with the following two solutions to a problem. Both students used the removeAll function from Exercise 16, which we abbreviate to r.

    Student A: $f(L) = \text{if } L = \langle\,\rangle \text{ then } \langle\,\rangle$
                        else head(L) :: r(head(L), f(tail(L))).
    Student B: $g(L) = \text{if } L = \langle\,\rangle \text{ then } \langle\,\rangle$
                        else head(L) :: g(r(head(L), tail(L))).

    a. Prove that $r(a, g(L)) = g(r(a, L))$ for all elements a and all lists L. *Hint:* Exercise 17 might be useful in the proof.

    b. Prove that $f(L) = g(L)$ for all lists L. *Hint:* Part (a) could be helpful.

    c. Can you find an appropriate name for f and g? Can you prove that the name you choose is correct?

Challenges

22. Prove that Condition 1 of (4.28) is a consequence of Condition 2 of (4.28).

23. Let G be the grammar $S \rightarrow a \mid abS$, and let $M = \{(ab)^n a \mid n \in \mathbb{N}\}$. Use (3.15) to construct an inductive definition for $L(G)$. Then use (4.31) to prove that $M = L(G)$.

24. A useful technique for recursively defined functions involves keeping—or accumulating—the results of function calls in *accumulating parameters*: The values in the accumulating parameters can then be used to compute subsequent values of the function that are then used to replace the old values in the accumulating parameters. We call the function by giving initial values to the accumulating parameters. Often these initial values are basis values for an inductively defined set of elements.

    For example, suppose we define the function f as follows:

    $$f(n,\, u,\, v) = \text{if } n = 0 \text{ then } u \text{ else } f(n-1,\, v,\, u+v).$$

    The second and third arguments to f are accumulating parameters because they always hold two possible values of the function. Prove each of the following statements.

    a. $f(n, 0, 1) = F_n$, the nth Fibonacci number.
    b. $f(n, 2, 1) = L_n$, the nth Lucas number.

    Hint: For Part (a), show that $f(n, 0, 1) = f(k, F_{n-k}, F_{n-k+1})$ for $0 \leq k \leq n$. A similar hint applies to Part (b).

25. A *derangement* of a string is an arrangement of the letters of the string such that no letter remains in the same position. In terms of bijections, a derangement of a set S is a bijection f on S such that $f(x) \neq x$ for all x in S. The number of derangements of an n-element set can be given by the following recursively defined function:

    $$d(1) = 0,$$
    $$d(2) = 1,$$
    $$d(n) = (n-1)(d(n-1) + d(n-2)) \quad (n \geq 3).$$

    Give an inductive proof that $d(n) = nd(n-1) + (-1)^n$ for $n \geq 2$.

4.5 Chapter Summary

Binary relations are common denominators for describing the ideas of equivalence, order, and inductive proof. The basic properties that a binary relation may or may not possess are reflexive, symmetric, transitive, irreflexive, and antisymmetric. Binary relations can be constructed from other binary relations by composition and closure, and by the usual set operations. Transitive closure plays an important part in algorithms for solving path problems—Warshall's algorithm, Floyd's algorithm, and the modification of Floyd's algorithm to find shortest paths.

Equivalence relations are characterized by being reflexive, symmetric, and transitive. These relations generalize the idea of basic equality by partitioning a set into classes of equivalent elements. Any set has a hierarchy of partitions ranging from fine to coarse. Equivalence relations can be generated from other relations by taking the transitive symmetric reflexive closure. They can also be generated from functions by the kernel relation. The equivalence problem can be solved by a novel tree structure. Kruskal's algorithm uses an equivalence relation to find a minimal spanning tree for a weighted undirected graph.

Order relations are characterized by being transitive and antisymmetric. Sets with these properties are called posets—for partially ordered sets—because it may be the case that not all pairs of elements are related. The ideas of successor and predecessor apply to posets. Posets can also be topologically sorted. A well-founded poset is characterized by the condition that no descending chain of elements can go on forever. This is equivalent to the condition that any nonempty subset has a minimal element. Well-founded sets can be constructed by mapping objects into a known well-founded set such as the natural numbers. Inductively defined sets are well-founded.

Inductive proof is a powerful technique that can be used to prove infinitely many statements. The most basic inductive proof technique is the principle of mathematical induction. Another useful inductive proof technique is well-founded induction. The important thing to remember about applying inductive proof techniques is to *make an assumption* and then *use the assumption* that you made. Inductive proof techniques can be used to prove properties of recursively defined functions and inductively defined sets.

chapter 5

Analysis
Techniques

Remember that time is money.

—Benjamin Franklin (1706–1790)

Time and space are important words in computer science because we want fast
algorithms and we want algorithms that don't use a lot of memory. The purpose
of this chapter is to study some fundamental techniques and tools that can be
used to analyze algorithms for the time and space that they require. Although the
study of algorithm analysis is beyond our scope, we'll give a brief introduction
to help show the need for the mathematical topics of the chapter.

The topics include techniques for finding closed forms or approximations
for sums, counting permutations and combinations, discrete probability, solving
recurrences, and comparing functions by rates of growth. Along the way we'll
give examples to show how the topics apply to analyzing algorithms.

chapter guide

5.1 (Analyzing Algorithms) introduces techniques for finding the worst-case performance of an algorithm and for finding an optimal algorithm.

5.2 (Summations and Closed Forms) introduces techniques for finding closed
forms for sums and for approximating sums that may not have a closed
form.

5.3 (Permutations and Combinations) introduces basic counting techniques for
permutations and combinations.

5.4 (Discrete Probability) introduces the ideas of discrete probability. We'll apply the results to the average-case performance of algorithms. We'll also
introduce Markov chains and apply the results to software development.

5.5 (Solving Recurrences) introduces techniques for solving recurrences that crop
up when analyzing an algorithm. We'll see solutions to recurrences that
arise from some divide-and-conquer algorithms. We'll introduce generating
functions and see how they can be used to solve recurrences.

5.6 (Comparing Rates of Growth) introduces approximation techniques for comparing the rates of growth of functions. We'll apply the results to functions that describe approximate running times of algorithms. We'll see some approximations for solutions to a variety of divide-and-conquer recurrences.

5.1 Analyzing Algorithms

An important question of computer science is: Can you convince another person that your algorithm is efficient? This takes some discussion. Let's start by stating the following problem.

The Optimal Algorithm Problem

Suppose algorithm A solves problem P. Is A the best solution to P?

What does "best" mean? Two typical meanings are *least time* and *least space*. In either case, we still need to clarify what it means for an algorithm to solve a problem in the least time or the least space. For example, an algorithm running on two different machines may take different amounts of time. Do we have to compare A to every possible solution of P on every type of machine? This is impossible. So we need to make a few assumptions in order to discuss the optimal algorithm problem. We'll concentrate on "least time" as the meaning of "best" because time is the most important factor in most computations.

5.1.1 Worst-Case Running Time

Instead of executing an algorithm on a real machine to find its running time, we'll analyze the algorithm by counting the number of certain operations that it will perform when executed on a real machine. In this way we can compare two algorithms by simply comparing the number of operations of the same type that each performs. If we make a good choice of the type of operations to count, we should get a good measure of an algorithm's performance. For example, we might count addition operations and multiplication operations for a numerical problem. On the other hand, we might choose to count comparison operations for a sorting problem.

The number of operations performed by an algorithm usually depends on the size or structure of the input. The size of the input again depends on the problem. For example, for a sorting problem, "size" usually means the number of items to be sorted. Sometimes inputs of the same size can have different structures that affect the number of operations performed. For example, some sorting algorithms perform very well on an input data set that is all mixed up but perform badly on an input set that is already sorted!

Because of these observations, we need to define the idea of a worst-case input for an algorithm A. An input of size n is a *worst-case input* if, when compared to all other inputs of size n, it causes A to execute the largest number of operations. Now let's get down to business. For any input I we'll denote its size by size(I), and we'll let time(I) denote the number of operations executed by A on I. Then the *worst-case function* for A is defined as follows:

$$W_A(n) = \max\{\text{time}(I) \mid I \text{ is an input and size}(I) = n\}.$$

Now let's discuss comparing different algorithms that solve a problem P. We'll always assume that the algorithms we compare use certain specified operations that we intend to count. If A and B are algorithms that solve P and if $W_A(n) \leq W_B(n)$ for all $n > 0$, then we know algorithm A has worst-case performance that is better than or equal to that of algorithm B. This gives us the proper tool to describe the idea of an optimal algorithm.

Definition of Optimal in the Worst Case
An algorithm A is *optimal in the worst case* for problem P, if for any algorithm B that exists, or ever will exist, the following relationship holds:

$$W_A(n) \leq W_B(n) \text{ for all } n > 0.$$

How in the world can we ever find an algorithm that is optimal in the worst case for a problem P? The answer involves the following three steps:

1. (Find an algorithm) Find or design an algorithm A to solve P. Then do an analysis of A to find the worst-case function W_A.

2. (Find a lower bound) Find a function F such that $F(n) \leq W_B(n)$ for all $n > 0$ and for all algorithms B that solve P.

3. Compare F and W_A. If $F = W_A$, then A is optimal in the worst case.

Suppose we know that $F \neq W_A$ in Step 3. This means that $F(n) < W_A(n)$ for some n. In this case there are two possible courses of action to consider:

1. Put on your construction hat and try to build a new algorithm C such that $W_C(n) \leq W_A(n)$ for all $n > 0$.

2. Put on your analysis hat and try to find a new function G such that $F(n) \leq G(n) \leq W_B(n)$ for all $n > 0$ and for all algorithms B that solve P.

We should note that zero is always a lower bound, but it's not very interesting because most algorithms take more than zero time. A few problems have optimal algorithms. For the vast majority of problems that have solutions, optimal algorithms have not yet been found. The examples contain both kinds of problems.

example **5.1** **Matrix Multiplication**

We can "multiply" two n by n matrices A and B to obtain the product AB, which is the n by n matrix defined by letting the element in the ith row and jth column of AB be the value of the expression

$$A_{i1}B_{1j} + A_{i2}B_{2j} + \cdots + A_{in}B_{nj}.$$

For example, let A and B be the following 2 by 2 matrices:

$$A = \begin{pmatrix} a & b \\ c & d \end{pmatrix}, \quad B = \begin{pmatrix} e & f \\ g & h \end{pmatrix}.$$

The product AB is the following 2 by 2 matrix:

$$AB = \begin{pmatrix} ae + bg & af + bh \\ ce + dg & cf + dh \end{pmatrix}.$$

Notice that the computation of AB takes eight multiplications and four additions. The definition of matrix multiplication of two n by n matrices uses n^3 multiplication operations and $n^2(n-1)$ addition operations.

A known lower bound for the number of multiplication operations needed to multiply two n by n matrices is n^2. Strassen [1969] showed how to multiply two matrices with about $n^{2.81}$ multiplication operations. The number 2.81 is an approximation to the value of $\log_2(7)$. It stems from the fact that a pair of 2 by 2 matrices can be multiplied by using seven multiplication operations.

Multiplication of larger-size matrices is broken down into multiplying many 2 by 2 matrices. Therefore, the number of multiplication operations becomes less than n^3. This revelation got research going in two camps. One camp is trying to find a better algorithm. The other camp is trying to raise the lower bound above n^2. Pan [1978] gave an algorithm to multiply two 70×70 matrices using 143,640 multiplications, which is less than $70^{2.81}$ multiplication operations. Coppersmith and Winograd [1987] gave an algorithm that, for large values of n, uses $n^{2.376}$ multiplication operations. So it goes.

end example

example **5.2** **Finding the Minimum**

Let's examine an optimal algorithm to find the minimum number in an unsorted list of n numbers. We'll count the number of comparison operations that an algorithm makes between elements of the list. To find the minimum number in a list of n numbers, the minimum number must be compared with the other $n-1$ numbers. Therefore, $n-1$ is a lower bound on the number of comparisons needed to find the minimum number in a list of n numbers.

If we represent the list as an array a indexed from 1 to n, then the following algorithm is optimal because the operation $\le$ is executed exactly $n-1$ times.

$$m := a[1];$$
$$\textbf{for } i := 2 \textbf{ to } n \textbf{ do}$$
$$\quad m := \text{if } m \leq a[i] \text{ then } m \text{ else } a[i]$$
$$\textbf{od}$$

end example

5.1.2 Decision Trees

We can often use a tree to represent the decision processes that take place in an algorithm. A *decision tree* for an algorithm is a tree whose nodes represent decision points in the algorithm and whose leaves represent possible outcomes. Decision trees can be useful in trying to construct an algorithm or trying to find properties of an algorithm. For example, lower bounds may equate to the depth of a decision tree.

If an algorithm makes decisions based on the comparison of two objects, then it can be represented by a *binary decision tree*. Each nonleaf node in the tree represents a pair of objects to be compared by the algorithm, and each branch from that node represents a path taken by the algorithm based on the comparison. Each leaf can represent an outcome of the algorithm. A *ternary decision tree* is similar except that each nonleaf node represents a comparison that has three possible outcomes.

Lower Bounds for Decision Tree Algorithms

Let's see whether we can compute lower bounds for decision tree algorithms. If a decision tree has depth d, then some path from the root to a leaf contains $d + 1$ nodes. Since the leaf is a possible outcome, it follows that there are d decisions made on the path. Since no other path from the root to a leaf can have more than $d + 1$ nodes, it follows that d is the worst-case number of decisions made by the algorithm.

Now, suppose that a problem has n possible outcomes and it can be solved by a binary decision tree algorithm. What is the best binary decision tree algorithm? We may not know the answer, but we can find a lower bound for the depth of any binary decision tree to solve the problem. Since the problem has n possible outcomes, it follows that any binary decision tree algorithm to solve the problem must have at least n leaves, one for each of the n possible outcomes. Recall that the number of leaves in a binary tree of depth d is at most 2^d.

So if d is the depth of a binary decision tree to solve a problem with n possible outcomes, then we must have $n \leq 2^d$. We can solve this inequality for d by taking $\log_2$ of both sides to obtain $\log_2 n \leq d$. Since d is a natural number, it follows that

$$\lceil \log_2 n \rceil \leq d.$$

In other words, any binary decision tree algorithm to solve a problem with n possible outcomes must have a depth of at least $\lceil \log_2 n \rceil$.

We can do the same analysis for ternary decision trees. The number of leaves in a ternary tree of depth d is at most 3^d. If d is the depth of a ternary decision tree to solve a problem with n possible outcomes, then we must have $n \leq 3^d$. Solve the inequality for d to obtain

$$\lceil \log_3 n \rceil \leq d.$$

In other words, any ternary decision tree algorithm to solve a problem with n possible outcomes must have a depth of at least $\lceil \log_3 n \rceil$.

Many sorting and searching algorithms can be analyzed with decision trees because they perform comparisons. Let's look at some examples to illustrate the idea.

example   **5.3   Binary Search**

Suppose we search a sorted list in a binary fashion. That is, we check the middle element of the list to see whether it's the key we are looking for. If not, then we perform the same operation on either the left half or the right half of the list, depending on the value of the key. This algorithm has a nice representation as a decision tree. For example, suppose we have the following sorted list of 15 numbers:

$$x_1, \ x_2, \ x_3, \ x_4, \ x_5, \ x_6, \ x_7, \ x_8, \ x_9, \ x_{10}, \ x_{11}, \ x_{12}, \ x_{13}, \ x_{14}, \ x_{15}.$$

Suppose we're given a number key K, and we must find whether it is in the list. The decision tree for a binary search of the list has the number x_8 at its root. This represents the comparison of K with x_8. If $K = x_8$, then we are successful in one comparison. If $K < x_8$, then we go to the left child of x_8; otherwise we go to the right child of x_8. The result is a ternary decision tree in which the leaves are labeled with either S, for successful search, or U, for unsuccessful search. The decision tree is pictured in Figure 5.1.

Since the depth of the tree is 4, it follows that there will be four comparisons in the worst case to find whether K is in the list. Is this an optimal algorithm? To see that the answer is yes, we can observe that there are 31 possible outcomes for the given problem: 15 leaves labeled with S to represent successful searches; and 16 leaves labeled with U to represent the gaps where $K < x_1$, $x_i < K < x_{i+1}$ for $1 \leq i < 15$, and $x_{15} < K$. Therefore, a worst-case lower bound for the number of comparisons is $\lceil \log_3 31 \rceil = 4$. Therefore, the algorithm is optimal.

end example

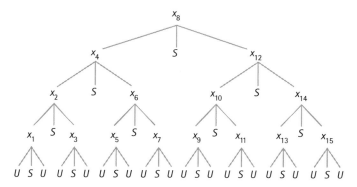

Figure 5.1   Decision tree for binary search.

example 5.4 **Finding a Bad Coin**

Suppose we are asked to use a pan balance to find the heavy coin among eight coins with the assumption that they all look alike and the other seven all have the same weight. One way to proceed is to always place coins in the two pans so that the bad coin is included and thus one pan will always go down.

This gives a binary decision tree, where each internal node of the tree represents the pan balance. If the left side goes down, then the heavy coin is on the left side of the balance. Otherwise, the heavy coin is on the right side of the balance. Each leaf represents one coin that is the heavy coin. Suppose we label the coins with the numbers $1, 2, \ldots, 8$.

The decision tree for one algorithm is shown in Figure 5.2, where the numbers on either side of a nonleaf node represent the coins on either side of the pan balance. This algorithm finds the heavy coin in three weighings. Can we do any better? Look at the next example.

end example

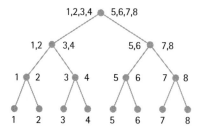

Figure 5.2   A binary decision tree.

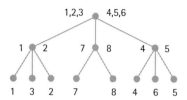

Figure 5.3 An optimal decision tree.

example 5.5 An Optimal Solution

The problem is the same as in Example 5.4. We are asked to use a pan balance to find the heavy coin among eight coins with the assumption that they all look alike and the other seven all have the same weight. In this case we'll weigh coins with the possibility that the two pans are balanced. So a decision tree can have nodes with three children.

We don't have to use all eight coins on the first weighing. For example, Figure 5.3 shows the decision tree for one algorithm. Notice that there is no middle branch on the middle subtree because, at this point, one of the coins, 7 or 8, must be the heavy one. This algorithm finds the heavy coin in two weighings.

This algorithm is an optimal pan-balance algorithm for the problem, where we are counting the number of weighings to find the heavy coin. To see this, notice that any one of the eight coins could be the heavy one. Therefore, there must be at least eight leaves on any algorithm's decision tree. Notice that a binary tree of depth d can have 2^d possible leaves. So to get eight leaves, we must have $2^d \geq 8$. This implies that $d \geq 3$. But a ternary tree of depth d can have 3^d possible leaves. So to get eight leaves, we must have $3^d \geq 8$, or $d \geq 2$. Therefore, 2 is a lower bound for the number of weighings. Since the algorithm solves the problem in two weighings, it is optimal.

end example

example 5.6 A Lower Bound Computation

Suppose we have a set of 13 coins in which at most one coin is bad and a bad coin may be heavier or lighter than the other coins. The problem is to use a pan balance to find the bad coin if it exists and say whether it is heavy or light. We'll find a lower bound on the heights of decision trees for pan-balance algorithms to solve the problem.

Any solution must tell whether a bad coin is heavy or light. Thus there are 27 possible outcomes: no bad coin and the 13 pairs of outcomes (ith coin light, ith coin heavy). Therefore, any decision tree for the problem must have at least 27 leaves. So a ternary decision tree of depth d must satisfy $3^d \geq 27$, or $d \geq 3$. This gives us a lower bound of 3.

Now the big question: Is there an algorithm to solve the problem, where the decision tree of the algorithm has depth 3? The answer is no. Just look at the cases of different initial weighings, and note in each case that the remaining possible outcomes cannot be distinguished with just two more weighings. Thus any decision tree for this problem must have depth 4 or more.

end example

Exercises

1. Draw a picture of the decision tree for an optimal algorithm to find the maximum number in the list x_1, x_2, x_3, x_4.

2. Suppose there are 95 possible answers to some problem. For each of the following types of decision tree, find a reasonable lower bound for the number of decisions necessary to solve the problem.

   a. Binary tree.   **b.** Ternary tree.   **c.** Four-way tree.

3. Find a nonzero lower bound on the number of weighings necessary for any ternary pan-balance algorithm to solve the following problem: A set of 30 coins contains at most one bad coin, which may be heavy or light. Is there a bad coin? If so, state whether it's heavy or light.

4. Find an optimal pan-balance algorithm to find a bad coin, if it exists, from 12 coins, where at most one coin is bad (i.e., heavier or lighter than the others). *Hint:* Once you've decided on the coins to weigh for the root of the tree, then the coins that you choose at the second level should be the same coins for all three branches of the tree.

5.2 Summations and Closed Forms

In trying to count things we often come up with expressions or relationships that need to be simplified to a form that can be easily computed with familiar operations.

Definition of Closed Form

A *closed form* is an expression that can be computed by applying a fixed number of familiar operations to the arguments. A closed form can't have an ellipsis because the number of operations to evaluate the form would not be fixed. For example, the expression $n(n+1)/2$ is a closed form, but the expression $1 + 2 + \ldots + n$ is not a closed form. In this section we'll see some ways to find closed forms for sums.

5.2.1 Basic Summations and Closed Forms

When we count things, we most often add things together. When evaluating a sum of two or more expressions, it sometimes helps to write the sum in a different form. Here are six examples to demonstrate some simple relationships.

$$c + c + c = 3c$$
$$ca_1 + ca_2 + ca_3 = c(a_1 + a_2 + a_3)$$
$$(a_1 - a_0) + (a_2 - a_1) + (a_3 - a_2) = (a_3 - a_0)$$
$$(a_0 - a_1) + (a_1 - a_2) + (a_2 - a_3) = (a_0 - a_3)$$
$$(a_1 + a_2 + a_3) + (b_1 + b_2 + b_3) = (a_1 + b_1) + (a_2 + b_2) + (a_3 + b_3)$$
$$(a_1x + a_2x^2 + a_3x^3) = x(a_1 + a_2x + a_3x^2)$$

To represent the sum of n terms $a_1, a_2, \ldots, a_n$ we use the following notation:

$$\sum_{k=1}^{n} a_k = a_1 + a_2 + \ldots + a_n.$$

The following list uses this notation to give some basic properties of sums, six of which generalize the preceding examples, and all of which are easily verified.

Summation Facts   **(5.1)**

a. $\displaystyle\sum_{k=m}^{n} c = (n - m + 1)c.$ (sum of a constant)

b. $\displaystyle\sum_{k=m}^{n} c\, a_k = c \sum_{k=m}^{n} a_k.$

c. $\displaystyle\sum_{k=1}^{n}(a_k - a_{k-1}) = a_n - a_0$ and $\displaystyle\sum_{k=1}^{n}(a_{k-1} - a_k) = a_0 - a_n.$ (collapsing sums)

d. $\displaystyle\sum_{k=m}^{n}(a_k + b_k) = \sum_{k=m}^{n} a_k + \sum_{k=m}^{n} b_k.$

e. $\displaystyle\sum_{k=m}^{n} a_k = \sum_{k=m}^{i} a_k + \sum_{k=i+1}^{n} a_k.$ $(m \le i < n)$

f. $\displaystyle\sum_{k=m}^{n} a_k x^{k+i} = x^i \sum_{k=m}^{n} a_k x^{k}.$

g. $\displaystyle\sum_{k=m}^{n} a_{k+i} = \sum_{k=m+i}^{n+i} a_k.$ (change index limits)

These facts are very useful in manipulating sums into simpler forms from which we might be able to find closed forms. In the following examples, we'll derive some closed forms that have many applications.

Sums of Powers

A closed form for the sum of powers $\sum_{k=1}^{n} k^m$ can be derived from closed forms for the sums of powers less than m. We'll use a technique that begins by considering the simple expression

$$k^{m+1} - (k-1)^{m+1}. \tag{5.2}$$

Expand this expression and collect terms to obtain a polynomial of the following form, where $c_m \neq 0$.

$$k^{m+1} - (k-1)^{m+1} = c_0 + c_1 k + c_2 k^2 + \cdots + c_m k^m. \tag{5.3}$$

Now comes the interesting part. We take the sum of both sides of (5.3) from 1 to n and observe two things about the resulting equation. The sum of the left side collapses to obtain

$$\sum_{k=1}^{n} \left(k^{m+1} - (k-1)^{m+1} \right) = n^{m+1}.$$

The sum on the right side becomes

$$\sum_{k=1}^{n} (c_0 + c_1 k + c_2 k^2 + \cdots + c_m k^m) =$$

$$c_0 \sum_{k=1}^{n} 1 + c_1 \sum_{k=1}^{n} k + c_2 \sum_{k=1}^{n} k^2 + \cdots + c_m \sum_{k=1}^{n} k^m.$$

So by summing both sides of (5.3) we obtain the equation

$$n^{m+1} = c_0 \sum_{k=1}^{n} 1 + c_1 \sum_{k=1}^{n} k + c_2 \sum_{k=1}^{n} k^2 + \cdots + c_m \sum_{k=1}^{n} k^m.$$

If we know the closed forms for the sums of powers less than m, we can substitute them for the sums in the equation, which results in an equation with just the sum $\sum_{k=1}^{n} k^m$ as the unknown quantity. Put it on one side of the equation by itself and its closed form will be sitting on the other side. We'll do some samples to demonstrate the method.

example **5.7 Closed Forms for Sums of Low Powers**

A closed form for the sum of the arithmetic progression 1, 2, ..., n is given in (4.25). We can write the sum and its closed form as

$$\sum_{k=1}^{n} k = \frac{n(n+1)}{2}. \tag{5.4}$$

To demonstrate this method we'll derive the closed form (5.4). So we begin with the expression

$$k^2 - (k-1)^2.$$

Expand the expression to obtain the equation

$$k^2 - (k-1)^2 = k^2 - k^2 + 2k - 1 = 2k - 1 = -1 + 2k.$$

Next we sum both sides from 1 to n and observe the resulting equation:

$$\sum_{k=1}^{n} (k^2 - (k-1)^2) = \sum_{k=1}^{n} (-1 + 2k). \tag{5.5}$$

The sum on the left side of (5.5) collapses to become

$$\sum_{k=1}^{n} (k^2 - (k-1)^2) = n^2 - 0^2 = n^2.$$

The sum on the right side of (5.5) becomes the following, where we use the known closed form (5.1a) for the sum of a constant.

$$\sum_{k=1}^{n} (-1 + 2k) = \sum_{k=1}^{n} (-1) + \sum_{k=1}^{n} 2k = -n + 2\sum_{k=1}^{n} k.$$

So Equation (5.5) becomes

$$n^2 = -n + 2\sum_{k=1}^{n} k.$$

Now solve for $\sum_{k=1}^{n} k$ to obtain the closed form in (5.4).

To make sure we have this method down, we'll find a closed form for $\sum_{k=1}^{n} k^2$. So we'll begin with the expression

$$k^3 - (k-1)^3.$$

Expand the expression to obtain the equation

$$k^3 - (k-1)^3 = k^3 - k^3 + 3k^2 - 3k + 1 = 1 - 3k + 3k^2.$$

Next we sum both sides from 1 to n and observe the resulting equation:

$$\sum_{k=1}^{n} (k^3 - (k-1)^3) = \sum_{k=1}^{n} (1 - 3k + 3k^2). \tag{5.6}$$

The sum on the left side of (5.6) collapses to become

$$\sum_{k=1}^{n} (k^3 - (k-1)^3) = n^3 - 0^3 = n^3.$$

The sum on the right side of (5.6) becomes the following, where we use the known closed forms (5.1a) and (5.4).

$$\sum_{k=1}^{n}(1 - 3k + 3k^2) = \sum_{k=1}^{n}1 - 3\sum_{k=1}^{n}k + 3\sum_{k=1}^{n}k^2$$

$$= n - 3\frac{n(n+1)}{2} + 3\sum_{k=1}^{n}k^2.$$

So Equation (5.6) becomes

$$n^3 = n - 3\frac{n(n+1)}{2} + 3\sum_{k=1}^{n}k^2.$$

We can solve this equation for $\sum_{k=1}^{n}k^2$ to obtain the closed form

$$\sum_{k=1}^{n}k^2 = \frac{n(n+1)(2n+1)}{6}. \tag{5.7}$$

end example

example **5.8 Finding the Sum of a Geometric Progression**

A closed form for the sum of the geometric progression $1, a, a^2, \ldots, a^n$, where $a \ne 1$, is given in (4.27). We can write the sum and its closed form as

$$\sum_{k=0}^{n}a^k = \frac{a^{n+1} - 1}{a - 1}. \tag{5.8}$$

We'll derive (5.8) to demonstrate a technique similar to that used for sums of powers. In this case we'll start with an expression that is the difference between two successive general terms of the sum. So we start with the expression

$$a^{k+1} - a^k.$$

Rewrite the expression to obtain

$$a^{k+1} - a^k = (a - 1)a^k.$$

Next we sum both sides of this equation from 1 to n to obtain the equation

$$\sum_{k=0}^{n}(a^{k+1} - a^k) = \sum_{k=0}^{n}(a - 1)a^k. \tag{5.9}$$

The sum on the left side of (5.9) collapses to

$$\sum_{k=0}^{n}(a^{k+1}-a^{k})=a^{n+1}-1.$$

The sum on the right side of (5.9) becomes

$$\sum_{k=0}^{n}(a-1)a^{k}=(a-1)\sum_{k=0}^{n}a^{k}.$$

So Equation (5.9) becomes

$$a^{n+1}-1=(a-1)\sum_{k=0}^{n}a^{k}.$$

Since $a \neq 1$, we can solve this equation for $\sum_{k=0}^{n}a^{k}$ to obtain the closed form in (5.8).

end example

example   **5.9   Closed Form for a Sum of Products**

We'll derive the closed form for the sum

$$\sum_{k=1}^{n}ka^{k}=a+2a^{2}+3a^{3}+\cdots+na^{n}.$$

We'll use a technique that is similar to those in the preceding examples. That is, we'll start with an expression whose sums will collapse and that expands to an expression whose sums will contain the sum we want. The expression is the difference between two successive general terms of the sum. So we start with the expression

$$(k+1)a^{k+1}-ka^{k}.$$

Rewrite the expression to

$$(k+1)a^{k+1}-ka^{k}=ka^{k+1}+a^{k+1}-ka^{k}=(a-1)ka^{k}+a^{k+1}.$$

Now take the sum of both sides from 1 to n to obtain

$$\sum_{k=1}^{n}\left((k+1)a^{k+1}-ka^{k}\right)=\sum_{k=1}^{n}\left((a-1)ka^{k}+a^{k+1}\right). \qquad (5.10)$$

The sum on the left side of (5.10) collapses to

$$\sum_{k=1}^{n}\left((k+1)a^{k+1}-ka^{k}\right)=(n+1)a^{n+1}-a.$$

The sum on the right side of (5.10) becomes the following, where we use the known closed form (5.8).

$$\sum_{k=1}^{n} \left((a-1)ka^k + a^{k+1}\right) = (a-1)\sum_{k=1}^{n} ka^k + a\sum_{k=1}^{n} a^k$$

$$= (a-1)\sum_{k=1}^{n} ka^k + a\left(\sum_{k=0}^{n} a^k - 1\right)$$

$$= (a-1)\sum_{k=1}^{n} ka^k + a\left(\frac{a^{n+1}-1}{a-1}\right) - a.$$

So Equation (5.10) becomes

$$(n+1)a^{n+1} - a = (a-1)\sum_{k=1}^{n} ka^k + a\left(\frac{a^{n+1}-1}{a-1}\right) - a.$$

Since $a \neq 1$, we can solve this equation for the sum to obtain the following closed form:

$$\sum_{k=1}^{n} ka^k = \frac{a-(n+1)a^{n+1} + na^{n+2}}{(a-1)^2}.$$

end example

The closed forms derived in the preceding examples are quite useful because they pop up in many situations when trying to count the number of operations performed by an algorithm. We'll list them here.

Closed Forms of Elementary Finite Sums (5.11)

a. $\displaystyle\sum_{k=1}^{n} k = \frac{n(n+1)}{2}.$

b. $\displaystyle\sum_{k=1}^{n} k^2 = \frac{n(n+1)(2n+1)}{6}.$

c. $\displaystyle\sum_{k=0}^{n} a^k = \frac{a^{n+1}-1}{a-1} \quad (a \neq 1).$

d. $\displaystyle\sum_{k=1}^{n} ka^k = \frac{a-(n+1)a^{n+1} + na^{a+2}}{(a-1)^2} \quad (a \neq 1).$

A Sum of Products Transformation

The summation facts listed in (5.1) are basic and useful, but not exhaustive. In fact, we will hardly scratch the surface of known summation facts. Let's look at an easily verifiable transformation for general sums of products that can be useful in breaking down a problem into simpler problems, somewhat akin to the idea of divide and conquer.

Abel's Summation Transformation

$$\sum_{k=0}^{n} a_k b_k = A_n b_n + \sum_{k=0}^{n-1} A_k (b_k - b_{k+1}), \quad \text{where } A_k = \sum_{i=0}^{k} a_k. \qquad (5.12)$$

Note that the formula also holds if the lower limit of 0 is replaced by any integer that is less than or equal to n.

This formula can be useful if either A_k or $b_k - b_{k+1}$ can be calculated. For example, we'll give an alternative derivation of the closed form in (5.11d). In this case we'll use the lower limit of 0 because (5.11d) has the same value whether we start at 1 or 0 and because the lower limit of 0 will simplify the closed form.

$$\sum_{k=1}^{n} ka^k = \sum_{k=0}^{n} ka^k = \sum_{k=0}^{n} a^k k = A_n n + \sum_{k=0}^{n-1} A_k(k - (k+1))$$

$$= A_n n - \sum_{k=0}^{n-1} A_k$$

$$= \left(\frac{a^{n+1} - 1}{a - 1} \right) n - \sum_{k=0}^{n-1} \left(\frac{a^{k+1} - 1}{a - 1} \right)$$

$$= \left(\frac{1}{a - 1} \right) \left(na^{n+1} - n - \sum_{k=0}^{n-1} (a^{k+1} - 1) \right)$$

$$= \left(\frac{1}{a - 1} \right) \left(na^{n+1} - n - a \sum_{k=0}^{n-1} a^k + \sum_{k=0}^{n-1} 1 \right)$$

$$= \left(\frac{1}{a - 1} \right) \left(na^{n+1} - n - a \left(\frac{a^n - 1}{a - 1} \right) + n \right)$$

$$= \frac{a - (n+1)a^{n+1} + na^{n+2}}{(a - 1)^2}.$$

Some Applications

Let's look at a few examples of problems that can be solved by finding closed forms for summations.

5.10 The Polynomial Problem

Suppose we're interested in the number of arithmetic operations needed to evaluate the following polynomial at some number x.

$$c_0 + c_1 x + c_2 x^2 + \cdots + c_n x^n.$$

The number of operations performed will depend on how we evaluate it. For example, suppose that we compute each term in isolation and then add up all the terms. There are n addition operations and each term of the form $c_i x^i$ takes i multiplication operations. So the total number of arithmetic operations is given by the following sum:

$$
\begin{aligned}
n + (0 + 1 + 2 + \cdots + n) &= n + \sum_{k=0}^{n} k \\
&= n + \frac{n(n+1)}{2} \\
&= \frac{n^2 + 3n}{2}.
\end{aligned}
$$

So for even small values of n there are many operations to perform. For example, if $n = 30$, then there are 495 arithmetic operations to perform. Can we do better? Sure, we can group terms so that we don't have to repeatedly compute the same powers of x. We'll continue the discussion after we've introduced recurrences in Section 5.5.

end example

 5.11 A Simple Sort

In this example we'll construct a simple sorting algorithm and analyze it to find the number of comparison operations. We'll sort an array a of numbers indexed from 1 to n as follows: Find the smallest element in a and exchange it with the first element. Then find the smallest element in positions 2 through n and exchange it with the element in position 2. Continue in this manner to obtain a sorted array. To write the algorithm, we'll use a function "min" and a procedure "exchange," which are defined as follows:

- min(a, i, n) is the index of the minimum number among the elements $a[i]$, $a[i+1]$, $\ldots$, $a[n]$. We can easily modify the algorithm in Example 5.2 to accomplish this task with $n - i$ comparisons.

- exchange$(a[i], a[j])$ is the usual operation of swapping elements and does not use any comparisons.

Now we can write the sorting algorithm as follows:

> **for** $i := 1$ **to** $n - 1$ **do**
> $\quad j := \min(a,\, i,\, n);$
> $\quad$ exchange$(a[i],\, a[j])$
> **od**

Now let's compute the number of comparison operations. The algorithm for $\min(a, i, n)$ makes $n - i$ comparisons. So as i moves from 1 to $n - 1$, the number of comparison operations moves from $n - 1$ to $n - (n - 1)$. Adding these comparisons gives the sum of an arithmetic progression,

$$(n - 1) + (n - 2) + \cdots + 1 = \frac{n(n - 1)}{2}.$$

The algorithm makes the same number of comparisons no matter what the form of the input array, even if it is sorted to begin with. So any arrangement of numbers is a worst-case input. For example, to sort 1,000 items it would take 499,500 comparisons, no matter how the items are arranged at the start.

There are many faster sorting algorithms. For example, an algorithm called "heapsort" takes no more than $2n\log_2 n$ comparisons for its worst-case performance. So for 1,000 items, heapsort would take a maximum of 20,000 comparisons—quite an improvement over our simple sort algorithm. In Section 5.3 we'll discover a good lower bound for the worst-case performance of comparison sorting algorithms.

> end example

example 5.12   A Problem of Loops

Let's look at a general problem of counting operations that occur within loops. We'll assume that we have an algorithm with a procedure P, where $P(j)$ executes $3j$ operations of a certain type. In the following algorithm we assume that n is a given positive integer.

> $i := 1;$
> **while** $i < n$ **do**
> $\quad i := 2i;$
> $\quad\quad$ **for** $j := 1$ **to** i **do** $P(j)$ **od**
> **od**

Our task will be to count the total number of operations executed by P as a function of n, which we'll denote by $T(n)$. Note, for example, that if $n = 1$, then the while-loop can't be entered and we must have $T(1) = 0$.

To find a formula for $T(n)$ we might start by observing the for-loop. For each i the for-loop calls $P(1)$, $P(2)$, $P(3)$, $\ldots$, $P(i)$. Since each call to $P(j)$ executes $3j$ operations, it follows that for each i the number of operations executed by the calls on P by the for-loop is $3(1+2+3+\ldots+i)$. Let $f(i)$ denote this expression. We can use (5.11a) to calculate $f(i) = 3i(i+1)/2$.

Now we must find the values of i that are used to enter the for-loop. Observe that the values of i to enter the while-loop are 1, 2, 4, 8, ..., 2^k, where $2^k < n \le 2^{k+1}$. So the values of i to enter the for-loop are $2, 4, 8, \ldots, 2^{k+1}$. Now we can write an expression for $T(n)$.

$$
\begin{aligned}
T(n) &= \sum_{m=1}^{k+1} f(2^m) = \sum_{m=1}^{k+1} (3/2)2^m(2^m + 1) \\
&= 3\sum_{m=1}^{k+1} 2^{m-1}(2^m + 1) \\
&= 3\sum_{m=0}^{k} 2^m(2^{m+1} + 1) \\
&= 3\sum_{m=0}^{k} (2^{2m+1} + 2^m) \\
&= 3\sum_{m=0}^{k} 2^{2m+1} + 3\sum_{m=0}^{k} 2^m \\
&= 6\sum_{m=0}^{k} 4^m + 3\sum_{m=0}^{k} 2^m \\
&= 2(4^{k+1} - 1) + 3(2^{k+1} - 1).
\end{aligned}
$$

Now, to obtain a function of n, recall that $2^k < n \le 2^{k+1}$. Apply $\log_2$ to the inequality to obtain $k < \log_2 n \le k + 1$. Therefore, $\lceil \log_2 n \rceil = k + 1$ and we can substitute for $k + 1$ to obtain the following expression for $T(n)$.

$$
\begin{aligned}
T(n) &= 2(4^{\lceil \log_2 n \rceil} - 1) + 3(2^{\lceil \log_2 n \rceil} - 1) \\
&= 2 \cdot 4^{\lceil \log_2 n \rceil} + 3 \cdot 2^{\lceil \log_2 n \rceil} - 5.
\end{aligned}
$$

end example

5.2.2 Approximating Sums

Sometimes a sum does not have a closed form or a closed form is hard to find. In such cases we can try to find an approximation for the sum.

Some Simple Approximation Techniques

If we have a sum that has a term of maximum value, then we can replace each term by that value to obtain an upper bound on the sum. Similarly, for a minimum-valued term. For example, suppose we have the following sum of logs:

$$\sum_{k=1}^{n} \log k = \log 1 + \log 2 + \cdots + \log n.$$

Since $\log n$ is the maximum value and $\log 1 = 0$ is the minimum value, we can say that

$$0 \leq \sum_{k=1}^{n} \log k \leq n \log n.$$

We can sometimes obtain closer bounds by splitting up the sum and bounding each part. For example, we'll split the sum up into two almost equal size sums as follows:

$$\sum_{k=1}^{n} \log k = (\log 1 + \log 2 + \cdots + \log \lfloor n/2 \rfloor) + (\log (\lfloor n/2 \rfloor + 1) + \cdots + \log n)$$

$$= \sum_{k=1}^{\lfloor n/2 \rfloor} \log k + \sum_{k=\lfloor n/2 \rfloor + 1}^{n} \log k.$$

To get a better lower bound we can replace each term of the first sum by 0 and each term of the second sum by $\log \lfloor n/2 \rfloor$. To get a better upper bound we can replace each term of the first sum by $\log \lfloor n/2 \rfloor$ and each term of the second sum by $\log n$. This gives us

$$\lfloor n/2 \rfloor \log \lfloor n/2 \rfloor \leq \sum_{k=1}^{n} \log k \leq \lfloor n/2 \rfloor \left(\log \lfloor n/2 \rfloor + \log n \right).$$

5.2.3   Approximations with Definite Integrals

A powerful technique for establishing upper and lower bounds for a sum comes from elementary calculus. If you don't have a background in calculus, you can safely skip the next couple of paragraphs.

We are interested in approximating a sum of the following form, where f is a continuous function with nonnegative values.

$$\sum_{k=1}^{n} f(k) = f(1) + f(2) + \cdots + f(n). \tag{5.13}$$

Each number $f(k)$ can be thought of as the area of a rectangle of width 1 and height $f(k)$. In fact we'll be more specific and let the base of the rectangle be the closed interval $[k, k+1]$. So the sum we want represents the area of n rectangles, whose bases consist of the following partition of the closed interval $[1, n+1]$.

$$[1, 2], [2, 3], \ldots, [n, n+1].$$

The area under the curve $f(x)$ above the x-axis for x in the closed interval $[1, n+1]$ is given by the definite integral

$$\int_1^{n+1} f(x)dx.$$

So this definite integral is an approximation for our sum:

$$\sum_{k=1}^n f(k) \approx \int_1^{n+1} f(x)dx. \tag{5.14}$$

Evaluating Definite Integrals

To evaluate a definite integral of f we need to find a function F that is an antiderivative of f (the derivative of $F(x)$ is $f(x)$) and then apply the fundamental theorem of calculus that relates the two functions as follows:

$$\int_a^b f(x)dx = F(x)|_a^b = F(b) - F(a).$$

Here is a listing of some useful functions and antiderivatives.

$f(x)$	$F(x)$
$x^r (r \neq -1)$	$x^{r+1}/(r+1)$
$1/x$	$\ln x$
$\ln x$	$x \ln x - x$
e^x	e^x
c^x	$c^x/\ln x$

$$(5.15)$$

Bounds for Monotonic Functions

Assume f is a monotonic increasing function, which means that $x < y$ implies $f(x) \le f(y)$. Then the area of each rectangle with base $[k, k+1]$ and height $f(k)$ is less than or equal to the area of region under the graph of f on the interval. So (5.14) gives us the following upper bound on the sum (5.13):

$$\sum_{k=1}^n f(k) \le \int_1^{n+1} f(x)dx. \tag{5.16}$$

We can obtain a lower bound for the sum (5.13) by noticing that the area of each rectangle with base $[k - 1, k]$ and height $f(k)$ is greater than or equal to the area of region under the graph of f on the interval. So in this case the sum (5.13) represents the area of n rectangles, whose bases consist of the following partition of the closed interval $[0, n]$.

$$[0, 1], [1, 2], \ldots, [n - 1, n].$$

So we obtain the following lower bound on the sum (5.13):

$$\int_0^n f(x)dx \leq \sum_{k=1}^n f(k). \tag{5.17}$$

Putting (5.16) and (5.17) together we obtain the following bounds on the sum (5.13):

$$\int_0^n f(x)dx \leq \sum_{k=1}^n f(k) \leq \int_1^{n+1} f(x)dx. \tag{5.18}$$

If f is monotonic decreasing (i.e., $x < y$ implies $f(x) \geq f(y)$), then we use entirely similar reasoning to obtain the following bounds of the sum:

$$\int_1^{n+1} f(x)dx \leq \sum_{k=1}^n f(k) \leq \int_0^n f(x)dx. \tag{5.19}$$

A Note on Lower Limits

For ease of presentation we used a specific lower limit of 1 for the summation, which gave rise to lower limits of 0 and 1 in the bounding definite integrals. Any natural number m could replace 1 in the sum, with corresponding replacements of 0 by $m - 1$ and 1 by m in the bounding definite integrals.

example 5.13 **Some Sample Approximations**

We'll find bounds for the following sum, where r is a real number and $r \neq -1$.

$$\sum_{k=1}^n k^r = 1^r + 2^r + \cdots + n^r.$$

If $r = 0$, then the sum becomes $1 + 1 + \ldots + 1 = n$. If $r > 0$, then x^r is increasing for $x \geq 0$. So we can obtain bounds by using (5.18) and the antiderivative from

(5.15) to obtain lower and upper bounds as follows:

$$\text{(lower bound)} \quad \sum_{k=1}^{n} k^r \geq \int_0^n x^r \, dx = \left. \frac{x^{r+1}}{r+1} \right|_0^n = \frac{n^{r+1}}{r+1}.$$

$$\text{(upper bound)} \quad \sum_{k=1}^{n} k^r \leq \int_1^{n+1} x^r \, dx = \left. \frac{x^{r+1}}{r+1} \right|_1^{n+1} = \frac{(n+1)^{r+1}}{r+1} - \frac{1}{r+1}.$$

If $r < 0$, then x^r is decreasing for $x > 0$, but is not defined at $x = 0$. So the upper bound from (5.19) does not exist. We can work around the problem by raising each lower limit of (5.19) by 1. This gives the inequality:

$$\int_2^{n+1} x^r \, dx \leq \sum_{k=2}^{n} k^r \leq \int_1^n x^r \, dx.$$

Notice that the middle sum is not what we want, but after we find the bounds, we can add the first term of the sum, which is 1 in this case, to the bounds. So we can evaluate the two definite integrals using the antiderivative from (5.15) to obtain bounds as follows:

$$\text{(upper bound)} \quad \sum_{k=2}^{n} k^r \leq \int_1^n x^r \, dx = \left. \frac{x^{r+1}}{r+1} \right|_1^n = \frac{n^{r+1}}{r+1} - \frac{1}{r+1}.$$

$$\text{(lower bound)} \quad \sum_{k=2}^{n} k^r \geq \int_2^{n+1} x^r \, dx = \left. \frac{x^{r+1}}{r+1} \right|_2^{n+1} = \frac{(n+1)^{r+1}}{r+1} - \frac{2^{r+1}}{r+1}.$$

Adding 1 to each bound gives us the bounds

$$1 + \frac{(n+1)^{r+1}}{r+1} - \frac{2^{r+1}}{r+1} \leq \sum_{k=1}^{n} k^r \leq 1 + \frac{n^{r+1}}{r+1} - \frac{1}{r+1}.$$

end example

5.2.4 Harmonic Numbers

For a positive integer n, the sum of the n numbers $1, 1/2, 1/3, \ldots, 1/n$ is called the nth *harmonic number* and it is usually denoted by H_n. In other words, the nth harmonic number is

$$H_n = \sum_{k=1}^{n} (1/k) = 1 + \frac{1}{2} + \frac{1}{3} + \cdots + \frac{1}{n}. \tag{5.20}$$

So $H_1 = 1$, $H_2 = 1 + \frac{1}{2}$, and so on. The first few harmonic numbers are

$$1, \frac{3}{2}, \frac{11}{6}, \frac{25}{12}, \frac{137}{60}, \frac{49}{20}, \cdots.$$

Harmonic numbers are interesting for two reasons. The first is that they appear in a wide variety of counting problems and the second is that there is no closed form for the sum. This brings up the question of approximating H_n without having to actually add up all the terms.

Notice that $1/x$ is decreasing for $x > 0$, but it is not defined for $x = 0$. So the upper bound from (5.19) does not exist. We can work around the problem as we did in Example 5.13 by raising each lower limit of (5.19) by 1. This gives the inequality

$$\int_2^{n+1} (1/x)dx \le \sum_{k=2}^{n} (1/k) \le \int_1^{n} (1/x)dx.$$

Notice that the middle sum is now $H_n - 1$. The table in (5.15) tells us that an antiderivative of $1/x$ is $\ln x$. So we can evaluate the two definite integrals and obtain the following inequality:

$$\ln(n + 1) - \ln 2 \le H_n - 1 \le \ln n.$$

Now add 1 to the terms of the inequality to obtain

$$\ln\left(\frac{n + 1}{2}\right) + 1 \le H_n \le \ln(n) + 1. \tag{5.21}$$

The difference between these two bounds is less than $\ln 2 = 0.693\ldots$. There are better bounds than those of (5.21). We'll see another one in the exercises.

Sums with Terms That Contain Harmonic Numbers

We'll derive two well-known sums that involve harmonic numbers. The first sum simply adds up the first n harmonic numbers.

$$\sum_{k=1}^{n} H_k = (n + 1)H_n - n. \tag{5.22}$$

We'll rearrange the terms of the sum to see whether we can discover a workable pattern.

$$\sum_{k=1}^{n} H_k = H_1 + H_2 + H_3 + \cdots + H_n$$

$$= (1) + (1 + 1/2) + (1 + 1/2 + 1/3) + \cdots + (1 + 1/2 + 1/3 + \cdots + 1/n).$$
$$= n(1) + (n - 1)(1/2) + (n - 2)(1/3) + \cdots + (1)(1/n)$$
$$= \sum_{k=1}^{n} (n - k + 1)(1/k)$$
$$= \sum_{k=1}^{n} (n + 1)(1/k) - \sum_{k=1}^{n} (k/k)$$
$$= (n + 1)\sum_{k=1}^{n} (1/k) - \sum_{k=1}^{n} 1$$
$$= (n + 1)H_n - n.$$

The second sum adds up the first n products of the form kH_k.

$$\sum_{k=1}^{n} kH_k = \frac{n(n + 1)}{2} H_n - \frac{n(n - 1)}{4}. \tag{5.23}$$

We can discover a workable pattern for this sum by rearranging terms as follows:

$$\sum_{k=1}^{n} kH_k = 1H_1 + 2H_2 + 3H_3 + \cdots + nH_n$$

$$= 1(1) + 2(1 + 1/2) + 3(1 + 1/2 + 1/3) + \cdots + n(1 + 1/2 + 1/3 + \cdots + 1/n).$$
$$= (1)(1 + 2 + 3 + \cdots + n) + (1/2)(2 + 3 + \cdots + n) + (1/3)(3 + \cdots + n) + \cdots + (1/n)(n).$$

We'll leave the remainder of the derivation as one of the exercises.

example 5.14 Sums within Sums

We'll consider the following algorithm with two loops:

```
k := 1;
while k < n do
      k := k + 1;
          for j := 1 to k do S(j) od
od
```

We will assume that each call to $S(j)$ executes about n/j operations of a type that we wish to count. For example, $S(j)$ might do some work on a collection of $\lfloor n/j \rfloor$ subsets of an n-element set, each of size j with the possibility of some subset of size less than j remaining. We should be using $\lfloor n/j \rfloor$ instead of n/j. But we'll work with n/j to keep the calculations simpler.

To start things off, we'll examine the for-loop for some value of k. This means that $S(j)$ is called k times with j taking values $1, 2, \ldots, k$. Since $S(j)$ executes n/j operations, the number of operations executed in each for-loop by S is

$$\sum_{j=1}^{k} (n/j) = n \sum_{j=1}^{k} (1/j) = nH_k.$$

Now we need to find the values of k at the for-loop. The values of k that enter the while-loop are $1, 2, \ldots, n-1$. Since k gets incremented by 1 upon entry, the values of k at the for-loop are $2, 3, \ldots, n$. So the number of operations by S is given by

$$\sum_{k=2}^{n} nH_k = n \sum_{k=2}^{n} H_k = n \left(\sum_{k=1}^{n} H_k - H_1 \right) = n \left(\sum_{k=1}^{n} H_k - 1 \right) = n(n+1)H_n - 2n.$$

Since we have an approximation for H_n, we can get a pretty good idea of the number of operations.

end example

We'll see some other ways to represent approximations in Section 5.6. To see other results about harmonic numbers, see Knuth [1968] or Graham, Knuth, and Patashnik [1989].

5.2.5 Polynomials and Partial Fractions

Before we proceed further with summations, let's recall a few facts about polynomials and partial fractions that we can use to simplify quotients of polynomials (also known as rational functions). So we will consider expressions of the following form, where $p(x)$ and $q(x)$ are two polynomials:

$$\frac{p(x)}{q(x)}.$$

To work with partial fractions the degree of $p(x)$ must be less than the degree of $q(x)$. If the degree of $p(x)$ is greater than or equal to the degree of $q(x)$, then we can transform the expression into the following form, where $s(x)$, $p_1(x)$, and $q_1(x)$ are polynomials and the degree of $p_1(x)$ is less than the degree of $q_1(x)$:

$$\frac{p(x)}{q(x)} = s(x) + \frac{p_1(x)}{q_1(x)}.$$

The transformation can be carried out by using long division for polynomials.

Dividing Polynomials

For example, suppose we have the following quotient of polynomials.

$$\frac{2x^3 + 1}{x^2 + 3x + 2}.$$

The degree of the numerator, which is 3, is greater than or equal to the degree of the denominator, which is 2. So we can divide the numerator by the denominator (the divisor) by using long division as follows, where division takes place between the terms of highest degree:

$$
\begin{array}{r}
2x \\
x^2 + 3x + 2 \overline{\big)\ 2x^3 \qquad\quad + 1} \\
2x^3 + 6x^2 + 4x \\
\hline
-6x^2 - 4x + 1
\end{array}
$$

Since the degree of the remainder, which is 2, is still greater than or equal to the degree of the divisor, we need to carry out the division one more step to obtain

$$
\begin{array}{r}
2x - 6 \\
x^2 + 3x + 2 \overline{\big)\ 2x^3 \qquad\quad + 1} \\
2x^3 + 6x^2 + 4x \\
\hline
-6x^2 - 4x + 1 \\
-6x^2 - 18x - 12 \\
\hline
14x + 13
\end{array}
$$

The degree of the remainder, which is 1, is now less than the degree of the divisor. So we can proceed just like the division algorithm for integers to write the dividend as the divisor times the quotient plus the remainder as follows:

$$2x^3 + 1 = (x^2 + 3x + 2)(2x - 6) + (14x + 13).$$

Now divide the equation by the divisor $x^2 + 3x + 2$ to obtain the following desired form:

$$\frac{2x^3 + 1}{x^2 + 3x + 2} = (2x - 6) + \frac{14x + 13}{x^2 + 3x + 2}.$$

Partial Fractions

Now assume that we have the following quotient of polynomials $p(x)$ and $q(x)$, where the degree of $p(x)$ is less than the degree of $q(x)$.

$$\frac{p(x)}{q(x)}.$$

Then the *partial fraction* representation (or expansion or decomposition) of the expression is a sum of terms that satisfy the following rules, where $q(x)$ has been factored into a product of linear and/or quadratic factors.

Partial Fractions

1. If the linear polynomial $ax + b$ is repeated k times as a factor of $q(x)$, then add the following terms to the partial fraction representation, where $A_1, ..., A_k$ are constants to be determined:

$$\frac{A_1}{ax + b} + \frac{A_2}{(ax + b)^2} + \cdots + \frac{A_k}{(ax + b)^k}.$$

2. If the quadratic polynomial $cx^2 + dx + e$ is repeated k times as a factor of $q(x)$, then add the following terms to the partial fraction representation, where A_i and B_i are constants to be determined:

$$\frac{A_1 x + B_1}{cx^2 + dx + e} + \frac{A_2 x + B_2}{(cx^2 + dx + e)^2} + \cdots + \frac{A_k x + B_k}{(cx^2 + dx + e)^k}.$$

Some Sample Partial Fractions

Here are a few samples of partial fractions that can be obtained from the two rules.

$$\frac{x - 1}{x(x - 2)(x + 1)} = \frac{A}{x} + \frac{B}{x - 2} + \frac{C}{x + 1}$$

$$\frac{x^3 - 1}{x^2(x - 2)^3} = \frac{A}{x} + \frac{B}{x^2} + \frac{C}{x - 2} + \frac{D}{(x - 2)^2} + \frac{E}{(x - 2)^3}$$

$$\frac{x^2}{(x - 1)(x^2 + x + 1)} = \frac{A}{x - 1} + \frac{Bx + C}{x^2 + x + 1}$$

$$\frac{x}{(x - 1)(x^2 + 1)^2} = \frac{A}{x - 1} + \frac{Bx + C}{x^2 + 1} + \frac{Dx + C}{(x^2 + 1)^2}.$$

Determining the Constants

To determine the constants in a partial fraction representation, we can solve simultaneous equations. If there are n constants to be found, then we need to create n equations. One way to accomplish this is to multiply the equation by the greatest common denominator, collect terms, and equate coefficients. Another was is to pick n values for x, with the restriction that no value of x makes any denominator zero.

For example, suppose we want to represent the following expression as the sum of partial fractions.

$$\frac{x+1}{(2x-1)(3x-1)}.$$

Since the degree of the numerator is less than the degree of the denominator, the rules tell us to write

$$\frac{x+1}{(2x-1)(3x-1)} = \frac{A}{2x-1} + \frac{B}{3x-1}.$$

Now we need to create two equations in A and B. One way to proceed is to multiply the equation by the greatest common denominator and then equate coefficients in the resulting equation. So we multiply both sides by $(2x-1)(3x-1)$ to obtain

$$x+1 = A(3x-1) + B(2x-1).$$

Collect terms on the right side to obtain

$$x+1 = (3A+2B)x + (-A-B).$$

Now equate coefficients to obtain the two equations

$$1 = -A - B$$
$$1 = 3A + 2B.$$

Solving for A and B, we get $A = 3$ and $B = -4$.

An alternative way to obtain two equations in A and B is to pick two values to substitute for x in the equation to obtain two equations. For example, let $x = 0$ and $x = 1$ to obtain the two equations

$$1 = -A - B$$
$$1 = A + (1/2)B.$$

Solving for A and B, we get $A = 3$ and $B = -4$, as shown. So the partial fraction representation of the expression is

$$\frac{x+1}{(2x-1)(3x-1)} = \frac{3}{2x-1} - \frac{4}{3x-1}.$$

example **5.15 Partial Fractions and Collapsing Sums**

Consider the following summation.

$$\sum_{k=1}^{n} \frac{1}{k^2+k}.$$

It's always fun to write out a few terms of the sum to see whether a pattern of some kind emerges. You might try it. But with new tools available, we might notice that the summand has a partial fraction representation as

$$\frac{1}{k^2 + k} = \frac{1}{k(k+1)} = \frac{A}{k} + \frac{B}{k+1} = \frac{1}{k} - \frac{1}{k+1}.$$

So we can write the sum as follows, where the sum of differences collapses to the answer.

$$\sum_{k=1}^{n} \frac{1}{k^2 + k} = \sum_{k=1}^{n} \left(\frac{1}{k} - \frac{1}{k+1} \right) = \frac{1}{1} - \frac{1}{n+1} = 1 - \frac{1}{n+1}.$$

end example

example **5.16 A Sum with a Harmonic Answer**

We'll evaluate the following sum:

$$\sum_{k=1}^{n} \frac{14k + 13}{k^2 + 3k + 2}.$$

We can use partial fractions to put the expression into the following workable form:

$$\frac{14k + 13}{k^2 + 3k + 2} = \frac{14k + 13}{(k+1)(k+2)} = \frac{-1}{k+1} + \frac{15}{k+2}.$$

So we can calculate the sum as follows:

$$\sum_{k=1}^{n} \frac{14k + 13}{k^2 + 3k + 2} = \sum_{k=1}^{n} \left(\frac{15}{k+2} - \frac{1}{k+1} \right)$$

$$= 15 \sum_{k=1}^{n} \frac{1}{k+2} - \sum_{k=1}^{n} \frac{1}{k+1}$$

$$= 15 \sum_{k=3}^{n+2} \frac{1}{k} - \sum_{k=2}^{n+1} \frac{1}{k}$$

$$= 15(H_{n+2} - H_2) - (H_{n+1} - H_1).$$

end example

Exercises

Closed Forms for Sums

1. Expand each expression into a sum of terms. Don't evaluate.

    a. $\displaystyle\sum_{k=1}^{5}(2k+3).$
    b. $\displaystyle\sum_{k=1}^{5}k3^{k}.$
    c. $\displaystyle\sum_{k=0}^{4}(5-k)3^{k}.$

2. (*Changing Limits of Summation*). Given the following summation expression:

$$\sum_{k=1}^{n}g(k-1)a_{k}x^{k+1}.$$

    For each of the following lower limits of summation, find an equivalent summation expression that starts with that lower limit:

    a. $k=0.$
    b. $k=2.$
    c. $k=-1.$
    d. $k=3.$
    e. $k=-2.$

3. Find a closed form for each of the following sums:

    a. $3+6+9+12+\cdots+3n.$
    b. $3+9+15+21+\cdots+(6n+3).$
    c. $3+6+12+24+\cdots+3(2^{n}).$
    d. $3+(2)3^{2}+(3)3^{3}+(4)3^{4}+\cdots+n3^{n}.$

4. Use summation facts and known closed forms to transform each of the following summations into a closed form:

    a. $\displaystyle\sum_{k=1}^{n}(2k+2).$
    b. $\displaystyle\sum_{k=1}^{n}(2k-1).$

    c. $\displaystyle\sum_{k=1}^{n}(2k+3).$
    d. $\displaystyle\sum_{k=1}^{n}(4k-1).$

    e. $\displaystyle\sum_{k=1}^{n}(4k-2).$
    f. $\displaystyle\sum_{k=1}^{n}k2^{k}.$

    g. $\displaystyle\sum_{k=1}^{n}k(k+1).$
    h. $\displaystyle\sum_{k=2}^{n}(k^{2}-k).$

5. Verify the two collapsing sum formulas in (5.1c) for the case $n=3$.

6. Verify Abel's summation transformation (5.12) for the case $n=2$.

7. Verify the following *Summation by Parts* formula for the case $n=2$.

$$\sum_{k=0}^{n}a_{k}(b_{k+1}-b_{k})=a_{n+1}b_{n+1}-a_{0}b_{0}+\sum_{k=0}^{n}b_{k+1}(a_{k+1}-a_{k}).$$

8. Use properties of sums and logs to calculate the given value for each of the following sums:

a. $\sum_{i=0}^{k-1} 2^i (1/5^i)^{\log_5 2} = k.$

b. $\sum_{i=0}^{k-1} 3^{2i} (1/5^i)^{\log_5 3} = (3^k - 1)/2.$

c. $\sum_{i=0}^{k-1} (1/5^i)^{\log_5 2} = 2 - (1/2)^{k-1}.$

9. Use properties of sums and logs to show that the following equation holds, where H_k is the kth harmonic number and $n = 2^k$.

$$\sum_{i=0}^{k-1} \frac{1}{\log_2(n/2^i)} = H_k.$$

10. In each case find a closed form for the sum in terms of harmonic numbers.

a. $\sum_{k=0}^{n} \frac{1}{2k+1}.$

b. $\sum_{k=1}^{n} \frac{1}{2k-1}.$

11. In each case find a closed form for the sum in terms of harmonic numbers by using division and partial fractions to simplify the summand.

a. $\sum_{k=1}^{n} \frac{k}{k+1}.$

b. $\sum_{k=1}^{n} \frac{k^2}{k+1}.$

12. Finish the derivation of the sum (5.23) starting from where the sum was rearranged.

Analyzing Algorithms

13. Given the following algorithm, find a formula in terms of n for each case:

$$\textbf{for } i := 1 \textbf{ to } n \textbf{ do}$$
$$\textbf{for } j := 1 \textbf{ to } i \textbf{ do } x := x + f(x) \textbf{ od};$$
$$x := x + g(x)$$
$$\textbf{od}$$

a. Find the number of times that the assignment statement ($:=$) is executed during the running of the program.

b. Find the number of times that the addition operation ($+$) is executed during the running of the program.

14. Given the following algorithm, find a formula in terms of n for each case:

$$i := 1;$$
while $i < n + 1$ **do**
    $i := i + 1;$
    for $j := 1$ **to** i **do** S **od**
od

   a. Find the number of times that the statement S is executed during the running of the program.

   b. Find the number of times that the assignment statement $(:=)$ is executed during the running of the program.

15. Given the following algorithm, find a formula in terms of n for each case:

$$i := 1;$$
while $i ¡ n + 1$ **do**
    $i := i + 2;$
    for $j := 1$ **to** i **do** S **od**
od

   a. Find the number of times that the statement S is executed during the running of the program.

   b. Find the number of times that the assignment statement $(:=)$ is executed during the running of the program.

Challenges

16. Use the collapsing sums technique from Example 5.7 to find a closed form for $\sum_{k=1}^{n} k^3$.

17. Use the technique introduced in Example 5.9 to derive a closed form for the sum

$$\sum_{k=1}^{n} k^2 a^k \text{ by using the known closed forms for } \sum_{k=1}^{n} k a^k \text{ and } \sum_{k=1}^{n} a^k.$$

18. In each case use (5.18) to find upper and lower bounds for the sum and compare the results with the actual closed form for the sum.

   a. $\sum_{k=1}^{n} k.$ b. $\sum_{k=1}^{n} k^2.$

19. Verify that the two bounds in (5.21) differ by less than ln 2.

20. Notice that the average value $(1/2)(1/k + 1/(k+ 1))$ of the areas of the two rectangles with base $[k, k + 1]$ and heights $1/k$ and $1/(k + 1)$, respectively (i.e., the area of the trapezoid with sides of length $1/k$ and $1/(k + 1)$) is larger than the definite integral of $1/x$ on the interval $[k, k + 1]$.

a. Use this fact to obtain the following lower bound for H_n.

$$\ln n + \frac{1}{2n} + \frac{1}{2}.$$

b. Show that the lower bound in Part (a) is better (i.e., greater) than the lower bound given in (5.21).

5.3   Permutations and Combinations

Whenever we need to count the number of ways that some things can be arranged or the number of subsets of things, there are some nice techniques that can help out. That's what our discussion of permutations and combinations will be about.

Two Counting Rules

Before we start, we should recall two useful rules of counting.

The *rule of sum* states that if there are m choices for some event to occur and n choices for another event to occur and the events are disjoint, then there are $m + n$ choices for either event to occur. This is just another way to say that the cardinality of the union of disjoint sets is the sum of the cardinalities of the two sets (1.11).

The *rule of product* states that if there are m choices for some event and n choices for another event, then there are mn choices for both events. This is just another way to say that the cardinality of the Cartesian product of two finite sets is the product of the cardinalities of the two sets (1.17).

For example, if a vending machine has 6 types of drink and 18 types of snack food, then there are $6 + 18 = 24$ possible choices in the machine and $6 \cdot 18 = 108$ possible choices of a drink and a snack.

Both rules are sometimes used to answer a question. For example, suppose we have a bookshelf with 5 technical books, 12 biographies, and 37 novels. We intend to take two books of different type to read while on vacation. There are $5 \cdot 12 = 60$ ways to choose a technical book and a biography, $5 \cdot 37 = 185$ ways to choose a technical book and a novel, and $12 \cdot 37 = 444$ ways to choose a biography and novel. So there are $50 + 185 + 444 = 689$ ways to choose two books of different type.

5.3.1   Permutations (Order Is Important)

An arrangement (i.e., an ordering) of distinct objects is called a *permutation* of the objects. For example, one permutation of the three letters in the set

$\{a, b, c\}$ is *bca*. We can count the number of permutations by noticing that there are 3 choices for the first letter. For each choice of a first letter there are 2 choices remaining for the second letter, and upon picking a second letter there is one letter remaining. Therefore there are $3 \cdot 2 \cdot 1 = 6$ permutations of a 3-element set. The six permutations of $\{a, b, c\}$ can be listed as

$$abc, acb, bac, bca, cab, cba.$$

The idea generalizes to permutations of an n-element set. There are n choices for the first element. For each of these choices there are $n - 1$ choices for the second element. Continuing in this way, we obtain $n \cdot (n - 1) \cdots 2 \cdot 1$ different permutations of n elements. The notation for this product is the symbol $n!$, which we read as "n factorial." So we have

$$n! = n \cdot (n - 1) \cdots 2 \cdot 1.$$

By convention, we set $0! = 1$. So here's the first rule of permutations.

Permutations (5.24)

The number of permutations of n distinct objects is n factorial.

Now suppose we want to count the number of permutations of r elements chosen from an n-element set, where $1 \leq r \leq n$. There are n choices for the first element. For each of these choices there are $n - 1$ choices for the second element. We continue this process r times to obtain the answer,

$$n(n - 1) \cdots (n - r + 1) = \frac{n!}{(n - r)!}. \qquad (5.25)$$

This number is denoted by $P(n, r)$. Here's the definition and the rule.

Permutations

An *r-permutation* of n distinct objects is a permutation of r of the objects. The number of r-permutations of n distinct objects is

$$P(n,\ r) = \frac{n!}{(n - r)!}. \qquad (5.26)$$

Notice that $P(n,\ 1) = n$ and $P(n,\ n) = n!$. If $S = \{a,\ b,\ c,\ d\}$, then there are 12 permutations of two elements from S, given by the formula $P(4, 2) = 4!/2! = 12$. The permutations are listed as follows:

$$ab,\ ba,\ ac,\ ca,\ ad,\ da,\ bc,\ cb,\ bd,\ db,\ cd,\ dc.$$

Permutations with Repeated Elements

Permutations can be thought of as arrangements of objects selected from a set *without replacement*. In other words, we can't pick an element from the set more than once. If we can pick an element more than once, then the objects are said to be selected *with replacement*. In this case the number of arrangements of r objects from an n-element set is just n^r. We can state this idea in terms of bags as follows: The number of distinct permutations of r objects taken from a bag containing n distinct objects, each occurring r times, is n^r. For example, consider the bag $B = [a, a, b, b, c, c]$. Then the number of distinct permutations of two objects chosen from B is 3^2, and they can be listed as follows:

$$aa, \ ab, \ ac, \ ba, \ bb, \ bc, \ ca, \ cb, \ cc.$$

Let's look now at permutations of all the elements in a bag. For example, suppose we have the bag $B = [a, a, b, b, b]$. We can write down the distinct permutations of B as follows:

$$aabbb, \ ababb, \ abbab, \ abbba, \ baabb, \ babab, \ babba, \ bbaab, \ bbaba, \ bbbaa.$$

There are 10 strings. Let's see how to compute the number 10 from the information we have about the bag B. One way to proceed is to place subscripts on the elements in the bag, obtaining the five distinct elements a_1, a_2, b_1, b_2, b_3. Then we get $5! = 120$ permutations of the five distinct elements. Now we remove all the subscripts on the elements, and we find that there are many repeated strings among the original 120 strings.

For example, suppose we remove the subscripts from the two strings,

$$a_1 b_1 b_2 a_2 b_3 \quad \text{and} \quad a_2 b_1 b_3 a_1 b_2.$$

Then we obtain two occurrences of the string *abbab*. If we wrote all occurrences down, we would find 12 strings, all of which reduce to the string *abbab* when subscripts are removed. This is because there are 2! permutations of the letters a_1 and a_2, and there are 3! permutations of the letters b_1, b_2, and b_3. So there are $2!3! = 12$ distinct ways to write the string *abbab* when we use subscripts. Of course, the number is the same for any string of two a's and three b's. Therefore, the number of distinct strings of two a's and three b's is found by dividing the total number of subscripted strings by $2!3!$ to obtain $5!/2!3! = 10$. This argument generalizes to obtain the following result about permutations that can contain repeated elements.

Permutations of a Bag (5.27)

Let B be an n-element bag with k distinct elements, where each of the numbers $m_1,\dots, m_k$ denotes the number of occurrences of each element. Then the number of permutations of the n elements of B is

$$\frac{n!}{m_1! \cdots m_k!}.$$

Now let's look at a few examples to see how permutations (5.24)–(5.27) can be used to solve a variety of problems. We'll start with an important result about sorting.

example 5.17 Worst-Case Lower Bound for Comparison Sorting

Let's find a lower bound for the number of comparisons performed by any algorithm that sorts by comparing elements in the list to be sorted. Assume that we have a set of n distinct numbers. Since there are $n!$ possible arrangements of these numbers, it follows that any algorithm to sort a list of n numbers has $n!$ possible input arrangements. Therefore, any decision tree for a comparison sorting algorithm must contain at least $n!$ leaves, one leaf for each possible outcome of sorting one arrangement.

We know that a binary tree of depth d has at most 2^d leaves. So the depth d of the decision tree for any comparison sort of n items must satisfy the inequality

$$n! \leq 2^d.$$

We can solve this inequality for the natural number d as follows:

$$\log_2 n! \leq d$$
$$\lceil \log_2 n! \rceil \leq d.$$

In other words, $\lceil \log_2 n! \rceil$ is a worst-case lower bound for the number of comparisons to sort n items. The number $\lceil \log_2 n! \rceil$ is hard to calculate for large values of n. We'll see in Section 5.6 that it is approximately $n \log_2 n$.

end example

example 5.18 People in a Circle

In how many ways can 20 people be arranged in a circle if we don't count a rotation of the circle as a different arrangement? There are 20! arrangements of 20 people in a line. We can form a circle by joining the two ends of a line. Since there are 20 distinct rotations of the same circle of people, it follows that there are

$$\frac{20!}{20} = 19!$$

distinct arrangements of 20 people in a circle. Another way to proceed is to put one person in a certain fixed position of the circle. Then fill in the remaining 19 people in all possible ways to get 19! arrangements.

end example

example 5.19 Rearranging a String

How many distinct strings can be made by rearranging the letters of the word *banana*? One letter is repeated twice, one letter is repeated three times, and one letter stands by itself. So we can answer the question by finding the number of permutations of the bag of letters $[b, a, n, a, n, a]$. Therefore, (5.27) gives us the result

$$\frac{6!}{1!2!3!} = 60.$$

end example

example 5.20 Strings with Restrictions

How many distinct strings of length 10 can be constructed from the two digits 0 and 1 with the restriction that five characters must be 0 and five must be 1? The answer is

$$\frac{10!}{5!5!} = 252$$

because we are looking for the number of permutations from a 10-element bag with five 1's and five 0's.

end examp

example 5.21 Constructing a Code

Suppose we want to build a code to represent each of 29 distinct objects with a binary string having the same minimal length n, where each string has the same number of 0's and 1's. Somehow we need to solve an inequality like

$$\frac{n!}{k!k!} \geq 29,$$

where $k = n/2$. We find by trial and error that $n = 8$. Try it.

end example

5.3.2 Combinations (Order Is Not Important)

When we choose, or select, some objects from a set of objects, without regard to order, the choice is called a *combination*. Since order does not matter we can represent combinations as sets. For example, one combination of two elements from the set $\{a, b, c, d\}$ is the subset $\{a, b\}$. There are six 2-element combinations from $\{a, b, c, d\}$, which we can list as

$$\{a, b\}, \{a, c\}, \{a, d\}, \{b, c\}, \{b, d\}, \{c, d\}.$$

We want to count combinations without listing them. We can get the idea from our little example. We'll start by observing that the number of 2-permutations of the set is $4 \cdot 3 = 12$. But we're not counting permutations. We've counted each 2-element combination twice. For example, $\{a, b\}$ has been counted twice, once for each of its permutations ab and ba. So we must divide 12 by 2 to obtain the correct number of combinations.

We can generalize this idea to find a formula for the number of r-element combinations of an n-element set. First count the number of r-permutations of n elements, which is given by the formula

$$P(n,\ r) = \frac{n!}{(n-r)!}.$$

Now observe that each r-element combination has been counted $r!$ times, once for each of its permutations. So we must divide $P(n, r)$ by $r!$ to obtain the number of r-element combinations of an n-element set

$$\frac{P(n,\ r)}{r!} = \frac{n!}{r!(n-r)!}.$$

This formula is denoted by the expression $C(n, r)$. Here's the definition and the rule.

Combinations (5.28)

An *r-combination* of n distinct objects is a combination of r of the objects. The number of *r-combinations* chosen from n distinct objects is

$$C(n,\ r) = \frac{n!}{r!(n-r)!}.$$

$C(n, r)$ is often read "n choose r."

---

example **5.22 Subsets of the Same Size**

Let $S = \{a, b, c, d, e\}$. We'll list all the three-element subsets of S:

$$\{a, b, c\}, \{a, b, d\}, \{a, b, e\}, \{a, c, d\}, \{a, c, e\},$$
$$\{a, d, e\}, \{b, c, d\}, \{b, c, e\}, \{b, d, e\}, \{c, d, e\}.$$

There are 10 such subsets, which we can verify by the formula

$$C(5,3) = \frac{5!}{3!2!} = 10.$$

end example

Binomial Coefficients

Notice how $C(n, r)$ crops up in the following binomial expansion of the expression $(x + y)^4$:

$$(x + y)^4 = x^4 + 4x^3y + 6x^2y^2 + 4xy^3 + y^4$$
$$= C(4, 0)\, x^4 + C(4, 1)\, x^3y + C(4, 2)\, x^2y^2 + C(4, 3)\, xy^3 + C(4, 4)\, y^4.$$

A useful way to represent $C(n, r)$ is with the *binomial coefficient symbol*:

$$\binom{n}{r} = C(n, r).$$

Using this symbol, we can write the expansion for $(x + y)^4$ as follows:

$$(x + y)^4 = x^4 + 4x^3y + 6x^2y^2 + xy^3 + y^4$$
$$= \binom{4}{0} x^4 + \binom{4}{1} x^3y + \binom{4}{2} x^2y^2 + \binom{4}{3} xy^3 + \binom{4}{4} y^4.$$

This is an instance of a well-known formula called the *binomial theorem*, which can be written as follows, where n is a natural number:

Binomial Theorem   (5.29)

$$(x + y)^n = \sum_{k=0}^{n} \binom{n}{k} x^{n-k} y^k.$$

Pascal's Triangle

The binomial coefficients for the expansion of $(x + y)^n$ can be read from the nth row of the table in Figure 5.4. The table is called *Pascal's triangle*—after the philosopher and mathematician Blaise Pascal (1623–1662). However, prior to the time of Pascal, the triangle was known in China, India, the Middle East, and Europe. Notice that any interior element is the sum of the two elements above and to its left.

But how do we really know that the following statement is correct?

Elements in Pascal's Triangle   (5.30)

The nth row kth column entry of Pascal's triangle is $\dbinom{n}{k}$.

	0	1	2	3	4	5	6	7	8	9	10
0	1										
1	1	1									
2	1	2	1								
3	1	3	3	1							
4	1	4	6	4	1						
5	1	5	10	10	5	1					
6	1	6	15	20	15	6	1				
7	1	7	21	35	35	21	7	1			
8	1	8	28	56	70	56	28	8	1		
9	1	9	36	84	126	126	84	36	9	1	
10	1	10	45	120	210	252	210	120	45	10	1

Figure 5.4   Pascal's triangle.

Proof: For convenience we will designate a position in the triangle by an ordered pair of the form (row, column). Notice that the edge elements of the triangle are all 1, and they occur at positions $(n, 0)$ or (n, n). Notice also that

$$\binom{n}{0} = 1 = \binom{n}{n}.$$

So (5.30) is true when $k = 0$ or $k = n$. Next, we need to consider the interior elements of the triangle. So let $n > 1$ and $0 < k < n$. We want to show that the element in position (n, k) is $\binom{n}{k}$. To do this, we need the following useful result about binomial coefficients:

$$\binom{n}{k} = \binom{n-1}{k} + \binom{n-1}{k-1}. \tag{5.31}$$

To prove (5.31), just expand each of the three terms and simplify. Continuing with the proof of (5.30), we'll use well-founded induction. To do this, we need to define a well-founded order on something. For our purposes we will let the something be the set of positions in the triangle. We agree that any position in row $n - 1$ precedes any position in row n. In other words, if $n' < n$, then (n', k') precedes (n, k) for any values of k' and k. Now we can use well-founded induction. We pick position (n, k) and assume that (5.30) is true for all pairs in row $n - 1$. In particular, we can assume that the elements in positions $(n - 1, k)$ and $(n - 1, k - 1)$ have values

$$\binom{n-1}{k} \quad \text{and} \quad \binom{n-1}{k-1}.$$

Now we use this assumption along with (5.31) to tell us that the value of the element in position (n, k) is $\binom{n}{k}$. QED.

Can you find some other interesting patterns in Pascal's triangle? There are lots of them. For example, look down the column labeled 2 and notice that, for each $n \geq 2$, the element in position $(n, 2)$ is the value of the arithmetic sum $1 + 2 + \cdots + (n - 1)$. In other words, we have the formula

$$\binom{n}{2} = \frac{n(n-1)}{2}.$$

Combinations with Repeated Elements

Let's continue our discussion about combinations by counting bags of things rather than sets of things. Suppose we have the set $A = \{a, b, c\}$. How many 3-element bags can we construct from the elements of A? We can list them as follows:

$$[a, a, a], [a, a, b], [a, a, c], [a, b, c], [a, b, b],$$
$$[a, c, c], [b, b, b], [b, b, c], [b, c, c], [c, c, c].$$

So there are ten 3-element bags constructed from the elements of $\{a, b, c\}$.

Let's see if we can find a general formula for the number of k-element bags that can be constructed from an n-element set. For convenience, we'll assume that the n-element set is $A = \{1, 2, \ldots, n\}$. Suppose that $b = [x_1, x_2, x_3, \ldots, x_k]$ is some k-element bag with elements chosen from A, where the elements of b are written so that $x_1 \leq x_2 \leq \cdots \leq x_k$. This allows us to construct the following k-element set:

$$B = \{x_1, x_2 + 1, x_3 + 2, \ldots, x_k + (k - 1)\}.$$

The numbers $x_i + (i - 1)$ are used to ensure that the elements of B are distinct elements in the set $C = \{1, 2, \ldots, n + (k - 1)\}$. So we've associated each k-element bag b over A with a k-element subset B of C. Conversely, suppose that $\{y_1, y_2, y_3, \ldots, y_k\}$ is some k-element subset of C, where the elements are written so that $y_1 \leq y_2 \leq \cdots \leq y_k$. This allows us to construct the k-element bag

$$[y_1, y_2 - 1, y_3 - 2, \ldots, y_k - (k - 1)],$$

whose elements come from the set A. So we've associated each k-element subset of C with a k-element bag over A.

Therefore, the number of k-element bags over an n-element set is exactly the same as the number of k-element subsets of a set with $n + (k - 1)$ elements. This gives us the following result.

Bag Combinations   (5.32)

The number of k-element bags whose distinct elements are chosen from an n-element set, where k and n are positive, is given by

$$\binom{n + k - 1}{k}.$$

---

example **5.23 Selecting Coins**

In how many ways can four coins be selected from a collection of pennies, nickels, and dimes? Let $S = \{$penny, nickel, dime$\}$. Then we need the number of 4-element bags chosen from S. The answer is

$$\binom{3+4-1}{4} = \binom{6}{4} = 15.$$

end example

example **5.24 Selecting a Committee**

In how many ways can five people be selected from a collection of Democrats, Republicans, and Independents? Here we are choosing five-element bags from a set of three characteristics $\{$Democrat, Republican, Independent$\}$. The answer is

$$\binom{3+5-1}{5} = \binom{7}{5} = 21.$$

end example

 Exercises

Permutations and Combinations

1. Evaluate each of the following expressions.
   a. $P(6, 6)$.   b. $P(6, 0)$.   **c.** $P(6, 2)$.
   d. $P(10, 4)$.   **e.** $C(5, 2)$.   f. $C(10, 4)$.

2. Let $S = \{a, b, c\}$. Write down the objects satisfying each of the following descriptions.
   a. All permutations of the three letters in S.
   b. All permutations consisting of two letters from S.
   c. All combinations of the three letters in S.
   d. All combinations consisting of two letters from S.
   e. All bag combinations consisting of two letters from S.

3. For each part of Exercise 2, write down the formula, in terms of P or C, for the number of objects requested.

4. Given the bag $B = [a, a, b, b]$, write down all the bag permutations of B, and verify with a formula that you wrote down the correct number.

5. Find the number of ways to arrange the letters in each of the following words. Assume all letters are lowercase.

   a. Computer.   **b.** Radar.   **c.** States.

   d. Mississippi.   **e.** Tennessee.

6. A *derangement* of a string is a permutation of the letters such that each letter changes its position. For example, a derangement of the string ABC is BCA. But ACB is not a derangement of ABC, since A does not change position. Write down all derangements for each of the following strings.

   a. A.   **b.** AB.   **c.** ABC.   **d.** $ABCD$.

7. Suppose we want to build a code to represent 29 objects in which each object is represented as a binary string of length n, which consists of k 0's and m 1's, and $n = k + m$. Find n, k, and m, where n has the smallest possible value.

8. We wish to form a committee of seven people chosen from five Democrats, four Republicans, and six Independents. The committee will contain two Democrats, two Republicans, and three Independents. In how many ways can we choose the committee?

9. Each of the following problems refers to a property of Pascal's triangle (Figure 5.4).

   a. Each row has a largest number. Find a formula to describe which column contains the largest number in row n.

   b. Show that the sum of the numbers in row n is 2^n.

10. A deck of 52 playing cards has four suits (Clubs, Diamonds, Hearts, Spades) with each suit having thirteen cards with ranks from high to low: Ace, King, Queen, Jack, 10, 9, 8, 7, 6, 5, 4, 3, and 2. In each case, find the number of possible 5-card hands with the given property.

   a. The total number of 5-card hands.

   b. Straight flush: A 5-card sequence of the same suit, such as 3, 4, 5, 6, 7 of Clubs. (Note that the Ace can also be used as the low card in a straight.)

   c. Four of a kind: Four cards of the same rank and one of another rank.

   d. Full house: Three of one rank and two of another rank such as 10, 10, 10, Jack, Jack.

   e. Flush: All five cards the same suit but not a straight flush.

   f. Straight: A 5-card sequence but not a straight flush.

   g. Three of a kind: Three of the same rank and two others not of the same rank, such as Queen, Queen, Queen, 4, 5.

   h. Two pair: Each pair has a different rank with a 5th card of neither rank, such as 4, 4, Queen, Queen, 5.

i. One pair: Two cards of the same rank and three other different rank cards, such as Ace, Ace, 4, 9, Jack.

j. High card: Five different ranks but not a flush and not a straight.

Challenges

11. Test the birthday problem on a group of people.

12. Suppose an operating system must schedule the execution of n processes, where each process consists of k separate actions that must be done in order. Assume that any action of one process may run before or after any action of another process. How many execution schedules are possible?

13. Count the number of strings consisting of n 0's and n 1's such that each string is subject to the following restriction: As we scan a string from left to right, the number of 0's is never greater than the number of 1's. For example, the string 110010 is OK, but the string 100110 is not. *Hint:* Count the total number of strings of length $2n$ with n 0's and n 1's. Then try to count the number that are not OK, and subtract this number from the total number.

14. Given a nonempty finite set S with n elements, prove that there are $n!$ bijections from S to S.

5.4 Discrete Probability

The founders of probability theory were Blaise Pascal (1623–1662) and Pierre Fermat (1601–1665). They developed the principles of the subject in 1654 during a correspondence about games of chance. It started when Pascal was asked about a gambling problem. The problem asked how the stakes of a "points" game should be divided up between two players if they quit before either had enough points to win.

Probability comes up whenever we ask about the chance of something happening. To answer such a question requires one to make some kind of assumption. For example, we might ask about the average behavior of an algorithm. That is, instead of the worst-case performance, we might be interested in the average-case performance. This can be a bit tricky because it usually forces us to make one or two assumptions. Some people hate to make assumptions. But it's not so bad. Let's do an example.

Suppose we have a sorted list of the first 15 prime numbers, and we want to know the average number of comparisons needed to find a number in the list, using a binary search. The decision tree for a binary search of the list is pictured in Figure 5.5.

After some thought, you might think it reasonable to add up all the path lengths from the root to a leaf marked with an S (for successful search) and

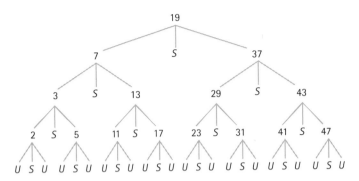

Figure 5.5   Binary search decision tree.

divide by the number of S leaves, which is 15. In this case there are eight paths of length 4, four paths of length 3, two paths of length 2, and one path of length 1. So we get

$$\text{Average path length} = \frac{32 + 12 + 4 + 1}{15} = \frac{49}{15} \approx 3.27.$$

This gives us the average number of comparisons needed to find a number in the list. Or does it? Have we made any assumptions here? Yes, we assumed that each path in the tree has the same chance of being traversed as any other path. Of course, this might not be the case. For example, suppose that we always wanted to look up the number 37. Then the average number of comparisons would be two. So our calculation was made under the assumption that each of the 15 numbers had the same chance of being picked.

5.4.1 Probability Terminology

Let's pause here and introduce some notions and notations for *discrete probability*, which gives us methods to calculate the likelihood of events that have a finite number of outcomes. If some operation or experiment has n possible outcomes and each outcome has the same chance of occurring, then we say that each outcome has *probability* $1/n$. In the preceding example we assumed that each number had probability $1/15$ of being picked. As another example, let's consider the coin-flipping problem. If we flip a fair coin, then there are two possible outcomes, assuming that the coin does not land on its edge. Thus the probability of a head is $1/2$, and the probability of a tail is $1/2$. If we flip the coin 1,000 times, we should expect about 500 heads and 500 tails. So probability has something to do with expectation.

Now for some terminology. The set of all possible outcomes of an experiment is called a *sample space* or *probability space*. The elements in a sample space are called *sample points* or simply *points*. Further, any subset of a sample space is called an *event*. For example, suppose we flip two coins and are interested in the set of possible outcomes. Let H and T stand for head and tail, respectively, and let the string HT mean that the first coin lands H and the second coin lands T. Then the sample space for this experiment is the set

$$\{HH,\ HT,\ TH,\ TT\}.$$

For example, the event that one coin lands as a head and the other coin lands as a tail can be represented by the subset $\{HT,\ TH\}$.

To discuss probability we need to make assumptions or observations about the probabilities of sample points. Here is the terminology.

Probability Distribution

A *probability distribution* on a sample space S is an assignment of probabilities to the points of S such that the sum of all the probabilities is 1.

Let's describe a probability distribution from a more formal point of view. Let $S = \{x_1,\ x_2,\ldots,\ x_n\}$ be a sample space. A probability distribution P on S is a function

$$P : S \to [0,\ 1]$$

such that

$$P(x_1) + P(x_2) + \cdots + P(x_n) = 1.$$

For example, in the two-coin-flip experiment it makes sense to define the following probability distribution on the sample space $S = \{HH,\ HT,\ TH,\ TT\}$:

$$P(HH) = P(HT) = P(TH) = P(TT) = \frac{1}{4}.$$

Probability of an Event

Once we have a probability distribution P defined on the points of a sample space S, we can use P to define the probability of any event E in S.

Probability of an Event

The *probability* of an event E is denoted by $P(E)$ and is defined by

$$P(E) = \sum_{x \in E} P(x).$$

In particular, we have $P(S) = 1$ and $P(\varnothing) = 0$. If A and B are two events, then the following formula follows directly from the definition and the inclusion-exclusion principle.

$$P(A \cup B) = P(A) + P(B) - P(A \cap B).$$

This formula has a very useful consequence. If E' is the complement of E in S, then $S = E \cup E'$ and $E \cap E' = \varnothing$. So it follows from the formula that

$$P(E') = 1 - P(E).$$

example   5.25   Complement of an Event

In our two-coin-flip example, let E be the event that at least one coin is a tail. Then $E = \{HT,\ TH,\ TT\}$. We can calculate $P(E)$ as follows:

$$P(E) = P(\{HT,\ TH,\ TT\}) = P(HT) + P(TH) + P(TT) = \tfrac{1}{4} + \tfrac{1}{4} + \tfrac{1}{4} = \tfrac{3}{4}.$$

But we also could observe that the complement E' is the event that both coins are heads. So we could calculate

$$P(E) = 1 - P(E') = 1 - P(HH) = 1 - \tfrac{1}{4} = \tfrac{3}{4}.$$

end example

Classic Example: The Birthday Problem

Suppose we ask 25 people, chosen at random, their birthday (month and day). Would you bet that they all have different birthdays? It seems a likely bet that no two have the same birthday since there are 365 birthdays in the year. But, in fact, the probability that two out of 25 people have the same birthday is greater than $1/2$. Again, we're assuming some things here, which we'll get to shortly. Let's see why this is the case. The question we want to ask is:

> Given n people in a room, what is the probability that at least two of the people have the same birthday (month and day)?

We'll neglect leap year and assume that there are 365 days in the year. So there are 365^n possible n-tuples of birthdays for n people. This set of n-tuples is our sample space S. We'll also assume that birthdays are equally distributed throughout the year. So for any n-tuple $(b_1, \ldots, b_n)$ of birthdays, we have $P(b_1, \ldots, b_n) = 1/365^n$. The event E that we are concerned with is the subset of S consisting of all n-tuples that contain two or more equal entries. So our question can be written as follows:

$$\text{What is } P(E)?$$

To answer the question, let's use the complement technique. That is, we'll compute the probability of the event $E' = S - E$, consisting of all n-tuples that have distinct entries. In other words, no two of the n people have the same birthday. Then the probability that we want is $P(E) = 1 - P(E')$. So let's concentrate on E'.

n	$P(E)$
10	0.117
20	0.411
23	0.507
30	0.706
40	0.891

Figure 5.6   Birthday table.

An n-tuple is in E' exactly when all its components are distinct. The cardinality of E' can be found in several ways. For example, there are 365 possible values for the first element of an n-tuple in E'. For each of these 365 values there are 364 values for the second element of an n-tuple in E'. Thus we obtain

$$365 \cdot 364 \cdot 363 \cdots (365 - n + 1)$$

n-tuples in E'. Since each n-tuple of E' is equally likely with probability $1/365^n$, it follows that

$$P(E') = \frac{365 \cdot 364 \cdot 363 \cdots (365 - n + 1)}{365^n}.$$

Thus the probability that we desire is

$$P(E) = 1 - P(E') = 1 - \frac{365 \cdot 364 \cdot 363 \cdots (365 - n + 1)}{365^n}.$$

The table in Figure 5.6 gives a few calculations for different values of n. Notice the case when $n = 23$. The probability is better than 0.5 that two people have the same birthday. Try this out the next time you're in a room full of people. It always seems like magic when two people have the same birthday.

example **5.26 Switching Pays**

Suppose there is a set of three numbers. One of the three numbers will be chosen as the winner of a three-number lottery. We pick one of the three numbers. Later, one of the two remaining numbers is identified and we are told that it is not the winner, and we are given the chance to keep the number that we picked or to switch and choose the remaining number. What should we do? We should switch.

To see this, notice that once we pick a number, the probability that we did not pick the winner is $2/3$. In other words, it is more likely that one of the other two numbers is a winner. So when we are told that one of the other numbers is not the winner, it follows that the remaining other number has probability $2/3$

of being the winner. So go ahead and switch. Try this experiment a few times with a friend to see that in the long run it's better to switch.

Another way to see that switching is the best policy is to modify the problem to a set of 50 numbers and a 50-number lottery. If we pick a number, then the probability that we did not pick a winner is 49/50. Later we are told that 48 of the remaining numbers are not winners, but we are given the chance to keep the number we picked or switch and choose the remaining number. What should we do? We should switch because the chance that the remaining number is the winner is 49/50.

end example

5.4.2   Conditional Probability

If we ask a question about the chance of something happening given that something else has happened, we are using conditional probability.

Conditional Probability
If A and B are events and $P(B) \neq 0$, then *the conditional probability of A given B* is denoted by $P(A|B)$ and defined by

$$P(A|B) = \frac{P(A \cap B)}{P(B)}.$$

We can think of $P(A|B)$ as the probability of the event $A \cap B$ when the sample space is restricted to B once we make an appropriate adjustment to the probability distribution.

example   5.27   Majoring in Two Subjects

In a university it is known that 1% of students major in mathematics and 2% major in computer science. Further, it is known that 0.1% of students major in both mathematics and computer science. If we learn that a student is a computer science major, what is the probability that the student is a mathematics major? If we learn that a student is a mathematics major, what is the probability that the student is a computer science major?

To answer the questions, let A and B be the sets of mathematics majors and computer science majors, respectively. Then we have the three probabilities

$$P(A) = 0.01, \ P(B) = 0.02, \text{ and } P(A \cap B) = 0.001.$$

The two questions are answered by $P(A \mid B)$ and $P(B \mid A)$, respectively. Here are the calculations.

$$P(A|B) = \frac{P(A \cap B)}{P(B)} = \frac{0.001}{0.02} = 0.05 \text{ and } P(B|A) = \frac{P(B \cap A)}{P(A)} = \frac{0.001}{0.01} = 0.10.$$

end example

A Useful Consequence of the Definition

From the definition of conditional probability we can obtain a simple relationship that has many applications. Given the definition

$$P(A|B) = \frac{P(A \cap B)}{P(B)}.$$

Multiply both sides by $P(B)$ to obtain the equation

$$P(A \cap B) = P(B)P(A|B).$$

Since intersection of sets is commutative, we also obtain the equation

$$P(A \cap B) = P(A)P(B|A).$$

example **5.28 Online Advertising**

A company advertises its product on a website. The company estimates that the ad will be read by 20% of the people who visit the site. It further estimates that if the ad is read, then the probability that the reader will buy the product is 0.005. What is the probability that a visitor to the website will read the ad and buy the product?

Let A be the event "read the ad" and let B be the event "buy the product." So we are given $P(A) = 0.20$ and $P(B|A) = 0.005$. We are asked to find the probability $P(A \cap B)$, which is

$$P(A \cap B) = P(A)P(B|A) = (0.20)(0.005) = 0.001.$$

end example

After-the-Fact (*a posteriori*) Probabilities

We usually use probability to give information regarding whether an event will happen in the future based on past experience. This is often called *a priori* probability. But probability can sometimes be used to answer questions about whether an event has happened based on new information. This is often called *a posteriori* probability (or after-the-fact probability).

example **5.29   Sports and Weather**

Suppose a sports team wins 75% of the games it plays in good weather, but the team wins only 50% of the games it plays in bad weather. The historic weather pattern for September has good weather two-thirds of the time and bad weather the rest of the time. If we read in the paper that the team has won a game on September 12, what is the probability that the weather was bad on that day?

Let W and L be the events win and lose, respectively. Let G and B be the events good weather and bad weather, respectively. We are given the conditional probabilities $P(W \mid G) = 3/4$ and $P(W \mid B) = 1/2$. The probabilities for the weather in September are $P(G) = 2/3$ and $P(B) = 1/3$. The question asks us to find $P(B \mid W)$. The definition gives us the formula

$$P(B|W) = \frac{P(B \cap W)}{P(W)}.$$

How do we find $P(W)$? Since G and B are mutually exclusive, we have

$$W = (G \cap W) \cup (B \cap W).$$

Therefore, we have

$$P(W) = P(G \cap W) + P(B \cap W).$$

So our calculation becomes

$$P(B|W) = \frac{P(B \cap W)}{P(W)} = \frac{P(B \cap W)}{P(G \cap W) + P(B \cap W)}.$$

Although we are not given $P(B \cap W)$ and $P(G \cap W)$, we can calculate the values using the given information together with the equations

$$P(B \cap W) = P(B)P(W|B) \text{ and } P(G \cap W) = P(G)P(W|G).$$

So our calculation becomes

$$P(B|W) = \frac{P(B \cap W)}{P(W)} = \frac{P(B \cap W)}{P(G \cap W) + P(B \cap W)}$$

$$= \frac{P(W|B)P(B)}{P(W|G)P(G) + P(W|B)P(B)}$$

$$= \frac{(1/2)(1/3)}{(3/4)(2/3) + (1/2)(1/3)} = 1/4.$$

end example

Bayes' Theorem

The formula we derived in the previous example is an instance of a well-known theorem called *Bayes' theorem*. Suppose a sample space is partitioned into disjoint (i.e., mutually exclusive) events $H_1, \ldots, H_n$ and E is another event such that $P(E) \neq 0$. Then we have the following formula for each $P(H_i \mid E)$.

$$P(H_i|E) = \frac{P(H_i \cap E)}{P(H_1 \cap E) + \cdots + P(H_n \cap E)}$$

$$= \frac{P(H_i)P(E|H_i)}{P(H_1)P(E|H_1) + \cdots + P(H_n)P(E|H_n)}.$$

Each $P(H_i)$ is the *a priori* probability of H_i and each $P(H_i|E)$ is the *a posteriori* probability of H_i given E.

Bayes' theorem is used in many contexts. For example, H_1, ..., H_n might be a set of mutually exclusive hypotheses that are possible causes of some outcome that can be tested by experiment and E could be an experiment that results in the outcome.

example 5.30 Software Errors

Suppose that the input data set for a program is partitioned into two types, one makes up 60% of the data and the other makes up 40%. Suppose further that inputs from the two types cause warning messages 15% of the time and 20% of the time, respectively. If a random warning message is received, what are the chances that it was caused by an input of each type?

To solve the problem we can use Bayes' theorem. Let E_1 and E_2 be the two sets of data and let B be the set of data that causes warning messages. Then we want to find $P(E_1|B)$ and $P(E_2|B)$. Now we are given the following probabilities:

$$P(E_1) = 0.6, \ P(E_2) = 0.4, \ P(B|E_1) = 0.15, \ P(B|E_2) = 0.2$$

So we can calculate $P(E_1 \mid B)$ as follows:

$$P(E_1|B) = \frac{P(E_1)P(B|E_1)}{P(E_1)P(B|E_1) + P(E_2)P(B|E_2)}$$

$$= \frac{(0.6)(0.15)}{(0.6)(0.15) + (0.4)(0.2)} = \frac{0.09}{0.17} \approx 0.53.$$

A similar calculation gives $P(E_2|B) \approx 0.47$.

end example

5.4.3   Independent Events

Informally, two events A and B are independent if they don't influence each other. If A and B don't influence each other and their probabilities are nonzero, we would like to say that $P(A|B) = P(A)$ and $P(B|A) = P(B)$. This condition will follow from the definition of independence. Two events A and B are *independent* if the following equation holds:

$$P(A \cap B) = P(A)\,P(B).$$

It's interesting to note that if A and B are independent events, then so are the three pairs of events A and B', A' and B, and A' and B'. We'll discuss this in the exercises.

The nice thing about independent events is that they simplify the task of assigning probabilities and computing probabilities.

example   5.31   Independence of Events

In the two-coin-flip example, let A be the event that the first coin is heads and let B be the event that the two coins come up different. Then $A = \{HT,\ HH\}$, $B = \{HT,\ TH\}$, and $A \cap B = \{HT\}$. If each coin is fair, then A and B are independent because $P(A) = P(B) = 1/2$ and $P(A \cap B) = 1/4$.

Of course many events are not independent. For example, if C is the event that at least one coin is tails, then $C = \{HT,\ TH,\ TT\}$. It follows that $A \cap C = \{HT\}$ and $B \cap C = \{HT,\ TH\}$. If the coins are fair, then it follows that A and C are dependent events and also that B and C are dependent events.

end example

Repeated Independent Trials

Independence is often used to assign probabilities for repeated trials of the same experiment. We'll be content here to discuss repeated trials of an experiment with two outcomes, where the trials are independent. For example, if we flip a coin n times, it's reasonable to assume that each flip is independent of the other flips. To make things a bit more general, we'll assume that a coin comes up either heads with probability p or tails with probability $1 - p$. Here is the question that we want to answer.

What is the probability that the coin comes up heads exactly k times?

To answer this question we need to consider the independence of the flips. For example, if we let A_i be the event that the ith flip comes up heads, then $P(A_i) = p$ and $P(A_i') = 1 - p$. Suppose now that we ask the probability that the first k flips come up heads and the last $n - k$ flips come up tails. Then we are asking about the probability of the event

$$A_1 \cap \cdots \cap A_k \cap A_{k+1}' \cap \cdots \cap A_n'.$$

Since each event in the intersection is independent of the other events, the probability of the intersection is the product of probabilities

$$p^k(1-p)^{n-k}.$$

We get the same answer for each arrangement of k heads and $n - k$ tails. So we'll have an answer to the question if we can find the number of different arrangements of k heads and $n - k$ tails. By (5.27) there are

$$\frac{n!}{k!\,(n-k)!}$$

such arrangements. This is also $C(n, k)$, which we can represent by the binomial coefficient symbol. So if a coin flip is repeated n times, then the probability of k successes is given by the expression

$$\binom{n}{k} p^k(1-p)^{n-k}.$$

This set of probabilities is called the *binomial distribution*. The name fits because by the binomial theorem, the sum of the probabilities as k goes from 0 to n is 1. We should note that although we used coin flipping to introduce the ideas, the binomial distribution applies to any experiment with two outcomes that has repeated trials.

example 5.32 A Good Golfer

A golfer wins 60% of the tournaments that s/he enters. What is the probability that the golfer will win exactly five of the next seven tournaments? There are two outcomes, win and lose, with probabilities 0.6 and 0.4, respectively. So the probability of exactly five wins in seven tournaments is given by

$$\binom{7}{5} (0.6)^5(0.4)^2 \approx 0.26.$$

end example

5.4.4 Expectation and Average Behavior

Let's get back to talking about averages and expectations. We all know that the average of a bunch of numbers is the sum of the numbers divided by the number of numbers. So what's the big deal? The deal is that we often assign numbers to each outcome in a sample space. For example, in our beginning discussion we assigned a path length to each of the first 15 prime numbers. We added up the 15 path lengths and divided by 15 to get the average. Makes sense, doesn't it? But remember, we assumed that each number was equally likely to occur. This is not always the case. So we also have to consider the probabilities assigned to the points in the sample space.

Let's look at another example to set the stage for a definition of expectation. Suppose we agree to flip a coin. If the coin comes up heads, we agree to pay 4 dollars; if it comes up tails, we agree to accept 5 dollars. Notice here that we have assigned a number to each of the two possible outcomes of this experiment. What is our expected take from this experiment? It depends on the coin. Suppose the coin is fair. After one flip we are either 4 dollars poorer or 5 dollars richer. Suppose we play the game 10 times. What then? Well, since the coin is fair, it seems likely that we can expect to win five times and lose five times. So we can expect to pay 20 dollars and receive 25 dollars. Thus our expectation from 10 flips is 5 dollars.

Suppose we knew that the coin was biased with $P(\text{head}) = 2/5$ and $P(\text{tail}) = 3/5$. What would our expectation be? Again, we can't say much for just one flip. But for 10 flips we can expect about four heads and six tails. Thus we can expect to pay out 16 dollars and receive 30 dollars, for a net profit of 14 dollars. An equation to represent our reasoning follows:

$$10P\,(\text{head})\,(-4) + 10P\,(\text{tail})\,(5) = 10\,(2/5)\,(-4) + 10\,(3/5)\,(5) = 70/5 = 14.$$

Can we learn anything from this equation? Yes, we can. The 14 dollars represents our take over 10 flips. What's the average profit? Just divide by 10 to get $1.40. This can be expressed by the following equation:

$$P\,(\text{head})\,(-4) + P\,(\text{tail})\,(5) = (2/5)\,(-4) + (3/5)\,(5) = 7/5 = 1.4.$$

So we can compute the average profit per flip without using the number of coin flips. The average profit per flip is $1.40 no matter how many flips there are. That's what probability gives us. It's called *expectation*, and we'll generalize from this example to define expectation for any sample space having an assignment of numbers to the sample points.

Definition of Expectation

Let S be a sample space, P a probability distribution on S, and $V : S \to \mathbb{R}$ an assignment of numbers to the points of S. Suppose $S = \{x_1, x_2, \ldots, x_n\}$. Then the *expected value* (or *expectation*) of V is defined by the following formula.

$$E(V) = V(x_1)P(x_1) + V(x_2)P(x_2) + \cdots + V(x_n)P(x_n).$$

So when we want the average behavior, we're really asking for the expectation. For example, in our little coin-flip example we have $S = \{\text{head}, \text{tail}\}$, $P(\text{head}) = 2/5$, $P(\text{tail}) = 3/5$, $V(\text{head}) = -4$, and $V(\text{tail}) = 5$. So the expectation of V is calculated by $E(V) = (-4)(2/5) + 5(3/5) = 1.4$.

We should note here that in probability theory the function V is called a *random variable*.

Average Performance of an Algorithm

To compute the average performance of an algorithm A, we must do several things: First, we must decide on a sample space to represent the possible inputs of size n. Suppose our sample space is $S = \{I_1, I_2, \ldots, I_k\}$. Second, we must define a probability distribution P on S that represents our idea of how likely it is that the inputs will occur. Third, we must count the number of operations required by A to process each sample point. We'll denote this count by the function $V : S \to \mathbb{N}$. Lastly, we'll let $\text{Avg}_A(n)$ denote the average number of operations to execute A as a function of input size n. Then $\text{Avg}_A(n)$ is just the expectation of V:

$$\text{Avg}_A(n) = E(V) = V(I_1)P(I_1) + V(I_2)P(I_2) + \cdots + V(I_k)P(I_k).$$

To show that an algorithm A is *optimal in the average case* for some problem, we need to specify a particular sample space and probability distribution. Then we need to show that $\text{Avg}_A(n) \leq \text{Avg}_B(n)$ for all $n > 0$ and for all algorithms B that solve the problem. The problem of finding lower bounds for the average case is just as difficult as finding lower bounds for the worst case. So we're often content to just compare known algorithms to find the best of the bunch.

Analysis of Sequential Search

Suppose we have the following algorithm to search for an element X in an array L, indexed from 1 to n. If X is in L, the algorithm returns the index of the rightmost occurrence of X. The index 0 is returned if X is not in L:

$$i := n;$$
$$\textbf{while } i \geq 1 \textbf{ and } X \neq L[i] \textbf{ do}$$
$$i := i - 1$$
$$\textbf{od}$$

We'll count the average number of comparisons $X \neq L[i]$ performed by the algorithm. First we need a sample space. Suppose we let I_i denote the input case where the rightmost occurrence of X is at the ith position of L. Let I_{n+1} denote the case in which X is not in L. So the sample space is the set

$$\{I_1, I_2, \ldots, I_{n+1}\}.$$

Let $V(I)$ denote the number of comparisons made by the algorithm when the input has the form I. Looking at the algorithm, we obtain

$$V(I_i) = n - i + 1 \quad \text{for } 1 \leq i \leq n,$$
$$V(I_{n+1}) = n.$$

Suppose we let q be the probability that X is in L. Thus $1 - q$ is the probability that X is not in L. Let's also assume that whenever X is in L, its position is random. This gives us the following probability distribution P over the sample space:

$$P(I_i) = \frac{q}{n} \quad \text{for } 1 \leq i \leq n,$$
$$P(I_{n+1}) = 1 - q.$$

Therefore, the expected number of comparisons made by the algorithm for this probability distribution is given by the expected value of V:

$$
\begin{aligned}
\text{Avg}_A(n) = E(V) &= V(I_1)P(I_1) + \cdots + V(I_{n+1})P(I_{n+1}) \\
&= \frac{q}{n}\left(n + (n-1) + \cdots + 1\right) + (1-q)n \\
&= q\left(\frac{n+1}{2}\right) + (1-q)n.
\end{aligned}
$$

Let's observe a few things about the expected number of comparisons. If we know that X is in L, then $q = 1$. So the expectation is $(n+1)/2$ comparisons. If we know that X is not in L, then $q = 0$, and the expectation is n comparisons. If X is in L and it occurs at the first position, then the algorithm takes n comparisons. So the worst case occurs for the two input cases I_{n+1} and I_1, and we have $W_A(n) = n$.

5.4.5 Finite Markov Chains

We mentioned before that probability comes up whenever we ask about the chance of something happening. To answer such a question requires one to make some kind of assumption. Now we'll look at processes that change state over time where each change of state depends only on the previous state and a given probability distribution about the chances of changing from any state to any other state. Such a process is called a *Markov chain*. The main property is that the next state depends only on the current state and the given probability for changing states.

We'll introduce the ideas of *finite Markov chains*, where the number of states is finite, with a two-state example to keep things manageable. For example, in weather forecasting, the two states might be sunny and cloudy. We must find a probability distribution about the chances of changing states. For example, the probability of a sunny day tomorrow given that today is sunny might be 0.6 and the probability of a sunny day tomorrow given that today is cloudy might be 0.3. From these conditional probabilities we can then forecast the weather. Of course, there can also be several states. For example, the weather example might have the three states rain, sunny, and partly cloudy.

So the process is very general, with almost unlimited applications. It applies to any process where a change of state depends only on the given state of affairs and a given (i.e., assumed) probability of moving from that state to another state. Now let's get down to business and introduce the ideas. The account we give is based on that given in the elegant little booklet on probability by Bates [1965].

Introduction with a Two–State Example

For ease of notation, we'll assume that there are two states, 0 and 1. We'll assume that the probability of moving from 0 to 1 is 0.6, so the probability of moving from 0 to 0 is 0.4. We'll assume that the probability of moving from 1 to 1 is 0.3, so the probability of moving from 1 to 0 is 0.7.

We can represent these four probabilities by the following matrix where the first row gives the probabilities of entering states 0 and 1 when starting from state 0 and the second row gives the probabilities of entering states 0 and 1 when starting from state 1.

$$P = \begin{pmatrix} 0.1 & 0.9 \\ 0.6 & 0.4 \end{pmatrix}.$$

The matrix P is called the *transition matrix* of the chain.

We can also picture the situation with a directed graph where the nodes are the states and each edge is labeled with the probability of traversing the edge to the next state.

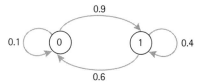

For example, suppose the process starts in state 0 and we are interested in the chance of entering state 1 after two stages (or transitions). There are two possible paths of length two from state 0 that end in state 1: the path 0, 1, 1 and the path 0, 0, 1. In other words, we can move to state 1 and then again to state 1 with probability $(0.9)(0.4)$ or we can move to state 0 and then to state 1 with probability $(0.1)(0.9)$. We'll represent these two possible paths by the event $\{011, 001\}$. So the desired probability is,

$$P(\{011, 001\}) = (0.9)(0.4) + (0.1)(0.9) = 0.45.$$

For example, if we start in state 0 the chance of entering state 0 after two stages is represented by the event $\{000, 010\}$. So the desired probability is,

$$P(\{000, 010\}) = (0.1)(0.1) + (0.9)(0.6) = 0.55.$$

Similarly, to calculate the two-stage probabilities for starting in state 1 we have the two events $\{100, 110\}$ and $\{101, 111\}$. So the desired probabilities are,

$$P(\{100, 110\}) = (0.6)(0.1) + (0.4)(0.6) = 0.30$$
$$P(\{101, 111\}) = (0.6)(0.9) + (0.4)(0.4) = 0.70.$$

We can represent these four probabilities by the following matrix, where the first row gives the probabilities for starting in state 0 and ending in states 0 and 1, and the second row gives the probabilities for starting in state 1 and ending in states 0 and 1:

$$\begin{pmatrix} 0.55 & 0.45 \\ 0.30 & 0.70 \end{pmatrix}.$$

Now, we come to an interesting relationship. The preceding matrix of two-stage probabilities is just the product of the matrix P with itself, which you should verify for yourself:

$$P^2 = PP = \begin{pmatrix} 0.1 & 0.9 \\ 0.6 & 0.4 \end{pmatrix} \begin{pmatrix} 0.1 & 0.9 \\ 0.6 & 0.4 \end{pmatrix} = \begin{pmatrix} 0.55 & 0.45 \\ 0.30 & 0.70 \end{pmatrix}.$$

The nice part is that this kind of relationship works for any number of stages. In other words, if we are interested in the probability of entering some state after n stages, we can find the answer in the matrix P^n.

Let's demonstrate this relationship for three stages of the process. For example, if we start in state 0 the chance of entering state 1 after three stages is represented by the event $\{0001, 0011, 0101, 0111\}$. So the desired probability is,

$$P(\{0001, 0011, 0101, 0111\})$$
$$= (0.1)(0.1)(0.9) + (0.1)(0.9)(0.4) + (0.9)(0.6)(0.9) + (0.9)(0.4)(0.4) = 0.675.$$

Now we could calculate the other probabilities in the same way. But instead, we'll calculate the matrix P^3.

$$P^3 = PP^2 = \begin{pmatrix} 0.1 & 0.9 \\ 0.6 & 0.4 \end{pmatrix} \begin{pmatrix} 0.55 & 0.45 \\ 0.30 & 0.70 \end{pmatrix} = \begin{pmatrix} 0.325 & 0.675 \\ 0.450 & 0.550 \end{pmatrix}.$$

Notice, for example, that the entry in the second row first column is 0.675. As expected, it is the probability of entering state 1 after three stages starting from state 0. As an exercise, you should compute the other three probabilities and confirm that they are the other three entries of P^3.

It can be shown that this process extends to any n. In other words, we have the following general property for any finite Markov chain.

Markov Chain Property

Given the transition matrix P, the probabilities after n stages of the process are given by P^n, the nth power P.

The Starting State

At first glance it appears that the process will give a different result depending on whether we start in state 0 or state 1. For example, suppose we flip a coin to choose the starting state. Then the probability of entering state 0 after one stage is

$$(0.5)(0.1) + (0.5)(0.6) = 0.35,$$

and the probability of entering state 1 after one stage is

$$(0.5)(0.9) + (0.5)(0.4) = 0.65.$$

Notice that we can calculate these one-stage results by multiplying the vector $(0.5, 0.5)$ times the matrix P to obtain

$$(0.5,\ 0.5)\, P = (0.5,\ 0.5) \begin{pmatrix} 0.1 & 0.9 \\ 0.6 & 0.4 \end{pmatrix} = (0.35,\ 0.65)\,.$$

In a similar way, we can obtain the two-stage probabilities by multiplying the initial vector $(0.5, 0.5)$ times P^2 to obtain

$$(0.5,\ 0.5)\, P^2 = (0.5,\ 0.5) \begin{pmatrix} 0.1 & 0.9 \\ 0.6 & 0.4 \end{pmatrix} \begin{pmatrix} 0.1 & 0.9 \\ 0.6 & 0.4 \end{pmatrix}$$

$$= (0.35,\ 0.65) \begin{pmatrix} 0.1 & 0.9 \\ 0.6 & 0.4 \end{pmatrix} = (0.425,\ 0.575)\,.$$

Let's observe what happens to these vectors if we continue this process for a few stages. Let X_n be the vector of probabilities after n stages of the process that starts with

$$X_0 = (0.5, 0.5).$$

X_0 is called the *initial probability vector* of the process. We've already calculated X_1 and X_2 as

$$X_1 = X_0 P = (0.5, 0.5)P = (0.35, 0.65),$$
$$X_2 = X_0 P^2 = (X_0 P)P = X_1 P = (0.35, 0.65)P = (0.425, 0.575).$$

You can verify that the next three values are

$$X_3 = X_0 P^3 = X_2 P = (0.3875, 0.6125),$$
$$X_4 = X_0 P^4 = X_3 P = (0.40625, 0.59375),$$
$$X_5 = X_0 P^5 = X_4 P = (0.396875, 0.603125).$$

Some Questions

Notice that the values of the vectors appear to be approaching some kind of constant value that looks like it might be (0.4, 0.6). Do we have to go on calculating to see the value? Does the value, if it exists, depend on the initial vector X_0? In other words, in the long run does it make a difference which state we start with?

Some Answers

These questions can be answered if we consider the following question that may seem a bit strange. Is there a probability vector $X = (p, q)$, where $p + q = 1$, such that $XP = X$? To see that the answer is yes, we'll solve the equation for the unknown p. The equation $XP = X$ becomes

$$(p,\ q) \begin{pmatrix} 0.1 & 0.9 \\ 0.6 & 0.4 \end{pmatrix} = (p,\ q).$$

This gives us the two equations

$$(0.1)p + (0.6)q = p,$$
$$(0.9)p + (0.4)q = q.$$

Each equation gives the relation

$$p = (2/3)q.$$

Since $p + q = 1$, we can solve for p and q to obtain $p = 0.4$ and $q = 0.6$, which tells us that the probability vector $X = (0.4, 0.6)$ satisfies the equation $XP = X$. So the answer to our strange question is yes.

Now, we noticed that the sequence of probability vectors X_0, X_1, X_2, X_3, X_4, X_5 appears to be approaching the vector (0.4, 0.6), which just happens to be the solution to the equation $XP = X$. Is there some connection? The following theorem tells all.

Markov Chain Theorem

If P is the transition matrix of a finite Markov chain such that some power of P has no zero entries, then the following statements hold:

a. There is a unique probability vector X such that $XP = X$ and X has no zero entries.

Continued ➡

➥ ➥

---

b. As n increases, matrix P^n approaches the matrix that has X in each row.

c. If X_0 is any initial probability vector, then as n increases, the vector $X_n = X_0 P^n$ approaches the unique probability vector X.

Now we can answer the questions we asked about the process. Part (a) answers the strange question that asks whether there is a probability vector X such that $XP = X$. Part (c) answers the question that asks about whether the values of X_n appear to be approaching some kind of constant value. The constant value is the X from Part (a). Parts (b) and (c) tell us that X does not depend on the initial vector X_0. In other words, in the long run, it makes no difference how the process starts. So in our example, the probability of entering state 0 in the long run is 0.4 and the probability of entering state 1 is 0.6, no matter how we chose the starting state.

Many–State Examples

We introduced Markov chains with a two-state example. But the ideas extend to any finite number of states. For example, in weather forecasting, we might have three or more conditions such as sunny, partly cloudy, rainy, etc. We might have a golf example, where a golfer on any hole has par, less than par, or greater than par. We could have a baseball example with the three hitting conditions for a player of single, double, and other hit (triple or home run). In business, the applications are endless. Similarly, in computer science, applications to performance of software/hardware abound. In the following example, we'll apply Markov chains to software development.

example **5.33 Software Development**

We'll assume that a set of instructions for a program is partitioned into three sequences, which we'll call A, B, and C, where the last instruction in each sequence is a jump/branch instruction. While testing the software, we find that once the execution enters one of these sequences, the jump/branch instructions at the end of the sequences have the following transfer pattern:

> A transfers back to A 60% of the time, to B 20% of the time, and to C 20% of the time.

> B transfers to A 20% of the time, back to B 50% of the time, and to C 30% of the time.

> C transfers to A 40% of the time, B 40% of the time, and back to C 20% of the time.

The transition matrix for this set of probabilities is

$$P = \begin{pmatrix} 0.6 & 0.2 & 0.2 \\ 0.2 & 0.5 & 0.3 \\ 0.4 & 0.4 & 0.2 \end{pmatrix}.$$

Now we can answer some questions, like, which code sequence will be used the most during long executions? Since P has no zero entries, we can find the probability vector X such that $XP = X$. To solve for X, we'll let $X = (x, y, z)$, where $x + y + z = 1$. Then the matrix equation $XP = X$ becomes

$$(x, y, z) \begin{pmatrix} 0.6 & 0.2 & 0.2 \\ 0.2 & 0.5 & 0.3 \\ 0.4 & 0.4 & 0.2 \end{pmatrix} = (x, y, z).$$

This gives us the three equations,

$$(0.6)x + (0.2)y + (0.4)z = x,$$
$$(0.2)x + (0.5)y + (0.4)z = y,$$
$$(0.2)x + (0.3)y + (0.2)z = z.$$

Now we can solve these equations subject to the probability restriction,

$$x + y + z = 1.$$

After collecting terms in the first three equations, we obtain the following set of four equations:

$$\begin{aligned} x + \quad y + \quad z &= 1, \\ -0.4x + \quad 0.2y + \quad 0.4z &= 0, \\ 0.2x + -0.5y + \quad 0.4z &= 0, \\ 0.2x + \quad 0.3y + -0.8z &= 0. \end{aligned}$$

At this point, let's recall two elementary operations on sets of linear equations that can be performed without changing the solutions:

a. Replace an equation by multiplying it by a nonzero number.

b. Replace an equation by adding it to a nonzero multiple of another equation.

By using these two elementary operations we can transform the set of equations to a simpler form that is easy to solve. The process is called *Gaussian elimination*.

For example, if we like to work with whole numbers, we can use operation (a) and multiply each of the last three equations by 10. This gives us the following set of equations:

$$
\begin{aligned}
x + \quad y + \quad z &= 1, \\
-4x + \quad 2y + \quad 4z &= 0, \\
2x + -5y + \quad 4z &= 0, \\
2x + \quad 3y + -8z &= 0.
\end{aligned}
$$

We can use operation (b) to eliminate x from the second equation by adding 4 times the first equation to the second equation. Similarly, we can eliminate x from the third and fourth equations by adding -2 times the first equation to each equation. The result is the following set of equations:

$$
\begin{aligned}
x + \quad y + \quad z &= 1, \\
6y + \quad 8z &= 4, \\
-7y + \quad 2z &= -2, \\
y + -10z &= -2.
\end{aligned}
$$

Now we can eliminate y from the second equation by adding -6 times the last equation to the second equation. Similarly, we can eliminate y from the third equation by adding 7 times the last equation to the third equation. The result is the following set of equations:

$$
\begin{aligned}
x + y + \quad z &= 1, \\
68z &= 16, \\
-68z &= -16, \\
y + -10z &= -2.
\end{aligned}
$$

Now we can back substitute to find $z = 4/17$, $y = -2 + 10z = 6/17$, and $x = 1 - y - z = 7/17$. So the probability vector is

$$
X = (x, y, z) = (7/17, 6/17, 4/17) \approx (0.41, 0.35, 0.24).
$$

We can conclude that no matter which sequence is entered initially, in executions that require many jump/branch instructions, sequence A will be executed about 41% of the time and those in sequences B and C will be executed 35% and 24% of the time, respectively. This information can be used to decide which instructions to place in cache memory. In a dynamic setting, when a jump/branch instruction is executed, the probability vector can be used to predict the next instruction to execute so that it can be pre-fetched and ready to execute.

end example

5.4.6 Approximations (Monte Carlo Method)

Sometimes it is not so easy to find a formula to solve a problem. In some of these cases we can find reasonable approximations by repeating some experiment many times and then observing the results. For example, suppose we have an irregular shape drawn on a piece of paper and we would like to know the area of the shape. The *Monte Carlo method* would have us randomly choose a large number of points on the paper. Then the area of the shape would be pretty close to the percentage of points that lie within the shape multiplied by the area of the paper.

The Monte Carlo method is useful in probability not only to check a calculated answer for a problem, but to find reasonable answers to problems for which we have no other answer. For example, a computer simulating thousands of repetitions of an experiment can give a pretty good approximation to the average outcome of the experiment.

 Exercises

Discrete Probability

1. Suppose three fair coins are flipped. Find the probability for each of the following events.

   a. Exactly one coin is a head.   b. Exactly two coins are tails.
   c. At least one coin is a head.   d. At most two coins are tails.

2. Suppose a pair of dice are flipped. Find the probability for each of the following events.

   a. The sum of the dots is 7.
   b. The sum of the dots is even.
   c. The sum of the dots is either 7 or 11.
   d. The sum of the dots is at least 5.

3. A team has probability 2/3 of winning whenever it plays. Find each of the following probabilities that the team will win.

   a. Exactly 4 out of 5 games.
   b. At most 4 out of 5 games.
   c. Exactly 4 out of 5 games given that it has already won the first 2 games of a 5-game series.

4. A baseball player's batting average is .250. Find each of the following probabilities that he will get hits.

   a. Exactly 2 hits in 4 times at bat.
   b. At least one hit in 4 times at bat.

5. A computer program uses one of three procedures for each piece of input. The procedures are used with probabilities 1/3, 1/2, and 1/6. Negative results are detected at rates of 10%, 20%, and 30% by the three procedures, respectively. Suppose a negative result is detected. Find the probabilities that each of the procedures was used.

6. A commuter crosses one of three bridges, A, B, or C, to go home from work, crossing A with probability 1/3, B with probability 1/6, and C with probability 1/2. The commuter arrives home by 6 p.m. 75%, 60%, and 80% of the time by crossing bridges A, B, and C, respectively. If the commuter arrives home by 6 p.m., find the probability that bridge A was used. Also find the probabilities for bridges B and C.

7. A student is chosen at random from a class of 80 students that has 20 honor students, 30 athletes, and 40 that are neither honor students nor athletes.

   a. What is the probability that the student selected is an athlete given that he or she is an honor student?
   b. What is the probability that the student selected is an honor student given that he or she is an athlete?
   c. Are the events "honor student" and "athlete" independent?

8. Suppose we have an algorithm that must perform 2,000 operations as follows: The first 1,000 operations are performed by a processor with a capacity of 100,000 operations per second. Then the second 1,000 operations are performed by a processor with a capacity of 200,000 operations per second. Find the average number of operations per second performed by the two processors to execute the 2,000 operations.

9. Consider each of the following lottery problems.

   a. Find the chances of winning a lottery that allows you to pick six numbers from the set $\{1, 2, \ldots, 49\}$.
   b. Suppose that a lottery consists of choosing a set of five numbers from the set $\{1, 2, \ldots, 49\}$. Suppose further that smaller prizes are given to people with four of the five winning numbers. What is the probability of winning a smaller prize?
   c. Suppose that a lottery consists of choosing a set of six numbers from the set $\{1, 2, \ldots, 49\}$. Suppose further that smaller prizes are given to people with four or five of the six winning numbers. What is the probability of winning a smaller prize?
   d. Find a formula for the probability of winning a smaller prize that goes with choosing k of the winning m numbers from the set $\{1, \ldots, n\}$, where $k < m < n$.

10. For each of the following problems, compute the expected value.

   a. The expected number of dots that show when a die is tossed.

   b. The expected score obtained by guessing all 100 questions of a true-false exam in which a correct answer is worth 1 point and an incorrect answer is worth $-1/2$ point.

11. Use the Monte Carlo method to answer each question about throwing darts.

   a. A dart lands inside a square of side length x. What is the probability that the dart landed outside the circle that is inscribed in the square?

   b. An irregular shape S is drawn within a square of side length x. A number of darts are thrown at the square with 70 landing inside the square but outside S and 45 landing inside S. What is the approximate area of S?

Markov Chains

12. This exercise will use the transition matrix P from the introductory example.

$$P = \begin{pmatrix} 0.1 & 0.9 \\ 0.6 & 0.4 \end{pmatrix}$$

   a. Calculate P^4, P^8, and P^{16}. Notice that it does not require a large value of n for P^n to get close to the matrix

$$\begin{pmatrix} 0.4 & 0.6 \\ 0.4 & 0.6 \end{pmatrix}$$

   described in Part (b) of the Markov chain theorem.

   b. For the initial probability vector $X_0 = (0.5, 0.5)$ from the introductory example, calculate the probability vectors $X_4 = X_0 P^4$, $X_8 = X_0 P^8$, and $X_{16} = X_0 P^{16}$.

   c. Suppose the starting state of the Markov chain is chosen by tossing a fair die, where state 0 is chosen if the top of the die is six, and otherwise state 1 is chosen. In other words, the initial probability vector is $X_0 = (1/6, 5/6)$. Calculate the probability vectors X_4, X_8, and X_{16}, and compare your results with those of Part (b).

13. A company has gathered statistics on three of its products A, B, and C. (You can think of A, B, and C as three breakfast cereals, or as three models of automobile, or as any three products that compete with each other for market share.) The statistics show that customers switch between products according to the following transition matrix:

$$P = \begin{pmatrix} 0 & 0.5 & 0.5 \\ 0.5 & 0.2 & 0.3 \\ 0.3 & 0 & 0.7 \end{pmatrix}.$$

a. Calculate P^2 and observe that it has no zero entries.

b. Since Part (a) shows that P^2 has no zero entries, the Markov theorem tells us that there is a unique probability vector X such that $XP = X$ and X has no zero entries. Find the probability vector X such that $XP = X$.

c. Calculate P^4 and P^8. Notice that the sequence P, P^2, P^4, P^8 gives good evidence of the fact that P^n approaches the matrix with X in each row.

d. Let $X_0 = (0.1, 0.8, 0.1)$ be the initial probability vector with respect to customers buying the products A, B, and C. Compute $X_1 = X_0P$, $X_2 = X_0P^2$, $X_4 = X_0P^4$, and $X_8 = X_0P^8$.

e. Let $X_0 = (0.3, 0.1, 0.6)$ be the initial probability vector with respect to customers buying the products A, B, and C. Compute $X_1 = X_0P$, $X_2 = X_0P^2$, $X_4 = X_0P^4$, and $X_8 = X_0P^8$. Compare the results with those of Part (d).

f. Suppose that the company manufactures the three products A, B, and C on the same assembly line that can produce 1200 items in a certain period of time. How many items of each type should be manufactured?

Challenges

14. Show that if S is a sample space and A is an event, then S and A are independent events. What about the independence of two events A and B that are disjoint?

15. Prove that if A and B are independent events, then so are the three pairs of events A and B', A' and B, and A' and B'.

16. (*Average-Case Analysis of Binary Search*).

   a. Assume that we have a sorted list of 15 elements, x_1, x_2, ..., x_{15}. Calculate the average number of comparisons made by a binary search algorithm to look for a key that may or may not be in the list. Assume that the key has probability $1/2$ of being in the list and that each of the events "key $= x_i$" is equally likely for $1 \leq i \leq 15$.

    b. Generalize the problem to find a formula for the average number of comparisons used to look for a key in a sorted list of size $n = 2^k - 1$, where k is a natural number. Assume that the key has probability p of being in the list and that each of the events "key $= x_i$" is equally likely for $1 \leq i \leq n$. Test your formula with $n = 15$ and $p = 1/2$ to see that you get the same answer as Part (a).

17. In each case find the number n of repetitions required to ensure the requested probability.

    a. How many flips of a fair coin will ensure that the probability of obtaining at least one head is greater than or equal to 0.99?

    b. How many flips of a fair coin will ensure that the probability of obtaining at least two heads is greater than or equal to 0.99?

5.5   Solving Recurrences

Many counting problems result in answers that are expressed in terms of recursively defined functions. For example, any program that contains recursively defined procedures or functions will give rise to such expressions. Many of these expressions have closed forms that can simplify the counting process. So let's discuss how to find closed forms for such expressions.

Definition of Recurrence Relation

Any recursively defined function f with domain $\mathbb{N}$ that computes numbers is called a *recurrence* or a *recurrence relation*. When working with recurrences we often write f_n in place of $f(n)$. For example, the following definition is a recurrence:

$$f(0) = 1$$
$$f(n) = 2f(n-1) + n.$$

We can also write this recurrence in the following useful form:

$$f_0 = 1$$
$$f_n = 2f_{n-1} + n.$$

To *solve a recurrence* f we must find an expression for the general term f_n that is not recursive.

5.5.1 Solving Simple Recurrences

Let's start with some simple recurrences that can be solved without much fanfare. The recurrences we'll be considering have the following general form, where a_n and b_n denote either constants or expressions involving n but not involving f.

$$f_0 = b_0, \tag{5.33}$$
$$f_n = a_n f_{n-1} + b_n.$$

We'll look at two similar techniques for solving these recurrences.

Solving by Substitution

One way to solve recurrences of the form (5.33) is by *substitution*, where we start with the definition for f_n and keep substituting for occurrences of f on the right side of the equation until we discover a pattern that allows us to skip ahead and eventually replace the basis f_0. We'll demonstrate the substitution technique with the following example.

example **5.34 Solving by Substitution**

We'll solve the following recurrence by substitution.

$$r_0 = 1,$$
$$r_n = 2r_{n-1} + n.$$

Notice in the following solutions that we never multiply numbers. Instead we keep track of products to help us discover general patterns. Once we find a pattern we emphasize it with parentheses and exponents. Each line represents a substitution and regrouping of terms.

$$r_n = 2r_{n-1} + n$$
$$= 2^2 r_{n-2} + 2(n-1) + n$$
$$= 2^3 r_{n-3} + 2^2(n-2) + 2(n-1) + n$$
$$\vdots$$
$$= 2^{n-1} r_1 + 2^{n-2}(2) + 2^{n-2}(2) + \cdots + 2^2(n-2) + 2^1(n-1) + 2^0(n)$$
$$= 2^n r_0 + 2^{n-1}(1) + 2^{n-2}(2) + \cdots + 2^2(n-2) + 2^1(n-1) + 2^0(n)$$
$$= 2^n(1) + 2^{n-1}(1) + 2^{n-2}(2) + \cdots + 2^2(n-2) + 2^1(n-1) + 2^0(n).$$

Now we'll put it into closed form using (5.1), (5.11c), and (5.11d). Be sure you can see the reason for each step. We'll start by keeping the first term where it is and reversing the rest of the sum to get it in a nicer form.

$$
\begin{aligned}
r_n &= 2^n\,(1) + n + 2\,(n-1) + 2^2\,(n-2) + \cdots + 2^{n-2}\,(2) + 2^{n-1}\,(1) \\
&= 2^n + \left[2^0\,(n) + 2^1\,(n-1) + 2^2\,(n-2) + \cdots + 2^{n-2}\,(2) + 2^{n-1}\,(1)\right] \\
&= 2^n + \sum_{i=0}^{n-1} 2^i\,(n-i) \\
&= 2^n + n\sum_{i=0}^{n-1} 2^i - \sum_{i=0}^{n-1} i2^i \\
&= 2^n + n\,(2^n - 1) - \left(2 - n2^n + (n-1)\,2^{n+1}\right) \\
&= 2^n\,(1 + n + n - 2n + 2) - n - 2 \\
&= 3\,(2^n) - n - 2.
\end{aligned}
$$

Now check a few values of r_n to make sure that the sequence of numbers for the closed form and the recurrence are the same: $1, 3, 8, 19, 42, \ldots$.

end example

Solving by Cancellation

An alternative technique to solve recurences of the form (5.33) is by *cancellation*, where we start with the general equation for f_n. The term on the left side of each succeeding equation is the same as the term that contains f on the right side of the preceding equation. We normally write a few terms until a pattern emerges. The last equation always contains the basis element f_0 on the right side. Now we add up the equations and observe that, except for f_n in the first equation, all terms on the left side of the remaining equations cancel with like terms on the right side of preceding equations. We'll demonstrate the cancellation technique with the following example.

example   **5.35   Solving by Cancellation**

We'll solve the recurrence in Example 5.34 by cancellation:

$$
\begin{aligned}
r_0 &= 1, \\
r_n &= 2r_{n-1} + n.
\end{aligned}
$$

Starting with the general term, we obtain the following sequence of equations, where the term on the left side of a new equation is always the term that contains

r from the right side of the preceding equation.

$$r_n = 2r_{n-1} + n$$
$$2r_{n-1} = 2^2 r_{n-2} + 2(n-1)$$
$$2^2 r_{n-2} = 2^3 r_{n-3} + 2^2 (n-2)$$
$$\vdots$$
$$2^{n-2} r_2 = 2^{n-1} r_1 + 2^{n-2}(2)$$
$$2^{n-1} r_1 = 2^n r_0 + 2^{n-1}(1).$$

Now add up all the equations, cancel the like terms, and replace r_0 by its value to get the following equation.

$$r_n = 2^n(1) + n + 2(n-1) + 2^2(n-2) + \cdots + 2^{n-2}(2) + 2^{n-1}(1).$$

Notice that, except for the ordering of terms, the solution is the same as the one obtained by substitution in Example 5.34.

end example

Since mistakes are easy to make, it is nice to know that you can always check your solution against the original recurrence by testing. You can also give an induction proof that your solution is correct.

The Polynomial Problem

In Example 5.10 we found the number of arithmetic operations in a polynomial of degree n. By grouping terms of the polynomial we can reduce repeated multiplications. For example, here is the grouping when $n = 3$:

$$c_0 + c_1 x + c_2 x^2 + c_3 x^3 = c_0 + x(c_1 + x(c_2 + x(c_3))).$$

Notice that the expression on the left uses nine operations while the expression on the right uses six. The following function will evaluate a polynomial with terms grouped in this way, where C is the list of coefficients:

$$\text{poly}(C, x) = \text{if } C = \langle\,\rangle \text{ then } 0 \text{ else } \text{head}(C) + x \cdot \text{poly}(\text{tail}(C), x).$$

For example, we'll evaluate the expression $\text{poly}(\langle\, a, b, c, d \rangle, x)$:

$$\begin{aligned}
\text{poly}(\langle a, b, c, d \rangle, x) &= a + x \cdot \text{poly}(\langle b, c, d \rangle, x)\\
&= a + x \cdot (b + x \cdot \text{poly}(\langle c, d \rangle, x))\\
&= a + x \cdot (b + x \cdot (c + x \cdot \text{poly}(\langle d \rangle, x)))\\
&= a + x \cdot (b + x \cdot (c + x \cdot (d + x \cdot \text{poly}(\langle\,\rangle, x))))\\
&= a + x \cdot (b + x \cdot (c + d \cdot 0)).
\end{aligned}$$

So there are six arithmetic operations performed by poly to evaluate a polynomial of degree 3. Let's figure out how many operations are performed to evaluate a polynomial of degree n. Let $T(n)$ denote the number of arithmetic operations performed by poly(C, x) when C has length n. If $n = 0$, then $C = \langle \ \rangle$ and

$$\text{poly}(C, x) = \text{poly}(\langle \ \rangle, x) = 0.$$

Therefore, $T(0) = 0$. If $n > 0$, then $C \neq \langle \ \rangle$ and

$$\text{poly}(C, x) = \text{head}(C) + x \cdot \text{poly}(\text{tail}(C), x).$$

This expression has two arithmetic operations plus the number of operations performed by poly(tail(C), x). Since tail(C) has $n - 1$ elements, it follows that poly(tail(C), x) performs $T(n - 1)$ operations. Therefore, for $n > 0$ we have $T(n) = T(n - 1) + 2$. So we have the following recursive definition:

$$T(0) = 0$$
$$T(n) = T(n - 1) + 2.$$

We'll solve this recurrence by cancellation.

$$T(n) = T(n - 1) + 2$$
$$T(n - 1) = T(n - 2) + 2$$

$$\vdots$$

$$T(2) = T(1) + 2$$
$$T(1) = T(0) + 2.$$

Now add the equations, cancel like terms, and replace $T(0)$ by 0 to obtain $T(n) = 2n$. This is quite a savings in the number of arithmetic operations to evaluate a polynomial of degree n. For example, if $n = 30$, then poly uses only 60 operations compared with 494 operations using the method discussed in Example 5.10.

The n-Ovals Problem

Suppose we are given the following sequence of three numbers:

$$2, 4, 8.$$

What is the next number in the sequence? The problem below might make you think about your answer.

The n-Ovals Problem

Suppose that n ovals (an oval is a closed curve that does not cross over itself) are drawn on the plane such that no three ovals meet in a point and each pair of ovals intersects in exactly two points. How many distinct regions of the plane are created by n ovals?

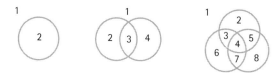

Figure 5.7   One, two, and three ovals.

For example, the diagrams in Figure 5.7 show the cases for one, two, and three ovals. If we let r_n denote the number of distinct regions of the plane for n ovals, then it's clear that the first three values are

$$r_1 = 2,$$
$$r_2 = 4,$$
$$r_3 = 8.$$

What is the value of r_4? Is it 16? Check it out. To find r_n, consider the following description: $n - 1$ ovals divide the region into r_{n-1} regions. The nth oval will meet each of the previous $n - 1$ ovals in $2(n - 1)$ points. So the nth oval will itself be divided into $2(n - 1)$ arcs. Each of these $2(n - 1)$ arcs splits some region in two. Therefore, we add $2(n - 1)$ regions to r_{n-1} to obtain r_n. This gives us the following recurrence.

$$r_1 = 2,$$
$$r_n = r_{n-1} + 2(n - 1).$$

We'll solve it by the substitution technique:

$$
\begin{aligned}
r_n &= r_{n-1} + 2(n - 1) \\
&= r_{n-2} + 2(n - 2) + 2(n - 1) \\
&\ \ \vdots \\
&= r_1 + 2(1) + \cdots + 2(n - 2) + 2(n - 1) \\
&= 2 + 2(1) + \cdots + 2(n - 2) + 2(n - 1).
\end{aligned}
$$

Now we can find a closed form for r_n.

$$
\begin{aligned}
r_n &= 2 + 2(1) + \cdots + 2(n - 2) + 2(n - 1) \\
&= 2 + 2(1 + 2 + \cdots (n - 2) + (n - 1)) \\
&= 2 + 2\sum_{i=1}^{n-1} i \\
&= 2 + 2\frac{(n - 1)(n)}{2} \\
&= n^2 - n + 2.
\end{aligned}
$$

For example, we can use this formula to calculate $r_4 = 14$. Therefore, the sequence of numbers 2, 4, 8 could very well be the first three numbers in the following sequence for the n-ovals problem.

$$2, 4, 8, 14, 22, 32, 44, 62, 74, 92, \dots.$$

5.5.2 Divide-and-Conquer Recurrences

Sometimes a problem can be solved by dividing it up into smaller problems, each of which has an easier solution. The solutions to the smaller problems are then used to construct the solution to the larger problem. This technique is called *divide and conquer*. Algorithms that solve problems by divide and conquer are naturally recursive in nature and have running times that are expressed as recurrences.

We'll introduce the idea by considering divide-and-conquer algorithms that split a problem of size n into a smaller problems, where each subproblem has size n/b. We'll let $f(n)$ denote the number of operations required to obtain the solution to the problem of size n from the a solutions to the subproblems of size n/b. Let $T(n)$ be the total number of operations to solve the problem with input size n. This gives us the following recurrence to describe $T(n)$:

$$T(n) = aT(n/b) + f(n). \tag{5.34}$$

To make sure that the size of each subproblem is an integer, we assume that $n = b^k$ for positive integers b and k with $b > 1$. This defines T for the values, $n = 1, b, b^2, b^3, \dots, b^k$. Now we can solve for $T(n)$. We'll use the cancellation method to obtain the following sequence of equations:

$$T(n) = aT(n/b) + f(n)$$
$$aT(n/b) = a^2 T(n/b^2) + af(n/b)$$
$$a^2 T(n/b^2) = a^3 T(n/b^3) + a^2 f(n/b^2)$$

$$\vdots$$

$$a^{k-1} T(n/b^{k-1}) = a^k T(n/b^k) + a^{k-1} f(n/b^{k-1}).$$

Now we can add the equations, cancel like terms, and use the fact that $n/b^k = 1$ (because $n = b^k$) to obtain

$$T(n) = a^k T(1) + a^{k-1} f(n/b^{k-1}) + \cdots + af(n/b) + f(n)$$
$$= a^k T(1) + \sum_{i=0}^{k-1} a^i f(n/b^i).$$

Therefore, we have the following formula for $T(n)$:

$$T(n) = a^k T(1) + \sum_{i=0}^{k-1} a^i f(n/b^i). \tag{5.35}$$

example **5.36 Some Specific Recurrences**

We'll evaluate (5.35) for some specific recurrences (5.34). For example, suppose we have the following recurrence, where $n = 2^k$:

$$T(1) = 1$$
$$T(n) = 3T(n/2) + n.$$

In this case we have $a = 3$, $b = 2$, and $f(n) = n$. So we can use (5.35) to obtain

$$T(n) = 3^k + \sum_{i=0}^{k-1} 3^i(n/2^i) = 3^k + n\sum_{i=0}^{k-1}(3/2)^i = 3^k + n\left(\frac{(3/2)^k - 1}{1/2}\right).$$

Since $k = \log_2 n$, we can replace k in the last expression and use properties of logs and exponents to obtain

$$T(n) = 3^{\log_2 n} + n\left(\frac{(3/2)^{\log_2 n} - 1}{1/2}\right) = n^{\log_2 3} + 2n(n^{\log_2 3 - 1} - 1)$$

$$= 3n^{\log_2 3} - 2n.$$

For a second example, suppose the recurrence (5.34) is

$$T(1) = 1$$
$$T(n) = 2T(n/2) + n.$$

In this case we have $a = 2$, $b = 2$, and $f(n) = n$. So we can use (5.35) to obtain

$$T(n) = 2^k + \sum_{i=0}^{k-1} 2^i(n/2^i) = 2^k + n\sum_{i=0}^{k-1}(1)^i = 2^k + nk = n + n\log_2 n.$$

end example

5.5.3 Generating Functions

For some recurrence problems we need to find new techniques. For example, suppose we wish to find a closed form for the nth Fibonacci number F_n, which is defined by the recurrence system

$$F_0 = 0,$$
$$F_1 = 1,$$
$$F_n = F_{n-1} + F_{n-2} \quad (n \geq 2).$$

We can't use substitution or cancellation with this system because F occurs twice on the right side of the general equation. This problem belongs to a large class of problems that need a more powerful technique.

The technique that we present comes from the simple idea of equating the coefficients of two polynomials. For example, suppose we have the following equation.

$$a + bx + cx^2 = 4 + 7x^2.$$

We can solve for a, b, and c by equating coefficients to yield $a = 4$, $b = 0$, and $c = 7$. We'll extend this idea to expressions that have infinitely many terms of the form $a_n x^n$ for each natural number n. Let's get to the definition.

Definition of a Generating Function

The *generating function* for the infinite sequence a_0, a_1, ..., a_n, ... is the following infinite expression, which is also called a formal power series or an infinite polynomial:

$$A(x) = a_0 + a_1 x + a_2 x^2 + \cdots + a_n x^n + \cdots$$

$$= \sum_{n=0}^{\infty} a_n x^n.$$

Two generating functions may be added by adding the corresponding coefficients. Similarly, two generating functions may be multiplied by extending the rule for multiplying regular polynomials. In other words, multiply each term of one generating function by every term of the other generating function, and then add up all the results. Two generating functions are equal if their corresponding coefficients are equal.

We'll be interested in those generating functions that have closed forms. For example, let's consider the following generating function for the infinite sequence 1, 1,..., 1, ...:

$$\sum_{n=0}^{\infty} x^n.$$

This generating function is often called a *geometric series*, and its closed form is given by the following formula.

Geometric Series Generating Function   (5.36)

$$\frac{1}{1-x} = \sum_{n=0}^{\infty} x^n.$$

To justify Equation (5.36), multiply both sides of the equation by $1 - x$.

Using a Generating Function Formula

But how can we use this formula to solve recurrences? The idea, as we shall see, is to create an equation in which $A(x)$ is the unknown, solve for $A(x)$, and hope that our solution has a nice closed form. For example, if we find that

$$A(x) = \frac{1}{1-2x},$$

then we can rewrite it using (5.36) in the following way.

$$A(x) = \frac{1}{1-2x} = \frac{1}{1-(2x)} = \sum_{n=0}^{\infty} (2x)^n = \sum_{n=0}^{\infty} 2^n x^n.$$

Now we can equate coefficients to obtain the solution $a_n = 2^n$. In other words, the solution sequence is $1, 2, 4, \ldots, 2^n, \ldots$.

Finding a Generating Function Formula

How do we obtain the closed form for $A(x)$? It's a four-step process, and we'll present it with an example. Suppose we want to solve the following recurrence:

$$a_0 = 0, \tag{5.37}$$
$$a_1 = 1,$$
$$a_n = 5a_{n-1} - 6a_{n-2} \quad (n \geq 2).$$

Step 1

Use the general equation in the recurrence to write an infinite polynomial with coefficients a_n. We start the index of summation at 2 because the general equation in (5.37) holds for $n \geq 2$. Thus we obtain the following equation:

$$\sum_{n=2}^{\infty} a_n x^n = \sum_{n=2}^{\infty} (5a_{n-1} - 6a_{n-2}) x^n$$

$$= \sum_{n=2}^{\infty} 5a_{n-1} x^n - \sum_{n=2}^{\infty} 6a_{n-2} x^n \tag{5.38}$$

$$= 5 \sum_{n=2}^{\infty} a_{n-1} x^n - 6 \sum_{n=2}^{\infty} a_{n-2} x^n.$$

We want to solve for $A(x)$ from this equation. Therefore, we need to transform each infinite polynomial in (5.38) into an expression containing $A(x)$. To do this, notice that the left-hand side of (5.38) can be written as

$$\sum_{n=2}^{\infty} a_n x^n = A(x) - a_0 - a_1 x$$

$$= A(x) - x \quad \text{(substitute for } a_0 \text{ and } a_1\text{)}.$$

The first infinite polynomial on the right side of (5.38) can be written as

$$\sum_{n=2}^{\infty} a_{n-1} x^n = \sum_{n=1}^{\infty} a_n x^{n+1} \quad \text{(by a change of indices)}$$

$$= x \sum_{n=1}^{\infty} a_n x^n$$

$$= x \left(A\left(x\right) - a_0 \right)$$

$$= x A\left(x\right).$$

The second infinite polynomial on the right side of (5.38) can be written as

$$\sum_{n=2}^{\infty} a_{n-2} x^n = \sum_{n=0}^{\infty} a_n x^{n+2} \quad \text{(by a change of indices)}$$

$$= x^2 \sum_{n=0}^{\infty} a_n x^n$$

$$= x^2 A\left(x\right).$$

Thus (5.38) can be rewritten in terms of $A(x)$ as

$$A(x) - x = 5x A(x) - 6x^2 A(x). \tag{5.39}$$

Step 1 can often be done equationally by starting with the definition of $A(x)$ and continuing until an equation involving $A(x)$ is obtained. For this example the process goes as follows:

$$A\left(x\right) = \sum_{n=0}^{\infty} a_n x^n$$

$$= a_0 + a_1 x + \sum_{n=2}^{\infty} a_n x^n$$

$$= x + \sum_{n=2}^{\infty} a_n x^n$$

$$= x + \sum_{n=2}^{\infty} \left(5a_{n-1} - 6a_{n-2} \right) x^n$$

$$= x + 5 \sum_{n=2}^{\infty} a_{n-1} x^n - 6 \sum_{n=2}^{\infty} a_{n-2} x^n$$

$$= x + 5x \left(A\left(x\right) - a_0 \right) - 6x^2 A\left(x\right)$$

$$= x + 5x A\left(x\right) - 6x^2 A\left(x\right).$$

Step 2

Solve the equation for $A(x)$ and try to transform the result into an expression containing closed forms of known generating functions. We solve (5.39) by isolating $A(x)$ as follows:

$$A(x)(1 - 5x + 6x^2) = x.$$

Therefore, we can solve for $A(x)$ and try to obtain known closed forms, which can then be replaced by generating functions:

$$
\begin{aligned}
A(x) &= \frac{x}{1 - 5x + 6x^2} \\
&= \frac{x}{(2x - 1)(3x - 1)} \\
&= \frac{1}{2x - 1} - \frac{1}{3x - 1} \qquad \text{(partial fractions)} \\
&= -\frac{1}{1 - 2x} + \frac{1}{1 - 3x} \qquad \text{(put into the form } \tfrac{1}{1-t}\text{)} \\
&= -\sum_{n=0}^{\infty} (2x)^n + \sum_{n=0}^{\infty} (3x)^n \\
&= -\sum_{n=0}^{\infty} 2^n x^n + \sum_{n=0}^{\infty} 3^n x^n \\
&= \sum_{n=0}^{\infty} (-2^n + 3^n) x^n.
\end{aligned}
$$

Step 3

Equate coefficients, and obtain the result. In other words, we equate the original definition for $A(x)$ and the form of $A(x)$ obtained in Step 2:

$$\sum_{n=0}^{\infty} a_n x^n = \sum_{n=0}^{\infty} (-2^n + 3^n) x^n.$$

These two infinite polynomials are equal if and only if the corresponding coefficients are equal. Equating the coefficients, we obtain the following closed form for a_n:

$$a_n = 3^n - 2^n \quad \text{for } n \geq 0. \tag{5.40}$$

Step 4 (Check the answer)

To make sure that no mistakes were made in Steps 1 to 3, we should check to see whether (5.40) is the correct answer to (5.37). Since the recurrence has two basis cases, we'll start by verifying the special cases for $n = 0$ and $n = 1$. These cases are verified below:

$$a_0 = 3^0 - 2^0 = 0,$$
$$a_1 = 3^1 - 2^1 = 1.$$

Now verify that (5.40) satisfies the general case of (5.37) for $n \geq 2$. We'll start on the right side of (5.37) and substitute (5.40) to obtain the left side of (5.37).

$$
\begin{aligned}
5a_{n-1} - 6a_{n-2} &= 5\left(3^{n-1} - 2^{n-1}\right) - 6\left(3^{n-2} - 2^{n-2}\right) &&\text{(substitution)} \\
&= 3^n - 2^n &&\text{(simplification)} \\
&= a_n.
\end{aligned}
$$

More Generating Functions

There are many useful generating functions. Since our treatment is not intended to be exhaustive, we'll settle for listing two more generating functions that have many applications.

Two More Useful Generating Functions

$$
\frac{1}{(1-x)^{k+1}} = \sum_{n=0}^{\infty} \binom{k+n}{n} x^n \quad \text{for } k \in \mathbb{N}. \tag{5.41}
$$

$$
(1+x)^r = \sum_{n=0}^{\infty} \left(\frac{r(r-1)\cdots(r-n+1)}{n!} \right) x^n \quad \text{for } r \in \mathbb{R}. \tag{5.42}
$$

The numerator of the coefficient expression for the nth term in (5.42) contains a product of n numbers. When $n = 0$, we use the convention that a vacuous product—of zero numbers—has the value 1. Therefore the 0th term of (5.42) is $1/0! = 1$. So the first few terms of (5.42) look like the following:

$$
(1+x)^r = 1 + rx + \frac{r(r-1)}{2}x^2 + \frac{r(r-1)(r-2)}{6}x^3 + \cdots .
$$

The Problem of Parentheses

Suppose we want to find the number of ways to parenthesize the expression

$$
t_1 + t_2 + \cdots + t_{n-1} + t_n \tag{5.43}
$$

so that a parenthesized form of the expression reflects the process of adding two terms. For example, the expression $t_1 + t_2 + t_3 + t_4$ has several different forms as shown in the following expressions:

$$((t_1 + t_2) + (t_3 + t_4))$$
$$(t_1 + (t_2 + (t_3 + t_4)))$$
$$(t_1 + ((t_2 + t_3) + t_4))$$
$$\vdots$$

To solve the problem, we'll let b_n denote the total number of possible parenthe-sizations for an n-term expression. Notice that if $1 \le k \le n - 1$, then we can split the expression (5.43) into two subexpressions as follows:

$$t_1 + \cdots + t_{n-k} \quad \text{and} \quad t_{n-k+1} + \cdots + t_n. \tag{5.44}$$

So there are $b_{n-k} b_k$ ways to parenthesize the expression (5.43) if the final $+$ is placed between the two subexpressions (5.44). If we let k range from 1 to $n - 1$, we obtain the following formula for b_n when $n \ge 2$:

$$b_n = b_{n-1} b_1 + b_{n-2} b_2 + \cdots + b_2 b_{n-2} + b_1 b_{n-1}. \tag{5.45}$$

But we need $b_1 = 1$ for (5.45) to make sense. It's OK to make this assumption because we're concerned only about expressions that contain at least two terms. Similarly, we can let $b_0 = 0$. So we can write down the recurrence to describe the solution as follows:

$$b_0 = 0, \tag{5.46}$$
$$b_1 = 1,$$
$$b_n = b_n b_0 + b_{n-1} b_1 + \cdots + b_1 b_{n-1} + b_0 b_n \quad (n \ge 2).$$

Notice that this system cannot be solved by substitution or cancellation. Let's try generating functions. Let $B(x)$ be the generating function for the sequence

$$b_0, b_1, \ldots, b_n, \ldots.$$

So $B(x) = \sum_{n=0}^{\infty} b_n x^n$. Now let's try to apply the four-step procedure for generating functions. First we use the general equation in the recurrence to introduce the partial (since $n \ge 2$) generating function

$$\sum_{n=2}^{\infty} b_n x^n = \sum_{n=2}^{\infty} (b_n b_0 + b_{n-1} b_1 + \cdots + b_1 b_{n-1} + b_0 b_n) x^n. \tag{5.47}$$

Now the left-hand side of (5.47) can be written in terms of $B(x)$:

$$\sum_{n=2}^{\infty} b_n x^n = B(x) - b_1 x - b_0$$
$$= B(x) - x \quad \text{(since } b_0 = 0 \text{ and } b_1 = 1\text{)}.$$

Before we discuss the right-hand side of (5.47), notice that we can write the product

$$B\left(x\right)B\left(x\right) = \left(\sum_{n=0}^{\infty} b_n x^n\right)\left(\sum_{n=0}^{\infty} b_n x^n\right)$$
$$= \sum_{n=0}^{\infty} c_n x^n,$$

where $c_0 = b_0 b_0$ and, for $n > 0$,

$$c_n = b_n b_0 + b_{n-1} b_1 + \cdots + b_1 b_{n-1} + b_0 b_n.$$

So the right-hand side of (5.47) can be written as

$$\sum_{n=2}^{\infty} \left(b_n b_0 + b_{n-1} b_1 + \cdots + b_1 b_{n-1} + b_0 b_n\right) x^n$$
$$= B\left(x\right)B\left(x\right) - b_0 b_0 - \left(b_1 b_0 + b_0 b_1\right) x$$
$$= B\left(x\right)B\left(x\right) \quad \text{(since } b_0 = 0\text{)}.$$

Now (5.47) can be written in simplified form as

$$B\left(x\right) - x = B\left(x\right)B\left(x\right) \quad \text{or} \quad B\left(x\right)^2 - B\left(x\right) + x = 0.$$

Now, thinking of $B(x)$ as the unknown, the equation is a quadratic equation with two solutions:

$$B\left(x\right) = \frac{1 \pm \sqrt{1 - 4x}}{2}.$$

Which solution should we choose? Notice that $\sqrt{1 - 4x}$ is the closed form for generating a function obtained from (5.42), where $r = \frac{1}{2}$. Thus we can write

$$\sqrt{1 - 4x} = \left(1 + \left(-4x\right)\right)^{\frac{1}{2}}$$
$$= \sum_{n=0}^{\infty} \frac{\frac{1}{2}\left(\frac{1}{2} - 1\right)\left(\frac{1}{2} - 2\right) \cdots \left(\frac{1}{2} - n + 1\right)}{n!} \left(-4x\right)^n$$
$$= \sum_{n=0}^{\infty} \frac{\frac{1}{2}\left(-\frac{1}{2}\right)\left(-\frac{3}{2}\right) \cdots \left(-\frac{2n-3}{2}\right)}{n!} \left(-2\right)^n 2^n x^n$$
$$= 1 + \sum_{n=1}^{\infty} \frac{\left(-1\right)\left(1\right)\left(3\right) \cdots \left(2n - 3\right)}{n!} 2^n x^n$$
$$= 1 + \sum_{n=1}^{\infty} \left(-\frac{2}{n}\right)\binom{2n - 2}{n - 1} x^n.$$

The transformation to obtain the last expression is left as an exercise. Notice that, for $n \geq 1$, the coefficient of x^n is negative in this generating function. In other words, the nth term $(n \geq 1)$ of the generating function for $\sqrt{1-4x}$ always has a negative coefficient. Since we need positive values for b_n, we must choose the following solution of our quadratic equation:

$$B\left(x\right) = \frac{1}{2} - \frac{1}{2}\sqrt{1-4x}.$$

Putting things together, we can write our desired generating function as follows:

$$\sum_{n=0}^{\infty} b_n x^n = B\left(x\right) = \frac{1}{2} - \frac{1}{2}\sqrt{1-4x}$$

$$= \frac{1}{2} - \frac{1}{2}\left\{1 + \sum_{n=1}^{\infty}\left(-\frac{2}{n}\right)\binom{2n-2}{n-1}x^n\right\}$$

$$= 0 + \sum_{n=1}^{\infty}\frac{1}{n}\binom{2n-2}{n-1}x^n.$$

Now we can finish the job by equating coefficients to obtain the following solution:

$$b_n = \text{if } n = 0 \text{ then } 0 \text{ else } \frac{1}{n}\binom{2n-2}{n-1}.$$

The Problem of Binary Trees

Suppose we want to find, for any natural number n, the number of structurally distinct binary trees with n nodes. Let b_n denote this number. We can figure out a few values by experiment. For example, since there is one empty binary tree and one binary tree with a single node, we have $b_0 = 1$ and $b_1 = 1$. It's also easy to see that $b_2 = 2$, and for $n = 3$ we see after a few minutes that $b_3 = 5$.

Let's consider b_n for $n \geq 1$. A tree with n nodes has a root and two subtrees whose combined nodes total $n - 1$. For each k in the interval $0 \leq k \leq n - 1$ there are b_k left subtrees of size k and b_{n-1-k} right subtrees of size $n - 1 - k$. So for each k there are $b_k b_{n-1-k}$ binary trees with n nodes. Therefore, the number b_n of binary trees can be given by the sum of these products as follows:

$$b_n = b_0 b_{n-1} + b_1 b_{n-2} + \cdots + b_k b_{n-k} + \cdots + b_{n-2} b_1 + b_{n-1} b_0.$$

Now we can write down the recurrence to describe the solution as follows:

$$b_0 = 1,$$
$$b_n = b_0 b_{n-1} + b_1 b_{n-2} + \cdots + b_k b_{n-k} + \cdots + b_{n-2} b_1 + b_{n-1} b_0 \quad (n \geq 1).$$

Notice that this system cannot be solved by cancellation or substitution. Let's try generating functions. Let $B(x)$ be the generating function for the sequence

$$b_0, \ b_1, \ldots, \ b_n, \ \ldots .$$

So $B(x) = \sum_{n=0}^{\infty} b_n x^n$. Now let's try to apply the four-step procedure for generating functions. First we use the general equation in the recurrence to introduce the partial (since $n \geq 1$) generating function

$$\sum_{n=1}^{\infty} b_n x^n = \sum_{n=1}^{\infty} (b_0 b_{n-1} + b_1 b_{n-2} + \cdots + b_{n-2} b_1 + b_{n-1} b_0) \, x^n. \qquad (5.48)$$

Now the left-hand side of (5.48) can be written in terms of $B(x)$.

$$\sum_{n=1}^{\infty} b_n x^n = B(x) - b_0$$
$$= B(x) - 1 \quad (\text{since } b_0 = 1).$$

The right-hand side of (5.48) can be written as follows:

$$\sum_{n=1}^{\infty} (b_0 b_{n-1} + b_1 b_{n-2} + \cdots + b_{n-2} b_1 + b_{n-1} b_0) \, x^n$$
$$= \sum_{n=0}^{\infty} (b_0 b_n + b_1 b_{n-1} + \cdots + b_{n-1} b_1 + b_n b_0) \, x^{n+1}$$
$$= x \sum_{n=0}^{\infty} (b_0 b_n + b_1 b_{n-1} + \cdots + b_{n-1} b_1 + b_n b_0) \, x^n$$
$$= x B(x) B(x)$$

So (5.48) can be written in simplified form as

$$B(x) - 1 = x B(x) B(x) \quad \text{or} \quad x B(x)^2 - B(x) + 1 = 0.$$

With $B(x)$ as the unknown, the quadratic equation has the following two solutions:

$$B(x) = \frac{1 \pm \sqrt{1 - 4x}}{2x}.$$

Which solution should we choose? In the previous problem on parentheses, we observed that

$$\sqrt{1-4x} = 1 + \sum_{n=1}^{\infty} \left(-\frac{2}{n}\right) \binom{2n-2}{n-1} x^n.$$

If we multiply both sides of the solution equation by $2x$ we obtain

$$2xB(x) = 1 \pm \sqrt{1-4x}.$$

Now substitute the generating functions for $B(x)$ and $\sqrt{1-4x}$ to obtain

$$\sum_{n=0}^{\infty} 2b_n x^{n+1} = 1 \pm \left(1 + \sum_{n=1}^{\infty} \left(-\frac{2}{n}\right) \binom{2n-2}{n-1} x^n\right).$$

Since the constant term on the left side is 0, it must be the case that the constant term on the right side is also 0. This can only happen if we choose – from $\pm$. This also ensures that the values of b_n are positive. So we must choose the following solution of our quadratic equation:

$$B(x) = \frac{1-\sqrt{1-4x}}{2x}.$$

Putting things together, we can write our desired generating function as follows:

$$\sum_{n=0}^{\infty} b_n x^n = B(x) = \frac{1-\sqrt{1-4x}}{2x} = \frac{1}{2x}\left(1 - \sqrt{1-4x}\right)$$

$$= \frac{1}{2x} \sum_{n=1}^{\infty} \left(\frac{2}{n}\right) \binom{2n-2}{n-1} x^n$$

$$= \sum_{n=1}^{\infty} \frac{1}{n} \binom{2n-2}{n-1} x^{n+1}$$

$$= \sum_{n=0}^{\infty} \frac{1}{n+1} \binom{2n}{n} x^n.$$

Now we can finish the job by equating coefficients to obtain

$$b_n = \frac{1}{n+1} \binom{2n}{n}.$$

Exercises

Simple Recurrences

1. Solve each of the following recurrences by the substitution technique and the cancellation technique. Put each answer in closed form (no ellipsis allowed).

a. $a_1 = 0$,   **b.** $a_1 = 0$,   **c.** $a_0 = 1$,

    $a_n = a_{n-1} + 4$. $a_n = a_{n-1} + 2n$. $a_n = 2a_{n-1} + 3$.

2. For each of the following definitions, find a recurrence to describe the number of times the cons operation :: is called. Solve each recurrence.

a. $\mathrm{cat}(L, M) =$ if $L = \langle\,\rangle$ then M else $\mathrm{head}(L) :: \mathrm{cat}(\mathrm{tail}(L), M)$.

b. $\mathrm{dist}(x, L) =$ if $L = \langle\,\rangle$ then $\langle\,\rangle$

                else $(x :: \mathrm{head}(L) :: \langle\,\rangle) :: \mathrm{dist}(x, \mathrm{tail}(L))$.

c. $\mathrm{power}(L) =$ **if** $L = \langle\,\rangle$ **then** return $\langle\,\rangle :: \langle\,\rangle$

           else

                $A := \mathrm{power}(\mathrm{tail}(L))$;

                $B := \mathrm{dist}(\mathrm{head}(L), A)$;

                $C := \mathrm{map}(::, B)$;

                return $\mathrm{cat}(A, C)$

          fi

3. (*Towers of Hanoi*). The *Towers of Hanoi* puzzle was invented by Lucas in 1883. It consists of three stationary pegs with one peg containing a stack of n disks that form a tower (each disk has a hole in the center for the peg) in which each disk has a smaller diameter than the disk below it. The problem is to move the tower to one of the other pegs by transferring one disk at a time from one peg to another peg, no disk ever being placed on a smaller disk. Find the minimum number of moves H_n to do the job.

      Hint: It takes 0 moves to transfer a tower of 0 disks and 1 move to transfer a tower of 1 disk. So $H_0 = 0$ and $H_1 = 1$. Try it out for $n = 2$ and $n = 3$ to get the idea. Then try to find a recurrence relation for the general term H_n as follows: Move the tower consisting of the top $n - 1$ disks to the nonchosen peg; then move the bottom disk to the chosen peg; then move the tower of $n - 1$ disks onto the chosen peg.

4. (*Diagonals in a Polygon*). A diagonal in a polygon is a line from one vertex to another nonadjacent vertex. For example, a triangle doesn't have any diagonals because each vertex is adjacent to the other vertices. Find the number of diagonals in an n-sided polygon, where $n \geq 3$.

5. (*The n-Lines Problem*). Find the number of regions in a plane that are created by n lines, where no two lines are parallel and where no more than two lines intersect at any point.

Generating Functions

6. Given the generating function $A(x) = \sum_{n=0}^{\infty} a_n x^n$, find a closed form for the general term a_n for each of the following representations of $A(x)$.

a. $A(x) = \dfrac{1}{x-2} - \dfrac{2}{3x+1}$.

b. $A(x) = \dfrac{1}{2x+1} + \dfrac{3}{x+6}$.

c. $A(x) = \dfrac{1}{3x-2} - \dfrac{1}{(1-x)^2}$.

7. Use generating functions to solve each of the following recurrences.

a. $a_0 = 0$,
   $a_1 = 4$,
   $a_n = 2a_{n-1} + 3a_{n-2}$ $(n \geq 2)$.

b. $a_0 = 0$,
   $a_1 = 1$,
   $a_n = 7a_{n-1} - 12a_{n-2}$ $(n \geq 2)$.

c. $a_0 = 0$,
   $a_1 = 1$,
   $a_2 = 1$,
   $a_n = 2a_{n-1} + a_{n-2} - 2a_{n-3}$ $(n \geq 3)$.

8. Use generating functions to solve each recurrence in Exercise 1. For those recurrences that do not have an a_0 term, assume that $a_0 = 0$.

Proofs and Challenges

9. Prove in two different ways that the following equation holds for all positive integers n, as indicated:

$$\frac{(1)(1)(3)\cdots(2n-3)}{n!}2^n = \frac{2}{n}\binom{2n-2}{n-1}.$$

a. Use induction.

b. Transform the left side into the right side by "inserting" the missing even numbers in the numerator.

10. Find a closed form for the nth Fibonacci number defined by the following recurrence system.

$$F_0 = 0,$$
$$F_1 = 1,$$
$$F_n = F_{n-1} + F_{n-2} \quad (n \geq 2).$$

5.6　Comparing Rates of Growth

Sometimes it makes sense to approximate the number of steps required to execute an algorithm because of the difficulty involved in finding a closed form for an expression or the difficulty in evaluating an expression. To approximate one function with another function, we need some way to compare them. That's where "rate of growth" comes in. We want to give some meaning to statements like "f has the same growth rate as g" and "f has a lower growth rate than g."

For our purposes we will consider functions whose domains and codomains are subsets of the real numbers. We'll examine the asymptotic behavior of two functions f and g by comparing $f(n)$ and $g(n)$ for large positive values of n (i.e., as n approaches infinity).

5.6.1　Big Oh

We'll begin by discussing the meaning of the statement "the growth rate of f is bounded above by the growth rate of g." Here is the definition.

Definition of Big Oh

We say *the growth rate of f is bounded above by the growth rate of g* if there are positive numbers c and m such that

$$|f(n)| \leq c|g(n)| \text{ for } n \geq m.$$

In this case we write $f(n) = O(g(n))$ and we say that $f(n)$ is *big oh* of $g(n)$.

There are several useful consequences of the definition. For example, if $0 \leq f(n) \leq g(n)$ for all $n \geq m$ for some m, then $f(n) = O(g(n))$ because we can let $c = 1$. For another example, suppose that $f_1(n) = O(g(n))$ and $f_2(n) = O(g(n))$. Then there are constants such that $|f_1(n)| \leq c_1|g(n)|$ for $n \geq m_1$, and $|f_2(n)| \leq c_2|g(n)|$ for $n \geq m_2$. It follows that

$$|f_1(n) + f_2(n)| \leq |f_1(n)| + |f_2(n)| \leq (c_1 + c_2)|g(n)| \text{ for } n \geq max\{m_1, m_2\}.$$

Therefore, $f_1(n) + f_2(n) = O(g(n))$.

The following list contains these properties and others that we'll leave as exercises.

Properties of Big Oh　　　　　　　　　　　　　　　　　　　　　　　(5.49)

　a. $f(n) = O(f(n))$.

　b. If $f(n) = O(g(n))$ and $g(n) = O(h(n))$, then $f(n) = O(h(n))$.

Continued ➡

c. If $0 \leq f(n) \leq g(n)$ for all $n \geq m$, then $f(n) = O(g(n))$.

d. If $f(n) = O(g(n))$ and a is any real number, then $af(n) = O(g(n))$.

e. If $f_1(n) = O(g(n))$ and $f_2(n) = O(g(n))$, then $f_1(n) + f_2(n) = O(g(n))$.

f. If f_1 and f_2 have nonnegative values and $f_1(n) = O(g_1(n))$ and $f_2(n) = O(g_2(n))$, then $f_1(n) + f_2(n) = O(g_1(n)) + g_2(n))$.

 5.37 Polynomials and Big Oh

We'll prove the following property of polynomials.

> If $p(n)$ is a polynomial of degree m or less, then $p(n) = O(n^m)$.

To prove the statement we'll let $p(n) = a_0 + a_1 n + \cdots + a_m n^m$. If k is an integer such that $0 \leq k \leq m$, then $n^k \leq n^m$ for $n \geq 1$. So by (5.49c) we have $n^k = O(n^m)$. Now we can apply (5.49d) to obtain $a_k n^k = O(n^m)$. Finally, we can apply (5.49e) repeatedly to the terms of $p(n)$ to obtain

$$p(n) = a_0 + a_1 n + \cdots + a_m n^m = O(n^m).$$

end example

Using Big Oh for Upper Bounds

We can use big oh as an approximate measuring stick when discussing different algorithms that solve the same problem. For example, suppose we have an algorithm to solve a problem P and the worst-case running time of the algorithm is a polynomial of degree m. Then we can say that an optimal algorithm in the worst case for P, if one exists, must have a worst-case running time of $O(n^m)$. In other words, an optimal algorithm for P must have a worst-case running time with growth rate bounded above by that of n^m.

5.6.2 Big Omega

Now let's go the other way and discuss the mirror image of big oh. We want a notation that gives meaning to the statement "the growth rate of f is bounded below by the growth rate of g." Here is the definition.

Definition of Big Omega

We say *the growth rate of f is bounded below by the growth rate of g* if there are positive numbers c and m such that

$$|f(n)| \geq c|g(n)| \text{ for } n \geq m.$$

In this case we write $f(n) = \Omega(g(n))$ and we say that $f(n)$ is *big omega* of $g(n)$.

As a consequence of the definition we have the following nice relationship between big omega and big oh.

$$f(n) = \Omega(g(n)) \text{ if and only if } g(n) = O(f(n)).$$

To see this, notice that using the constant c for one of the definitions corresponds to using the constant $1/c$ for the other definition. In other words, we have

$$|f(n)| \geq c|g(n)| \text{ if and only if } |g(n)| \leq (1/c)|f(n)|.$$

Many properties of big omega are similar to those of big oh. We'll list some of them in the exercises.

Using Big Omega for Lower Bounds

Although less useful than big oh, big omega can be used to describe rough lower bounds for optimal algorithms. For example, if we know that the lower bound for all algorithms to solve a particular problem is given by a polynomial of degree m, then we can say that an optimal algorithm in the worst case for P, if one exists, must have a worst-case running time of $\Omega(n^m)$. In other words, an optimal algorithm for P must have a worst-case running time with growth rate bounded below by that of n^m.

5.6.3   Big Theta

Now let's discuss the meaning of the statement "f has the same growth rate as g." Basically, this means that both big oh and big omega hold for f and g. Here is the definition.

Definition of Big Theta

A function f has the *same growth rate* as g (or f has the *same order* as g) if we can find a number m and two positive constants c and d such that

$$c|g(n)| \leq |f(n)| \leq d|g(n)| \text{ for } n \geq m.$$

In this case we write $f(n) = \Theta(g(n))$ and say that $f(n)$ is *big theta* of $g(n)$.

If $f(n) = \Theta(g(n))$ and we also know that $g(n) \neq 0$ for all $n \geq m$, then we can divide the inequality in the definition by $g(n)$ to obtain

$$c \leq \left| \frac{f(n)}{g(n)} \right| \leq d \text{ for all } n \geq m.$$

This inequality gives us a better way to think about "having the same growth rate." It tells us that the ratio of the two functions is always within a fixed bound beyond some point. We can always take this point of view for functions that count the steps of algorithms because they are positive valued.

It's easy to verify that the relation "has the same growth rate as" is an equivalence relation. In other words, the following three properties hold for all functions.

Properties of Big Theta (5.50)

   a. $f(n) = \Theta(f(n))$.

   b. If $f(n) = \Theta(g(n))$, then $g(n) = \Theta(f(n))$.

   c. If $f(n) = \Theta(g(n))$ and $g(n) = \Theta(h(n))$, then $f(n) = \Theta(h(n))$.

Now let's see whether we can find some functions that have the same growth rate. To start things off, suppose f and g are proportional. This means that there is a nonzero constant a such that $f(n) = ag(n)$ for all n. In this case, the definition of big theta is satisfied by letting $d = c = |a|$. Thus we have the following statement.

Proportionality (5.51)

If two functions f and g are proportional, then $f(n) = \Theta(g(n))$.

---

example 5.38 **The Log Function**

Recall that log functions with different bases are proportional. In other words, if we have two bases $a > 1$ and $b > 1$, then

$$\log_a n = (\log_a b)(\log_b n) \quad \text{for all } n > 0.$$

So we can disregard the base of the log function when considering rates of growth. In other words, we have

$$\log_a n = \Theta(\log_b n). \tag{5.52}$$

end example

example 5.39 **Harmonic Numbers**

Recall from Section 5.2 that the nth harmonic number H_n is defined as the sum

$$H_n = \sum_{k=1}^{n} (1/k) = 1 + \frac{1}{2} + \frac{1}{3} + \cdots + \frac{1}{n}.$$

We also found the following bounds for H_n:

$$\ln\left(\frac{n+1}{2}\right) + 1 \le H_n \le \ln(n) + 1.$$

We'll show that $H_n = \Theta(\ln n)$ by working on the two bounds. With the upper bound we have the following inequality for $n \ge 3$:

$$\ln(n) + 1 \le \ln n + \ln n = 2 \ln n.$$

With the lower bound we have the following inequality:

$$\ln\left(\frac{n+1}{2}\right) + 1 = \ln\left(\frac{n+1}{2}\right) + \ln e = \ln\left((e/2)(n+1)\right) \ge \ln(n+1) > \ln n.$$

So we have the following inequality for $n > 3$:

$$\ln n < H_n < 2 \ln n.$$

Therefore, $H_n = \Theta(\ln n)$. By (5.52) we have $\ln n = \Theta(\log_b n)$ for any base b. So we can disregard the base and write

$$H_n = \Theta(\log n). \tag{5.53}$$

end example

The following theorem gives us a nice tool for showing that two functions have the same growth rate.

Theorem (5.54)

If $\displaystyle\lim_{n \to \infty} \frac{f(n)}{g(n)} = c$ where $c \ne 0$ and $c \ne \infty$, then $f(n) = \Theta(g(n))$.

---

example **5.40 Polynomials and Big Theta**

We'll show that if $p(n)$ is a polynomial of degree m, then $p(n) = \Theta(n^m)$. Let

$$p(n) = a_0 + a_1 n + \cdots + a_m n^m.$$

Then we have the following limit:

$$\lim_{n \to \infty} \frac{p(n)}{n^m} = \lim_{n \to \infty} \frac{a_0 + a_1 n + a_2 n^2 + \cdots + a_{m-1} n^{m-1} + a_m n^m}{n^m}$$

$$= \lim_{n \to \infty} \left(\frac{a_0}{n^m} + \frac{a_1}{n^{m-1}} + \frac{a_2}{n^{m-2}} + \cdots + \frac{a_{m-1}}{n} + \frac{a_m}{1}\right) = a_m.$$

Since $p(n)$ has degree m, $a_m \neq 0$. So by (5.54) we have $p(n) = \Theta(n^m)$.

end example

We should note that the limit in (5.54) is not a necessary condition for $f(n) = \Theta(g(n))$. For example, suppose we let f and g be the two functions

$$f(n) = \text{if } n \text{ is odd then 2 else 4,}$$
$$g(n) = 2.$$

We can write $1 \cdot g(n) \leq f(n) \leq 2 \cdot g(n)$ for all $n \geq 1$. Therefore, $f(n) = \Theta(g(n))$. But the quotient $f(n)/g(n)$ alternates between the two values 1 and 2. Therefore, the limit of the quotient does not exist. Still the limit test (5.54) will work for the majority of functions that occur in analyzing algorithms.

Approximations

Approximations can be quite useful for those of us who can't remember formulas that we don't use all the time. For example, the first four of the following approximations are the summation formulas from (5.11) written in terms of Θ.

Some Approximations

$$\sum_{k=1}^{n} k = \Theta(n^2). \tag{5.55}$$

$$\sum_{k=1}^{n} k^2 = \Theta(n^3). \tag{5.56}$$

$$\sum_{k=0}^{n} a^k = \Theta(a^{n+1}) \quad (a \neq 1) \tag{5.57}$$

$$\sum_{k=1}^{n} ka^k = \Theta(na^{n+1}) \quad (a \neq 1) \tag{5.58}$$

$$\sum_{k=1}^{n} k^r = \Theta(n^{r+1}) \quad (\text{for any real number } r \neq -1) \tag{5.59}$$

Notice that (5.56) and (5.57) are special cases of (5.59). Formula (5.59) follows from Example 5.13, where upper and lower bounds were calculated for the sum. It is an easy exercise to show that both bounds are $\Theta(n^{r+1})$.

example **5.41 A Worst-Case Lower Bound for Sorting**

Let's clarify a statement that we made in Example 5.17. We showed that $\lceil \log_2 n! \rceil$ is the worst-case lower bound for comparison sorting algorithms. But $\log n!$ is hard to calculate for even modest values of n. We stated that $\lceil \log_2 n! \rceil$ is approximately equal to $n \log_2 n$. Now we can make the following statement:

$$\log n! = \Theta(n \log n). \tag{5.60}$$

To prove this statement, we'll find some bounds on $\log n!$ as follows:

$$
\begin{aligned}
\log n! &= \log n + \log(n-1) + \ldots + \log 1 \\
&\leq \log n + \log n + \ldots + \log n \quad &(n \text{ terms}) \\
&= n \log n.
\end{aligned}
$$

$$
\begin{aligned}
\log n! &= \log n + \log(n-1) + \ldots + \log 1 \\
&\geq \log n + \log(n-1) + \ldots + \log(\lceil n/2 \rceil) \quad &(\lceil n/2 \rceil \text{ terms}) \\
&\geq \log\lceil n/2 \rceil + \ldots + \log\lceil n/2 \rceil \quad &(\lceil n/2 \rceil \text{ terms}) \\
&= \lceil n/2 \rceil \log\lceil n/2 \rceil \\
&\geq (n/2) \log(n/2).
\end{aligned}
$$

So we have the inequality:

$$(n/2) \log(n/2) \leq \log n! \leq n \log n.$$

It's easy to see (i.e., as an exercise) that if $n > 4$, then $(1/2) \log n < \log (n/2)$. Therefore, we have the following inequality for $n > 4$:

$$(1/4)(n \log n) \leq (n/2) \log(n/2) \leq \log n! \leq n \log n.$$

So there are nonzero constants $1/4$ and 1 and the number 4 such that

$$(1/4)(n \log n) \leq \log n! \leq (1)(n \log n) \quad \text{for all } n > 4.$$

This tells us that $\log n! = \Theta(n \log n)$.

 end example

Big Theta and Factorial

An important approximation to $n!$ is *Stirling's formula*—named for the mathematician James Stirling (1692–1770)—which is written as

$$n! = \Theta\left(\sqrt{2\pi n}\left(\frac{n}{e}\right)^n\right). \tag{5.61}$$

Using Big Theta to Discuss Algorithms

Let's see how we can use big theta to discuss the approximate performance of algorithms. For example, the worst-case performance of the binary search algorithm is $\Theta(\log n)$ because the actual value is $1 + \lfloor \log_2 n \rfloor$. Both the average and worst-case performances of a linear sequential search are $\Theta(n)$ because the average number of comparisons is $(n + 1)/2$ and the worst-case number of comparisons is n.

For sorting algorithms that sort by comparison, the worst-case lower bound is $\lceil \log_2 n! \rceil = \Theta(n \log n)$. Many sorting algorithms, like the simple sort algorithm in Example 5.11, have worst-case performance of $\Theta(n^2)$. The "dumbSort" algorithm, which constructs a permutation of the given list and then checks to see whether it is sorted, may have to construct all possible permutations before it gets the right one. Thus dumbSort has worst-case performance of $\Theta(n!)$. An algorithm called "heapsort" will sort any list of n items using at most $2n \log_2 n$ comparisons. So heapsort is a $\Theta(n \log n)$ algorithm in the worst case.

5.6.4 Little Oh

Now let's discuss the meaning of the statement "f has a lower growth rate than g." Here is the definition.

Definition of Little Oh

A function f has a *lower growth rate* than g (or f has *lower order* than g) if

$$\lim_{n \to \infty} \frac{f(n)}{g(n)} = 0.$$

In this case we write $f(n) = o(g(n))$ and we say that $f(n)$ is *little oh* of $g(n)$.

An equivalent definition states that $f(n) = o(g(n))$ if and only if for every $\epsilon > 0$ there is $m > 0$ such that $|f(n)| \leq \epsilon |g(n)|$ for all $n \geq m$.

For example, the quotient n/n^2 approaches 0 as n goes to infinity. Therefore, $n = o(n^2)$, and we can say that n has lower order than n^2. Equivalently, for any $\epsilon > 0$, we have $n \leq \epsilon n^2$ for $n \geq 1/\epsilon$. For another example, if a and b are positive numbers such that $a < b$, then $a^n = o(b^n)$. To see this, notice that the quotient $a^n/b^n = (a/b)^n$ approaches 0 as n approaches infinity because $0 < a/b < 1$.

For those readers familiar with derivatives, the evaluation of limits can often be accomplished by using L'Hôpital's rule.

Theorem (5.62)

If $\lim\limits_{n \to \infty} f(n) = \lim\limits_{n \to \infty} g(n) = \infty$ or $\lim\limits_{n \to \infty} f(n) = \lim\limits_{n \to \infty} g(n) = 0$ and f and g are differentiable beyond some point, then

$$\lim_{n \to \infty} \frac{f(n)}{g(n)} = \lim_{n \to \infty} \frac{f'(n)}{g'(n)}.$$

---

example **5.42 Powers of Log Grow Slow**

We'll start by showing that $\log n$ has lower order than n. In other words, we'll show that

$$\log n = o(n).$$

Since both n and $\log n$ approach infinity as n approaches infinity, we can apply (5.62) to $(\log n)/n$. We can write $\log n$ in terms of the natural log as $\log n = (\log e)(\ln n)$. So the derivative of $\log n$ is $(\log e)(1/n)$ and we obtain

$$\lim_{n \to \infty} \frac{\log n}{n} = (\log e) \lim_{n \to \infty} \frac{\ln n}{n} = (\log e) \lim_{n \to \infty} \frac{(1/n)}{1} = 0.$$

So $\log n$ has lower order than n, and we can write $\log n = o(n)$.

   This is actually an example of the more general result that any power of log has lower order than any positive power of n. In other words, for any real numbers k and m with $m > 0$, it follows that

$$(\log n)^k = o(n^m). \tag{5.63}$$

For example, $(\log n)^{100}$ has lower order than $n^{0.001}$. For the proof we'll consider several cases. If $k \leq 0$, then $-k \geq 0$. So we have

$$\lim_{n \to \infty} \frac{(\log n)^k}{n^m} = \lim_{n \to \infty} \frac{1}{(\log n)^{-k} n^m} = 0.$$

So assume $k > 0$. If $k < m$, let $\epsilon = m - k$ so that $m = k + \epsilon$ and $\epsilon \geq 0$. So we have

$$\lim_{n \to \infty} \frac{(\log n)^k}{n^m} = \lim_{n \to \infty} \frac{(\log n)^k}{n^{k+\varepsilon}} = \left(\lim_{n \to \infty} \frac{\log n}{n} \right)^k \left(\lim_{n \to \infty} \frac{1}{n^\varepsilon} \right) = 0 \cdot 0 = 0.$$

If $k = m$, then

$$\lim_{n \to \infty} \frac{(\log n)^m}{n^m} = \left(\lim_{n \to \infty} \frac{\log n}{n} \right)^m = 0.$$

If $k > m$, then the proof uses (5.62) repeatedly until $\log n$ disappears from the numerator. For example, if $k = 1.5$ and $m = 0.3$, then we have

$$\lim_{n\to\infty} \frac{(\log n)^{1.5}}{n^{0.3}} = (\log e)^{1.5} \lim_{n\to\infty} \frac{(\ln n)^{1.5}}{n^{0.3}} = (\log e)^{1.5} \lim_{n\to\infty} \frac{1.5(\ln n)^{0.5}(1/n)}{0.3n^{-0.7}}$$

$$= (\log e)^{1.5}\left(\frac{1.5}{0.3}\right) \lim_{n\to\infty} \frac{(\ln n)^{0.5}}{n^{0.3}} = (\log e)^{1.5}\left(\frac{1.5}{0.3}\right)$$

$$\lim_{n\to\infty} \frac{0.5(\ln n)^{-0.5}(1/n)}{0.3n^{-0.7}}$$

$$= (\log e)^{1.5}\left(\frac{0.75}{0.09}\right) \lim_{n\to\infty} \frac{(\ln n)^{-0.5}}{n^{0.3}} = (\log e)^{1.5}\left(\frac{0.75}{0.09}\right)$$

$$\lim_{n\to\infty} \frac{1}{(\ln n)^{0.5}n^{0.3}} = 0.$$

end example

A Hierarchy of Familiar Functions

Let's list a hierarchy of some familiar functions according to their growth rates, where $f(n) \prec g(n)$ means that $f(n) = o(g(n))$:

$$1 \prec \log n \prec n \prec n \log n \prec n^2 \prec n^3 \prec 2^n \prec 3^n \prec n! \prec n^n. \qquad (5.64)$$

This hierarchy can help us compare different algorithms. For example, we would certainly choose an algorithm with running time $\Theta(\log n)$ over an algorithm with running time $\Theta(n)$.

5.6.5 Using the Symbols

Let's see how we can use the symbols that we've defined so far to discuss algorithms. For example, suppose we have constructed an algorithm A to solve some problem P. Suppose further that we've analyzed A and found that it takes $5n^2$ operations in the worst case for an input of size n. This allows us to make a few general statements. First, we can say that the worst-case performance of A is $\Theta(n^2)$. Second, we can say that an optimal algorithm for P, if one exists, must have a worst-case performance of $O(n^2)$. In other words, an optimal algorithm for P must do no worse than our algorithm A.

Continuing with our example, suppose some good soul has computed a worst-case theoretical lower bound of $\Theta(n \log n)$ for any algorithm that solves P. Then we can say that an optimal algorithm, if one exists, must have a worst-case performance of $\Omega(n \log n)$. In other words, an optimal algorithm for P can do no better than the given lower bound of $\Theta(n \log n)$.

Alternative Ways to Use the Symbols

Before we leave our discussion of approximate optimality, let's look at some other ways to use the symbols. The four symbols Θ, o, O, and Ω can also be used to represent terms within an expression. For example, the equation

$$h(n) = 4n^3 + O(n^2)$$

means that $h(n)$ equals $4n^3$ plus some term of order at most n^2. When used as part of an expression, big oh is the most popular of the four symbols because it gives a nice way to concentrate on those terms that contribute the most muscle.

We should also note that the four symbols Θ, o, O, and Ω can be formally defined to represent sets of functions. In other words, for a function g we define the following four sets:

O(g) is the set of functions of order bounded above by that of g.

$\Omega(g)$ is the set of functions of order bounded below by that of g.

$\Theta(g)$ is the set of functions with the same order as g.

o(g) is the set of functions with lower order than g.

When set representations are used, we can use an expression like $f(n) \in \Theta(g(n))$ to mean that f has the same order as g. However, convention is strong to use $f(n) = \Theta(g(n))$.

The symbols can be used on both sides of an equation under the assumption that the left side of the equation represents a set of functions that is a subset of the set of functions on the right side. For example, consider the following equation:

$$3n^2 + O(g(n)) = O(h(n)).$$

We must interpret the equation as a subset relation with the property that for every function $f(n)$ in $O(g(n))$, the function $3n^2 + f(n)$ is in $O(h(n))$. In other words, with the conventional notation, if $f(n) = O(g(n))$, then $3n^2 + f(n) = O(h(n))$. Remember that an equal sign means subset. For example the following property holds for this convention.

If f and g are nonnegative, then $O(f(n)) + O(g(n)) = O(f(n) + g(n))$.

It tells us that if $h_1(n) = O(f(n))$ and $h_2(n) = O(g(n))$, then $h_1(n) + h_2(n) = O(f(n) + g(n))$. This property extends to any finite sum of nonnegative functions.

There are many mixed properties too. For example, here are three properties that follow directly from the definitions of the symbols:

$$\Theta(f(n)) = O(f(n)), \ \Theta(f(n)) = \Omega(f(n)), \ \text{and} \ o(f(n)) = O(f(n)).$$

Here are a few more properties:

$$\Theta(f(n)) + O(f(n)) = O(f(n)).$$
$$o(f(n)) + O(f(n)) = O(f(n)).$$
If $f(n) = O(g(n))$, then $\Theta(f(n)) = O(g(n))$.
If $f(n) = O(g(n))$, then $\Theta(f(n)h(n)) = O(g(n)h(n))$.

These properties along with some others are included in the exercises. They are all consequences of the definitions of the symbols.

Divide-and-Conquer Recurrences

In Section 5.5.2 we discussed the following recurrence (5.34) for divide-and-conquer algorithms that split problems into equal size subproblems.

$$T(n) = aT(n/b) + f(n).$$

We obtained the following solution (5.35) under the assumption that $n = b^k$.

$$T(n) = a^k T(1) + \sum_{i=0}^{k-1} a^i f(n/b^i). \tag{5.65}$$

It's nice to know that there are some general methods that give approximations for $T(n)$ for various kinds of functions f. We'll illustrate the idea with the following theorem that allows f to be a polynomial or a nonnegative real power of n, or any real power of $\log n$, or a product of the two. For example, $f(n)$ could be any of the following expressions.

$$n^5, \quad 3n^2 + 2n + 5, \quad \sqrt{n}, \quad \log n, \quad \sqrt{\log n}, \quad n^2 \log n, \quad n/\sqrt{\log n}.$$

Simple Tests for Some Divide-and-Conquer Recurrences (5.66)

Let $T(n)$ be defined by one of the following recurrences, where $a \geq 1$ is a real number and $b \geq 2$ is an integer.

$$T(n) = aT(n/b) + f(n), \text{ where } n \text{ is a power of } b,$$
$$T(n) = aT(\lfloor n/b \rfloor) + f(n),$$
$$T(n) = aT(\lceil n/b \rceil) + f(n).$$

For the base cases let $T(n)$ be a positive real number for $n < b$. Let

$$f(n) = \Theta(n^\alpha (\log n)^\beta),$$

Continued ➧

⇒ ➡

where $f(n)$ is a nonnegative real number for $n \geq b$, α is a nonnegative real number, and β is any real number. Then $T(n)$ has the following approximations.

1. If $\alpha < \log_b a$, then $T(n) = \Theta(n^{\log_b a})$.

2. If $\alpha = \log_b a$, then three cases occur as follows:

$$T(n) = \begin{cases} \Theta(n^{\log_b a}(\log n)^{\beta+1}) & \text{if } \beta > -1 \\ \Theta(n^{\log_b a} \log \log n) & \text{if } \beta = -1 \\ \Theta(n^{\log_b a}) & \text{if } \beta < -1 \end{cases}.$$

3. If $\alpha > \log_b a$, then two cases occur as follows:

$$T(n) = \begin{cases} \Theta(f(n)) & \text{if } \beta \geq 0 \\ O(n^{\alpha}) & \text{if } \beta < 0 \end{cases}.$$

The next example gives a variety of divide and conquer recurrences for which (5.66) can be applied to approximate $T(n)$. We'll follow the example with a proof.

example **5.43 Divide and Conquer Recurrences**

1. $T(n) = 4T(n/2) + n \log n$. We have $a = 4$ and $b = 2$, which gives $\log_b a = 2$. Since $\alpha = 1$ it follows that $\alpha < \log_b a$. So $T(n) = \Theta(n^2)$.

2. $T(n) = 4T(n/2) + n/\log n$. We have $a = 4$ and $b = 2$, which gives $\log_b a = 2$. Since $\alpha = 1$ it follows that $\alpha < \log_b a$. So $T(n) = \Theta(n^2)$.

3. $T(n) = 2T(n/2) + n \log n$. We have $a = 2$ and $b = 2$, which gives $\log_b a = 1$. Since $\alpha = 1$ it follows that $\alpha = \log_b a$. Since $\beta > -1$ it follows that $T(n) = \Theta(n(\log n)^2)$.

4. $T(n) = 2T(n/2) + n/\log n$. We have $a = 2$ and $b = 2$, which gives $\log_b a = 1$. Since $\alpha = 1$ it follows that $\alpha = \log_b a$. Since $\beta = -1$ it follows that $T(n) = \Theta(n \log \log n)$.

5. $T(n) = 2T(n/2) + n/(\log n)^2$. We have $a = 2$ and $b = 2$, which gives $\log_b a = 1$. Since $\alpha = 1$ it follows that $\alpha = \log_b a$. Since $\beta < -1$ it follows that $T(n) = \Theta(n)$.

6. $T(n) = 2T(n/2) + n^2 \log n$. We have $a = 2$ and $b = 2$, which gives $\log_b a = 1$. Since $\alpha = 2$ it follows that $\alpha > \log_b a$. Since $\beta \geq 0$ it follows that $T(n) = \Theta(n^2 \log n)$.

7. $T(n) = 2T(n/2) + n^2/\log n$. We have $a = 2$ and $b = 2$, which gives $\log_b a = 1$. Since $\alpha = 2$ it follows that $\alpha > \log_b a$. Since $\beta < 0$ it follows that $T(n) = O(n^2)$.

end example

The proof illustrates the use not only of the approximation symbols, but also basic properties of logs and summations. First we'll prove the result for the case where n is a power of b. Then we'll see how the floor and ceiling cases give the same results.

So assume that $n = b^k$. Then $k = \log_b n$ and we have $a^k = a^{\log_b n} = n^{\log_b a}$. Since $T(1)$ is a positive constant, we can write the first term of (5.65) as

$$a^k T(1) = n^{\log_b a} T(1) = \Theta(n^{\log_b a}).$$

We'll examine the summation term of (5.65) for each of the three cases.

1. Case $\alpha < \log_b a$. Let $\epsilon = (1/2)(\log_b a - \alpha)$. Then $2\epsilon = \log_b a - \alpha$ and $\alpha = \log_b a - 2\epsilon$. Since $\epsilon > 0$ it follows from (5.63) that $(\log n)^\beta = o(n^\varepsilon)$. Therefore,

$$f(n) = \Theta(n^\alpha (\log n)^\beta) = O(n^{\log_b a - 2\varepsilon} n^\varepsilon) = O(n^{\log_b a - \varepsilon}).$$

So by using properties of big oh the general term of the sum becomes

$$a^i f(n/b^i) = O(a^i (n/b^i)^{\log_b a - \varepsilon}).$$

By using properties of logs and exponents we obtain

$$a^i (n/b^i)^{\log_b a - \varepsilon} = n^{\log_b a - \varepsilon} (b^\varepsilon)^i.$$

So we can write the sum as follows.

$$\sum_{i=0}^{k-1} a^i f(n/b^i) = \sum_{i=0}^{k-1} O(n^{\log_b a - \varepsilon} (b^\varepsilon)^i) = O\left(\sum_{i=0}^{k-1} n^{\log_b a - \varepsilon} (b^\varepsilon)^i\right)$$

$$= O\left(n^{\log_b a - \varepsilon} \sum_{i=0}^{k-1} (b^\varepsilon)^i\right)$$

$$= O\left(n^{\log_b a - \varepsilon} \frac{(b^\varepsilon)^k - 1}{b^\varepsilon - 1}\right) = O\left(n^{\log_b a - \varepsilon} \frac{n^\varepsilon - 1}{b^\varepsilon - 1}\right)$$

$$= O\left(n^{\log_b a - \varepsilon} \frac{n^\varepsilon}{b^\varepsilon - 1}\right) = O(n^{\log_b a - \varepsilon} n^\varepsilon) = O(n^{\log_b a}).$$

So the two terms of $T(n)$ in (5.65) have the following approximations:

$$a^k T(1) = \Theta(n^{\log_b a}) \text{ and } \sum_{i=0}^{k-1} a^i f(n/b^i) = O(n^{\log_b a}).$$

Therefore, since the terms of $T(n)$ are nonnegative, it follows from Exercise 1b that

$$T(n) = \Theta(n^{\log_b a}).$$

2. **Case $\alpha = \log_b a$.** In this case we can write

$$f(n) = \Theta(n^{\log_b a}(\log n)^{\beta}).$$

By using properties of big theta, logs, and exponents, we obtain

$$a^i f(n/b^i) = \Theta(a^i(n/b^i)^{\log_b a}(\log(n/b^i))^{\beta}) = \Theta(n^{\log_b a}(\log n - i \log b)^{\beta})$$
$$= \Theta(n^{\log_b a}(\log_b n - i)^{\beta}) = \Theta(n^{\log_b a}(k - i)^{\beta}).$$

The second line above follows from the first line because logs of different bases are proportional. Now we can use a big theta property from Exercise 18d and some summation properties to obtain

$$\sum_{i=0}^{k-1} a^i f(n/b^i) = \Theta\left(\sum_{i=0}^{k-1} n^{\log_b a}(k-i)^{\beta}\right) = \Theta\left(n^{\log_b a}\sum_{i=0}^{k-1}(k-i)^{\beta}\right)$$
$$= \Theta\left(n^{\log_b a}\sum_{i=1}^{k}(i)^{\beta}\right). \qquad (5.67)$$

The last expression of (5.67) has different values depending on β. If $\beta \neq -1$, then we can use (5.59) to obtain

$$\Theta\left(n^{\log_b a}\sum_{i=1}^{k}(i)^{\beta}\right) = \Theta\left(n^{\log_b a}k^{\beta+1}\right) = \Theta\left(n^{\log_b a}(\log_b n)^{\beta+1}\right)$$
$$= \Theta(n^{\log_b a}(\log n)^{\beta+1}).$$

So we can approximate the summation term of $T(n)$ as follows:

$$\sum_{i=0}^{k-1} a^i f(n/b^i) = \Theta(n^{\log_b a}(\log n)^{\beta+1}).$$

Recall that the first term of $T(n)$ approximates to $a^k T(1) = \Theta(n^{\log_b a})$. If $\beta > -1$, then we have $n^{\log_b a} \leq n^{\log_b a}(\log n)^{\beta+1}$, and it follows that

$$T(n) = \Theta(n^{\log_b a}(\log n)^{\beta+1}).$$

On the other hand, If $\beta < -1$, then we have $n^{\log_b a}(\log n)^{\beta+1} \leq n^{\log_b a}$, and it follows that

$$T(n) = \Theta(n^{\log_b a}).$$

If $\beta = -1$, then the sum in the last expression of (5.67) is H_k, the kth harmonic number. We also know from (5.53) that $H_k = \Theta(\log k)$. So we obtain the following for the last expression of (5.67):

$$\Theta\left(n^{\log_b a}\sum_{i=1}^{k}(i)^{-1}\right) = \Theta\left(n^{\log_b a}\sum_{i=1}^{k}\frac{1}{i}\right) = \Theta\left(n^{\log_b a}H_k\right)$$
$$= \Theta\left(n^{\log_b a}\log k\right) = \Theta(n^{\log_b a}\log\log n).$$

So we have

$$\sum_{i=0}^{k-1}a^i f(n/b^i) = \Theta(n^{\log_b a}\log\log n).$$

Since the first term of $T(n)$ is $a^k T(1) = \Theta(n^{\log_b a})$ and $n^{\log_b a} \leq n^{\log_b a}\log\log n$, it follows that

$$T(n) = \Theta(n^{\log_b a}\log\log n).$$

3. **Case $\alpha > \log_b a$.** In this case we have $n^{\log_b a} < n^{\alpha}$. So the first term of $T(n)$ can be written as

$$a^k T(1) = n^{\log_b a}T(1) = o(n^{\alpha}).$$

Let $\epsilon = \alpha - \log_b a$. Then $\epsilon > 0$ and $\alpha = \log_b a + \epsilon$. Now we'll take the two cases for the value of β.

Assume that $\beta \geq 0$. It follows that $n^{\alpha} \leq n^{\alpha}(\log n)^{\beta}$. So we have $a^k T(1) = o(f(n))$. For simplicity, we'll let $g(n) = n^{\alpha}(\log n)^{\beta}$ so that $f(n) = \Theta(g(n))$. Then it is an interesting exercise to show that

$$ag(n/b) \leq (1/b^{\varepsilon})g(n).$$

It follows by induction that

$$a^i g(n/b^i) \leq (1/b^{\varepsilon})^i g(n).$$

Since $f(n) = \Theta(g(n))$, we can use properties of big theta and big oh to conclude that

$$a^i f(n/b^i) = \Theta(a^i g(n/b^i)) = O((1/b^{\varepsilon})^i g(n)).$$

Now take the sum of these terms and use the formula for the sum of a geometric progression to obtain

$$\sum_{i=0}^{k-1} a^i f(n/b^i) = \sum_{i=0}^{k-1} O((1/b^\varepsilon)^i g(n)) = O\left(g(n) \sum_{i=0}^{k-1} (1/b^\varepsilon)^i\right)$$

$$= O\left(g(n)\left(\frac{(1/b^\varepsilon)^k - 1}{(1/b^\varepsilon) - 1}\right)\right)$$

$$= O\left(g(n)\left(\frac{1}{1 - (1/b^\varepsilon)}\right)\right) = O\left(g(n)\right).$$

So the sum is $O(g(n))$. Since $f(n) = \Theta(g(n))$ if and only if $g(n) = \Theta(f(n))$, it follows that the sum is $O(f(n))$. But we know that $f(n/b^i)$ occurs in each term of the given sum and each term is nonnegative. So the sum is $\Omega(f(n))$. Since the sum is both $O(f(n))$ and $\Omega(f(n))$, it must be $\Theta(f(n))$. Since the first term of $T(n)$ is $o(f(n))$, it follows that $T(n) = \Theta(f(n))$.

Now assume that $\beta < 0$. In this case we have

$$a f(n/b) = a(n/b)^\alpha (\log n)^\beta \le a(n/b)^\alpha = a(n/b)^{\log_b a + \varepsilon} = \cdots = (1/b^\varepsilon) n^\alpha.$$

It follows by induction that

$$a^i f(n/b^i) \le (1/b^\varepsilon)^i n^\alpha.$$

Now take the sum of these terms and use the formula for the sum of a geometric progression to obtain

$$\sum_{i=0}^{k-1} a^i f(n/b^i) = \sum_{i=0}^{k-1} O((1/b^\varepsilon)^i n^\alpha) = O\left(n^\alpha \sum_{i=0}^{k-1} (1/b^\varepsilon)^i\right)$$

$$= O\left(n^\alpha\left(\frac{(1/b^\varepsilon)^k - 1}{(1/b^\varepsilon) - 1}\right)\right)$$

$$= O\left(n^\alpha\left(\frac{1}{1 - (1/b^\varepsilon)}\right)\right) = O\left(n^\alpha\right).$$

Since the first term of $T(n)$ is $o(n^\alpha)$, it follows that $T(n) = O(n^\alpha)$.

Extension to Floor or Ceiling

The solutions in (5.66) also hold for floor and ceilings. In other words when n is not a power of b, the term $T(n/b)$ can be replaced in the recurrence by either $T(\lfloor n/b \rfloor)$ or $T(\lceil n/b \rceil)$ so that the arguments are integers. We'll work with the floor first. So we have the recurrence

$$T(n) = aT(\lfloor n/b \rfloor) + f(n). \tag{5.68}$$

The following property of floor is given in Example 2.6, where p is a positive integer:

$$\lfloor \lfloor x \rfloor / p \rfloor = \lfloor x/p \rfloor.$$

For example, since b is a positive integer, then we have the following equation for any m:

$$\lfloor \lfloor n/b^m \rfloor / b \rfloor = \lfloor (n/b^m)/b \rfloor = \lfloor n/b^{m+1} \rfloor.$$

We'll use this property to solve the recurrence in the form (5.68). By using the substitution method we obtain

$$\begin{aligned}
T(n) &= aT\left(\lfloor n/b \rfloor\right) + f(n) \\
&= a^2 T\left(\lfloor n/b^2 \rfloor\right) + af\left(\lfloor n/b \rfloor\right) + f(n) \\
&= a^3 T\left(\lfloor n/b^3 \rfloor\right) + a^2 f\left(\lfloor n/b^2 \rfloor\right) + af\left(\lfloor n/b \rfloor\right) + f(n) \\
&\vdots
\end{aligned}$$

Continuing in this way, we obtain

$$T(n) = a^k T\left(\lfloor n/b^k \rfloor\right) + \sum_{i=0}^{k-1} a^i f\left(\lfloor n/b^i \rfloor\right). \tag{5.69}$$

Since $T(n)$ is a positive constant for $n < b$ and since $\lfloor n/b^k \rfloor < n/b^k < b$, it follows that $T(\lfloor n/b^k \rfloor)$ is a positive constant. So the first term of (5.69) becomes

$$a^k T(\lfloor n/b^k \rfloor) = n^{\log_b a} T(\lfloor n/b^k \rfloor) = \Theta(n^{\log_b a}).$$

Now we'll look at the summation in (5.69). The goal is to verify that it has the same growth rate as the summation without the floor symbols. Now we can tackle f, where $f(n) = \Theta(n^\alpha (\log n)^\beta)$. The floor property gives $1 < x/2 < \lfloor x \rfloor \le x$ for sufficiently large values of x (e.g., $x > 4$ works). So if $\alpha > 0$, it follows that

$$(1/2)^\alpha x^\alpha < \lfloor x \rfloor^\alpha < x^\alpha.$$

Therefore, $\lfloor x \rfloor^\alpha = \Theta(x^\alpha)$. Letting $x = n/b^i$, we obtain

$$\lfloor n/b^i \rfloor^\alpha = \Theta\left((n/b^i)^\alpha\right).$$

We can also observe that for sufficiently large values of x we have $\sqrt{x} < \lfloor x \rfloor < x$. Now apply log to each term to obtain

$$(1/2)\log x < \log \lfloor x \rfloor < \log x.$$

For sufficiently large x we have $1 < (1/2)\log x$. So if $\beta > 0$, it follows that

$$(1/2)^\beta (\log x)^\beta < (\log \lfloor x \rfloor)^\beta < (\log x)^\beta.$$

Therefore, $(\log \lfloor x \rfloor)^\beta = \Theta\left((\log x)^\beta\right)$. Letting $x = n/b^i$, we obtain

$$\left(\log \lfloor n/b^i \rfloor\right)^\beta = \Theta\left(\left(\log(n/b^i)\right)^\beta\right).$$

Since $f(n) = \Theta(n^\alpha (\log n)^\beta)$, we can put things together to obtain

$$
\begin{aligned}
f\left(\lfloor n/b^i \rfloor\right) &= \Theta\left(\lfloor n/b^i \rfloor^\alpha \left(\log \lfloor n/b^i \rfloor\right)^\beta\right) \\
&= \Theta\left((n/b^i)^\alpha \left(\log(n/b^i)\right)^\beta\right) = \Theta\left(f(n/b^i)\right).
\end{aligned}
$$

So the second term of (5.69) becomes

$$
\sum_{i=0}^{k-1} a^i f\left(\lfloor n/b^i \rfloor\right) = \sum_{i=0}^{k-1} a^i \Theta\left(f(n/b^i)\right)
$$
$$
= \Theta\left(\sum_{i=0}^{k-1} a^i f(n/b^i)\right).
$$

So the two terms of (5.69) yield results with the same growth rate as the results we obtained by assuming n is a power of b. So the theorem holds for floor. The argument for ceiling is similar and we will omit it.

Master Theorems

The simple test (5.66) is actually a special case of an important general result of Akra and Bazzi [1998]. The Akra–Bazzi method works for a much larger class of functions that satisfy certain conditions. For example, the function f must be nonnegative and it must satisfy conditions that can be met if the derivative of f is big oh of n^c for some constant c. Then the solution to the recurrence

$$T(n) = aT(n/b) + f(n)$$

is given by

$$T(n) = \Theta\left(n^{\log_b a}\left(1 + \int_b^n \frac{f(x)}{x^{1+\log_b a}} dx\right)\right).$$

The definite integral in the expression for $T(n)$ can often be evaluated by checking a table of integrals. Otherwise, approximation techniques are required. For example, any function of the form $f(n) = \Theta(n^\alpha (\log n)^\beta)$ satisfies the Akra–Bazzi conditions. So the solution is given by

$$T(n) = \Theta\left(n^{\log_b a}\left(1 + \int_b^n x^{\alpha - \log_b a - 1}(\log x)^\beta dx\right)\right).$$

This formula calculates the results given in (5.66). For example, if $\alpha = \log_b a$ and $\beta \ne -1$, then the expression is evaluated as follows:

$$T(n) = \Theta \left(n^{\log_b a} \left(1 + \int_b^n x^{-1} (\log x)^\beta dx \right) \right)$$

$$= \Theta \left(n^{\log_b a} \left(1 + \frac{(\log x)^{\beta+1}}{\beta + 1} \bigg|_b^n \right) \right) = \Theta \left(n^{\log_b a} (\log n)^{\beta+1} \right).$$

The Akra–Bazzi method also works for divide and conquer recurrences that arise from splitting up problems into subproblems of different sizes. Such recurrences have the form

$$T(n) = a_1 T(n/b_1) + a_2 T(n/b_2) + \cdots + a_m T(n/b_m) + f(n).$$

In this case, $T(n)$ has the following value, where p is a constant that must be calculated:

$$T(n) = \Theta \left(n^p \left(1 + \int_1^n \frac{f(x)}{x^{1+p}} dx \right) \right).$$

The constant p is the unique real number satisfying the equation

$$a_1/b_1^p + a_2/b_2^p + \cdots + a_m/b_m^p = 1.$$

For example, in the recurrence of (5.66) the solution to the equation $a/b^p = 1$ is $p = \log_b a$.

From a historical point of view, the Akra–Bazzi method generalized a theorem, often called the master theorem for divide and conquer recurrences, presented in the popular book on algorithms by Cormen, Leiserson, and Rivest [1990]. The master theorem is similar to (5.66) but allows for a wider variety of functions f because some need only satisfy big oh or big omega bounds rather than big theta bounds.

 Exercises

Using the Definitions

1. Use the definitions to prove each statement.

   a. If $f(n) \le g(n) \le 0$ for all n, then $g(n) = O(f(n))$.

   b. If f and g are nonnegative and $f(n) = \Theta(h(n))$ and $g(n) = O(h(n))$, then $f(n) + g(n) = \Theta(h(n))$.

2. Use the definition of big theta to prove each statement.

   a. $1 + 1/n = \Theta(1)$.

   b. $1 - 1/n = \Theta(1)$.

3. Show that $(n + 1)^r = \Theta(n^r)$ for any real number r.

4. Prove each statement, where $k > 0$ is a real number.

    a. $\log(kn) = \Theta(\log n)$.
    b. $\log(k + n) = \Theta(\log n)$.

5. Use the definitions of big oh and little oh to prove each statement.

    a. If $\epsilon > 0$ and $f(n) = O(n^{k-\varepsilon})$, then $f(n) = o(n^k)$.
    b. If $\epsilon > 0$ and $f(n) = \Omega(n^{k+\varepsilon})$, then $n^k = o(f(n))$.

6. Find a place to insert $\log \log n$ in the sequence (5.64).

7. Prove the following sequence of orders: $n \prec n \log n \prec n^2$.

8. For any constant k, show that $n^k \prec 2^n$.

9. Prove the following sequence of orders: $2^n \prec n! \prec n^n$.

10. For each of the following values of n, calculate the following three numbers: the exact value of $n!$, Stirling's approximation (5.61) for the value of $n!$, and the difference between the two values.

    a. $n = 5$.   **b.** $n = 10$.

11. Find an appropriate place in the sequence (5.64) for each function.

    a. $f(n) = \log 1 + \log 2 + \log 3 + \ldots + \log n$.
    b. $f(n) = \log 1 + \log 2 + \log 4 + \ldots + \log 2^n$.

12. Use (5.66) to approximate $T(n)$ for each of the following recurrences. Then compare your answers to the exact answers given in Example 5.36.

    a. $T(n) = 3T(n/2) + n$.   **b.** $T(n) = 2T(n/2) + n$.

13. Use (5.66) to approximate $T(n)$ for each of the following recurrences:

    a. $T(n) = 2T(n/2) + \sqrt{n}$.
    b. $T(n) = 2T(n/4) + \sqrt{n/\log n}$.
    c. $T(n) = 2T(n/4) + \sqrt{n}/\log n$.
    d. $T(n) = 2T(n/4) + \sqrt{n}/(\log n)^2$.
    e. $T(n) = T(n/2) + \sqrt{n}$.
    f. $T(n) = 3T(n/2) + n^2 \log n$.
    g. $T(n) = 3T(n/2) + n^2/\log n$.

General Properties

14. Prove each of the following properties of big oh:

    a. $f(n) = O(f(n))$.

    b. If $f(n) = O(g(n))$ and $g(n) = O(h(n))$, then $f(n) = O(h(n))$.

    c. If $f(n) = O(g(n))$, then $af(n) = O(g(n))$ for any real number a.

    d. If $f(n) = O(g(n)$, then $af(n) = O(ag(n))$ for any number a.

    e. If $f(n) = O(g(n)$, then $f(n/b) = O(g(n/b))$ for $b > 0$.

    f. If $f_1(n) = O(g_1(n))$ and $f_2(n) = O(g_2(n))$, then $f_1(n)f_2(n) = O(g_1(n)g_2(n))$.

    g. If f_1 and f_2 have nonnegative values and $f_1(n) = O(g_1(n))$ and $f_2(n) = O(g_2(n))$, then $f_1(n) + f_2(n) = O(g_1(n) + g_2(n))$.

15. Prove each of the following properties of big oh:

    a. $O(O(f(n))) = O(f(n))$.

    b. $f(n)O(g(n)) = O(f(n))O(g(n))$.

    c. $f(n)O(g(n)) = O(f(n)g(n))$.

    d. $O(af(n)) = O(f(n))$ for any real number a.

    e. If f and g are nonnegative, then $O(f(n)) + O(g(n)) = O(f(n) + g(n))$.

    f. If f is nonnegative, then $\sum_{k=1}^{n} O(f(k)) = O\left(\sum_{k=1}^{n} f(k)\right)$.

16. Prove each of the following properties of big omega:

    a. $f(n) = \Omega(f(n))$.

    b. If $f(n) = \Omega(g(n))$ and $g(n) = \Omega(h(n))$, then $f(n) = \Omega(h(n))$.

    c. If f and g satisfy the relation $0 \leq f(n) \leq g(n)$ for $n \geq m$, then $g(n) = \Omega(f(n))$.

    d. If $f(n) = \Omega(g(n))$, then $af(n) = \Omega(g(n))$ for any real number $a \neq 0$.

    e. If f_1 and f_2 have nonnegative values and $f_1(n) = \Omega(g(n))$ and $f_2(n) = \Omega(g(n))$, then $f_1(n) + f_2(n) = \Omega(g(n))$.

    f. If f_1 and f_2 have nonnegative values and $f_1(n) = \Omega(g_1(n))$ and $f_2(n) = \Omega(g_2(n))$, then $f_1(n) + f_2(n) = \Omega(g_1(n) + g_2(n))$.

17. Prove each of the following properties of big theta:

    a. $f(n) = \Theta(f(n))$.

    b. If $f(n) = \Theta(g(n))$, then $g(n) = \Theta(f(n))$.

    c. If $f(n) = \Theta(g(n))$ and $g(n) = \Theta(h(n))$, then $f(n) = \Theta(h(n))$.

    d. If $f(n) = \Theta(g(n)$, then $af(n) = \Theta(ag(n))$ for any real number a.

    e. If $f(n) = \Theta(g(n)$, then $f(n/b) = \Theta(g(n/b))$ for $b > 0$.

18. Prove each of the following properties of big theta:

    a. $\Theta(\Theta(f(n))) = \Theta(f(n))$.

    b. $\Theta(af(n)) = \Theta(f(n))$ for $a \neq 0$.

    c. If $f_1(n) = \Theta(g_1(n))$ and $f_2(n) = \Theta(g_2(n))$ and f_1, f_2, g_1, and g_2, are nonnegative, then $f_1(n) + f_2(n) = \Theta(g_1(n) + g_2(n))$.

    d. If $f(k) = \Theta(g(k))$ for $1 \leq k \leq n$ and f and g are nonnegative, then
$$\sum_{k=1}^{n} f(k) = \Theta\left(\sum_{k=1}^{n} g(k)\right).$$

19. Prove each of the following properties of little oh:

    a. if $f(n) = o(g(n))$ and $g(n) = o(h(n))$, then $f(n) = o(h(n))$.

    b. $o(f(n)) + o(f(n)) = o(f(n))$.

    c. $o(f(n))o(g(n)) = o(f(n)g(n))$.

    d. $o(o(f(n))) = o(f(n))$.

20. Prove each of the following mixed properties:

    a. $\Theta(f(n)) + O(f(n)) = O(f(n))$.

    b. $o(f(n)) + O(f(n)) = O(f(n))$.

    c. If $f(n) = O(g(n))$, then $\Theta(f(n)) = O(g(n))$.

    d. If $f(n) = O(g(n))$, then $\Theta(f(n)h(n)) = O(g(n)h(n))$.

    e. If $f(n) = o(g(n)$, then $O(f(n)) = o(g(n))$.

    f. If $f(n) = o(g(n)$, then $O(f(n)h(n)) = o(g(n)h(n))$.

    g. If $f(n) = o(g(n)$, then $\Theta(f(n)) = o(g(n))$.

5.7 Chapter Summary

This chapter introduces some basic tools and techniques that are used to analyze algorithms. Analysis by worst-case running time is discussed. A lower bound is a value that can't be beat by any algorithm in a particular class. An algorithm is optimal if its performance matches the lower bound.

    Counting problems often give rise to finite sums that need closed form solutions. Properties of sums together with summation notation provide us with techniques to find closed forms for many finite sums. Approximation techniques can be used when closed forms are difficult to evaluate or do not exist, as in the case of harmonic numbers.

    Two useful things to count are permutations, in which order is important, and combinations, in which order is not important. Pascal's triangle contains formulas for combinations, which are the same as binomial coefficients. There are formulas to count permutations and combinations that allow repeated elements.

Discrete probability gives us tools to calculate the chance of events that have a finite number of outcomes. We can use expectation together with counting techniques to analyze the average-case performance of an algorithm. Markov chains can be used to predict the future state of systems, where each change of state depends only on the previous state and a given probability distribution about the chances of changing from any state to any other state.

Counting problems often give rise to recurrences. Some simple recurrences can be solved by either substitution or cancellation to obtain a finite sum, which can then be transformed into a closed form. The use of generating functions provides a powerful technique for solving recurrences.

Often it makes sense to find approximations for functions that describe the number of operations performed by an algorithm. The rates of growth of two functions can be compared in various ways—big theta, little oh, big oh, and big omega. There are theorems that can be used to obtain approximations for recurrences that arise from the analysis of divide-and-conquer algorithms.

Notes

In this chapter we've just scratched the surface of techniques for manipulating expressions that crop up in counting things while analyzing algorithms. The book by Knuth [1968] contains the first account of a collection of techniques for the analysis of algorithms. The book by Graham, Knuth, and Patashnik [1989] contains a host of techniques, formulas, anecdotes, and further references to the literature. The book also introduces an alternative notation for working with sums, which often makes it easier to manipulate them without having to change the expressions for the upper and lower limits of summation. The notation is called Iverson's convention, and it is also described in the article by Knuth [1992].

chapter 6

Elementary Logic

... if it was so, it might be; and if it were so, it would be:
but as it isn't, it ain't. That's logic.

—Tweedledee in *Through the Looking-Glass*
by Lewis Carroll (1832–1898)

Why is it important to study logic? Two things that we continually try to accomplish are to understand and to be understood. We attempt to understand an argument given by someone so that we can agree with the conclusion or, possibly, so that we can say that the reasoning does not make sense. We also attempt to express arguments to others without making a mistake. A formal study of logic will help improve these fundamental communication skills.

Why should a student of computer science study logic? A computer scientist needs logical skills to argue whether a problem can be solved on a machine, to transform statements from everyday language to a variety of computer languages, to argue that a program is correct, and to argue that a program is efficient. Computers are constructed from logic devices and are programmed in a logical fashion. Computer scientists must be able to understand and apply new ideas and techniques for programming, many of which require a knowledge of the formal aspects of logic.

In this chapter we'll discuss the formal character of sentences that contain words like "and," "or," and "not" or a phrase like "if *A* then *B*."

chapter guide

contingency. We'll introduce the idea of equivalence, and we'll use it to find disjunctive and conjunctive normal forms for formulas.

6.3 (Formal Reasoning) introduces basic techniques of formal reasoning. We'll introduce rules that will allow us to write proofs in a formal manner that still reflect the way we do informal reasoning.

6.4 (Formal Axiom Systems) introduces axiom systems. We'll look at a specific set of three axioms and a single rule that are sufficient to prove any required statement of the propositional calculus.

6.1   How Do We Reason?

How do we reason with each other in our daily lives? We probably make arguments, where an *argument* is a finite sequence of statements called *premises* followed by a single statement called the *conclusion*. The conclusion of the argument normally begins with a word or phrase such as one of the following:

Therefore, So, Thus, Hence, It follows that.

For example, here's an argument that might have been made by Descartes before he made the famous statement, "I exist."

If I think, then I exist.

If I do not think, then I think.

Therefore, I exist.

An argument is said to be *valid* if upon assuming that the premises are true it follows that the conclusion is true. So the validity of an argument does not depend on whether the premises are true, but only on whether the conclusion follows from the assumption that the premises are true. We reason by trying to make valid arguments and by trying to decide whether arguments made by others are valid. But how do we tell whether an argument is valid? It seems that each of us has a personal reasoning system that includes some rules of logic that we don't think about and we don't know where they came from.

For example, the most common rule of logic is called *modus ponens* (Latin for "mode that affirms"), and it works like this: Suppose A and B are two statements and we assume that A and "If A then B" are both true. We can then conclude that B is true. For example, consider the following three sentences:

If it is raining, then there are clouds in the sky.

It is raining.

Therefore, there are clouds in the sky.

We use the modus ponens rule without thinking about it. We certainly learned it when we were children, probably by testing a parent. For example, if a child receives a hug from a parent after performing some action, it might dawn on the child that the hug follows after the action. The parent might reinforce the situation by saying, "If you do that again, then you will be rewarded." Parents often make statements such as: "If you touch that stove burner, then you will burn your finger." After touching the burner, the child probably knows a little bit more about modus ponens. A parent might say, "If you do that again, then you are going to be punished." The normal child probably will do it again and notice that punishment follows. Eventually, in the child's mind, the statement "If... then... punishment" is accepted as a true statement, and the modus ponens rule has taken root.

Most of us are also familiar with the false reasoning exhibited by the *non sequitur* (Latin for "It does not follow"). For example, someone might make several true statements and then conclude that some other statement is true, even though it does not follow from the preceding statements. The hope is that we can recognize this kind of false reasoning so that we never use it. For example, we can probably agree that the following four sentences form a non sequitur:

> You squandered the money entrusted to you.
> You did not keep required records.
> You incurred more debt than your department is worth.
> Therefore, you deserve a promotion.

When two people disagree on what they assume to be true or on how they reason about things, then they have problems trying to reason with each other. Some people call this "lack of communication." Other people call it something worse, especially when things like non sequiturs are part of a person's reasoning system. Can common ground be found? Are there any reasoning systems that are, or should be, contained in everyone's personal reasoning system? The answer is yes. The study of logic helps us understand and describe the fundamental parts of all reasoning systems.

What Is a Calculus?

The Romans used small beads called "calculi" to perform counting tasks. The word "calculi" is the plural of the word "calculus." So it makes sense to think that "calculus" has something to do with calculating. Since there are many kinds of calculation, it shouldn't surprise us that "calculus" is used in many different contexts. Let's give a definition.

A *calculus* is a language of expressions of some kind, with definite rules for forming the expressions. There are values, or meanings, associated with the expressions, and there are definite rules to transform one expression into another expression having the same value.

The English language is something like a calculus, where the expressions are sentences formed by English grammar rules. Certainly, we associate meanings with English sentences. But there are no definite rules for transforming one sentence into another. So our definition of a calculus is not quite satisfied. Let's try again with a programming language X. We'll let the expressions be the programs written in the X language. Is this a calculus? Well, there are certainly rules for forming the expressions, and the expressions certainly have meaning. Are there definite rules for transforming one X language program into another X language program? For most modern programming languages the answer is no. So we don't quite have a calculus. We should note that compilers transform X language programs into Y language programs, where X and Y are different languages. Thus a compiler does not qualify as a calculus transformation rule.

In mathematics the word "calculus" usually means the calculus of real functions. For example, the two expressions

$$D_x[f(x)g(x)] \quad \text{and} \quad f(x)D_xg(x) + g(x)D_xf(x)$$

are equivalent in this calculus. The calculus of real functions satisfies our definition of a calculus because there are definite rules for forming the expressions and there are definite rules for transforming expressions into equivalent expressions.

We'll be studying some different kinds of "logical" calculi. In a logical calculus the expressions are defined by rules, the values of the expressions are related to the concepts of true and false, and there are rules for transforming one expression into another. We'll start with a question.

How Can We Tell Whether Something Is a Proof?

When we reason with each other, we normally use informal proof techniques from our personal reasoning systems. This brings up a few questions:

> What is an informal proof?
> What is necessary to call something a proof?
> How can I tell whether an informal proof is correct?
> Is there a proof system to learn for each subject of discussion?
> Can I live my life without all this?

A formal study of logic will provide us with some answers to these questions. We'll find general methods for reasoning that can be applied informally in many different situations. We'll introduce a precise language for expressing arguments formally, and we'll discuss ways to translate an informal argument into a formal argument. This is especially important in computer science, in which formal solutions (programs) are required for informally stated problems.

6.2   Propositional Calculus

To discuss reasoning, we need to agree on some rules and notation about the truth of sentences. A sentence that is either true or false is called a *proposition*.

P	Q	¬P	P ∨ Q	P ∧ Q	P → Q
T	T	F	T	T	T
T	F	F	T	F	F
F	T	T	T	F	T
F	F	T	F	F	T

Figure 6.1   Truth tables.

For example, each of the following lines contains a proposition:

> Winter begins in June in the Southern Hemisphere.
> $2 + 2 = 4$.
> If it is raining, then there are clouds in the sky.
> I may or may not go to a movie tonight.
> All integers are even.
> There is a prime number greater than a googol.

For this discussion we'll denote propositions by the letters P, Q, and R, possibly subscripted. Propositions can be combined to form more complicated propositions, just the way we combine sentences, using the words "not," "and," "or," and the phrase "if... then...". These combining operations are often called *connectives*. We'll denote them by the following symbols and words:

> ¬   not, negation.
> ∧   and, conjunction.
> ∨   or, disjunction.
> →   conditional, implication.

Two common ways to read the expression $P \rightarrow Q$ are "if P then Q" and "P implies Q." Other less common readings are "Q if P," "P is a sufficient condition for Q," and "Q is a necessary condition for P." P is called the *antecedent* and Q is called the *consequent* of $P \rightarrow Q$.

Now that we have some symbols, we can denote propositions in symbolic form. For example, if P denotes the proposition "It is raining" and Q denotes the proposition "There are clouds in the sky," then $P \rightarrow Q$ denotes the proposition "If it is raining, then there are clouds in the sky." Similarly, $\neg P$ denotes the proposition "It is not raining."

The four logical operators are defined to reflect their usage in everyday English. Figure 6.1 is a *truth table* that defines the operators for the possible truth values of their operands.

6.2.1 Well-Formed Formulas and Semantics

Like any programming language or any natural language, whenever we deal with symbols, at least two questions always arise. The first deals with syntax: Is an expression grammatically (or syntactically) correct? The second deals with semantics: What is the meaning of an expression? Let's look at the first question first.

A grammatically correct expression is called a *well-formed formula,* or *wff* for short, which can be pronounced "woof." To decide whether an expression is a wff, we need to precisely define the syntax (or grammar) rules for the formation of wffs in our language. So let's do it.

Syntax

As with any language, we must agree on a set of symbols to use as the alphabet. For our discussion we will use the following sets of symbols:

Truth symbols:	T (or True), F (or False)
Connectives:	$\neg$, $\rightarrow$, $\wedge$, $\vee$
Propositional variables:	Uppercase letters like P, Q, and R
Punctuation symbols:	(,)

Next we need to define those expressions (strings) that form the wffs of our language. We do this by giving the following informal inductive definition for the set of propositional wffs.

The Definition of a Wff

A wff is either a truth symbol, or a propositional variable, or the negation of a wff, or the conjunction of two wffs, or the disjunction of two wffs, or the implication of one wff from another, or a wff surrounded by parentheses.

For example, the following expressions are wffs:

$$\text{True, False, } P, \neg\, Q, P \wedge Q, P \rightarrow Q, (P \vee Q) \wedge R, P \wedge Q \rightarrow R.$$

If we need to justify that some expression is a wff, we can apply the inductive definition. Let's look at an example.

example 6.1 Analyzing a Wff

We'll show that the expression $P \wedge Q \vee R$ is a wff. First, we know that P, Q, and R are wffs because they are propositional variables. Therefore, $Q \vee R$ is a wff because it's a disjunction of two wffs. It follows that $P \wedge Q \vee R$ is a wff because it's a conjunction of two wffs. We could have arrived at the same conclusion by saying that $P \wedge Q$ is a wff and then stating that $P \wedge Q \vee R$ is a wff, since it is the disjunction of two wffs.

end example

Semantics

Can we associate a truth table with each wff? Yes we can, once we agree on a hierarchy of precedence among the connectives. For example, $P \wedge Q \vee R$ is a perfectly good wff. But to find a truth table, we need to agree on which connective to evaluate first. We will define the following hierarchy of evaluation for the connectives of the propositional calculus:

$$\neg \qquad \text{(highest, do first)}$$
$$\wedge$$
$$\vee$$
$$\rightarrow \qquad \text{(lowest, do last)}$$

We also agree that the operations $\wedge$, $\vee$, and $\rightarrow$ are left associative. In other words, if the same operation occurs two or more times in succession, without parentheses, then evaluate the operations from left to right. Be sure you can tell the reason for each of the following lines, where each line contains a wff together with a parenthesized wff with the same meaning:

$P \vee Q \wedge R$	means	$P \vee (Q \wedge R)$.
$P \rightarrow Q \rightarrow R$	means	$(P \rightarrow Q) \rightarrow R$.
$\neg P \vee Q$	means	$(\neg P) \vee Q$.
$\neg P \rightarrow P \wedge Q \vee R$	means	$(\neg P) \rightarrow ((P \wedge Q) \vee R)$.
$\neg \neg P$	means	$\neg (\neg P)$.

Any wff has a natural syntax tree that clearly displays the hierarchy of the connectives. For example, the syntax tree for the wff $P \wedge (Q \vee \neg R)$ is given by the diagram in Figure 6.2.

Now we can say that any wff has a unique truth table. For example, suppose we want to find the truth table for the wff

$$\neg P \rightarrow Q \wedge R.$$

From the hierarchy of evaluation we know that this wff has the following parenthesized form:

$$(\neg P) \rightarrow (Q \wedge R).$$

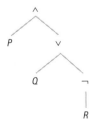

Figure 6.2   Syntax tree.

P	Q	R	¬P	Q ∧ R	¬P → Q ∧ R
T	T	T	F	T	T
T	T	F	F	F	T
T	F	T	F	F	T
T	F	F	F	F	T
F	T	T	T	T	T
F	T	F	T	F	F
F	F	T	T	F	F
F	F	F	T	F	F

Figure 6.3   Truth table.

So we can construct the truth table as follows: Begin by writing down all possible truth values for the three variables P, Q, and R. This gives us a table with eight lines. Next, compute a column of values for $\neg P$. Then compute a column of values for $Q \wedge R$. Finally, use these two columns to compute the column of values for $\neg P \to Q \wedge R$. Figure 6.3 gives the result.

Although we've talked some about meaning, we haven't specifically defined the *meaning*, or *semantics*, of a wff. Let's do it now.

The Meaning of a Wff

The meaning of the truth symbol T (or True) is true and the meaning of the truth symbol F (or False) is false. In other words, T (or True) is a truth and F (or False) is a falsity. Otherwise, the meaning of a wff is its truth table.

Tautology, Contradiction, and Contingency

A wff is a *tautology* if its truth table values are all T.

A wff is a *contradiction* if its truth table values are all F.

A wff is a *contingency* if its truth table has at least one T and at least one F.

We have the following fundamental examples:

T (or True) is a tautology.

$P \vee \neg P$ is a tautology.

F (or False) is a contradiction.

$P \wedge \neg P$ is a contradiction.

P is a contingency.

Notational Convenience

We will often use uppercase letters to refer to arbitrary propositional wffs. For example, if we say, "A is a wff," we mean that A represents some arbitrary wff. We also use uppercase letters to denote specific propositional wffs. For example, if we want to talk about the wff $P \wedge (Q \vee \neg R)$ several times in a discussion, we might let $W = P \wedge (Q \vee \neg R)$. Then we can refer to W instead of always writing down the symbols $P \wedge (Q \vee \neg R)$.

6.2.2 Logical Equivalence

In our normal discourse we often try to understand a sentence by rephrasing it in some way. Of course, we always want to make sure that the two sentences have the same meaning. This idea carries over to formal logic too, where we want to describe the idea of equivalence between two wffs.

Definition of Logical Equivalence

Two wffs A and B are *logically equivalent* (or *equivalent*) if they have the same truth value for each assignment of truth values to the set of all propositional variables occurring in the wffs. In this case we write

$$A \equiv B.$$

If two wffs contain the same propositional variables, then they will be equivalent if and only if they have the same truth tables. For example, the wffs $\neg P \vee Q$ and $P \rightarrow Q$ both contain the propositional variables P and Q. The truth tables for the two wffs are shown in Figure 6.4. Since the tables are the same, we have $\neg P \vee Q \equiv P \rightarrow Q$.

Two wffs that do not share the same propositional variables can still be equivalent. For example, the wffs $\neg P$ and $\neg P \vee (Q \wedge \neg P)$ don't share Q. Since the truth table for $\neg P$ has two lines and the truth table for $\neg P \vee (Q \wedge \neg P)$ has four lines, the two truth tables can't be the same. But we can still compare the truth values of the wffs for each truth assignment to the variables that occur in both wffs. We can do this with a truth table using the variables P and Q as shown in Figure 6.5. Since the columns agree, we know the wffs are equivalent. So we have $\neg P \equiv \neg P \vee (Q \wedge \neg P)$.

P	Q	$\neg P \vee Q$	$P \rightarrow Q$
T	T	T	T
T	F	F	F
F	T	T	T
F	F	T	T

Figure 6.4 Equivalent wffs.

P	Q	$\neg P$	$\neg P \vee (Q \wedge \neg P)$
T	T	F	F
T	F	F	F
F	T	T	T
F	F	T	T

Figure 6.5 Equivalent wffs.

Basic Equivalences (6.1)

Negation	Disjunction	Conjunction	Implication
$\neg\neg A \equiv A$	$A \vee \text{True} \equiv \text{True}$	$A \wedge \text{True} \equiv A$	$A \rightarrow \text{True} \equiv \text{True}$
	$A \vee \text{False} \equiv A$	$A \wedge \text{False} \equiv \text{False}$	$A \rightarrow \text{False} \equiv \neg A$
	$A \vee A \equiv A$	$A \wedge A \equiv A$	$\text{True} \rightarrow A \equiv A$
	$A \vee \neg A \equiv \text{True}$	$A \wedge \neg A \equiv \text{False}$	$\text{False} \rightarrow A \equiv \text{True}$
			$A \rightarrow A \equiv \text{True}$

Some Conversions	Absorption laws
$A \rightarrow B \equiv \neg A \vee B$	$A \wedge (A \vee B) \equiv A$
$\neg(A \rightarrow B) \equiv A \wedge \neg B$	$A \vee (A \wedge B) \equiv A$
$A \rightarrow B \equiv A \wedge \neg B \rightarrow \text{False}$	$A \wedge (\neg A \vee B) \equiv A \wedge B$
$\wedge$ and $\vee$ are associative.	$A \vee (\neg A \wedge B) \equiv A \vee B$
$\wedge$ and $\vee$ are commutative.	
$\wedge$ and $\vee$ distribute over each other:	**De Morgan's Laws**
$A \wedge (B \vee C) \equiv (A \wedge B) \vee (A \wedge C)$	$\neg(A \wedge B) \equiv \neg A \vee \neg B$
$A \vee (B \wedge C) \equiv (A \vee B) \wedge (A \vee C)$	$\neg(A \vee B) \equiv \neg A \wedge \neg B$

Figure 6.6   Equivalences, conversions, basic laws.

When two wffs don't have any propositional variables in common, the only way for them to be equivalent is that they are either both tautologies or both contradictions. Can you see why? For example, $P \vee \neg P \equiv Q \rightarrow Q \equiv \text{True}$.

The definition of equivalence also allows us to make the following useful formulation in terms of conditionals and tautologies.

Equivalence

$A \equiv B$ if and only if $(A \rightarrow B) \wedge (B \rightarrow A)$ is a tautology

if and only if $A \rightarrow B$ and $B \rightarrow A$ are tautologies.

Before we go much further, let's list a few basic equivalences. Figure 6.6 shows a collection of equivalences, all of which are easily verified by truth tables, so we'll leave them as exercises.

Reasoning with Equivalences

Can we do anything with the basic equivalences? Sure. We can use them to show that other wffs are equivalent without checking truth tables. But first we need to observe two general properties of equivalence.

The first thing to observe is that equivalence is an "equivalence" relation. In other words, $\equiv$ satisfies the reflexive, symmetric, and transitive properties. The transitive property is the most important property for our purposes. It can be stated as follows for any wffs W, X, and Y:

$$\text{If } W \equiv X \text{ and } X \equiv Y, \text{ then } W \equiv Y.$$

This property allows us to write a sequence of equivalences and then conclude that the first wff is equivalent to the last wff, just the way we do it with ordinary equality of algebraic expressions.

The next thing to observe is the *replacement rule* for equivalences, which is similar to the old rule: "You can always replace equals for equals."

Replacement Rule

If a wff W is changed by replacing a subwff (i.e., a wff that is part of W) by an equivalent wff, then the wff obtained in this way is equivalent to W.

Can you see why this is OK for equivalences? For example, suppose we want to simplify the wff $B \rightarrow (A \vee (A \wedge B))$. We might notice that one of the laws from (6.1) gives $A \vee (A \wedge B) \equiv A$. Therefore, we can apply the replacement rule and write the following equivalence:

$$B \rightarrow (A \vee (A \wedge B)) \equiv B \rightarrow A.$$

Let's do an example to illustrate the process of showing that two wffs are equivalent without checking truth tables.

example 6.2 A Conditional Relationship

The following equivalence shows an interesting relationship involving the connective $\rightarrow$.

$$A \rightarrow (B \rightarrow C) \equiv B \rightarrow (A \rightarrow C).$$

We'll prove it using equivalences that we already know. Make sure you can give the reason for each line of the proof.

$$
\begin{aligned}
\text{Proof:} \quad A \rightarrow (B \rightarrow C) &\equiv A \rightarrow (\neg B \vee C) \\
&\equiv \neg A \vee (\neg B \vee C) \\
&\equiv (\neg A \vee \neg B) \vee C \\
&\equiv (\neg B \vee \neg A) \vee C \\
&\equiv \neg B \vee (\neg A \vee C) \\
&\equiv B \rightarrow (\neg A \vee C) \\
&\equiv B \rightarrow (A \rightarrow C). \quad \text{QED.}
\end{aligned}
$$

end example

This example illustrates that we can use known equivalences like (6.1) as rules to transform wffs into other wffs that have the same meaning. This justifies the word "calculus" in the name "propositional calculus."

We can also use known equivalences to prove that a wff is a tautology. Here's an example.

example 6.3 An Equivalence with Descartes

We'll prove the validity of the following argument that, as noted earlier, might have been made by Descartes.

If I think, then I exist.

If I do not think, then I think.

Therefore, I exist.

We can formalize the argument by letting A and B be "I think" and "I exist," respectively. Then the argument can be represented by saying that the conclusion B follows from the premises $A \to B$ and $\neg A \to A$. A truth table for the following wff can then be used to show the argument is valid.

$$(A \to B) \land (\neg A \to A) \to B.$$

But we'll prove that the wff is a tautology by using basic equivalences to show it is equivalent to True. Make sure you can give the reason for each line of the proof.

Proof:
$$(A \to B) \land (\neg A \to A) \to B \equiv (\neg A \lor B) \land (\neg \neg A \lor A) \to B$$
$$\equiv (\neg A \lor B) \land (A \lor A) \to B$$
$$\equiv (\neg A \lor B) \land A \to B$$
$$\equiv B \land A \to B$$
$$\equiv \neg (B \land A) \lor B$$
$$\equiv \neg B \lor \neg A \lor B$$
$$\equiv \text{True} \lor \neg A$$
$$\equiv \text{True}. \qquad \text{QED.}$$

end example

Is It a Tautology, a Contradiction, or a Contingency?

Suppose our task is to find whether a wff W is a tautology, a contradiction, or a contingency. If W contains n variables, then there are 2^n different assignments of truth values to the variables of W. Building a truth table with 2^n rows can be tedious when n is moderately large.

Are there any other ways to determine the meaning of a wff? Yes. One way is to use equivalences to transform the wff into a wff that we recognize as a tautology, a contradiction, or a contingency. But another way, called Quine's method, combines the substitution of variables with the use of equivalences. To describe the method we need a definition.

Definition

If A is a variable in the wff W, then the expression $W(A/\text{True})$ denotes the wff obtained from W by replacing all occurrences of A by True. Similarly, we define $W(A/\text{False})$ to be the wff obtained from W by replacing all occurrences of A by False.

For example, if $W = (A \rightarrow B) \wedge (A \rightarrow C)$, then $W(A/\text{True})$ and $W(A/\text{False})$ have the following values, where we've continued in each case with some basic equivalences.

$$W\,(A/\text{True}) = (\text{True} \rightarrow B) \wedge (\text{True} \rightarrow C) \equiv B \wedge C.$$
$$W\,(A/\text{False}) = (\text{False} \rightarrow B) \wedge (\text{False} \rightarrow C) \equiv \text{True} \wedge \text{True} \equiv \text{True}.$$

Now comes the key observation that allows us to use these ideas to decide the truth value of a wff.

Substitution Properties

1. W is a tautology iff $W(A/\text{True})$ and $W(A/\text{False})$ are tautologies.

2. W is a contradiction iff $W(A/\text{True})$ and $W(A/\text{False})$ are contradictions.

For example, in our little example we found that $W(A/\text{True}) \equiv B \wedge C$, which is a contingency, and $W(A/\text{False}) \equiv \text{True}$, which is a tautology. Therefore, W is a contingency.

The idea of Quine's method is to construct $W(A/\text{True})$ and $W(A/\text{False})$ and then to simplify these wffs by using the basic equivalences. If we can't tell the truth values, then choose another variable and apply the method to each of these wffs. A complete example is in order.

example **6.4   Quine's Method**

Suppose that we want to check the meaning of the following wff W:

$$[(A \wedge B \rightarrow C) \wedge (A \rightarrow B)] \rightarrow (A \rightarrow C).$$

First we compute the two wffs $W(A/\text{True})$ and $W(A/\text{False})$ and simplify them using basic equivalences where appropriate.

$$
\begin{aligned}
W(A/\text{True}) &= [(\text{True} \wedge B \rightarrow C) \wedge (\text{True} \rightarrow B)] \rightarrow (\text{True} \rightarrow C) \\
&\equiv [(B \rightarrow C) \wedge (\text{True} \rightarrow B)] \rightarrow (\text{True} \rightarrow C) \\
&\equiv [(B \rightarrow C) \wedge B] \rightarrow C.
\end{aligned}
$$

$$
\begin{aligned}
W(A/\text{False}) &= [(\text{False} \wedge B \rightarrow C) \wedge (\text{False} \rightarrow B)] \rightarrow (\text{False} \rightarrow C) \\
&\equiv [(\text{False} \rightarrow C) \wedge \text{True}] \rightarrow \text{True} \\
&\equiv \text{True}.
\end{aligned}
$$

Therefore, $W(A/\text{False})$ is a tautology. Now we need to check the simplification of $W(A/\text{True})$. Call it X. We continue the process by constructing the two wffs $X(B/\text{True})$ and $X(B/\text{False})$:

$$
\begin{aligned}
X(B/\text{True}) &= [(\text{True} \rightarrow C) \wedge \text{True}] \rightarrow C \\
&\equiv [C \wedge \text{True}] \rightarrow C \\
&\equiv C \rightarrow C \\
&\equiv \text{True}.
\end{aligned}
$$

So $X(B/\text{True})$ is a tautology. Now let's look at $X(B/\text{False})$.

$$
\begin{aligned}
X(B/\text{False}) &= [(\text{False} \rightarrow C) \wedge \text{False}] \rightarrow C \\
&\equiv [\text{True} \wedge \text{False}] \rightarrow C \\
&\equiv \text{False} \rightarrow C \\
&\equiv \text{True}.
\end{aligned}
$$

So $X(B/\text{False})$ is also a tautology. Therefore, X is a tautology and it follows that W is a tautology.

end example

Quine's method can also be described graphically with a binary tree. Let W be the root. If N is any node, pick one of its variables, say V, and let the two children of N be $N(V/\text{True})$ and $N(V/\text{False})$. Each node should be simplified as much as possible. Then W is a tautology if all leaves are True, a contradiction if all leaves are False, and a contingency otherwise. Let's illustrate the idea with the wff $P \rightarrow Q \wedge P$. The binary tree in Figure 6.7 shows that the wff $P \rightarrow Q \wedge P$ is a contingency because Quine's method gives one False leaf and two True leaves.

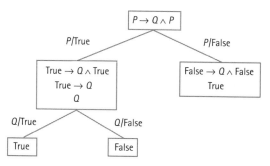

Figure 6.7   Quine's method.

6.2.3 Truth Functions and Normal Forms

A *truth function* is a function with a finite number of arguments, where the argument values and function values are from the set {True, False}. So any wff defines a truth function. For example, the function g defined by

$$g(P, Q) = P \wedge Q$$

is a truth function. Is the converse true? In other words, is every truth function a wff? The answer is yes. To see why this is true, we'll present a technique to construct a wff for any truth function.

For example, suppose we define a truth function f by saying that $f(P, Q)$ is true exactly when P and Q have opposite truth values. Is there a wff that has the same truth table as f? We'll introduce the technique with this example. Figure 6.8 is the truth table for f. We'll explain the statements on the right side of this table.

We've written the two wffs $P \wedge \neg Q$ and $\neg P \wedge Q$ on the second and third lines of the table because f has the value T on these lines. Each wff is a conjunction of argument variables or their negations according to their values on the same line subject to the following two rules:

If P has the value T, then put P in the conjunction.
If P has the value F, then put $\neg P$ in the conjunction.

Let's see why we want to follow these rules. Notice that the truth table for $P \wedge \neg Q$ in Figure 6.9 has exactly one T and it occurs on the second line.

P	Q	$f(P, Q)$	Action
T	T	F	
T	F	T	Create $P \wedge \neg Q$
F	T	T	Create $\neg P \wedge Q$
F	F	F	

Figure 6.8   A truth function.

P	Q	f(P, Q)	P ∧ ¬ Q	¬ P ∧ Q
T	T	F	F	F
T	F	T	T	F
F	T	T	F	T
F	F	F	F	F

Figure 6.9   A truth function.

Similarly, the truth table for $\neg P \wedge Q$ has exactly one T and it occurs on the third line of the table.

So each of the tables for $P \wedge \neg Q$ and $\neg P \wedge Q$ has exactly one T per column, and they occur on the same lines where f has the value T. Since there is one conjunctive wff for each occurrence of T in the table for f, it follows that the table for f can be obtained by taking the disjunction of the tables for $P \wedge \neg Q$ and $\neg P \wedge Q$. Thus we obtain the following equivalence.

$$f(P,\ Q) \equiv (P \wedge \neg Q) \vee (\neg P \wedge Q).$$

Let's do another example to get the idea. Then we'll discuss the special forms that we obtain by using this technique.

example   **6.5   Converting a Truth Function**

Let f be the truth function defined as follows:

$f(P,\ Q,\ R) =$ True if and only if either $P = Q =$ False or $Q = R =$ True.

Then f is true in exactly the following four cases:

f(F, F, T),
f(F, F, F),
f(T, T, T),
f(F, T, T).

So we can construct a wff equivalent to f by taking the disjunction of the four wffs that correspond to these four cases. The disjunction follows.

$$(\neg P \wedge \neg Q \wedge R) \vee (\neg P \wedge \neg Q \wedge \neg R) \vee (P \wedge Q \wedge R) \vee (\neg P \wedge Q \wedge R).$$

end example

The method we have described can be generalized to construct an equivalent wff for any truth function with True for at least one if its values. If a truth function doesn't have any True values, then it is a contradiction and is equivalent to False. So every truth function is equivalent to some propositional wff. We'll state this as the following theorem:

Truth Functions (6.2)

Every truth function is equivalent to a propositional wff defined in terms of the connectives ¬, ∧, and ∨.

Now we're going to discuss some useful forms for propositional wffs. But first we need a little terminology. A *literal* is a propositional variable or its negation. For example, P, Q, $\neg P$, and $\neg Q$ are literals.

Disjunctive Normal Form

A *fundamental conjunction* is either a literal or a conjunction of two or more literals. For example, P and $P \wedge \neg Q$ are fundamental conjunctions. A *disjunctive normal form* (DNF) is either a fundamental conjunction or a disjunction of two or more fundamental conjunctions. For example, the following wffs are DNFs:

$$P \vee (\neg P \wedge Q),$$
$$(P \wedge Q) \vee (\neg Q \wedge P),$$
$$(P \wedge Q \wedge R) \vee (\neg P \wedge Q \wedge R).$$

Sometimes the trivial cases are hardest to see. For example, try to explain why the following four wffs are DNFs: P, $\neg P$, $P \vee \neg P$, and $\neg P \wedge Q$. The propositions that we constructed for truth functions are DNFs.

It is often the case that a DNF is equivalent to a simpler DNF. For example, the DNF $P \vee (P \wedge Q)$ is equivalent to the simpler DNF P by using (6.1). For another example, consider the following DNF:

$$(P \wedge Q \wedge R) \vee (\neg P \wedge Q \wedge R) \vee (P \wedge R).$$

The first fundamental conjunction is equivalent to $(P \wedge R) \wedge Q$, which we see contains the third fundamental conjunction $P \wedge R$ as a subexpression. Thus the first term of the DNF can be absorbed by (6.1) into the third term, which gives the following simpler equivalent DNF:

$$(\neg P \wedge Q \wedge R) \vee (P \wedge R).$$

For any wff W we can always construct an equivalent DNF. If W is a contradiction, then it is equivalent to the single term DNF $P \wedge \neg P$. If W is not a contradiction, then we can write down its truth table and use the technique that we used for truth functions to construct a DNF. So we can make the following statement.

Disjunctive Normal Form (6.3)

Every wff is equivalent to a DNF.

Another way to construct a DNF for a wff is to transform it into a DNF by using the equivalences of (6.1). In fact we'll outline a short method that will always do the job:

First, remove all occurrences (if there are any) of the connective $\to$ by using the equivalence

$$A \to B \equiv \neg A \vee B.$$

Next, move all negations inside to create literals by using De Morgan's equivalences

$$\neg (A \wedge B) \equiv \neg A \vee \neg B \text{ and } \neg (A \vee B) \equiv \neg A \wedge \neg B.$$

Finally, apply the distributive equivalences to obtain a DNF. Let's look at an example.

example **6.6 A DNF Construction**

We'll construct a DNF for the wff $((P \wedge Q) \to R) \wedge S$.

$$\begin{aligned}((P \wedge Q) \to R) \wedge S &\equiv (\neg (P \wedge Q) \vee R) \wedge S \\ &\equiv (\neg P \vee \neg Q \vee R) \wedge S \\ &\equiv (\neg P \wedge S) \vee (\neg Q \wedge S) \vee (R \wedge S).\end{aligned}$$

end example

Suppose W is a wff having n distinct propositional variables. A DNF for W is called a *full disjunctive normal form* if each fundamental conjunction has exactly n literals, one for each of the n variables appearing in W. For example, the following wff is a full DNF:

$$(P \wedge Q \wedge R) \vee (\neg P \wedge Q \wedge R).$$

The wff $P \vee (\neg P \wedge Q)$ is a DNF but not a full DNF because the variable Q does not occur in the first fundamental conjunction.

The truth table technique to construct a DNF for a truth function automatically builds a full DNF because all of the variables in a wff occur in each fundamental conjunction. So we can state the following result.

Full Disjunctive Normal Form   **(6.4)**

Every wff that is not a contradiction is equivalent to a full DNF.

Conjunctive Normal Form

In a manner entirely analogous to the previous discussion we can define a *fundamental disjunction* to be either a literal or the disjunction of two or more literals. A *conjunctive normal form* (CNF) is either a fundamental disjunction or a conjunction of two or more fundamental disjunctions. For example, the following wffs are CNFs:

$$P \wedge (\neg P \vee Q),$$
$$(P \vee Q) \wedge (\neg Q \vee P),$$
$$(P \vee Q \vee R) \wedge (\neg P \vee Q \vee R).$$

Let's look at some trivial examples. Notice that the following four wffs are CNFs: P, $\neg P$, $P \wedge \neg P$, and $\neg P \vee Q$. As in the case for DNFs, some CNFs are equivalent to simpler CNFs. For example, the CNF $P \wedge (P \vee Q)$ is equivalent to the simpler CNF P by (6.1).

Suppose some wff W has n distinct propositional letters. A CNF for W is called a *full conjunctive normal form* if each fundamental disjunction has exactly n literals, one for each of the n variables that appear in W. For example, the following wff is a full CNF:

$$(P \vee Q \vee R) \wedge (\neg P \vee Q \vee R).$$

On the other hand, the wff $P \wedge (\neg P \vee Q)$ is a CNF but not a full CNF.

It's possible to write any truth function f that is not a tautology as a full CNF. In this case we associate a fundamental disjunction with each line of the truth table in which f has the value False. Let's return to our original example, in which $f(P, Q) =$ True exactly when P and Q have opposite truth values. Figure 6.10 shows the values for f together with a fundamental disjunction created for each line where f has the value False.

In this case, $\neg P$ is added to the disjunction if $P =$ True, and P is added to the disjunction if $P =$ False. Then we take the conjunction of these disjunctions to obtain the following conjunctive normal form of f:

$$f(P, Q) \equiv (\neg P \vee \neg Q) \wedge (P \vee Q).$$

P	Q	$f(P, Q)$	Action
T	T	F	Create $\neg P \vee \neg Q$
T	F	T	
F	T	T	
F	F	F	Create $P \vee Q$

Figure 6.10　A truth function.

Of course, any tautology is equivalent to the single term CNF $P \vee \neg P$. Now we can state the following results for CNFs, which correspond to statements (6.3) and (6.4) for DNFs:

Conjunctive Normal Form

Every wff is equivalent to a CNF. $\hspace{4cm}$ (6.5)

Every wff that is not a tautology is equivalent to a full CNF. $\hspace{1.5cm}$ (6.6)

We should note that some authors use the terms "disjunctive normal form" and "conjunctive normal form" to describe the expressions that we have called "full disjunctive normal forms" and "full conjunctive normal forms." For example, they do not consider $P \vee (\neg P \wedge Q)$ to be a DNF. We use the more general definitions of DNF and CNF because they are useful in describing methods for automatic reasoning and they are useful in describing methods for simplifying digital logic circuits.

Constructing Full Normal Forms Using Equivalences

We can construct full normal forms for wffs without resorting to truth table techniques. Let's start with the full disjunctive normal form. To find a full DNF for a wff, we first convert it to a DNF by the usual actions: eliminate conditionals, move negations inside, and distribute $\wedge$ over $\vee$. For example, the wff $P \wedge (Q \rightarrow R)$ can be converted to a DNF in two steps, as follows:

$$P \wedge (Q \rightarrow R) \equiv P \wedge (\neg Q \vee R)$$
$$\equiv (P \wedge \neg Q) \vee (P \wedge R).$$

The right side of the equivalence is a DNF. However, it's not a full DNF because the two fundamental conjunctions don't contain all three variables. The trick to add the extra variables can be described as follows:

Adding a Variable to a Fundamental Conjunction

To add a variable, say R, to a fundamental conjunction C without changing the value of C, write the following equivalences:

$$C \equiv C \wedge \text{True} \equiv C \wedge (R \vee \neg R) \equiv (C \wedge R) \vee (C \wedge \neg R).$$

Let's continue with our example. First, we'll add the letter R to the fundamental conjunction $P \wedge \neg Q$. Be sure to justify each step of the following calculation:

$$P \wedge \neg Q \equiv (P \wedge \neg Q) \wedge \text{True}$$
$$\equiv (P \wedge \neg Q) \wedge (R \vee \neg R)$$
$$\equiv (P \wedge \neg Q \wedge R) \vee (P \wedge \neg Q \wedge \neg R).$$

Next, we'll add the variable Q to the fundamental conjunction $P \wedge R$:

$$
\begin{aligned}
P \wedge R &\equiv (P \wedge R) \wedge \text{True} \\
&\equiv (P \wedge R) \wedge (Q \vee \neg Q) \\
&\equiv (P \wedge R \wedge Q) \vee (P \wedge R \wedge \neg Q).
\end{aligned}
$$

Lastly, we put the two wffs together to obtain a full DNF for $P \wedge (Q \to R)$:

$$
(P \wedge \neg Q \wedge R) \vee (P \wedge \neg Q \wedge \neg R) \vee (P \wedge R \wedge Q) \vee (P \wedge R \wedge \neg Q).
$$

Notice that the wff can be simplified to a full DNF with three fundamental conjunctions.

example 6.7 Constructing a Full DNF

We'll construct a full DNF for the wff $P \to Q$. Make sure to justify each line of the following calculation.

$$
\begin{aligned}
P \to Q &\equiv \neg P \vee Q \\
&\equiv (\neg P \wedge \text{True}) \vee Q \\
&\equiv (\neg P \wedge (Q \vee \neg Q)) \vee Q \\
&\equiv (\neg P \wedge Q) \vee (\neg P \wedge \neg Q) \vee Q \\
&\equiv (\neg P \wedge Q) \vee (\neg P \wedge \neg Q) \vee (Q \wedge \text{True}) \\
&\equiv (\neg P \wedge Q) \vee (\neg P \wedge \neg Q) \vee (Q \wedge (P \vee \neg P)) \\
&\equiv (\neg P \wedge Q) \vee (\neg P \wedge \neg Q) \vee (Q \wedge P) \vee (Q \wedge \neg P) \\
&\equiv (\neg P \wedge Q) \vee (\neg P \wedge \neg Q) \vee (Q \wedge P).
\end{aligned}
$$

end example

We can proceed in an entirely analogous manner to find a full CNF for a wff. The trick in this case is to add variables to a fundamental disjunction without changing its truth value. It goes as follows:

Adding a Variable to a Fundamental Disjunction

To add a variable, say R, to a fundamental disjunction D without changing the value of D, write the following equivalences:

$$
D \equiv D \vee \text{False} \equiv D \vee (R \wedge \neg R) \equiv (D \vee R) \wedge (D \vee \neg R).
$$

For example, let's find a full CNF for the wff $P \wedge (P \to Q)$. To start off, we put the wff in conjunctive normal form as follows:

$$
P \wedge (P \to Q) \equiv P \wedge (\neg P \vee Q).
$$

The right side is not a full CNF because the variable Q does not occur in the fundamental disjunction P. So we'll apply the trick to add the variable Q. Make sure you can justify each step in the following calculation:

$$\begin{aligned} P \wedge (P \to Q) &\equiv P \wedge (\neg P \vee Q) \\ &\equiv (P \vee \text{False}) \wedge (\neg P \vee Q) \\ &\equiv (P \vee (Q \wedge \neg Q)) \wedge (\neg P \vee Q) \\ &\equiv (P \vee Q) \wedge (P \vee \neg Q) \wedge (\neg P \vee Q). \end{aligned}$$

The result is a full CNF that is equivalent to the original wff. Let's do another example.

example 6.8 Constructing a Full CNF

We'll construct a full CNF for $(P \to (Q \vee R)) \wedge (P \vee Q)$. After converting the wff to conjunctive normal form, all we need to do is add the variable R to the fundamental disjunction $P \vee Q$. Here's the transformation:

$$\begin{aligned} (P \to (Q \vee R)) \wedge (P \vee Q) &\equiv (\neg P \vee Q \vee R) \wedge (P \vee Q) \\ &\equiv (\neg P \vee Q \vee R) \wedge ((P \vee Q) \vee \text{False}) \\ &\equiv (\neg P \vee Q \vee R) \wedge ((P \vee Q) \vee (R \wedge \neg R)) \\ &\equiv (\neg P \vee Q \vee R) \wedge ((P \vee Q \vee R) \wedge (P \vee Q \vee \neg R)) \\ &\equiv (\neg P \vee Q \vee R) \wedge (P \vee Q \vee R) \wedge (P \vee Q \vee \neg R). \end{aligned}$$

end example

6.2.4 Adequate Sets of Connectives

A set of connectives is *adequate* (also *functionally complete*) if every truth function is equivalent to a wff defined in terms of the connectives. We've already seen in (6.2) that every truth function is equivalent to a propositional wff defined in terms of the connectives $\neg$, $\wedge$, and $\vee$. Therefore, $\{\neg, \wedge, \vee\}$ is an adequate set of connectives.

Are there any 2-element adequate sets of connectives? The answer is yes. Consider the two connectives $\neg$ and $\vee$. Since $\{\neg, \wedge, \vee\}$ is adequate, we'll have an adequate set if we can show that $\wedge$ can be written in terms of $\neg$ and $\vee$. This can be seen by the equivalence

$$A \wedge B \equiv \neg(\neg A \vee \neg B).$$

So $\{\neg, \vee\}$ is an adequate set of connectives. Other adequate sets of connectives are $\{\neg, \wedge\}$ and $\{\neg, \to\}$. We'll leave these as exercises.

Are there any single connectives that are adequate? The answer is yes, but we won't find one among the four basic connectives. There is a connective called the

P	Q	NAND (P, Q)
T	T	F
T	F	T
F	T	T
F	F	T

Figure 6.11   A truth table.

P	Q	NOR (P, Q)
T	T	F
T	F	F
F	T	F
F	F	T

Figure 6.12   A truth table.

NAND operator, which is short for the "Negation of AND." We'll write NAND in functional form $\text{NAND}(P, Q)$, since there is no well-established symbol for it. Figure 6.11 shows the truth table for NAND.

To see that NAND is adequate, we have to show that the connectives in some adequate set can be defined in terms of it. For example, we can write negation in terms of NAND as follows:

$$\neg P \equiv \text{NAND}(P, P).$$

We'll leave it as an exercise to show that the other connectives can be written in terms of NAND.

Another single adequate connective is the NOR operator. NOR is short for the "Negation of OR." Figure 6.12 shows the truth table for NOR. We'll leave it as an exercise to show that NOR is an adequate connective. NAND and NOR are important because they represent the behavior of two important building blocks for logic circuits.

 Exercises

Syntax and Semantics

1. Write down the parenthesized version of each of the following expressions.

   a. $\neg P \wedge Q \rightarrow P \vee R.$
   b. $P \vee \neg Q \wedge R \rightarrow P \vee R \rightarrow \neg Q.$
   c. $A \rightarrow B \vee \neg C \wedge D \wedge E \rightarrow F.$

2. Remove as many parentheses as possible from each of the following wffs.

   a. $(((P \vee Q) \rightarrow (\neg R)) \vee (((\neg Q) \wedge R) \wedge P)).$
   b. $((A \rightarrow (B \vee C)) \rightarrow (A \vee (\neg (\neg B)))).$

3. Let A, B, and C be propositional wffs. Find two different wffs, where the statement "If A then B else C" reflects the meaning of each wff.

Equivalence

4. Use truth tables to verify the equivalences in (6.1).

5. Use other equivalences to prove the equivalence

$$A \to B \equiv A \wedge \neg B \to \text{False}.$$

   Hint: Start with the right side.

6. Show that $\to$ is not associative. That is, show that $(A \to B) \to C$ is not equivalent to $A \to (B \to C)$.

7. Use Quine's method to show that each wff is a contingency.

   a. $A \vee B \to B$.
   b. $(A \to B) \wedge (B \to \neg A) \to A$.
   c. $(A \to B) \wedge (B \to C) \to (C \to A)$.
   d. $(A \vee B \to C) \wedge A \to (C \to B)$.
   e. $(A \to B) \vee ((C \to \neg B) \wedge \neg C)$.
   f. $(A \vee B) \to (C \vee A) \wedge (\neg C \vee B)$.

8. Use Quine's method to show that each wff is a tautology.

   a. $(A \to B) \wedge (B \to C) \to (A \to C)$.
   b. $(A \vee B) \wedge (A \to C) \wedge (B \to D) \to (C \vee D)$.
   c. $(A \to C) \wedge (B \to D) \wedge (\neg C \vee \neg D) \to (\neg A \vee \neg B)$.
   d. $(A \to (B \to C)) \to ((A \to B) \to (A \to C))$.
   e. $(\neg B \to \neg A) \to ((\neg B \to A) \to B)$.
   f. $(A \to B) \to (C \vee A \to C \vee B)$.
   g. $(A \to C) \to ((B \to C) \to (A \vee B \to C))$.
   h. $(A \to B) \to (\neg (B \wedge C) \to \neg (C \wedge A))$.

9. Verify each of the following equivalences by writing an equivalence proof. That is, start on one side and use known equivalences to get to the other side.

   a. $(A \to B) \wedge (A \vee B) \equiv B$.
   b. $A \wedge B \to C \equiv (A \to C) \vee (B \to C)$.
   c. $A \wedge B \to C \equiv A \to (B \to C)$.
   d. $A \vee B \to C \equiv (A \to C) \wedge (B \to C)$.
   e. $A \to B \wedge C \equiv (A \to B) \wedge (A \to C)$.
   f. $A \to B \vee C \equiv (A \to B) \vee (A \to C)$.

10. Show that each wff is a tautology by using equivalences to show that each wff is equivalent to True.

    a. $A \rightarrow A \vee B$.

    b. $A \wedge B \rightarrow A$.

    c. $(A \vee B) \wedge \neg A \rightarrow B$.

    d. $A \rightarrow (B \rightarrow A)$.

    e. $(A \rightarrow B) \wedge \neg B \rightarrow \neg A$.

    f. $(A \rightarrow B) \wedge A \rightarrow B$.

    g. $A \rightarrow (B \rightarrow (A \wedge B))$.

    h. $(A \rightarrow B) \rightarrow ((A \rightarrow \neg B) \rightarrow \neg A)$.

Normal Forms

11. Use equivalences to transform each of the following wffs into a DNF.

    a. $(P \rightarrow Q) \rightarrow P$.

    b. $P \rightarrow (Q \rightarrow P)$.

    c. $Q \wedge \neg P \rightarrow P$.

    d. $(P \vee Q) \wedge R$.

    e. $P \rightarrow Q \wedge R$.

    f. $(A \vee B) \wedge (C \rightarrow D)$.

12. Use equivalences to transform each of the following wffs into a CNF.

    a. $(P \rightarrow Q) \rightarrow P$.

    b. $P \rightarrow (Q \rightarrow P)$.

    c. $Q \wedge \neg P \rightarrow P$.

    d. $(P \vee Q) \wedge R$.

    e. $P \rightarrow Q \wedge R$.

    f. $(A \wedge B) \vee E \vee F$.

    g. $(A \wedge B) \vee (C \wedge D) \vee (E \rightarrow F)$.

13. For each of the following functions, write down the full DNF and full CNF representations.

    a. $f(P, Q) = $ True if and only if P is True.

    b. $f(P, Q, R) = $ True if and only if either Q is True or R is False.

14. Transform each of the following wffs into a full DNF if possible.

    a. $(P \rightarrow Q) \rightarrow P$.

    b. $Q \wedge \neg P \rightarrow P$.

    c. $P \rightarrow (Q \rightarrow P)$.

    d. $(P \vee Q) \wedge R$.

    e. $P \rightarrow Q \wedge R$.

15. Transform each of the following wffs into a full CNF if possible.

    a. $(P \to Q) \to P$.

    b. $P \to (Q \to P)$.

    c. $Q \wedge \neg P \to P$.

    d. $P \to Q \wedge R$.

    e. $(P \vee Q) \wedge R$.

Challenges

16. Show that each of the following sets of operations is an adequate set of connectives.

    a. $\{\neg, \wedge\}$.   b. $\{\neg, \to\}$.   **c.** $\{\text{false}, \to\}$.

    d. $\{\text{NAND}\}$.   **e.** $\{\text{NOR}\}$.

17. Show that there are no adequate single binary connectives other than NAND and NOR. *Hint:* Let f be the truth function for an adequate binary connective. Show that $f(\text{True}, \text{True}) = \text{False}$ and $f(\text{False}, \text{False}) = \text{True}$ because the negation operation must be represented in terms of f. Then consider the remaining cases in the truth table for f.

6.3   Formal Reasoning

We have seen that truth tables are sufficient to find the truth of any proposition. However, if a proposition contains several connectives, then a truth table can become quite complicated. When we use an equivalence proof rather than truth tables to decide the equivalence of two wffs, it seems a bit closer to the way we communicate with each other. But the usual way we reason is a bit more informal.

We reason informally by writing sentences in English mixed with symbols and expressions from some domain of discourse and try to make conclusions based on assumptions. Now we're going to focus on the formal structure of proof, where the domain of discourse consists of propositions. Although there is no need to formally reason about the truth of propositions, many parts of logic need tools other than truth tables to determine the truth of wffs. Formal reasoning tools for propositional calculus also carry over to other areas of discourse that use logic. So let's get started.

An *argument* is a finite sequence of wffs called *premises* followed by a single wff called the *conclusion*. An argument is *valid* if upon assuming that the premises are true it follows that the conclusion is true. For example, the following argument is valid.

Premises: $A \vee B$,

                 $\neg A$,

                 $B \rightarrow C$,

Conclusion: $B \wedge C$.

This argument is valid because if the premises $A \vee B$, $\neg A$, and $B \rightarrow C$ are true, then the conclusion $B \wedge C$ is true, which can be verified by a truth table.

Our goal is to show that an argument is valid by using a formal proof process that uses a specific set of rules. There are various ways to choose the rules depending on how close we want the formal proof process to approximate the way we reason informally. We'll introduce two approaches to formal proof. In this section we'll study the *natural deduction* approach, where the rules reflect the informal (i.e., natural) way that we reason. In Section 6.4 we'll introduce the *axiomatic* approach, where the premises for arguments are limited to a fixed set of wffs, called *axioms*, and where there is only one rule, modus ponens. When we apply logic to a particular subject, an axiom might also be a statement about the subject under study. For example, "On any two distinct points there is always a line," is an axiom for Euclidean geometry.

6.3.1   Proof Rules

A *proof* (or *derivation*) is a finite sequence of wffs, where each wff is either a premise or the result of applying a rule to certain previous wffs in the sequence. The rules are often called *proof rules* or *inference* rules.

A proof rule gives a condition for which a new wff can be added to a proof. The condition can be a listing of one or more specific wffs or it can be a derivation from a specific premise to a specific conclusion. We'll represent a proof rule by an expression of the form

$$\frac{\text{condition}}{\text{wff}}.$$

We say that the condition of the rule *infers* the wff.

For example, let's look at the modus ponens rule. If we find A and $A \rightarrow B$ in a proof and we apply the rule, then B gets added to the proof. Since the order of occurrence of A and $A \rightarrow B$ does not matter, we can represent the modus ponens rule by either

$$\frac{A, A \rightarrow B}{B} \quad \text{or} \quad \frac{A \rightarrow B, A}{B}.$$

Before we go on, let's list the proof rules that we'll be using to construct proofs. We'll discuss their properties as we go along.

The Proof Rules   **(6.7)**

Conjunction (Conj)	Simplification (Simp)
$$\frac{A, B}{A \land B}.$$	$$\frac{A \land B}{A} \quad \text{and} \quad \frac{A \land B}{B}.$$
Addition (Add)	Disjunction Syllogism (DS)
$$\frac{A}{A \lor B} \quad \text{and} \quad \frac{A}{B \lor A}.$$	$$\frac{A \lor B, \neg A}{B} \quad \text{and} \quad \frac{A \lor B, \neg B}{A}.$$
Modus Ponens (MP)	Conditional Proof (CP)
$$\frac{A, A \to B}{B}.$$	From A, derive B <hr> $A \to B$

Double Negation (DN)	Contradiction (Contr)	Indirect Proof (IP)
$$\frac{\neg \neg A}{A} \quad \text{and} \quad \frac{A}{\neg \neg A}.$$	$$\frac{A, \neg A}{\text{False}}.$$	From $\neg A$, derive False <hr> A

The seven rules Conj, Simp, Add, DS, MP, DN, and Contr are all valid arguments that come from simple tautologies. For example, the modus ponens rule, MP, infers B from A and $A \to B$. It's easy to check (say, with a truth table) that $A \land (A \to B) \to B$ is a tautology. It follows that the truth of the premises A and $A \to B$ implies the truth of the conclusion B. Therefore, MP is a valid argument. The same reasoning applies to the other six rules.

The condition for the CP rule is "From A, derive B," which is short for "There is a derivation from the premise A that ends with B." If we find such a derivation in a proof and apply the CP rule, then $A \to B$ is added to the proof. Similarly, the condition for the IP rule is "From $\neg W$, derive False," which is short for "There is a derivation from the premise $\neg W$ that ends with a contradiction." If we find such a derivation in a proof and apply the IP rule, then W is added to the proof. We'll have more to say about CP and IP shortly.

The proof rules (6.7) will allow us to prove any valid argument involving wffs of the propositional calculus. In fact, we could get by without some of the rules. For example, the contradiction rule (Contr) is not required because the IP rule could be rephrased as "From $\neg A$, derive B and $\neg B$ for some wff B." But we include Contr because it helps make indirect proofs easier to understand and closer to the way we reason informally. Some rules (i.e., DN and the extra rules in DS and Add) can be derived from the other rules. But we include them because they clarify how the rules can be used.

6.3.2 Proofs

Now that we have some proof rules, we can start doing proofs. As we said earlier, a proof (or derivation) is a finite sequence of wffs, where each wff is either a premise or the result of applying a rule to certain previous wffs in the sequence.

Note that there are some restrictions on which wffs can be used by the proof rules. We'll get to them when we discuss the CP and IP rules.

We'll write proofs in table format, where each line is numbered and contains a wff together with the reason it's there. For example, a proof sequence

$$W_1, \ldots, W_n$$

will be written in the following form:

1. W_1 Reason for W_1
2. W_2 Reason for W_2

$\vdots$ $\vdots$

$n.$ W_n Reason for W_n

The reason column for each line always contains a short indication of why the wff is on the line. If the line depends on previous lines because of a rule, then we'll always include the line numbers of those previous lines and the name of the rule.

Notation for Premises

For each line that contains a premise, we'll write the letter P in the reason column of the line.

example 6.9 A Formal Proof

Earlier we considered the argument with premises $A \vee B$, $\neg A$, and $B \rightarrow C$, and conclusion $B \wedge C$. We "proved" that the argument is valid by saying that the truth of the three premises implies the truth of the conclusion, which can be verified by a truth table. Our goal now is to give a formal proof that the argument is valid by using the rules of our proof system.

We start the proof by writing down the premises on the first three lines and then leaving some space before writing the conclusion as follows:

1. $A \vee B$ P
2. $\neg A$ P
3. $B \rightarrow C$ P

   ...

   $B \wedge C$?

After examining the rules in our proof system, we see that no rule can be used to infer the conclusion from the premises. So a proof will have to include one or more new wffs inserted between the premises and the conclusion in such a way that each wff is obtained by a rule.

How do we proceed? It's a good idea to examine the conclusion to see what might be needed to obtain it. For example, we might observe that the conclusion $B \wedge C$ is the conjunction of B and C. So if we can somehow get B and C on two lines, then the Conj rule would give us the conclusion $B \wedge C$. Then we might notice that the DS rule applied to the premises $A \vee B$ and $\neg A$ gives us B. Finally, we might see that the MP rule applied to B and the premise $B \rightarrow C$ gives us C. Here's a finished proof.

1. $A \vee B$ P
2. $\neg A$ P
3. $B \rightarrow C$ P
4. B 1, 2, DS
5. C 3, 4, MP
6. $B \wedge C$ 4, 5, Conj
    QED.

end example

Proofs with CP or IP

Suppose a proof consists of a derivation from the premise A to B followed by the application of CP to obtain $A \rightarrow B$. If no other lines of the proof use CP or IP, then the derivation from the premise A to B uses only rules that are valid arguments. So the truth of A implies the truth of B. Therefore, $A \rightarrow B$ is a tautology. For IP we get a similar result. Suppose a proof consists of a derivation from the premise $\neg W$ to False followed by the application of IP to obtain W. If no other lines of the proof use CP or IP, then the derivation from $\neg W$ to False uses only rules that are valid arguments. So the truth of $\neg W$ leads to a contradiction. Therefore, W is a tautology.

When CP is applied to a derivation from the premise A to B to obtain $A \rightarrow B$, the premise A is said to be *discharged*. Similarly, when IP is applied to a derivation from the premise $\neg W$ to False to obtain W, the premise $\neg W$ is said to be *discharged*.

When proofs contain more than one application of CP or IP, then there are restrictions on which lines of a proof can be used to justify later lines. Basically, if a premise has been discharged by the use of CP or IP, then the premise together with the wffs derived from it may not be used to justify any subsequent lines of the proof. We'll get to that shortly. For now, we'll look at some simple examples that have just one application of CP at the end of the proof.

Multiple Premises for CP

The condition for the CP rule is a derivation from a premise A to B. If A is a conjunction of wffs, such as $A = C \wedge D$, then we usually need to work with C

and D separately. We can do this by using Simp as follows:

1. $C \wedge D$ P
2. C 1, Simp
3. D 2, Simp

But we can save time and space by writing C and D as premises to begin with as follows:

1. C P
2. D P

So any derivation that results in B in one proof can also be done in the other proof. If this is the case, then CP can be applied to obtain $C \wedge D \rightarrow B$. In the first proof CP discharges the premise $C \wedge D$ and in the second proof CP discharges the two premises C and D.

The last line of a CP or IP Proof

If a proof ends by applying CP or IP to a derivation, then instead of writing the result of CP or IP on the last line, we will write QED along with the line numbers of the derivation followed by CP or IP. We omit the writing of the result because it can be quite lengthy in some cases. The next examples demonstrate the idea.

example 6.10 A CP Proof

We'll prove that the following conditional wff is a tautology.

$$(A \vee B) \wedge \neg A \wedge (B \rightarrow C) \rightarrow (B \wedge C).$$

Since the antecedent of the wff is a conjunction of three wffs, it suffices to construct a proof of $B \wedge C$ from the three premises $A \vee B$, $\neg A$, and $B \rightarrow C$, and then apply the CP rule. Here's a proof.

1. $A \vee B$ P
2. $\neg A$ P
3. $B \rightarrow C$ P
4. B 2, 3, DS
5. C 4, 5, MP
6. $B \wedge C$ 5, 6, Conj
    QED   1–7, CP.

The last line of the proof indicates that CP has been applied to the derivation on lines 1 thru 6 that starts with the three premises and ends with $B \wedge C$. Therefore, the result of CP applied to the derivation is the wff

$$(A \vee B) \wedge \neg A \wedge (B \to C) \to (B \wedge C).$$

Our convention is to not write out the wff, but rather to put QED in its place. Since the only use of CP or IP is at the end of the proof, it follows that the wff is a tautology.

end example

example 6.11   A Simple CP Proof of $A \to A$

We'll prove the simple tautology $A \to A$. In other words, we'll show that A follows from the premise A.

1.	A	P
2.	$A \vee A$	1, Add
3.	$A \wedge (A \vee A)$	1, 2, Conj
4.	A	2, Simp
	QED	1–4, CP.

end example

Subproofs and Discharged Premises

It is often the case that a proof contains another proof, called a *subproof*, which proves a statement that is needed for the proof to continue. A subproof is a derivation that always starts with a new premise and always ends by applying CP or IP to the derivation. When a subproof ends, the premise is discharged and the wffs of the derivation become inactive, which means that they may not be used to justify any subsequent line of the proof.

A subproof is denoted by indenting the wffs of the derivation. We do not indent the wff on the CP or IP line because it is needed for the proof to continue. Here's an example of a proof structure that contains a subproof.

1.	W_1		P
2.	W_2		Reason (May use line 1)
3.	W_3		Reason (May use lines 1–2)
4.		W_4	P (New premise for subproof)
5.		W_5	Reason (May use lines 1–4)
6.		W_6	Reason (May use lines 1–5)
7.		W_7	Reason (May use lines 1–6)
8.	W_8		CP or IP from lines 4–7
9.	W_9		Reason (May use lines 1–3 and 8)
10.	W_{10}		Reason (May use lines 1–3 and 8–9)
	QED		CP or IP from lines 1–3 and 8–10.

The premise on line 4 is discharged by the use of CP or IP on line 8. So lines 4 thru 7 become inactive and may not be used to justify subsequent lines of the proof.

If a subproof is nested within another subproof, then the premise of the innermost subproof must be discharged before the premise of the outermost subproof. In other words, finish the innermost subproof first with CP or IP so the result is available for use as part of the outermost subproof.

Subproofs usually depend on prior premises that have not yet been discharged. For example, suppose that we have a derivation from the premise A to B and we apply CP to obtain $A \to B$. Suppose further that the derivation from A to B depends on a prior premise C that is not discharged. Then B depends on both C and A. In other words, the truth of C and A implies the truth of B. So we have the tautology

$$(C \wedge A) \to B.$$

It follows from Exercise 9c of Section 6.2 that

$$(C \wedge A) \to B \equiv C \to (A \to B).$$

Since C is not discharged, we can apply MP to C and $C \to (A \to B)$ to obtain $A \to B$. This result generalizes to the case where a subproof depends on two or more prior undischarged premises.

So we've found that if a derivation from the premise A to B depends on prior undischarged premises, then the truth of $A \to B$ depends on the truth of those prior premises. This means that $A \to B$ is the conclusion of a valid argument from those prior premises. So $A \to B$ can be used to justify subsequent lines of the proof until one of those prior premises becomes discharged.

The argument is the same for the IP rule. In other words, if a derivation from the premise $\neg W$ to False depends on prior undischarged premises, then W is the conclusion of a valid argument from those premises. So W can be used to justify subsequent lines of the proof until one of those prior premises becomes discharged.

It follows from the preceding paragraphs that if a proof satisfies the restrictions we've described and all premises in the proof are discharged, then the last wff of the proof is a tautology. Here's a listing of the restrictions with CP and IP.

Restrictions with CP and IP

After CP or IP has been applied to a derivation, the lines of the derivation become inactive and may not be used to justify subsequent lines of the proof.

For nested subproofs, finish the innermost subproof first with CP or IP so the result is available for use as part of the outermost subproof.

Let's do some examples to cement the ideas. We'll start with some CP examples and then introduce the use of IP.

example **6.12 A Proof with a Subproof**

We'll give a proof to show that the following wff is a tautology.

$$((A \lor B) \to (B \land C)) \to (B \to C).$$

This wff is a conditional and the conclusion is also a conditional. So the proof will contain a subproof of the conditional $B \to C$.

1.	$(A \lor B) \to (B \land C)$	P
2.	B	P [for $B \to C$]
3.	$A \lor B$	2, Add
4.	$B \land C$	1, 3, MP
5.	C	4, Simp
6.	$B \to C$	2–5, CP
	QED	1, 6, CP.

end example

example **6.13 A Simple Tautology: $A \to (B \to A)$**

We'll prove the simple tautology $A \to (B \to A)$.

1.	A	P
2.	B	P [for $B \to A$]
3.	$A \land B$	1, 2, Conj
4.	A	3, Simp
5.	$B \to A$	2–4, CP
	QED	1, 5, CP.

end example

example 6.14 Another Simple Tautology: $\neg A \rightarrow (A \rightarrow B)$

We'll prove the simple tautology $\neg A \rightarrow (A \rightarrow B)$.

1.	$\neg A$	P
2.	A	P [for $A \rightarrow B$]
3.	$A \vee B$	2, Add
4.	B	1, 3, DS
5.	$A \rightarrow B$	2–4, CP
	QED	1, 5, CP.

end example

example 6.15 A Proof with a Subproof with a Subproof

We'll give a proof that the following wff is a tautology.

$$(A \rightarrow (B \rightarrow C)) \rightarrow ((A \rightarrow B) \rightarrow (A \rightarrow C)).$$

This wff is a conditional and its conclusion is also a conditional with a conditional in its conclusion. So the proof will contain a subproof of $(A \rightarrow B) \rightarrow (A \rightarrow C)$, which will start with $A \rightarrow B$ as a new premise. This subproof will contain a subproof of $A \rightarrow C$, which will start with A as a new premise. Here's the proof.

1.	$A \rightarrow (B \rightarrow C)$	P
2.	$A \rightarrow B$	P [for $(A \rightarrow B) \rightarrow (A \rightarrow C)$]
3.	A	P [for $A \rightarrow C$]
4.	$B \rightarrow C$	1, 3, MP
5.	B	2, 3, MP
6.	C	4, 5, MP
7.	$A \rightarrow C$	3–6, CP
8.	$(A \rightarrow B) \rightarrow (A \rightarrow C)$	2, 7, CP
	QED	1, 8, CP.

end example

Using the IP Rule

Suppose we want to prove a statement, but we just can't seem to find a way to get going. We might try *proof by contradiction* (i.e., *reductio ad absurdum*). In other words, assume that the statement to be proven is false and argue from there until a contradiction of some kind is reached. Then conclude that the original statement is true. The idea is based on the following simple equivalence for any wff W:

$$W \equiv \neg W \rightarrow \text{False}.$$

This equivalence together with the CP rule gives us the IP rule. Here's why. If we have a derivation from the premise $\neg W$ to False, then we can apply CP to obtain $\neg W \rightarrow$ False, which is equivalent to W. So we have the IP rule.

Since False is equivalent to $B \wedge \neg B$ for any wff B, there can be different ways to find a contradiction. You might try proof by contradiction whenever there doesn't seem to be enough information from the given premises or when you run out of ideas.

Using IP to Prove a Conditional

When proving a conditional of the form $V \rightarrow W$, it is often easier to use IP as part of a CP proof as follows: Start the CP proof by listing the antecedent V as a premise for the CP proof. Then start an IP subproof by listing $\neg W$ as a new premise. Once a contradiction is reached the IP rule can be used to infer W. Then CP gives the desired result.

example   **6.16   A CP Proof with an IP Subproof**

We'll prove that the following conditional is a tautology.

$$(A \vee B) \wedge (B \rightarrow C) \wedge \neg C \rightarrow A.$$

The antecedent is a conjunction of the three wffs $A \vee B$, $B \rightarrow C$, and $\neg C$. So we'll start the proof by writing down these three wffs as premises for CP. The goal is to obtain A and apply CP. We'll obtain A with an IP subproof by assuming $\neg A$ as a new premise.

1.	$A \vee B$		P
2.	$B \rightarrow C$		P
3.	$\neg C$		P
4.		$\neg A$	P [for A]
5.		B	1, 4, DS
6.		C	2, 5, MP
7.		False	3, 6, Contr
8.	A		4–7, IP
	QED		1–3, 6, CP.

example **6.17 An IP Proof with an IP Subproof of $A \lor \neg A$**

We'll give an IP proof of the simple tautology

$$A \lor \neg A.$$

The proof starts with the negation of the wff as a new premise. Then in an attempt to find a contradiction, it starts an IP subproof to obtain A by assuming $\neg A$.

1.	$\neg (A \lor \neg A)$		P [for $A \lor \neg A$]
2.		$\neg A$	P [for A]
3.		$A \lor \neg A$	2, Add
4.		False	1, 3, Contr
5.	A		2–4, IP
6.	$A \lor \neg A$		5, Add
7.	False		1, 6, Contr
	QED		1, 5–7, IP.

end example

example **6.18 Using IP Twice with Descartes**

We'll prove the validity of the following argument noted earlier, which might have been made by Descartes.

If I think, then I exist.
If I do not think, then I think.
Therefore, I exist.

We can formalize the argument by letting A and B be "I think" and "I exist," respectively. Then the argument can be represented by saying that the conclusion B follows from the premises $A \to B$ and $\neg A \to A$. Here's a proof.

1.	$A \to B$			P
2.	$\neg A \to A$			P
3.		$\neg B$		P [for B]
4.			$\neg \neg A$	P [for $\neg A$]
5.			A	4, DN
6.			B	1, 5, MP
7.			False	3, 6, Contr
8.		$\neg A$		4–7, IP
9.		A		2, 8, MP
10.		False		8, 9, Contr
11.	B			3, 8–10, IP
	QED.			

end example

6.3.3 Derived Rules

There are other useful rules that can be derived from the rules in our list (6.7). In fact, some rules in the list can be derived from the other rules in the list (i.e., DN and the additional rules in DS and Add). But we include them because they make the rules easier to use. The next three examples show how to derive these three rules. After that, we'll derive four new proof rules that have many uses.

example **6.19** The Double Negation (DN) Rules

We'll show that the double negation rules can be derived from the other original rules. In other words, we'll derive the following rules.

$$\frac{\neg\,\neg\,A}{A} \text{ and } \frac{A}{\neg\,\neg\,A}.$$

Here's a proof that the first rule represents a valid argument.

1.	$\neg\,\neg\,A$	P
2.	$\neg\,A$	P [for A]
3.	False	1, 2, Contr
4.	A	2, 3, IP
	QED.	

Here's a proof that the second rule represents a vaild argument.

1.	A	P
2.	$\neg\,\neg\,\neg\,A$	P [for $\neg\,\neg\,A$]
3.	$\neg\,A$	2, DN [proved above]
4.	False	1, 3, Contr
5.	$\neg\,\neg\,A$	2–4, IP
	QED.	

end example

example **6.20** The Other Disjunctive Syllogism (DS) Rule

We'll show that one disjunctive syllogism rule can be derived from the other original rules. In other words, we'll derive the following rule.

$$\frac{A \lor B,\ \neg\,B}{A}.$$

Here's a proof that the rule represents a valid argument.

1. $A \vee B$ P
2. $\neg B$ P
3. $\neg A$ P [for A]
4. B 1, 3, DS [other DS]
5.   False   2, 4, Contr
6. A 3–5, IP
     QED.

end example

example **6.21 The Other Addition (Add) Rule**

We'll show that one addition rule can be derived from the other original rules. In other words, we'll derive the following rule.

$$\frac{A}{B \vee A}.$$

Here's a proof that the rule represents a valid argument.

1. A P
2. $A \vee B$ 1, Add
3. $\neg (B \vee A)$ P [for $B \vee A$]
4. $\neg B$ P [for B]
5. $\neg A$ 2, 4, DS
6.   False   1, 5, Contr
7. B 4–6, IP
8. $B \vee A$ 7, Add [other Add]
9.   False   3, 8, Contr
10. $B \vee A$ 3, 7–9, IP
     QED.

end example

Four New Derived Rules

Here are four derived rules that can be very useful. We'll list them next and then give proofs that they are derivable from the original rules.

Derived Rules   (6.8)

Modus Tollens (MT) (Latin for "mode that denies") $$\dfrac{A \to B, \quad \neg\, B}{\neg\, A}.$$	Proof by Cases (Cases) $$\dfrac{A \lor B, \quad A \to C, \quad B \to C}{C}.$$
Hypothetical Syllogism (HS) $$\dfrac{A \to B, \quad B \to C}{A \to C}.$$	Constructive Dilemma (CD) $$\dfrac{A \lor B, \quad A \to C, \quad B \to D}{C \lor D}.$$

> example **6.22   Proof of Modus Tollens (MT)**

We'll show that the modus tollens rule can be derived from the original rules. In other words, we'll derive the following rule.

$$\frac{A \to B, \quad \neg\, B}{\neg\, A}.$$

Here's a proof that the rule represents a valid argument.

1.	$A \to B$	P
2.	$\neg\, B$	P
3.	$\qquad \neg\,\neg\, A$	P [for $\neg$ A]
4.	$\qquad A$	3, DN
5.	$\qquad B$	1, 4, MP
6.	$\qquad$ False	2, 5, Contr
7.	$\neg\, A$	3–6, IP
	QED.	

> end example

> example **6.23   Proof of Proof by Cases (Cases)**

We'll show that the proof by cases rule can be derived from the original rules. In other words, we'll derive the following rule.

$$\frac{A \lor B, \quad A \to C, \quad B \to C}{C}.$$

Here's a proof that the rule represents a valid argument.

1. $A \lor B$ P
2. $A \to C$ P
3. $B \to C$ P
4. $\neg C$ P [for C]
5. $\neg A$ 4, MT
6. B 1, 5, DS
7. C 3, 6, MP
8.   False   4, 7, Contr
9. C 4–8, IP
     QED.

end example

example   **6.24**   **Proof of Hypothetical Syllogism (HS)**

We'll show that the hypothetical syllogism rule can be derived from the original rules. In other words, we'll derive the following rule.

$$\frac{A \to B, \quad B \to C}{A \to C}.$$

Here's a proof that the rule represents a valid argument.

1. $A \to B$ P
2. $B \to C$ P
3. A P [for $A \to C$]
4. B 1, 3, MP
5. C 2, 4, MP
6. $A \to C$ 3–5, CP
     QED.

end example

example   **6.25**   **Proof of Constructive Dilemma (CD)**

We'll show that the constructive dilemma rule can be derived from the original rules. In other words, we'll derive the following rule.

$$\frac{A \lor B, \quad A \to C, \quad B \to D}{C \lor D}.$$

Here's a proof that the rule represents a valid argument.

1.	$A \lor B$	P
2.	$A \to C$	P
3.	$B \to D$	P
4.	$\quad A$	P [for $A \to C \lor D$]
5.	$\quad C$	2, 4, MP
6.	$\quad C \lor D$	5, Add
7.	$A \to C \lor D$	4–6, CP
8.	$\quad B$	P [for $B \to C \lor D$]
9.	$\quad D$	3, 8, MP
10.	$\quad C \lor D$	9, Add
11.	$B \to C \lor D$	8–10, CP
12.	$C \lor D$	1, 7, 11, Cases
	QED.	

6.3.4   Theorems, Soundness, and Completeness

We'll begin by giving the definition of a theorem for our proof system. Then we'll discuss how theorems are related to tautologies.

Definition of Theorem

A *theorem* is the last wff in a proof for which all premises have been discharged.

A proof system for propositional calculus is said to be *sound* if every theorem is a tautology. It's nice to know that our proof system is sound. This follows from our previous discussion about the fact that the last wff of a proof is a tautology if all premises have been discharged.

    A proof system for propositional calculus is said to be *complete* if every tautology is a theorem. It's nice to know that our proof system is complete. This fact follows from a completeness result in the next section for a system that depends on three wffs that it uses as axioms. The completeness of our system follows because each of three wffs used as axioms has a proof in our system.

Using Previously Proved Results (Theorems)

If we know that some wff is a theorem, then we can use it in another proof. That is, we can place the theorem on some line of the proof. If the theorem is a conditional of the form $V \to W$, then we can use it as a derived proof rule. In other words, if we find V on some line of a proof, we can write W on a subsequent line.

The proof rules can be used to prove the basic equivalences listed in (6.1). So each equivalence $V \equiv W$ gives us two derived rules, one for $V \to W$ and the other for $V \to W$.

When we place a theorem on a proof line or when we use a theorem as a derived proof rule, we'll indicate it by writing

$$T$$

in the reason column. Think of T as a theorem.

example **6.26** **Proving a Theorem Using Three Theorems**

We'll prove the simple tautology $(A \to B) \vee (B \to A)$ by using three theorems and the constructive dilemma rule.

1.	$A \vee \neg A$	T [Example 6.17]
2.	$A \to (B \to A)$	T [Example 6.13]
3.	$\neg A \to (A \to B)$	T [Example 6.14]
4.	$(A \to B) \vee (B \to A)$	1, 2, 3, CD
	QED.	

end example

6.3.5 Practice Makes Perfect

Some proofs are straightforward, while others can be brain busters. Remember, when you construct a proof, it may take several false starts before you come up with a correct proof sequence. Study the examples and then do lots of exercises.

Basic Equivalences

A good way to practice formal proof techniques is to write proofs for the basic equivalences. We've already seen proofs for the tautologies $A \to A$ and $A \vee \neg A$ in Example 6.11 and Example 6.17, respectively. In the following examples we'll give proofs for a few more of the basic equivalences. Proofs for other basic equivalences are in the exercises.

example **6.27** $A \vee B \equiv B \vee A$

Proof of $A \vee B \rightarrow B \vee A$:

1.	$A \vee B$	P
2.	$\neg (B \vee A)$	P [for $B \vee A$]
3.	$\neg \neg A$	P [for $\neg A$]
4.	A	3, DN
5.	$B \vee A$	4, Add
6.	False	2, 5, Contr
7.	$\neg A$	3–6, IP
8.	B	1, 7, DS
9.	$B \vee A$	8, Add
10.	False	2, 9, Contr
11.	$B \vee A$	2, 7–10, IP
	QED	1, 11, CP.

The proof of $B \vee A \rightarrow A \vee B$ is similar.

end example

example **6.28** $A \rightarrow B \equiv \neg A \vee B$

Proof of $(A \rightarrow B) \rightarrow (\neg A \vee B)$:

1.	$A \rightarrow B$	P
2.	$\neg (\neg A \vee B)$	P [for $\neg A \vee B$]
3.	$\neg A$	P [for A]
4.	$\neg A \vee B$	3, Add
5.	False	2, 4, Contr
6.	A	3–5, IP
7.	B	1, 6, MP
8.	$\neg A \vee B$	7, Add
9.	False	2, 8, Contr
10.	$\neg A \vee B$	2, 6–9, IP
	QED	1, 10, CP.

Proof of $(\neg A \vee B) \to (A \to B)$:

1. $\neg A \vee B$ P
2. A P [for $A \to B$]
3. $\neg \neg A$ 2, DN
4. B 1, 3, DS
5. $A \to B$ 2–4, CP
     QED   1, 5, CP.

example   **6.29** $\neg (A \to B) \equiv A \wedge \neg B$

Proof of $\neg (A \to B) \to (A \wedge \neg B)$:

1. $\neg (A \to B)$ P
2. $\neg A$ P [for A]
3. $\neg A \vee B$ 2, Add
4. $A \to B$ 3, T [Example 6.28]
5.   False   1, 4, Contr
6. A 2–5, IP
7. $\neg \neg B$ P [for $\neg B$]
8. B 7, DN
9. $B \to (A \to B)$ T, [Example 6. 13]
10. $A \to B$ 8, 9, MP
11.   False   1, 10, Contr
12. $\neg B$ 7–11, IP
13. $A \wedge \neg B$ 6, 12, Conj
     QED   1, 6, 12, 13, CP.

Proof of $(A \wedge \neg B) \to \neg (A \to B)$:

1. $A \wedge \neg B$ P
2. $\neg \neg (A \to B)$ P [for $\neg (A \to B)$]
3. $A \to B$ 2, DN
4. A 1, Simp
5. B 3, 4, MP
6. $\neg B$ 1, Simp
7.   False   5, 6, Contr
8. $\neg (A \to B)$ 2–7, IP
     QED   1, 6, CP.

example **6.30** $\neg (A \vee B) \equiv \neg A \wedge \neg B$

Proof of $\neg (A \vee B) \rightarrow (\neg A \wedge \neg B)$:

1.	$\neg (A \vee B)$	P
2.	$\neg \neg A$	P [for $\neg A$]
3.	A	2, DN
4.	$A \vee B$	3, Add
5.	False	1, 4, Contr
6.	$\neg A$	2–5, IP
7.	$\neg \neg B$	P [for $\neg B$]
8.	B	7, DN
9.	$A \vee B$	8, Add
10.	False	1, 9, Contr
11.	$\neg B$	7–10, IP
12.	$\neg A \wedge \neg B$	6, 11, Conj
	QED	1, 6, 11, 12, CP.

Proof of $(\neg A \wedge \neg B) \rightarrow \neg (A \vee B)$:

1.	$\neg A \wedge \neg B$	P
2.	$\neg \neg (A \vee B)$	P [for $\neg (A \vee B)$]
3.	$A \vee B$	2, DN
4.	$\neg A$	1, Simp
5.	B	3, 4, DS
6.	$\neg B$	1, Simp
7.	False	5, 6, Contr
8.	$\neg (A \vee B)$	2–7, IP
	QED	1, 8, CP.

end example

■ Exercises

Proof Structures

1. Let W denote the wff $(A \vee B \rightarrow C) \rightarrow (B \vee C)$. It's easy to see that W is not a tautology. For example, let $A = B = C = $ False. Suppose someone claims that the following proof shows that W is a tautology.

1.	$A \vee B \rightarrow C$	P
2.	A	P
3.	$A \vee B$	2, Add
4.	C	1, 3, MP
5.	$B \vee C$	4, Add
	QED	1–5, CP.

What is wrong with the claim?

2. Let W denote the wff $(A \rightarrow (B \wedge C)) \rightarrow (A \rightarrow B) \wedge C$. It's easy to see that W is not a tautology. Suppose someone claims that the following sequence of statements is a "proof" of W:

1.	$A \rightarrow (B \wedge C)$	P
2.	A	P [for $A \rightarrow B$]
3.	$B \wedge C$	1, 2, MP
4.	B	3, Simp
5.	$A \rightarrow B$	2–4, CP
6.	C	3, Simp
7.	$(A \rightarrow B) \wedge C$	5, 6, Conj
	QED	1, 5–7, CP.

What is wrong with this "proof" of W?

3. Find the number of premises required for a proof of each of the following wffs. Assume that the letters stand for other wffs.

    a. $A \rightarrow (B \rightarrow (C \rightarrow D))$.

    b. $((A \rightarrow B) \rightarrow C) \rightarrow D$.

4. Give a formalized version of the following proof.

If I am dancing, then I am happy. There is a mouse in the house or I am happy. I am sad. Therefore, there is a mouse in the house and I am not dancing.

Formal Proofs

5. Give a formal proof for each of the following tautologies by using the CP rule. Do not use the IP rule.

    a. $A \rightarrow (B \rightarrow (A \wedge B))$.

    b. $A \rightarrow (\neg B \rightarrow (A \wedge \neg B))$.

    c. $(A \vee B \rightarrow C) \wedge A \rightarrow C$.

    d. $(B \rightarrow C) \rightarrow (A \wedge B \rightarrow A \wedge C)$.

    e. $(A \vee B \rightarrow C \wedge D) \rightarrow (B \rightarrow D)$.

    f. $(A \vee B \rightarrow C) \wedge (C \rightarrow D \wedge E) \rightarrow (A \rightarrow D)$.

    g. $(\neg A \vee \neg B) \wedge (B \vee C) \wedge (C \rightarrow D) \rightarrow (A \rightarrow D)$.

    h. $(A \rightarrow (B \rightarrow C)) \rightarrow (B \rightarrow (A \rightarrow C))$.

    i. $(A \rightarrow C) \rightarrow (A \wedge B \rightarrow C)$.

    j. $(A \rightarrow C) \rightarrow (A \rightarrow B \vee C)$.

    k. $(A \rightarrow B) \rightarrow (C \vee A \rightarrow C \vee B)$.

6. Give a formal proof for each of the following tautologies by using the CP rule and by using the IP rule at least once in each proof.

  a. $A \rightarrow (B \rightarrow A)$.

  b. $(A \rightarrow B) \wedge (A \vee B) \rightarrow B$.

  c. $\neg B \rightarrow (B \rightarrow C)$.

  d. $(A \rightarrow C) \rightarrow (A \rightarrow B \vee C)$.

  e. $(A \rightarrow B) \rightarrow ((A \rightarrow \neg B) \rightarrow \neg A)$.

  f. $(A \rightarrow B) \rightarrow ((B \rightarrow C) \rightarrow (A \vee B \rightarrow C))$.

  g. $(A \rightarrow B) \rightarrow (C \vee A \rightarrow C \vee B)$.

  h. $(C \rightarrow A) \wedge (\neg C \rightarrow B) \rightarrow (A \vee B)$.

7. Give a formal proof for each of the following tautologies by using the CP rule and by using the IP rule at least once in each proof.

  a. $(A \vee B \rightarrow C) \wedge A \rightarrow C$.

  b. $(B \rightarrow C) \rightarrow (A \wedge B \rightarrow A \wedge C)$.

  c. $(A \vee B \rightarrow C \wedge D) \rightarrow (B \rightarrow D)$.

  d. $(A \vee B \rightarrow C) \wedge (C \rightarrow D \wedge E) \rightarrow (A \rightarrow D)$.

  e. $\neg (A \wedge B) \wedge (B \vee C) \wedge (C \rightarrow D) \rightarrow (A \rightarrow D)$.

  f. $(A \rightarrow B) \rightarrow ((B \rightarrow C) \rightarrow (A \vee B \rightarrow C))$.

  g. $(A \rightarrow (B \rightarrow C)) \rightarrow (B \rightarrow (A \rightarrow C))$.

  h. $(A \rightarrow C) \rightarrow (A \wedge B \rightarrow C)$.

Challenges

8. Prove that the following rule, called the Destructive Dilemma rule, can be derived from the original and derived proof rules.

$$\frac{\neg C \vee \neg D, \quad A \rightarrow C, \quad B \rightarrow D}{\neg A \vee \neg B}.$$

9. Two students came up with the following different wffs to formalize the statement "If A then B else C."

$(A \wedge B) \vee (\neg A \wedge C)$.

$(A \rightarrow B) \wedge (\neg A \rightarrow C)$.

Prove that the two wffs are equivalent by finding formal proofs for the following two statements.

  a. $((A \wedge B) \vee (\neg A \wedge C)) \rightarrow ((A \rightarrow B) \wedge (\neg A \rightarrow C))$.

  b. $((A \rightarrow B) \wedge (\neg A \rightarrow C)) \rightarrow ((A \wedge B) \vee (\neg A \wedge C))$.

Basic Properties

For each of the following exercises try to use only the proof rules and derived proof rules. If you must use T, then use it only if the theorem has already been proved.

10. Prove the following basic properties of conjunction.

    a. $A \wedge \neg \text{ False} \equiv A$.

    b. $A \wedge \text{False} \equiv \text{False}$.

    c. $A \wedge A \equiv A$.

    d. $A \wedge \neg A \equiv \text{False}$.

    e. $A \wedge B \equiv B \wedge A$.

    f. $A \wedge (B \wedge C) \equiv (A \wedge B) \wedge C$.

11. Prove the following basic properties of disjunction.

    a. $A \vee \neg \text{ False} \equiv \neg \text{ False}$.

    b. $A \vee \text{False} \equiv A$.

    c. $A \vee A \equiv A$.

    d. $A \vee (B \vee C) \equiv (A \vee B) \vee C$.

12. Prove the following basic properties of implication.

    a. $A \rightarrow \neg \text{ False}$.

    b. $A \rightarrow \text{False} \equiv \neg A$.

    c. $\neg \text{ False} \rightarrow A \equiv A$.

    d. $\text{False} \rightarrow A$.

13. Prove the following basic conversions.

    a. $A \rightarrow B \equiv \neg B \rightarrow \neg A$.

    b. $A \rightarrow B \equiv A \wedge \neg B \rightarrow \text{False}$.

    c. $\neg (A \wedge B) \equiv \neg A \vee \neg B$.

    d. $A \wedge (B \vee C) \equiv (A \wedge B) \vee (A \wedge C)$.

    e. $A \vee (B \wedge C) \equiv (A \vee B) \wedge (A \vee C)$.

14. Prove the following absorption equivalences.

    a. $A \wedge (A \vee B) \equiv A$.

    b. $A \vee (A \wedge B) \equiv A$.

    c. $A \wedge (\neg A \vee B) \equiv A \wedge B$.

    d. $A \vee (\neg A \wedge B) \equiv A \vee B$.

6.4   Formal Axiom Systems

Although truth tables are sufficient to decide the truth of a propositional wff, most of us do not reason by truth tables. We reason in a way that is similar to the natural deduction approach of Section 6.3, where we used the proof rules listed in (6.7) and (6.8). In this section we'll introduce the *axiomatic* approach, where the premises for arguments are limited to a fixed set of wffs, called *axioms*.

   Our goal is to have a formal proof system with two properties. We want our proofs to yield theorems that are tautologies, and we want any tautology to be provable as a theorem. In other words, we want soundness and completeness.

Soundness: All proofs yield theorems that are tautologies.

Completeness: All tautologies are provable as theorems.

6.4.1   An Example Axiom System

Is there a simple formal system for which we can show that the propositional calculus is both sound and complete? Yes, there is. In fact, there are many of them. Each one specifies a small fixed set of axioms and inference rules. The pioneer in this area was the mathematician Gottlob Frege (1848–1925). He formulated the first such axiom system [1879]. We'll discuss it further in the exercises. Later, in 1930, J. Lukasiewicz showed that Frege's system, which has six axioms, could be replaced by the following three axioms, where A, B, and C can represent any wff generated by propositional variables and the two connectives $\neg$ and $\rightarrow$.

Frege-Lukasiewicz Axioms   (6.9)

   1. $A \rightarrow (B \rightarrow A)$.

   2. $(A \rightarrow (B \rightarrow C)) \rightarrow ((A \rightarrow B) \rightarrow (A \rightarrow C))$.

   3. $(\neg A \rightarrow \neg B) \rightarrow (B \rightarrow A)$.

The only inference rule used is modus ponens. Although the axioms may appear a bit strange, they can all be verified by truth tables. Also note that conjunction and disjunction are missing. But this is no problem, since we know that they can be written in terms of implication and negation.

   We'll use this system to prove the CP rule (i. e., the deduction theorem). But first we must prove a result that we'll need, namely that $A \rightarrow A$ is a theorem provable from the axioms. Notice that the proof uses only the given axioms and modus ponens.

Lemma 1. $A \rightarrow A$ **is provable from the axioms.**

In the following proof B can be any wff, including A.

1. $(A \rightarrow ((B \rightarrow A) \rightarrow A)) \rightarrow ((A \rightarrow (B \rightarrow A)) \rightarrow (A \rightarrow A))$ Axiom 2
2. $A \rightarrow ((B \rightarrow A) \rightarrow A)$ Axiom 1
3. $(A \rightarrow (B \rightarrow A)) \rightarrow (A \rightarrow A)$ 1, 2, MP
4. $A \rightarrow (B \rightarrow A)$ Axiom 1
5. $A \rightarrow A$ 3, 4, MP

   QED.

Now we're in position to prove the CP rule, which is also called the deduction theorem. The proof, which is due to Herbrand (1930), uses only the given axioms, modus ponens, and Lemma 1.

Deduction Theorem (Conditional Proof Rule, CP)

If A is a premise in a proof of B, then there is a proof of $A \rightarrow B$ that does not use A as a premise.

Proof: Assume that A is a premise in a proof of B. We must show that there is a proof of $A \rightarrow B$ that does not use A as a premise. Suppose that $B_1,..., B_n$ is a proof of B that uses the premise A. We'll show by induction that for each k in the interval $1 \leq k \leq n$, there is a proof of $A \rightarrow B_k$ that does not use A as a premise. Since $B = B_n$, the result will be proved. If $k = 1$, then either $B_1 = A$ or B_1 is an axiom or a premise other than A. If $B_1 = A$, then $A \rightarrow B_1 = A \rightarrow A$, which by Lemma 1 has a proof that does not use A as a premise. If B_1 is an axiom or a premise other than A, then the following proof of $A \rightarrow B_1$ does not use A as a premise.

1. B_1 An axiom or premise other than A
2. $B_1 \rightarrow (A \rightarrow B_1)$ Axiom 1
3. $A \rightarrow B_1$ 1, 2, MP

Now assume that for each $i < k$ there is a proof of $A \rightarrow B_i$ that does not use A as a premise. We must show that there is a proof of $A \rightarrow B_k$ that does not use A as a premise. If $B_k = A$ or B_k is an axiom or a premise other than A, then we can use the same argument for the case when $k = 1$ to conclude that there is a proof of $A \rightarrow B_k$ that does not use A as a premise. If B_k is not an axiom or a premise, then it is inferred by MP from two wffs in the proof of the form B_i and $B_j = B_i \rightarrow B_k$, where $i < k$ and $j < k$. Since by $i < k$ and $j < k$, the induction hypothesis tells us that there are proofs of $A \rightarrow B_i$ and $A \rightarrow (B_i \rightarrow B_k)$, neither of which contains A as a premise. Now consider the following proof, where lines 1 and 2 represent the proofs of $A \rightarrow B_i$ and $A \rightarrow (B_i \rightarrow B_k)$.

1.	Proof of $A \to B_i$	Induction
2.	Proof of $A \to (B_i \to B_k)$	Induction
3.	$(A \to (B_i \to B_k)) \to ((A \to B_i) \to (A \to B_k))$	Axiom 2
4.	$(A \to B_i) \to (A \to B_k)$	2, 3, MP
5.	$A \to B_k$	1, 4, MP

So there is a proof of $A \to B_k$ that does not contain A as a premise. Let $k = n$ to obtain the desired result because $B_n = B$. QED.

Once we have the CP rule, proofs become much easier since we can have premises in our proofs. But still, everything that we do must be based only on the axioms, MP, CP, and any results we prove along the way. The system is sound because the axioms are tautologies and MP maps tautologies to a tautology. In other words, every proof yields a theorem that is a tautology.

The remarkable thing is that this little axiom system is complete in the sense that there is a proof within the system for every tautology of the propositional calculus. We'll give a few more examples to get the flavor of reasoning from a very small set of axioms.

example 6.31 **Hypothetical Syllogism**

We'll use CP to prove the hypothetical syllogism proof rule.

Hypothetical Syllogism (HS) (6.10)

From the premises $A \to B$ and $B \to C$, there is a proof of $A \to C$.

Proof:
1.	$A \to B$	P
2.	$B \to C$	P
3.	A	P [for $A \to C$]
4.	B	1, 3, MP
5.	C	2, 4, MP
6.	$A \to C$	3–5, CP
	QED.	

We can apply the CP rule to HS to say that from the premise $A \to B$ there is a proof of $(B \to C) \to (A \to C)$. One more application of the CP rule tells us that the following wff has a proof with no premises:

$$(A \to B) \to ((B \to C) \to (A \to C)).$$

This is just another way to represent the HS rule as a tautology.

end example

Six Sample Proofs

Now that we have CP and HS rules, it should be easier to prove statements within the axiom system. Let's prove the following six statements.

Six Tautologies (6.11)

a. $\neg A \rightarrow (A \rightarrow B)$.

b. $\neg \neg A \rightarrow A$.

c. $A \rightarrow \neg \neg A$.

d. $(A \rightarrow B) \rightarrow (\neg B \rightarrow \neg A)$.

e. $A \rightarrow (\neg B \rightarrow \neg (A \rightarrow B))$.

f. $(A \rightarrow B) \rightarrow ((\neg A \rightarrow B) \rightarrow B)$.

In the next six examples we'll prove these six statements using only the axioms, MP, CP, HS, and previously proven results.

example 6.32 $\neg A \rightarrow (A \rightarrow B)$

Proof:

1.	$\neg A$	P
2.	$\neg A \rightarrow (\neg B \rightarrow \neg A)$	Axiom 1
3.	$\neg B \rightarrow \neg A$	1, 2, MP
4.	$(\neg B \rightarrow \neg A) \rightarrow (A \rightarrow B)$	Axiom 3
5.	$A \rightarrow B$	3, 4, MP
	QED	1–5, CP.

end example

example 6.33 $\neg \neg A \rightarrow A$

Proof:

1.	$\neg \neg A$	P
2.	$\neg \neg A \rightarrow (\neg A \rightarrow \neg \neg \neg A)$	T [Example 6.32]
3.	$\neg A \rightarrow \neg \neg \neg A$	1, 2, MP
4.	$(\neg A \rightarrow \neg \neg \neg A) \rightarrow (\neg \neg A \rightarrow A)$	Axiom 3
5.	$\neg \neg A \rightarrow A$	3, 4, MP
6.	A	1, 5, MP
	QED	1–6, CP.

end example

example   **6.34** $A \rightarrow \neg \neg A$

Proof: 1. $\neg \neg \neg A \rightarrow \neg A$ T [Example 6.33]
         2. $(\neg \neg \neg A \rightarrow \neg A) \rightarrow (A \rightarrow \neg \neg A)$ Axiom 3
         3. $A \rightarrow \neg \neg A$ 1, 2, MP
              QED.

end example

example   **6.35** $(A \rightarrow B) \rightarrow (\neg B \rightarrow \neg A)$

Proof: 1. $A \rightarrow B$ P
         2. $\neg \neg A \rightarrow A$ T [Example 6.33]
         3. $\neg \neg A \rightarrow B$ 1, 2, HS
         4. $B \rightarrow \neg \neg B$ T [Example 6.34]
         5. $\neg \neg A \rightarrow \neg \neg B$ 1, 3, 4, HS
         6. $(\neg \neg A \rightarrow \neg \neg B) \rightarrow (\neg B \rightarrow \neg A)$ Axiom 3
         7. $\neg B \rightarrow \neg A$ 5, 6, MP
              QED   1–7, CP.

end example

example   **6.36** $A \rightarrow (\neg B \rightarrow \neg (A \rightarrow B))$

Proof: 1. A P
         2. $A \rightarrow B$ P [for $(A \rightarrow B) \rightarrow B$]
         3. B 1, 2, MP
         4. $(A \rightarrow B) \rightarrow B$ 2, 3, CP
         5. $((A \rightarrow B) \rightarrow B) \rightarrow (\neg B \rightarrow \neg (A \rightarrow B))$ T [Example 6.35]
         6. $\neg B \rightarrow \neg (A \rightarrow B)$ 4, 5, MP
              QED   1, 4–6, CP.

end example

example **6.37** $(A \rightarrow B) \rightarrow ((\neg A \rightarrow B) \rightarrow B)$

Proof:
1. $A \rightarrow B$ .... P
2. $(A \rightarrow B) \rightarrow (\neg B \rightarrow \neg A)$ .... T [Example 6.35]
3. $\neg B \rightarrow \neg A$ .... 1, 2, MP
4. $\neg A \rightarrow B$ .... P [for $(\neg A \rightarrow B) \rightarrow B$]
5. $\neg B \rightarrow B$ .... 3, 4, HS
6. $\neg B \rightarrow (B \rightarrow \neg (A \rightarrow B))$ .... T [Example 6.32]
7. $(\neg B \rightarrow (B \rightarrow \neg (A \rightarrow B)))$
          $\rightarrow ((\neg B \rightarrow B) \rightarrow (\neg B \rightarrow \neg (A \rightarrow B)))$ Axiom 2
8. $(\neg B \rightarrow B) \rightarrow (\neg B \rightarrow \neg (A \rightarrow B))$ .... 6, 7, MP
9. $\neg B \rightarrow \neg (A \rightarrow B)$ .... 5, 8, MP
10. $(\neg B \rightarrow \neg (A \rightarrow B)) \rightarrow ((A \rightarrow B) \rightarrow B))$ ..... Axiom 3
11. $(A \rightarrow B) \rightarrow B)$ .... 9, 10, MP
12. B .... 1, 11, MP
13. $(\neg A \rightarrow B) \rightarrow B$ .... 4–12, CP
    QED .... 1–3, 13, CP.

end example

Completeness of the Axiom System

As we mentioned earlier, it is a remarkable result that this axiom system is complete. The interesting thing is that we now have enough tools to find a proof of completeness.

Lemma 2

Let W be a wff with propositional variables $P_1, ..., P_m$. For any truth assignment to the variables, let $Q_1, ..., Q_m$ be defined by letting each Q_k be either P_k or $\neg P_k$ depending on whether P_k is assigned True or False, respectively. Then from the premises $Q_1, ..., Q_m$ there is either a proof of W or a proof of $\neg W$ depending on whether the assignment makes W true or false, respectively.

Proof: The proof is by induction on the number n of connectives that occcur in W. If $n = 0$, then W is just a propositional variable P. If P is assigned True, then we must find a proof of P using P as its own premise.

1. P Premise
2. $P \rightarrow P$ Lemma 1
3. P 1, 2, MP
    QED   1–3, CP.

If P is assigned False, then we must find a proof of $\neg P$ from premise $\neg P$.

1.	$\neg P$	Premise
2.	$\neg P \to \neg P$	Lemma 1
3.	$\neg P$	1, 2, MP
	QED	1–3, CP.

So assume W is a wff with n connectives where $n > 0$ and assume that the lemma is true for all wffs with less than n connectives. Now W has one of two forms, $W = \neg A$ or $W = A \to B$. It follows that A and B each have less than n connectives, so by induction there are proofs from the premises $Q_1, \ldots, Q_m$ of either A or $\neg A$ and of either B or $\neg B$ depending on whether they are made true or false by the truth assignment. We'll argue for each form of W.

Let $W = \neg A$. If W is true, then A is false, so there is a proof from the premises of $\neg A = W$. If W is false, then A is true, so there is a proof from the premises of A. Now (6.11c) gives us $A \to \neg \neg A$. So by MP there is a proof from the premises of $\neg \neg A = \neg W$.

Let $W = A \to B$. Assume W is true. Then either A is false or B is true. If A is false, then there is a proof from the premises of $\neg A$. Now (6.11a) gives us $\neg A \to (A \to B)$. So by MP there is a proof from the premises of $(A \to B) = W$. If B is true, then there is a proof from the premises of B. Now Axiom 1 gives us $B \to (A \to B)$. So by MP there is a proof from the premises of $(A \to B) = W$.

Assume W is false. Then A is true and B is false. So there is a proof from the premises of A and there is a proof from the premises of $\neg B$. Now (6.11e) gives us $A \to (\neg B \to \neg (A \to B))$. So by two applications of MP there is a proof from the premises of $\neg (A \to B) = \neg W$. QED.

Theorem (Completeness)

Any tautology can be proven as a theorem in the axiom system.

Proof: Let W be a tautology with propositional variables $P_1, \ldots, P_m$. Since W is always true, it follows from Lemma 2 that for any truth assignment to the propositional variables $P_1, \ldots, P_m$ that occur in W, there is a proof of W from the premises $Q_1, \ldots, Q_m$, where each Q_k is either P_k or $\neg P_k$ according to whether the truth assignment to P_k is True or False, respectively. Now if P_m is assigned True, then $Q_m = P_m$ and if P_m is assigned False, then $Q_m = \neg P_m$. So there are two proofs of W, one with premises $Q_1, \ldots, Q_{m-1}, P_m$ and one with premises $Q_1, \ldots, Q_{m-1}, \neg P_m$. Now apply the CP rule to both proofs to obtain the following two proofs.

A proof from premises $Q_1, \ldots, Q_{m-1}$ of $(P_m \to W)$.

A proof from premises $Q_1, \ldots, Q_{m-1}$ of $(\neg P_m \to W)$.

We combine the two proofs into one proof. Now (6.11f) gives us the following statement, which we add to the proof:

$$(P_m \rightarrow W) \rightarrow ((\neg\, P_m \rightarrow W) \rightarrow W).$$

Now with two applications of MP, we obtain W. So there is a proof of W from premises $Q_1, \ldots, Q_{m-1}$. Now use the same procedure for $m-1$ more steps to obtain a proof of W with no premises. QED.

6.4.2 Other Axiom Systems

There are many other small axiom systems for the propositional calculus that are sound and complete. For example, Frege's original axiom system consists of the following six axioms together with the modus ponens inference rule, where A, B, and C can represent any wff generated by propositional variables and the two connectives $\neg$ and $\rightarrow$.

Frege's Axioms (6.12)

1. $A \rightarrow (B \rightarrow A)$.

2. $(A \rightarrow (B \rightarrow C)) \rightarrow ((A \rightarrow B) \rightarrow (A \rightarrow C))$.

3. $(A \rightarrow (B \rightarrow C)) \rightarrow (B \rightarrow (A \rightarrow C))$.

4. $(A \rightarrow B) \rightarrow (\neg\, B \rightarrow \neg\, A)$.

5. $\neg\, \neg\, A \rightarrow A$.

6. $A \rightarrow \neg\, \neg\, A$.

If we examine the proof of the CP rule, we see that it depends only on Axioms 1 and 2 of (6.9), which are the same as Frege's first two axioms. Also, HS is proved with CP. So we can continue reasoning from these axioms with CP and HS in our tool kit. We'll discuss this system in the exercises.

Another small axiom system, due to Hilbert and Ackermann [1938], consists of the following four axioms together with the modus ponens inference rule, where $A \rightarrow B$ is used as an abbreviation for $\neg\, A \vee B$, and where A, B, and C can represent any wff generated by propositional variables and the two connectives $\neg$ and $\vee$.

Hilbert-Ackermann Axioms (6.13)

1. $A \vee A \rightarrow A$.

2. $A \rightarrow A \vee B$.

3. $A \vee B \rightarrow B \vee A$.

4. $(A \rightarrow B) \rightarrow (C \vee A \rightarrow C \vee B)$.

We'll discuss this system in the exercises too.

There are important reasons for studying small formal systems like the axiom systems we've been discussing. Small systems are easier to test and easier to compare with other systems because there are only a few basic operations to worry about. For example, if we build a program to do automatic reasoning, it may be easier to implement a small set of axioms and inference rules. This also applies to computers with small instruction sets and to programming languages with a small number of basic operations.

 Exercises

Axiom Systems

1. In the Frege-Lukasiewicz axiom system (6.9), prove the following version of the IP rule: From the premises $\neg A \rightarrow B$ and $\neg A \rightarrow \neg B$, there is a proof of A. Use only the axioms (6.9), MP, CP, HS, and the six tautologies (6.11).

2. In Frege's axiom system (6.12), prove that Axiom 3 follows from Axioms 1 and 2. That is, prove the following statement from Axioms 1 and 2:

$$(A \rightarrow (B \rightarrow C)) \rightarrow (B \rightarrow (A \rightarrow C)).$$

3. In Frege's axiom system (6.12), prove that $(\neg A \rightarrow \neg B) \rightarrow (B \rightarrow A)$.

4. In the Hilbert-Ackermann axiom system (6.13), prove each of the following statements. Do not use the CP rule. You can use any previous statement in the list to prove a subsequent statement.

   a. $(A \rightarrow B) \rightarrow ((C \rightarrow A) \rightarrow (C \rightarrow B))$.
   b. (HS) If $A \rightarrow B$ and $B \rightarrow C$ are theorems, then $A \rightarrow C$ is a theorem.
   c. $A \rightarrow A$ (i.e., $\neg A \vee A$).

    d. $A \vee \neg A$.

    e. $A \rightarrow \neg \neg A$.

    f. $\neg \neg A \rightarrow A$.

    g. $\neg A \rightarrow (A \rightarrow B)$.

    h. $(A \rightarrow B) \rightarrow (\neg B \rightarrow \neg A)$.

    i. $(\neg B \rightarrow \neg A) \rightarrow (A \rightarrow B)$.

Logic Puzzles

5. The county jail is full. The sheriff, Anne Oakley, brings in a newly caught criminal and decides to make some space for the criminal by letting one of the current inmates go free. She picks prisoners A, B, and C to choose from. She puts blindfolds on A and B because C is already blind. Next she selects three hats from five hats hanging on the hat rack, two of which are red and three of which are white, and places the three hats on the prisoner's heads. She hides the remaining two hats. Then she takes the blindfolds off A and B and tells them what she has done, including the fact that there were three white hats and two red hats to choose from. Sheriff Oakley then says, "If you can tell me the color of the hat you are wearing, without looking at your own hat, then you can go free." The following things happen:

   1. A says that he can't tell the color of his hat. So the sheriff has him returned to his cell.

   2. Then B says that he can't tell the color of his hat. So he is also returned to his cell.

   3. Then C, the blind prisoner, says that he knows the color of his hat. He tells the sheriff, and she sets him free.

   What color was C's hat, and how did C do his reasoning?

6. Four men and four women were nominated for two positions on the school board. One man and one woman were elected to the positions. Suppose the men are named A, B, C, and D and the women are named E, F, G, and H. Further, suppose that the following four statements are true:

   1. If neither A nor E won a position, then G won a position.

   2. If neither A nor F won a position, then B won a position.

   3. If neither B nor G won a position, then C won a position.

   4. If neither C nor F won a position, then E won a position.

   Who were the two people elected to the school board?

6.5 Chapter Summary

Propositional calculus is the basic building block of formal logic. Each wff represents a statement that can be checked by truth tables to determine whether it is a tautology, a contradiction, or a contingency. There are basic equivalences (6.1) that allow us to simplify and transform wffs into other wffs. We can use these equivalences with Quine's method to determine whether a wff is a tautology, a contradiction, or a contingency. We can also use the equivalences to transform any wff into a DNF or a CNF. Any truth function has one of these forms.

Propositional calculus also provides us with formal techniques for proving properties of wffs without using truth tables. The rules for constructing formal proofs reflect the natural way that we reason informally about the validity of arguments.

We want formal reasoning systems to be sound—proofs yield theorems that are tautologies—and complete—all tautologies can be proven as theorems. The system of antural deduction presented in Section 6.3 is sound and complete. There are even small axiomatic systems for the propositional calculus that are sound and complete. We presented one in Section 6.4 that has three axioms and one inference rule.

Notes

The logical symbols that we've used in this chapter are not universal. So you should be flexible in your reading of the literature. From a historical point of view, Whitehead and Russell [1910] introduced the symbols $\supset$, $\vee$, $\cdot$, $\sim$, and $\equiv$ to stand for implication, disjunction, conjunction, negation, and equivalence, respectively. A prefix notation for the logical operations was introduced by Lukasiewicz [1929], where the letters C, A, K, N, and E stand for implication, disjunction, conjunction, negation, and equivalence, respectively. So in terms of our notation we have $Cpq = p \rightarrow q$, $Apq = p \vee q$, $Kpq = p \wedge q$, $Nq = \neg q$, and $Epq = p \equiv q$. This notation is called Polish notation, and its advantage is that each expression has a unique meaning without using parentheses and precedence. For example, $(p \rightarrow q) \rightarrow r$ and $p \rightarrow (q \rightarrow r)$ are represented by the expressions $CCpqr$ and $CpCqr$, respectively. The disadvantage of the notation is that it's harder to read. For example, $CCpqKsNr = (p \rightarrow q) \rightarrow (s \wedge \neg r)$.

The fact that a wff W is a theorem is often denoted by placing a turnstile in front of it as follows:

$$\vdash W.$$

Turnstiles are also used in discussing proofs that have premises. For example, the notation

$$A_1, A_2, \ldots, A_n \vdash B$$

means that there is a proof of B from the premises $A_1, A_2, \ldots, A_n$.

We should again emphasize that the logic we are studying in this book deals with statements that are either true or false. This is sometimes called the *Law of the Excluded Middle*: Every statement is either true or false. If our logic does not assume the law of the excluded middle, then we can no longer use indirect proof because we can't conclude that a statement is false from the assumption that it is not true. A logic called *intuitionist logic* omits this law and thus forces all proofs to be direct. Intuitionists like to construct things in a direct manner.

Logics that assume the law of the excluded middle are called *two-valued logics*. Some logics take a more general approach and consider statements that may not be true or false. For example, in a *three-valued logic* we might use the numbers 0, 0.5, and 1 to stand for the truth values False, Unknown, and True, respectively. We can build truth tables for this logic by defining $\neg A = 1 - A$, $A \vee B = \max(A, B)$, and $A \wedge B = \min(A, B)$. We still use the equivalence $A \rightarrow B \equiv \neg A \vee B$. So we can discuss three-valued logic.

In a similar manner we can discuss *n-valued logic* for any natural number $n \geq 2$, where statements have one of n truth values that can be represented by n numbers in the range 0 to 1. Some logics, called *fuzzy logics*, assign truth values over an infinite set such as the closed unit interval $[0, 1]$.

There are also logics that examine the behavior of "is necessary" and "is possible." Such logics are called *modal logics*, but the term is also used to refer to other logics such as *deontic logic*, which examines the behavior of "is obligated" and "is permitted," and *temporal logic*, which examines the behavior of statements with respect to time such as "has been," "will be," "has always been," and "will always be." For example, the quote at the beginning of this chapter could be represented and analyzed in a modal logic.

All these logics have applications in computer science but they are beyond our scope and purpose. However, it's nice to know that they all depend on a good knowledge of two-valued logic. In this chapter we've covered the fundamental parts of two-valued logic—the properties and reasoning rules of the propositional calculus. We'll see that these ideas occur in all the logics and applications that we cover in the next three chapters.

chapter 7

Predicate Logic

Error of opinion may be tolerated
where reason is left free to combat it.

— Thomas Jefferson (1743–1826)

We need a new logic if we want to describe arguments that deal with all instances or with some instance. In this chapter we'll introduce the notions and notations of first-order predicate calculus. This logic will allow us to analyze and symbolize a wider variety of statements and arguments than can be done with propositional logic.

chapter guide

7.1 First-Order Predicate Calculus

Propositional calculus provides adequate tools for reasoning about propositional wffs, which are combinations of propositions. But a proposition is a sentence taken as a whole. With this restrictive definition, propositional calculus doesn't provide the tools to do everyday reasoning. For example, in the following argument it is impossible to find a formal way to test the correctness of the inference without further analysis of each sentence.

> All computer science majors own a personal computer.
> Socrates does not own a personal computer.
> Therefore, Socrates is not a computer science major.

To discuss such an argument, we need to break up the sentences into parts. The words in the set {All, own, not} are important to understand the argument. Somehow we need to symbolize a sentence so that the information needed for reasoning is characterized in some way. Therefore, we will study the inner structure of sentences.

7.1.1   Predicates and Quantifiers

The statement "x owns a personal computer" is not a proposition because its truth value depends on x. If we give x a value, like $x =$ Socrates, then the statement becomes a proposition because it is either true or false. From a grammatical point of view, the property "owns a personal computer" is a predicate, where a predicate is the part of a sentence that gives a property of the subject. A predicate usually contains a verb, like "owns" in our example. The word predicate comes from the Latin word *praedicare*, which means to proclaim.

From the logic point of view, a *predicate* is a relation, which of course we can also think of as a property. For example, suppose we let $p(x)$ mean "x owns a personal computer." Then p is a predicate that describes the relation (i.e., property) of owning a personal computer. Sometimes it's convenient to call $p(x)$ a predicate, although p is the actual predicate. If we replace the variable x by some definite value such as Socrates, then we obtain the proposition p(Socrates). For another example, suppose that for any two natural numbers x and y we let $q(x, y)$ mean "$x < y$." Then q is the predicate that we all know of as the "less than" relation. For example, the proposition $q(1, 5)$ is true, and the proposition $q(8, 3)$ is false.

The Existential Quantifier

Let $p(x)$ mean that x is an odd integer. Then the proposition $p(9)$ is true, and the proposition $p(20)$ is false. Similarly, the following proposition is true:

$$p(2) \lor p(3) \lor p(4) \lor p(5).$$

We can describe this proposition by saying,

"There exists an element x in the set {2, 3, 4, 5} such that $p(x)$ is true."

If we let $D = \{2, 3, 4, 5\}$, the statement can be shortened to

"There exists $x \in D$ such that $p(x)$ is true."

If we don't care where x comes from and we don't care about the meaning of $p(x)$, then we can still describe the preceding disjunction by saying:

"There exists an x such that $p(x)$."

This expression has the following symbolic representation:

$$\exists x \ p(x).$$

This expression is not a proposition because we don't have a specific set of elements over which x can vary. So $\exists x \ p(x)$ does not have a truth value. The symbol $\exists x$ is called an *existential quantifier*.

The Universal Quantifier

Now let's look at conjunctions rather than disjunctions. As before, we'll start by letting $p(x)$ mean that x is an odd integer. Suppose we have the following proposition:

$$p(1) \land p(3) \land p(5) \land p(7).$$

This conjunction is true and we can represent it by the following statement, where $D = \{1, 3, 5, 7\}$.

"For every x in D, $p(x)$ is true."

If we don't care where x comes from and we don't care about the meaning of $p(x)$, then we can still describe the preceding conjunction by saying:

"For every x, $p(x)$."

This expression has the following symbolic representation:

$$\forall x \ p(x).$$

This expression is not a proposition because we don't have a specific set of elements over which x can vary. So $\forall x \ p(x)$ does not have a truth value. The symbol $\forall x$ is called a *universal quantifier*.

Using Quantifiers

Let's see how the quantifiers can be used together to represent certain statements. If $p(x, y)$ is a predicate and we let the variables x and y vary over the set $D = \{0, 1\}$, then the proposition

$$[p(0, 0) \lor p(0, 1)] \land [p(1, 0) \lor p(1, 1)]$$

can be represented by the following quantified expression:

$$\forall x \ \exists y \ p(x, y).$$

To see this, notice that the two disjunctions can be written as follows:

$$p(0,0) \vee p(0,1) = \exists y\ p(0,y) \quad \text{and} \quad p(1,0) \vee p(1,1) = \exists y\ p(1,y).$$

So we can write the proposition as follows:

$$[p(0,0) \vee p(0,1)] \wedge [p(1,0) \vee p(1,1)] = \exists y\ p(0,y) \wedge \exists y\ p(1,y)$$
$$= \forall x\ \exists y\ p(x,y).$$

Here's an alternative way to get the same result:

$$[p(0,0) \vee p(0,1)] \wedge [p(1,0) \vee p(1,1)] = \forall x\ [p(x,0) \vee p(x,1)] = \forall x\ \exists y\ p(x,y).$$

Now let's go the other way. We'll start with an expression containing different quantifiers and try to write it as a proposition. For example, if we use the same set of values $D = \{0, 1\}$, then the quantified expression

$$\exists y\ \forall x\ p(x,\ y)$$

denotes the proposition

$$[p(0,\ 0) \wedge p(1,\ 0)] \vee [p(0,\ 1) \wedge p(1,\ 1)]\ .$$

To see this, notice that we can evaluate the quantified expression in either of two ways as follows:

$$\exists y\ \forall x\ p(x,y) = \forall x\ p(x,0) \vee \forall x\ p(x,1) = [p(0,0) \wedge p(1,0)] \vee [p(0,1) \wedge p(1,1)].$$
$$\exists y\ \forall x\ p(x,y) = \exists y\ [p(0,y) \wedge p(1,y)] = [p(0,0) \wedge p(1,0)] \vee [p(0,1) \wedge p(1,1)].$$

Of course, not every expression containing quantifiers results in a proposition. For example, if $D = \{0, 1\}$, then the expression $\forall x\ p(x,\ y)$ can be written as follows:

$$\forall x\ p(x,\ y) = p(0,\ y) \wedge p(1,\ y).$$

To obtain a proposition, each variable of the expression must be quantified or assigned some value in D. We'll discuss this shortly when we talk about semantics.

The next two examples will introduce us to the important process of formalizing English sentences with quantifiers.

example 7.1 Formalizing Sentences

We'll formalize the three sentences about Socrates that we listed at the beginning of the section. Over the domain of all people, let $C(x)$ mean that x is a computer science major, and let $P(x)$ mean that x owns a personal computer. Then the sentence "All computer science majors own a personal computer" can be formalized as

$$\forall x\ (C(x) \to P(x)).$$

The sentence "Socrates does not own a personal computer" becomes

$$\neg\ P(\text{Socrates}).$$

The sentence "Socrates is not a computer science major" becomes

$$\neg\ C(\text{Socrates}).$$

end example

example 7.2 Formalizing Sentences

Suppose we consider the following two elementary facts about the set $\mathbb{N}$ of natural numbers.

1. Every natural number has a successor.

2. There is no natural number whose successor is 0.

Let's formalize these sentences. We'll begin by writing down a semiformal version of the first sentence:

For each $x \in \mathbb{N}$ there exists $y \in \mathbb{N}$ such that y is a successor of x.

If we let $s(x, y)$ mean that y is a successor of x, then the formal version of the sentence can be written as follows:

$$\forall x\ \exists y\ s(x,\ y).$$

Now let's look at the second sentence. It can be written in a semiformal version as follows:

There does not exist $x \in \mathbb{N}$ such that 0 is a successor of x.

The formal version of this sentence is the following sentence, where $a = 0$:

$$\neg\ \exists x\ s(x,\ a).$$

end example

These notions of quantification belong to a logic called *first-order predicate calculus*. The term "first-order" refers to the fact that quantifiers apply only to variables that range over the domain of discourse. In Chapter 8 we'll discuss "higher-order" logics in which quantifiers can quantify additional things. To discuss first-order predicate calculus, we need to give a precise description of its well-formed formulas and their meanings. That's the task of this section.

7.1.2   Well-Formed Formulas

To give a precise description of a first-order predicate calculus, we need an alphabet of symbols. For this discussion we'll use several kinds of letters and symbols, described as follows:

Individual variables:	x, y, z
Individual constants:	a, b, c
Function letters:	f, g, h
Predicate letters:	p, q, r
Connective symbols:	$\neg$, $\rightarrow$, $\wedge$, $\vee$
Quantifier symbols:	$\exists$, $\forall$
Punctuation symbols:	$(\,,)$, ",,"

From time to time we may use other letters, strings of letters, or subscripted letters. The number of arguments for a predicate or function will normally be clear from the context. A predicate with no arguments is considered to be a proposition.

A *term* is either a variable, a constant, or a function applied to arguments that are terms. For example, x, a, and $f(x, g(b))$ are terms. An *atomic formula* (or simply *atom*) is a predicate applied to arguments that are terms. For example, $p(x, a)$ and $q(y, f(c))$ are atoms.

We can define the wffs—the well-formed formulas—of first-order predicate calculus inductively as follows:

Definition of a Wff (Well-Formed Formula)

1. Any atom is a wff.

2. If W and V are wffs and x is a variable, then the following expressions are also wffs:

$$(W),\ \neg\ W,\ W \vee V,\ W \wedge V,\ W \rightarrow V,\ \exists x\ W,\ \text{and}\ \forall x\ W.$$

To write formulas without too many parentheses and still maintain a unique meaning, we'll agree that the quantifiers have the same precedence as the negation symbol. We'll continue to use the same hierarchy of precedence for the operators $\neg$, $\wedge$, $\vee$, and $\rightarrow$. Therefore, the hierarchy of precedence now looks like the following:

$$\neg,\ \exists x,\ \forall y \quad \text{(highest, do first)}$$
$$\wedge$$
$$\vee$$
$$\rightarrow \qquad \text{(lowest, do last)}$$

If any of the quantifiers or the negation symbol appear next to each other, then the rightmost symbol is grouped with the smallest wff to its right. Here are a few wffs in both unparenthesized form and parenthesized form:

Unparenthesized Form	Parenthesized Form
$\forall x \, \neg \, \exists y \, \forall z \, p(x, y, z)$	$\forall x \, (\neg \, (\exists y \, (\forall z \, p(x, y, z))))$
$\exists x \, p(x) \vee q(x)$	$(\exists x \, p(x)) \vee q(x)$
$\forall x \, p(x) \rightarrow q(x)$	$(\forall x \, p(x)) \rightarrow q(x)$
$\exists x \, \neg \, p(x, y) \rightarrow q(x) \wedge r(y)$	$(\exists x \, (\neg \, p(x, y))) \rightarrow (q(x) \wedge r(y))$
$\exists x \, p(x) \rightarrow \forall x \, q(x) \vee p(x) \wedge r(x)$	$(\exists x \, p(x)) \rightarrow ((\forall x \, q(x)) \vee (p(x) \wedge r(x)))$

Scope, Bound, and Free

Now let's discuss the relationship between the quantifiers and the variables that appear in a wff. When a quantifier occurs in a wff, it influences some occurrences of the quantified variable. The extent of this influence is called the scope of the quantifier, which we define as follows:

Definition of Scope

The *scope* of $\exists x$ in $(\exists x \, W)$ is W. Similarly, the scope of $\forall x$ in $(\forall x \, W)$ is W. In the absence of parentheses, the *scope* of a quantifier is the smallest wff immediately to its right.

For example, the scope of $\exists x$ in the wff

$$\exists x \, p(x, \, y) \rightarrow q(x)$$

is $p(x, \, y)$ because the parenthesized version of the wff is $(\exists x \, p(x, \, y)) \rightarrow q(x)$. On the other hand, the scope of $\exists x$ in the wff

$$\exists x \, (p(x, \, y) \rightarrow q(x))$$

is the conditional $p(x, \, y) \rightarrow q(x)$. Now let's classify the occurrences of variables that occur in a wff.

Bound and Free Variables

An occurrence of a variable x in a wff is said to be *bound* if it lies within the scope of either $\exists x$ or $\forall x$ or if it is the quantifier variable x itself. Otherwise, an occurrence of x is said to be *free* in the wff.

For example, consider the following wff:

$$\exists x \; p(x, \, y) \rightarrow q(x).$$

The first two occurrences of x are bound because the scope of $\exists x$ is $p(x, \, y)$. The only occurrence of y is free, and the third occurrence of x is free.

So every occurrence of a variable in a wff can be classified as either bound or free, and this classification is determined by the scope of the quantifiers in the wff. Now we're in a position to discuss the meaning of wffs.

7.1.3   Interpretations and Semantics

Up to this point a wff was just a string of symbols. For a wff to have meaning, an interpretation must be given to its symbols so that the wff becomes a statement that has a truth value.

Interpretations

We'll look at some examples to introduce the properties of an interpretation for a wff. For the first example, let $p(x)$ be the statement "x is an even," where x takes values from the set of integers. Since x is a free variable in $p(x)$, the statement is neither true nor false. If we let x be the number 236, then $p(236)$ is the statement "236 is an even integer," which is true. If we let x be –5, then $p(-5)$ is the statement "–5 is an even integer," which is false.

For the second example, we'll give an interpretation for the wff

$$\forall x \; p(x).$$

Let $p(x)$ be "x has a parent," where the variable x takes values from the set of people. With this interpretation, the wff becomes the statement "Every person has a parent," which is true. If we let $p(x)$ be "x has a child," again over the set of people, then the wff becomes the statement "Every person has a child," which is false.

For the third example, we'll give an interpretation for the wff

$$\exists x \; \forall y \; p(x, y).$$

Let $p(x, \, y)$ be "$x \leq y$," where the variables x and y take values from the set of natural numbers. With this interpretation, the wff becomes the statement "Some natural number is less than or equal to every natural number," which is true because we know that zero is the smallest natural number. However, if we let x and y take values from the set of integers, then the wff becomes the statement "Some integer is less than or equal to every integer," which is false because there is no such integer.

Here is a description of the properties of an interpretation that allow a wff to have a truth value.

Properties of an Interpretation

An *interpretation* for a wff consists of a nonempty set D, called the *domain*, together with the following assignments of symbols that occur in the wff:

1. Each predicate letter is assigned a relation over D.

2. Each function letter is assigned a function over D.

3. Each free variable is assigned a value in D. All occurrences of the same variable are assigned the same value in D.

4. Each individual constant is assigned a value in D. All occurrences of the same constant are assigned the same value in D.

We will always make assignments transparent by identifying the predicate symbols and function symbols of a wff with the assigned relations and functions over the domain. We will also identify the constants and free variables in a wff with the assigned elements of the domain.

example **7.3** Interpreting a Constant and a Function

We will give two interpretations for the following wff.

$$\forall x \ (p(x, c) \ \rightarrow \ p(f(x, x), x)).$$

For the first interpretation let D be the set of positive rational numbers, let $p(x, y)$ be "$x < y$," let $c = 2$, and let $f(x, y)$ be the product $x \cdot y$. Then the interpreted wff becomes the statement "For every positive rational number x, if $x < 2$ then $x \cdot x < x$," which is false. For example, if we let $x = 3/2$, then $3/2 \cdot 3/2 = 9/4$, which is not less than $3/2$.

For a second interpretation let $c = 1$ and keep everything else the same. We know that if $0 < x < 1$, then multiplying through by x gives $x \cdot x < x$. So the interpreted wff is true.

end example

Substitutions for Free Variables

Suppose W is a wff, x is a free variable in W, and t is a term. Then the wff obtained from W by replacing all free occurrences of x by t is denoted by

$$W(x/t).$$

The expression x/t is called a *substitution* (or *binding*). For example, if $W = \forall y \ p(x, y)$, then we have $W(x/t) = \forall y \ p(t, y)$. Notice that $W(y/t) = W$ because y does not occur free in W.

Whenever we want to emphasize the fact that a wff W might contain a free variable x, we'll represent W by the expression

$$W(x).$$

When this is the case, we often write $W(t)$ to denote the wff $W(x/t)$.

example **7.4 Substituting for a Variable**

Let $W = p(x, y) \vee \exists y\ q(x, y)$. Notice that x occurs free twice in W, so for any term t we have

$$W(x/t) = p(t, y) \vee \exists y\ q(t, y).$$

For example, here are some results for four different values of t.

$$W\,(x/a) = p\,(a, y) \vee \exists y\ q\,(a, y)\,,$$
$$W\,(x/y) = p\,(y, y) \vee \exists y\ q\,(y, y)\,,$$
$$W\,(x/z) = p\,(z, y) \vee \exists y\ q\,(z, y)\,,$$
$$W\,(x/f(x, y, z)) = p\,(f\,(x, y, z)\,, y) \vee \exists y\ q\,(f(x, y, z)\,, y)\,.$$

Notice that y occurs free once in W, so for any term t we have

$$W(y/t) = p(x, t) \vee \exists y q(x, y).$$

We can also apply one substitution and then another. For example,

$$W(x/a)(y/b) = (p(a, y) \vee \exists y\ q(a, y))(y/b) = p(a, b) \vee \exists y\ q(a, y).$$

end example

Let's record some simple yet useful facts about substitutions, all of which follow directly from the definition.

Properties of Substitutions   **(7.1)**

1. x/t distributes over the connectives $\neg$, $\wedge$, $\vee$, $\rightarrow$. For example,

$$(\neg A)\,(x/t) = \neg\,A\,(x/t)\,.$$
$$(A \wedge B)\,(x/t) = A\,(x/t) \wedge B\,(x/t)\,.$$
$$(A \vee B)\,(x/t) = A\,(x/t) \vee B\,(x/t)\,.$$
$$(A \rightarrow B)\,(x/t) = A\,(x/t) \rightarrow B\,(x/t)\,.$$

2. If $x \neq y$, then x/t distributes over $\forall y$ and $\exists y$. For example,

$$(\forall y\ W)\,(x/t) = \forall y\,(W(x/t))\,.$$
$$(\exists y\ W)\,(x/t) = \exists y\,(W(x/t))\,.$$

Continued ➡

➡ ➡

---

3. If x is not free in W, then $W(x/t) = W$. For example,

$$(\forall x \ W)(x/t) = \forall x \ W.$$
$$(\exists x \ W)(x/t) = \exists x \ W.$$

Semantics

We have an intuitive understanding of truth for sentences that use the quantifiers "for every" and "there exists" when we are familiar with the domain of discourse. For an arbitrary interpretation, the following definition shows how to find the truth value of a wff.

Truth Value of a Wff

The truth value of a wff with respect to an interpretation with domain D is obtained by recursively applying the following rules:

1. A wff with no quantifiers has the truth value of the proposition obtained by applying the interpretation to it.

2. $\forall x \ W$ is true if and only if $W(x/d)$ is true for every $d \in D$.

3. $\exists x \ W$ is true if and only if $W(x/d)$ is true for some $d \in D$.

When a wff is true with respect to an interpretation I, we say that the wff *is true for I*. Otherwise the wff *is false for I*.

---

example 7.5 Finding the Truth of a Wff

We'll describe the typical process to follow when trying to find the truth value of a wff with respect to some interpretation I with domain D. Suppose the wff has the form

$$\forall x \ \exists y \ W,$$

where W does not contain any quantifiers. The wff will be true if $(\exists y \ W)(x/d)$ is true for every $d \in D$. A substitution property tells us that

$$(\exists y \ W)(x/d) = \exists y \ (W(x/d)).$$

So we must consider the truth value of $\exists y \ (W(x/d))$ for every $d \in D$. We know that $\exists y \ (W(x/d))$ will be true if there is some element $e \in D$ such that

$W(x/d)(y/e)$ is true. Since our assumption is that W does not contain any quantifiers, the truth value of $W(x/d)(y/e)$ is the truth value of the proposition obtained by applying the interpretation to this wff.

end example

Models and Countermodels

An interpretation that makes a wff true is called a *model*. An interpretation that makes a wff false is called a *countermodel*.

7.1.4 Validity

Can any wff be true for every possible interpretation? Although it may seem unlikely, this property holds for many wffs. The property is important enough to introduce some terminology. A wff is *valid* if it's true for all possible interpretations. So a wff is valid if every interpretation is a model. Otherwise, the wff is *invalid*. A wff is *unsatisfiable* if it's false for all possible interpretations. So a wff is unsatisfiable if all of its interpretations are countermodels. Otherwise, it is *satisfiable*. From these definitions we see that every wff satisfies exactly one of the following pairs of properties:

valid and thus also satisfiable,

invalid and satisfiable,

unsatisfiable and thus also invalid.

In propositional calculus the words *tautology*, *contingency*, and *contradiction* correspond, respectively, to the preceding three pairs of properties.

example **7.6 A Satisfiable and Invalid Wff**

The wff $\exists x \, \forall y \, (p(y) \to q(x, y))$ is satisfiable and invalid. To see that the wff is satisfiable, notice that the wff is true with respect to the following interpretation: The domain is the singleton $\{3\}$, and we define $p(3) = \text{True}$ and $q(3, 3) = \text{True}$. To see that the wff is invalid, notice that it is false with respect to the following interpretation: The domain is still the singleton $\{3\}$, but now we define $p(3) = \text{True}$ and $q(3, 3) = \text{False}$.

end example

Proving Validity

In propositional calculus we can use truth tables to decide whether any propositional wff is a tautology. But how can we show that a wff of predicate calculus is valid? We can't check the infinitely many interpretations of the wff to see whether each one is a model. So we are forced to use some kind of reasoning to show that a wff is valid. Here are two strategies to prove validity.

Direct approach: If the wff has the form $A \rightarrow B$, then assume that there is an arbitrary interpretation for $A \rightarrow B$ that is a model for A. Show that the interpretation is a model for B. This proves that any interpretation for $A \rightarrow B$ is a model for $A \rightarrow B$. So $A \rightarrow B$ is valid.

Indirect approach: Assume that the wff is invalid, and try to obtain a contradiction. Start by assuming the existence of a countermodel for the wff. Then try to argue toward a contradiction of some kind. For example, if the wff has the form $A \rightarrow B$, then a countermodel for $A \rightarrow B$ makes A true and B false. This information should be used to find a contradiction.

We'll demonstrate these proof strategies in the next example. But first we'll list a few valid conditionals to have something to talk about.

Some Valid Conditionals   (7.2)

   a. $\forall x \, A(x) \rightarrow \exists x \, A(x)$.

   b. $\exists x \, (A(x) \wedge B(x)) \rightarrow \exists x \, A(x) \wedge \exists x \, B(x)$.

   c. $\forall x \, A(x) \vee \forall x \, B(x) \rightarrow \forall x \, (A(x) \vee B(x))$.

   d. $\forall x \, (A(x) \rightarrow B(x)) \rightarrow (\forall x \, A(x) \rightarrow \forall x \, B(x))$.

   e. $\exists y \, \forall x \, P(x, y) \rightarrow \forall x \, \exists y \, P(x, y)$.

We should note that the converses of these wffs are invalid. We'll leave this to the exercises. In the following example we'll use the direct approach and the indirect approach to prove the validity of (7.2e). The proofs of (7.2a–7.2d) are left as exercises.

example 7.7 Proving Validity

Let W denote the following wff:

$$\exists y \, \forall x \, P(x, y) \rightarrow \forall x \, \exists y \, P(x, y).$$

We'll give two proofs to show that W is valid—one direct and one indirect. In both proofs we'll let A be the antecedent and B be the consequent of W.

Direct approach: Let M be an interpretation with domain D for W such that M is a model for A. Then there is an element $d \in D$ such that $\forall x\, P(x, d)$ is true. Therefore, $P(e, d)$ is true for all $e \in D$, which says that $\exists y\, P(e, y)$ is true for all $e \in D$. This says that M is also a model for B. Therefore, W is valid. QED.

Indirect approach: Assume that W is invalid. Then it has a countermodel with domain D that makes A true and B false. Therefore, there is an element $d \in D$ such that $\exists y\, P(d, y)$ is false. Thus $P(d, e)$ is false for all $e \in D$. Now we are assuming that A is true. Therefore, there is an element $c \in D$ such that $\forall x\, P(x, c)$ is true. In other words, $P(b, c)$ is true for all $b \in D$. In particular, this says that $P(d, c)$ is true. But this contradicts the fact that $P(d, e)$ is false for all elements $e \in D$. Therefore, W is valid. QED.

end example

Closures

There are two interesting transformations that we can apply to any wff containing free variables. One is to universally quantify each free variable, and the other is to existentially quantify each free variable. It seems reasonable to expect that these transformations will change the meaning of the original wff, as the following examples show:

$p(x) \wedge \neg\, p(y)$ is satisfiable, but $\forall x\, \forall y\, (p(x) \wedge \neg\, p(y))$ is unsatisfiable.

$p(x) \rightarrow p(y)$ is invalid, but $\exists x\, \exists y\, (p(x) \rightarrow p(y))$ is valid.

The interesting thing about the process is that validity is preserved if we universally quantify the free variables and unsatisfiability is preserved if we existentially quantify the free variables. To make this more precise, we need a little terminology.

Suppose W is a wff with free variables $x_1, \ldots, x_n$. The *universal closure* of W is the wff

$$\forall x_1 \cdots \forall x_n\ W.$$

The *existential closure* of W is the wff

$$\exists x_1 \cdots \exists x_n\ W.$$

For example, suppose $W = \forall x\, p(x, y)$. W has y as its only free variable. So the universal closure of W is

$$\forall y\, \forall x\, p(x, y),$$

and the existential closure of W is

$$\exists y \; \forall x \; p(x, \, y).$$

As we have seen, the meaning of a wff may change by taking either of the closures. But there are two properties that don't change, and we'll state them for the record as follows:

Closure Properties   (7.3)

1. A wff is valid if and only if its universal closure is valid.

2. A wff is unsatisfiable if and only if its existential closure is unsatisfiable.

Proof: We'll prove Part (1) first. To start things off we'll show that if x is a free variable in a wff W, then

$$W \text{ is valid if and only if } \forall x \; W \text{ is valid.}$$

Suppose that W is valid. Let I be an interpretation with domain D for $\forall x \; W$. If I is not a model for $\forall x \; W$, then there is some element $d \in D$ such that $W(x/d)$ is false for I. This being the case, we can define an interpretation J for W by letting J be I, with the free variable x assigned to the element $d \in D$. Since W is valid, it follows that $W(x/d)$ is true for J. But $W(x/d)$ with respect to J is the same as $W(x/d)$ with respect to I, which is false. This contradiction shows that I is a model for $\forall x \; W$. Therefore, $\forall x \; W$ is valid.

Suppose $\forall x \; W$ is valid. Let I be an interpretation with domain D for W, where x is assigned the value $d \in D$. Now define an interpretation J for $\forall x \; W$ by letting J be obtained from I by removing the assignment of x to d. Then J is an interpretation for the valid wff $\forall x \; W$. So $W(x/e)$ is true for J for all elements $e \in D$. In particular, $W(x/d)$ is true for J, and thus also for I. Therefore, I is a model for W. Therefore, W is valid.

The preceding two paragraphs tell us that if x is free in W, then W is valid if and only if $\forall x \; W$ is valid. The proof now follows by induction on the number n of free variables in a wff W. If $n = 0$, then W does not have any free variables, so W is its own universal closure. So assume that $n > 0$ and assume that Part (1) is true for any wff with k free variables, where $k < n$. If x is a free variable, then $\forall x \; W$ contains $n - 1$ free variables, and it follows by induction that $\forall x \; W$ is valid if and only if its universal closure is valid. But the universal closure of $\forall x \; W$ is the same as the universal closure of W. So it follows that W is valid if and only if the universal closure of W is valid. This proves Part (1).

The proof of Part (2) is similar to that of Part (1) and we'll leave it as an exercise. QED.

7.1.5 The Validity Problem

We'll end this section with a short discussion about deciding the validity of wffs. First we need to introduce the general notion of decidability. Any problem that can be stated as a question with a yes or no answer is called a *decision problem*. Practically every problem can be stated as a decision problem, perhaps after some work. A decision problem is called *decidable* if there is an algorithm that halts with the answer to the problem. Otherwise, the problem is called *undecidable*. A decision problem is called *partially decidable* if there is an algorithm that halts with the answer yes if the problem has a yes answer but may not halt if the problem has a no answer.

Now let's get back to logic. The *validity problem* can be stated as follows:

Given a wff, is it valid?

The validity problem for propositional calculus can be stated as follows: Given a wff, is it a tautology? This problem is decidable by Quine's method. Another algorithm would be to build a truth table for the wff and then check it.

Although the validity problem for first-order predicate calculus is undecidable, it is partially decidable. There are two partial decision procedures for first-order predicate calculus that are of interest: *natural deduction* (due to Gentzen [1935]) and *resolution* (due to Robinson [1965]). Natural deduction is a formal reasoning system that models the natural way we reason about the validity of wffs by using proof rules, as we did in Chapter 6 and as we'll discuss in Section 7.3. Resolution is a mechanical way to reason, which is not easily adaptable to people. It is, however, adaptable to machines. Resolution is an important ingredient in logic programming and automatic reasoning, which we'll discuss in Chapter 9.

 Exercises

Quantified Expressions

1. Write down the proposition denoted by each of the following expressions, where the variables take values in the domain {0, 1}.

    a. $\exists x \; \forall y \; p(x, y)$.

    b. $\forall y \; \exists x \; p(x, y)$.

2. Write down a quantified expression over some domain to denote each of the following propositions or predicates.

    a. $q(0) \wedge q(1)$.

    b. $q(0) \vee q(1)$.

    c. $p(x, 0) \wedge p(x, 1)$.

    d. $p(0, x) \vee p(1, x)$.

    e. $p(1) \vee p(3) \vee p(5) \vee \cdots$.

    f. $p(2) \wedge p(4) \wedge p(6) \wedge \cdots$.

Syntax, Scope, Bound, and Free

3. Explain why each of the following expressions is a wff.

    a. $\exists x \, p(x) \rightarrow \forall x \, p(x)$.
                    b. $\exists x \, \forall y \, (p(y) \rightarrow q(f(x), y))$.

4. Explain why the expression $\forall y \, (p(y) \rightarrow q(f(x), p(x)))$ is not a wff.

5. For each of the following wffs, label each occurrence of the variables as either bound or free.

    a. $p(x, y) \vee (\forall y \, q(y) \rightarrow \exists x \, r(x, y))$.

    b. $\forall y \, q(y) \wedge \neg \, p(x, y)$.

    c. $\neg \, q(x, y) \vee \exists x \, p(x, y)$.

6. Write down a single wff containing three variables x, y, and z, with the following properties: x occurs twice as a bound variable; y occurs once as a free variable; z occurs three times, once as a free variable and twice as a bound variable.

Interpretations

7. Let isFatherOf(x, y) be "x is the father of y," where the domain is the set of all people now living or who have lived. Find the truth value for each of the following wffs.

    a. $\forall x \, \exists y \,$ isFatherOf(x, y).

    b. $\forall y \, \exists x \,$ isFatherOf(x, y).

    c. $\exists x \, \forall y \,$ isFatherOf(x, y).

    d. $\exists y \, \forall x \,$ isFatherOf(x, y).

    e. $\exists x \, \exists y \,$ isFatherOf(x, y).

    f. $\exists y \, \exists x \,$ isFatherOf(x, y).

    g. $\forall x \, \forall y \,$ isFatherOf(x, y).

    h. $\forall y \, \forall x \,$ isFatherOf(x, y).

8. Let $W = \exists x\ \forall y\ (p(y) \rightarrow q(x,\ y))$. Find the truth value of W with respect to each of the following interpretations, where $q(x,\ y)$ is "$x = y$."

   a. The domain is $\{a\}$ and $p(a) = $ True.
   b. The domain is $\{a\}$ and $p(a) = $ False.
   c. The domain is $\{a,\ b\}$ and $p(a) = p(b) = $ True.
   d. Any domain D for which $p(d) = $ True for at most one element $d \in D$.

9. Let $W = \forall x\ (p(f(x,\ x),\ x) \rightarrow p(x,\ y))$. Find the truth value of W with respect to each of the following interpretations.

   a. The domain is the set $\{a, b\}$, p is equality, $y = a$, and f is defined by $f(a,\ a) = a$ and $f(b,\ b) = b$.
   b. The domain is the set of natural numbers, p is equality, $y = 0$, and f is the function defined by $f(a,b) = (a + b) \bmod 3$.

10. Let $B(x)$ mean x is a bird, let $W(x)$ mean x is a worm, and let $E(x,\ y)$ mean x eats y. Find an English sentence to describe each of the following statements.

   a. $\forall x\ \forall y\ (B(x) \wedge W(y) \rightarrow E(x,\ y))$.
   b. $\forall x\ \forall y\ (E(x,\ y) \rightarrow B(x) \wedge W(y))$.

11. Let $p(x)$ mean that x is a person, let $c(x)$ mean that x is a chocolate bar, and let $e(x,\ y)$ mean that x eats y. For each of the following wffs, write down an English sentence that reflects the interpretation of the wff.

   a. $\exists x\ (p(\mathrm{x}) \wedge \forall y\ (c(y) \rightarrow e(x,\ y)))$.
   b. $\forall y\ (c(y) \wedge \exists x\ (p(\mathrm{x}) \wedge e(x,\ y)))$.

12. Let $e(x,\ y)$ mean that $x = y$, let $p(x,\ y)$ mean that $x < y$, and let $d(x,\ y)$ mean that x divides y. For each of the following statements about the natural numbers, find a formal quantified expression.

   a. Every natural number other than 0 has a predecessor.
   b. Any two nonzero natural numbers have a common divisor.

13. Given the wff $W = \exists x\ p(x) \rightarrow \forall x\ p(x)$.

   a. Find all possible interpretations of W over the domain $D = \{a\}$. Also give the truth value of W over each of the interpretations.
   b. Find all possible interpretations of W over the domain $D = \{a,\ b\}$. Also give the truth value of W over each of the interpretations.

14. Find a model for each of the following wffs.

    a. $p(c) \land \exists x \neg p(x)$.

    b. $\exists x \, p(x) \rightarrow \forall x \, p(x)$.

    c. $\exists y \, \forall x \, p(x, y) \rightarrow \forall x \, \exists y \, p(x, y)$.

    d. $\forall x \, \exists y \, p(x, y) \rightarrow \exists y \, \forall x \, p(x, y)$.

    e. $\forall x \, (p(x, f(x)) \rightarrow p(x, y))$.

15. Find a countermodel for each of the following wffs.

    a. $p(c) \land \exists x \neg p(x)$.

    b. $\exists x \, p(x) \rightarrow \forall x \, p(x)$.

    c. $\forall x \, (p(x) \lor q(x)) \rightarrow \forall x \, p(x) \lor \forall x \, q(x)$.

    d. $\exists x \, p(x) \land \exists x \, q(x) \rightarrow \exists x \, (p(x) \land q(x))$.

    e. $\forall x \, \exists y \, p(x, y) \rightarrow \exists y \, \forall x \, p(x, y)$.

    f. $\forall x \, (p(x, f(x)) \rightarrow p(x, y))$.

    g. $(\forall x \, p(x) \rightarrow \forall x \, q(x)) \rightarrow \forall x \, (p(x) \rightarrow q(x))$.

Validity

16. Given the wff $W = \forall x \, \forall y \, (p(x) \rightarrow p(y))$.

    a. Show that W is true for any interpretation whose domain is a singleton.

    b. Show that W is not valid.

17. Given the wff $W = \forall x \, p(x, x) \rightarrow \forall x \, \forall y \, \forall z \, (p(x, y) \lor p(x, z) \lor p(y, z))$.

    a. Show that W is true for any interpretation whose domain is a singleton.

    b. Show that W is true for any interpretation whose domain has two elements.

    c. Show that W is not valid.

18. Find an example of a wff that is true for any interpretation having a domain with three or fewer elements but is not valid. *Hint:* Look at the structure of the wff in Exercise 17.

19. Prove that each of the following wffs is valid. *Hint:* Either show that every interpretation is a model or assume that the wff is invalid and find a contradiction.

    a. $\forall x \, (p(x) \rightarrow p(x))$.

    b. $p(c) \rightarrow \exists x \, p(x)$.

    c. $\forall x \, p(x) \rightarrow \exists x \, p(x)$.

    d. $\exists x \, (A(x) \land B(x)) \rightarrow \exists x \, A(x) \land \exists x \, B(x)$.

e. $\forall x \ A(x) \lor \forall x \ B(x) \rightarrow \forall x \ (A(x) \lor B(x))$.

f. $\forall x \ (A(x) \rightarrow B(x)) \rightarrow (\exists x \ A(x) \rightarrow \exists x \ B(x))$.

g. $\forall x \ (A(x) \rightarrow B(x)) \rightarrow (\forall x \ A(x) \rightarrow \exists x \ B(x))$.

h. $\forall x \ (A(x) \rightarrow B(x)) \rightarrow (\forall x \ A(x) \rightarrow \forall x \ B(x))$.

20. Prove that each of the following wffs is unsatisfiable. *Hint:* Either show that every interpretation is a countermodel or assume that the wff is satisfiable and find a contradiction.

   a. $p(c) \land \neg \ p(c)$.   b. $\exists x \ (p(x) \land \neg \ p(x))$.

   c. $\exists x \ \forall y \ (p(x, \ y) \land \neg \ p(x, \ y))$.

Further Thoughts

21. For a wff W, let $c(W)$ denote the wff obtained from W by replacing the free variables of W by distinct constants. Prove that W has a model if and only if $c(W)$ has a model.

22. Prove that any wff of the form $A \rightarrow B$ is valid if and only if whenever A is valid, then B is valid.

23. Prove Part (2) of (7.3) by using a proof similar to that of Part (1). A wff is unsatisfiable if and only if its existential closure is unsatisfiable.

7.2   Equivalent Formulas

In our normal discourse we often try to understand a sentence by rephrasing it in some way such that the meaning remains the same. We've seen how this idea of equivalence carries over to propositional calculus. Now we'll extend the idea to predicate calculus.

7.2.1   Logical Equivalence

Two wffs A and B are said to be *logically equivalent* (or *equivalent*) if they both have the same truth value with respect to every interpretation of both A and B. By an interpretation of both A and B, we mean that all free variables, constants, functions, and predicates that occur in either A or B are interpreted with respect to a single domain. If A and B are equivalent, then we write

$$A \equiv B.$$

We should note that any two valid wffs are equivalent because they are both true for any interpretation. Similarly, any two unsatisfiable wffs are equivalent because they are both false for any interpretation. The definition of equivalence also allows us to make the following useful formulation in terms of conditionals and validity.

Equivalence

$A \equiv B$ if and only if $(A \to B) \land (B \to A)$ is valid

if and only if $A \to B$ and $B \to A$ are both valid.

Instances of Propositional Wffs

To start things off, let's see how propositional equivalences give rise to predicate calculus equivalences. A wff W is an *instance* of a propositional wff V if W is obtained from V by replacing each propositional variable of V by a wff, where all occurrences of each propositional variable in V are replaced by the same wff. For example, the wff

$$\forall x \, p(x) \to \forall x \, p(x) \lor q(x)$$

is an instance of $P \to P \lor Q$ because Q is replaced by $q(x)$ and both occurrences of P are replaced by $\forall x \, p(x)$.

If W is an instance of a propositional wff V, then the truth value of W for any interpretation can be obtained by assigning truth values to the propositional variables of V. For example, suppose we define an interpretation with domain $D = \{a, b\}$ and we set $p(a) = p(b) = $ True and $q(a) = q(b) = $ False. For this interpretation, the truth value of the wff $\forall x \, p(x) \to \forall x \, p(x) \lor q(x)$ is the same as the truth value of the propositional wff $P \to P \lor Q$, where $P = $ True and $Q = $ False.

So we can say that two wffs are equivalent if they are instances of two equivalent propositional wffs, where both instances are obtained by using the same replacement of propositional variables. For example, we have

$$\forall x \, p(x) \to q(x) \equiv \neg \, \forall x \, p(x) \lor q(x)$$

because the left and right sides are instances of the left and right sides of the propositional equivalence $P \to Q \equiv \neg \, P \lor Q$, where both occurrences of P are replaced by $\forall x \, p(x)$ and both occurrences of Q are replaced by $q(x)$. We'll state the result again for emphasis:

Equivalent Instances

Two wffs are equivalent whenever they are instances of two equivalent propositional wffs, where both instances are obtained by using the same replacement of propositional variables.

Equivalences involving Quantifiers

Let's see whether we can find some more equivalences to make our logical life easier. We'll start by listing equivalences that relate the two quantifiers by negation. For any wff W we have the following two equivalences.

Quantifiers and Negation   (7.4)

$$\neg\,(\forall x\ W) \equiv \exists x\ \neg\ W \quad \text{and} \quad \neg\,(\exists x\ W) \equiv \forall x\ \neg\ W.$$

It's easy to believe these two equivalences. For example, we can illustrate the equivalence $\neg\,(\forall x\ W) \equiv \exists x\ \neg\ W$ by observing that the negation of the statement "Something is true for all possible cases" has the same meaning as the statement "There is some case for which the something is false." Similarly, we can illustrate the equivalence $\neg\,(\exists x\ W) \equiv \forall x\ \neg\ W$ by observing that the negation of the statement "There is some case for which something is true" has the same meaning as the statement "Every case of the something is false."

Another way to demonstrate these equivalences is to use De Morgan's laws. For example, let $W = p(x)$ and suppose that we have an interpretation with domain $D = \{0, 1, 2, 3\}$. Then no matter what values we assign to p, we can apply De Morgan's laws to obtain the following propositional equivalence:

$$\begin{aligned}
\neg\,(\forall x\ p\,(x)) &\equiv \neg\,(p\,(0) \wedge p\,(1) \wedge p\,(2) \wedge p\,(3)) \\
&\equiv \neg\,p\,(0) \vee \neg\,p\,(1) \vee \neg\,p\,(2) \vee \neg\,p\,(3) \\
&\equiv \exists x\ \neg\,p\,(x).
\end{aligned}$$

We also get the following equivalence:

$$\begin{aligned}
\neg\,(\exists x\ p\,(x)) &\equiv \neg\,(p\,(0) \vee p\,(1) \vee p\,(2) \vee p\,(3)) \\
&\equiv \neg\,p\,(0) \wedge \neg\,p\,(1) \wedge \neg\,p\,(2) \wedge \neg\,p\,(3) \\
&\equiv \forall x\ \neg\,p\,(x).
\end{aligned}$$

These examples are nice, but they don't prove (7.4). Let's give an actual proof, using validity, of the equivalences (7.4). We'll prove the first equivalence, $\neg\,(\forall x\ W) \equiv \exists x\ \neg\ W$, and then use it to prove the second equivalence.

Proof: Let I be an interpretation with domain D for the wffs $\neg\,(\forall x\ W)$ and $\exists x\ \neg\ W$. We want to show that I is a model for one of the wffs if and only if I is a model for the other wff. The following equivalent statements do the job:

$$I \text{ is a model for } \neg\,(\forall x W) \quad \text{iff} \quad \neg\,(\forall x W) \text{ is true for } I$$
$$\text{iff} \quad \forall x W \text{ is false for } I$$
$$\text{iff} \quad W(x/d) \text{ is false for some } d \in D$$
$$\text{iff} \quad \neg\, W(x/d) \text{ is true for some } d \in D$$
$$\text{iff} \quad \exists x \,\neg\, W \text{ is true for } I$$
$$\text{iff} \quad I \text{ is a model for } \exists x \,\neg\, W.$$

This proves the equivalence $\neg\,(\forall\,x\,W) \equiv \exists x \,\neg\, W$. Now, since W is arbitrary, we can replace W by the wff $\neg\, W$ to obtain the following equivalence:

$$\neg\,(\forall x \,\neg\, W) \equiv \exists x \,\neg\,\neg\, W.$$

Now take the negation of both sides of this equivalence, and simplify the double negations to obtain the second equivalence of (7.4):

$$\forall x \,\neg\, W \equiv \neg\,(\,\exists x\; W).\; \text{QED.}$$

Now let's look at two equivalences that allow us to interchange universal quantifiers if they are next to each other; the same holds for existential quantifiers.

Interchanging Quantifers of the Same Type $\hspace{3em}$ (7.5)

$$\forall x\; \forall y\; W \equiv \forall y\; \forall x\; W \quad \text{and} \quad \exists x\; \exists y\; W \equiv \exists y\; \exists x\; W.$$

Again, this is easy to believe. For example, suppose that $W = p(x,\,y)$ and we have an interpretation with domain $D = \{0,\,1\}$. Then we have the following equivalences.

$$\forall x\; \forall y\; p\,(x,y) \equiv \forall y\; p\,(0,y) \wedge \forall y\; p\,(1,y)$$
$$\equiv (p\,(0,0) \wedge p\,(0,1)) \wedge (p\,(1,0) \wedge p\,(1,1))$$
$$\equiv (p\,(0,0) \wedge p\,(1,0)) \wedge (p\,(0,1) \wedge p\,(1,1))$$
$$\equiv \forall x\; p\,(x,0) \wedge \forall x\; p\,(x,1)$$
$$\equiv \forall y\; \forall x\; p\,(x,y).$$

We also have the following equivalences.

$$\exists x\; \exists y\; p\,(x,y) \equiv \exists y\; p\,(0,y) \vee \exists y\; p\,(1,y)$$
$$\equiv (p\,(0,0) \vee p\,(0,1)) \vee (p\,(1,0) \vee p\,(1,1))$$
$$\equiv (p\,(0,0) \vee p\,(1,0)) \vee (p\,(0,1) \vee p\,(1,1))$$
$$\equiv \exists x\; p\,(x,0) \vee \exists x\; p\,(x,1)$$
$$\equiv \exists y\; \exists x\; p\,(x,y).$$

We'll leave the proofs of equivalences (7.5) as exercises.

Equivalences Containing Quantifiers and Connectives

It's time to start looking at some equivalences that involve quantifiers and connectives. For example, the following equivalence involves both quantifiers and the conditional connective. It shows that we can't always distribute a quantifier over a conditional.

An Equivalence to be Careful With (7.6)

$$\exists x \ (p(x) \rightarrow q(x)) \equiv \forall x \ p(x) \rightarrow \exists x \ q(x).$$

---

example **7.8 Proof of an Equivalence**

We'll give a proof of (7.6) consisting of two subproofs showing that each side implies the other. First we'll prove the validity of

$$\exists x \ (p(x) \rightarrow q(x)) \rightarrow (\forall x \ p(x) \rightarrow \exists x \ q(x)).$$

Proof: Let I be a model for $\exists x \ (p(x) \rightarrow q(x))$ with domain D. Then $\exists x \ (p(x) \rightarrow q(x))$ is true for I, which means that $p(d) \rightarrow q(d)$ is true for some $d \in D$. Therefore, either $p(d)$ is false or both $p(d)$ and $q(d)$ are true for some $d \in D$. If $p(d)$ is false, then $\forall x \ p(x)$ is false for I; if both $p(d)$ and $q(d)$ are true, then $\exists x \ q(x)$ is true for I. In either case $\forall x \ p(x) \rightarrow \exists x \ q(x)$ is true for I. Therefore, I is a model for $\forall x \ p(x) \rightarrow \exists x \ q(x)$. QED.

Now we'll prove the validity of

$$(\forall x \ p(x) \rightarrow \exists x \ q(x)) \rightarrow \exists x \ (p(x) \rightarrow q(x)).$$

Proof: Let I be a model for $\forall x \ p(x) \rightarrow \exists x \ q(x)$ with domain D. Then $\forall x \ p(x) \rightarrow \exists x \ q(x)$ is true for I. Therefore, either $\forall x \ p(x)$ is false for I or both $\forall x \ p(x)$ and $\exists x \ q(x)$ are true for I. If $\forall x \ p(x)$ is false for I, then $p(d)$ is false for some $d \in D$. Therefore, $p(d) \rightarrow q(d)$ is true. If both $\forall x \ p(x)$ and $\exists x \ q(x)$ are true for I, then there is some $c \in D$ such that both $p(c)$ and $q(c)$ are true. Thus $p(c) \rightarrow q(c)$ is true. So in either case, $\exists x \ (p(x) \rightarrow q(x))$ is true for I. Thus I is a model for $\exists x \ (p(x) \rightarrow q(x))$. QED.

end example

Of course, once we know some equivalences, we can use them to prove other equivalences. For example, let's see how previous results can be used to prove the following equivalences.

Distributing the Quantifiers (7.7)

a. $\exists x \ (p(x) \lor q(x)) \equiv \exists x \ p(x) \lor \exists x \ q(x).$

b. $\forall x \ (p(x) \land q(x)) \equiv \forall x \ p(x) \land \forall x \ q(x).$

Proof of (7.7a): $\exists x\,(p\,(x) \vee q\,(x)) \equiv \exists x\,(\neg\,p\,(x) \rightarrow q\,(x))$

$$\equiv \forall x\,\neg\,p\,(x) \rightarrow \exists x\,q\,(x) \qquad \text{(by 7.6)}$$

$$\equiv \neg\,\exists x\,p\,(x) \rightarrow \exists x\,q\,(x) \qquad \text{(by 7.4)}$$

$$\equiv \exists x\,p\,(x) \vee \exists x\,q\,(x) \qquad \text{QED.}$$

Proof of (7.7b): Use the fact that $\forall x\,(p(x) \wedge q(x)) \equiv \neg\,\exists x\,(\neg\,p(x) \vee \neg\,q(x))$ and then apply (7.7a). QED.

Restricted Equivalences

Some interesting and useful equivalences can occur when certain restrictions are placed on the variables. To start things off, we'll see how to change the name of a quantified variable in a wff without changing the meaning of the wff.

We'll illustrate the renaming problem with the following interpreted wff to represent the fact over the integers that for every integer x there is an integer y greater than x:

$$\forall x\,\exists y\,x < y.$$

Can we replace all occurrences of the quantifier variable x with some other variable? If we choose a variable different from x and y, say z, we obtain

$$\forall z\,\exists y\,z < y.$$

This looks perfectly fine. But if we choose y to replace x, then we obtain

$$\forall y\,\exists y\,y < y.$$

This looks bad. Not only has $\forall y$ lost its influence, but the statement says there is an integer y such that $y < y$, which is false. So we have to be careful when renaming quantified variables. We'll always be on solid ground if we pick a new variable that does not occur anywhere in the wff. Here's the rule.

Renaming Rule (7.8)

If y is a new variable that does not occur in $W(x)$, then the following equivalences hold:

a. $\exists x\,W(x) \equiv \exists y\,W(x/y)$.

b. $\forall x\,W(x) \equiv \forall y\,W(x/y)$.

---

example **7.9 Renaming Variables**

We'll rename the quantified variables in the following wff so that they are all distinct:

$$\forall x \; \exists y \; (p(x, y) \rightarrow \exists x \; q(x, y) \vee \forall y \; r(x, y)).$$

Since there are four quantifiers using just the two variables x and y, we need two new variables, say z and w, which don't occur in the wff. We can replace any of the quantified wffs. So we'll start by replacing $\forall x$ by $\forall z$ and each x bound to $\forall x$ by z to obtain the following equivalent wff:

$$\forall z \; \exists y \; (p(z, y) \rightarrow \exists x \; q(x, y) \vee \forall y \; r(z, y)).$$

Notice that $p(x, y)$ and $r(x, y)$ have changed to $p(z, y)$ and $r(z, y)$ because the scope of $\forall x$ is the entire wff while the scope of $\exists x$ is just $q(x, y)$. Now let's replace $\forall y \; r(z, y)$ by $\forall w \; r(z, w)$ to obtain the following equivalent wff:

$$\forall z \; \exists y \; (p(z, y) \rightarrow \exists x \; q(x, y) \vee \forall w \; r(z, w)).$$

We end up with an equivalent wff with distinct quantified variables.

end example

Now we'll look at some restricted equivalences that allow us to move a quantifier past a wff that doesn't contain a free occurrence of the quantified variable.

Equivalences with Restrictions

If x does not occur free in C, *then* the following equivalences hold.

Simplification (7.9)

$\forall x \; C \equiv C$ and $\exists x \; C \equiv C$.

Disjunction (7.10)

a. $\forall x \; (C \vee A(x)) \equiv C \vee \forall x \; A(x)$.

b. $\exists x \; (C \vee A(x)) \equiv C \vee \exists x \; A(x)$.

Conjunction (7.11)

a. $\forall x \; (C \wedge A(x)) \equiv C \wedge \forall x \; A(x)$.

b. $\exists x \; (C \wedge A(x)) \equiv C \wedge \exists x \; A(x)$.

Continued ➡

➡ ➡

Implication   (7.12)

    a. $\forall x\ (C \rightarrow A(x)) \equiv C \rightarrow \forall x\ A(x).$

    b. $\exists x\ (C \rightarrow A(x)) \equiv C \rightarrow \exists x\ A(x).$

    c. $\forall x\ (A(x) \rightarrow C) \equiv \exists x\ A(x) \rightarrow C.$

    d. $\exists x\ (A(x) \rightarrow C) \equiv \forall x\ A(x) \rightarrow C.$

Proof: We'll prove (7.10a). The important point in this proof is the assumption that x is not free in C. This means that any substitution x/t does not change C. In other words, $C(x/t) = C$ for all possible terms t. We'll assume that I is an interpretation with domain D. With these assumptions we can start.

    If I is a model for $\forall x\ (C \vee A(x))$, then $(C \vee A(x))(x/d)$ is true with respect to I for all d in D. Now write $(C \vee A(x))(x/d)$ as

$$(C \vee A(x))(x/d) = C(x/d) \vee A(x)(x/d) \quad \text{(substitution property)}$$
$$= C \vee A(x)(x/d) \quad \text{(because } x \text{ is not free in } C).$$

So $C \vee A(x)(x/d)$ is true for I for all d in D. Since the truth of C is not affected by any substitution for x, it follows that either C is true for I or $A(x)(x/d)$ is true for I for all d in D. So either I is a model for C or I is a model for $\forall x\ A(x)$. Therefore, I is a model for $C \vee \forall x\ A(x)$.

    Conversely, if I is a model for $C \vee \forall x\ A(x)$, then $C \vee \forall x\ A(x)$ is true for I. Therefore, either C is true for I or $\forall x\ A(x)$ is true for I. Suppose that C is true for I. Then, since x is not free in C, we have $C = C(x/d)$ for any d in D. So $C(x/d)$ is true for I for all d in D. Therefore, $C(x/d) \vee A(x)(x/d)$ is also true for I for all d in D. Substitution gives $C(x/d) \vee A(x)(x/d) = (C \vee A(x))(x/d)$. So $(C \vee A(x))(x/d)$ is true for I for all d in D. This means I is a model for $\forall x\ (C \vee A(x))$. Suppose $\forall x A(x)$ is true for I. Then $A(x)(x/d)$ is true for I for all d in D. So $C(x/d) \vee A(x)(x/d)$ is true for I for all d in D, and thus $(C \vee A(x))(x/d)$ is true for I for all d in D. So I is a model for $C \vee \forall x\ A(x)$. QED.

    The proof of (7.10b) is similar and we'll leave it as an exercise. Once we have the equivalences (7.10), the other equivalences are simple consequences. For example, we'll prove (7.11b):

$$\exists x\ (C \wedge A(x)) \equiv \neg\ \forall x\ \neg\ (C \wedge A(x)) \qquad \text{(by 7.4)}$$
$$\equiv \neg\ \forall x\ (\neg\ C \vee \neg\ A(x))$$
$$\equiv \neg\ (\neg\ C \vee \forall x\ \neg\ A(x)) \qquad \text{(by 7.10a)}$$
$$\equiv \neg\ (\neg\ C \vee \neg\ \exists x\ A(x)) \qquad \text{(by 7.4)}$$
$$\equiv C \wedge \exists x\ A(x).$$

The implication equivalences (7.12) are also easily derived from the other equivalences. For example, we'll prove (7.12c):

$$\forall x \, (A\,(x) \rightarrow C) \equiv \forall x \, (\neg \, A\,(x) \vee C)$$
$$\equiv \forall x \, \neg \, A\,(x) \vee C \qquad \text{(by 7.10a)}$$
$$\equiv \neg \, \exists x \, A\,(x) \vee C \qquad \text{(by 7.4)}$$
$$\equiv \exists x \, A\,(x) \rightarrow C.$$

Now that we have some equivalences on hand, we can use them to prove other equivalences. In other words, we have a set of rules to transform wffs into other wffs having the same meaning. This justifies the word "calculus" in the name "predicate calculus."

7.2.2 Normal Forms

In propositional calculus we know that any wff is equivalent to a wff in conjunctive normal form and to a wff in disjunctive normal form. Let's see whether we can do something similar with the wffs of predicate calculus. We'll start with a definition. A wff W is in *prenex normal form* if it has the form

$$W = Q_1 x_1 \ldots Q_n x_n \, M,$$

where each $Q_i x_i$ is a quantifier, each x_i is distinct, and M is a wff with no quantifiers. For example, the following wffs are in prenex normal form:

$$p\,(x)\,,$$
$$\exists x \, p\,(x)\,,$$
$$\forall x \, p\,(x, y)\,,$$
$$\forall x \, \exists y \, (p\,(x, y) \rightarrow q\,(x))\,,$$
$$\forall x \, \exists y \, \forall z \, (p\,(x) \vee q\,(y) \wedge r\,(x, z))\,.$$

Is any wff equivalent to some wff in prenex normal form? Yes. In fact there's an easy algorithm to obtain the desired form. The idea is to make sure that variables have distinct names and then apply equivalences that send all quantifiers to the left end of the wff. Here's the algorithm:

Prenex Normal Form Algorithm   **(7.13)**

Any wff W has an equivalent prenex normal form, which can be constructed as follows:

1. Rename the variables of W so that no quantifiers use the same variable name and such that the quantified variable names are distinct from the free variable names.

2. Move quantifiers to the left by using equivalences (7.4), (7.10), (7.11), and (7.12).

The renaming of variables is important to the success of the algorithm. For example, we can't replace $p(x) \vee \forall x\, q(x)$ by $\forall x\, (p(x) \vee q(x))$ because they aren't equivalent. But we can rename variables to obtain the following equivalence:

$$p(x) \vee \forall x\, q(x) \equiv p(x) \vee \forall y\, q(y) \equiv \forall y\, (p(x) \vee q(y)).$$

example 7.10 **Prenex Normal Form**

We'll put the following wff W into prenex normal form:

$$A(x) \wedge \forall x\, (B(x) \rightarrow \exists y\, C(x,\, y) \vee \neg\, \exists y\, A(y)).$$

First notice that y is used in two quantifiers and x occurs both free and in a quantifier. After changing names, we obtain the following equivalent wff:

$$A(x) \wedge \forall z\, (B(z) \rightarrow \exists y\, C(z,\, y) \vee \neg\, \exists w\, A(w)).$$

Now each quantified variable is distinct, and the quantified variables are distinct from the free variable x. We'll apply equivalences to move all the quantifiers to the left:

$$
\begin{aligned}
W &\equiv A\,(x) \wedge \forall z\, (B\,(z) \rightarrow \exists y\, C\,(z,y) \vee \neg\, \exists w\, A\,(w)) \\
&\equiv \forall z\, (A\,(x) \wedge (B\,(z) \rightarrow \exists y\, C\,(z,y) \vee \neg\, \exists w\, A\,(w))) && \text{(by 7.11)} \\
&\equiv \forall z\, (A\,(x) \wedge (B\,(z) \rightarrow \exists y\, (C\,(z,y) \vee \neg\, \exists w\, A\,(w)))) && \text{(by 7.10)} \\
&\equiv \forall z\, (A\,(x) \wedge \exists y\, (B\,(z) \rightarrow C\,(z,y) \vee \neg\, \exists w\, A\,(w))) && \text{(by 7.12)} \\
&\equiv \forall z\, \exists y\, (A\,(x) \wedge (B\,(z) \rightarrow C\,(z,y) \vee \neg\, \exists w\, A\,(w))) && \text{(by 7.11)} \\
&\equiv \forall z\, \exists y\, (A\,(x) \wedge (B\,(z) \rightarrow C\,(z,y) \vee \forall w\, \neg\, A\,(w))) && \text{(by 7.4)} \\
&\equiv \forall z\, \exists y\, (A\,(x) \wedge (B\,(z) \rightarrow \forall w\, (C\,(z,y) \vee \neg\, A\,(w)))) && \text{(by 7.10)} \\
&\equiv \forall z\, \exists y\, (A\,(x) \wedge \forall w\, (B\,(z) \rightarrow C\,(z,y) \vee \neg\, A\,(w))) && \text{(by 7.12)} \\
&\equiv \forall z\, \exists y\, \forall w\, (A\,(x) \wedge (B\,(z) \rightarrow C\,(z,y) \vee \neg\, A\,(w))) && \text{(by 7.11)}
\end{aligned}
$$

This wff is in the desired prenex normal form.

end example

There are two special prenex normal forms that correspond to the disjunctive normal form and the conjunctive normal form for propositional calculus. We define a *literal* in predicate calculus to be an atom or the negation of an atom. For example, $p(x)$ and $\neg\, q(x,\, y)$ are literals. A prenex normal form is called a *prenex disjunctive normal form* if it has the form

$$Q_1 x_1\, \ldots\, Q_n x_n\, (C_1 \vee \cdots \vee C_k),$$

where each C_i is a conjunction of one or more literals. Similarly, a prenex normal form is called a *prenex conjunctive normal form* if it has the form

$$Q_1 x_1 \ \ldots \ Q_n x_n \ (D_1 \wedge \cdots \wedge D_k),$$

where each D_i is a disjunction of one or more literals.

It's easy to construct either of these normal forms from a prenex normal form. Just eliminate conditionals, move $\neg$ inward, and distribute either $\wedge$ over $\vee$ or $\vee$ over $\wedge$. If we want to start with an arbitrary wff, then we can put everything together in a nice little algorithm. We can save some thinking by removing all conditionals at an early stage of the process. Then we won't have to remember the formulas (7.12). The algorithm can be stated as follows:

Prenex Disjunctive/Conjunctive Normal Form Algorithm   **(7.14)**

Any wff W has an equivalent prenex disjunctive/conjunctive normal form, which can be constructed as follows:

1. Rename the variables of W so that no quantifiers use the same variable name and such that the quantified variable names are distinct from the free variable names.

2. Remove implications by using the equivalence $A \rightarrow B \equiv \neg A \vee B$.

3. Move negations to the right to form literals by using the equivalences (7.4) and the equivalences $\neg (A \wedge B) \equiv \neg A \vee \neg B$, $\neg (A \vee B) \equiv \neg A \wedge \neg B$, and $\neg \neg A \equiv A$.

4. Move quantifiers to the left by using equivalences (7.10) and (7.11).

5. To obtain the disjunctive normal form, distribute $\wedge$ over $\vee$. To obtain the conjunctive normal form, distribute $\vee$ over $\wedge$.

Now let's do an example that uses (7.14) to transform a wff into prenex normal form.

> example   **7.11   Prenex CNF and DNF**

Let W be the following wff, which is the same wff from Example 7.10:

$$A(x) \wedge \forall x \ (B(x) \rightarrow \exists y \ C(x, y) \vee \neg \exists y \ A(y)).$$

We'll use algorithm (7.14) to construct a prenex conjunctive normal form and a prenex disjunctive normal form of W.

$$W = A(x) \wedge \forall x \, (B(x) \to \exists y \, C(x,y) \vee \neg \, \exists y \, A(y))$$
$$\equiv A(x) \wedge \forall z \, (B(z) \to \exists y \, C(z,y) \vee \neg \, \exists w \, A(w)) \qquad \text{(rename variables)}$$
$$\equiv A(x) \wedge \forall z \, (\neg \, B(z) \vee \exists y \, C(z,y) \vee \neg \, \exists w \, A(w)) \qquad \text{(remove } \to \text{)}$$
$$\equiv A(x) \wedge \forall z \, (\neg \, B(z) \vee \exists y \, C(z,y) \vee \forall w \, \neg \, A(w)) \qquad \text{(by 7.4)}$$
$$\equiv \forall z \, (A(x) \wedge (\neg \, B(z) \vee \exists y \, C(z,y) \vee \forall w \, \neg \, A(w))) \qquad \text{(by 7.11)}$$
$$\equiv \forall z \, (A(x) \wedge \exists y \, (\neg \, B(z) \vee \, C(z,y) \vee \forall w \, \neg \, A(w))) \qquad \text{(by 7.10)}$$
$$\equiv \forall z \, \exists y \, (A(x) \wedge (\neg \, B(z) \vee C(z,y) \vee \forall w \, \neg \, A(w))) \qquad \text{(by 7.11)}$$
$$\equiv \forall z \, \exists y \, (A(x) \wedge \forall w \, (\neg \, B(z) \vee C(z,y) \vee \neg \, A(w))) \qquad \text{(by 7.10)}$$
$$\equiv \forall z \, \exists y \, \forall w \, (A(x) \wedge (\neg \, B(z) \vee C(z,y) \vee \neg \, A(w))) \qquad \text{(by 7.11)}$$

This wff is in prenex conjunctive normal form. We'll distribute $\wedge$ over $\vee$ to obtain the following prenex disjunctive normal form:

$$\equiv \forall z \, \exists y \, \forall w \, ((A(x) \wedge \neg \, B(z)) \vee (A(x) \wedge C(z,\,y)) \vee (A(x) \wedge \neg \, A(w))).$$

end example

7.2.3 Formalizing English Sentences

Now that we have a few tools at hand, let's see whether we can find some heuristics for formalizing English sentences. We'll look at several sentences dealing with people and the characteristics of being a politician and being crooked. Let $p(x)$ denote the statement "x is a politician," and let $q(x)$ denote the statement "x is crooked." For each of the following sentences we've listed a formalization with quantifiers. Before you look at each formalization, try to find one of your own. It may be correct, even though it doesn't look like the listed answer.

"Some politician is crooked."	$\exists x \, (p(x) \wedge q(x))$.
"No politician is crooked."	$\forall x \, (p(x) \to \neg \, q(x))$.
"All politicians are crooked."	$\forall x \, (p(x) \to q(x))$.
"Not all politicians are crooked."	$\exists x \, (p(x) \wedge \neg \, q(x))$.
"Every politician is crooked."	$\forall x \, (p(x) \to q(x))$.
"There is an honest politician."	$\exists x \, (p(x) \wedge \neg \, q(x))$.
"No politician is honest."	$\forall x \, (p(x) \to q(x))$.
"All politicians are honest."	$\forall x \, (p(x) \to \neg \, q(x))$.

Can we notice anything interesting about the formalizations of these sentences? Yes, we can. Notice that each formalization satisfies one of the following two properties:

The universal quantifier $\forall x$ quantifies a conditional.

The existential quantifier $\exists x$ quantifies a conjunction.

To see why this happens, let's look at the statement "Some politician is crooked." We came up with the wff $\exists x \ (p(x) \wedge q(x))$. Someone might argue that the answer could also be the wff $\exists x \ (p(x) \rightarrow q(x))$. Notice that the second wff is true even if there are no politicians, while the first wff is false in this case, as it should be. For another example, notice that the second wff is true in the case that some computer scientist is crooked while all politicians are honest, but the first wff is false in this case, as it should be. Another way to see the difference is to look at equivalent wffs. From (7.6) we have the equivalence $\exists x \ (p(x) \rightarrow q(x)) \equiv \forall x \ p(x) \rightarrow \exists x \ q(x)$. Let's see how the wff $\forall x \ p(x) \rightarrow \exists x \ q(x)$ reads when applied to our example. It says, "If everyone is a politician, then someone is crooked." This doesn't seem to convey the same thing as our original sentence.

Another thing to notice is that people come up with different answers. For example, the second sentence, "No politician is crooked," might also be written as follows:

$$\neg \ \exists x \ (p(x) \wedge q(x)).$$

It's nice to know that this answer is OK too because it's equivalent to the listed answer, $\forall x \ (p(x) \rightarrow \neg \ q(x))$. We'll prove the equivalence of the two wffs by applying (7.4) as follows:

$$\neg \ \exists x \ (p\left(x\right) \wedge q\left(x\right)) \equiv \forall x \ \neg \ (p\left(x\right) \wedge q\left(x\right))$$
$$\equiv \forall x \ (\neg \ p\left(x\right) \vee \neg \ q\left(x\right))$$
$$\equiv \forall x \ (p\left(x\right) \rightarrow \neg \ q\left(x\right)).$$

Of course, not all sentences are easy to formalize. For example, suppose we want to formalize the following sentence:

It is not the case that not every widget has no defects.

Suppose we let $w(x)$ mean "x is a widget" and let $d(x)$ mean "x has a defect." We might look at the latter portion of the sentence, which says, "every widget has no defects." We can formalize this statement as $\forall x \ (w(x) \rightarrow \neg \ d(x))$. Now the beginning part of the sentence says, "It is not the case that not." This is a double negation. So the formalization of the entire sentence is

$$\neg \ \neg \ \forall x \ (w(x) \rightarrow \neg \ d(x)),$$

which of course is equivalent to $\forall x \ (w(x) \rightarrow \neg \ d(x))$.

Let's discuss the little words "is" and "are." Their usage can lead to quite different formalizations. For example, the three statements

"4 is 2 + 2," "x is a widget," and "widgets are defective"

have the three formalizations $4 = 2 + 2$, $w(x)$, and $\forall x \ (w(x) \rightarrow d(x))$. So we have to be careful when we try to formalize English sentences.

As a final example, which we won't discuss, consider the following sentence taken from Section 2, Article I, of the Constitution of the United States of America.

No person shall be a Representative who shall not have attained to the Age of twenty-five Years, and been seven Years a Citizen of the United States, and who shall not, when elected, be an Inhabitant of that State in which he shall be chosen.

7.2.4 Summary

Here, all in one place, are some equivalences and restricted equivalences.

Equivalences

1. $\neg \, \forall x \; W(x) \equiv \exists x \, \neg \, W(x).$

2. $\neg \, \exists x \; W(x) \equiv \forall x \, \neg \, W(x).$

3. $\forall x \, \forall y \; W(x, y) \equiv \forall y \, \forall x \; W(x, y).$

4. $\exists x \, \exists y \; W(x, y) \equiv \exists y \, \exists x \; W(x, y).$

5. $\exists x \; (A(x) \to B(x)) \equiv \forall x \, A(x) \to \exists x \, B(x).$

6. $\exists x \; (A(x) \lor B(x)) \equiv \exists x \, A(x) \lor \exists x \, B(x).$

7. $\forall x \; (A(x) \land B(x)) \equiv \forall x \, A(x) \land \forall x \, B(x).$

Restricted Equivalences

The following equivalences hold if x does not occur free in the wff C:

Simplification

$\forall x \; C \equiv C$ and $\exists x \; C \equiv C.$

Disjunction

$\forall x \; (C \lor A(x)) \equiv C \lor \forall x \, A(x).$
$\exists x \; (C \lor A(x)) \equiv C \lor \exists x \, A(x).$

Conjunction

$\forall x \; (C \land A(x)) \equiv C \land \forall x \, A(x).$
$\exists x \; (C \land A(x)) \equiv C \land \exists x \, A(x).$

Implication

$\forall x \; (C \to A(x)) \equiv C \to \forall x \, A(x).$
$\exists x \; (C \to A(x)) \equiv C \to \exists x \, A(x).$
$\forall x \; (A(x) \to C) \equiv \exists x \, A(x) \to C.$
$\exists x \; (A(x) \to C) \equiv \forall x \, A(x) \to C.$

Exercises

Proving Equivalences with Validity

1. Prove each of the following equivalences with validity arguments (i.e., use interpretations and models).

    a. $\forall x\ (A(x) \wedge B(x)) \equiv \forall x\ A(x) \wedge \forall x\ B(x)$.

    b. $\exists x\ (A(x) \vee B(x)) \equiv \exists x\ A(x) \vee \exists x\ B(x)$.

    c. $\exists x\ (A(x) \rightarrow B(x)) \equiv \forall x\ A(x) \rightarrow \exists x\ B(x)$.

    d. $\forall x\ \forall y\ W(x,\ y) \equiv \forall y\ \forall x\ W(x,\ y)$.

    e. $\exists x\ \exists y\ W(x,\ y) \equiv \exists y\ \exists x\ W(x,\ y)$.

2. Assume that x does not occur free in the wff C. With this assumption, prove each of the following equivalences with validity arguments (i.e., use interpretations and models).

    a. $\forall x\ C \equiv C$.

    b. $\exists x\ C \equiv C$.

    c. $\exists x\ (C \vee A(x)) \equiv C \vee \exists x\ A(x)$.

Proving Equivalences with Equivalence

3. Assume that x does not occur free in the wff C. With this assumption, prove each of the following statements with an equivalence proof that uses the equivalence listed in parentheses.

    a. $\forall x\ (C \rightarrow A(x)) \equiv C \rightarrow \forall x\ A(x)$.   (use 7.10a)

    b. $\exists x\ (C \rightarrow A(x)) \equiv C \rightarrow \exists x\ A(x)$.   (use 7.10b)

    c. $\exists x\ (A(x) \rightarrow C) \equiv \forall x\ A(x) \rightarrow C$.   (use 7.10b)

    d. $\forall x\ (C \wedge A(x)) \equiv C \wedge \forall x\ A(x)$.   (use 7.10b)

Prenex Normal Forms

4. Use equivalences to construct a prenex conjunctive normal form for each of the following wffs.

    a. $\forall x\ (p(x) \vee q(x)) \rightarrow \forall x\ p(x) \vee \forall x\ q(x)$.

    b. $\exists x\ p(x) \wedge \exists x\ q(x) \rightarrow \exists x\ (p(x) \wedge q(x))$.

    c. $\forall x\ \exists y\ p(x,\ y) \rightarrow \exists y\ \forall x\ p(x,\ y)$.

    d. $\forall x\ (p(x,\ f(x)) \rightarrow p(x,\ y))$.

    e. $\forall x\ \forall y\ (p(x,\ y) \rightarrow \exists z\ (p(x,\ z) \wedge p(y,\ z)))$.

    f. $\forall x\ \forall y\ \forall z\ (p(x,\ y) \wedge p(y,\ z) \rightarrow p(x,\ z)) \wedge \forall x\ \neg\ p(x,\ x)$
        $\rightarrow \forall x\ \forall y\ (p(x,\ y) \rightarrow \neg\ p(y,\ x))$.

5. Use equivalences to construct a prenex disjunctive normal form for each of the following wffs.

    a. $\forall x\ (p(x) \lor q(x)) \to \forall x\ p(x) \lor \forall x\ q(x)$.

    b. $\exists x\ p(x) \land \exists x\ q(x) \to \exists x\ (p(x) \land q(x))$.

    c. $\forall x\ \exists y\ p(x,\ y) \to \exists y\ \forall x\ p(x,\ y)$.

    d. $\forall x\ (p(x,\ f(x)) \to p(x,\ y))$.

    e. $\forall x\ \forall y\ (p(x,\ y) \to \exists z\ (p(x,\ z) \land p(y,\ z)))$.

    f. $\forall x\ \forall y\ \forall z\ (p(x,\ y) \land p(y,\ z) \to p(x,\ z)) \land \forall x\ \neg\ p(x,\ x)$
       $\to \forall x\ \forall y\ (p(x,\ y) \to \neg\ p(y,\ x))$.

6. Recall that an equivalence $A \equiv B$ stands for the wff $(A \to B) \land (B \to A)$. Let C be a wff that does not contain the variable x.

    a. Find a countermodel to show that the following statement is invalid: $(\forall x\ W(x) \equiv C) \equiv \forall x\ (W(x) \equiv C)$.

    b. Find a prenex normal form for the statement $(\forall x\ W(x) \equiv C)$.

Formalizing English Sentences

7. Formalize each of the following English sentences, where the domain of discourse is the set of all people, where $C(x)$ means x is a committee member, $G(x)$ means x is a college graduate, $R(x)$ means x is rich, $S(x)$ means x is smart, $O(x)$ means x is old, and $F(x)$ means x is famous.

    a. Every committee member is rich and famous.

    b. Some committee members are old.

    c. All college graduates are smart.

    d. No college graduate is dumb.

    e. Not all college graduates are smart.

8. Formalize each of the following statements, where $B(x)$ means x is a bird, $W(x)$ means x is a worm, and $E(x,\ y)$ means x eats y.

    a. Every bird eats worms.

    b. Some birds eat worms.

    c. Only birds eat worms.

    d. Not all birds eat worms.

    e. Birds only eat worms.

    f. No bird eats only worms.

    g. Not only birds eat worms.

9. Formalize each argument as a wff, where $P(x)$ means x is a person, $S(x)$ means x can swim, and $F(x)$ means x is a fish.

    a. All fish can swim. John can't swim. Therefore, John is not a fish.

    b. Some people can't swim. All fish can swim. Therefore, there is some person who is not a fish.

10. Formalize each statement, where $P(x)$ means x is a person, $B(x)$ means x is a bully, $K(x, y)$ means x is kind to y, $C(x)$ means x is a child, $A(x)$ means x is an animal, $G(x)$ means x plays golf, and $N(x, y)$ means x knows y.

   a. All people except bullies are kind to children.

   b. Bullies are not kind to children.

   c. Bullies are not kind to themselves.

   d. Not everyone plays golf.

   e. Everyone knows someone who plays golf.

   f. People who play golf are kind to animals.

   g. People who are not kind to animals do not play golf.

7.3   Formal Proofs in Predicate Calculus

To reason formally about wffs in predicate calculus, we need some proof rules. It's nice to know that all the proof rules of propositional calculus can still be used for predicate calculus.

For example, let's take the modus ponens proof rule of propositional calculus and prove that it also works for predicate calculus. In other words, we'll show that modus ponens maps valid wffs to a valid wff.

Proof: Let A and $A \rightarrow B$ be valid wffs. We need to show that B is valid. Suppose we have an interpretation for B with domain D. We can use D to give an interpretation to A by assigning values to all the predicates, functions, free variables, and constants that occur in A but not B. This gives us interpretations for A, B, and $A \rightarrow B$ over the domain D. Since we are assuming that A and $A \rightarrow B$ are valid, it follows that A and $A \rightarrow B$ are true for these interpretations over D. Now we can apply the modus ponens rule for propositions to conclude that B is true with respect to the given interpretation over D. Since the given interpretation of B was arbitrary, it follows that every interpretation of B is a model. Therefore, B is valid. QED.

The arguments are similar for the rules Conj, Simp, Add, DS, MP, DN, Contr, MT, Cases, HS, and CD. So we have a built-in collection of rules to do formal reasoning in predicate calculus. But we need more.

Sometimes it's hard to reason about statements that contain quantifiers. The natural approach is to remove quantifiers from statements, do some reasoning with the unquantified statements, and then restore any needed quantifiers. We might call this the RRR method of reasoning with quantifiers—*remove, reason,* and *restore.*

But quantifiers cannot be removed and restored at will. So we'll spend a little time discussing restrictions that govern their use. Then we'll show that CP, and thus also IP, can be used as proof rules of predicate calculus.

7.3.1 Universal Instantiation (UI)

Let's start by using our intuition and see how far we can get. It seems reasonable to say that if a property holds for everything, then it holds for any particular thing. In other words, we should be able to infer $W(x)$ from $\forall x\, W(x)$. Similarly, we should be able to infer $W(c)$ from $\forall x\, W(x)$ for any constant c.

Can we infer $W(t)$ from $\forall x\, W(x)$ for any term t? This seems OK too, but there may be a problem if $W(x)$ contains a free occurrence of x that lies within the scope of a quantifier. For example, suppose we let

$$W(x) = \exists y\, p(x,\, y).$$

Now if we let $t = y$, then we obtain

$$W(t) = W(y) = \exists y\, p(y,\, y).$$

But we can't always infer $\exists y\, p(y,\, y)$ from $\forall x\, \exists y\, p(x,\, y)$. For example, let $p(x,\, y)$ mean "x is a child of y." Then the statement $\forall x\, \exists y\, p(x,\, y)$ is true because every person x is a child of some person y. But the statement $\exists y\, p(y,\, y)$ is false because no person is their own child.

Trouble arises when we try to infer $W(t)$ from $\forall x\, W(x)$ in situations where t contains an occurrence of a variable that is quantified in $W(x)$ and x occurs free within the scope of that quantifier. We must restrict our inferences so that this does not happen. To make things precise, we'll make the following definition.

Definition of Free to Replace

We say that a term t *is free to replace* x in $W(x)$ if no free occurrence of x in $W(x)$ is in the scope of a quantifier that binds a variable in t. Equivalently we can say that a term t is free to replace x in $W(x)$ if both $W(t)$ and $W(x)$ have the same bound occurrences of variables.

example **7.12 Free to Replace**

A term t is free to replace x in $W(x)$ under any of the following conditions.

  a. $t = x$.

  b. t is a constant.

c. The variables of t do not occur in $W(x)$.

d. The variables of t do not occur within the scope of a quantifier in $W(x)$.

e. x does not occur free within the scope of a quantifier in $W(x)$.

f. $W(x)$ does not have any quantifiers.

end example

Going the other way, we can say that a term t is *not* free to replace x in $W(x)$ if t contains a variable that is quantified in $W(x)$ *and* x occurs free within the scope of that quantifier. We can also observe that when t is not free to replace x in $W(x)$, then $W(t)$ has more bound variables than $W(x)$, while if t is free to replace x in $W(x)$, then $W(t)$ and $W(x)$ have the same bound variables.

example **7.13 Not Free to Replace**

We'll examine some terms t that are not free to replace x in $W(x)$, where $W(x)$ is the following wff:

$$W(x) = q(x) \land \exists y\; p(x, y).$$

We'll examine the following two terms:

$$t = y \text{ and } t = f(x, y).$$

Notice that both terms contain the variable y, which is quantified in $W(x)$ and there is a free occurrence of x within the scope of the quantifier $\exists y$. So each t is not free to replace x in $W(x)$. We can also note the difference in the number of bound variables between $W(t)$ and $W(x)$. For example, for $t = y$ we have

$$W(t) = W(y) = q(y) \land \exists y\; p(y, y).$$

So $W(t)$ has one more bound occurrence of y than $W(x)$. The same property holds for $t = f(x, y)$.

end example

Now we're in a position to state the *universal instantiation* rule along with the restriction on its use.

Universal Instantiation Rule (UI) (7.15)

$$\frac{\forall x\; W(x)}{W(t)} \quad \text{\textit{Restriction:} } t \text{ is free to replace } x \text{ in } W(x).$$

Special cases where the restriction is always satisfied:

$$\frac{\forall x\; W(x)}{W(x)} \quad \text{and} \quad \frac{\forall x\; W(x)}{W(c)} \quad \text{(for any constant } c\text{).}$$

Proof: The key point in the proof comes from the observation that if t is free to replace x in $W(x)$, then for any interpretation, the interpretation of $W(t)$ is the same as the interpretation of $W(d)$, where d is the interpreted value of t. To state this more concisely for an interpretation I, let tI be the interpreted value of t by I, let $W(t)I$ be the interpretation of $W(t)$ by I, and let $W(tI)I$ be the interpretation of $W(tI)$ by I. Now we can state the key point as an equation. If t is free to replace x in $W(x)$, then for any interpretation I, the following equation holds:

$$W(t)I = W(tI)I.$$

This can be proved by induction on the number of quantifiers and the number of connectives that occur in $W(x)$ and we'll leave it as an exercise. The UI rule follows easily from this. Let I be an interpretation with domain D that is a model for $\forall x \ W(x)$. Then $W(x)I$ is true for all $x \in D$. Since $tI \in D$, it follows that $W(tI)I$ is true. But we have the equation $W(t)I = W(tI)I$, so it also follows that $W(t)I$ is true. So I is a model for $W(t)$. Therefore, the wff $\forall x \ W(x) \rightarrow W(t)$ is valid. QED.

example **7.14 Truth Value Preserved**

We'll give some examples to show that $W(t)I = W(tI)I$ holds when t is free to replace x. Let $W(x)$ be the following wff:

$$W(x) = \exists y \ p(x, y, z).$$

Then, for any term t, we have

$$W(t) = \exists y \ p(t, y, z).$$

Let I be an interpretation with domain $D = \{a, b, c\}$ that assigns any free occurrences of x, y, and z to a, b, and c. In each of the following examples, t is free to replace x in $W(x)$.

1. $t = b$: $W(t)I = \exists y \ p(b, y, z)I = \exists y \ p(b, y, c) = W(b)I = W(tI)I.$
2. $t = x$: $W(t)I = \exists y \ p(x, y, z)I = \exists y \ p(a, y, c) = \exists y \ p(a, y, z)I = W(a)I = W(tI)I.$
3. $t = z$: $W(t)I = \exists y \ p(z, y, z)I = \exists y \ p(c, y, c) = \exists y \ p(c, y, z)I = W(c)I = W(tI)I.$
4. $t = f(x, z)$: $W(t)I = \exists y \ p(f(x, z), y, z)I$
   $= \exists y \ p(f(a, c), y, c)$
   $= \exists y \ p(f(a, c), y, z)I$
   $= W(f(a, c))I = W(tI)I.$

end example

example   **7.15** **Truth Value Not Preserved**

We'll give some examples to show that $W(t)I \neq W(tI)I$ when t is not free to replace x in $W(x)$. We'll use the wff from Example 7.14.

$$W(x) = \exists y \; p(x, \, y, \, z).$$

Notice that each of the following terms t is not free to replace x in $W(x)$ because each one contains the quantified variable y and x occurs free in the scope of the quantifier.

$$t = y \quad \text{and} \quad t = f(x, \, y).$$

Let I be the interpretation with domain $D = \{1, \, 2, \, 3\}$ that assigns any free occurrences of x, y, and z to 1, 2, and 3, respectively. Let $p(u, \, v, \, w)$ mean that u, v, and w are all distinct and let $f(u, \, v)$ be the maximum of u and v.

1. $t = y$: $W(t)\,I = W(y)\,I$
$$= \exists y \; p\,(y, y, z)\,I$$
$$= \exists y \; p\,(y, y, 3)\,, \; \text{which is false.}$$

         $W(tI)\,I = W(yI)\,I$
$$= W(2)\,I$$
$$= \exists y \; p\,(2, y, z)\,I$$
$$= \exists y \; p\,(2, y, 3)\,, \; \text{which is true.}$$

2. $t = f(x, y)$: $W(t)\,I = W(f\,(x, y))\,I$
$$= \exists y \; p\,(f\,(x, y)\,, y, z)\,I$$
$$= \exists y \; p\,(f\,(1, y)\,, y, 3)\,, \; \text{which is false (try different y's).}$$

         $W(tI)\,I = W(f\,(x, y)\,I)\,I$
$$= W(f\,(1, 2)\,I)\,I$$
$$= W(2)\,I$$
$$= \exists y \; p\,(2, y, z)\,I$$
$$= \exists y \; p\,(2, y, 3)\,, \; \text{which is true.}$$

So in each case $W(t)I$ and $W(tI)I$ have different truth values.

end example

7.3.2 Existential Generalization (EG)

It seems to make sense that if a property holds for a particular thing, then the property holds for some thing. For example, we know that 5 is a prime number, so it makes sense to conclude that there is some prime number. In other words, if we let $p(x)$ mean "x is a prime number," then from $p(5)$ we can infer $\exists x\, p(x)$. So far, so good. If a wff can be written in the form $W(t)$ for some term t, can we infer $\exists x\, W(x)$?

After a little thought, this appears to be related to the UI rule in its contrapositive form. In other words, notice the following equivalences:

$$W(t) \rightarrow \exists x\, W(x) \equiv \neg\, \exists x\, W(x) \rightarrow \neg\, W(t)$$
$$\equiv \forall x\, \neg\, W(x) \rightarrow \neg\, W(t)\,.$$

The last wff is an instance of the UI rule, which tells us that the wff is valid if t is free to replace x in $\neg\, W(x)$. Since $W(x)$ and $\neg\, W(x)$ differ only by negation, it follows that t is free to replace x in $\neg\, W(x)$ if and only if t is free to replace x in $W(x)$. Therefore, we can say that

$$W(t) \rightarrow \exists x\, W(x) \text{ is valid if } t \text{ is free to replace } x \text{ in } W(x).$$

So we have the *existential generalization* rule along with the restriction on its use.

Existential Generalization Rule (EG)   **(7.16)**

$$\frac{W(t)}{\exists x\, W(x)} \quad \textit{Restriction: } t \text{ is free to replace } x \text{ in } W(x)\,.$$

Special cases where the restriction is always satisfied:

$$\frac{W(x)}{\exists x\, W(x)} \quad \text{and} \quad \frac{W(c)}{\exists x\, W(x)} \quad \text{(for any constant } c\text{).}$$

Usage Note

There is a kind of forward-backward reasoning to keep in mind when using the EG rule. If we want to apply EG to a wff, then we must be able to write the wff in the form $W(t)$ for some term t such that $W(t)$ is obtained from $W(x)$ by replacing all free occurrences of x by t. In other words, we must have $W(t) = W(x)(x/t)$. Once this is done, we check to see whether t is free to replace x in $W(x)$.

example **7.16 Using the EG Rule**

Let's examine the use of EG on some sample wffs. We'll put each wff into the form $W(t)$ for some term t, where $W(t)$ is obtained from a wff $W(x)$ for some variable x, and t is free to replace x in $W(x)$. Then we'll use EG to infer $\exists x\ W(x)$.

1. $\forall y\ p(c, y)$.

   We can write $\forall y\ p(c, y) = W(c)$, where $W(x) = \forall y\ p(x, y)$. Now since c is a constant, we can use EG to infer

   $$\exists x\ \forall y\ p(x, y).$$

   For example, over the domain of natural numbers, let $p(x, y)$ mean $x \leq y$ and let $c = 0$. Then from $\forall y\ (0 \leq y)$ we can use EG to infer $\exists x\ \forall y\ (x \leq y)$.

2. $p(x, y, c)$.

   We can write $p(x, y, c) = W(c)$, where $W(z) = p(x, y, z)$. Since c is a constant, the EG rule can be used to infer

   $$\exists z\ p(x, y, z).$$

   Notice that we can also write $p(x, y, c)$ as either $W(x)$ or $W(y)$ with no substitutions. So EG can also be used to infer $\exists x\ p(x, y, c)$ and $\exists y\ p(x, y, c)$.

3. $\forall y\ p(f(x, z), y)$.

   We can write $\forall y\ p(f(x, z), y) = W(f(x, z))$, where $W(x) = \forall y\ p(x, y)$. Notice that the term $f(x, z)$ is free to replace x in $W(x)$. So we can use EG to infer

   $$\exists x\ \forall y\ p(x, y).$$

   We can also write $\forall y\ p(f(x, z), y)$ as either of the forms $W(x)$ or $W(z)$ with no substitutions. Therefore, we can use EG to infer $\exists x\ \forall y\ p(f(x, z), y)$ and $\exists z\ \forall y\ p(f(x, z), y)$.

end example

7.3.3 Existential Instantiation (EI)

It seems reasonable to say that if a property holds for some thing, then it holds for a particular thing. This type of reasoning is used quite often in proofs that proceed in the following way. Assume that we are proving some statement and during the proof we have the wff

$$\exists x\ W(x).$$

We then say that $W(c)$ holds for a particular object c. From this point the proof proceeds with more deductions and finally reaches a conclusion that does not contain any occurrence of the object c.

Difficulty (Choice of the Constant)

We have to be careful about our choice of the constant. For example, if the wff $\exists x\, p(x, b)$ occurs in a proof, then we can't say that $p(b, b)$ holds. To see this, suppose we let $p(x, b)$ mean "x is a parent of b." Then $\exists x\, p(x, b)$ is true, but $p(b, b)$ is false because b is not a parent of b. So we can't pick a constant that is already in the wff.

But we need to restrict the choice of constant further. Suppose, for example, that we have the following partial "attempted" proof.

1.	$\exists x\, p(x)$	P
2.	$\exists x\, q(x)$	P
3.	$p(c)$	1, proposed EI rule
4.	$q(c)$	2, proposed EI rule
5.	$p(c) \wedge q(c)$	3, 4, Conj

This can't continue because line 5 does not follow from the premises. For example, suppose over the domain of integers we let $p(x)$ mean "x is odd" and let $q(x)$ mean "x is even." Then the premises are true because there is an odd number and there is an even number. But line 5 says that c is even and c is odd, which is a false statement. So we have the following restriction on the choice of constant.

Restriction: Choose a new constant that does not occur on any previous line of the proof.

Difficulty (Constant and Conclusion)

There is one more restriction on the choice of constant. Namely, the constant cannot appear in any conclusion. For example, suppose starting with the premise $\exists x\, p(x)$ we deduce $p(c)$ and then claim by conditional proof that we have proven the validity of the wff $\exists x\, p(x) \rightarrow p(c)$. But this wff is not valid. For example, consider the interpretation with domain $\{0, 1\}$, where we assign the constant $c = 1$ and let $p(0) = $ True and $p(1) = $ False. Then the interpreted wff has a true antecedent and a false consequent, which makes the interpreted wff false. So we have the following additional restriction on the choice of constant.

Restriction: Choose a constant that does not occur in the statement to be proved.

Now we're in position to state the *existential instantiation* rule along with the restrictions on its use.

Existential Instantiation Rule (EI)   (7.17)

If $\exists x \; W(x)$ occurs on some line of a proof, then $W(c)$ may be placed on any subsequent line of the proof (subject to the following restrictions).

Restrictions: Choose c to be a new constant in the proof and such that c does not occur in the statement to be proven.

Proof: We'll give an idea of the proof. Suppose that the EI rule is used in a proof of the wff A and the constant c does not occur in A. We'll show that the proof of A does not need the EI rule. (So there is no harm in using EI.) Let P be the conjunction of the premises in the proof, except $\exists x \; W(x)$ if it happens to be a premise. Therefore, the wff $P \wedge \exists x \; W(x) \wedge W(c) \to A$ is valid. So it follows that the wff $P \wedge \exists x \; W(x) \to (W(c) \to A)$ is also valid. Let y be a variable that does not occur in the proof. Then $P \wedge \exists x \; W(x) \to (W(y) \to A)$ is also valid because any interpretation assigning y a value yields the same wff by assigning c that value. It follows from the proof of (7.3) that $\forall y \; (P \wedge \exists x \; W(x) \to (W(y) \to A))$ is valid. Since y does not occur in $P \wedge \exists x \; W(x)$ or A, we have the following equivalences.

$$\forall y \; (P \wedge \; \exists x \; W(x) \to W(y) \to A))$$
$$\equiv P \wedge \exists x \; W(x) \to \forall y \; W(y) \to A \qquad \text{(by 7.12)}$$
$$\equiv P \wedge \exists x \; W(x) \to (\exists y \; W(y) \to A) \qquad \text{(by 7.12)}$$
$$\equiv P \wedge \exists x \; W(x) \wedge \exists y \; W(y) \to A$$
$$\equiv P \wedge \exists x \; W(x) \to A.$$

So A can be proved without the use of EI. QED.

7.3.4 Universal Generalization (UG)

It seems reasonable to say that if some property holds for an arbitrary thing, then the property holds for all things. This type of reasoning is used quite often in proofs that proceed in the following way. We prove that some property holds for an arbitrary element x and then conclude that the property holds for all x. Here's a more detailed description of the technique in terms of a wff $W(x)$ over some domain D.

> We let x be an arbitrary but fixed element of the domain D. Next, we construct a proof that $W(x)$ is true. Then we say that since x was arbitrary, it follows that $W(x)$ is true for all x in D. So from a proof of $W(x)$, we have proved $\forall x \; W(x)$.

So if we find $W(x)$ on some line in a proof, when can we put $\forall x \, W(x)$ on a subsequent line of the proof? As we'll see, there are a couple of restrictions that must be met.

Difficulty (Free Variables in a Premise)

Over the domain of natural numbers let $p(x)$ mean that x is a prime number. Then $\forall x \, p(x)$ means that every natural number x is a prime number. Since there are natural numbers that are not prime, we can't infer $\forall x \, p(x)$ from $p(x)$. In other words, $p(x) \rightarrow \forall x \, p(x)$ is not valid.

This illustrates a problem that can occur in a proof when there is a premise containing a free variable x and we later try to generalize with respect to the variable.

1. $p(x)$ P
2. $\forall x \, p(x)$ 1, *Do not use the UG rule*. It doesn't work!

Restriction: Among the wffs used to infer $W(x)$, x is not free in any premise.

Difficulty (Free Variables and EI)

Another problem can occur when we try to generalize with respect to a variable that occurs free in a wff constructed with EI. For example, consider the following attempted proof, which starts with the premise that for any natural number x there is some natural number y greater than x.

1. $\forall x \, \exists y \, (x < y)$ P
2. $\exists y \, (x < y)$ 1, UI
3. $x < c$ 2, EI
4. $\forall x \, (x < c)$ 3, *Do not use the UG rule*. It doesn't work.
5. $\exists y \, \forall x \, (x < y)$ 4, EG.
    Not QED.

We better not allow line 4 because the conclusion on line 5 says that there is a natural number y greater than every natural number x, which we know to be false.

Restriction: Among wffs used to infer $W(x)$, x is not free in any wff inferred by EI.

Now we're in position to state the *universal generalization* rule the restrictions on its use.

Universal Generalization Rule (UG)   **(7.18)**

If $W(x)$ occurs on some line of a proof, then $\forall x\ W(x)$ may be placed on any subsequent line of the proof (subject to the following restrictions).

Restrictions: Among the wffs used to obtain $W(x)$, x is not free in any premise and x is not free in any wff obtained by EI.

Proof: We'll give a general outline that works if the restrictions are satisfied. Suppose we have a proof of $W(x)$. Let P be the conjunction of premises in the proof and any wffs obtained by EI that contain a free occurrence of x. Then $P \rightarrow W(x)$ is valid. We claim that $P \rightarrow \forall x\ W(x)$ is valid. For if not, then $P \rightarrow \forall x\ W(x)$ has some interpretation I with domain D such that P is true with respect to I and $\forall x\ W(x)$ is false with respect to I. So there is an element $d \in D$ such that $W(d)$ is false with respect to I. Now let J be the interpretation of $P \rightarrow W(x)$ with the same domain D and with all assignments the same as I but with the additional assignment of the free variable x to d. Since x is not free in P, it follows that P with respect to J is the same as P with respect to I. So P is true with respect to J. But since $P \rightarrow W(x)$ is valid, it follows that $W(d)$ is true with respect to J. But x is not free in $W(d)$. So $W(d)$ with respect to J is the same as $W(d)$ with respect to I, which contradicts $W(d)$ being false with respect to I. Therefore, $P \rightarrow \forall x\ W(x)$ is valid. So we can place $\forall x\ W(x)$ on any subsequent line of the proof. QED.

It's nice to know that the restrictions of the UG rule are almost always satisfied. For example, if the premises in the proof don't contain any free variables and if the proof doesn't use the EI rule, then use the UG rule with abandon.

Conditional Proof Rule

Now that we've discussed the quantifier proof rules, let's take a minute to discuss the extension of the conditional proof rule from propositional calculus to predicate calculus. Recall that it allows us to prove a conditional $A \rightarrow B$ with a conditional proof of B from the premise A. The result is called the *conditional proof rule* (CP). It is also known as the deduction theorem.

Conditional Proof Rule (CP)   **(7.19)**

If A is a premise in a proof of B, then there is a proof of $A \rightarrow B$ that does not use A as a premise.

Proof: Let $W_1, \ldots, W_n = B$ be a proof of B that contains A as a premise. We'll show by induction that for each k in the interval $1 \leq k \leq n$, there is a proof of $A \rightarrow W_k$ that does not use A as a premise. Since $B = W_n$, the result will be proved. For the case $k = 1$, the argument is the same as that given in the proof of the CP rule for propositional calculus. Let $k > 1$ and assume that for each $i < k$ there is a proof of $A \to W_i$ that does not use A as a premise. If W_k is not obtained by a quantifier rule, then the argument in the proof of the CP rule for propositional calculus constructs a proof of $A \to W_k$ that does not use A as a premise. The construction also guarantees that the premises needed in the proof of $A \to W_k$ are the premises other than A that are needed in the original proof of W_k.

Suppose that W_k is obtained by a quantifier rule from W_i, where $i < k$. First, notice that if A is not needed to prove W_i, then A is not needed to prove W_k. So we can remove A from the given proof of W_k. Now add the valid wff $W_k \to (A \to W_k)$ to the proof and then use MP to infer $A \to W_k$. This gives us a proof of $A \to W_k$ that does not use A as a premise. Second, notice from the proof of the EI rule (7.17) that for any proof that uses EI, there is an alternative proof that does not use EI. So we can assume that A is needed in the proof of W_i and EI is not used in the given proof.

If W_k is obtained from W_i by UG, then $W_k = \forall x \, W_i$ where x is not free in any premise needed to prove W_i. So x is not free in A. Induction gives a proof of $A \to W_i$ that does not use A as a premise and x is not free in any premise needed to prove $A \to W_i$. So we can use UG to obtain $\forall x \, (A \to W_i)$. Since x is not free in A, it follows from (7.12a) that $\forall x \, (A \to W_i) \to (A \to \forall x \, W_i)$ is valid. Now use MP to infer $A \to \forall x \, W_i$. So we have a proof of $A \to W_k$ that does not use A as a premise.

If W_k is obtained from W_i by UI, then there is a wff $C \, (x)$ such that $W_i = \forall x \, C \, (x)$ and $W_k = C \, (t)$, where t is free to replace x in $C \, (x)$. The proof of the UI rule tells us that $\forall x \, C \, (x) \to C \, (t)$ is valid. Induction gives a proof of $A \to \forall x \, C \, (x)$ that does not use A as a premise. Now use HS to infer $A \to C \, (t)$. So we have a proof of $A \to W_k$ that does not use A as a premise.

If W_k is obtained from W_i by EG, then there is a wff $C \, (x)$ such that $W_k = C \, (t)$ and $W_k = \exists x \, C \, (x)$, where t is free to replace x in $C \, (x)$. The proof of the EG rule tells us that $C \, (t) \to \exists x \, C \, (x)$ is valid. By induction there is a proof of $A \to C \, (t)$ that does not use A as a premise. Now use HS to infer $A \to \exists x \, C \, (x)$. So we have a proof of $A \to W_k$ that does not use A as a premise. QED.

7.3.5 Examples of Formal Proofs

Finally we can get down to business and do some proofs. The following examples show the usefulness of the four quantifier rules. Notice in most cases that we can use the less restrictive forms of the rules.

example **7.17 Part of an Equivalence**

We'll give an indirect formal proof of the following statement:

$$\forall x \, \neg \, W(x) \to \neg \, \exists x \, W(x).$$

1. $\forall x \, \neg \, W(x)$ P
2. $\neg \, \neg \, \exists x \, W(x)$ P [for $\neg \, \exists x \, W(x)$]
3. $\exists x \, W(x)$ 2, DN
4. $W(c)$ 3, EI
5. $\neg \, W(c)$ 1, UI
6.   False   4, 5, Contr
7. $\neg \, \exists x \, W(x)$ 2–6, IP
      QED   1, 7, CP.

We'll prove the converse of $\forall x \, \neg \, W(x) \rightarrow \neg \, \exists x \, W(x)$ in Example 7.26.

end example

example **7.18** **Using Hypothetical Syllogism**

We'll prove the following statement:

$$\forall x \, (A(x) \rightarrow B(x)) \wedge \forall x \, (B(x) \rightarrow C(x)) \rightarrow \forall x \, (A(x) \rightarrow C(x)).$$

1. $\forall x \, (A(x) \rightarrow B(x))$ P
2. $\forall x \, (B(x) \rightarrow C(x))$ P
3. $A(x) \rightarrow B(x)$ 1, UI
4. $B(x) \rightarrow C(x)$ 2, UI
5. $A(x) \rightarrow C(x)$ 3, 4, HS
6. $\forall x \, (A(x) \rightarrow C(x))$ 5, UG
      QED   1–6, CP.

end example

example **7.19** **Lewis Carroll's Logic**

The following argument is from *Symbolic Logic* by Lewis Carroll.

Babies are illogical. Nobody is despised who can manage a crocodile. Illogical persons are despised. Therefore babies cannot manage crocodiles.

We'll formalize the argument over the domain of people. Let $B(x)$ mean "x is a baby," $L(x)$ mean "x is logical," $D(x)$ mean "x is despised," and $C(x)$ mean "x can manage a crocodile." Then the four sentences become

$$\forall x \, (B\,(x) \rightarrow \neg \, L\,(x)) \, .$$
$$\forall x \, (C\,(x) \rightarrow \neg \, D\,(x)) \, .$$
$$\forall x \, (\neg \, L\,(x) \rightarrow D\,(x)) \, .$$
$$\text{Therefore, } \forall x \, (B\,(x) \rightarrow \neg \, C\,(x)) \, .$$

Here is a formal proof that the argument is correct.

1.	$\forall x\,(B(x) \to \neg\,L(x))$	P
2.	$\forall x\,(C(x) \to \neg\,D(x))$	P
3.	$\forall x\,(\neg\,L(x) \to D(x))$	P
4.	$B(x) \to \neg\,L(x)$	1, UI
5.	$C(x) \to \neg\,D(x)$	2, UI
6.	$\neg\,L(x) \to D(x)$	3, UI
7.	$\qquad B(x)$	P [for $B(x) \to \neg\,C(x)$]
8.	$\qquad \neg\,L(x)$	4, 7, MP
9.	$\qquad D(x)$	6, 8, MP
10	$\qquad \neg\,\neg\,D(x)$	9, DN
11.	$\qquad \neg\,C(x)$	5, 10, MT
12.	$B(x) \to \neg\,C(x)$	7–11, CP
13.	$\forall x\,(B(x) \to \neg\,C(x))$	12, UG
	QED	1–6, 12, 13, CP.

Note that this argument holds for any interpretation. In other words, we've shown that the wff $A \to B$ is valid, where A and B are defined as follows:

$$A = \forall x\,(B\,(x) \to \neg\,L\,(x)) \land \forall x\,(C\,(x) \to \neg\,D\,(x)) \land \forall x\,(\neg\,L\,(x) \to D\,(x)),$$
$$B = \forall x\,(B\,(x) \to \neg\,C\,(x)).$$

end example

example **7.20 Swapping Universal Quantifiers**

We'll prove the following general statement about swapping universal quantifiers:

$$\forall x\,\forall y\;W \to \forall y\,\forall x\;W.$$

1.	$\forall x\,\forall y\;W$	P
2.	$\forall y\;W$	1, UI
3.	W	2 , UI
4.	$\forall x\;W$	3, UG
5.	$\forall y\,\forall x\;W$	4, UG
	QED	1–5, CP.

The converse of the statement can be proved in the same way. Therefore, we have a formal proof of the following equivalence in (7.5).

$$\forall x\,\forall y\;W \equiv \forall y\,\forall x\;W.$$

end example

example **7.21 Renaming Variables**

We'll give formal proofs of the equivalences that rename variables (7.8): Let $W(x)$ be a wff, and let y be a variable that does not occur in $W(x)$. Then the following renaming equivalences hold:

$$\exists x \; W(x) \equiv \exists y \; W(y),$$
$$\forall x \; W(x) \equiv \forall y \; W(y).$$

First we'll prove $\exists x \; W(x) \equiv \exists y \; W(y)$, which will require proofs of the two statements

$$\exists x \; W(x) \to \exists y \; W(y) \quad \text{and} \quad \exists y \; W(y) \to \exists x \; W(x).$$

Proof of $\exists x \; W(x) \to \exists y \; W(y)$:

1. $\exists x \; W(x)$ P
2. $W(c)$ 1, EI
3. $\exists y \; W(y)$ 2, EG
     QED   1–3, CP.

The proof of $\exists y \; W(y) \to \exists x \; W(x)$ is similar.

Next, we'll prove $\forall x \; W(x) \equiv \forall y \; W(y)$ by proving the two statements

$$\forall x \; W(x) \to \forall y \; W(y) \quad \text{and} \quad \forall y \; W(y) \to \forall x \; W(x).$$

Proof of $\forall x \; W(x) \to \forall y \; W(y)$:

1. $\forall x \; W(x)$ P
2. $W(y)$ 1, UI, y is free to replace x
3. $\forall y \; W(y)$ 2, UG
     QED   1–3, CP.

The proof of $\forall y \; W(y) \to \forall x \; W(x)$ is similar.

end example

example **7.22 Using EI, UI, and EG**

We'll prove the statement

$$\forall x \; p(x) \land \exists x \; q(x) \to \exists x \; (p(x) \land q(x)).$$

1. $\forall x \; p(x)$ P
2. $\exists x \; q(x)$ P
3. $q(c)$ 2, EI
4. $p(c)$ 1, UI
5. $p(c) \land q(c)$ 3, 4, Conj
6. $\exists x \; (p(x) \land q(x))$ 5, EG
     QED   1–6, CP.

end example

example **7.23 Formalizing an Argument**

Consider the following three statements:

> Every computer science major is a logical thinker.
>
> John is a computer science major.
>
> Therefore, there is some logical thinker.

We'll formalize these statements as follows: Let $C(x)$ mean "x is a computer science major," let $L(x)$ mean "x is a logical thinker," and let the constant b mean "John." Then the three statements can be written more concisely as follows, over the domain of people:

$$\forall x\, (C(x) \to L(x))$$
$$C(b)$$
$$\text{Therefore, } \exists x\, L(x).$$

These statements can be written as the following conditional wff:

$$\forall x\, (C(x) \to L(x)) \wedge C(b) \to \exists x\, L(x).$$

Although we started with a specific set of English sentences, we now have a first-order wff. We'll prove that this conditional wff is valid as follows:

1.	$\forall x\, (C(x) \to L(x))$	P
2.	$C(b)$	P
3.	$C(b) \to L(b)$	1, UI
4.	$L(b)$	2, 3, MP
5.	$\exists x\, L(x)$	4, EG
	QED	1–5, CP.

end example

example **7.24 Formalizing an Argument**

Let's consider the following argument:

> All computer science majors are people.
>
> Some computer science majors are logical thinkers.
>
> Therefore, some people are logical thinkers.

We'll give a formalization of this argument. Let $C(x)$ mean "x is a computer science major," $P(x)$ mean "x is a person," and $L(x)$ mean "x is a logical thinker." Now the statements can be represented by the following wff:

$$\forall x\, (C(x) \to P(x)) \wedge \exists x\, (C(x) \wedge L(x)) \to \exists x\, (P(x) \wedge L(x)).$$

We'll prove that this wff is valid as follows:

1. $\forall x \ (C(x) \rightarrow P(x))$ P
2. $\exists x \ (C(x) \wedge L(x))$ P
3. $C(c) \wedge L(c)$ 2, EI
4. $C(c) \rightarrow P(c)$ 1, UI
5. $C(c)$ 3, Simp
6. $P(c)$ 4, 5, MP
7. $L(c)$ 3, Simp
8. $P(c) \wedge L(c)$ 6, 7, Conj
9. $\exists x \ (P(x) \wedge L(x))$ 8, EG
   QED   1–9, CP.

end example

example **7.25 Move Quantifiers with Care**

We'll give a proof of the validity of the following wff:

$$\forall x \ A(x) \vee \forall x \ B(x) \rightarrow \forall x \ (A(x) \vee B(x)).$$

1. $\forall x \ A(x) \vee \forall x \ B(x)$ P
2. $\quad \forall x \ A(x)$ P [for $\forall x \ A(x) \rightarrow \forall x \ (A(x) \vee B(x))$]
3. $\quad A(x)$ 2, UI
4. $\quad A(x) \vee B(x)$ 3, Add
5. $\quad \forall x \ (A(x) \vee B(x))$ 4, UG
6. $\forall x \ A(x) \rightarrow \forall x \ (A(x) \vee B(x))$ 2–5, CP
7. $\quad \forall x \ B(x)$ P [for $\forall x \ B(x) \rightarrow \forall x \ (A(x) \vee B(x))$]
8. $\quad B(x)$ 7, UI
9. $\quad A(x) \vee B(x)$ 8, Add
10. $\quad \forall x \ (A(x) \vee B(x))$ 9, UG
11. $\forall x \ B(x) \rightarrow \forall x \ (A(x) \vee B(x))$ 7–10, CP
12. $\forall x \ (A(x) \vee B(x))$ 1, 6, 11, CD
    QED   1–6, 11, 12, CP.

end example

example **7.26 An Equivalence**

In Example 7.17 we gave a formal proof of the statement

$$\forall x \ \neg \ W(x) \rightarrow \ \neg \ \exists x \ W(x).$$

Now we're in a position to give a formal proof of its converse. Thus we'll have a formal proof of the following equivalence (7.4):

$$\forall x \neg W(x) \equiv \neg \exists x \; W(x).$$

The converse that we want to prove is the wff $\neg \exists x \; W(x) \rightarrow \forall x \neg W(x)$. To prove this statement, we'll divide the proof into two parts. First, we'll prove the statement $\neg \exists x \; W(x) \rightarrow \neg W(x)$.

1.	$\neg \exists x \; W(x)$	P
2.	$\quad W(x)$	P [for $\neg$ W(x)]
3.	$\quad \exists x \; W(x)$	2, EG
4.	$\quad$ False	1, 4, Contr
5.	$\neg W(x)$	2–4, IP
	QED	1, 5, CP.

Now we can easily prove the statement $\neg \exists x \; W(x) \rightarrow \forall x \neg W(x)$.

1.	$\neg \exists x \; W(x)$	P
2.	$\neg \exists x \; W(x) \rightarrow \neg W(x)$	T [proved above]
3.	$\neg W(x)$	1, 2, MP
4.	$\forall x \neg W(x)$	3, UG
	QED	1–4, CP.

end example

example 7.27 An Incorrect Proof

Suppose we're given the following wff.

$$\exists x \; P(x) \wedge \exists x \; Q(x) \rightarrow \exists x \; (P(x) \wedge Q(x)).$$

This wff is not valid! For example, over the integers, if $P(x)$ means that x is even and $Q(x)$ means that x is odd, then the antecedent is true and the consequent is false. We'll give an incorrect proof sequence that claims to show that the wff is valid.

1.	$\exists x \; P(x)$	P
2.	$\exists x \; Q(x)$	P
3.	$P(c)$	1, EI
4.	$Q(c)$	2, EI [No: c already occurs in line 3]
5.	$P(c) \wedge Q(c)$	3, 4, Conj
6.	$\exists x \; (P(x) \wedge Q(x))$	5, EG
	Not QED	1–6, CP.

end example

example   **7.28** **Formalizing a Numerical Argument**

We'll formalize the following informal proof that the sum of any two odd integers is even. Proof: Let x and y be arbitrary odd integers. Then there exist integers m and n such that $x = 2m + 1$ and $y = 2n + 1$. Now add x and y to obtain

$$x + y = 2m + 1 + 2n + 1 = 2(m + n + 1).$$

Therefore, $x + y$ is an even integer. Since x and y are arbitrary integers, it follows that the sum of any two odd integers is even. QED.

Now we'll write a more formal version of this proof, where $odd(x)$ means x is odd and $even(x)$ means x is even. We'll start the proof with an indented subproof of the statement $odd(x) \land odd(y) \rightarrow even(x + y)$. Then we'll apply UG to obtain the desired result.

1.	$odd(x)$	P
2.	$odd(y)$	P
3.	$\exists z\ (x = 2z + 1)$	1, definition of odd
4.	$\exists z\ (y = 2z + 1)$	2, definition of odd
5.	$x = 2m + 1$	3, EI
6.	$y = 2n + 1$	4, EI
7.	$x + y = 2(m + n + 1)$	5, 6, algebra
8.	$\exists z\ (x + y = 2z)$	7, EG
9.	$even(x + y)$	8, definition of even
10.	$odd(x) \land odd(y) \rightarrow even(x + y)$	1–9, CP
11.	$\forall y\ (odd(x) \land odd(y) \rightarrow even(x + y))$	10, UG
12.	$\forall x\ \forall y\ (odd(x) \land odd(y) \rightarrow even(x + y))$	11, UG
	QED.	

end example

 Exercises

Restrictions Using Quantifiers

1. Each of the following proof segments contains an invalid use of a quantifier proof rule. In each case, state why the proof rule cannot be used.

    a. 1. $x < 4$ P

         2. $\forall x\ (x < 4)$ 1, UG.

    b. 1. $\exists x\ (y < x)$ P

         2. $y < c$ 1, EI

         3. $\forall y\ (y < c)$ 2, UG.

   c. 1. $\forall y \; (y < f(y))$ P

       2. $\exists x \; \forall y \; (y < x)$ 1, EG.

   d. 1. $q(x, \; c)$ P

       2. $\exists x \; q(x, \; x)$ 1, EG.

   e. 1. $\exists x \; p(x)$ P

       2. $\exists x \; q(x)$ P

       3. $p(c)$ 1, EI

       4. $q(c)$ 2, EI.

   f. 1. $\forall x \; \exists y \; (x < y)$ P

       2. $\exists y \; (y < y)$ 1, UI.

2. Let W be the wff $\forall x \; (p(x) \lor q(x)) \to \forall x \; p(x) \lor \forall x \; q(x)$. It's easy to see that W is not valid. For example, let $p(x)$ mean "x is odd" and $q(x)$ mean "x is even" over the domain of integers. Then the antecedent is true, and the consequent is false. Suppose someone claims that the following sequence of statements is a "proof" of W:

    1. $\forall x \; (p(x) \lor q(x))$ P

    2. $p(x) \lor q(x)$ 1, UI

    3. $\forall x \; p(x) \lor q(x)$ 2, UG

    4. $\forall x \; p(x) \lor \forall x \; q(x)$ 3, UG

        QED   1–4, CP.

What is wrong with this "proof" of W?

3. Let W be the wff

$$\exists x \; P(x) \land \exists x \; (P(x) \to Q(x)) \to \exists x \; Q(x).$$

   a. Find a countermodel to show that W is not valid.

   b. The following argument attempts to prove that W is valid. Find an error in the argument.

    1. $\exists x \; P(x)$ P

    2. $P(d)$ 1, EI

    3. $\exists x \; (P(x) \to Q(x))$ P

    4. $P(d) \to Q(d)$ 3, EI

    5. $Q(d)$ 2, 4, MP

    6. $\exists x \; Q(x)$ 5, EG.

4. Explain what is wrong with the following attempted proof.

    1. $p(x)$ P

    2. $\forall x \; q(x)$ P

    3. $q(x)$ 2, UI

4. $p(x) \wedge q(x)$ 1, 3, Conj
5. $\forall x \ (p(x) \wedge q(x))$ 4, UG.

5. We'll give a formal proof of the following statement.

$$\forall x \ (p(x) \rightarrow q(x) \vee p(x)).$$

1. $p(x)$ $P \ [\text{for } \forall x \ (p(x) \rightarrow q(x) \vee p(x))]$
2. $q(x) \vee p(x)$ 1, Add
3. $p(x) \rightarrow q(x) \vee p(x)$ 1, 2, CP
4. $\forall x \ (p(x) \rightarrow q(x) \vee p(x))$ 3, UG
      QED.

Suppose someone argues against this proof as follows: The variable x is free in the premise on line 1, which is used to infer line 3, so we can't use UG to generalize the wff on line 3. What is wrong with this argument?

Direct Proofs

6. Give a formal proof that each of the following wffs is valid by using the CP rule. Do not use the IP rule.

a. $\forall x \ p(x) \rightarrow \exists x \ p(x).$

b. $\forall x \ (p(x) \rightarrow q(x)) \wedge \exists x \ p(x) \rightarrow \exists x \ q(x).$

c. $\exists x \ (p(x) \wedge q(x)) \rightarrow \exists x \ p(x) \wedge \exists x \ q(x).$

d. $\forall x \ (p(x) \rightarrow q(x)) \rightarrow (\exists x \ p(x) \rightarrow \exists x \ q(x)).$

e. $\forall x \ (p(x) \rightarrow q(x)) \rightarrow (\forall x \ p(x) \rightarrow \exists x \ q(x)).$

f. $\forall x \ (p(x) \rightarrow q(x)) \rightarrow (\forall x \ p(x) \rightarrow \forall x \ q(x)).$

g. $\exists y \ \forall x \ p(x, \ y) \rightarrow \forall x \ \exists y \ p(x, \ y).$

h. $\exists x \ \forall y \ p(x, \ y) \wedge \forall x \ (p(x, \ x) \rightarrow \exists y \ q(y, \ x)) \rightarrow \exists y \ \exists x \ q(x, \ y).$

Indirect Proofs

7. Give a formal proof that each of the following wffs is valid by using the CP rule and by using the IP rule in each proof.

a. $\forall x \ p(x) \rightarrow \exists x \ p(x).$

b. $\forall x \ (p(x) \rightarrow q(x)) \wedge \exists x \ p(x) \rightarrow \exists x \ q(x).$

c. $\exists y \ \forall x \ p(x, \ y) \rightarrow \forall x \ \exists y \ p(x, \ y).$

d. $\exists x \ \forall y \ p(x, \ y) \wedge \forall x \ (p(x, \ x) \rightarrow \exists y \ q(y, \ x)) \rightarrow \exists y \ \exists x \ q(x, \ y).$

e. $\forall x \ p(x) \vee \forall x \ q(x) \rightarrow \forall x \ (p(x) \vee q(x)).$

Transforming English Arguments

8. Transform each informal argument into a formalized wff. Then give a formal proof of the wff.

a. Every dog either likes people or hates cats. Rover is a dog. Rover loves cats. Therefore, some dog likes people.

b. Every committee member is rich and famous. Some committee members are old. Therefore, some committee members are old and famous.

c. No human beings are quadrupeds. All men are human beings. Therefore, no man is a quadruped.

d. Every rational number is a real number. There is a rational number. Therefore, there is a real number.

e. Some freshmen like all sophomores. No freshman likes any junior. Therefore, no sophomore is a junior.

Equivalences

9. Give a formal proof for each of the following equivalences as follows: To prove $W \equiv V$, prove the two statements $W \to V$ and $V \to W$.

a. $\exists x\ \exists y\ W(x,\ y) \equiv \exists y\ \exists x\ W(x,\ y)$.

b. $\forall x\ (A(x) \wedge B(x)) \equiv \forall x\ A(x) \wedge \forall x\ B(x)$.

c. $\exists x\ (A(x) \vee B(x)) \equiv \exists x\ A(x) \vee \exists x\ B(x)$.

d. $\exists x\ (A(x) \to B(x)) \equiv \forall x\ A(x) \to \exists x\ B(x)$.

Challenges

10. Give a formal proof of $A \to B$, where A and B are defined as follows:

$$A = \forall x\ (\exists y\ (q\ (x,y) \wedge s\ (y)) \to \exists y\ (p\ (y) \wedge r\ (x,y))),$$
$$B = \neg\ \exists x\ p\ (x) \to \forall x\ \forall y\ (q\ (x,y) \to \neg\ s\ (y)).$$

11. Give a formal proof of $A \to B$, where A and B are defined as follows:

$A = \exists x\ (r(x) \wedge \forall y\ (p(y) \to q(x,\ y))) \wedge \forall x\ (r(x) \to \forall y\ (s(y) \to \neg\ q(x,\ y)))$,

$B = \forall x\ (p(x) \to \neg\ s(x))$.

12. Each of the following wffs is *invalid*. Nevertheless, for each wff you are to construct a proof sequence that claims to be a proof of the wff but that fails because of the improper use of one or more proof rules. Also indicate which rules you use improperly and why the use is improper.

a. $\exists x\ A(x) \to \forall x\ A(x)$.

b. $\exists x\ A(x) \wedge \exists x\ B(x) \to \exists x\ (A(x) \wedge B(x))$.

c. $\forall x\ (A(x) \vee B(x)) \to \forall x\ A(x) \vee \forall x\ B(x)$.

d. $(\forall x\ A(x) \to \forall x\ B(x)) \to \forall x\ (A(x) \to B(x))$.

e. $\forall x\ \exists y\ W(x,\ y) \to \exists y\ \forall x\ W(x,\ y)$.

13. Assume that x does not occur free in the wff C. Use either CP or IP to give a formal proof for each of the following equivalences.

   a. $\forall x \ (C \wedge A(x)) \equiv C \wedge \forall x \ A(x)$.
   b. $\exists x \ (C \wedge A(x)) \equiv C \wedge \exists x \ A(x)$.
   c. $\forall x \ (C \vee A(x)) \equiv C \vee \forall x \ A(x)$.
   d. $\exists x \ (C \vee A(x)) \equiv C \vee \exists x \ A(x)$.
   e. $\forall x \ (C \rightarrow A(x)) \equiv C \rightarrow \forall x \ A(x)$.
   f. $\exists x \ (C \rightarrow A(x)) \equiv C \rightarrow \exists x \ A(x)$.
   g. $\forall x \ (A(x) \rightarrow C) \equiv \exists x \ A(x) \rightarrow C$.
   h. $\exists x \ (A(x) \rightarrow C) \equiv \forall x \ A(x) \rightarrow C$.

14. Prove that each of the following proof rules of propositional calculus can be used in predicate calculus.

   a. Modus tollens.
   b. Hypothetical syllogism.

15. Let $W(x)$ be a wff and t a term to be substituted for x in $W(x)$.

   a. Suppose t is free to replace x in $W(x)$ and there is a variable in t that is bound in $W(x)$. What can you say?
   b. Suppose no variable of t is bound in $W(x)$. What can you say?

16. If the term t is free to replace x in $W(x)$ and I is an interpretation, then $W(t)I = W(tI)I$. Prove the statement by induction on the number of connectives and quantifiers.

17. Any binary relation that is irreflexive and transitive is also asymmetric. Here is an informal proof. Let p be a binary relation on a set A such that p is irreflexive and transitive. Suppose, by way of contradiction, that p is not asymmetric. Then there are elements $a, b \in A$ such that $p(a, b)$ and $p(b, a)$. Since p is transitive, it follows that $p(a, a)$. But this contradicts the fact that p is irreflexive. Therefore, p is asymmetric. Give a formal proof of the statement, where the following wffs represent the three properties:

   Irreflexive: $\forall x \ \neg \ p(x, x)$.

   Transitive: $\forall x \ \forall y \ \forall z \ (p(x, y) \wedge p(y, z) \rightarrow p(x, z))$.

   Asymmetric: $\forall x \ \forall \ y \ (p(x, y) \rightarrow \neg \ p(y, x))$.

7.4 Chapter Summary

First-order predicate calculus extends propositional calculus by allowing wffs to contain predicates and quantifiers of variables. Meanings for these wffs are defined in terms of interpretations over nonempty sets called domains. A wff is valid if it's true for all possible interpretations. A wff is unsatisfiable if it's false for all possible interpretations.

There are basic equivalences that allow us to simplify and transform wffs into other wffs. We can use equivalences to transform any wff into a prenex DNF or prenex CNF. Equivalences can also be used to compare different formalizations of the same English sentence.

To decide whether a wff is valid, we can try to transform it into an equivalent wff that we know to be valid. But, in general, we must rely on some type of informal or formal reasoning. A formal reasoning system for first-order predicate calculus can use all the rules and proof techniques of propositional calculus. But we need four additional proof rules for the quantifiers: universal instantiation, existential instantiation, universal generalization, and existential generalization.

Notes

Now we have the basics of logic—propositional calculus and first-order predicate calculus. In Section 6.4 we introduced a formal axiom system for propositional calculus and we observed that the system is complete, which means that every tautology can be proven as a theorem within the system.

It's nice to know that there is a similar statement for predicate calculus, which is due to the logician and mathematician Kurt Gödel (1906–1978). Gödel showed that first-order predicate calculus is complete. In other words, there are formal systems for first-order predicate calculus such that every valid wff can be proven as a theorem. The formal system presented by Gödel [1930] used fewer axioms and fewer inference rules than the system that we've been using in this chapter.

chapter 8

Applied Logic

Once the people begin to reason,
all is lost.

— Voltaire (1694–1778)

When we reason, we usually do it in a particular domain of discourse. For example, we might reason about computer science, politics, mathematics, physics, automobiles, or cooking. But these domains are usually too large to do much reasoning. So we normally narrow our scope of thought and reason in domains such as imperative programming languages, international trade, plane geometry, optics, suspension systems, or pasta recipes.

No matter what the domain of discussion, we usually try to correctly apply inferences while we are reasoning. Since each of us has our own personal reasoning system, we sometimes find it difficult to understand one another. In an attempt to find common ground among the various ways that people reason, we introduced propositional calculus and first-order predicate calculus. So we've looked at some formalizations of logic.

Can we go a step further and formalize the things that we talk about? Many subjects can be formalized by giving some axioms that define the properties of the objects being discussed. For example, when we reason about geometry, we make assumptions about points and lines. When we reason about automobile engines, we make certain assumptions about how they work. When we combine first-order predicate calculus with the formalization of some subject, we obtain a reasoning system called a *first-order theory*.

chapter guide

8.3 (Higher-Order Logics) introduces logics that are beyond the first order. We'll give some examples to show how higher-order logics can be used to formalize much of our natural discourse.

8.1   Equality

Equality is a familiar notion to most of us. For example, we might compare two things to see whether they are equal, or we might replace a thing by an equal thing during some calculation. In fact, equality is so familiar that we might think that it does not need to be discussed further. But we are going to discuss it further because different domains of discourse often use equality in different ways. If we want to formalize some subject that uses the notion of equality, then it should be helpful to know basic properties that are common to all equalities.

A first-order theory is called a *first-order theory with equality* if it contains a two-argument predicate, say e, that captures the properties of equality required by the theory. We usually denote $e(x, y)$ by the familiar

$$x = y.$$

Similarly, we let $x \neq y$ denote $\neg\, e(x, y)$.

Let's examine how we use equality in our daily discourse. We always assume that any term is equal to itself. For example, $x = x$ and $f(c) = f(c)$. We might call this "syntactic equality."

Another familiar use of equality might be called "semantic equality." For example, although the expressions $2 + 3$ and $1 + 4$ are not syntactically equal, we still write $2 + 3 = 1 + 4$ because they both represent the same number.

Another important use of equality is to replace equals for equals in an expression. The following examples should get the point across.

$$\text{If } x + y = 2z, \text{ then } (x + y) + w = 2z + w.$$
$$\text{If } x = y, \text{ then } f(x) = f(y).$$
$$\text{If } f(x) = f(y), \text{ then } g(f(x)) = g(f(y)).$$
$$\text{If } x = y + z, \text{ then } (8 < x) \equiv (8 < y + z).$$
$$\text{If } x = y, \text{ then } p(x) \vee q(w) \equiv p(y) \vee q(w).$$

8.1.1   Describing Equality

Let's try to describe some fundamental properties that all first-order theories with equality should satisfy. Of course, we want equality to satisfy the basic property that each term is equal to itself. The following axiom will suffice for this purpose.

Equality Axiom (EA) (8.1)

$$\forall x \ (x = x).$$

This axiom tells us that $x = x$ for all variables x. The axiom is sometimes called the *law of identity*. But we also want to say that $t = t$ for any term t. For example, if a theory contains a term such as $f(x)$, we certainly want to say that $f(x) = f(x)$. Do we need another axiom to tell us that each term is equal to itself? No. All we need is a little proof sequence as follows:

1. $\forall x \ (x = x)$ EA
2. $t = t$ 1, UI.

So for any term t we have $t = t$. Because this is such a useful result, we'll also refer to it as EA. In other words, we have

---

Equality Axiom (EA) (8.2)

$$t = t \text{ for all terms } t.$$

Now let's try to describe that well-known piece of folklore, *equals can replace equals*. Since this idea has such a wide variety of uses, it's hard to tell where to begin. So we'll start with a rule that describes the process of replacing some occurrence of a term in a predicate by an equal term. In this rule, p denotes an arbitrary predicate with one or more arguments. The letters t and u represent arbitrary terms.

Equals-for-Equals (EE) (8.3)

$$(t = u) \wedge p(\ldots \ t \ \ldots) \rightarrow p(\ldots \ u \ \ldots).$$

The notations $\ldots \ t \ \ldots$ and $\ldots \ u \ \ldots$ indicate that t and u occur in the same argument place of p. In other words, u replaces the indicated occurrence of t. Since (8.3) is an implication, we can use it as a proof rule in the following equivalent form.

Equals-for-Equals (EE) (8.4)

$$\frac{t = u, \ p(\ldots t \ldots)}{p(\ldots u \ldots)}.$$

The EE rule is sometimes called the *principle of extensionality.* Let's see what we can conclude from EE. Whenever we discuss equality of terms, we usually want the following two properties to hold for all terms:

Symmetric: $(t = u) \rightarrow (u = t)$.

Transitive: $(t = u) \wedge (u = v) \rightarrow (t = v)$.

We'll use the EE rule to prove the symmetric property in the next example and leave the transitive property as an exercise.

example   8.1   A Proof of Symmetry

We'll prove the symmetric property $(t = u) \rightarrow (u = t)$.

1. $t = u$ P
2. $t = t$ EA
3. $u = t$ 1, 2, EE
   QED   1–3, CP.

To see why the statement on line 3 follows from the EE rule, we'll let $p(x, y)$ mean "$x = y$." Then the proof can be rewritten in terms of p as follows:

1. $t = u$ P
2. $p(t, t)$ EA
3. $p(u, t)$ 1, 2, EE
   QED   1–3, CP.

end example

Another thing we would like to conclude from EE is that equals can replace equals in a term like $f(\ldots \ t \ldots)$. In other words, we would like the following wff to be valid:

$$(t = u) \rightarrow f(\ldots \ t \ldots) = f(\ldots \ u \ldots).$$

To prove that this wff is valid, we'll let $p(t, u)$ mean "$f(\ldots \ t \ldots) = f(\ldots \ u \ldots)$." Then the proof goes as follows:

1. $t = u$ P
2. $p(t, t)$ EA
3. $p(t, u)$ 1, 2, EE
   QED   1–3, CP.

When we're dealing with axioms for a theory, we sometimes write down more axioms than we really need. For example, some axiom might be deducible as a theorem from the other axioms. The practical purpose for this is to have a listing of the useful properties all in one place. For example, to describe equality for terms, we might write down the following five statements as axioms.

Equality Axioms for Terms (8.5)

In these axioms the letters t, u, and v denote arbitrary terms, f is an arbitrary function, and p is an arbitrary predicate.

EA:	$t = t$.
Symmetric:	$(t = u) \to (u = t)$.
Transitive:	$(t = u) \wedge (u = v) \to (t = v)$.
EE (functional form):	$(t = u) \to f(\ldots t \ldots) = f(\ldots u \ldots)$.
EE (predicate form):	$(t = u) \wedge p(\ldots t \ldots) \to p(\ldots u \ldots)$.

The EE axioms in (8.5) allow only a single occurrence of t to be replaced by u. We may want to substitute more than one "equals for equals" at the same time. For example, if $x = a$ and $y = b$, we would like to say that $f(x, y) = f(a, b)$. It's nice to know that simultaneous use of equals for equals can be deduced from the axioms. For example, we'll prove the following statement:

$$(x = a) \wedge (y = b) \to f(x, y) = f(a, b).$$

1.	$x = a$	P
2.	$y = b$	P
3.	$f(x, y) = f(a, y)$	1, EE
4.	$f(a, y) = f(a, b)$	2, EE
5.	$f(x, y) = f(a, b)$	3, 4, Transitive
	QED	1–5, CP.

This proof can be extended to substitute any number of equals for equals simultaneously in a function or in a predicate. In other words, we could have written the two EE axioms of (8.5) in the following form.

Multiple Replacement EE (8.6)

In these axioms the letters t and u denote arbitrary terms, f is an arbitrary function, and p is an arbitrary predicate.

EE (function): $(t_1 = u_1) \wedge \cdots \wedge (t_k = u_k) \to f(t_1, \ldots, t_k) = f(u_1, \ldots, u_k)$.

EE (predicate): $(t_1 = u_1) \wedge \cdots \wedge (t_k = u_k) \wedge p(t_1, \ldots, t_k) \to p(u_1, \ldots, u_k)$.

So the two axioms (8.1) and (8.3) are sufficient for us to deduce all the axioms in (8.5) together with those of (8.6).

Working with Numeric Expressions

Let's consider the set of arithmetic expressions over the domain of integers, together with the usual arithmetic operations. The terms in this theory are arithmetic expressions, such as

$$35, \quad x, \quad 2 + 8, \quad x + y, \quad 6x - 5 + y.$$

Equality of terms comes into play when we write statements such as

$$3 + 6 = 2 + 7, \quad 4 \neq 2 + 3.$$

Axioms give us the algebraic properties of the operations. For example, we know that the $+$ operation is associative, and we know that $x + 0 = x$ and $x + -x = 0$. We can reason in such a theory by using predicate calculus with equality. In the next two examples we'll give an informal proof and a formal proof of the following well-known statement:

$$\forall x \, ((x + x = x) \rightarrow (x = 0)).$$

example 8.2 An Informal Proof

First we'll do an informal equational type proof. Let x be any number such that $x + x = x$. Then we have the following equations:

$x = x + 0$	Algebra
$ = x + (x + -x)$	Algebra
$ = (x + x) + -x$	Algebra
$ = x + -x$	Hypothesis
$ = 0$	Algebra

Since x was arbitrary, the statement is true for all x. QED.

end example

example 8.3 A Formal Proof

In the informal proof we used several instances of equals for equals. Now let's look at a formal proof in all its glory. We'll start the proof with an indented subproof of the statement $(x + x = x) \rightarrow (x = 0)$. Then we'll apply UG to obtain the desired result.

1.	$x + x = x$	P [for $(x + x = x) \rightarrow (x = 0)$]
2.	$-x = -x$	EA
3.	$(x + x) + -x = x + -x$	1, 2, EE
4.	$x + (x + -x) = (x + x) + -x$	Algebra
5.	$x + (x + -x) = x + -x$	3, 4, Transitivity

6. $\quad x + -x = 0$ $\qquad\qquad$ Algebra
7. $\quad x + 0 = 0$ $\qquad\qquad$ 5, 6, EE
8. $\quad x = x + 0$ $\qquad\qquad$ Algebra
9. $\quad x = 0$ $\qquad\qquad$ 7, 8, Transitivity
10. $(x + x = x) \to (x = 0)$ $\qquad$ 1–9, CP
11. $\forall x \, ((x + x = x) \to (x = 0))$ $\quad$ 10, UG
      QED.

Let's explain the two uses of EE. For line 3, let $f(u, v) = u + v$. Then the wff on line 3 results from lines 1 and 2 together with the following instance of EE in functional form:

$$(x + x = x) \to f(x + x, -x) = f(x, -x).$$

For line 7, let $p(u, v)$ denote the statement "$u + v = v$." Then the wff on line 7 results from lines 5 and 6 together with the following instance of EE in predicate form:

$$(x + -x = 0) \land p(x, x + -x) \to p(x, 0).$$

end example

Partial Order Theories

A *partial order theory* is a first-order theory with equality that also contains an ordering predicate that is antisymmetric and transitive. If the ordering predicate is reflexive, we denote it by $\leq$. If it is irreflexive, we denote it by $<$.

For example, the antisymmetric and transitive properties for $\leq$ can be written as follows, where x, y, and z are arbitrary elements.

Antisymmetric: $(x \leq y) \land (y \leq x) \to (x = y).$
  Transitive: $(x \leq y) \land (y \leq z) \to (x \leq z).$

We can use equality to define either one of the relations $<$ and $\leq$ in terms of the other in the following way.

$$x < y \quad \text{means} \quad (x \leq y) \land (x \neq y),$$
$$x \leq y \quad \text{means} \quad (x < y) \lor (x = y).$$

We can do formal reasoning in such a first-order theory in much the same way that we reason informally.

example 8.4 **An Obvious Statement**

Most of us use the following statement without even thinking:

$$(x < y) \to (x \leq y).$$

We'll give two different proofs of the statement. Here's the first proof:

1. $x < y$ $\qquad\qquad\quad$ P
2. $(x \le y) \land (x \ne y)$ $\quad$ 1, T [definition]
3. $x \le y$ $\qquad\qquad$ 2, Simp
    QED $\qquad\qquad\quad$ 1–3, CP.

Here's an alternative proof:

1. $x < y$ $\qquad\qquad\quad$ P
2. $(x < y) \lor (x = y)$ $\quad$ 1, Add
3. $x \le y$ $\qquad\qquad$ 2, T [definition]
    QED $\qquad\qquad\quad$ 1–3, CP.

end example

8.1.2 Extending Equals for Equals

The EE rule for replacing equals for equals in a predicate can be extended to other wffs. For example, we can use the EE rule to prove the following more general statement about wffs without quantifiers.

EE for Wffs with no Quantifiers $\qquad\qquad\qquad\qquad\qquad$ (8.7)

If $W(x)$ has no quantifiers, then the following wff is valid:

$$(t = u) \land W(t) \to W(u).$$

We assume that $W(t)$ is obtained from $W(x)$ by replacing one or more occurrences of x by t and that $W(u)$ is obtained from $W(t)$ by replacing one or more occurrences of t by u.

For example, if $W(x) = p(x, y) \land q(x, x)$, then we might have $W(t) = p(t, y) \land q(x, t)$, where only two of the three occurrences of x are replaced by t. In this case we might have $W(u) = p(u, y) \land q(x, t)$, where only one occurrence of t is replaced by u. In other words, the following wff is valid:

$$(t = u) \land p(t, y) \land q(x, t) \to p(u, y) \land q(x, t).$$

What about wffs that contain quantifiers? Even when a wff has quantifiers, we can use the EE rule if we are careful not to introduce new bound occurrences of variables. Here is the full-blown version of EE.

EE for Wffs with Quantifiers (8.8)

If $W(x)$ is a wff and t and u are terms that are free to replace x in $W(x)$, then the following wff is valid:

$$(t = u) \land W(t) \rightarrow W(u).$$

We assume that $W(t)$ is obtained from $W(x)$ by replacing one or more free occurrences of x by t and that $W(u)$ is obtained from $W(x)$ by replacing one or more free occurrences of x that were used to obtain $W(t)$ by u.

For example, suppose $W(x) = \exists y \, p(x, y)$. Then for any terms t and u that do not contain occurrences of y, the following wff is valid:

$$(t = u) \land \exists y \, p(t, y) \rightarrow \exists y \, p(u, y).$$

The exercises contain some samples to show how EE for predicates (8.3) can be used to prove some simple extensions of EE to more general wffs.

 Exercises

Equals for Equals

1. Use the EE rule to prove the double replacement rule:

$$(s = v) \land (t = w) \land p(s, t) \rightarrow p(v, w).$$

2. Show that the transitive property $(t = u) \land (u = v) \rightarrow (t = v)$ can be deduced from the other axioms for equality (8.5).

3. Give a formal proof of the following statement about the integers:

$$(c = a^i) \land (i \leq b) \land \neg (i < b) \rightarrow (c = a^b).$$

4. Use the equality axioms (8.5) to prove each of the following versions of EE, where p and q are predicates, t and u are terms, and x, y, and z are variables.

   a. $(t = u) \land \neg \, p(\ldots \, t \ldots) \rightarrow \neg \, p(\ldots \, u \ldots).$
   b. $(t = u) \land p(\ldots \, t \ldots) \land q(\ldots \, t \ldots) \rightarrow p(\ldots \, u \ldots) \land q(\ldots \, u \ldots).$
   c. $(t = u) \land (p(\ldots \, t \ldots) \lor q(\ldots \, t \ldots)) \rightarrow p(\ldots \, u \ldots) \lor q(\ldots \, u \ldots).$
   d. $(x = y) \land \exists z \, p(\ldots \, x \, \ldots) \rightarrow \exists z \, p(\ldots \, y \, \ldots).$
   e. $(x = y) \land \forall z \, p(\ldots \, x \, \ldots) \rightarrow \forall z \, p(\ldots \, y \, \ldots).$

5. Prove the validity of the wff $\forall x \, \exists y \, (x = y).$

6. Prove each of the following equivalences.

    a. $p(x) \equiv \exists y \, ((x = y) \wedge p(y))$.

    b. $p(x) \equiv \forall y \, ((x = y) \to p(y))$.

Formalizing English Sentences

7. Formalize the definition for each statement about the integers.

    a. $\mathrm{odd}(x)$ means x is odd.

    b. $\mathrm{even}(x)$ means x is even.

    c. $\mathrm{div}(a, b)$ means a divides b.

    d. $r = a \bmod b$.

    e. $d = \gcd(a, b)$.

8. Formalize each of the following statements.

    a. There is at most one x such that $A(x)$ is true.

    b. There are exactly two x and y such that $A(x)$ and $A(y)$ are true.

    c. There are at most two x and y such that $A(x)$ and $A(y)$ are true.

9. Students were asked to formalize the statement "There is a unique x such that $A(x)$ is true." The following wffs were given as answers.

$$\exists x \, (A(x) \wedge \forall y \, (A(y) \to (x = y))),$$
$$\exists x \, A(x) \wedge \forall x \, \forall y \, (A(x) \wedge A(y) \to (x = y)).$$

Prove that the wffs are equivalent by performing the following tasks.

    a. Prove that the first wff implies the second.

    b. Prove that the second wff implies the first.

Hint: You might want to include the use of indirect proof in your proofs.

8.2 Program Correctness

An important and difficult problem of computer science can be stated as

$$\text{"Prove that a program is correct."} \tag{8.9}$$

This takes some discussion. One major question to ask before we can prove that a program is correct is "What is the program supposed to do?" If we can state in English what a program is supposed to do, and English is the programming language, then the statement of the problem may itself be a proof of its correctness.

    Normally, a problem is stated in some language X, and its solution is given in some language Y. For example, the statement of the problem might use English mixed with some symbolic notation, while the solution might be in a programming language. How do we prove correctness in cases like this? Often the answer depends on the programming language. As an example, we'll look at a formal system for proving the correctness of imperative programs.

8.2.1 Imperative Program Correctness

An *imperative program* consists of a sequence of statements that represent commands. The most important statement is the assignment statement. Other statements are used for control, such as looping and taking alternate paths.

Suppose we want to prove that a program does some particular thing. We must represent the thing that we want to prove in terms of a precondition P, which states what is supposed to be true before the program starts, and a postcondition Q, which states what is supposed to be true after the program halts. If S denotes the program, then we will describe this informal situation with the following expression, which is called a *Hoare triple*:

$$\{P\}\ S\ \{Q\}.$$

The letters P and Q denote statements that describe properties of the variables that occur in S. P is called a *precondition* for S, and Q is called a *postcondition* for S. We assume that P and Q are wffs from a first-order theory with equality that depends on the program S. For example, if the program manipulates numbers, then the first-order theory must include the numerical operations and properties that are required to describe the problem at hand. If the program processes strings, then the first-order theory must include the string operations.

The wffs for the formal proof system are the Hoare triples. For example, suppose S is the single assignment statement $x := x + 1$. Then the following expression is a wff in our logic:

$$\{x > 4\}\ x := x + 1\ \{x > 5\}.$$

Now let's associate a truth value with each Hoare triple.

The Truth Value of $\{P\}\ S\ \{Q\}$

The truth value of $\{P\}\ S\ \{Q\}$ is the truth value of the following statement:

> If P is true before S is executed and the execution of S terminates, then Q is true after the execution of S.

If $\{P\}\ S\ \{Q\}$ is true, we say S is *correct* with respect to precondition P and postcondition Q. Strictly speaking, we should say that S is *partially correct* because the truth of Q is required only when S terminates. If we also know that S terminates, then we say S is *totally correct*. We'll discuss termination at the end of the section.

Sometimes it's easy to observe whether $\{P\}\ S\ \{Q\}$ is true. For example, from our knowledge of the assignment statement, most of us will agree that the following wff is true:

$$\{x > 4\}\ x := x + 1\ \{x > 5\}.$$

On the other hand, most of us will also agree that the following wff is false:

$$\{x > 4\}\ x := x + 1\ \{x > 6\}.$$

But we need some proof methods to verify our intuition. We'll start with an axiom and then give some proof rules.

The Assignment Axiom

The axioms depend on the types of assignments allowed by the assignment statement. The proof rules depend on the control structures of the language. So we had better agree on a language before we go any further in our discussion. To keep things simple, we'll assume that the assignment statement has the following form, where x is a variable and t is a term:

$$x := t.$$

So the only thing we can do is assign a value to a variable. This effectively restricts the language so that it cannot use other structures, such as arrays and records. In other words, we can't make assignments like $a[i] := t$ or $a.b := t$.

Since our assignment statement is restricted to the form $x := t$, we need only one axiom. It's called the *assignment axiom*, and we'll motivate the discovery of the axiom by an example. Suppose we're told that the following wff is correct:

$$\{P\}\ x := 4\ \{y > x\}.$$

In other words, if P is true before the execution of the assignment statement, then after its execution the statement $y > x$ is true. What should P be? From our knowledge of the assignment statement we might guess that P has the following definition:

$$P = (y > 4).$$

This is about the most general statement we can make. Notice that P can be obtained from the postcondition $y > x$ by replacing x by 4. The assignment axiom generalizes this idea in the following way.

Assignment Axiom (AA)   (8.10)

$$\{Q(x/t)\}\ x := t\ \{Q\},$$

where t is free to replace x in Q.

The notation $Q(x/t)$ denotes the wff obtained from Q by replacing all free occurrences of x by t. The axiom is often called the "backward" assignment axiom because the precondition is constructed from the postcondition.

Let's see how the assignment axiom works in a backward manner. When using AA, always start by writing down the form of (8.10) with an empty precondition as follows:

$$\{ \quad \} \, x := t \, \{Q\}.$$

Now the task is to construct the precondition by replacing all free occurrences of x in Q by t.

For example, suppose we know that $x < 5$ is the postcondition for the assignment statement $x := x + 1$. We start by writing down the following partially completed version of AA:

$$\{ \quad \} \, x := x + 1 \, \{x < 5\}.$$

Then we use AA to construct the precondition. In this case we replace the x by $x + 1$ in the postcondition $x < 5$. This gives us the precondition $x + 1 < 5$, and we can write down the completed instance of the assignment axiom:

$$\{x + 1 < 5\} \, x := x + 1 \, \{x < 5\}.$$

The Consequence Rule

It happens quite often that the precondition constructed by AA doesn't quite match what we're looking for. For example, most of us will agree that the following wff is correct.

$$\{x < 3\} \, x := x + 1 \, \{x < 5\}.$$

But we've already seen that AA applied to this assignment statement gives

$$\{x + 1 < 5\} \, x := x + 1 \, \{x < 5\}.$$

Since the two preconditions don't match, we have some more work to do. In this case we know that for any number x we have $(x < 3) \rightarrow (x + 1 < 5)$.

Let's see why this is enough to prove that $\{x < 3\} \, x := x + 1 \, \{x < 5\}$ is correct. If $x < 3$ before the execution of $x := x + 1$, then we also know that $x + 1 < 5$ before execution of $x := x + 1$. Now AA tells us that $x < 5$ is true after execution of $x := x + 1$. So $\{x < 3\} \, x := x + 1 \, \{x < 5\}$ is correct.

This kind of argument happens so often that we have a proof rule to describe the situation for any program S. It's called the *consequence rule*:

Consequence Rules (8.11)

$$\frac{P \rightarrow R \text{ and } \{R\} \, S \, \{Q\}}{\{P\} \, S \, \{Q\}} \quad \text{and} \quad \frac{\{P\} \, S \, \{T\} \text{ and } T \rightarrow Q}{\{P\} \, S \, \{Q\}}.$$

Notice that each consequence rule requires two proofs: a proof of correctness and a proof of an implication. Let's do an example.

 example   8.5 The Assignment Axiom and the Consequence Rule

We'll prove the correctness of the following wff:

$$\{x < 5\}\ x := x + 1\ \{x < 7\}.$$

To start things off, we'll apply (8.10) to the assignment statement and the post-condition to obtain the following wff:

$$\{x + 1 < 7\}\ x := x + 1\ \{x < 7\}.$$

This isn't what we want. We got the precondition $x + 1 < 7$, but we need the precondition $x < 5$. Let's see whether we can apply (8.11) to the problem. In other words, let's see whether we can prove the following statement:

$$(x < 5) \rightarrow (x + 1 < 7).$$

This statement is certainly true, and we'll include its proof in the following formal proof of correctness of the original wff.

1.	$\{x + 1 < 7\}\ x := x + 1\ \{x < 7\}$	AA
2.	$\quad\quad x < 5$	P [for $(x < 5) \rightarrow (x + 1 < 7)$]
3.	$\quad\quad x + 1 < 6$	2, T
4.	$\quad\quad 6 < 7$	T
5.	$\quad\quad x + 1 < 7$	3, 4, Transitive
6.	$(x < 5) \rightarrow (x + 1 < 7)$	2–5, CP
	QED	1, 6, Consequence.

 end example

Although assignment statements are the core of imperative programming, we can't do much programming without control structures. So let's look at a few fundamental control structures together with their corresponding proof rules.

The Composition Rule

The most basic control structure is the composition of two statements S_1 and S_2, which we denote by $S_1; S_2$. This means execute S_1 and then execute S_2. The *composition rule* can be used to prove the correctness of the composition of two statements.

Composition Rule   (8.12)

$$\frac{\{P\}\, S_1\, \{R\}\ \text{and}\ \{R\}\, S_2\, \{Q\}}{\{P\}\, S_1;\, S_2\, \{Q\}}.$$

The composition rule extends naturally to any number of program statements in a sequence. For example, suppose we prove that the following three wffs are correct.

$$\{P\}\ S_1\ \{R\},\quad \{R\}\ S_2\ \{T\},\quad \{T\}\ S_3\ \{Q\}.$$

Then we can infer that $\{P\}\ S_1;\ S_2;\ S_3\ \{Q\}$ is correct.

For (8.12) to work, we need an intermediate condition R to place between the two statements. Intermediate conditions often appear naturally during a proof, as the next example shows.

---

example **8.6 The Composition Rule**

We'll show the correctness of the following wff:

$$\{(x > 2) \wedge (y > 3)\}\ x := x + 1;\ y := y + x\ \{y > 6\}.$$

This wff matches the bottom of the composition proof rule (8.12). Since the program statements are assignments, we can use the AA rule to move backward from the postcondition to find an intermediate condition to place between the two assignments. Then we can use AA again to move backward from the intermediate condition. Here's the proof.

Proof: First we'll use AA to work backward from the postcondition through the second assignment statement:

1. $\{y + x > 6\}\ y := y + x\ \{y > 6\}$ AA

Now we can take the new precondition and use AA to work backward from it through the first assignment statement:

2. $\{y + x + 1 > 6\}\ x := x + 1\ \{y + x > 6\}$ AA

Now we can use the composition rule (8.12) together with lines 1 and 2 to obtain line 3 as follows:

3. $\{y + x + 1 > 6\}\ x := x + 1;\ y := y + x\ \{y > 6\}$ 1, 2, Comp

At this point the precondition on line 3 does not match the precondition for the wff that we are trying to prove correct. Let's try to apply the consequence rule (8.11) to the situation.

4.	$(x > 2) \land (y > 3)$	P
5.	$x > 2$	4, Simp
6.	$y > 3$	4, Simp
7.	$x + y > 2 + y$	5, T
8.	$2 + y > 2 + 3$	6, T
9.	$x + y > 2 + 3$	7, 8, Transitive
10.	$x + y + 1 > 6$	9, T
11.	$(x > 2) \land (y > 3) \to (x + y + 1 > 6)$	4–10, CP

Now we're in position to apply the consequence rule to lines 3 and 11:

12. $\{(x > 2) \land (y > 3)\}\ x := x + 1;\ y := y + x\ \{y > 6\}$

$\qquad\qquad\qquad\qquad\qquad\qquad\qquad\qquad$ 3, 11, Consequence.

$\qquad$ QED

end example

The If–Then Rule

The statement **if** C **then** S means that S is executed if C is true and S is bypassed if C is false. For statements of this form we have the following *if-then* proof rule.

If–Then Rule $\hfill$ (8.13)

$$\frac{\{P \land C\}\, S\, \{Q\} \text{ and } P \land \neg\, C \to Q}{\{P\}\, \textbf{if } C \textbf{ then } S\, \{Q\}}.$$

The wff $P \land \neg\, C \to Q$ is required in the premise of (8.13) because if C is false, then S does not execute. But we still need Q to be true after C has been determined to be false during the execution of the if-then statement. Let's do an example.

example **8.7   The If–Then Rule**

We'll show that the following wff is correct:

$$\{\text{True}\}\ \textbf{if } x < 0 \textbf{ then } x := -x\ \{x \geq 0\}.$$

Proof: Since the wff fits the pattern of (8.13), all we need to do is prove the following two statements:

1. $\{\text{True} \land (x < 0)\}\ x := -x\ \{x \geq 0\}$.
2. $\text{True} \land \neg\, (x < 0) \to (x \geq 0)$.

The proofs are easy. We'll combine them into one formal proof:

1.	$\{-x \geq 0\} \; x := -x \; \{x \geq 0\}$	AA
2.	$\text{True} \wedge (x < 0)$	P
3.	$x < 0$	2, Simp
4.	$-x > 0$	3, T
5.	$-x \geq 0$	4, Add
6.	$\text{True} \wedge (x < 0) \rightarrow (-x \geq 0)$	2–5, CP
7.	$\{\text{True} \wedge (x < 0)\} \; x := -x \; \{x \geq 0\}$	1, 6, Consequence
8.	$\text{True} \wedge \neg\,(x < 0)$	P
9.	$\neg\,(x < 0)$	8, Simp
10.	$x \geq 0$	9, T
11.	$\text{True} \wedge \neg\,(x < 0) \rightarrow (x \geq 0)$	8–10, CP
	QED	7, 11, If-then.

end example

The If–Then–Else Rule

The statement **if** C **then** S_1 **else** S_2 means that S_1 is executed if C is true and S_2 is executed if C is false. For statements of this form we have the following *if-then-else* proof rule.

If–Then–Else Rule (8.14)

$$\frac{\{P \wedge C\} \, S_1 \, \{Q\} \;\; \text{and} \;\; \{P \wedge \neg\, C\} \, S_2 \, \{Q\}}{\{P\} \; \textbf{if} \; C \; \textbf{then} \; S_1 \; \textbf{else} \; S_2 \, \{Q\}} \; .$$

---

example **8.8** The If-Then-Else Rule

Suppose we're given the following wff, where even(x) means that x is an even integer:

$$\{\text{True}\} \; \textbf{if} \; \text{even}(x) \; \textbf{then} \; y := x \; \textbf{else} \; y := x + 1 \; \{\text{even}(y)\}.$$

We'll give a formal proof that this wff is correct. The wff matches the bottom of rule (8.14). Therefore, the wff will be correct by (8.14) if we can show that the following two wffs are correct:

1. $\{\text{True} \wedge \text{even}(x)\} \; y := x \; \{\text{even}(y)\}.$
2. $\{\text{True} \wedge \text{odd}(x)\} \; y := x + 1 \; \{\text{even}(y)\}.$

To make the proof formal, we need to give formal descriptions of even(x) and odd(x). This is easy to do over the domain of integers.

$$\text{even}\,(x) = \exists k\,(x = 2k)\,,$$
$$\text{odd}\,(x) = \exists k\,(x = 2k + 1)\,.$$

To avoid clutter, we'll use even(x) and odd(x) in place of the formal expressions. If you want to see why a particular line holds, you might make the substitution for even or odd and then see whether the statement makes sense. We'll combine the two proofs into the following formal proof:

1.	$\{\text{even}(x)\}\ y := x\ \{\text{even}(y)\}$	AA
2.	$\text{True} \wedge \text{even}(x)$	P
3.	$\text{even}(x)$	2, Simp
4.	$\text{True} \wedge \text{even}(x) \rightarrow \text{even}(x)$	2, 3, CP
5.	$\{\text{True} \wedge \text{even}(x)\}\ y := x\ \{\text{even}(y)\}$	1, 4, Consequence
6.	$\{\text{even}(x + 1)\}\ y := x + 1\ \{\text{even}(y)\}$	AA
7.	$\text{True} \wedge \text{odd}(x)$	P
8.	$\text{odd}(x)$	7, Simp
9.	$\text{even}(x + 1)$	8, T
10.	$\text{True} \wedge \text{odd}(x) \rightarrow \text{even}(x + 1)$	7–9, CP
11.	$\{\text{True} \wedge \text{odd}(x)\}\ y := x + 1\ \{\text{even}(y)\}$	6, 10, Consequence
	QED	5, 11, If-then-else.

end example

The While Rule

The last proof rule that we will consider is the *while rule*. The statement **while** C **do** S means that S is executed if C is true, and if C is still true after S has executed, then the process is started over again. Since the body S may execute more than once, there must be a close connection between the precondition and postcondition for S. This can be seen by the appearance of P in all preconditions and postconditions of the rule.

While Rule	**(8.15)**

$$\frac{\{P \wedge C\}\,S\,\{P\}}{\{P\}\,\textbf{while}\ C\ \textbf{do}\ \ S\,\{P \wedge \neg\,C\}}.$$

The wff P is called a *loop invariant* because it must be true before and after each execution of the body S. Loop invariants can be tough to find in programs with no documentation. On the other hand, in writing a program, a loop invariant can be a helpful tool for specifying the actions of while loops.

example 8.9 The While Rule

We'll prove the correctness of the following wff, where $\text{int}(x)$ means x is an integer.

$$\{\text{int}(x) \wedge (x > 0)\}$$
$$\textbf{while } \text{even}(x) \textbf{ do } x := x/2$$
$$\{\text{int}(x) \wedge (x > 0) \wedge \text{ odd}(x)\}.$$

This wff matches the conclusion of the while rule, where

$$P = \text{int}(x) \wedge (x > 0), \ C = \text{even}(x), \text{ and } S = \text{“}x := x/2\text{”}.$$

So by the while rule the wff will be correct if we can prove the correctness of

$$\{P \wedge C\} \ x := x/2 \ \{P\}.$$

Here's a proof:

1. $\{\text{int}(x/2) \wedge (x/2 > 0)\} \ x := x/2 \ \{\text{int}(x) \wedge (x > 0)\}$ AA
2. $\quad \text{int}(x) \wedge (x > 0) \wedge \text{even}(x)$ $P[\text{for } P \wedge C \rightarrow \ \text{int}(x/2) \wedge (x/2 > 0)]$
3. $\quad x > 0$ 2, Simp
4. $\quad x/2 > 0$ 3, T
5. $\quad \text{int}(x) \wedge \text{even}(x)$ 2, Simp
6. $\quad \text{int}(x) \wedge \text{even}(x) \rightarrow \ \text{int}(x/2)$ T
7. $\quad \text{int}(x/2)$ 5, 6, MP
8. $\quad \text{int}(x/2) \wedge (x/2 > 0)$ 4, 7, Conj
9. $\text{int}(x) \wedge (x > 0) \wedge \text{even}(x) \rightarrow \ \text{int}(x/2) \wedge (x/2 > 0)$ 2–8, CP
10. $\{P \wedge C\} \ x := x/2 \ \{P\}.$ 1, 9, Consequence
    QED.

end example

Discovering a Loop Invariant

To further illustrate the idea of working with while loops, we'll work our way through an example that will force us to discover a loop invariant in order to prove the correctness of a wff. Suppose we want to prove the correctness of the following program to compute the power a^b of two natural numbers a and b, where $a > 0$ and $b \geq 0$:

$$\{(a > 0) \wedge (b \geq 0)\} \qquad\qquad (8.16)$$

$$i := 0;$$

$$p := 1;$$

while $i < b$ **do**

$$\qquad p := p \cdot a;$$

$$\qquad i := i + 1$$

od

$$\{p = a^b\}$$

The program consists of three statements. So we can represent the program and its precondition and postcondition in the following form:

$$\{(a > 0) \wedge (b \geq 0)\}\, S_1;\, S_2;\, S_3\, \{p = a^b\}.$$

In this form, S_1 and S_2 are the first two assignment statements, and S_3 represents the while statement. The composition rule (8.12) tells us that we can prove the wff is correct if we can find proofs of the following three statements for some wffs P and Q.

$$\{(a > 0) \wedge (b \geq 0)\}\, S_1\, \{Q\}\,,$$

$$\{Q\}\, S_2\, \{P\}\,,$$

$$\{P\}\, S_3\, \{p = a^b\}\,.$$

Where do P and Q come from? If we know P, then we can use AA to work backward through S_2 to find Q. But how do we find P? Since S_3 is a while statement, P should be a loop invariant. So we need to do a little work.

From (8.15) we know that a loop invariant P for the while statement S_3 must satisfy the following form:

$$\{P\}\ \text{\textbf{while}}\ i < b\ \text{\textbf{do}}\ p := p \cdot a;\ i := i + 1\ \text{\textbf{od}}\ \{P \wedge \neg\, (i < b)\}.$$

Let's try some possibilities for P. Suppose we set $P \wedge \neg\, (i < b)$ equivalent to the program's postcondition $p = a^b$ and try to solve for P. This won't work because $p = a^b$ does not contain the letter i. So we need to be more flexible in our thinking. Since we have the consequence rule, all we really need is an invariant P such that $P \wedge \neg\, (i < b)$ implies $p = a^b$.

After staring at the program, we might notice that the equation $p = a^i$ holds both before and after the execution of the two assignment statements in the body of the while statement. It's also easy to see that the inequality $i \leq b$ holds before and after the execution of the body. So let's try the following definition for P:

$$(p = a^i) \wedge (i \leq b).$$

This P has more promise. Notice that $P \wedge \neg (i < b)$ implies $i = b$, which gives us the desired postcondition $p = a^b$. Next, by working backward from P through the two assignment statements, we wind up with the statement

$$(1 = a^0) \wedge (0 \leq b).$$

This statement can certainly be derived from the precondition $(a > 0) \wedge (b \geq 0)$. So P does OK from the start of the program down to the beginning of the while loop. All that remains is to prove the following statement:

$$\{P\} \textbf{ while } i < b \textbf{ do } p := p \cdot a;\ i := i + 1 \textbf{ od } \{P \wedge \neg (i < b)\}.$$

By (8.15), all we need to prove is the following statement:

$$\{P \wedge (i < b)\}\ p := p \cdot a;\ i := i + 1\ \{P\}.$$

This can be done easily, working backward from P through the two assignment statements. We'll put everything together in the following example.

example **8.10 A Correctness Proof**

We'll prove the correctness of program (8.16) to compute the power a^b of two natural numbers a and b, where $a > 0$ and $b \geq 0$. We'll use the loop invariant $P = (p = a^i) \wedge (i \leq b)$ for the while statement. To keep things straight, we'll insert $\{P\}$ as the precondition for the while loop and $\{P \wedge \neg (i < b)\}$ as the postcondition for the while loop to obtain the following representation of the program.

$$\{(a > 0) \wedge (b \geq 0)\}$$
$$i := 0;$$
$$p := 1;$$
$$\{P\} = \{(p = a^i) \wedge (i \leq b)\}$$
$$\textbf{while } i < b \textbf{ do}$$
$$\qquad p := p \cdot a;$$
$$\qquad i := i + 1$$
$$\textbf{od}$$
$$\{P \wedge \neg C\} = \{(p = a^i) \wedge (i \leq b) \wedge \neg (i < b)\}$$
$$\{p = a^b\}$$

We'll start by proving that $P \wedge \neg C \rightarrow (p = a^b)$.

1. $(p = a^i) \wedge (i \leq b) \wedge \neg (i < b)$ P
2. $p = a^i$ 1, Simp
3. $(i \leq b) \wedge \neg (i < b)$ 1, Simp
4. $i = b$ 3, T
5. $p = a^b$ 2, 4, EE
6. $(p = a^i) \wedge (i \leq b) \wedge \neg (i < b) \rightarrow (p = a^b)$ 1–5, CP

Next, we'll prove the correctness of $\{P\}$ **while** $i < b$ **do** S $\{P \wedge \neg (i < b)\}$. The while proof rule tells us to prove the correctness of $\{P \wedge (i < b)\}$ S $\{P\}$.

7. $\{(p = a^{i+1}) \wedge (i + 1 \leq b)\}$
        $i := i + 1 \{(p = a^i) \wedge (i \leq b)\}$ AA
8. $\{(p \cdot a = a^{i+1}) \wedge (i + 1 \leq b)\}$
        $p := p \cdot a \{(p = a^{i+1}) \wedge (i + 1 \leq b)\}$ AA
9. $(p = a^i) \wedge (i \leq b) \wedge (i < b)$ P
10. $p = a^i$ 9, Simp
11. $i < b$ 9, Simp
12. $i + 1 \leq b$ 11, T [for integers]
13. $a = a$ EA
14. $p \cdot a = a^{i+1}$ 10, 13, EE
15. $(p \cdot a = a^{i+1}) \wedge (i + 1 \leq b)$ 12, 14, Conj
16. $P \wedge (i < b) \rightarrow (p \cdot a = a^{i+1}) \wedge (i + 1 \leq b)$ 9–15, CP
17. $\{P \wedge (i < b)\}$ $p := p \cdot a;$ $i := i + 1$ $\{P\}$ 7, 8, 16, Comp, Conseq
18. $\{P\}$ **while** $i < b$ **do** $p := p \cdot a;$
        $i := i + 1$ **od** $\{P \wedge \neg (i < b)$ 17, While

Now let's work on the two assignment statements that begin the program. So we'll prove the correctness of $\{(a > 0) \wedge (b \geq 0)\}$ $i := 0;$ $p := 1$ $\{P\}$.

19. $\{(1 = a^i) \wedge (i \leq b)\}$ $p := 1$ $\{(p = a^i) \wedge (i \leq b)\}$ AA
20. $\{(1 = a^0) \wedge (0 \leq b)\}$ $i := 0$ $\{(1 = a^i) \wedge (i \leq b)\}$ AA
21. $(a > 0) \wedge (b \geq 0)$ P
22. $a > 0$ 21, Simp
23. $b \geq 0$ 21, Simp
24. $1 = a^0$ 22, T
25. $(1 = a^0) \wedge (0 \leq b)$ 23, 24, Conj
26. $(a > 0) \wedge (b \geq 0) \rightarrow (1 = a^0) \wedge (0 \leq b)$ 21–25, CP
27. $\{(a > 0) \wedge (b \geq 0)\}$ $i := 0;$ $p := 1$ $\{P\}$ 19, 20, 26, Comp,
                                                           Conseq

The proof is finished by using the Composition and Consequence rules:

QED   27, 18, 6, Comp,

Conseq.

8.2.2 Array Assignment

Since arrays are fundamental structures in imperative languages, we'll modify our proof system so that we can handle assignment statements like $a[i] := t$. In other words, we want to be able to construct a precondition for the following partial wff:

$$\{\ \}\ a[i] := t\ \{Q\}.$$

What do we do? We might try to work backward, as with AA, and replace all occurrences of $a[i]$ in Q by t. Let's try it and see what happens. Let $Q(a[i]/t)$ denote the wff obtained from Q by replacing all occurrences of $a[i]$ by t. We'll call the following statement the "attempted" array assignment axiom:

$$\text{Attempted AAA:}\ \{Q\,(a\,[i]\,/t)\}\,a\,[i] := t\,\{Q\}. \tag{8.17}$$

Since we're calling (8.17) the Attempted AAA, let's see whether we can find something wrong with it. For example, suppose we have the following wff, where the letter i is a variable:

$$\{\text{true}\}\ a[i] := 4\ \{a[i] = 4\}.$$

This wff is clearly correct, and we can prove it with (8.17).

1. $\{4 = 4\}\ a[i] := 4\ \{a[i] = 4\}$ Attempted AAA
2. $\text{true} \rightarrow (4 = 4)$ T
     QED   1, 2, Consequence.

At this point, things seem OK. But let's try another example. Suppose we have the following wff, where i and j are variables:

$$\{(i = j) \wedge (a[i] = 3)\}\ a[i] := 4\ \{a[j] = 4\}.$$

This wff is also clearly correct because $a[i]$ and $a[j]$ both represent the same indexed array variable. Let's try to prove that the wff is correct by using (8.17). The first line of the proof looks like

1. $\{a[j] = 4\}\ a[i] := 4\ \{a[j] = 4\}$ Attempted AAA

Since the precondition on line 1 is not the precondition of the wff, we need to use the consequence rule, which states that we must prove the following wff:

$$(i = j) \wedge (a[i] = 3) \rightarrow (a[j] = 4).$$

But this wff is invalid because a single array element can't have two distinct values.

So we now have an example of an array assignment statement that we "know" is correct, but we don't have the proper tools to prove that it is correct. What went wrong? Well, since the expression $a[i]$ does not appear in the postcondition $\{a[j] = 4\}$, the attempted AAA (8.17) just gives us back the postcondition as the precondition. This stops us in our tracks because we are now forced to prove an invalid conditional wff.

The problem is that (8.17) does not address the possibility that i and j might be equal. So we need a more sophisticated assignment axiom for arrays. Let's start again and try to incorporate the preceding remarks. We want an axiom to fill in the precondition of the following partial wff:

$$\{\ \}\ a[i] := t\ \{Q\}.$$

Of course, we need to replace all occurrences of $a[i]$ in Q by t. But we also need to replace all occurrences of $a[j]$ in Q, where j is any arithmetic expression, by an expression that allows the possibility that $j = i$. We can do this by replacing each occurrence of $a[j]$ in Q by the following if-then-else statement:

$$\text{"if } j = i \text{ then } t \text{ else } a[j].\text{"}$$

For example, if the equation $a[j] = s$ occurs in Q, then the precondition will contain the following equation:

$$(\text{if } j = i \text{ then } t \text{ else } a[j]) = s.$$

When an equation contains an if-then-else statement, we can write it without if-then-else as a conjunction of two wffs. For example, the following two statements are equivalent for terms s, t, and u:

$$(\text{if } C \text{ then } t \text{ else } u) = s,$$
$$(C \rightarrow (t = s)) \wedge (\neg\, C \rightarrow (u = s)).$$

For example, the following two statements are equivalent:

$$(\text{if } j = i \text{ then } t \text{ else } a\,[j]\,) = s,$$
$$((j = i) \rightarrow (t = s)) \wedge ((j \neq i) \rightarrow (a\,[j] = s)).$$

So when we use the if-then-else form in a wff, we are still within the bounds of a first-order theory with equality. Now let's put things together and state the correct axiom for array assignment.

Array Assignment Axiom (AAA)   (8.18)

$$\{P\}\ a[i] := t\ \{Q\},$$

where P is constructed from Q by the following rules:

1. Replace all occurrences of $a[i]$ in Q by t.
2. Replace all occurrences of $a[j]$ in Q by "if $j = i$ then t else $a[j]$".

Note: i and j may be any arithmetic expressions that do not contain a.

It is very important that the index expressions i and j don't contain the array name. For example, $a[a[k]]$ is *not* OK, but $a[k + 1]$ is OK. To see why we can't use arrays within arrays when applying AAA, consider the following wff:

$$\{(a[1] = 2) \wedge (a[2] = 2)\}\ a[a[2]] := 1\ \{a[a[2]] = 1\}.$$

This wff is false because the assignment statement sets $a[2] = 1$, which makes the postcondition into the equation $a[1] = 1$, contradicting the fact that $a[1] = 2$. But we can use AAA to improperly "prove" that the wff is correct, as the following sequence shows.

1. $\{1 = 1\}\ a[a[2]] := 1\ \{a[a[2]] = 1\}$ AAA attempt with $a[a[\ldots]]$
2. $(a[1] = 2) \wedge (a[2] = 2) \rightarrow (1 = 1)$ T
     Not QED   1, 2, Conseq.

The exclusion of arrays within arrays is not a real handicap because an assignment statement like $a[a[i]] := t$ can be rewritten as the following sequence of two assignment statements:

$$j := a[i];\ a[j] := t.$$

Similarly, a logical statement like $a[a[i]] = t$ appearing in a precondition or postcondition can be rewritten as

$$\exists x\ ((x = a[i]) \wedge (a[x] = t)).$$

Now let's see whether we can use (8.18) to prove the correctness of the wff that we could not prove before.

example **8.11 Using the Array Assignment Axiom**

We want to prove the correctness of the following wff:

$$\{(i = j) \wedge (a[i] = 3)\}\ a[i] := 4\ \{a[j] = 4\}.$$

This wff represents a simple reassignment of an array element, where the index of the array element is represented by two variable names. We'll include all the details of the consequence part of the proof, which uses the conjunction form of an if-then-else equation.

1.	$\{(\text{if } j = i \text{ then } 4 \text{ else } a[j]) = 4\}\ a[i] := 4\ \{a[j] = 4\}$	AAA
2.	$(i = j) \wedge (a[i] = 3)$	P
3.	$i = j$	2, Simp
4.	$j = i$	3, Symmetry
5.	$4 = 4$	EA
6.	$(j = i) \rightarrow (4 = 4)$	5, T [trivial]
7.	$(j \neq i) \rightarrow (a[j] = 4)$	4, T [vacuous]
8.	$((j = i) \rightarrow (4 = 4)) \wedge ((j \neq i) \rightarrow (a[j] = 4))$	6, 7, Conj
9.	$(\text{if } j = i \text{ then } 4 \text{ else } a[j]) = 4$	8, T
10.	$(i = j) \wedge (a[i] = 3) \rightarrow ((\text{if } j = i \text{ then } 4 \text{ else } a[j]) = 4)$	2–9, CP
	QED	1, 10, Conseq.

end example

8.2.3 Termination

Program correctness as we have been discussing it does not consider whether loops terminate. In other words, the correctness of the wff $\{P\}\ S\ \{Q\}$ includes the assumption that S halts. That's why this kind of correctness is called *partial correctness*. For *total correctness* we can't assume that loops terminate. We must prove that they terminate.

Introductory Example

For example, suppose we're presented with the following while loop, and the only information we know is that the variables take integer values:

$$\textbf{while } x \neq y \textbf{ do} \qquad (8.19)$$
$$x := x - 1;$$
$$y := y + 1;$$
$$\textbf{od}$$

We don't have enough information to be able to tell for certain whether the loop terminates. For example, if we initialize $x = 4$ and $y = 5$, then the loop will run forever. In fact, the loop will run forever whenever $x < y$. If we initialize $x = 6$ and $y = 3$, the loop will also run forever. After a little study and thought, we can see that the loop will terminate if initially we have $x \geq y$ and $x - y$ is an even number.

This example shows that the precondition (i.e., the loop invariant) must contain enough information to decide whether the loop terminates. We're going to discuss a general method for proving termination of a loop. But first we need to discuss a few preliminary ideas.

The State of a Computation

The *state* of a computation at some point is a tuple that represents the values of the variables at that point in the computation. For example, the tuple (x, y) denotes an arbitrary state of program (8.19). For our purposes the only time a state will change is when an assignment statement is executed.

For example, let the initial state of a computation for (8.19) be $(10, 6)$. For this state the loop condition is true because $10 \neq 6$. After the execution of the first assignment statement, the state becomes $(9, 6)$. Then after the execution of the second assignment statement, the state becomes $(9, 7)$. So the state changes from $(10, 6)$ to $(9, 7)$ after one iteration of the loop. For this state the loop condition is true because $9 \neq 7$. So a second iteration of the loop can begin. We can see that the state changes from $(9, 7)$ to $(8, 8)$ after the second iteration of the loop. For this state the loop condition is $8 \neq 8$, which is false, so the loop terminates.

The Termination Theorem

Program (8.19) terminates for the initial state $(10, 6)$ because with each iteration of the loop the value $x - y$ gets smaller, eventually equaling zero. In other words, $x - y$ takes on the sequence of values 4, 2, 0. This is the key point in showing loop termination. There must be some decreasing sequence of numbers that stops at some point. In more general terms, the numbers must form a decreasing sequence in some well-founded set. For program (8.19) the well-founded set is the set $\mathbb{N}$ of natural numbers.

To show loop termination, we need to find a well-founded set $\langle W, \prec \rangle$ together with a way to associate the state of the ith iteration of the loop with an element $x_i \in W$ such that the elements form a decreasing sequence

$$x_1 \succ x_2 \succ x_3 \cdots .$$

Since W is well-founded, the sequence must stop. Thus the loop must halt.

Let's put things together and describe the general process to prove termination of a program **while** C **do** S with respect to a loop invariant P. We'll assume that we already know, or we have already proven, that the body S terminates. This reflects the normal process of working from the inside out when doing termination proofs.

Termination Theorem   **(8.20)**

The program **while** C **do** S terminates with respect to the loop invariant P if the following conditions are met, where s is the program state before the execution of S and t is the program state after the execution of S.

1. Find a well-founded set $\langle W, \prec \rangle$.

2. Find an expression f in terms of the program variables.

3. Prove that if P and C are true for state s, then

$$f(s), f(t) \in W \text{ and } f(s) \succ f(t).$$

Proof: Notice that (8.20) requires that $f(s)$, $f(t) \in W$. The reason for this is that f is an expression that may very well not even be defined for certain states or it may be defined but not a member of W. So we must check that $f(s)$ and $f(t)$ are defined and members of W are in good standing. Then the statement $f(s) \succ f(t)$ will ensure that the loop terminates. So assume we have met all the conditions of (8.20). Let s_i represent the state prior to the ith execution of S. Then s_{i+1} represents the state after the ith execution of S. Therefore, we have the following decreasing sequence of elements in W:

$$f(s_1) \succ f(s_2) \succ f(s_3) \succ \cdots .$$

Since W is a well-founded set, this sequence must stop because all descending chains must be finite. Therefore the loop terminates. QED.

example 8.12 A Termination Proof

Let's show that the following program terminates with respect to the loop invariant $P = (x \geq y) \wedge \text{even}(x - y)$ where all variables take integer values:

$$\begin{aligned}
&\textbf{while } x \neq y \textbf{ do}\\
&\qquad x := x - 1;\\
&\qquad y := y + 1;\\
&\textbf{od}
\end{aligned}$$

We'll leave the correctness proof with respect to P as an exercise. For a well-founded set we'll choose $\mathbb{N}$ with the usual ordering, and for the program variables (x, y) we'll define

$$f(x, y) = x - y.$$

If $s = (x, y)$ is the state before the execution of the loop's body and t is the state after execution of the loop's body, then

$$t = (x - 1, y + 1).$$

So the expressions $f(s)$ and $f(t)$ become

$$f(s) = f(x, y) = x - y,$$
$$f(t) = f(x - 1, y + 1) = (x - 1) - (y + 1) = x - y - 2.$$

To prove that the program terminates with respect to P, we must prove the following statement.

If P and C are true for state s, then $f(s), f(t) \in \mathbb{N}$ and $f(s) > f(t)$.

So assume that P and C are true for state s. This means that the following two statements are true.

$$(x \geq y) \wedge \text{even}(x - y) \quad \text{and} \quad (x \neq y).$$

Since $x \geq y$ and $x \neq y$, it follows that $x > y$. So we have $x - y > 0$. Therefore, $f(s) \in \mathbb{N}$. Since $x - y$ is even and positive, it follows that $x - y - 2 \geq 0$. So $f(t) \in \mathbb{N}$. Finally, since $x - y > x - y - 2$, it follows that $f(s) > f(t)$. Therefore, the program terminates with respect to P.

end example

example 8.13 A Termination Proof

Let's look at a popular example of termination that needs a well-founded set other than the natural numbers. Suppose we have the following while loop where integer(x) means x is an integer and random( ) is a random number generator that returns a natural number.

$$\{\text{integer}(x)\}$$
while $x \neq 0$ **do**
       if $x < 0$ **then** $x :=$ random( ) **else** $x := x - 1$ **fi**
od
$$\{\text{integer}(x) \wedge x = 0\}$$

After some study it becomes clear that the program terminates because if x is initially a negative integer, then it is assigned a random natural number. So after at most one iteration of the loop, x is a natural number. Subsequent iterations decrement x to zero, which terminates the loop.

To prove termination from a formal point of view we need to find a well-founded set W and an expression f such that $f(s) \succ f(t)$ where s represents x at the beginning of the loop body ($s = x$) and t represents x at the end of the loop body (either t is a random natural number or $t = x - 1$). Since we don't know in advance whether x is negative, we don't know how many times the loop will execute because it depends on the random natural number that is generated. So we can't define $W = \mathbb{N}$ and $f(x) = x$ because $f(x)$ may not be in $\mathbb{N}$.

But we can get the job done with the well-founded set $W = \mathbb{N} \times \mathbb{N}$ with the lexicographic ordering. Then we can define

$$f(x) = \text{if } x < 0 \text{ then } (-x, 0) \text{ else } (0, x).$$

Notice, for example, that

$$f(0) \prec f(1) \prec f(2) \prec \cdots \prec f(-1) \prec f(-2) \prec f(-3) \prec \cdots.$$

Thus we have $f(s), f(t) \in W$ and $f(s) \succ f(t)$. Therefore, (8.20) tells us that the loop terminates.

end example

As a final remark to this short discussion, we should remember the fundamental requirement that programs with loops need loop invariants that contain enough restrictions to ensure that the loops terminate.

Note

Hopefully, this introduction has given you the flavor of proving properties of programs. There are many mechanical aspects to the process. For example, the backward application of the AA and AAA rules is a simple substitution problem that can be automated. We've omitted many important results. For example, if the programming language has other control structures, such as for-loops and repeat-loops, then new proof rules must be constructed. The original papers in these areas are by Hoare [1969] and Floyd [1967]. A good place to start reading more about this subject is the survey paper by Apt [1981].

Different languages usually require different proof systems to handle the program correctness problem. For example, declarative languages, in which programs can consist of recursive definitions, require methods of inductive proof for proving program correctness.

 Exercises

Assignment Statements

1. Prove that the following wff is correct over the domain of integers:

$$\{\text{True} \wedge \text{even}(x)\} \ y := x + 1 \ \{\text{odd}(y)\}.$$

2. Prove that each of the following wffs is correct. Assume that the domain is the set of integers.

   a. $\{(a > 0) \wedge (b > 0)\} \ x := a; \ y := b \ \{x + y > 0\}.$

   b. $\{a > b\} \ x := -a; \ y := -b \ \{x < y\}.$

3. Both of the following wffs claim to correctly perform the swapping process. The first one uses a temporary variable. The second does not. Prove that each wff is correct. Assume that the domain is the real numbers.

   a. $\{x < y\}$ temp $:= x; x := y; y :=$ temp $\{y < x\}$.

   b. $\{x < y\}$ $y := y + x; x := y - x; y := y - x$ $\{y < x\}$.

If-Then and If-Then-Else Statements

4. Prove that each of the following wffs is correct. Assume that the domain is the set of integers.

   a. $\{x < 10\}$ **if** $x \geq 5$ **then** $x := 4$ $\{x < 5\}$.
   b. $\{\text{True}\}$ **if** $x \neq y$ **then** $x := y$ $\{x = y\}$.
   c. $\{\text{True}\}$ **if** $x < y$ **then** $x := y$ $\{x \geq y\}$.
   d. $\{\text{True}\}$ **if** $x > y$ **then** $x := y + 1; y := x + 1$ **fi** $\{x \leq y\}$.

5. Prove that each of the following wffs is correct. Assume that the domain is the set of integers.

   a. $\{\text{True}\}$ **if** $x < y$ **then** max $:= y$ **else** max $:= x$ $\{(\max \geq x) \land (\max \geq y)\}$.
   b. $\{\text{True}\}$ **if** $x < y$ **then** $y := y - 1$ **else** $x := -x; y := -y$ **fi** $\{x \leq y\}$.

6. Show that each of the following wffs is *not* correct over the domain of integers.

   a. $\{x < 5\}$ **if** $x \geq 2$ **then** $x := 5$ $\{x = 5\}$.
   b. $\{\text{True}\}$ **if** $x < y$ **then** $y := y - x$ $\{y > 0\}$.

While Statements

7. Prove that the following wff is correct, where x and y are integers.

$$\{(x \geq y) \land \text{ even } (x - y)\}$$
$$\textbf{while } x \neq y \textbf{ do}$$
$$x := x - 1;$$
$$y := y + 1;$$
$$\textbf{od}$$
$$\{(x \geq y) \land \text{ even } (x - y) \land (x = y)\}.$$

8. Prove that each of the following wffs is correct.

    a. The program computes the floor of a nonnegative real number x. *Hint:* Let the loop invariant be $(i \leq x)$.

$$\{x \geq 0\}$$
$$i := 0;$$
$$\textbf{while } i \leq x - 1 \textbf{ do } i := i + 1 \textbf{ od}$$
$$\{i = \text{floor}(x)\}.$$

    b. The program computes the floor of a negative real number x. *Hint:* Let the loop invariant be $(x < i + 1)$.

$$\{x < 0\}$$
$$i := -1;$$
$$\textbf{while } x < i \textbf{ do } i := i - 1 \textbf{ od}$$
$$\{i = \text{floor}(x)\}.$$

    c. The program computes the floor of an arbitrary real number x, where the statements S_1 and S_2 are the two programs from Parts (a) and (b).

$$\{\text{True}\} \textbf{ if } x \geq 0 \textbf{ then } S_1 \textbf{ else } S_2 \; \{i = \text{floor}(x)\}.$$

9. Given a natural number n, the following program computes the sum of the first n natural numbers. Prove that the wff is correct. *Hint:* Let the loop invariant be $(s = i(i + 1)/2) \wedge (i \leq n)$.

$$\{n \geq 0\}$$
$$i := 0;$$
$$s := 0;$$
$$\textbf{while } i < n \textbf{ do}$$
$$\quad i := i + 1;$$
$$\quad s := s + i$$
$$\textbf{od}$$
$$\{s = n(n+1)/2\}.$$

10. The following program implements the division algorithm for natural numbers. It computes the quotient and the remainder of the division of a natural number by a positive natural number. Prove that the wff is correct. *Hint:* Let the loop invariant be $(a = yb + x) \land (0 \le x)$.

$$\{(a \ge 0) \land (b > 0)\}$$
$$x := a;$$
$$y := 0;$$
$$\textbf{while } b \le x \textbf{ do}$$
$$\qquad x := x - b;$$
$$\qquad y := y + 1$$
$$\textbf{od};$$
$$r := x;$$
$$q := y$$
$$\{(a = qb + r) \land (0 \le r < b)\}.$$

11. (*Greatest Common Divisor*). The following program claims to find the greatest common divisor gcd(a, b) of two positive integers a and b. Prove that the wff is correct.

$$\{(a > 0) \land (b > 0)\}$$
$$x := a;$$
$$y := b;$$
$$\textbf{while } x \ne y \textbf{ do}$$
$$\qquad \textbf{if } x > y \textbf{ then } x := x - y \textbf{ else } y := y - x$$
$$\textbf{od};$$
$$\text{great} := x$$
$$\{\text{gcd}(a, b) = \text{great}\}.$$

Hints: Use gcd$(a, b) =$ gcd(x, y) as the loop invariant. You may use the following useful fact derived from (2.2) for any integers w and z: gcd$(w, z) =$ gcd$(w - z, z)$.

12. Write a program to compute the ceiling of an arbitrary real number. Give the program a precondition and a postcondition, and prove that the resulting wff is correct. *Hint:* Look at Exercise 8.

Array Assignment

13. For each of the following partial wffs, fill in the precondition that results by applying the array assignment axiom (8.18).

a. $\{ \ \} \ a[i - 1] := 24 \ \{a[j] = 24\}.$
b. $\{ \ \} \ a[i] := 16 \ \{(a[i] = 16) \land (a[j + 1] = 33)\}.$
c. $\{ \ \} \ a[i + 1] := 25; \ a[j - 1] := 12 \ \{(a[i] = 12) \land (a[j] = 25)\}.$

14. Prove that each of the following wffs is correct.

   a. $\{(i = j + 1) \wedge (a[j] = 39)\}\ a[i - 1] := 24\ \{a[j] = 24\}$.
   b. $\{\text{even}(a[i]) \wedge (i = j + 1)\}\ a[j] := a[i] + 1\ \{\text{odd}(a[i - 1])\}$.
   c. $\{(i = j - 1) \wedge (a[i] = 25) \wedge (a[j] = 12)\}$
      $a[i + 1] := 25;\ a[j - 1] := 12$
      $\{(a[i] = 12) \wedge (a[j] = 25)\}$.

15. The following wffs are not correct. For each wff, apply the array assignment axiom to the postcondition and assignment statements to obtain a condition Q. Show that the precondition does not imply Q.

   a. $\{\text{even}(a[i])\}\ a[i + 1] := a[i] + 1\ \{\text{even}(a[i + 1])\}$.
   b. $\{a[2] = 2\}\ i := a[2];\ a[i] := 1\ \{(a[i] = 1) \wedge (i = a[2])\}$.
   c. $\{\forall j\ ((1 \le j \le 5) \rightarrow (a[j] = 23))\}\ i := 3;\ a[i] := 355$
      $\{\forall j\ ((1 \le j \le 5) \rightarrow (a[j] = 23))\}$.
   d. $\{(a[1] = 2) \wedge (a[2] = 2)\}\ a[a[2]] := 1\ \{\exists x\ ((x = a[2]) \wedge (a[x] = 1))\}$.

Termination

16. Prove that each of the following loops terminates with respect to the given loop invariant P, where $\text{int}(x)$ means x is an integer and $\text{even}(x)$ means x is even. *Hint:* In each case use the well-founded set $\mathbb{N}$ with the usual ordering.

   a. **while** $i < x$ **do** $i := i + 1$ **od** with $P = \text{int}(i) \wedge \text{int}(x) \wedge (i \le x)$.
   b. **while** $i < x$ **do** $x := x - 1$ **od** with $P = \text{int}(i) \wedge \text{int}(x) \wedge (i \le x)$.
   c. **while** $\text{even}(x) \wedge (x \ne 0)$ **do** $x := x/2$ **od** with $P = \text{int}(x)$.

17. Given the following while loop:

   while $x \ne y$ **do if** $x < y$ **then** $y := y - x$ **else** $x := x - y$ **od**.

   Let $P = \text{pos}(x) \wedge \text{pos}(y)$ be the loop invariant, where $\text{pos}(z)$ means z is a positive integer. Prove that the loop terminates for each of the following choices of f and well-founded set W.

   a. $f(x, y) = x + y$ and $W = \mathbb{N}$.
   b. $f(x, y) = \max(x, y)$ and $W = \mathbb{N}$.
   c. $f(x, y) = (x, y)$ and $W = \mathbb{N} \times \mathbb{N}$ with the lexicographic ordering.

18. Exercise 17 demonstrates that the following loop terminates for the loop invariant $P = \text{pos}(x) \wedge \text{pos}(y)$, where $\text{pos}(z)$ means z is a positive integer.

   while $x \ne y$ **do if** $x < y$ **then** $y := y - x$ **else** $x := x - y$ **od**.

Show that if we let $W = \mathbb{N}$ with the usual ordering, then each of the following choices for f cannot be used to prove termination of the loop.

a. $f(x, y) = |x - y|$.

b. $f(x, y) = \min(x, y)$.

Challenges

19. Prove the total correctness of the following program to compute $a \bmod b$, where a and b are natural numbers with $b > 0$. *Hint:* Let the loop invariant be $\exists x \ (a = xb + r) \land (0 \leq r)$.

$$\{(a \geq 0) \land (b > 0)\}$$
$$r := a;$$
$$\textbf{while } r \geq b \textbf{ do } r := r - b \textbf{ od}$$
$$\{r = a \bmod b\}.$$

20. Let member(a, L) be the test for membership of a in list L. Prove the total correctness of the following program to compute member(a, L). *Hint:* Let the loop invariant be "member(a, x) = member(a, L)".

$$\{\text{True}\}$$
$$x := L;$$
$$\textbf{while } x \neq \langle \ \rangle \textbf{ and } a \neq \text{ head}\,(L) \textbf{ do } x := \text{ tail}\,(x) \textbf{ od;}$$
$$r := (x \neq \langle \ \rangle)$$
$$\{r = \text{member}\,(a, L)\}.$$

8.3 Higher–Order Logics

In first-order predicate calculus the only things that can be quantified are individual variables, and the only things that can be arguments for predicates are terms (i.e., constants, variables, or functional expressions with terms as arguments). If we loosen up a little and allow our wffs to quantify other things like predicates or functions, or if we allow our predicates to take arguments that are predicates or functions, then we move to a *higher-order logic*.

Is higher-order logic necessary? The purpose of this section is to convince you that the answer is yes. After some examples we'll give a general definition that will allow us to discuss nth-order logic for any natural number n.

Some Introductory Examples

We often need higher-order logic to express simple statements about the things that interest us. We'll do a few examples to demonstrate.

example 8.14 **Formalizing a Statement**

Let's try to formalize the following statement:

"There is a function that is larger than the log function."

This statement asserts the existence of a function. So if we want to formalize the statement, we'll need to use higher-order logic to quantify a function. We might formalize the statement as

$$\exists f \; \forall x \; (f(x) > \log x).$$

This wff is an instance of the following more general wff, where $>$ is an instance of p and log is an instance of g.

$$\exists f \; \forall x \; p(f(x), g(x)).$$

end example

example 8.15 **Equality**

Let's formalize the notion of equality. Suppose we agree to say that x and y are identical if all their properties are the same. We'll signify this by writing $x = y$. Can we express this thought in formal logic? Sure. If P is some property, then we can think of P as a predicate, and we'll agree that $P(x)$ means that x has property P. Then we can define $x = y$ as the following higher-order wff:

$$\forall P \; ((P(x) \rightarrow P(y)) \wedge (P(y) \rightarrow P(x))).$$

This wff is higher-order because the predicate P is quantified.

end example

example 8.16 **Mathematical Induction**

Suppose that we want to formalize the following version of the principle of mathematical induction:

For any predicate P, if $P(0)$ is true and if for all natural numbers n, $P(n)$ implies $P(n + 1)$, then $P(n)$ is true for all n.

We can represent the statement with the following higher-order wff:

$$\forall P \; (P(0) \wedge \forall n \; (P(n) \rightarrow P(n + 1)) \rightarrow \forall n \; P(n)).$$

This wff is an instance of the following more general higher-order wff, where $c = 0$ and $s(n) = n + 1$:

$$\forall P \; (P(c) \wedge \forall n \; (P(n) \rightarrow P(s(n))) \rightarrow \forall n \; P(n)).$$

end example

Now that we have some examples, let's get down to business and discuss higher-order logic in a general setting that allows us to classify the different orders of logic.

8.3.1 Classifying Higher-Order Logics

To classify higher-order logics, we need to make an assumption about the relationship between predicates and sets.

Identifying Sets with Predicates

We'll assume that predicates are sets and that sets are predicates. Let's see why we can think of predicates and sets as the same thing. For example, if P is a predicate with one argument, we can think of P as a set in which $x \in P$ if and only if $P(x)$ is true. Similarly, if S is a set of 3-tuples, we can think of S as a predicate in which $S(x, y, z)$ is true if and only if $(x, y, z) \in S$.

The relationship between sets and predicates allows us to look at some wffs in a new light. For example, consider the following wff:

$$\forall x \ (A(x) \to B(x)).$$

In addition to the usual reading of this wff as "For every x, if $A(x)$ is true, then $B(x)$ is true," we can now read it in terms of sets by saying, "For every x, if $x \in A$, then $x \in B$." In other words, we have a wff that represents the statement "A is a subset of B."

Definition of Higher-Order Logic

The identification of predicates and sets puts us in position to define higher-order logics.

Higher-Order Logic and Higher-Order Wff

A logic is called *higher-order* if it allows sets to be quantified or if it allows sets to be elements of other sets.

A wff that quantifies a set or has a set as an argument to a predicate is called a *higher-order wff*.

For example, the following two wffs are higher-order wffs:

$$\exists S \ S(x) \qquad \text{The set } S \text{ is quantified.}$$
$$S(x) \land T(S) \qquad \text{The set } S \text{ is an element of the set } T.$$

Functions are Sets

Let's see how functions fit into the picture. Recall that a function can be thought of as a set of 2-tuples. For example, if $f(x) = 3x$ for all $x \in \mathbb{N}$, then we can think of f as the set

$$f = \{(x, 3x) \mid x \in \mathbb{N}\}.$$

So whenever a wff contains a quantified function name, the wff is actually quantifying a set and thus is a higher-order wff by our definition. Similarly, if a wff contains a function name as an argument to a predicate, then the wff is higher-order. For example, the following two wffs are higher-order wffs:

$\exists f \; \forall x \; p(f(x), \, g(x))$ The function f is a set and is quantified.

$p(f(x)) \land q(f)$ The function f is a set and is an element of the set q.

Since we can think of a function as a set and we are identifying sets with predicates, we can also think of a function as a predicate. For example, let f be the function

$$f = \{(x, \, 3x) \mid x \in \mathbb{N}\}.$$

We can think of f as a predicate with two arguments. In other words, we can write the wff $f(x, 3x)$ and let it mean "x is mapped by f to $3x$," which of course we usually write as $f(x) = 3x$.

Classifying Orders of Logic

Now let's see whether we can classify the different orders of logic. We'll start with the two logics that we know best. A propositional calculus is called a *zero-order logic* and a first-order predicate calculus is called a *first-order logic*. We want to continue the process by classifying higher-order logics as second-order, third-order, and so on. To do this, we need to attach an order to each predicate and each quantifier that occurs in a wff. We'll start with the *order of a predicate*.

The Order of a Predicate

A predicate has *order* 1 if its arguments are terms (i.e., constants, individual variables, or functional expressions with terms as arguments). Otherwise, the predicate has *order* $n + 1$, where n is the highest order among its arguments that are not terms.

For example, for each of the following wffs we've given the order of its predicates (i.e., sets):

$S(x) \land T(S)$ S has order 1, and T has order 2.

$p(f(x)) \land q(f)$ p has order 1, f has order 1, and q has order 2.

The reason that the function f has order 1 is that any function, when thought of as a predicate, takes only terms for arguments. Thus any function name has order 1. Remember to distinguish between $f(x)$ and f; $f(x)$ is a term, and f is a function (i.e., a set or a predicate).

We can also relate the order of a predicate to the level of nesting of its arguments, where we think of a predicate as a set. For example, if a wff contains the three statements $S(x)$, $T(S)$, and $P(T)$, then we have $x \in S$, $S \in T$, and $T \in P$. The orders of S, T, and P are 1, 2, and 3. So the order of a predicate (or set) is the maximum number of times the symbol $\in$ is used to get from the set down to its most basic elements.

Now we'll define the *order of a quantifier*.

The Order of a Quantifier

A quantifer has *order* 1 if it quantifies an individual variable. Otherwise, the quantifier has *order* $n + 1$, where n is the order of the predicate being quantified.

For example, let's find the orders of the quantifiers in the wff that follows. Try your luck before you read the answers.

$$\forall x \; \exists S \; \exists T \; \exists f \; (S(x, f(x)) \land T(S)).$$

The quantifier $\forall x$ has order 1 because x is an individual variable. $\exists S$ has order 2 because S has order 1. $\exists T$ has order 3 because T has order 2. $\exists f$ has order 2 because f is a function name, and all function names have order 1.

Now we can make a simple definition for the order of a wff.

The Order of a Wff

The *order of a wff* is the highest order of any of its predicates and quantifiers.

---

example **8.17 Orders of Wffs**

Here are a few sample wffs and their orders.

First-Order Wffs

$S(x)$	S has order 1.
$\forall x \; S(x)$	Both S and $\forall x$ have order 1.

Second-Order Wffs

$S(x) \land T(S)$	S has order 1, and T has order 2.
$\exists S \; S(x)$	S has order 1, and $\exists S$ has order 2.
$\exists S \; (S(x) \land T(S))$	S has order 1, and $\exists S$ and T have order 2.
$P(x, f, f(x))$	P has order 2 because f has order 1.

Third-Order Wffs

$S(x) \land T(S) \land P(T)$	S, T, and P have orders 1, 2, and 3.
$\forall T \; (S(x) \land T(S))$	S, T, $\forall T$ have orders 1, 2, and 3.
$\exists T \; (S(x) \land T(S) \land P(T))$	S, T, P, and $\exists T$ have orders 1, 2, 3, and 3.

end example

Now we can make the definition of an nth-order logic.

The Order of a Logic
An *nth-order logic* is a logic whose wffs have order n or less.

Let's do some examples that transform sentences into higher-order wffs.

example **8.18 Subsets**

Suppose we want to represent the following statement in formal logic.

"There is a set of natural numbers that doesn't contain 4."

Since the statement asserts the existence of a set, we'll need an existential quantifier. The set must be a subset of the natural numbers, and it must not contain the number 4. Putting these ideas together, we can write a mixed version (informal and formal) as follows:

$$\exists S \ (S \text{ is a subset of } \mathbb{N} \text{ and } \neg \ S(4)).$$

Let's see whether we can finish the formalization. We've seen that the general statement "A is a subset of B" can be formalized as follows:

$$\forall x \ (A(x) \rightarrow B(x)).$$

Therefore, we can write the following formal version of our statement:

$$\exists S \ (\forall x \ (S(x) \rightarrow \mathbb{N}(x)) \wedge \neg \ S(4)).$$

This wff is second-order because S has order 1, so $\exists S$ has order 2.

end example

example **8.19 Cities, Streets, and Addresses**

Suppose we think of a city as a set of streets and a street as a set of house addresses. We'll try to formalize the following statement:

"There is a city with a street named Main, and
there is an address 1140 on Main Street."

Suppose C is a variable representing a city and S is a variable representing a street. If x is a name, then we'll let $N(S, x)$ mean that the name of S is x. A third-order logic formalization of the sentence can be written as follows:

$$\exists C \ \exists S \ (C(S) \wedge N(S, \text{Main}) \wedge S(1140)).$$

This wff is third-order because S has order 1, so C has order 2 and $\exists C$ has order 3.

end example

8.3.2 Semantics

How do we attach a meaning to a higher-order wff? The answer is that we construct an interpretation for the wff. The usual approach (called *standard semantics*) is to specify a domain D (a nonempty set) of individuals that we use to give meaning to the constants, the free variables, and the functions and predicates that are not quantified. The quantified individual variables, functions, and predicates are allowed to vary over all possible meanings in terms of D.

Let's try to make the idea of an interpretation clear with some examples.

example 8.20 A Second-Order Interpretation

We'll give an interpretation for the following second-order wff:

$$\exists S \ \exists T \ \forall x \ (S(x) \rightarrow \neg \ T(x)).$$

Suppose we let the domain be $D = \{a, b\}$. We observe that S and T are predicates of order 1, and they are both quantified. So S and T can vary over all possible single-argument predicates over D. For example, the following list shows the four possible predicate definitions for S together with the corresponding set definitions for S:

Predicate Definitions for S	*Set Definitions for S*
$S(a)$ and $S(b)$ are both true	$S = \{a, b\}$
$S(a)$ is true and $S(b)$ is false	$S = \{a\}$
$S(a)$ is false and $S(b)$ is true	$S = \{b\}$
$S(a)$ and $S(b)$ are both false	$S = \varnothing$

We can see from this list that there are as many possibilities for S as there are subsets of D. A similar statement holds for T. Now it's easy to see that our example wff is true for our interpretation. For example, if we choose $S = \{a, b\}$ and $T = \varnothing$, then S is always true and T is always false. Thus

$$S(a) \rightarrow \neg \ T(a) \text{ and } S(b) \rightarrow \neg \ T(b) \text{ are both true.}$$

Therefore, $\exists S \ \exists T \ \forall x \ (S(x) \rightarrow \neg \ T(x))$ is true for the interpretation.

end example

example 8.21 A Second-Order Interpretation

We'll give an interpretation for the following second-order wff:

$$\exists S \ \forall x \ \exists y \ S(x, y).$$

Let $D = \{a, b\}$. Since S takes two arguments, it has 16 possible definitions, one corresponding to each subset of 2-tuples over D. For example, if $S = \{(a, a), (b, a)\}$, then $S(a, a)$ and $S(b, a)$ are both true, and $S(a, b)$ and $S(b, b)$ are both false. Thus the wff $\exists S \; \forall x \; \exists y \; S(x, y)$ is true for our interpretation.

end example

example   **8.22   A Third-Order Interpretation**

We'll give an interpretation for the following third-order wff:

$$\exists T \; \forall x \; (T(S) \rightarrow S(x)).$$

We'll let $D = \{a, b\}$. Since S is not quantified, it is a normal predicate and we must give it a meaning. Suppose we let $S(a)$ be true and $S(b)$ be false. This is represented by $S = \{a\}$. Now T is an order 2 predicate because it takes an order 1 predicate as its argument. T is also quantified, so it is allowed to vary over all possible predicates that take arguments like S.

From the viewpoint of sets, the arguments to T can be any of the four subsets of D. Therefore, T can vary over any of the 16 subsets of $\{\varnothing, \{a\}, \{b\}, \{a, b\}\}$. For example, one possible value for T is $T = \{\varnothing, \{b\}\}$. If we think of T as a predicate, this means that $T(\varnothing)$ and $T(\{b\})$ are both true, while $T(\{a\})$ and $T(\{a, b\})$ are both false. This value of T makes the wff $\forall x \; (T(S) \rightarrow S(x))$ true. Thus the wff $\exists T \; \forall x \; (T(S) \rightarrow S(x))$ is true for our interpretation.

end example

As we have shown in the examples, we can give interpretations to higher-order wffs. This means that we can also use the following familiar terms in our discussions about higher-order wffs.

model, countermodel, valid, invalid, satisfiable, and unsatisfiable.

What about formal reasoning with higher-order wffs? That's next.

8.3.3   Higher–Order Reasoning

Gödel proved a remarkable result in 1931. He proved that if a formal system is powerful enough to describe all the arithmetic formulas over the natural numbers and the system is consistent, then it is not complete. In other words, there is a valid arithmetic formula that can't be proven as a theorem in the system. Even if additional axioms were added to make the formula provable, then there would exist a new valid formula that is not provable in the larger system. A very readable account of Gödel's proof is given by Nagel and Newman [1958].

It is known that the formulas of arithmetic over the natural numbers can be described in second-order logic. So it follows as a corollary of Gödel's result that second-order logic is not complete.

What does it really mean when we have a formal system that is not complete? It means that we might have to leave the formalism to prove that some wffs are valid. In other words, we may need to argue informally—using only our wits and imaginations—to prove some statements. In some sense this is nice because it justifies our existence as reasoning beings and there will always be enough creative work for us to do—perhaps aided by computers.

Formal Proofs

Even though higher-order logics are not complete, we can still do formal reasoning to prove the validity of many higher-order wffs. Before we give some formal proofs, we need to say something about proof rules for quantifiers in higher-order logic. We'll use the same rules that we used for first-order logic, but we'll apply them to higher-order wffs when the need arises. Basically, the rules are a reflection of our natural discourse. To keep things clear, when we use an uppercase letter like P as a variable, we'll sometimes use the lowercase letter p to represent an instantiation of P. Here is an example to demonstrate the ideas.

example 8.23 A Valid Second-Order Wff

We'll give an informal proof and a formal proof that the following second-order wff is valid:

$$\exists P \; \forall Q \; \forall x \; (Q(x) \to P(x)).$$

Proof: Let I be an interpretation with domain D. Then the wff has the following meaning with respect to I: There is a subset P of D such that for every subset Q of D it follows that $x \in Q$ implies $x \in P$. This statement is true because we can choose P to be D. So I is a model for the wff. Since I was an arbitrary interpretation, it follows that the wff is valid. QED.

In the formal proof, we'll use EI to instantiate Q to a particular value represented by lowercase q. Also, since P is universally quantified, we can use UI to instantiate P to any predicate. We'll instantiate P to True, so that $P(x)$ becomes True(x), which means that True$(x) = $ True.

1.	$\neg \, \exists P \; \forall Q \; \forall x \; (Q(x) \to P(x))$	P [for an IP proof]
2.	$\forall P \; \exists Q \; \exists x \; (Q(x) \wedge \neg \, P(x))$	1, T
3.	$\exists Q \; \exists x \; (Q(x) \wedge \neg \, \text{True}(x))$	2, UI [instantiate P to True]
4.	$\exists x \; (q(x) \wedge \neg \, \text{True}(x))$	3, EI
5.	$q(c) \wedge \neg \, \text{True}(c)$	4, EI
6.	$q(c) \wedge \text{False}$	5, T
7.	False	6, T
	QED	1–7, IP.

end example

A Euclidean Geometry Example

Let's see how higher-order logic comes into play when we discuss elementary geometry. From an informal viewpoint, the wffs of Euclidean geometry are English sentences. For example, the following four statements describe part of Hilbert's axioms for Euclidean plane geometry.

1. On any two distinct points there is always a line.

2. On any two distinct points there is not more than one line.

3. Every line has at least two distinct points.

4. There are at least three points not on the same line.

Can we formalize these axioms? Let's assume that a line is a set of points. So two lines are equal if they have the same set of points. We'll also assume that arbitrary points are denoted by the variables x, y, and z and individual points by the constants a, b, and c. We'll denote arbitrary lines by the variables L, M, and N and we'll denote individual lines by the constants l, m, and n. We'll let the predicate $L(x)$ denote the fact that x is a point on line L or, equivalently, L is a line on the point x. Now we can write the four axioms as second-order wffs as follows:

1. $\forall x \, \forall y \, ((x \neq y) \to \exists L \, (L(x) \wedge L(y)))$.

2. $\forall x \, \forall y \, ((x \neq y) \to \forall L \, \forall M \, (L(x) \wedge L(y) \wedge M(x) \wedge M(y) \to (L = M)))$.

3. $\forall L \, \exists x \, \exists y \, ((x \neq y) \wedge L(x) \wedge L(y))$.

4. $\exists x \, \exists y \, \exists z \, ((x \neq y) \wedge (x \neq z) \wedge (y \neq z) \wedge \forall L \, (L(x) \wedge L(y) \to \neg \, L(z)))$.

In the following examples, we'll prove that there are at least two distinct lines. In the first example, we'll give an informal proof of the statement. In the second example, we'll formalize the statement and give a formal proof.

example **8.24 An Informal Theorem and Proof**

Let's prove the following theorem.

There are at least two distinct lines.

Proof: Axiom 4 tells us that there are three distinct points a, b, and c not on the same line. By Axiom 1 there is a line l on a and b, and again by Axiom 1 there is a line m on a and c. By Axiom 4, c is not on line l. Therefore, it follows that $l \neq m$. QED.

end example

example   **8.25 A Formal Theorem and Proof**

Now we'll formalize the theorem from Example 8.24 and give a formal proof. A formalized version of the theorem can be written as

$$\exists L \; \exists M \; \exists x \; (\neg \, L(x) \wedge M(x)).$$

Here's the proof:

1.	$\exists x \; \exists y \; \exists z \; ((x \neq y) \wedge (x \neq z) \wedge (y \neq z)$ $\wedge \; \forall L \; (L(x) \wedge L(y) \rightarrow \neg \, L(z)))$	Axiom 4
2.	$(a \neq b) \wedge (a \neq c) \wedge (b \neq c) \wedge \forall L \; (L(a) \wedge L(b) \rightarrow \neg \, L(c))$	1, EI, EI, EI
3.	$\forall x \; \forall y \; ((x \neq y) \rightarrow \exists L \; (L(x) \wedge L(y)))$	Axiom 1
4.	$(a \neq b) \rightarrow \exists L \; (L(a) \wedge L(b))$	3, UI, UI
5.	$a \neq b$	2, Simp
6.	$\exists L \; (L(a) \wedge L(b))$	4, 5, MP
7.	$l(a) \wedge l(b)$	6, EI
8.	$(a \neq c) \rightarrow \exists L \; (L(a) \wedge L(c))$	3, UI, UI
9.	$a \neq c$	2, Simp
10.	$\exists L \; (L(a) \wedge L(c))$	8, 9, MP
11.	$m(a) \wedge m(c)$	10, EI
12.	$\forall L \; (L(a) \wedge L(b) \rightarrow \neg \, L(c))$	2, Simp
13.	$l(a) \wedge l(b) \rightarrow \neg \, l(c)$	12, UI
14.	$\neg \, l(c)$	7, 13, MP
15.	$m(c)$	11, Simp
16.	$\neg \, l(c) \wedge m(c)$	14, 15, Conj
17.	$\exists x \; (\neg \, l(x) \wedge m(x))$	16, EG
18.	$\exists M \; \exists x \; (\neg \, l(x) \wedge M(x))$	17, EG
19.	$\exists L \; \exists M \; \exists x \; (\neg \, L(x) \wedge M(x)).$	18, EG
	QED.	

end example

A few more Euclidean geometry problems (both informal and formal) are included in the exercises.

◀ Exercises

Orders of Logic

1. State the minimal order of logic needed to describe each of the following wffs.

   a. $\forall x \ (Q(x) \rightarrow P(Q))$.

   b. $\exists x \ \forall g \ \exists p \ (q(c, g(x)) \land p(g(x)))$.

   c. $A(B) \land B(C) \land C(D) \land D(E) \land E(F)$.

   d. $\exists P \ (A(B) \land B(C) \land C(D) \land P(A))$.

   e. $S(x) \land T(S, x) \rightarrow U(T, S, x)$.

   f. $\forall x \ (S(x) \land T(S, x) \rightarrow U(T, S, x))$.

   g. $\forall x \ \exists S \ (S(x) \land T(S, x) \rightarrow U(T, S, x))$.

   h. $\forall x \ \exists S \ \exists T \ (S(x) \land T(S, x) \rightarrow U(T, S, x))$.

   i. $\forall x \ \exists S \ \exists T \ \exists U \ (S(x) \land T(S, x) \rightarrow U(T, S, x))$.

Formalizing English Sentences

2. Formalize each of the following sentences as a wff in second-order logic.

   a. There are sets A and B such that $A \cap B = \varnothing$.

   b. There is a set S with two subsets A and B such that $S = A \cup B$.

3. Formalize each of the following sentences as a wff in an appropriate higher-order logic. Also figure out the order of the logic that you use in each case.

   a. Every state has a city named Springfield.

   b. There is a nation with a state that has a county named Washington.

   c. A house has a room with a bookshelf containing a book by Thoreau.

   d. There is a continent with a nation containing a state with a county named Lincoln, which contains a city named Central City that has a street named Broadway.

   e. Some set has a partition consisting of two subsets.

4. Formalize the basis of mathematical induction: If S is a subset of $\mathbb{N}$ and $0 \in S$ and $x \in S$ implies $succ(x) \in S$, then $S = \mathbb{N}$.

5. If R is a relation, let $B(R)$ mean that R is a binary relation. Formalize the following statement about relations: Every binary relation that is irreflexive and transitive is asymmetric.

Validity

6. Show that the following wff is satisfiable and invalid:

$$\exists x \ \exists y \ (p(x, y) \rightarrow \forall Q \ (Q(x) \rightarrow \neg \ Q(y))).$$

7. Show that each of the following wffs is valid with an informal validity argument.

  a. $\forall S \; \exists x \; S(x) \rightarrow \exists x \; \forall S \; S(x)$.
  b. $\forall x \; \exists S \; S(x) \rightarrow \exists S \; \forall x \; S(x)$.
  c. $\exists S \; \forall x \; S(x) \rightarrow \forall x \; \exists S \; S(x)$.
  d. $\exists x \; \forall S \; S(x) \rightarrow \forall S \; \exists x \; S(x)$.

8. Give an informal proof and a formal proof that the following second-order wff is valid:

$$\forall P \; \exists Q \; \forall x \; (Q(x) \rightarrow P(x)).$$

More Geometry

9. Use the facts from the Euclidean geometry example to give an informal proof for each of the following statements. You may use any of these statements to prove a subsequent statement.

  a. For each line there is a point not on the line.
  b. Two lines cannot intersect in more than one point.
  c. Through each point there exist at least two lines.
  d. Not all lines pass through the same point.

10. Formalize each of the following statements as a wff in second-order logic, using the variable names from the Euclidean geometry example. Then provide a formal proof for each wff.

  a. Not all points lie on the same line.
  b. Two lines cannot intersect in more than one point.
  c. Through each point there exist at least two lines.
  d. Not all lines pass through the same point.

8.4 Chapter Summary

A first-order theory is a formal treatment of some subject that uses first-order predicate calculus. We often need the idea of equality when applying logic in a formal manner to a particular subject. Equality can be added to first-order logic in such a way that the following familiar notion is included: Equals can replace—be substituted for—equals.

We can prove elementary statements about imperative programs with a formal proof system where each program is bounded by two conditions—a precondition and a postcondition. The system uses only one axiom—the assignment axiom. Some useful proof rules are the consequence rule and the rules for composition, if-then, if-then-else, and while statements. The system can be extended by adding axioms and proof rules for items that are normally found in imperative languages, such as arrays and other loop forms. We also have a procedure for proving the termination of while loops.

When formalizing a subject, we often need higher forms of logic to express statements. Higher-order logic extends first-order logic by allowing predicates and functions to be quantified and to be arguments in predicates. We can classify the order of a logic if we make the association that a predicate is a set. Even though higher-order logics are not complete, we can still reason formally within these logics just as we do in propositional logic and first-order logic.

chapter 9

Computational Logic

Let us not dream that reason can ever be popular.
Passions, emotions, may be made popular, but
reason remains ever the property of the few.

— Johann Wolfgang von Goethe (1749–1832)

Can reasoning be automated? The answer is yes, for some logics. In this chapter we'll discuss how to automate the reasoning process for first-order logic. We might start by automating the "natural deduction" proof techniques that we introduced in Chapters 6, 7, and 8. A problem with this approach is that there are many proof rules that can be applied in many different ways. In this chapter we'll look at a more mechanical way to perform deduction. We'll introduce a single proof rule, called resolution, which can be applied automatically by a computer. We'll also see that the resolution rule is used for the execution of logic programs.

chapter guide

9.1 (Automatic Reasoning) introduces the resolution proof rule. To understand the rule, we'll need to discuss clauses, clausal forms, substitution, and unification. We'll see how the rule can be applied in a mechanical fashion to prove theorems.

9.2 (Logic Programming) introduces logic programming and shows how resolution is applied to perform the computation of a logic program. We'll also give some elementary techniques for constructing logic programs.

9.1 Automatic Reasoning

Let's look at the mechanical side of logic. We're going to introduce a proof rule that can be applied automatically. As fate would have it, the rule must be applied while trying to prove that a wff is unsatisfiable. This is not really a problem, because we know that a wff is valid if and only if its negation is unsatisfiable. In other words, if we want to prove that the wff W is valid, then we can do so by trying to prove that $\neg\, W$ is unsatisfiable. For example, if we want to prove the validity of the conditional $A \to B$, then we can try to prove the unsatisfiability of its negation $A \land \neg\, B$.

The new proof rule, which is called the *resolution rule*, can be applied over and over again in an attempt to show unsatisfiability. We can't present the resolution rule yet because it can be applied only to wffs that are written in a special form, called *clausal form*. So let's get to it.

9.1.1   Clauses and Clausal Forms

We need to introduce a little terminology before we can describe a clausal form. Recall that a *literal* is either an atom or the negation of an atom. For example, $p(x)$ and $\neg\, q(x,\, b)$ are literals. To distinguish whether a literal has a negation sign, we may use the terms *positive literal* and *negative literal*. $p(x)$ is a positive literal, and $\neg\, q(x,\, b)$ is a negative literal.

A *clause* is a disjunction of zero or more literals. For example, the following wffs are clauses:

$$p(x),$$
$$\neg\, q(x, b),$$
$$\neg\, p(a) \lor p(b),$$
$$p(x) \lor \neg\, q(a, y) \lor p(a).$$

The clause that is a disjunction of zero literals is called the *empty clause*, and it's denoted by the following special box symbol:

$$\square.$$

The empty clause is assigned the value false. We'll soon see why this makes sense when we discuss resolution.

A *clausal form* is the universal closure of a conjunction of clauses. In other words, a clausal form is a prenex conjunctive normal form, in which all quantifiers are universal and there are no free variables. For ease of notation we'll often represent a clausal form by the set consisting of its clauses. For example, each line in the following list shows a clausal form together with its representation as a set of clauses.

Clausal Form	*Set of Clauses*
$\forall x\; p(x)$	$\{p(x)\}$
$\forall x\; \neg\, q(x, b)$	$\{\neg\, q(x, b)\}$
$\forall x\; \forall y\; (p(x) \land \neg\, q(y, b))$	$\{p(x), \neg\, q(y, b)\}$
$\forall x\; \forall y\; (p(y, f(x)) \land (q(y) \lor \neg\, q(a)))$	$\{p(y, f(x)), q(y) \lor \neg\, q(a)\}$
$(p(a) \lor p(b)) \land q(a, b)$	$\{p(a) \lor p(b), q(a, b)\}$

Notice that the last clausal form does not need quantifiers because it doesn't have any variables. In other words, it's a proposition. In fact, for propositions a clausal form is just a conjunctive normal form (CNF).

Usage Note

When we use the words "valid," "invalid," "satisfiable," or "unsatisfiable" to describe a set S of clauses we are always referring to the clausal form that S denotes. For example, we can say that the set of clauses $\{p(x),\ \neg\ p(x)\}$ is unsatisfiable because $\forall x\ (p(x) \wedge \neg\ p(x))$ is unsatisfiable.

Constructing Clausal Forms

It's easy to see that some wffs are not equivalent to any clausal form. For example, let's consider the following wff:

$$\forall x\ \exists y\ p(x,\ y).$$

This wff is not a clausal form, and it isn't equivalent to any clausal form because it has an existential quantifier. Since clausal forms are the things that resolution needs to work on, it's nice to know that we can associate a clausal form with each wff in such a way that the clausal form is unsatisfiable if and only if the wff is unsatisfiable. Let's see how to find such a clausal form for each wff.

To construct a clausal form for a wff, we can start by constructing a prenex conjunctive normal form for the wff. If there are no free variables and all the quantifiers are universal, then we have a clausal form. Otherwise, we need to get rid of the free variables and the existential quantifiers and still retain enough information to be able to detect whether the original wff is unsatisfiable. Luckily, there's a way to do this. The technique is due to the mathematician Thoralf Skolem (1887–1963), and it appears in his paper [1928].

Let's introduce Skolem's idea by considering the following example wff:

$$\forall x\ \exists y\ p(x,\ y).$$

In this case the quantifier $\exists y$ is inside the scope of the quantifier $\forall x$. So it may be that y depends on x. For example, if we let $p(x,\ y)$ mean "x has a successor y," then y certainly depends on x. If we're going to remove the quantifier $\exists y$ from $\forall x\ \exists y\ p(x,\ y)$, then we'd better leave some information about the fact that y may depend on x. Skolem's idea was to use a new function symbol, say f, and replace each occurrence of y within the scope of $\exists y$ by the term $f(x)$. After performing this operation, we obtain the following wff, which is now in clausal form:

$$\forall x\ p(x, f(x)).$$

We can describe the general method for eliminating existential quantifiers as follows:

Skolem's Rule (9.1)

Let $\exists x\ W(x)$ be a wff or part of a larger wff. If $\exists x$ is not inside the scope of a universal quantifier, then pick a new constant c, and

$$\text{replace } \exists x\ W(x) \text{ by } W(c).$$

If $\exists x$ is inside the scope of universal quantifiers $\forall x_1, ..., \forall x_n$, then pick a new function symbol f, and

$$\text{replace } \exists x\ W(x) \text{ by } W(f(x_1, ..., x_n)).$$

The constants and functions introduced by the rule are called *Skolem functions*.

---

example 9.1   Applying Skolem's Rule

Let's apply Skolem's rule to the following wff:

$$\exists x\ \forall y\ \forall z\ \exists u\ \forall v\ \exists w\ p(x, y, z, u, v, w).$$

Since the wff contains three existential quantifiers, we'll use (9.1) to create three Skolem functions to replace the existentially quantified variables as follows:

Replace x by b because $\exists x$ is not in the scope of a universal quantifier.

Replace u by $f(y, z)$ because $\exists u$ is in the scope of $\forall y$ and $\forall z$.

Replace w by $g(y, z, v)$ because $\exists w$ is in the scope of $\forall y$, $\forall z$, and $\forall v$.

Now we can apply (9.1) to eliminate the existential quantifiers by making the above replacements to obtain the following clausal form:

$$\forall y\ \forall z\ \forall v\ p(b, y, z, f(y, z), v, g(y, z, v)).$$

end example

Now we have the ingredients necessary to construct clausal forms with the property that a wff and its clausal form are either both unsatisfiable or both satisfiable.

Skolem's Algorithm (9.2)

Given a wff W, there exists a clausal form such that W and the clausal form are either both unsatisfiable or both satisfiable. In other words, W has a

---

Continued ➡

➡ ➡

> model if and only if the clausal form has a model. The clausal form can be constructed from W by the following steps:
>
> **1.** Construct the prenex conjunctive normal form of W.
>
> **2.** Replace all occurrences of each free variable by a new constant.
>
> **3.** Use Skolem's rule (9.1) to eliminate the existential quantifiers.

Proof: We already know that any wff is equivalent to its prenex conjunctive normal form, and we also know that any wff W with free variables has a model if and only if the wff obtained from W by replacing each occurrence of a free variable by a new constant has a model. So we may assume that W is already in prenex conjunctive normal form with no free variables.

Let $s(W)$ denote the clausal form constructed from W by Skolem's algorithm. We must show that W has a model if and only if $s(W)$ has a model. One direction is easy because $s(W) \rightarrow W$ is valid. To see this, start with the premise $s(W)$ and use UI to remove all universal quantifiers. Then use EG and UG to restore the quantifiers of W in proper order. So if $s(W)$ has a model, then W has a model. We'll prove the converse by induction on the number n of existential quantifiers in W.

If $n = 0$, then $W = s(W)$, so any model for W is also a model for $s(W)$. Now assume that $n > 0$ and assume that, for any wff A of the same form with less than n existential quantifiers, if A has a model, then $s(A)$ has a model. Since $n > 0$, let $\exists y$ be the leftmost existential quantifier in W. There may be universal quantifiers to the left of $\exists y$ in the prenex form. So, for some natural number k, we can write W in the form

$$W = \forall x_1 \ldots \forall x_k \exists y \ C(x_1, \ldots, x_k, y).$$

Now apply Skolem's rule to W by replacing y by $f(x_1, \ldots, x_k)$ to obtain the wff

$$A = \forall x_1 \ldots \forall x_k \ C(x_1, \ldots, x_k, f(x_1, \ldots, x_k)).$$

Suppose that W has a model I with domain D. We will use I to construct a model for A. The main task is to define f as a function over the domain D. Let $d_1, \ldots, d_k \in D$. Since W is true for I, there exists $e \in D$ such that $C(d_1, \ldots, d_k, e)$ is true for I. But there might be more than one e such that $C(d_1, \ldots, d_k, e)$ is true for I. Choose any one such e and define $f(d_1, \ldots, d_k) = e$. This defines f as a function over D. Let J be the interpretation with domain D, where C is the same relation used for I and f is the function just defined. It follows that J is a model for A. So if W has a model, then A has a model. Now A has fewer than n existential quantifiers. Induction tells us that if A has a model, then $s(A)$ has a model. So if W has a model, then $s(A)$ has a model. But since A is obtained from W by applying Skolem's rule, it follows that $s(A) = s(W)$. So if W has a model, then $s(W)$ has a model. QED.

Remarks About the Algorithm

Before we do some examples, let's make a couple of remarks about the steps of the algorithm. Step 2 could be replaced by the statement "Take the existential closure." But then Step 3 would remove these same quantifiers by replacing each of the newly quantified variables with a new constant name. So we saved time and did it all in one step. Step 2 can be done at any time during the process. We need Step 2 because we know that a wff and its existential closure are either both unsatisfiable or both satisfiable.

Step 3 can be applied during Step 1 after all implications have been eliminated and after all negations have been pushed to the right, but before all quantifiers have been pushed to the left. Often this will reduce the number of variables in the Skolem function. Another way to simplify the Skolem function is to push all quantifiers to the right as far as possible before applying Skolem's rule.

example **9.2 Applying Skolem's Algorithm**

Suppose W is the following wff:

$$\forall x \, \neg \, p(x) \wedge \forall y \, \exists z \, q(y, z).$$

First we'll apply (9.2) as stated. In other words, we calculate the prenex form of W by moving the quantifiers to the left to obtain

$$\forall x \, \forall y \, \exists z \, (\neg p(x) \wedge q(y, z)).$$

Then we apply Skolem's rule (9.1), which says that we replace z by $f(x, y)$ to obtain the following clausal form for W.

$$\forall x \, \forall y \, (\neg p(x) \wedge q(y, f(x, y))).$$

Now we'll start again with W, but we'll apply (9.1) during Step 1 after all implications have been eliminated and after all negations have been pushed to the right. There is nothing to do in this regard. So, before we move the quantifiers to the left, we'll apply (9.1). In this case the quantifier $\exists z$ is only within the scope of $\forall y$, so we replace z by $f(y)$ to obtain

$$\forall x \, \neg p(x) \wedge \forall y \, q(y, f(y)).$$

Now finish constructing the prenex form by moving the universal quantifiers to the left to obtain the following clausal form for W:

$$\forall x \, \forall y \, (\neg p(x) \wedge q(y, f(y))).$$

So we get a simpler clausal form for W in this case.

end example

example **9.3 A Simple Clausal Form**

Suppose we have a wff with no variables (i.e., a propositional wff). For example, let W be the wff

$$(p(a) \to q) \land ((q \land s(b)) \to r).$$

To find the clausal form for W, we need only apply equivalences from propositional calculus to find a CNF as follows:

$$(p(a) \to q) \land ((q \land s(b)) \to r) \equiv (\neg\, p(a) \lor q) \land (\neg(q \land s(b)) \lor r)$$
$$\equiv (\neg\, p(a) \lor q) \land (\neg\, q \lor \neg\, s(b) \lor r).$$

end example

example **9.4 Finding a Clausal Form**

We'll use (9.2) to find a clausal form for the following wff:

$$\exists y\ \forall x\ (p(x) \to q(x,\, y)) \land \forall x\ \exists y\ (q(x,\, x) \land s(y) \to r(x)).$$

The first step is to find the prenex conjunctive normal form. Since there are two quantifiers with the same name, we'll do some renaming to obtain the following wff:

$$\exists y\ \forall x\ (p(x) \to q(x,\, y)) \land \forall w\ \exists z\ ((q(w,\, w) \land s(z)) \to r(w)).$$

Next, we eliminate the conditionals to obtain the following wff:

$$\exists y\ \forall x\ (\neg\, p(x) \lor q(x,\, y)) \land \forall w\ \exists z\ (\neg\, (q(w,\, w) \land s(z)) \lor r(w)).$$

Now, push negation to the right to obtain the following wff:

$$\exists y\ \forall x\ (\neg\, p(x) \lor q(x,\, y)) \land \forall w\ \exists z\ (\neg\, q(w,\, w) \lor \neg\, s(z) \lor r(w)).$$

Next, we'll apply Skolem's rule (9.1) to eliminate the existential quantifiers and obtain the following wff:

$$\forall x\ (\neg\, p(x) \lor q(x,\, a)) \land \forall w\ (\neg\, q(w,\, w) \lor \neg\, s(f(w)) \lor r(w)).$$

Lastly, we push the universal quantifiers to the left and obtain the desired clausal form:

$$\forall x\ \forall w\ ((\neg\, p(x) \lor q(x,\, a)) \land (\neg\, q(w,\, w) \lor \neg\, s(f(w)) \lor r(w))).$$

end example

example **9.5 Finding a Clausal Form**

We'll construct a clausal form for the following wff:

$$\forall x\ (p(x) \to \exists y\ \forall z\ ((p(w) \lor q(x,\, y)) \to \forall w\ r(x,\, w))).$$

The free variable w is also used in the quantifier $\forall w$, and the quantifier $\forall z$ is superfluous. So we'll do some renaming, and we'll remove $\forall z$ to obtain the following wff:

$$\forall x \ (p(x) \rightarrow \exists y \ ((p(w) \vee q(x, \, y)) \rightarrow \forall z \ r(x, \, z))).$$

We remove the conditionals in the usual way to obtain the following wff:

$$\forall x \ (\neg \, p(x) \vee \exists y \ (\neg \ (p(w) \vee q(x, \, y)) \vee \forall z \ r(x, \, z))).$$

Next, we move negation inward to obtain the following wff:

$$\forall x \ (\neg \, p(x) \vee \exists y \ ((\neg \ p(w) \wedge \neg \ q(x, \, y)) \vee \forall z \ r(x, \, z))).$$

Now we can apply Skolem's rule (9.1) to eliminate $\exists y$ and replace the free variable w by b to get the following wff:

$$\forall x \ (\neg \, p(x) \vee ((\neg \ p(b) \wedge \neg \ q(x, \, f(x))) \vee \forall z \ r(x, \, z))).$$

Next, we push the universal quantifier $\forall z$ to the left, obtaining the following wff:

$$\forall x \ \forall z \ (\neg \ p(x) \vee ((\neg \ p(b) \wedge \neg \ q(x, \, f(x))) \vee r(x, \, z))).$$

Lastly, we distribute $\vee$ over $\wedge$ to obtain the following clausal form:

$$\forall x \ \forall z \ ((\neg \ p(x) \vee \neg \ p(b) \vee r(x, \, z)) \wedge (\neg \ p(x) \vee \neg \ q(x, \, f(x)) \vee r(x, \, z))).$$

end example

So we can transform any wff into a wff in clausal form in which the two wffs are either both unsatisfiable or both satisfiable. Since the resolution rule tests clausal forms for unsatisfiability, we're a step closer to describing the idea of resolution. Before we introduce the general idea of resolution, we're going to pause and discuss resolution for the simple case of propositions.

9.1.2 Resolution for Propositions

It's easy to see how resolution works for propositional clauses (i.e., clauses with no variables). The resolution proof rule works something like a cancellation process. It takes two clauses and constructs a new clause from them by deleting all occurrences of a positive literal p from one clause and all occurrences of $\neg \, p$ from the other clause. For example, suppose we are given the following two propositional clauses:

$$p \vee q,$$
$$\neg \, p \vee r \vee \neg \, p.$$

We obtain a new clause by first eliminating p from the first clause and eliminating the two occurrences of $\neg \, p$ from the second clause. Then we take the disjunction of the leftover clauses to form the new clause:

$$q \vee r.$$

Let's write down the resolution rule in a more general way. Suppose we have two propositional clauses of the following forms:

$$p \vee A,$$
$$\neg\, p \vee B.$$

Let $A - p$ denote the disjunction obtained from A by deleting all occurrences of p. Similarly, let $B - \neg\, p$ denote the disjunction obtained from B by deleting all occurrences of $\neg\, p$. The resolution rule allows us to infer the propositional clause

$$(A - p) \vee (B - \neg\; p).$$

Here's the rule.

Resolution Rule for Propositions   (9.3)

$$\frac{p \vee A, \; \neg\, p \vee B}{(A - p) \vee (B - \neg\, p)}.$$

Although the rule may look strange, it's a good rule. That is, it preserves truth. To see this, we can suppose that $(p \vee A) \wedge (\neg\, p \vee B) = \text{True}$. If p is true, then the equation reduces to $B = \text{True}$. Since $\neg\, p$ is false, we can remove all occurrences of $\neg\, p$ from B and still have $B - \neg\, p = \text{True}$. Therefore, $(A - p) \vee (B - \neg\; p) = \text{True}$. We obtain the same result if p is false. So the resolution rule does its job.

A proof by resolution uses only the resolution rule. So we can define a *resolution proof* as a sequence of clauses, ending with the empty clause, in which each clause in the sequence either is a premise or is inferred by the resolution rule from two preceding clauses in the sequence. Notice that the empty clause is obtained from (9.3) when A is either empty or contains only copies of p and when B is either empty or contains only copies of $\neg\, p$. For example, the simplest version of (9.3) can be stated as follows:

$$\frac{p, \neg\, p}{\square}.$$

In other words, we obtain the well-known tautology $p \wedge \neg\, p \rightarrow \text{False}$.

example **9.6 A Resolution Proof**

Let's prove that the following clausal form is unsatisfiable:

$$(\neg\, p \vee q) \wedge (p \vee q) \wedge (\neg\, q \vee p) \wedge (\neg\, p \vee \neg\, q).$$

In other words, we'll prove that the following set of clauses is unsatisfiable:

$$\{\neg\, p \vee q, \, p \vee q, \, \neg\, q \vee p, \, \neg\, p \vee \neg\, q\}.$$

The following resolution proof does the job:

1. $\neg p \vee q$ P
2. $p \vee q$ P
3. $\neg q \vee p$ P
4. $\neg p \vee \neg q$ P
5. $q \vee q$ 1, 2, Resolution
6. p 3, 5, Resolution
7. $\neg p$ 4, 5, Resolution
8. $\square$ 6, 7, Resolution
   QED.

end example

Now let's get back on our original track, which is to describe the resolution rule for clauses of first-order predicate calculus.

9.1.3   Substitution and Unification

When we discuss the resolution rule for clauses that contain variables, we'll see that a certain kind of matching is required. For example, suppose we are given the following two clauses:

$$p(x, y) \vee q(y),$$
$$r(z) \vee \neg q(b).$$

The matching that we will discuss allows us to replace all occurrences of the variable y by the constant b, thus obtaining the following two clauses:

$$p(x, b) \vee q(b),$$
$$r(z) \vee \neg q(b).$$

Notice that one clause contains $q(b)$ and the other contains its negation $\neg q(b)$. Resolution will allow us to cancel them and construct the disjunction of the remaining parts, which is the clause $p(x, b) \vee r(z)$.

We need to spend a little time to discuss the process of replacing variables by terms. If x is a variable and t is a term, then the expression x/t is called a *binding* of x to t and can be read as "x gets t" or "x is bound to t" or "x has value t" or "x is replaced by t." For example, three typical bindings are written as follows:

$$x/a, \quad y/z, \quad w/f(b, v).$$

Definition of Substitution

A *substitution* is a finite set of bindings $\{x_1/t_1, \ldots, x_n/t_n\}$, where variables $x_1, \ldots, x_n$ are all distinct and $x_i \neq t_i$ for each i. We use lowercase Greek letters to denote substitutions. The *empty substitution*, which is just the empty set, is denoted by the Greek letter ϵ.

What do we do with substitutions? We apply them to expressions, an *expression* being a finite string of symbols. Let E be an expression, and let θ be the following substitution:

$$\theta = \{x_1/t_1, \ldots, x_n/t_n\}.$$

Then the *instance* of E by θ, denoted $E\theta$, is the expression obtained from E by simultaneously replacing all occurrences of the variables $x_1, \ldots, x_n$ in E by the terms $t_1, \ldots, t_n$, respectively. We say that $E\theta$ is obtained from E by *applying* the substitution θ to the expression E. For example, if $E = p(x, y, f(x))$ and $\theta = \{x/a, y/f(b)\}$, then $E\theta$ has the following form:

$$E\theta = p(x, y, f(x))\{x/a, y/f(b)\} = p(a, f(b), f(a)).$$

If S is a set of expressions, then the *instance* of S by θ, denoted $S\theta$, is the set of all instances of expressions in S by θ. For example, if $S = \{p(x, y), q(a, y)\}$ and $\theta = \{x/a, y/f(b)\}$, then $S\theta$ has the following form:

$$S\theta = \{p(x, y), q(a, y)\}\{x/a, y/f(b)\} = \{p(a, f(b)), q(a, f(b)\}.$$

Now let's see how we can combine two substitutions θ and σ into a single substitution that has the same effect as applying θ and then applying σ to any expression.

Compostion of Substitutions

The *composition* of two substitutions θ and σ is the substitution denoted by $\theta\sigma$ that satisfies the following property for any expression E:

$$E(\theta\sigma) = (E\theta)\sigma.$$

Although we have described the composition in terms of how it acts on all expressions, we can compute $\theta\sigma$ without any reference to an expression as follows:

Computing the Composition (9.4)

Given the two substitutions $\theta = \{x_1/t_1, \ldots, x_n/t_n\}$ and $\sigma = \{y_1/s_1, \ldots, y_m/s_m\}$. The composition $\theta\sigma$ is constructed as follows:

1. Apply σ to the denominators of θ to get $\{x_1/t_1\sigma, \ldots, x_n/t_n\sigma\}$.

2. Delete any bindings of the form x_i/x_i from the set on line 1.

3. Delete any y_i/s_i from σ if y_i is a variable in $\{x_1, \ldots, x_n\}$.

4. $\theta\sigma$ is the union of the sets constructed on lines 2 and 3.

The process looks complicated, but it's really quite simple. It's just a formalization of the following construction: For each distinct variable v occurring in the numerators of θ and σ, apply θ and then σ to v, obtaining the expression $(v\theta)\sigma$. The composition $\theta\sigma$ consists of all bindings $v/(v\theta)\sigma$ such that $v \neq (v\theta)\sigma$.

It's also nice to know that we can always check whether we constructed a composition correctly. Just make up an example atom containing the distinct variables in the numerators of θ and σ, say, $p(v_1, \ldots, v_k)$, and then check to make sure the following equation holds:

$$(p(v_1, \ldots, v_k)\theta)\sigma = p(v_1, \ldots, v_k)(\theta\sigma).$$

example 9.7 Finding a Composition

Let $\theta = \{x/f(y),\ y/z\}$ and $\sigma = \{x/a,\ y/b,\ z/y\}$. To find the composition $\theta\sigma$, we first apply σ to the denominators of θ to form the following set:

$$\{x/f(y)\sigma,\ y/z\sigma\} = \{x/f(b),\ y/y\}.$$

Now remove the binding y/y to obtain $\{x/f(b)\}$. Next, delete the bindings x/a and y/b from σ to obtain $\{z/y\}$. Finally, compute $\theta\sigma$ as the union of these two sets $\theta\sigma = \{x/f(b),\ z/y\}$.

Let's check to see whether the answer is correct. For our example atom we'll pick $p(x,\ y,\ z)$ because x, y, and z are the distinct variables occurring in the numerators of θ and σ. We'll make the following two calculations to see whether we get the same answer.

$$(p(x,y,z)\,\theta)\,\sigma = p(f(y),z,z)\,\sigma = p(f(b),y,y),$$
$$p(x,y,z)\,(\theta\sigma) = p(f(b),y,y).$$

end example

Three simple, but useful, properties of composition are listed next. The proofs are left as exercises.

Properties of Composition   (9.5)

For any substitutions θ and σ and any expression E the following statements hold.

1. $E(\theta\sigma) = (E\theta)\sigma$.

2. $E\epsilon = E$.

3. $\theta\epsilon = \epsilon\theta = \theta$.

Definition of Unifier

A substitution θ is called a *unifier* of a finite set S of literals if $S\theta$ is a singleton set. For example, if we let $S = \{p(x,\ b),\ p(a,\ y)\}$, then the substitution $\theta = \{x/a,\ y/b\}$ is a unifier of S because

$$S\theta = \{p(a,\ b)\},$$

which is a singleton set.

Some sets of literals don't have a unifier, while other sets have infinitely many unifiers. The range of possibilities can be shown by the following four simple examples.

1. $\{p(x),\ q(y)\}$ doesn't have a unifier.

2. $\{p(x),\ \neg\ p(x)\}$ doesn't have a unifier.

3. $\{p(x),\ p(a)\}$ has a unifier. Any unifier must contain the binding x/a and yield the singleton $\{p(a)\}$. E.g., $\{x/a\}$ and $\{x/a,\ y/z\}$ are unifiers of the set.

4. $\{p(x),\ p(y)\}$ has infinitely many unifiers that can yield different singletons. E.g., $\{x/y\}$, $\{y/x\}$, and $\{x/t,\ y/t\}$ for any term t are all unifers of the set.

Among the unifiers of a set there is always at least one unifier that can be used to construct every other unifier. To be specific, a unifier θ for S is called a *most general unifier* (mgu) for S if for every unifier α of S there exists a substitution σ such that $\alpha = \theta\sigma$. In other words, an mgu for S is a factor of every other unifier of S. Let's look at an example.

example **9.8 A Most General Unifier**

As we have noted, the set $S = \{p(x),\ p(y)\}$ has infinitely many unifiers that we can describe as follows:

$$\{x/y\},\ \{y/x\},\ \text{and}\ \{x/t,\ y/t\}\ \text{for any term}\ t.$$

The unifier $\{x/y\}$ is an mgu for S because we can write the other unifiers in terms of $\{x/y\}$ as follows: $\{y/x\} = \{x/y\}\{y/x\}$, and $\{x/t,\ y/t\} = \{x/y\}\{y/t\}$ for any term t. Similarly, $\{y/x\}$ is an mgu for S.

end example

Unification Algorithms

We want to find a way to construct an mgu for any set of literals. Before we do this, we need a little terminology to describe the set of terms that cause two or more literals in a set to be distinct.

Disagreement Set

If S is a set of literals, then the *disagreement set* of S is constructed in the following way.

1. Find the longest common substring that starts at the left end of each literal of S.

2. The disagreement set of S is the set of all the terms that occur in the literals of S that are immediately to the right of the longest common substring.

For example, we'll construct the disagreement set for the following set of three literals.

$$S = \{p(x, f(x), y),\ p(x, y, z),\ p(x, f(a), b)\}.$$

The longest common substring for the literals in S is the string

$$\text{``}p(x,\text{''}$$

of length four. The terms in the literals of S that occur immediately to the right of this string are $f(x)$, y, and $f(a)$. Thus the disagreement set of S is

$$\{f(x),\ y,\ f(a)\}.$$

Now we have the tools to describe a very important algorithm by Robinson [1965]. The algorithm computes, for a set of atoms, a most general unifier, if one exists.

Unification Algorithm (Robinson) (9.6)

Input: A finite set S of atoms.

Output: Either a most general unifier for S or a statement that S is not unifiable.

1. Set $k = 0$ and $\theta_0 = \epsilon$, and go to Step 2.

2. Calculate $S\theta_k$. If it's a singleton set, then stop (θ_k is the mgu for S). Otherwise, let D_k be the disagreement set of $S\theta_k$, and go to Step 3.

3. If D_k contains a variable v and a term t, such that v does not occur in t, then calculate the composition $\theta_{k+1} = \theta_k\{v/t\}$, set $k := k + 1$, and go to Step 2. Otherwise, stop (S is not unifiable).

The composition $\theta_k\{v/t\}$ in Step 3 is easy to compute for two reasons. The variable v doesn't occur in t, and v will never occur in any numerator of

θ_k. Therefore, the middle two steps of the composition construction (9.4) don't change anything. In other words, the composition $\theta_k \{v/t\}$ is constructed by applying $\{v/t\}$ to each denominator of θ_k and then adding the binding v/t to the result.

example 9.9 Finding a Most General Unifier

Let's try the algorithm on the set $S = \{p(x, f(y)), p(g(y), z)\}$. We'll list each step of the algorithm as we go.

1. Set $\theta_0 = \epsilon$.

2. $S\theta_0 = S\epsilon = S$ is not a singleton. $D_0 = \{x, g(y)\}$.

3. Variable x doesn't occur in term $g(y)$ of D_0.
   Put $\theta_1 = \theta_0 \{x/g(y)\} = \{x/g(y)\}$.

2. $S\theta_1 = \{p(g(y), f(y)), p(g(y), z)\}$ is not a singleton. $D_1 = \{f(y), z\}$.

3. Variable z does not occur in term $f(y)$ of D_1.
   Put $\theta_2 = \theta_1 \{z/f(y)\} = \{x/g(y), z/f(y)\}$.

2. $S\theta_2 = \{p(g(y), f(y))\}$ is a singleton. Therefore, the algorithm terminates with the mgu $\{x/g(y), z/f(y)\}$ for the set S.

end example

example 9.10 No Most General Unifier

Let's trace the algorithm on the set $S = \{p(x), p(g(x))\}$. We'll list each step of the algorithm as we go:

1. Set $\theta_0 = \epsilon$.

2. $S\theta_0 = S\epsilon = S$, which is not a singleton. $D_0 = \{x, g(x)\}$.

3. The only choices for a variable and a term in D_0 are x and $g(x)$. But the variable x occurs in $g(x)$. So the algorithm stops, and S is not unifiable.

This makes sense too. For example, if we were to apply the substitution $\{x/g(x)\}$ to S, we would obtain the set $\{p(g(x)), p(g(g(x)))\}$, which in turn gives us the same disagreement set $\{x, g(x)\}$. So the process would go on forever. Notice that a change of variables makes a big difference. For example, if we change the second atom in S to $p(g(y))$, then the algorithm unifies the set $\{p(x), p(g(y))\}$, obtaining the mgu $\{x/g(y)\}$.

end example

The following alternative algorithm for unification is due to Martelli and Montanari [1982]. It can be used on pairs of atoms.

Unification Algorithm (Martelli-Montanari)   (9.7)

Input: A singleton set $\{A = B\}$ where A and B are atoms or terms.

Output: Either a most general unifier of A and B or a statement that they are not unifiable.

Perform the following nondeterministic actions until no action can be performed or a halt with failure occurs. If there is no failure then the output is a set of equations of the form $\{x_1 = t_1, \ldots, x_n = t_n\}$ and the mgu is $\{x_1/t_1, \ldots, x_n/t_n\}$. Note: f and g represent function or predicate symbols.

Equation	Action
1. $f(s_1, \ldots, s_n) = f(t_1, \ldots, t_n)$.	Replace the equation with the equations $s_1 = t_1, \ldots, s_n = t_n$.
2. $f(s_1, \ldots, s_m) = g(t_1, \ldots, t_m)$ and either $f \neq g$ or $m \neq n$.	Halt with failure.
3. $x = x$	Delete the equation.
4. $t = x$ and t is not a variable.	Replace $t = x$ with $x = t$.
5. $x = t$, x does not occur in t, and x occurs in another equation.	Apply the substitution $\{x/t\}$ to all other equations.
6. $x = t$, t is not a variable, and x occurs in t.	Halt with failure.

---

example **9.11 Finding a Most General Unifier**

Let's try the algorithm on the two atoms $p(x, f(x))$ and $p(y, f(b))$. We'll list each set of equations generated by the algorithm together with the reason for each step.

$$\{p(x, f(x)) = p(y, f(b))\} \quad \text{Input}$$
$$\{x = y, f(x) = f(b)\} \quad \text{Equation (1)}$$
$$\{x = y, f(y) = f(b)\} \quad \text{Equation (5)}$$
$$\{x = y, y = b\} \quad \text{Equation (1)}$$
$$\{x = b, y = b\} \quad \text{Equation (5)}$$

Therefore, the mgu is $\{x/b, y/b\}$.

end example

9.1.4 Resolution: The General Case

Now we've got the tools to discuss resolution of clauses that contain variables. Let's look at a simple example to help us see how unification comes into play. Suppose we're given the following two clauses:

$$p(x, a) \lor \neg q(x),$$
$$\neg p(b, y) \lor \neg q(a).$$

We want to cancel $p(x, a)$ from the first clause and $\neg p(b, y)$ from the second clause. But they won't cancel until we unify the two atoms $p(x, a)$ and $p(b, y)$. An mgu for these two atoms is $\{x/b, y/a\}$. If we apply this unifier to the original two clauses, we obtain the following two clauses:

$$p(b, a) \lor \neg q(b),$$
$$\neg p(b, a) \lor \neg q(a).$$

Now we can cancel $p(b, a)$ from the first clause and $\neg p(b, a)$ from the second clause and take the disjunction of what's left to obtain the following clause:

$$\neg q(b) \lor \neg q(a).$$

That's the way the resolution rule works when variables are present. Now let's give a detailed description of the rule.

The Resolution Rule

The resolution rule takes two clauses and constructs a new clause. But the rule can be applied only to clauses that possess the following two properties.

Two Requirements for Resolution

1. The two clauses have no variables in common.

2. There are one or more atoms, $L_1, \ldots, L_k$, in one of the clauses and one or more literals, $\neg M_1, \ldots, \neg M_n$, in the other clause such that the set $\{L_1, \ldots, L_k, M_1, \ldots, M_n\}$ is unifiable.

The first property can always be obtained by renaming any common variables. For example, the variable x is used in both of the following clauses:

$$q(b, x) \lor p(x), \quad \neg q(x, a) \lor p(y).$$

We can replace x in the second clause with a new variable z to obtain the following two clauses that satisfy the first property:

$$q(b, x) \lor p(x), \quad \neg q(z, a) \lor p(y).$$

This action does not change the satisfiability or unsatisfiability of the clausal form containing the clauses.

Suppose we have two clauses that satisfy properties 1 and 2. Then they can be written in the following form, where C and D represent the other parts of each clause:

$$L_1 \vee \cdots \vee L_k \vee C \quad \text{and} \quad \neg\, M_1 \vee \cdots \vee \neg\, M_n \vee D.$$

Since the clauses satisfy the second property, we know that there is an mgu θ that unifies the set of atoms $\{L_1, \dots, L_k, M_1, \dots, M_n\}$. In other words, there is a unique atom N such that $N = L_i\theta = M_j\theta$ for any i and j. To be specific, we'll set

$$N = L_1\theta.$$

Now we're ready to do our canceling. Since there could be additional occurrences of N in $C\theta$, we'll let $C\theta - N$ denote the clause obtained from $C\theta$ by deleting all occurrences of the atom N. Similarly, let $D\theta - \neg\, N$ denote the clause obtained from $D\theta$ by deleting all occurrences of $\neg\, N$. The clause that we construct is the disjunction of any literals that are left after the cancellation:

$$(C\theta - N) \vee (D\theta - \neg\, N).$$

Summing all this up, we can state the resolution rule as follows:

Resolution Rule (R) $\hfill$ **(9.8)**

$$\frac{L_1 \vee \cdots \vee L_k \vee C, \ \neg\, M_1 \vee \cdots \vee \ \neg\, M_n \vee D}{(C\theta - N) \vee (D\theta - \neg\, N)}.$$

The clause constructed in the denominator of (9.8) is called a *resolvant* of the two clauses in the numerator. Let's describe how to use (9.8) to find a resolvant of the two clauses.

1. Check the two clauses for distinct variables (rename if necessary).

2. Find an mgu θ for the set of atoms $\{L_1, \dots, L_k, M_1, \dots, M_n\}$.

3. Apply θ to both clauses C and D.

4. Set $N = L_1\theta$.

5. Remove all occurrences of N from $C\theta$.

6. Remove all occurrences of $\neg\, N$ from $D\theta$.

7. Form the disjunction of the clauses in Steps 5 and 6. This is the resolvant.

Let's do some examples to get the look and feel of resolution before we forget everything.

example 9.12 **Resolving Two Clauses**

We'll try to find a resolvent of the following two clauses:

$$q\,(b,x) \lor p\,(x) \lor q\,(b,a)\,,$$
$$\neg\,q\,(y,a) \lor p\,(y).$$

We'll cancel the atom $q(b, x)$ in the first clause with the literal $\neg\,q(y, a)$ in the second clause. So we'll write the first clause in the form $L \lor C$, where L and C have the following values:

$$L = q\,(b,x) \quad \text{and} \quad C = p\,(x) \lor q\,(b,a)\,.$$

The second clause can be written in the form $\neg\,M \lor D$, where M and D have the following values:

$$M = q\,(y,a) \quad \text{and} \quad D = p\,(y)\,.$$

Now L and M, namely $q(b, x)$ and $q(y, a)$, can be unified by the mgu $\theta = \{y/b, x/a\}$. We can apply θ to either atom to obtain the common value $N = L\theta = M\theta = q(b, a)$. Now we can apply (9.8) to find the resolvent of the two clauses. First, compute the clauses $C\theta$ and $D\theta$:

$$C\theta = (p\,(x) \lor q\,(b,a))\,\{y/b, x/a\} = p\,(a) \lor q\,(b,a)\,,$$
$$D\theta = p\,(y)\,\{y/b, x/a\} = p\,(b)\,.$$

Next we'll remove all occurrences of $N = q(b, a)$ from $C\theta$ and remove all occurrences of $\neg\,N = \neg\,q(b, a)$ from $D\theta$:

$$C\theta - N = p\,(a) \lor q\,(b,a) - q\,(b,a) = p\,(a)\,,$$
$$D\theta - \neg\,N = p\,(b) - \neg\,q\,(b,a) = p\,(b)\,.$$

Lastly, we'll take the disjunction of the remaining clauses to obtain the resolvent $p(a) \lor p(b)$.

end example

example 9.13 **Resolving Two Clauses**

In this example we'll consider canceling two literals from one of the clauses. Suppose we have the following two clauses.

$$p\,(f\,(x)) \lor p\,(y) \lor \neg\,q\,(x)\,,$$
$$\neg\,p\,(z) \lor q\,(w).$$

We'll pick the disjunction $p(f(x)) \vee p(y)$ from the first clause to cancel with the literal $\neg\, p(z)$ in the second clause. So we need to unify the set of atoms $\{p(f(x)),$ $p(y),\ p(z)\}$. An mgu for this set is $\theta = \{y/f(x),\ z/f(x)\}$. The common value N obtained by applying θ to any of the atoms in the set is $N = p(f(x))$. To see how the cancellation takes place, we'll apply θ to both of the original clauses to obtain the clauses

$$p\left(f\left(x\right)\right) \vee p\left(f\left(x\right)\right) \vee \neg\, q\left(x\right),$$
$$\neg\, p\left(f\left(x\right)\right) \vee q\left(w\right).$$

We'll cancel $p(f(x)) \vee p(f(x))$ from the first clause and $\neg\, p(f(x))$ from the second clause, with no other deletions possible. So we take the disjunction of the remaining parts after cancellation to obtain the resolvent $\neg\, q(x) \vee q(w)$.

end example

What's so great about finding resolvents? Two things are great. One great thing is that the process is mechanical—it can be programmed. The other great thing is that the process preserves unsatisfiability. In other words, we have the following result.

Theorem   (9.9)

Let G be a resolvent of the clauses E and F. Then $\{E,\ F\}$ is unsatisfiable if and only if $\{E,\ F,\ G\}$ is unsatisfiable.

Now we're almost in position to describe how to prove that a set of clauses is unsatisfiable. Let S be a set of clauses where—after possibly renaming some variables—distinct clauses of S have disjoint sets of variables. We define the *resolution* of S, denoted by $R(S)$, to be the set

$$R(S) = S \cup \{G \mid G \text{ is a resolvent of a pair of clauses in } S\}.$$

We can conclude from (9.9) that S is unsatisfiable if and only if $R(S)$ is unsatisfiable. Similarly, $R(S)$ is unsatisfiable if and only if $R(R(S))$ is unsatisfiable. We can continue on in this way. To simplify the notation, we'll define $R^0(S) = S$ and $R^{n+1}(S) = R(R^n(S))$ for $n > 0$. So for any n we can say that

$$S \text{ is unsatisfiable if and only if } R^n(S) \text{ is unsatisfiable.}$$

Let's look at some examples to demonstrate the calculation of the sequence of sets $S,\ R(S),\ R^2(S), \ldots$.

example   **9.14 Calculating Resolutions**

Suppose we start with the following set of clauses:

$$S = \{p(x),\ \neg\, p(a)\}.$$

To compute $R(S)$, we must add to S all possible resolvants of pairs of clauses. There is only one pair of clauses in S, and the resolvant of $p(x)$ and $\neg\, p(a)$ is the empty clause. Thus $R(S)$ is the following set.

$$R(S) = \{p(x), \neg\, p(a), \Box\}.$$

Now let's compute $R(R(S))$. The only two clauses in $R(S)$ that can be resolved are $p(x)$ and $\neg\, p(a)$. Since their resolvant is already in $R(S)$, there's nothing new to add. So the process stops, and we have $R(R(S)) = R(S)$.

end example

example **9.15 Calculating Resolutions**

Consider the following set of three clauses.

$$S = \{p(x),\ q(y) \vee \neg\, p(y),\ \neg\, q(a)\}.$$

Let's compute $R(S)$. There are two pairs of clauses in S that have resolvants. The two clauses $p(x)$ and $q(y) \vee \neg\, p(y)$ resolve to $q(y)$. The clauses $q(y) \vee \neg\, p(y)$ and $\neg\, q(a)$ resolve to $\neg\, p(a)$. Thus $R(S)$ is the following set:

$$R(S) = \{p(x),\ q(y) \vee \neg\, p(y),\ \neg\, q(a),\ q(y),\ \neg\, p(\text{a})\}.$$

Now let's compute $R(R(S))$. The two clauses $p(x)$ and $\neg\, p(a)$ resolve to the empty clause, and nothing new is added by resolving any other pairs from $R(S)$. Thus $R(R(S))$ is the following set:

$$R(R(S)) = \{p(x), q(y) \vee \neg\, p(y), \neg\, q(a), q(y), \neg\, p(a), \Box\}.$$

It's easy to see that we can't get anything new by resolving pairs of clauses in $R(R(S))$. Thus we have $R^3(S) = R^2(S)$.

end example

These two examples have something very important in common. In each case the set S is unsatisfiable, and the empty clause occurs in $R^n(S)$ for some n. This is no coincidence as we can see by the following result of Robinson [1965].

Resolution Theorem   (9.10)

A finite set S of clauses is unsatisfiable iff $\Box \in R^n(S)$ for some $n \geq 0$.

The theorem provides us with an algorithm to prove that a wff is unsatisfiable. Let S be the set of clauses that make up the clausal form of the wff. Start by calculating all the resolvants of pairs of clauses from S. The new resolvants are added to S to form the larger set of clauses $R(S)$. If the empty clause has been calculated, then we are done. Otherwise, calculate resolvants of pairs of clauses in the set $R(S)$. Continue the process until we find a pair of clauses whose resolvant is the empty clause.

If we get to a point at which no new clauses are being created and we have not found the empty clause, then the process stops, and we conclude that the wff that we started with is satisfiable.

9.1.5 Theorem Proving with Resolution

Recall that a resolution proof is a sequence of clauses that ends with the empty clause, in which each clause is either a premise or can be inferred from two preceding clauses by the resolution rule. Recall also that a resolution proof is a proof of unsatisfiability. Since we normally want to prove that some wff is valid, we must first take the negation of the wff, then find a clausal form, and then attempt to do a resolution proof. We'll summarize the steps.

Steps to Prove that W is Valid

1. Form the negation $\neg\, W$. For example, if W is a conditional of the form $A \wedge B \wedge C \to D$, then $\neg\, W$ has the form $A \wedge B \wedge C \wedge \neg\, D$.

2. Use Skolem's algorithm (9.2) to convert line 1 into clausal form.

3. Take the clauses from line 2 as premises in the proof.

4. Apply the resolution rule (9.8) to derive the empty clause.

---

example **9.16 Binary Relations**

We'll prove that if a binary relation is irreflexive and transitive, then it is asymmetric. If p denotes a binary relation, then the three properties can be represented as follows.

Irreflexive: $\forall x \; \neg\, p(x,\, x)$.

Transitive: $\forall x \; \forall y \; \forall z \; (p(x,\, y) \wedge p(y,\, z) \to p(x,\, z))$.

Asymmetric: $\forall x \; \forall y \; (p(x,\, y) \to \neg\, p(y,\, x))$.

So we must prove that the wff W is valid, where

$$W = \text{Irreflexive} \wedge \text{Transitive} \to \text{Asymmetric}.$$

To use resolution, we must prove that $\neg\, W$ is unsatisfiable, where

$$\neg\, W = \text{Irreflexive} \wedge \text{Transitive} \wedge \neg\, \text{Asymmetric}.$$

Notice that ¬ Asymmetric has the following form:

$$\neg \text{ Asymmetric} = \neg\, \forall x\, \forall y\, (p(x,\,y) \rightarrow \neg\, p(y,\,x)) \equiv \exists x\, \exists y\, (p(x,\,y) \wedge p(y,\,x)).$$

First we put ¬ W into clausal form. The following table shows the clauses in the clausal forms for Irreflexive, Transitive, and ¬ Asymmetric.

Wff	Clauses
$\forall x \, \neg\, p\,(x,\,x)$	$\neg\, p\,(x,\,x)$
$\forall x\, \forall y\, \forall z\, (p\,(x,\,y) \wedge p\,(y,\,z) \rightarrow p\,(x,\,z))$	$\neg\, p\,(u,\,v) \vee \neg\, p\,(v,\,w) \vee p\,(u,\,w)$
$\exists x\, \exists y\, (p\,(x,\,y) \wedge p\,(y,\,x))$	$p\,(a,\,b)$ and $p\,(b,\,a)$

To do a resolution proof, we start with the four clauses as premises. Our goal is to construct resolvants to obtain the empty clause. Each resolution step includes the most general unifier used for that application of resolution.

1.	$\neg\, p(x,\,x)$	P
2.	$\neg\, p(u,\,v) \vee \neg\, p(v,\,w) \vee p(u,\,w)$	P
3.	$p(a,\,b)$	P
4.	$p(b,\,a)$	P
5.	$\neg\, p(x,\,v) \vee \neg\, p(v,\,x)$	1, 2, R, $\{u/x,\,w/x\}$
6.	$\neg\, p(b,\,a)$	3, 5, R, $\{x/a,\,v/b\}$
7.	$\square$	4, 6, R, { }
	QED.	

Therefore, we conclude that the properties of irreflexive and transitive imply the asymmetric property.

end example

example 9.17 The Family Tree Problem

Suppose we let p stand for the isParentOf relation and let g stand for the isGrandParentOf relation. Let G denote the relationship between g and p as follows:

$$G = \forall x\, \forall y\, \forall z\, (p(x,\,z) \wedge p(z,\,y) \rightarrow g(x,\,y)).$$

In other words, if x is a parent of z and z is a parent of y, then we conclude that x is a grandparent of y. Suppose we have the following facts about parents, where the letters a, b, c, d, and e denote the names of people:

$$p(a,\,b),\ p(c,\,b),\ p(b,\,d),\ p(a,\,e).$$

Now, suppose someone claims that $g(a,\,d)$ is implied by the given facts. Let P denote the conjunction of parent facts as follows:

$$P = p(a,\,b) \wedge p(c,\,b) \wedge p(b,\,d) \wedge p(a,\,e).$$

So the claim is that the wff W is valid, where

$$W = P \wedge G \rightarrow g(a, d).$$

To prove the claim using resolution, we must prove that $\neg\ W$ is unsatisfiable. We can observe that $\neg\ W$ has the following form:

$$\neg\ W = \neg\ (P \wedge G \rightarrow g(a, d)) \equiv P \wedge G \wedge \neg\ g(a, d).$$

We need to put $\neg\ W$ into clausal form. Since P is a conjunction of atoms, it is already in clausal form. So we need only work on G, which will be in clausal form if we replace the conditional. The result is the clause

$$\neg\ p(x, z) \vee \neg\ p(z, y) \vee g(x, y).$$

So the clausal form of $\neg\ W$ consists of the following six clauses.

$$p(a, b),\ p(c, b),\ p(b, d),\ p(a, e),\ \neg\ p(x, z) \vee \neg\ p(z, y) \vee g(x, y),\ \text{and}\ \neg\ g(a, d).$$

To do a resolution proof, we start with the six clauses as premises. Our goal is to construct resolvants to obtain the empty clause. Each resolution step includes the most general unifier used for that application of resolution. Here's the proof.

1.	$p(a, b)$	P
2.	$p(c, b)$	P
3.	$p(b, d)$	P
4.	$p(a, e)$	P
5.	$\neg\ p(x, z) \vee \neg\ p(z, y) \vee g(x, y)$	P
6.	$\neg\ g(a, d)$	P
7.	$\neg\ p(a, z) \vee \neg\ p(z, d)$	5, 6, R, $\{x/a,\ y/d\}$
8.	$\neg\ p(b, d)$	1, 7, R, $\{z/b\}$
9.	$\square$	3, 8, R, $\{\ \}$
	QED.	

Therefore, we conclude that $g(a, d)$ is implied from the given facts.

end example

example **9.18 Diagonals of a Trapezoid**

We'll give a resolution proof that the alternate interior angles formed by a diagonal of a trapezoid are equal. This problem is from Chang and Lee [1973]. Let $t(x, y, u, v)$ mean that x, y, u, and v are the four corner points of a trapezoid in clockwise order. Let $p(x, y, u, v)$ mean that edges xy and uv are parallel lines. Let $e(x, y, z, u, v, w)$ mean that angle xyz is equal to angle uvw. We'll assume the following two axioms about trapezoids.

Axiom 1: $\forall x \, \forall y \, \forall u \, \forall v \, (t(x, y, u, v) \rightarrow p(x, y, u, v))$.

Axiom 2: $\forall x \, \forall y \, \forall u \, \forall v \, (p(x, y, u, v) \rightarrow e(x, y, v, u, v, y))$.

To prove: $t(a, b, c, d) \rightarrow e(a, b, d, c, d, b)$.

To prepare for a resolution proof, we need to write each axiom in its clausal form. This gives us the following two clauses:

Axiom 1: $\neg \, t(x, y, u, v) \lor p(x, y, u, v)$.

Axiom 2: $\neg \, p(x, y, u, v) \lor e(x, y, v, u, v, y)$.

Next, we need to negate the statement to be proved and put the result in clausal form, which gives us the following two clauses:

$$t(a, b, c, d),$$
$$\neg \, e(a, b, d, c, d, b).$$

To do a resolution proof, we start with the four clauses as premises. Our goal is to construct resolvants to obtain the empty clause. Each resolution step includes the most general unifier used for that application of resolution.

1.	$\neg \, t(x, y, u, v) \lor p(x, y, u, v)$	P
2.	$\neg \, p(x, y, u, v) \lor e(x, y, v, u, v, y)$	P
3.	$t(a, b, c, d)$	P
4.	$\neg \, e(a, b, d, c, d, b)$	P
5.	$\neg \, p(a, b, c, d)$	2, 4, R, $\{x/a, y/b, v/d, u/c\}$
6.	$\neg \, t(a, b, c, d)$	1, 5, R, $\{x/a, y/b, u/c, v/d\}$
7.	$\square$	3, 6, R, $\{\ \}$

QED.

end example

9.1.6 Remarks

In the example proofs we didn't follow a specific strategy to help us choose which clauses to resolve. Strategies are important because they may help reduce the searching required to find a proof. Although a general discussion of strategy is beyond our scope, we'll present a strategy in the next section for the special case of logic programming.

The unification algorithm (9.6) is the original version given by Robinson [1965]. Other researchers have found algorithms that can be implemented more efficiently. For example, the paper by Paterson and Wegman [1978] presents a linear algorithm for unification.

There are also other versions of resolution. One approach uses two simple rules, called *binary resolution* and *factoring*, which can be used together to do the same job as resolution. Another rule, called *paramodulation*, is used when the equality predicate is present to take advantage of substituting equals for equals. An excellent introduction to automatic reasoning is contained in the book by Wos, Overbeek, Lusk, and Boyle [1984].

Another subject that we haven't discussed is automatic reasoning in higher-order logic. In higher-order logic it's undecidable whether a set of atoms can be unified. Still there are many interesting results about higher-order unification and there are automatic reasoning systems for some higher-order logics. For example, in second-order monadic logic (*monadic logic* restricts predicates to at most one argument) there is an algorithm to decide whether two atoms can be unified. For example, if F is a variable that represents a function, then the two atoms $F(a)$ and a can be unified by letting F be the constant function that returns the value a or by letting F be the identity function. The paper by Snyder and Gallier [1989] contains many results on higher-order unification.

Automatic theorem-proving techniques are an important and interesting part of computer science, with applications to almost every area of endeavor. Probably the most successful applications of automatic theorem proving will be interactive in nature, with the proof system acting as an assistant to the person using it. Typical tasks involve such things as finding ways to represent problems and information to be processed by an automatic theorem prover, finding algorithms that make proper choices in performing resolution, and finding algorithms to efficiently perform unification. We'll look at the programming side of theorem proving in the next section.

 Exercises

Constructing Clausal Forms

1. Use Skolem's algorithm, if necessary, to transform each of the following wffs into a clausal form.

    a. $(A \wedge B) \vee C \vee D$.

    b. $(A \wedge B) \vee (C \wedge D) \vee (E \rightarrow F)$.

    c. $\exists y \, \forall x \, (p(x, y) \rightarrow q(x))$.

    d. $\exists y \, \forall x \, p(x, y) \rightarrow q(x)$.

    e. $\forall x \, \forall y \, (p(x, y) \vee \exists z \, q(x, y, z))$.

    f. $\forall x \, \exists y \, \exists z \, [(\neg \, p(x, y) \wedge q(x, z)) \vee r(x, y, z)]$.

Resolution with Propositions

2. What is the resolvent of the propositional clause $p \vee \neg p$ with itself? What is the resolvent of $p \vee \neg \, p \vee q$ with itself?

3. Find a resolution proof to show that each of the following sets of propositional clauses is unsatisfiable.

    a. $\{A \lor B, \lnot A, \lnot B \lor C, \lnot C\}$.

    b. $\{p \lor q, \lnot p \lor r, \lnot r \lor \lnot p, \lnot q\}$.

    c. $\{A \lor B, A \lor \lnot C, \lnot A \lor C, \lnot A \lor \lnot B, C \lor \lnot B, \lnot C \lor B\}$.

Substitutions and Unification

4. Compute the composition $\theta\sigma$ of each of the following pairs of substitutions.

    a. $\theta = \{x/y\}, \sigma = \{y/x\}$.

    b. $\theta = \{x/y\}, \sigma = \{y/x, x/a\}$.

    c. $\theta = \{x/y, y/a\}, \sigma = \{y/x\}$.

    d. $\theta = \{x/f(z), y/a\}, \sigma = \{z/b\}$.

    e. $\theta = \{x/y, y/f(z)\}, \sigma = \{y/f(a), z/b\}$.

5. Use Robinson's unification algorithm to find a most general unifier for each of the following sets of atoms.

    a. $\{p(x, f(y, a), y), p(f(a, b), v, z)\}$.

    b. $\{q(x, f(x)), q(f(x), x)\}$.

    c. $\{p(f(x, g(y)), y), p(f(g(a), z), b)\}$.

    d. $\{p(x, f(x), y), p(x, y, z), p(w, f(a), b)\}$.

6. Use the Martelli-Montanari unification algorithm to find a most general unifier for each of the following sets of atoms.

    a. $\{p(x, f(y, a), y), p(f(a, b), v, z)\}$.

    b. $\{q(x, f(x)), q(f(x), x)\}$.

    c. $\{p(f(x, g(y)), y), p(f(g(a), z), b)\}$.

Resolution in First-Order Logic

7. What is the resolvent of the clause $p(x) \lor \lnot p(f(a))$ with itself? What is the resolvent of $p(x) \lor \lnot p(f(a)) \lor q(x)$ with itself?

8. Use resolution to show that each of the following sets of clauses is unsatisfiable.

    a. $\{p(x), q(y, a) \lor \lnot p(a), \lnot q(a, a)\}$.

    b. $\{p(u, v), q(w, z), \lnot p(y, f(x, y)) \lor \lnot p(f(x, y), f(x, y)) \lor \lnot q(x, f(x, y))\}$.

    c. $\{p(a) \lor p(x), \lnot p(a) \lor \lnot p(y)\}$.

    d. $\{p(x) \lor p(f(a)), \lnot p(y) \lor \lnot p(f(z))\}$.

    e. $\{q(x) \lor q(a), \lnot p(y) \lor \lnot p(g(a)) \lor \lnot q(a), p(z) \lor p(g(w)) \lor \lnot q(w)\}$.

Proving Theorems

9. Prove that each of the following propositional statements is a tautology by using resolution to prove that its negation is a contradiction.

   a. $(A \lor B) \land \neg A \rightarrow B$.

   b. $(p \rightarrow q) \land (q \rightarrow r) \rightarrow (p \rightarrow r)$.

   c. $(p \lor q) \land (q \rightarrow r) \land (r \rightarrow s) \rightarrow (p \lor s)$.

   d. $[(A \land B \rightarrow C) \land (A \rightarrow B)] \rightarrow (A \rightarrow C)$.

10. Prove that each of the following statements is valid by using resolution to prove that its negation is unsatisfiable.

   a. $\forall x \ p(x) \rightarrow \exists x \ p(x)$.

   b. $\forall x \ (p(x) \rightarrow q(x)) \land \exists x \ p(x) \rightarrow \exists x \ q(x)$.

   c. $\exists y \ \forall x \ p(x, y) \rightarrow \forall x \ \exists y \ p(x, y)$.

   d. $\exists x \ \forall y \ p(x, y) \land \forall x \ (p(x, x) \rightarrow \exists y \ q(y, x)) \rightarrow \exists y \ \exists x \ q(x, y)$.

   e. $\forall x \ p(x) \lor \forall x \ q(x) \rightarrow \forall x \ (p(x) \lor q(x))$.

Challenges

11. Let E be any expression, A and B two sets of expressions, and θ, σ, α any substitutions. Prove each of the following statements about composing substitutions.

   a. $E(\theta\sigma) = (E\theta)\sigma$.

   b. $E\epsilon = E$.

   c. $\theta\epsilon = \epsilon\theta = \theta$.

   d. $(\theta\sigma)\alpha = \theta(\sigma\alpha)$.

   e. $(A \cup B)\theta = A\theta \cup B\theta$.

12. Translate each of the following arguments into first-order predicate calculus. Then use resolution to prove that the resulting wffs are valid by proving that the negations are unsatisfiable.

   a. All computer science majors are people. Some computer science majors are logical thinkers. Therefore, some people are logical thinkers.

   b. Babies are illogical. Nobody is despised who can manage a crocodile. Illogical persons are despised. Therefore, babies cannot manage crocodiles.

13. Translate each of the following arguments into first-order predicate calculus. Then use resolution to prove that the resulting wffs are valid by proving the negations are unsatisfiable.

   a. Every dog either likes people or hates cats. Rover is a dog. Rover loves cats. Therefore, some dog likes people.

   b. Every committee member is rich and famous. Some committee members are old. Therefore, some committee members are old and famous.

c. No human beings are quadrupeds. All men are human beings. Therefore, no man is a quadruped.

d. Every rational number is a real number. There is a rational number. Therefore, there is a real number.

e. Some freshmen like all sophomores. No freshman likes any junior. Therefore, no sophomore is a junior.

14. In the proof of Skolem's algorithm it was shown that $s(W) \rightarrow W$ is valid, where $s(W)$ is the clausal form of W. Show that the converse is false by considering the following wff:

$$W = \forall x \, \exists y \, (p(x, y) \vee \neg \, p(y, y)).$$

a. Show that W is valid.

b. Find the clausal form of W and show that it is invalid.

15. It was noted after Skolem's algorithm that Skolem's rule could not be used to remove existential quantifiers until after all implications were eliminated and all negations were moved inward. To confirm this, do each of the following exercises, where W is the following wff and C is any wff that does not contain x or y:

$$W = (\exists x \, p(x) \rightarrow C) \wedge \exists y \, (p(y) \wedge \neg \, C).$$

a. Show that W is unsatisfiable.

b. Remove the two existential quantifiers from W with Skolem's rule (9.1). Show that the resulting wff is satisfiable.

c. Eliminate $\rightarrow$ from W and then apply (9.1) to the wff obtained. Show that the resulting wff is satisfiable.

d. Apply Skolem's algorithm correctly to W and show that the resulting clausal form is unsatisfiable.

9.2 Logic Programming

In this section we'll see how logic programming is related to logic. To start things off we'll give a gentle introduction to the Prolog language with a discussion about family trees. Then we'll see how resolution comes into the picture as a computation device for logic programs. Finally, we'll study a few basic techniques of logic programming.

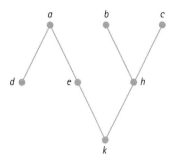

Figure 9.1   A family.

9.2.1 Family Trees

We'll start with a set of parent-child relations. If we let $p(x, y)$ mean "x is a parent of y," then we should start with some parent-child facts. For example, suppose the graph in Figure 9.1 represents a set of parent-child relations, where each node represents the root of a partial family tree with children directly below it and parents directly above it.

For example, d and e are the children of a. We can represent this tree with 6 parent-child facts as follows, where in Prolog each fact ends with a period:

$$p(a, d).$$
$$p(a, e).$$
$$p(b, h).$$
$$p(c, h).$$
$$p(e, k).$$
$$p(h, k).$$

Finding the Parents of a Person

Suppose that we want to find the parents of k. In Prolog we can find them by typing the following goal or query in response to the interactive prompt |?–, where the uppercase letter X stands for a variable.

$$|?–\ p(X,\ k).$$

The computation will search the program facts trying to match $p(X, k)$ with some fact in the program. In this case, the goal matches $p(e, k)$ and the system responds with

$$X = e\ ?$$

At this point, we either stop the computation or ask it to continue. If we stop it, then most systems simply answer yes, indicating that an answer was found.

If we continue, then the system will search for another match for the goal atom $p(X, k)$. In this case, the goal matches $p(h, k)$ and the system responds with

$$X = h \; ?$$

If we ask it to continue, it will search for another match. But there are none and the system responds with the answer no.

Finding the Grandparents of a Person

Now suppose we want to find some grandparent relations. To do this we can let $g(x, y)$ mean "x is a grandparent of y." From our knowledge of family relations we know that x is a grandparent of y if there is some z such that x is a parent of z and z is a parent of y. So g can be described in terms of p as follows:

$$g(x, y) \text{ if } p(x, z) \text{ and } p(z, y).$$

In Prolog, we represent this relationship as the following expression, where the variables start with uppercase letters:

$$g(X, Y) :\!- p(X, Z), p(Z, Y).$$

With the addition of this clause, the Prolog program now looks like the following collection of statements.

$$p\,(a, d)\,.$$
$$p\,(a, e)\,.$$
$$p\,(b, h)\,.$$
$$p\,(c, h)\,.$$
$$p\,(e, k)\,.$$
$$p\,(h, k)\,.$$
$$g\,(X, Y) : -p\,(X, Z),\; p\,(Z, Y)\,.$$

Suppose that we want to find the grandparents of k. We can find them by typing the following goal or query.

$$|?\!- g(U, k).$$

The system will search the program statements trying to match $g(U, k)$ with a fact or the left part of a clause. In this case, it matches $g(X, Y)$ with the unifier $\{U/X, Y/k\}$. Now this unifier is applied to the antecedents $p(X, Z)$ and $p(Z, Y)$ to obtain the two new goals

$$p(X, Z), p(Z, k).$$

These atoms have to be unified with some facts before the answer yes can be returned. So the system searches for two facts to match the two atoms. One such match to be found is the pair

$$p(a, e), p(e, k)$$

with the unifier $\{X/a,\ Z/e\}$. Then the composition of the two unifiers is applied to U:

$$U\ \{U/X,\ Y/k\}\ \{X/a,\ Z/e\} = X\ \{X/a,\ Z/e\} = a.$$

So the system responds with

$$U = a\ ?$$

If we continue, the system will find a match for the pair $p(b, h),\ p(h, k)$ with the unifier $\{X/b, Z/h\}$. So the computation responds with

$$U = b\ ?$$

If we continue, the system will find a match for the pair $p(c, h),\ p(h, k)$ with the unifier $\{X/c, Z/h\}$. So the computation responds with

$$U = c\ ?$$

If we continue the computation, the system will return the answer no because there are no other grandparents listed for k.

Of course we're only touching the surface of the kind of information that we can extract from the parent-child relations. For example, we might want to know answers to questions regarding ancestors, descendants, cousins, and so on. We'll see later that it is an easy matter to define rules to allow us to answer many such questions.

Now that we have a little knowledge of Prolog, in the next few paragraphs we'll define what a logic program is, and we'll show how logic program computations are performed.

9.2.2 Definition of a Logic Program

A *logic program* is a set of clauses, where each clause has exactly one positive literal (i.e., an atom) and zero or more negative literals. Such clauses have one of the following two forms, where A, B_1, ..., B_n are atoms:

A (one positive and no negative literals)

$A \vee \neg B_1 \vee \cdots \vee \neg B_n$ (one positive and some negative literals).

Notice how we can use equivalences to write a clause as a conditional:

$$A \vee \neg B_1 \vee \cdots \vee \neg B_n \equiv A \vee \neg (B_1 \wedge \cdots \wedge B_n) \equiv B_1 \wedge \cdots \wedge B_n \to A.$$

Such a clause is denoted by writing it backward as the following expression, where the conjunctions are replaced by commas.

$$A \leftarrow B_1, \ldots, B_n.$$

We can read this clause as "A is true if B_1, ..., B_n are all true." The atom A on the left side of the arrow is called the *head* of the clause and the atoms on the right form the *body* of the clause. A clause consisting of a single atom A can be read as "A is true." Such a clause is called a *fact* or a *unit* clause. It is a clause without a body.

Goals or Queries

Since a logic program is a set of clauses, we can ask whether anything can be inferred from the program by letting the clauses be premises. In fact, to execute a logic program we must give it a *goal* or *query*, which is a clause of the following form, where $B_1, \ldots, B_n$ are atoms.

$$\leftarrow B_1, \ldots, B_n.$$

It is a clause without a head. We can read the goal as "Are $B_1, \ldots, B_n$ inferred by the program?"

We should note that the clauses in logic programs are often called *Horn* clauses.

example   **9.19 The Result of an Execution**

Let P be the logic program consisting of the following three clauses:

$$q\,(a).$$
$$r\,(a).$$
$$p\,(x) \leftarrow q\,(x)\,, r\,(x).$$

Suppose we execute P with the following goal or query:

$$\leftarrow p(a).$$

We can read the query as "Is $p(a)$ true?" or "Is $p(a)$ inferred from P?" The answer is yes. We can argue informally. The two program facts tell us that $q(a)$ and $r(a)$ are both true. The program clause tells us that $p(x)$ is true if $q(x)$ and $r(x)$ are both true. So we infer that $p(a)$ is true by modus ponens. In what follows we'll see how the answer follows from resolution.

end example

9.2.3 Resolution and Logic Programming

Let's see why things are set up to use resolution. First of all, a logic program is a set of clauses, so it is already in the proper form for using resolution. To execute a logic program we need a goal. So suppose P is a logic program and we execute P with the following goal:

$$\leftarrow B_1, \ldots, B_n.$$

For this discussion we'll denote the existential closure of $B_1 \wedge \cdots \wedge B_n$ by

$$\exists\,(B_1 \wedge \cdots \wedge B_n).$$

With this notation, we read the goal $\leftarrow B_1, \ldots, B_n$ as

"Is $\exists\,(B_1 \wedge \cdots \wedge B_n)$ inferred by program P?"

In other words, the goal asks if there is a proof of $\exists\,(B_1 \wedge \cdots \wedge B_n)$ using the clauses of P as premises. Equivalently, the goal asks if there is a proof that the set $P \cup \{\neg\,\exists\,(B_1 \wedge \cdots \wedge B_n)\}$ is unsatisfiable.

Now we're getting somewhere because resolution is used to prove unsatisfiability. Now let's notice that $\neg\,\exists\,(B_1 \wedge \cdots \wedge B_n)$ can be written in the following form:

$$\neg\,\exists\,(B_1 \wedge \cdots \wedge B_n) \equiv \forall\,\neg\,(B_1 \wedge \cdots \wedge B_n) \equiv \forall\,(\neg\,B_1 \vee \cdots \vee \neg\,B_n).$$

Now, as luck would have it, the wff $\forall\,(\neg\,B_1 \vee \cdots \vee \neg\,B_n)$ is none other than a clause, where $\forall$ denotes universal closure. So the goal $\leftarrow B_1, \ldots, B_n$ becomes the following statement about a set of clauses:

"Is there a proof that the set $P \cup \{\forall\,(\neg\,B_1 \vee \cdots \vee \neg\,B_n)\}$ is unsatisfiable?"

As with the other clauses, we'll delete $\forall$ from the notation for the clause. So the goal $\leftarrow B_1, \ldots, B_n$ becomes the following statement about a set of clauses.

"Is there a proof that the set $P \cup \{(\neg\,B_1 \vee \cdots \vee \neg\,B_n)\}$ is unsatisfiable?"

Now the goal $\leftarrow B_1, \ldots, B_n$ is just notation for the clause $(\neg\,B_1 \vee \cdots \vee \neg\,B_n)$. So the statement can be phrased strictly in terms of logic program notation as

"Is there a proof that the set $P \cup \{\leftarrow B_1, \ldots, B_n\}$ is unsatisfiable?"

Answers to Goals

When we give a goal to a logic program, we usually want more than just the answer yes or no to the goal question. If the answer is yes, we may want to know the values of any variables that appear in the goal. The nice thing about resolution is that the unifiers constructed during the proof provide values for the variables. So we really can read the goal question as "Is there a substitution θ for the variables of $B_1, \ldots, B_n$ such that $(B_1 \wedge \cdots \wedge B_n)\theta$ is inferred by the program P?"

Let's look at an example to see how the notation for logic program clauses makes it easy to find answers to goal questions.

example **9.20** **Answering a Goal Question**

Let P be the following logic program.

$$q\,(a)\,.$$
$$p\,(f\,(x)) \leftarrow q\,(x)\,.$$

Suppose also that we have the following goal question:

$$\leftarrow p(y).$$

This means that we want an answer to the question, "Is there a substitution θ such that $p(y)\theta$ is inferred from P?" Let's give the answer first and then see how we got it. The answer is yes. Letting $\theta = \{y/f(a)\}$, we can evaluate $p(y)\theta$ as follows:

$$p(y)\theta = p(y)\{y/f(a)\} = p(f(a)).$$

We claim that $p(f(a))$ is inferred from P. Let's give a resolution proof showing that $P \cup \{\neg\, p(y)\}$ is unsatisfiable. First we'll write the two program clauses and the goal clause in the clausal form needed for resolution. Then we'll write them as premises. Here's the proof.

1. $q(a)$ P [program clause $q(a)$]
2. $p(f(x)) \vee \neg\, q(x)$ P [program clause $p(f(x)) \leftarrow q(x)$]
3. $\neg\, p(y)$ P [goal clause $\leftarrow p(y)$]
4. $\neg\, q(x)$ 2, 3, R, $\{y/f(x)\}$
5. $\square$ 1, 4, R, $\{x/a\}$
    QED.

Therefore, by the resolution theorem, $P \cup \{\neg\, p(y)\}$ is unsatisfiable. So the answer to the goal question is yes. But what value of y does the job? We can find it by applying the composition of the mgu's to y as follows:

$$y\, \{y/f(x)\}\, \{x/a\} = f(x)\, \{x/a\} = f(a).$$

Therefore, $p(f(a))$ is a logical consequence of program P.

end example

SLD–Resolution

There are three advantages to the notation used for logic programs.

1. The notation is easy to write down because we don't have to use the symbols $\neg$, $\wedge$, and $\vee$.

2. The notation allows us to interpret a program in two different ways. For example, suppose we have the clause $A \leftarrow B_1, \ldots, B_n$. This clause has the usual logical interpretation "A is true if $B_1, \ldots, B_n$ are all true." The clause also has the procedural interpretation "A is a procedure that is executed by executing the procedures $B_1, \ldots, B_n$ in the order they are written." Most logic programming systems allow this procedural interpretation.

3. The notation makes it easy to apply the resolution rule. We'll discuss this next.

Whenever we apply the resolution rule, we have to do a lot of choosing. We have to choose two clauses to resolve, and we have to choose literals to "cancel" from each clause. Since there are many choices, it's easy to understand why we can come up with many different proof sequences. When resolution is used with logic program clauses, we can specialize the rule.

The specialized rule always picks one clause to be the most recent line of the proof, which is always a goal clause. Start the proof by picking the initial goal. Select the leftmost atom in the goal clause as the literal to "cancel." For the second clause, pick a program clause whose head unifies with the atom selected from the goal clause. The resolvent of these two clauses is created by first replacing the leftmost atom in the goal clause by the body atoms of the program clause and then applying the unifier to the resulting goal. Here is a formal description of the rule, which is called the *SLD-resolution* rule:[1]

SLD–Resolution Rule $\hspace{6cm}$ (9.11)

To resolve the goal $\leftarrow B_1, \ldots, B_k$ with the clause $A \leftarrow A_1, \ldots, A_n$, perform the following steps:

1. Unify B_1 and A and obtain a most general unifier θ.

2. Replace B_1 by the body atoms $A_1, \ldots, A_n$.

3. Apply θ to the result to obtain the resolvent

$$\leftarrow (A_1, \ldots, A_n, B_2, \ldots, B_k)\theta.$$

Constructing a Logic Program Proof

To construct a logic program proof, we start by listing each program clause as a premise. Then we write the goal clause as a premise. Now we use (9.11) repeatedly to add new resolvents to the proof, each new resolvent being constructed from the goal on the previous line together with some program clause. We can summarize the application of (9.11) with the following four-step procedure:

1. Pick the goal clause on the last line of the partial proof, and select its leftmost atom, say B_1.

2. Find a program clause whose head unifies with B_1, say by θ. Be sure the two clauses have distinct sets of variables (rename if necessary).

3. Replace B_1 in the goal clause with the body of the program clause.

4. Apply θ to the goal constructed on line 3 to get the resolvent, which is placed on a new line of the proof.

---

[1]SLD-resolution means selective linear resolution of definite clauses. In our case we always "select" the leftmost atom of the goal clause.

The Family Tree Revisited

We'll introduce the use of the SLD-resolution rule by revisiting family tree relations. Suppose we are given the following logic program, where p means isParentOf and g means isGrandparentOf.

$$p(a, b).$$
$$p(d, b).$$
$$p(b, c).$$
$$g(x, y) \leftarrow p(x, z),\ p(z, y).$$

We'll execute the program by giving it the following goal question:

$$\leftarrow g(w,\ c).$$

Since there is a variable w in this goal, we can read the goal as the question

"Is there a grandparent for c?"

The resolution proof starts by letting the program clauses and the goal clause be premises. For this example we have the following five lines:

1. $p(a,\ b)$ P
2. $p(d,\ b)$ P
3. $p(b,\ c)$ P
4. $g(x,\ y) \leftarrow p(x,\ z),\ p(z,\ y)$ P
5. $\leftarrow g(w,\ c)$ P Initial goal

The proof starts by resolving the initial goal on line 5 with some program clause. The atom $g(w,\ c)$ from the initial goal unifies with $g(x,\ y)$, the head of the program clause on line 4, by the mgu

$$\theta_1 = \{w/x,\ y/c\}.$$

Therefore, we can use (9.11) to resolve the two clauses on lines 4 and 5. So we replace the goal atom $g(w,\ c)$ on line 5 with the body of the clause on line 4 and then apply the mgu θ_1 to the result to obtain the following resolvent.

$$p(x,\ z),\ p(z,\ c).$$

Let's compare what we've just done for logic program clauses using (9.11) to the case for first-order clauses using (9.8). The following two lines are copies of lines 4 and 5 in which we've included the clausal notation for each logic program clause:

Logic Program Notation	Clausal Notation
4. $g(x,\ y) \leftarrow p(x,\ z),\ p(z,\ y)$	$g(x,\ y) \vee \neg\, p(x,\ z) \vee \neg\, p(z,\ y)$
5. $\leftarrow g(w,\ c)$	$\neg\, g(w,\ c)$

We apply (9.11) to the logic program notation clauses, and we apply (9.8) to the clauses in clausal notation. This gives the following resolvant.

Logic Program Notation	Clausal Notation
$\leftarrow p(x, z), p(z, c)$	$\neg\, p(x, z) \vee \neg\, p(z, c)$

So we get the same answer with either method.

Now let's continue the proof. We'll write down the new resolvant on line 6 of our proof, in which we've added the mgu to the reason column:

$$6. \; \leftarrow p(x, z), p(z, c) \qquad 4, 5, R, \theta_1 = \{w/x, y/c\}$$

To continue the proof according to (9.11), we must choose this new goal on line 6 for one of the clauses, and we must choose its leftmost atom $p(x, z)$ for "cancellation." For the second clause we'll choose the clause on line 1 because its head $p(a, b)$ unifies with our chosen atom by the mgu

$$\theta_2 = \{x/a, z/b\}.$$

To apply (9.11), we must replace $p(x, z)$ on line 6 by the body of the clause on line 1 and then apply θ_2 to the result. Since the clause on line 1 does not have a body, we simply delete $p(x, z)$ from line 6 and apply θ_2 to the result, obtaining the resolvant $\leftarrow p(b, c)$. So we have a new goal:

$$\leftarrow p(b, c).$$

Let's compute this result in terms of both (9.11) and (9.8). The clauses on lines 1 and 6 take the following forms, in which we've added the regular clausal notation for each clause.

Logic Program Notation	Clausal Notation
1. $p(a, b)$	$p(a, b)$
6. $\leftarrow p(x, z), p(z, c)$	$\neg\, p(x, z) \vee \neg\, p(z, c)$

After applying (9.11) and (9.8) to the respective notations on lines 1 and 6, we obtain the following resolvant:

Logic Program Notation	Clausal Notation
$\leftarrow p(b, c)$	$\neg\, p(b, c)$

So we can continue the proof by writing down the new resolvant on line 7.

$$7. \; \leftarrow p(b, c) \qquad 1, 6, R, \theta_2 = \{x/a, z/b\}$$

To continue the proof using (9.11), we must choose the goal on line 7 together with its only atom $p(b, c)$. It unifies with the head $p(b, c)$ of the clause on line 3 by the empty unifier

$$\theta_3 = \{\ \}.$$

Since there is only one atom in the goal clause of line 7 and there is no body in the clause on line 3, it follows that the resolvant of the clauses on these two lines is just the empty clause. So our proof is completed by writing this information on line 8.

8.   □   3, 7, R, $\theta_3 = \{\ \}$
  QED.

To finish things off, we'll collect the eight steps of the proof and rewrite them as a single unit:

1. $p(a, b)$ P
2. $p(d, b)$ P
3. $p(b, c)$ P
4. $g(x, y) \leftarrow p(x, z), p(z, y)$ P
5. $\leftarrow g(w, c)$ P [initial goal]
6. $\leftarrow p(x, z), p(z, c)$ 4, 5, R, $\theta_1 = \{w/x, y/c\}$
7. $\leftarrow p(b, c)$ 1, 6, R, $\theta_2 = \{x/a, z/b\}$
8.   □   3, 7, R, $\theta_3 = \{\ \}$
  QED.

Since □ was obtained, the answer is yes to the goal

$$\leftarrow g(w, c).$$

In other words, "Yes, there is a grandparent w of c." But we want to know more. We want to know the value of w that gives a yes answer. We can recover the value by composing the three unifiers θ_1, θ_2, and θ_3 and then applying the result to w:

$$w\theta_1\theta_2\theta_3 = a.$$

So the goal question "Is there a grandparent w of c?" is answered

Yes,
$$w = a.$$

What about other possibilities? By looking at the facts, we notice for this example that d is also a grandparent of c. Can this answer be computed? Sure. Keep the first six lines of the proof as they are. Then resolve the goal on line 6 with the clause on line 2. The goal atom $p(x, z)$ on line 6 unifies with $p(d, b)$ on line 2 by mgu

$$\theta_2 = \{x/d, z/b\}.$$

This θ_2 is different from the previous θ_2. So we get a new line 7 and the same line 8 to obtain the following alternative proof.

7.	$\leftarrow p(b,\ c)$	2, 6, R, $\theta_2 = \{x/d,\ z/b\}$
8.	$\square$	3, 7, R, $\theta_3 = \{\ \}$
	QED.	

With this proof we obtain the following new value for w:

$$w\theta_1\theta_2\theta_3 = d.$$

So the goal question "Is there a grandparent w of c?" can also be answered

Yes,

$w = d.$

We can observe from the facts that a and d are the only grandparents of c, and we've come up with proofs to calculate them. So it's time to see whether a computation can actually come up with both answers.

9.2.4 Computation Trees

Now that we have an example under our belts, let's look again at the general picture. The preceding proof had two possible yes answers. We would like to find a way to represent all possible answers (i.e., proof sequences) for a goal. For our purposes a tree will do the job.

A *computation tree* for a goal is an ordered tree whose root is the goal. The children of any parent node are all the possible goals (i.e., resolvants) that can be obtained by resolving the parent goal with a program clause. We agree to order the children of each node from left to right in terms of the top-to-bottom ordering of the program clauses that are used with the parent to create the children. Each parent-child branch is labeled with the mgu obtained to create the child. A leaf may be the empty clause or a goal. If the empty clause occurs as a leaf, we write "yes" together with the values of any variables that occur in the original goal at the root of the tree. If a goal occurs as a leaf, this means that it can't be resolved with any program clause, so we write "failure." The computation tree will always show all possible answers for the given goal at its root.

For example, the computation tree for the goal $g(w,\ c)$ with respect to our example program can be pictured as shown in Figure 9.2. Notice that the tree contains all possible answers to the goal question.

Searching the Computation Tree

A logic programming system needs a strategy to search the computation tree for a leaf with a yes answer. The strategy used by most Prolog systems is the *depth-first* search strategy, which starts by traversing the tree down to the leftmost leaf. If the leaf is the empty clause, then the yes answer is reported. If the leaf is a failure leaf, then the search returns to the parent of the leaf. At this point a depth-first search is started at the next child to the right. If there is no next child, then the search returns to the parent of the parent, and a depth-first search starts with its next child to the right, and so on. If this process eventually

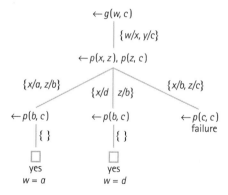

Figure 9.2 Computation tree.

returns to the root of the tree and there are no more paths to search, then failure is reported.

It might be desirable for a logic programming system to attempt to find all possible answers to a goal question. One strategy for attempting to find all possible answers is called *backtracking*. For example, with depth-first search we perform backtracking by continuing the depth-first search process from the point at which the last yes answer was found. In other words, when a yes answer is found, the system reports the answer and then continues just as though a failure leaf was encountered.

In the next few examples we'll construct some computation trees and discuss the problems that can arise in trying to find all possible answers to a goal question.

example **9.21 Many Possible Answers**

Let's consider the following two-clause program:

$$p(a).$$
$$p(\text{succ}(x)) \leftarrow p(x).$$

Suppose we give the following goal to the program:

$$\leftarrow p(x).$$

This goal will resolve with either one of the program clauses. So the root of the computation tree has two children. One child, the empty clause, results from the resolution of goal $\leftarrow p(x)$ with the fact $p(a)$. The other child results from the resolution of goal $\leftarrow p(x)$ with the clause $p(\text{succ}(x)) \leftarrow p(x)$.

But before this happens, we need to change variables. We'll replace x by x_1 in the program clause to obtain $p(\text{succ}(x_1)) \leftarrow p(x_1)$. Now resolving the goal $\leftarrow p(x)$ with this clause produces the goal $\leftarrow p(x_1)$, which becomes the second child of the root.

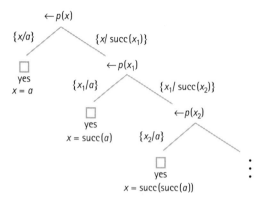

Figure 9.3 Infinitely many answers.

The process starts all over again with the goal $\leftarrow p(x_1)$. To keep track of variable names, we'll replace x by x_2 in the second program clause. Then resolve the goal $\leftarrow p(x_1)$ with $p(\mathrm{succ}(x_2)) \leftarrow p(x_2)$ to obtain the goal $\leftarrow p(x_2)$. This process continues forever.

The computation tree for this example is shown in Figure 9.3. It is an infinite tree that continues the indicated pattern forever. If we use the depth-first search rule, the first answer is "yes, $x = a$." If we force backtracking, the next answer we'll get is "yes, $x = \mathrm{succ}(a)$." If we force backtracking again, we'll get the answer "yes, $x = \mathrm{succ}(\mathrm{succ}(a))$." Continuing in this way, we can generate the following infinite sequence of possible values for x:

$$a, \ \mathrm{succ}(a), \ \mathrm{succ}(\mathrm{succ}(a)), \ldots, \ \mathrm{succ}^k(a), \ldots.$$

end example

example **9.22 Two Possible Answers**

Consider the following three-clause program, in which the third clause has more than one atom in its body.

$$q(a).$$
$$p(a).$$
$$p(f(x)) \leftarrow p(x), \ q(x).$$

Figure 9.4 shows a few levels of the computation tree for the goal $\leftarrow p(x)$. Notice that as we travel down the rightmost path from the root, the number of goal atoms at each node is increased by one for each new level.

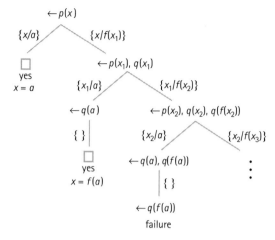

Figure 9.4 Only two answers.

Using the depth-first search rule, we obtain the answer "yes, $x = a$." Backtracking works one time to give the answer "yes, $x = f(a)$." If we force backtracking again, the computation takes an infinite walk down the tree, failing at each leaf.

end example

example **9.23 How to Miss Many Answers**

Suppose we're given the following three-clause program.

$$p\left(f\left(x\right)\right) \leftarrow p\left(x\right).$$
$$p\left(a\right).$$
$$p\left(b\right).$$

We'll start with the goal

$$\leftarrow p(x).$$

The computation tree will be a ternary tree because there are three "p" clauses that match each goal. Figure 9.5 shows the first few levels of the tree. The tree is infinite, and there are infinitely many yes answers to the goal question. The infinite sequence of possible values for x is

$$a,\ b,\ f(a),\ f(b),\ f(f(a)),\ f(f(b)),\ldots,\ f^{k}(a),\ f^{k}(b),\ldots.$$

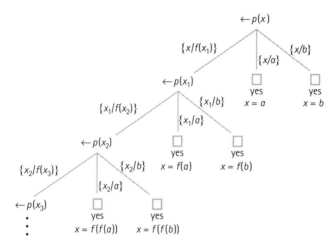

Figure 9.5   Many answers to miss.

Notice that if we used the depth-first search strategy, then the computation would take an infinite walk down the left branch of the tree. So although there are infinitely many answers, the depth-first search strategy won't find even one of them.

end example

   In the preceding example, depth-first search did not find any answers to the goal $\leftarrow p(x)$. Suppose we reordered the three program clauses as follows:

$$p(a).$$
$$p(b).$$
$$p(f(x)) \leftarrow p(x).$$

The computation tree corresponding to these three clauses can be searched in a depth-first fashion with backtracking to generate all the answers to the goal $\leftarrow p(x)$. Suppose we write the three clauses in the following order:

$$p(a).$$
$$p(f(x)) \leftarrow p(x).$$
$$p(b).$$

The computation tree for these three clauses, when searched with depth-first and backtracking, will yield some, but not all, of the possible answers.

   So when a logic programming language uses depth-first search, two problems can occur when the computation tree for a goal is infinite:

1. The yes answers found may depend on the order of the clauses.

2. Backtracking might not find all possible yes answers to a goal.

Many logic programming systems use the depth-first search strategy because it's efficient to implement and because it reflects the procedural interpretation of a clause. For example, the clause $A \leftarrow B, C$ represents a procedure named A that is executed by first calling procedure B and then calling procedure C.

Another search strategy is called *breadth-first search*. It looks for a yes answer by examining all the children of the root. Then it looks at all nodes at the next level of the tree, and so on. This strategy will find all possible answers to a goal question.

Some implementation strategies for searching the computation tree use breadth-first search with a twist. All children of a node are searched in parallel. A search at a particular node is started only when the goal atom has not already occurred at a higher level in the tree. If the goal atom matches a goal at a higher level in the tree, then the process waits for the answer to the other goal. When it receives the answer, then it continues with its search. This technique requires a table containing previous goal atoms and answers. It has proved useful in detecting certain kinds of loops that give rise to infinite computation trees. In some cases the search process won't take an infinite walk. An introduction to these ideas is given in Warren [1992].

9.2.5 Logic Programming Techniques

Let's spend some time discussing a few elementary techniques to construct logic programs. First we'll see how to construct logic programs that process relations. Then we'll discuss logic programs that process functions. The clauses in our examples are ordered to take advantage of the depth-first search strategy. This strategy is used by most Prolog systems.

A Technique for Relations

Logic programming allows us to easily process many relations because relations are just predicates. For example, suppose that we want to write the isAncestorOf relation in terms of the isParentOf relation, where an ancestor is either a parent, or a grandparent, or a great-grandparent, and so on. The next example discusses a technique for solving this type of problem.

example **9.24 Acyclic Transitive Closure**

The isAncestorOf relation is the transitive closure of the isParentOf relation. In general terms, suppose we're given a binary relation r whose graph is acyclic (i.e., there are no cycles) and we need to compute the transitive closure of r. If we let tc denote the transitive closure of r, the following two-clause program does the job:

$$\text{tc}\,(x, y) \leftarrow r\,(x, y).$$
$$\text{tc}\,(x, y) \leftarrow r\,(x, z),\ \text{tc}\,(z, y).$$

For example, suppose r is the isParentOf relation. Then tc is the isAncestorOf relation. The first clause can be read as "x is an ancestor of y if x is a parent of y," and the second clause can be read as "x is an ancestor of y if x is a parent of z and z is an ancestor of y." Here's the Prolog version of the program:

$$tc(X, Y) : -r(X, Y).$$
$$tc(X, Y) : -r(X, Z), tc(Z, Y).$$

end example

A Technique for Computing Functions

Now let's see whether we can find a technique to construct logic programs to compute functions. Actually, it's pretty easy. The major thing to remember in translating a function definition to a logic definition is that a functional equation like

$$f(x) = y$$

can be represented by a predicate expression such as

$$pf(x, y).$$

The predicate name "pf" can remind us that we have a "predicate for f." The predicate expression $pf(x, y)$ can still be read as "f of x is y."

Now let's discuss a technique to construct a logic program for a recursively defined function. If f is defined recursively, then there is at least one part of the definition that defines $f(x)$ in terms of some $f(y)$. In other words, some part of the definition of f has the following form, where $E(f(y))$ denotes an expression containing $f(y)$:

$$f(x) = E(f(y)).$$

Using our technique to create a predicate for this functional equation, we get the following expression:

$$pf(x, E(f(y))).$$

But we aren't done yet because the recursive definition of f causes $f(y)$ to occur as an argument in the predicate. Since we're trying to compute f by the predicate pf, we need to get rid of $f(y)$. The solution is to replace $f(y)$ by a new variable z. We can represent this replacement by writing down the following version of the expression:

$$pf(x, E(z)) \text{ where } z = f(y).$$

Now we have a functional equation $z = f(y)$, which we can replace by $pf(y, z)$. So we obtain the following expression:

$$pf(x, E(z)) \text{ where } pf(y, z).$$

The transformation to a logic program is now simple: Replace the word "where" by the symbol ← to obtain a logic program clause as follows:

$$\text{pf}(x,\ E(z)) \leftarrow \text{pf}(y,\ z).$$

So we have a general technique to transform a functional equation into a logic program. Here are the steps, all in one place:

$f(x) = E\left(f\left(y\right)\right)$	The given functional equation.
$\text{pf}(x, E\left(f\left(y\right)\right))$	Create a predicate expression.
$\text{pf}(x, E\left(z\right))$ where $z = f\left(y\right)$	Let $z = f\left(y\right)$.
$\text{pf}(x, E\left(z\right))$ where $\text{pf}(y, z)$	Create a predicate expression.
$\text{pf}(x, E\left(z\right)) \leftarrow \text{pf}(y, z)$	Create a clause.

Of course, there may be more work to do depending on the complexity of the expression $E(z)$. Let's do some examples to help get the look and feel of this process.

example **9.25 The Length Function**

Suppose we want to write a logic program to compute the length of a list. A recursively defined function f that does the job can be written as follows:

$$f(L) = \text{if } L = \langle\ \rangle \text{ then } 0 \text{ else } 1 + f(\text{tail}(L)).$$

It is easier to construct a logic program if we rewrite f in equational form as follows:

$$f(\langle\rangle) = 0$$
$$f(x :: T) = 1 + f(T).$$

We can start by writing down two predicate expressions to represent these two functional equations. We'll use the predicate name "length" as follows:

$$\text{length}(\langle\rangle, 0)$$
$$\text{length}(x :: T, 1 + f(T)).$$

The second expression contains an occurrence of the function f, which we're trying to define. So we'll replace $f(T)$ by a new variable z to obtain

$$\text{length}(x :: T, 1 + z) \text{ where } z = f(T).$$

Now replace the functional equation $z = f(T)$ by the predicate expression $\text{length}(T, z)$ to obtain

$$\text{length}(x :: T, 1 + z) \text{ where } \text{length}(T, z).$$

Now convert this expression to a logic program clause to obtain

$$\text{length}(x :: T, 1 + z) \leftarrow \text{length}(T, z).$$

So we have the following two-clause logic program:

$$\text{length}(\langle \rangle, 0).$$
$$\text{length}(x :: T, 1 + z) \leftarrow \text{length}(T, z).$$

Here's the Prolog version of the program:

$$\text{length}([\,], 0).$$
$$\text{length}([X|T], N) : -\text{length}(T, Z), N \text{ is } Z + 1.$$

9.26 Deleting an Element

Suppose we want to write a logic program to delete the first occurrence of an element from a list. A recursively defined function to do the job can be written as follows:

$$\begin{aligned}
\text{delete}\,(x, L) = \ &\text{if } L = \langle \rangle \text{ then } \langle \rangle \\
&\text{else if head}\,(L) = x \text{ then tail}\,(L) \\
&\text{else head}\,(L) :: \text{delete}\,(x, \text{tail}\,(L)).
\end{aligned}$$

It is easier to construct a logic program if we rewrite the function in equational form as follows, where the equations are applied in the order they are written:

$$\text{delete}(x, \langle \rangle) = \langle \rangle$$
$$\text{delete}(x, x :: T) = T$$
$$\text{delete}(x, y :: T) = y :: \text{delete}(x, T).$$

First we'll convert each equation to a predicate expression using the predicate named "remove" as follows:

$$\text{remove}(x, \langle \rangle, \langle \rangle)$$
$$\text{remove}(x, x :: T, T)$$
$$\text{remove}(x, y :: T, y :: \text{delete}(x, T)).$$

Since the functional value $\text{delete}(x, T)$ occurs in the third expression, we'll replace it by a new variable U to obtain

$$\text{remove}(x, y :: T, y :: U) \text{ where } U = \text{delete}(x, T).$$

Now we can replace the equation $U = \text{delete}(x, T)$ by the predicate expression remove(x, T, U) to obtain

$$\text{remove}(x, y :: T, y :: U) \text{ where } \text{remove}(x, T, U).$$

Now convert this expression to a logic program clause to obtain

$$\text{remove}(x, y :: T, y :: U) \leftarrow \text{remove}(x, T, U).$$

So we have the following three-clause logic program:

$$\text{remove}(x, \langle \rangle, \langle \rangle).$$
$$\text{remove}(x, x :: T, T).$$
$$\text{remove}(x, y :: T, y :: U) \leftarrow \text{remove}(x, T, U).$$

Here's the Prolog version of the program:

$$\text{remove}(X, [\,], [\,]).$$
$$\text{remove}(X, [X|T], T).$$
$$\text{remove}(X, [Y|T], [Y|U]) : -\text{remove}(X, T, U).$$

end example

9.2.6 A Note on Not Provable

Suppose the execution of a logic program P with goal atom A fails. Then we can say that A is not a logical consequence of P or that A is not provable from P. Can we say anything about $\neg A$? Since SLD-resolution only processes atoms, the only logical consequences of a program are atoms. In other words, if A is an atom, then $\neg A$ can never be a logical consequence of P.

However, the notion of "not provable" is useful in programming. Prolog uses the two-character symbol $\backslash +$ to denote "not provable," where a goal of the form $\backslash + A$ is processed by trying to satisfy the goal A. If A succeeds, then $\backslash + A$ fails. If A fails, then $\backslash + A$ succeeds.

example **9.27 The Blocks World**

Consider the following Prolog program to find out information about the blocks world. This world consists of stacks of building blocks, where we represent the fact that block a is on block b with the fact on(a, b). For example, a stack of three blocks can be represented with the facts

$$\text{on}(a, b).$$
$$\text{on}(b, c).$$

How can we tell if some block is on top of a stack? Let onTop(X) mean that X is on the top of a stack. Here is an implementation of the predicate:

$$\text{onTop}(X) :- \ \backslash + \text{on}(Y,\,X).$$

For our example, the goal onTop(a) executes by calling the goal $\backslash +$ on(Y, a). So the computation calls the goal on(Y, a), which fails. Therefore, $\backslash +$ on(Y, a) succeeds, so onTop(a) succeeds. Thus a is on top of a stack. Similarly, the goal onTop(b) fails because on(Y, b) succeeds. So b is not on top of the stack.

Notice that if d is any object not listed as one of the blocks, then the goal on(Y, d) fails. So $\backslash +$ on(Y, d) succeeds and thus onTop(d) also succeeds, which does not make any sense because d is not on top of any stack represented by the program. So care must be taken when using "not provable."

end example

The "not provable" rule has a formal name in logic programming, where it is called *negation as failure*. The name comes from a theorem that, roughly speaking, says that if the computation tree for P and A is finite with failure at each leaf, then $\neg \ A$ is a logical consequence of a set of first-order wffs that can be constructed from P.

 Exercises

Family Trees

1. Suppose you are given an isParentOf relation. Find a definition for each of the following relations.

   a. isChildOf.
   b. isGrandchildOf.
   c. isGreatGrandparentOf.

2. Suppose you are given an isParentOf relation. Try to find a definition for each of the following relations. *Hint:* You might want to consider some kind of test for equality.

   a. isSiblingOf.
   b. isCousinOf.
   c. isSecondCousinOf.
   d. isFirstCousinOnceRemovedOf.

Computation by SLD-Resolution

3. Suppose we're given the following logic program:

$$p\,(a, b)\,.$$
$$p\,(a, c)\,.$$
$$p\,(b, d)\,.$$
$$p\,(c, e)\,.$$
$$g\,(x, y) \leftarrow p\,(x, z)\,,\ p\,(z, y)\,.$$

   a. Find a resolution proof for the goal $g(a,\ w)$.

   b. Draw a picture of the computation tree for the goal $g(a,\ w)$.

4. Suppose we're given the following logic program:

$$p\,(a)\,.$$
$$p\,(g\,(x)) \leftarrow p\,(x)\,.$$
$$p\,(b)\,.$$

   a. Draw at least three levels of the computation tree for the goal $p(x)$.

   b. What are the possible yes answers for the goal $p(x)$?

   c. Describe the values of x that are generated by backtracking with the depth-first search strategy for the goal $p(x)$.

5. The following logic program claims to test an integer to see whether it is a natural number, where pred$(x,\ y)$ means that the predecessor of x is y:

$$\text{isNat}\,(0)\,.$$
$$\text{isNat}\,(x) \leftarrow \text{isNat}\,(y)\,,\ \text{pred}\,(x, y)\,.$$

   a. What happens when the goal is isNat(2)?

   b. What happens when the goal is isNat(–1)?

Logic Programming

6. Let r denote a binary relation. Write logic programs to compute each of the following relations.

   a. The symmetric closure of r.

   b. The reflexive closure of r.

7. Translate each of the following functional definitions into a logic program. *Hint:* First, translate the if-then-else definitions into equational definitions.

   a. The function f computes the nth Fibonacci number:

$$f(n) = \text{if } n = 0 \text{ then } 0 \text{ else if } n = 1 \text{ then } 1 \text{ else } f(n - 1) + f(n - 2).$$

b. The function "cat" computes the concatenation of two lists:

$$\text{cat}(x,\, y) = \text{if } x = \langle\,\rangle \text{ then } y \text{ else head}(x) :: \text{cat(tail}(x),\, y).$$

c. The function "nodes" computes the number of nodes in a binary tree:

$$\text{nodes}(t) = \text{if } t = \langle\,\rangle \text{ then } 0 \text{ else } 1 + \text{nodes(left}(t)) + \text{nodes(right}(t)).$$

8. Find a logic program to implement each of the following functions, where the variables represent elements or lists.

    a. equalLists(x, y) tests whether the lists x and y are equal.

    b. member(x, y) tests whether x is an element of the list y.

    c. all(x, y) is the list obtained from y by removing all occurrences of x.

    d. makeSet(x) is the list obtained from x by deleting repeated elements.

    e. subset(x, y) tests whether x, considered as a set, is a subset of y.

    f. equalSets(x, y) tests whether x and y, considered as sets, are equal.

    g. subBag(x, y) tests whether x, considered as a bag, is a subbag of y.

    h. equalBags(x, y) tests whether the bags x and y are equal.

Challenges

9. Suppose we have a schedule of classes with each entry having the form class$(i,\, s,\, t,\, p)$, which means that class i section s meets at time t in place p. Find a logic program to compute the possible schedules available for a given list of classes.

10. Write a logic program to test whether a propositional wff is a tautology. Assume that the wffs use the four operators in the set $\{\neg,\, \wedge,\, \vee,\, \rightarrow\}$. *Hint:* Use the method of Quine together with the fact that if A is a wff containing a propsitional variable p, then A is a tautology iff $A(p/\text{True})$ and $A(p/\text{False})$ are both tautologies. To assist in finding the propositional variables, assume that the predicate atom(x) means that x is a propositional variable.

9.3 Chapter Summary

The major component of automatic reasoning for first-order predicate calculus is the resolution proof rule. Resolution proofs work by showing that a wff is unsatisfiable. So to prove that a wff is valid, we can use resolution to show that its negation is unsatisfiable. Resolution requires wffs to be represented as sets of clauses, which can be constructed by Skolem's algorithm. Before each step of a resolution proof involving predicates, the unification algorithm must calculate a substitution—a most general unifier—that will unify a set of atoms. The process of applying the resolution rule can be programmed to perform automatic reasoning.

Logic programs consist of clauses that have one positive literal and zero or more negative literals. A logic program goal is a clause consisting of one or more negative literals. Logic program goals are computed by a modification of resolution called SLD-resolution. Each goal of a logic program has an associated computation tree that can be searched in a variety of ways. The depth-first search strategy is used by most logic programming languages. Elementary techniques for logic programming include the implementation of relations and recursively defined functions.

chapter 10

Algebraic Structures and Techniques

Algebraic rules of procedure were proclaimed as if they were divine revelations. . . .

— From *The History of Mathematics*
by David M. Burton

The word "algebra" comes from the word "al-jabr" in the title of the textbook *Hisâb al-jabr w'al-muqâbala*, which was written around 820 by the mathematician and astronomer al-Khowârizmî. The title translates roughly to "calculations by restoration and reduction," where restoration—al-jabr—refers to adding or subtracting a number on both sides of an equation, and reduction refers to simplification. We should also note that the word "algorithm" has been traced back to al-Khowârizmî because people used his name—mispronounced—when referring to a method of calculating with Hindu numerals that was contained in another of his books.

Having studied high school algebra, most of us probably agree that algebra has something to do with equations and simplification. In high school algebra we simplified a lot. In fact, we were often given the one word command "simplify" in the exercises. So we tried to somehow manipulate a given expression into one that was simpler than the given one, although this direction was a bit vague, and there always seemed to be a question about what "simplify" meant. We also tried to describe word problems in terms of algebraic equations and then to apply our simplification methods to extract solutions. Everything we did dealt with numbers and expressions for numbers.

In this chapter we'll clarify and broaden the idea of an algebra. The chapter introduces the notions and notations of algebra with special emphasis on the techniques and applications of algebra in computer science.

chapter guide

10.1 (What Is an Algebra?) introduces the idea of an algebra. We'll see that high school algebra is just one kind of algebra.

10.2 (Boolean Algebra) introduces Boolean algebra together with some techniques to simplify Boolean expressions and digital circuits.

10.3 (Abstract Data Types as Algebras) introduces the idea of an abstract data type as an algebra. As examples, we'll discuss some properties of the natural numbers, lists, strings, stacks, queues, binary trees, and priority queues.

10.4 (Computational Algebras) introduces relational algebras and functional algebras. We'll see how these ideas are applied to databases and functional programming.

10.5 (Other Algebraic Ideas) introduces a collection of algebraic ideas that are useful for computational problems. We'll introduce congruences and see how some results are applied in cryptology. We'll also introduce subalgebras and morphisms.

10.1   What Is an Algebra?

Before we say just what an algebra is, let's see how an algebra is used in the problem-solving process. An important part of problem solving is the process of transforming informal word problems into formal things like equations, expressions, or algorithms. Another important part of problem solving is the process of transforming these formal things into solutions by solving equations, simplifying expressions, or implementing algorithms. For example, in high school algebra we tried to describe certain word problems in terms of algebraic equations, and then we tried to solve the equations. An algebra should provide tools and techniques to help us describe informal problems in formal terms and to help us solve the resulting formal problems.

The Description Problem

How can we describe something to another person in such a way that the person understands exactly what we mean? One way is to use examples. But sometimes examples may not be enough for a proper understanding. It is often useful at some point to try to describe an object by describing some properties that it possesses. So we state the following general problem:

The Description Problem: Describe an object.

Whatever form a description takes, it should be communicated in a clear and concise manner so that examples or instances of the object can be easily checked for correctness. Try to describe one of the following things to a friend:

> A car.
> The left side of a person.
> The number zero.
> The concept of area.

Most likely, you'll notice that the description of an object often depends on the knowledge level of the audience.

We need some tools to help us describe properties of the things we are talking about, so we can check not only the correctness of examples, but also the correctness of the descriptions. Algebras provide us with natural notations that can help us give precise descriptions for many things, particularly those structures and ideas that are used in computer science.

High School Algebra

A natural example of an algebra that we all know and love is the algebra of numbers. We learned about it in school, and we probably had different ideas about what it was. First, we learned about arithmetic of the natural numbers $\mathbb{N}$, using the operation of addition. We came eventually to believe things like

$$7 + 12 = 19, \quad 3 + 5 = 5 + 3, \quad \text{and} \quad 4 + (6 + 2) = (4 + 6) + 2.$$

Soon we learned about multiplication, negative numbers, and the integers $\mathbb{Z}$. It seemed that certain numbers like 0 and 1 had special properties such as

$$14 + 0 = 14, \quad 1 \cdot 47 = 47, \quad \text{and} \quad 0 = 9 + (-9).$$

Somewhere along the line, we learned about division, the rational numbers $\mathbb{Q}$, and the fact that we could not divide by zero.

Then came the big leap. We learned to denote numbers by symbols like the letters x and y and by expressions like $x^2 + y$. We spent much time transforming one expression into another, such as $x^2 + 4x + 4 = (x + 2)(x + 2)$. All this had something to do with algebra, perhaps because that was the name of the class.

There are two main ingredients to the algebra that we studied in high school. The first is a set of numbers to work with, such as the real numbers $\mathbb{R}$. The second is a set of operations on the numbers, such as $-$ and $+$. We learned about the general properties of the operations, such as $x + y = y + x$ and $x + 0 = x$. And we learned to use these properties to simplify expressions and solve equations.

Now we are in position to discuss algebra from a more general point of view. We will see that high school algebra is just one of many different kinds of algebra.

10.1.1 Definition of an Algebra

An *algebra* is a structure consisting of one or more sets together with one or more operations on the sets. The sets are often called *carriers* of the algebra. This is a very general definition. If this is the definition of an algebra, how can it help us solve problems? As we will see, the utility of an algebra comes from knowing how to use the operations.

For example, high school algebra is an algebra with the single carrier $\mathbb{R}$, or maybe $\mathbb{Q}$. The operators of the algebra are $+$, $-$, $\cdot$, and $\div$. The constants 0 and 1 are also important to consider because they have special properties. Note that a constant can be thought of as a nullary operation (having arity zero). Many familiar properties hold among the operations, such as the fact that multiplication distributes over addition: $a \cdot (b + c) = a \cdot b + a \cdot c$; and the fact that we can cancel: If $a \neq 0$, then $a \cdot b = a \cdot c$ implies $b = c$.

Algebraic Expressions

An *algebraic expression* is a string of symbols used to represent an element in a carrier of an algebra. For example, 3, $8 \div x$, and $x^2 + y$ are algebraic expressions in high school algebra. But $x + y +$ is not an algebraic expression. The set of algebraic expressions is a language. The symbols in the alphabet are the operators and constants from the algebra together with variable names and grouping symbols, like parentheses and commas. The language of algebraic expressions over an algebra can be defined inductively as follows:

Algebraic Expressions

1. Constants and variables are algebraic expressions.

2. An operator applied to its arguments is an algebraic expression if the arguments are algebraic expressions.

For example, suppose x and y are variables and c is a constant. If g is a ternary operator, then the following five strings are algebraic expressions:

$$x, \quad y, \quad c, \quad g(x, y, c), \quad g(x, g(c, y, x), x).$$

Different algebraic expressions often mean the same thing. For example, the equation $2x = x + x$ makes sense to us because we look beyond the two strings $2x$ and $x + x$, which are not equal strings. Instead, we look at the possible values of the two expressions and conclude that they always have the same value, no matter what value x has. Two algebraic expressions are *equivalent* if they always evaluate to the same element in a carrier of the algebra. So the expressions $2x$ and $x + x$ are equivalent in high school algebra. We can make the idea of equivalence precise by giving an inductive definition. Assume that C is a carrier of an algebra.

Equivalent Algebraic Expressions

1. Any element in C is equivalent to itself.

2. Suppose E and E' are two algebraic expressions and x is a variable such that $E(x/b)$ and $E'(x/b)$ are equivalent for all elements b in C. Then E is equivalent to E'.

For example, the two expressions $(x + 2)^2$ and $x^2 + 4x + 4$ are equivalent in high school algebra. But $x + y$ is not equivalent to $5x$ because we can let $x = 1$ and $y = 2$, which makes $x + y = 3$ and $5x = 5$.

Describing an Algebra

The set of operators in an algebra is called the *signature* of the algebra. When describing an algebra, we need to decide which operators to put in the signature. For example, we may wish to list only the primitive operators (the constructors) that are used to build all other operators. On the other hand, we might want to list all the operators that we know about.

Let's look at a convenient way to denote an algebra. We'll list the carrier or carriers first, followed by a semicolon. The operators in the signature are listed next. For example, this notation is used to denote the following algebras.

$$\langle \mathbb{N}; \text{succ}, 0 \rangle \qquad \langle \mathbb{N}; +, \cdot, 0, 1 \rangle$$
$$\langle \mathbb{N}; +, 0 \rangle \qquad \langle \mathbb{Z}; +, \cdot, -, 0, 1 \rangle$$
$$\langle \mathbb{N}; \cdot, 1 \rangle \qquad \langle \mathbb{Q}; +, \cdot, -, \div, 0, 1 \rangle$$
$$\langle \mathbb{N}; \text{succ}, +, 0 \rangle \quad \langle \mathbb{R}; +, \cdot, -, \div, 0, 1 \rangle$$

The constants 0 and 1 are listed as operations to emphasize the fact that they have special properties, such as $x + 0 = x$ and $x \cdot 1 = x$.

It may also be convenient to use a picture to describe an algebra. The diagram in Figure 10.1 represents the algebra $\langle \mathbb{N}; +, 0 \rangle$.

The circle represents the carrier $\mathbb{N}$, of natural numbers. The two arrows coming out of $\mathbb{N}$ represent two arguments to the $+$ operator. The arrow from $+$ to $\mathbb{N}$ indicates that the result of $+$ is an element of $\mathbb{N}$. The fact that there are no arrows pointing at 0 means that 0 is a constant (an operator with no arguments), and the arrow from 0 to $\mathbb{N}$ means that 0 is an element of $\mathbb{N}$.

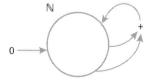

Figure 10.1 An algebra.

10.1.2   Concrete Versus Abstract

An algebra is called *concrete* if its carriers are specific sets of elements so that its operators are defined by rules applied to the carrier elements. High school algebra is a concrete algebra. In fact, all the examples that we have seen so far are concrete algebras.

An algebra is called *abstract* if it is not concrete. In other words, its carriers don't have any specific set interpretation. Thus its operators cannot be defined in terms of rules applied to the carrier elements because we don't have a description of them. Therefore the general properties of the operators in an abstract algebra must be given by axioms. An abstract algebra is a powerful description tool because it represents all the concrete algebras that satisfy its axioms.

So when we talk about an abstract algebra, we are really talking about all possible examples of the algebra. Is this a useful activity? Sure. Many times we are overwhelmed with important concepts, but we aren't given any tools to make sense of them. Abstraction can help to classify things and thus make sense of things that act in similar ways.

If an algebra is abstract, then we must be more explicit when trying to describe it. For example, suppose we write down the following algebra:

$$\langle S;\ s,\ a \rangle.$$

All we know at this point is that S is a carrier and there are two operators s and a. We don't even know the arity of s or a.

Suppose we're told that a is a constant of S and s is a unary operator on S. Now we know something, but not very much, about the algebra. We can use the operators s and a to construct the following algebraic expressions for elements of S:

$$a,\ s(a),\ s(s(a)),\ \ldots,\ s^n(a),\ \ldots.$$

This is the most we can say about the elements of S. There might be other elements in S, but we have no way of knowing it. The elements of S that we know about can be represented by all possible algebraic expressions made up from the operator symbols in the signature together with left and right parentheses.

example   **10.1**   **Induction Algebra**

An algebra $\langle S;\ s,\ a \rangle$ is called an *induction algebra* if s is a unary operator on S and a is a constant of S such that

$$S = \{a,\ s(a),\ s(s(a)),\ \ldots,\ s^n(a),\ \ldots\}.$$

The word "induction" is used because of the natural ordering on the carrier that can be used in inductive proofs.

The algebra $\langle \mathbb{N}; \text{succ}, 0 \rangle$ is a concrete example of an induction algebra, where $\text{succ}(x) = x + 1$. The algebraic expressions for the elements are

$$0, \text{succ}(0), \text{succ}(\text{succ}(0)), \ldots, \text{succ}^n(0), \ldots,$$

where $\text{succ}(0) = 1$, $\text{succ}(\text{succ}(0)) = 2$, and so on. So every natural number is represented by one of the algebraic expressions.

There are many concrete examples of induction algebras. For example, let $A = \{2, 1, 0, -1, -2, -3, \ldots\}$. Then the algebra $\langle A; \text{pred}, 2 \rangle$ is an induction algebra, where $\text{pred}(x) = x - 1$. The expressions for the elements are

$$2, \text{pred}(2), \text{pred}(\text{pred}(2)), \ldots, \text{pred}^n(2), \ldots,$$

where we have $\text{pred}(2) = 1$, $\text{pred}(\text{pred}(2)) = 0$, and so on. So every number in A is represented by one of the algebraic expressions.

end example

Axioms for Abstractions

Interesting things can happen when we add axioms to an abstract algebra. For example, the algebra $\langle S; s, a \rangle$ changes its character when we add the single axiom $s^6(x) = x$ for all $x \in S$. In this case we can say that the algebraic expressions define a finite set of elements, which can be represented by the following six expressions:

$$a, s(a), s^2(a), s^3(a), s^4(a), s^5(a).$$

A complete definition of an abstract algebra can be given by listing the carriers, operations, and axioms. For example, the abstract algebra that we've just been discussing can be defined as follows.

Carrier: S

Operations: $a \in S$

$\quad\quad\quad\quad s : S \to S$

Axiom: $s^6(x) = x$.

We'll always assume that the variable x is universally quantified over S.

example **10.2 A Finite Algebra**

The algebra $\langle \mathbb{N}_6; \text{succ}_6, 0 \rangle$, where $\text{succ}_6(x) = (x + 1) \bmod 6$, is a concrete example of the abstract algebra $\langle S; s, a \rangle$ with axiom $s^6(x) = x$. To see this, observe that the algebraic expressions for the carrier elements are

$$0, \text{succ}_6(0), \text{succ}_6(\text{succ}_6(0)), \ldots, \text{succ}_6^n(x), \ldots.$$

But we have $\text{succ}_6(x) = (x + 1) \bmod 6$. So the preceding algebraic expressions evaluate to an infinite repetition of six numbers in $\mathbb{N}_6$.

$$0, 1, 2, 3, 4, 5, 0, 1, 2, 3, 4, 5, \ldots.$$

In other words, we have $\text{succ}_6^6(x) = x$ for all $x \in \mathbb{N}_6$, which has the same form as the axiom in the abstract algebra.

end example

10.1.3 Working in Algebras

The goal of the following paragraphs is to get familiar with some elementary properties of algebraic operations. We'll see more examples of algebras and we'll observe whether they have any of the properties we've discussed.

Properties of the Operations

Let's look at some fundamental properties that may be associated with a binary operation. If $\circ$ is a binary operator on a set C, then an element $z \in C$ is called a *zero* for $\circ$ if the following condition holds:

$$z \circ x = x \circ z = z \quad \text{for all } x \in C.$$

For example, the number 0 is a zero for the multiply operation over the real numbers because $0 \cdot x = x \cdot 0 = 0$ for all real numbers x.

Continuing with the same binary operator $\circ$ and carrier C, we call an element $u \in C$ an *identity*, or *unit*, for $\circ$ if the following condition holds:

$$u \circ x = x \circ u = x \quad \text{for all } x \in C.$$

For example, the number 1 is an identity for the multiply operation over the real numbers because $1 \cdot x = x \cdot 1 = x$ for all numbers x. Similarly, the number 0 is an identity for the addition operation over real numbers because $0 + x = x + 0 = x$ for all numbers x.

Suppose u is an identity element for $\circ$, and $x \in C$. An element y in C is called an *inverse* of x if the following equation holds:

$$x \circ y = y \circ x = u.$$

For example, in the algebra $\langle \mathbb{Q} ; \cdot, 1 \rangle$ the number 1 is an identity element. We also know that if $x \neq 0$, then

$$x \cdot \frac{1}{x} = \frac{1}{x} \cdot x = 1.$$

In other words, all nonzero rational numbers have inverses.

Each of the following examples presents an algebra together with some observations about its operators.

example 10.3 Algebra of Sets

Let S be a set. Then the power set of S is the carrier for an algebra described as follows:

$$\langle \text{power}\,(S)\,;\cup,\cap,\varnothing,S\rangle.$$

Notice that if $A \in \text{power}(S)$, then $A \cup \varnothing = A$, and $A \cap S = A$. So $\varnothing$ is an identity for $\cup$, and S is an identity for $\cap$. Similarly, $A \cap \varnothing = \varnothing$, and $A \cup S = S$. Thus $\varnothing$ is a zero for $\cap$, and S is a zero for $\cup$. This algebra has many well-known properties. For example, $A \cup A = A$ and $A \cap A = A$ for any $A \in \text{power}(S)$. We also know that $\cap$ and $\cup$ are commutative and associative and that they distribute over each other.

end example

example 10.4 A Finite Algebra

Let $\mathbb{N}_n$ denote the set $\{0, 1, \ldots, n-1\}$, and let "max" be the function that returns the maximum of its two arguments. Consider the following algebra with carrier $\mathbb{N}_n$:

$$\langle \mathbb{N}_n;\ \text{max},\ 0,\ n-1\rangle.$$

Notice that max is commutative and associative. Notice also that for any $x \in \mathbb{N}_n$ it follows that $\max(x, 0) = \max(0, x) = x$. So 0 is an identity for max. It's also easy to see that for any $x \in \mathbb{N}_n$,

$$\max(x,\ n-1) = \max(n-1,\ x) = n-1.$$

So $n-1$ is a zero for the operator max.

end example

example 10.5 An Algebra of Functions

Let S be a set, and let F be the set of all functions of type $S \to S$. If we let $\circ$ denote the operation of composition of functions, then F is the carrier of an algebra $\langle F;\ \circ,\ \text{id}\rangle$. The function "id" denotes the identity function. In other words, we have the equation $\text{id} \circ f = f \circ \text{id} = f$ for all functions f in F. Therefore, id is an identity for $\circ$.

end example

Notice that we used the equality symbol "=" in the previous examples without explicitly defining it as a relation. The first example uses equality of sets, the second uses equality of numbers, and the third uses equality of functions. In our discussions we will usually assume an implicit equality theory on each carrier of an algebra. But, as we have said before, equality relations are operations that may need to be implemented when needed as part of a programming activity.

∘	a	b	c	d
a	a	b	c	d
b	b	c	d	a
c	c	d	a	b
d	d	a	b	c

Figure 10.2   Binary operation table.

Operation Tables

Any binary operation on a finite set can be represented by a table, called an *operation table*. For example, if ∘ is a binary operation on the set $\{a, b, c, d\}$, then the operation table for ∘ might look like Figure 10.2, where the elements of the set are used as row labels and column labels.

If x is a row label and y is a column label, then the element in the table at row x and column y represents the element $x \circ y$. For example, we have $c \circ d = b$.

We can often find out many things about a binary operation by observing its operation table. For example, notice in Figure 10.2 that the row labeled a and the column labeled a are copies of the row label and column label sequence $a\ b\ c\ d$. This tells us that a is an identity for ∘. It's also easy to see that ∘ is commutative and that each element has an inverse. Does ∘ have a zero? It's easy to see that the answer is no. Is ∘ associative? The answer is yes, but it's not very easy to check. We'll leave these problems as exercises.

It's also easy to see that there cannot be more than one identity for a binary operation. Can you see why from Figure 10.2? We'll prove the following general fact about identities for any binary operation.

Uniqueness of an Identity   (10.1)

Any binary operation has at most one identity.

Proof: Let ∘ be a binary operation on a set S. To show that ∘ has at most one identity, we'll assume that u and e are identities for ∘. Then we'll show that $u = e$. Remember, since u and e are identities, we know that $u \circ x = x \circ u = x$ and $e \circ x = x \circ e = x$ for all x in S. Thus we have the following equality:

$$e = e \circ u \qquad (u \text{ is an identity for } \circ)$$
$$ = u \qquad (e \text{ is an identity for } \circ). \quad \text{QED.}$$

Algebras with One Binary Operation

Some algebras are used so frequently that they have been given names. For example, any algebra of the form $\langle A; \circ \rangle$, where ∘ is a binary operation, is called a *groupoid*. If we know that the binary operation is associative, then the algebra is called a *semigroup*. If we know that the binary operation is associative and also has an identity, then the algebra is called a *monoid*. If we know that the binary operation is associative, has an identity, and each element has an inverse, then the algebra is called a *group*. So these words are used to denote certain properties of the binary operation. Here's a synopsis.

Groupoid: ∘ is a binary operation.

Semigroup: ∘ is an associative binary operation.

Monoid: ∘ is an associative binary operation with an identity.

Group: ∘ is an associative binary operation with an identity and every element has an inverse.

It's clear from the listing of properties that a group is a monoid, which is a semigroup, which is a groupoid. But things don't go the other way. For example, the algebra in Example 10.5 is a monoid but not a group, since not every function has an inverse.

We can have some fun with these names. For example, we can describe a group as a monoid with inverses, and we can describe a monoid as a semigroup with identity. When an algebra contains an operation that satisfies some special property beyond the axioms of the algebra, we often modify the name of the algebra with the name of the property. For example, the algebra $\langle \mathbb{Z}; +, 0 \rangle$ is a group. But we know that the operation $+$ is commutative. Therefore, we can call the algebra a "commutative" group.

Now let's discuss a few elementary results. To get our feet wet, we'll prove the following simple property that holds in any monoid.

Uniqueness of Inverses   (10.2)

In a monoid, if an element has an inverse, the inverse is unique.

Proof: Let $\langle M; \circ, u \rangle$ be a monoid. We will show that if an element x in M has an inverse, then the inverse is unique. In other words, if y and z are both inverses of x, then $y = z$. We can prove this result as follows:

$$
\begin{aligned}
y &= y \circ u && (u \text{ is the identity for } \circ) \\
  &= y \circ (x \circ z) && (z \text{ is an inverse of } x) \\
  &= (y \circ x) \circ z && (\circ \text{ is associative}) \\
  &= u \circ z && (y \text{ is an inverse of } x) \\
  &= z && (u \text{ is the identity for } \circ). \quad \text{QED.}
\end{aligned}
$$

If we have a group, then we know that every element has an inverse. Thus we can conclude from (10.2) that every element x in a group has a unique inverse, which is usually denoted by writing the symbol

$$x^{-1}.$$

example 10.6 Working with Groups

We can use the elementary properties of a group to obtain other properties. For example, let $\langle G; \circ, e \rangle$ be a group. This means that we know that $\circ$ is associative, e is an identity, and every element of G has an inverse. We'll prove two properties as examples. The first property is stated as follows:

$$\text{If } x \circ x = x \text{ holds for some element } x \in G, \text{ then } x = e. \qquad (10.3)$$

Proof: $\begin{aligned} x &= x \circ e & & (e \text{ is an identity for } \circ) \\ &= x \circ (x \circ x^{-1}) & & (x^{-1} \text{ is the inverse of } x) \\ &= (x \circ x) \circ x^{-1} & & (\circ \text{ is associative}) \\ &= x \circ x^{-1} & & (x \circ x = x \text{ is the hypothesis}) \\ &= e & & (x^{-1} \text{ is the inverse of } x). \text{ QED.} \end{aligned}$

A second property of groups is cancellation on the left. This property can be stated as follows:

$$\text{If } x \circ y = x \circ z \text{ then } y = z. \qquad (10.4)$$

Proof: $\begin{aligned} y &= e \circ y & & (e \text{ is an identity}) \\ &= (x^{-1} \circ x) \circ y & & (x^{-1} \text{ is the inverse of } x) \\ &= x^{-1} \circ (x \circ y) & & (\circ \text{ is associative}) \\ &= x^{-1} \circ (x \circ z) & & (\text{hypothesis}) \\ &= (x^{-1} \circ x) \circ z & & (\circ \text{ is associative}) \\ &= e \circ z & & (x^{-1} \text{ is the inverse of } x) \\ &= z & & (e \text{ is an identity}). \text{ QED.} \end{aligned}$

end example

Algebras with Several Operations

A natural example of an algebra with two binary operations is the integers together with the usual operations of addition and multiplication. We can denote this algebra by the structure $\langle \mathbb{Z}; +, \cdot, 0, 1 \rangle$. This algebra is a concrete example of an algebra called a ring, which we'll now define. A *ring* is an algebra with the structure

$$\langle A; +, \cdot, 0, 1 \rangle,$$

$+_5$	0	1	2	3	4
0	0	1	2	3	4
1	1	2	3	4	0
2	2	3	4	0	1
3	3	4	0	1	2
4	4	0	1	2	3

$\cdot_5$	0	1	2	3	4
0	0	0	0	0	0
1	0	1	2	3	4
2	0	2	4	1	3
3	0	3	1	4	2
4	0	4	3	2	1

Figure 10.3   Mod 5 addition and multiplication tables.

where $\langle A; +, 0 \rangle$ is a commutative group, $\langle A; \cdot, 1 \rangle$ is a monoid, and the operation $\cdot$ distributes over $+$ from the left and the right. This means that

$$a \cdot (b + c) = a \cdot b + a \cdot c \quad \text{and} \quad (b + c) \cdot a = b \cdot a + c \cdot a.$$

Check to see that $\langle \mathbb{Z}; +, \cdot, 0, 1 \rangle$ is indeed a ring.

If $\langle A; +, \cdot, 0, 1 \rangle$ is a ring with the additional property that $\langle A - \{0\}; \cdot, 1 \rangle$ is a commutative group, then it's called a *field*. The ring $\langle \mathbb{Z}; +, \cdot, 0, 1 \rangle$ is not a field because, for example, 3 does not have an inverse for multiplication. On the other hand, if we replace $\mathbb{Z}$ by $\mathbb{Q}$, the rational numbers, then $\langle \mathbb{Q}; +, \cdot, 0, 1 \rangle$ is a field. For example, 3 has inverse $1/3$ in $\mathbb{Q} - \{0\}$.

For another example of a field, let $\mathbb{N}_5 = \{0, 1, 2, 3, 4\}$ and let $+_5$ and $\cdot_5$ be addition mod 5 and multiplication mod 5, respectively. Then $\langle \mathbb{N}_5; +_5, \cdot_5, 0, 1 \rangle$ is a field. Figure 10.3 shows the operation tables for $+_5$ and $\cdot_5$. We'll leave the verification of the field properties as an exercise.

The next examples show some algebras that might be familiar to you.

example **10.7 Polynomial Algebras**

Let $\mathbb{R}[x]$ denote the set of all polynomials over x with real numbers as coefficients. It's a natural process to add and multiply two polynomials. So we have an algebra $\langle \mathbb{R}[x]; +, \cdot, 0, 1 \rangle$, where $+$ and $\cdot$ represent addition and multiplication of polynomials and 0 and 1 represent themselves. This algebra is a ring. Why isn't it a field?

end example

example **10.8 Matrix Algebras**

Suppose we let $M_n(\mathbb{R})$ denote the set of all n by n matrices with elements in $\mathbb{R}$. We can add two matrices A and B by letting $A_{ij} + B_{ij}$ be the element in the ith row and jth column of the sum. We can multiply A and B by letting $\sum_{k=1}^{n} A_{ik}B_{kj}$ be the element in the ith row and jth column of the product. Thus we have an algebra $\langle M_n(\mathbb{R}); +, \cdot, 0, 1 \rangle$, where $+$ and $\cdot$ represent matrix addition and multiplication, 0 represents the matrix with all entries zero, and 1 represents the matrix with 1's along the main diagonal and 0's elsewhere. This algebra is a ring. Why isn't it a field?

end example

example **10.9   Vector Algebras**

The algebra of n-dimensional vectors, with real numbers as components, can be described by listing two carriers $\mathbb{R}$ and $\mathbb{R}^n$. We can multiply a vector $(x_1, \ldots, x_n) \in \mathbb{R}^n$ by number $b \in \mathbb{R}$ to obtain a new vector by multiplying each component of the vector by b, obtaining

$$(bx_1, \ldots, bx_n).$$

If we let $\cdot$ denote this operation, then we have

$$b \cdot (x_1, \ldots, x_n) = (bx_1, \ldots, bx_n).$$

We can add vectors by adding corresponding components. For example,

$$(x_1, \ldots, x_n) + (y_1, \ldots, y_n) = (x_1 + y_1, \ldots, x_n + y_n).$$

Thus we have an algebra $\langle \mathbb{R}, \mathbb{R}^n; \cdot, + \rangle$ of n-dimensional vectors. Notice that the algebra has two carriers, $\mathbb{R}$ and $\mathbb{R}^n$. This is because they are both necessary to define the $\cdot$ operation, which has type $\mathbb{R} \times \mathbb{R}^n \to \mathbb{R}^n$.

end example

example **10.10   Power Series Algebras**

If we extend polynomials over x to allow infinitely many terms, then we obtain what are called *power series* (we also know them as generating functions). Letting $\mathbb{R}[[x]]$ denote the set of power series with real numbers as coefficients, we obtain the algebra $\langle \mathbb{R}[[x]]; +, \cdot, 0, 1 \rangle$, where $+$ and $\cdot$ represent addition and multiplication of power series and 0 and 1 represent themselves. This algebra is a ring. Why isn't it a field?

end example

 Exercises

Algebraic Properties

1. Let m and n be two integers with $m < n$. Let $A = \{m, m + 1, \ldots, n\}$, and let "min" be the function that returns the smaller of its two arguments. Does min have a zero? Identity? Inverses? If so, describe them.

2. Let $A = \{\text{True, False}\}$. For each of the following binary operations on A, answer the three questions: Does the operation have a zero? Does the operation have an identity? What about inverses?

   a. Conditional, $\rightarrow$.

   b. Conjunction, $\wedge$.

   c. Disjunction, $\vee$.

3. Given the algebra $\langle S; f, a \rangle$, where f is a unary operation and a is a constant of S and $f^5(x) = f^3(x)$ for all $x \in S$, find a finite set of algebraic expressions that will represent the distinct elements of S.

4. Given a binary operation on a finite set in table form, for each of the following parts, describe an easy way to detect whether the binary operation has the listed property.

   a. There is a zero.

   b. The operation is commutative.

   c. Inverses exist for each element of the set (assume there is an identity).

5. Let $A = \{a, b, c, d\}$, and let $\circ$ be a binary operation on A. For each of the following problems, write down a table for $\circ$ that satisfies the given properties.

   a. a is an identity, but no other element of A has an inverse.

   b. a is an identity, and every element of A has an inverse.

   c. a is a zero, and $\circ$ is not associative.

   d. a is an identity, and exactly two elements have inverses.

   e. a is an identity, and $\circ$ is commutative but not associative.

6. Let $A = \{a, b\}$. For each of the following problems, find an operation table satisfying the given condition for a binary operation $\circ$ on A.

   a. $\langle A; \circ \rangle$ is a group.

   b. $\langle A; \circ \rangle$ is a monoid but not a group.

   c. $\langle A; \circ \rangle$ is a semigroup but not a monoid.

   d. $\langle A; \circ \rangle$ is a groupoid but not a semigroup.

7. Write an algorithm to check a binary operation table for associativity.

Challenges

8. Given the algebra $\langle S; f, g, a \rangle$, where f and g are unary operations and a is a constant of S, suppose that $f(f(x)) = g(x)$ and $g(g(x)) = x$ for all $x \in S$.

    a. Show that $f(g(x)) = g(f(x))$ for all $x \in S$.

    b. Show that $f(f(f(f(x)))) = x$ for all $x \in S$.

    c. Find a finite set of algebraic expressions to represent the distinct elements of S.

9. Prove each of the following facts about a group $\langle G; \circ, e \rangle$.

    a. Cancellation on the right: If $y \circ x = z \circ x$, then $y = z$.

    b. The inverse of $x \circ y$ is $y^{-1} \circ x^{-1}$. In other words, $(x \circ y)^{-1} = y^{-1} \circ x^{-1}$.

10. Let $\mathbb{N}_5 = \{0, 1, 2, 3, 4\}$, and let $+_5$ and $\cdot_5$ be the two operations of addition mod 5 and multiplication mod 5, respectively. Show that $\langle \mathbb{N}_5; +_5, \cdot_5, 0, 1 \rangle$ is a field.

10.2 Boolean Algebra

Do the techniques of set theory and the techniques of logic have anything in common? Let's do an example to see that the answer is yes. When working with sets, we know that the following equation holds for all sets A, B, and C:

$$A \cup (B \cap C) = (A \cup B) \cap (A \cup C).$$

When working with propositions, we know that the following equivalence holds for all propositions A, B, and C:

$$A \vee (B \wedge C) \equiv (A \vee B) \wedge (A \vee C).$$

Certainly these two examples have a similar pattern. As we'll see shortly, sets and logic have a lot in common. They can both be described as concrete examples of a Boolean algebra. The name "Boolean" comes from the mathematician George Boole (1815–1864), who studied relationships between set theory and logic. Let's get to the definition.

Definition of Boolean Algebra

A *Boolean algebra* is an algebra with the structure $\langle B; +, \cdot, {}^-, 0, 1 \rangle$, where the following properties hold.

Defining Properties of a Boolean Algebra

1. $\langle B; +, 0 \rangle$ and $\langle B; \cdot, 1 \rangle$ are commutative monoids. In other words, the following properties hold for all x, y, $z \in B$:

$$(x + y) + z = x + (y + z), \quad (x \cdot y) \cdot z = x \cdot (y \cdot z),$$
$$x + y = y + x, \qquad\qquad x \cdot y = y \cdot x,$$
$$x + 0 = x, \qquad\qquad x \cdot 1 = x.$$

2. The operations $+$ and $\cdot$ distribute over each other. In other words, the following properties hold for all x, y, $z \in B$:

$$x \cdot (y + z) = (x \cdot y) + (x \cdot z) \quad \text{and} \quad x + (y \cdot z) = (x + y) \cdot (x + z).$$

3. $x + \overline{x} = 1$ and $x \cdot \overline{x} = 0$ for all elements $x \in B$. The element $\overline{x}$ is called the *complement* of x or the *negation* of x.

We often drop the dot and write xy in place of $x \cdot y$. We'll also reduce the need for parentheses by agreeing to the following precedence hierarchy:

$$\overline{} \qquad \text{highest (do it first),}$$

$$\cdot$$

$$+ \qquad \text{lowest (do it last).}$$

For example, the expression $x + y\overline{z}$ means the same thing as $(x + (y(\overline{z})))$.

---

example 10.11 **Sets**

Suppose $B = \text{power}(S)$ for some set S. Then B is the carrier of a Boolean algebra if we let union and intersection act as the operations $+$ and $\cdot$, let X' be the complement of X, let $\varnothing$ act as 0, and let S act as 1. For example, the two properties in Part 3 of the definition are represented by the following equations, where X is any subset of S:

$$X \cup X' = S \quad \text{and} \quad X \cap X' = \varnothing.$$

end example

---

example 10.12 **Logic**

Suppose we let B be the set of all propositional wffs of the propositional calculus. Then B is the carrier of a Boolean algebra if we let disjunction and conjunction act as the operations $+$ and $\cdot$, let $\neg\, X$ be the complement of X, let False act as 0, let True act as 1, and let logical equivalence act as equality. For example, the two properties in Part 3 of the definition are represented by the following equivalences, where X is any proposition:

$$X \vee \neg X \equiv \text{True} \quad \text{and} \quad X \wedge \neg X \equiv \text{False}.$$

We can also obtain a very simple Boolean algebra by using just the carrier {False, True} together with the operations $\vee$, $\wedge$, and $\neg$.

end example

example 10.13   Divisors

Let n be a product of distinct prime numbers. For example, n could be 30 because $30 = 2 \cdot 3 \cdot 5$, but n cannot be 12 because $12 = 2 \cdot 2 \cdot 3$, which is not a product of distinct primes. Let B_n be the set of positive divisors of n. Then B_n is the carrier of a Boolean algebra if we let "least common multiple" and "greatest common divisor" be the operations $+$ and $\cdot$, respectively. Let n/x be the complement of x, let 1 act as the zero, and let n act as the one.

With these definitions, all the properties of a Boolean algebra are satisfied. For example, the two properties in Part 3 of the definition are represented by the following equations, where $x \in B_n$:

$$\text{lcm}(x, \, n/x) = n \quad \text{and} \quad \gcd(x, \, n/x) = 1.$$

For an example, let $n = 10 = 2 \cdot 5$. Then $B_{10} = \{1, 2, 5, 10\}$, 1 is the zero, and 10 is the one. Thus, for example, the complement of 2 is 5, $\text{lcm}(2, 5) = 10$ (the one), and $\gcd(2, 5) = 1$ (the zero).

Notice what happens if we let $n = 12$. We get $B_{12} = \{1, 2, 3, 4, 6, 12\}$. The reason B_{12} does not yield a Boolean algebra with our definition is because 2 and its complement 6 don't satisfy the properties in Part 3 of the definition. Notice that $\text{lcm}(2, 6) = 6$, which is not the one, and $\gcd(2, 6) = 2$, which is not the zero.

end example

10.2.1   Simplifying Boolean Expressions

A fundamental problem of Boolean algebra, with applications to such areas as logic design and theorem-proving systems, is to simplify Boolean expressions so that they contain a small number of operations. Let's see how the axioms of Boolean algebra can help us obtain some useful simplification properties.

For example, in the Boolean algebra of propositions we have $P \wedge P \equiv P$ for any proposition P. Similarly, in the Boolean algebra of sets we have $S \cap S = S$ for any set S. Can we generalize these properties to all Boolean algebras? In other words, can we say $xx = x$ for every element x in the carrier of a Boolean algebra? The answer is yes. Let's prove it with equational reasoning. Be sure you can provide a reason for each step of the following proof:

$$x = x \cdot 1 = x \cdot (x + \overline{x}) = x \cdot x + x \cdot \overline{x} = x \cdot x + 0 = x \cdot x.$$

A related statement is $x + x = x$ for all elements x. Can you provide the proof? We'll state these two properties for the record.

Idempotent Properties (10.5)

$$x \cdot x = x \quad \text{and} \quad x + x = x.$$

A nice property of Boolean algebras is that results come in pairs. This is because the axioms come in pairs. In other words, $\langle B; +, 0 \rangle$ and $\langle B; \cdot, 1 \rangle$ are both commutative monoids; $+$ and $\cdot$ distribute over each other; and $x + \overline{x} = 1$ and $x \cdot \overline{x} = 0$ for all elements $x \in B$. The *duality principle* states that whenever a result A is true for a Boolean algebra, then a dual result A' is also true, where A' is obtained from the A by simultaneously replacing all occurrences of $\cdot$ by $+$, all occurrences of $+$ by $\cdot$, all occurrences of 1 by 0, and all occurrences of 0 by 1. A proof for the result A' can be obtained by making these same changes in the proof of A.

There are lots of properties that we can discover. For example, if S is a set, then $\varnothing \cap A = \varnothing$ for any subset A of S. This is an instance of a general property that holds for any Boolean algebra: $0 \cdot x = 0$ for every element x. This follows readily from (10.5) as follows:

$$0 \cdot x = (\overline{x} \cdot x) \cdot x = \overline{x} \cdot (x \cdot x) = \overline{x} \cdot x = 0.$$

Again, there is a dual result: $1 + x = 1$ for every x. Can you prove this result? We'll also state these two properties for the record:

Zero and One Properties (10.6)

$$0 \cdot x = 0 \quad \text{and} \quad 1 + x = 1.$$

Let's do an example to see how we can put our new knowledge to use in simplifying a Boolean expression. Suppose the function f is defined over a Boolean algebra by

$$f(x, y, z) = x + yz + z\overline{x}y + \overline{y}xz.$$

To evaluate f, we need to perform three $+$ operations, five $\cdot$ operations, and two $-$ operations. Can we do any better? Sure we can. We can simplify the expression for $f(x, y, z)$ as follows—make sure you can state a reason for each line:

$$
\begin{aligned}
f(x, y, z) &= x + yz + z\overline{x}y + \overline{y}xz \\
&= x + yz\left(1 + \overline{x}\right) + \overline{y}xz \\
&= x + yz1 + \overline{y}xz \\
&= x\left(1 + \overline{y}z\right) + yz \\
&= x1 + yz \\
&= x + yz.
\end{aligned}
$$

So f can be evaluated with only one $+$ operation and one $\cdot$ operation.

To simplify Boolean expressions, it's important to have a good knowledge of Boolean algebra together with some luck and ingenuity. We'll give a few more general properties that are very useful simplification tools. The following properties can be used to simplify an expression by reducing the number of operations by two. We'll leave the proofs as exercises.

Absorption Laws (10.7)

$$x + xy = x \quad \text{and} \quad x\left(x + y\right) = x.$$
$$x + \overline{x}y = x + y \quad \text{and} \quad x\left(\overline{x} + y\right) = xy.$$

In a Boolean algebra, complements are unique in the following sense: If an element acts like a complement of some element, then it is in fact the only complement of the element. Using symbols, we can state the result as follows:

Uniqueness of the Complement (10.8)

$$\text{If } x + y = 1 \text{ and } xy = 0, \text{ then } y = \overline{x}.$$

Proof: To prove this statement, we write the following equations:

$$
\begin{aligned}
y &= y1 \\
&= y\left(x + \overline{x}\right) \\
&= yx + y\overline{x} \\
&= 0 + y\overline{x} && (\text{since } xy = 0) \\
&= x\overline{x} + y\overline{x} \\
&= (x + y)\overline{x} \\
&= 1\overline{x} && (\text{since } x + y = 1) \\
&= \overline{x} && \text{QED.}
\end{aligned}
$$

Complements are quite useful in Boolean algebra. As a consequence of the uniqueness of complements (10.8), we have the following property:

Involution Law (10.9)

$$\bar{\bar{x}} = x.$$

Proof: Notice that $\bar{x} + x = 1$ and $\bar{x}x = 0$. Therefore, x acts like the complement of $\bar{x}$. Thus x is indeed equal to the complement of $\bar{x}$. That is, $x = \bar{\bar{x}}$. QED.

Recall from the propositional calculus that we have the following logical equivalence $\neg (p \wedge q) \equiv \neg p \vee \neg q$. This is an example of one of De Morgan's laws, which have the following forms in Boolean algebra.

De Morgan's Laws (10.10)

$$\overline{x + y} = \bar{x}\,\bar{y} \quad \text{and} \quad \overline{xy} = \bar{x} + \bar{y}.$$

Proof: We'll prove the first of the two laws and leave the second as an exercise. We'll show that $\bar{x}\,\bar{y}$ acts like the complement of $x + y$. Then we'll use (10.8) to conclude the result. First, we'll show that $(x + y) + \bar{x}\,\bar{y} = 1$ as follows:

$$
\begin{aligned}
(x + y) + \bar{x}\,\bar{y} &= (x + y + \bar{x})(x + y + \bar{y}) \\
&= (x + \bar{x} + y)(x + y + \bar{y}) \\
&= (1 + y)(x + 1) \\
&= 1 \cdot 1 \\
&= 1.
\end{aligned}
$$

Next we'll show that $(x + y)\bar{x}\,\bar{y} = 0$ as follows:

$$
\begin{aligned}
(x + y)\bar{x}\,\bar{y} &= x\bar{x}\,\bar{y} + y\bar{x}\,\bar{y} \\
&= x\bar{x}\,\bar{y} + \bar{x}y\bar{y} \\
&= 0\bar{y} + \bar{x}0 \\
&= 0 + 0 \\
&= 0.
\end{aligned}
$$

Thus $\bar{x}\,\bar{y}$ acts like a complement of $x + y$. So we can apply (10.8) to conclude that $\bar{x}\,\bar{y}$ is the complement of $x + y$. QED.

In the propositional calculus we can find a disjunctive normal form (DNF) and a conjunctive normal form (CNF) for any wff. These ideas carry over to any Boolean algebra, where $+$ corresponds to disjunction and $\cdot$ corresponds to conjunction. So we can make the following statement:

Any Boolean expression has a DNF and a CNF.

example   **10.14   DNF and CNF Constructions**

We'll construct a DNF and CNF for the expression

$$\overline{x + y} + z.$$

The following transformations do the job:

$$\overline{x+y} + z = \overline{x}\,\overline{y} + z \qquad \text{(DNF)}$$
$$= (\overline{x} + z)\,(\overline{y} + z) \qquad \text{(CNF)}.$$

end example

10.2.2   Digital Circuits

Now let's see what Boolean algebra has to do with digital circuits. A *digital circuit* (also called a *logic circuit*) is an electronic representation of a function whose input values are either high or low voltages and whose output value is either a high or low voltage. Digital circuits are used to represent and process information in digital computers. The high- and low-voltage values are normally represented by the two digits 1 and 0.

The basic electronic components used to build digital circuits are called *gates.* The three basic "logic" gates are the AND gate, the OR gate, and the NOT gate. These gates work just like the corresponding logical operations, where 1 means true and 0 means false. So we can represent digital circuits as Boolean expressions with values in the Boolean algebra whose carrier is $\{0, 1\}$, where 0 means false, 1 means true, and the operations $+$, $\cdot$, and $^-$ stand for $\vee$, $\wedge$, and $\neg$, respectively.

The three logic gates are represented graphically as shown in Figure 10.4, where the inputs are on the left and the outputs are on the right.

Arithmetic Circuits

These gates can be combined in various ways to form digital circuits to do all basic arithmetic operations. For example, suppose we want to add two binary digits x and y. The first thing to notice is that the result has a summand digit and a carry digit. We'll consider two functions, "carry" and "summand." Let's look at the carry function first. Notice that $\text{carry}(x, y) = 1$ if and only if $x = 1$ and $y = 1$. Thus we can define the carry function as follows:

$$\text{carry}(x, y) = xy.$$

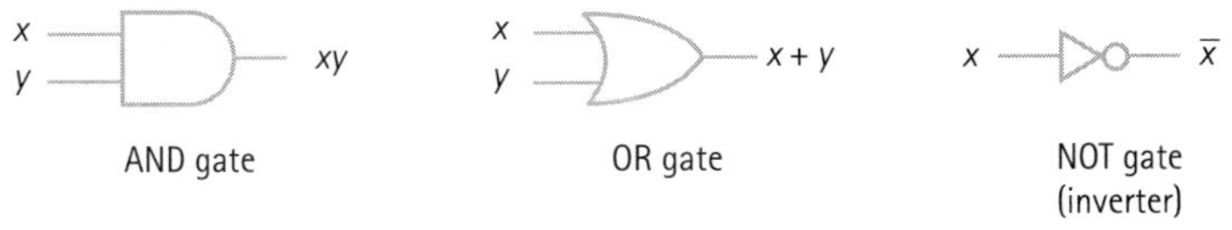

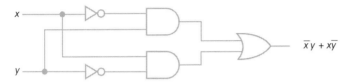

Figure 10.4   Logical gates.

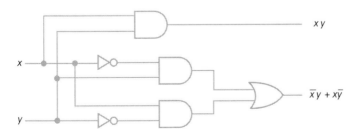

Figure 10.5   Summand circuit.

A circuit to implement the carry function consists of the simple AND gate shown in Figure 10.4.

Now let's look at the summand. It's clear that summand$(x, y) = 1$ if and only if either $x = 0$ and $y = 1$ or $x = 1$ and $y = 0$. Thus we can define the summand function as follows:

$$\text{summand } (x, y) = \overline{x}y + x\overline{y}.$$

A circuit to implement the summand function is shown in Figure 10.5.

We can combine the two circuits for the carry and the summand into one circuit that gives both outputs. The circuit is shown in Figure 10.6. Such a circuit is called a *half-adder*, and it's a fundamental building block in all arithmetic circuits.

example 10.15 A Simple Half-Adder

Let's see whether we can simplify the circuit for a half-adder. The preceding circuit for a half-adder has six gates. But we can do better. First, notice that the expression for the summand, $\overline{x}y + x\overline{y}$ has five operations: two negations, two

Figure 10.6   Half-adder circuit.

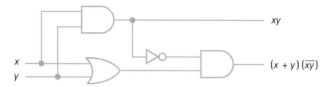

Figure 10.7   Simpler half-adder circuit.

conjunctions, and one disjunction. Let's rewrite it as follows (be sure to fill in the reasons for each step).

$$\overline{x}y + x\overline{y} = (\overline{x}y + x)(\overline{x}y + \overline{y})$$
$$= (x + y)(\overline{x} + \overline{y})$$
$$= (x + y)\overline{xy}.$$

This latter expression has four operations: two conjunctions, one disjunction, and one negation. Also, note that the expression xy is computed before the negation is applied. Therefore, we can also use this expression for the carry. So we have a simpler version of the half-adder, as shown in Figure 10.7.

end example

Constructing a Full Adder

We'll describe a circuit to add two binary numbers. For example, suppose we want to add the two binary numbers 1 0 1 1 and 1 1 1 0. The school method can be pictured as follows:

$$
\begin{array}{ccccc}
  & 1 & 1 & 0 & \leftarrow \text{ (carry bits)} \\
  & 1 & 0 & 1 & 1 \\
  & 1 & 1 & 1 & 0 \\
\hline
1 & 1 & 0 & 0 & 1
\end{array}
$$

So if we want to add two binary numbers, then we can start by using a half-adder on the two rightmost digits of each number. After that, we must be able to handle the addition of three binary digits: two binary digits and a carry from the preceding addition. A digital circuit to accomplish this latter feat is called a *full adder*.

We can build a full adder by using half-adders as components. Let's see how it goes. First, to get the big picture, we will denote a half-adder by a box with two input lines and two output lines, as shown in Figure 10.8.

To get an idea about the kind of circuit we need, let's look at a table of values for the outputs sum and carry. Figure 10.9 shows the values of sum and carry that are obtained by adding three binary digits.

Let's use Figure 10.9 to find DNFs for the sum and carry functions in terms of x, y, and z. Notice that the value of the sum is 1 in four places, on lines 2,

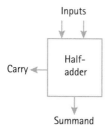

Figure 10.8   Half adder.

x	y	z	Sum	Carry
0	0	0	0	0
0	0	1	1	0
0	1	0	1	0
0	1	1	0	1
1	0	0	1	0
1	0	1	0	1
1	1	0	0	1
1	1	1	1	1

Figure 10.9   Adding three binary digits.

3, 5, and 8 of the table. So the DNF for the sum function will consist of the disjunction of four terms, with each conjunction constructed from the values of x, y, and z on the four lines. Similarly, the value of the carry is 1 in four places, on lines 4, 6, 7, and 8. So the DNF for the carry function depends on conjunctions that depend on these latter four lines. We obtain the following forms for sum and carry.

$$\text{sum}\,(x, y, z) = \overline{x}\,\overline{y}z + \overline{x}y\overline{z} + x\overline{y}\,\overline{z} + xyz$$
$$= \overline{x}\,(\overline{y}z + y\overline{z}) + x\,(\overline{y}\,\overline{z} + yz)\,.$$

$$\text{carry}\,(x, y, z) = \overline{x}yz + x\overline{y}z + xy\overline{z} + xyz$$
$$= yz + x\,(\overline{y}z + y\overline{z})\,.$$

At this point we could build a circuit for sum and carry. But let's study the expressions that we obtained. Notice first that the expression

$$\overline{y}z + y\overline{z}$$

occurs in both the sum formula and the carry formula. Recall also that this is the expression for the summand output of the half-adder. It can be shown that the expression $\overline{y}\,\overline{z} + yz$, in the sum function, is equal to the negation of the expression

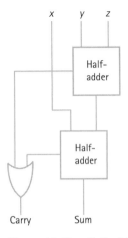

Figure 10.10   Full adder.

$\overline{y}z + y\overline{z}$ (the proof is left as an exercise). In other words, if we let $e = \overline{y}z + y\overline{z}$, then we can write the sum in the following form:

$$\mathrm{sum}\,(x, y, z) = \overline{x}e + x\overline{e}.$$

This shows us that sum(x, y, z) is just the summand output of a half-adder. So we can let y and z be inputs to a half-adder and then feed the summand output along with x into another half-adder to obtain the desired sum(x, y, z). Before we draw the diagram, we need to look at the carry function. We have written carry in the following form:

$$\mathrm{carry}\,(x, y, z) = yz + x\left(\overline{y}z + y\overline{z}\right).$$

Notice that the term yz is the carry output of a half-adder with input values y and z. Further, the term $x\left(\overline{y}z + y\overline{z}\right)$ is the carry output of a half-adder with inputs x and $\overline{y}z + y\overline{z}$, where $\overline{y}z + y\overline{z}$ is the output of the half-adder with inputs y and z. So we can draw a picture of the circuit for a full adder as shown in Figure 10.10.

Minimization

As we've seen in the previous two examples, we can get simpler digital circuits if we spend some time simplifying the corresponding Boolean expressions. Often a digital circuit must be built with the minimum number of components, where the components correspond to DNFs or CNFs.

This brings up the question of finding a *minimal* DNF for a Boolean expression. Here the word "minimal" is usually defined to mean the fewest number of fundamental conjunctions in a DNF, and if two DNFs have the same number of fundamental conjunctions, then the one with the fewest literals is minimal. The term *minimal* CNF is defined analogously. It's not always easy to find a minimal DNF for a Boolean expression.

example 10.16 A Minimal DNF

The Boolean expression $yz + yx$ is a minimal DNF for the expression

$$\overline{x}yz + xyz + xy\overline{z}.$$

We can show that these two expressions are equivalent as follows:

$$
\begin{aligned}
\overline{x}yz + xyz + xy\overline{z} &= (\overline{x} + x)\,yz + xy\overline{z} \\
&= yz + xy\overline{z} \\
&= y\,(z + x\overline{z}) \\
&= y\,(z + x) \\
&= yz + yx.
\end{aligned}
$$

But it takes some work to see that $yz + xy$ is a minimal DNF. First we need to argue that there is no equivalent DNF with just a single fundamental conjunction. Then we need to argue that there is no equivalent DNF with two fundamental conjunctions with fewer than four literals.

end example

There are formal methods that can be applied to the problem of finding minimal DNFs and minimal CNFs. We'll leave them to more specialized texts.

 Exercises

The Axioms

1. Let S be a set and $B = \text{power}(S)$. Suppose someone claims that B is a Boolean algebra with the following definitions for the operators.

$$
\begin{aligned}
+ \quad &\text{is union,} \\
\cdot \quad &\text{is difference,} \\
- \quad &\text{is complement with respect to } S, \\
0 \quad &\text{is } S, \\
1 \quad &\text{is } \varnothing.
\end{aligned}
$$

Is the result a Boolean algebra? Why or why not?

Boolean Expressions

2. Prove each of the following four absorption laws (10.7).

    a. $x + xy = x.$ **b.** $x\,(x + y) = x.$

    c. $x + \overline{x}y = x + y.$ **d.** $x\,(\overline{x} + y) = xy.$

3. Let $e = \overline{y}z + y\overline{z}$. Prove that $\overline{e} = \overline{y}\,\overline{z} + yz$.

4. Use Boolean algebra properties to prove each of the following equalities.

    a. $\overline{x} + \overline{y} + xyz = \overline{x} + \overline{y} + z.$

    b. $\overline{x} + \overline{y} + xy\overline{z} = \overline{x} + \overline{y} + \overline{xyz}.$

5. Simplify each of the following Boolean expressions.

    a. $x + \overline{x}y.$ **b.** $x\overline{y}\,\overline{x} + xy\overline{x}.$

    c. $\overline{x}\,\overline{y}z + x\overline{y}\,\overline{z} + x\overline{y}z.$ **d.** $xy + x\overline{y} + \overline{x}y.$

    e. $x\,(y + \overline{y}z) + \overline{y}z + yz.$ **f.** $x\,(y + z)\,(\overline{x} + y)\,(\overline{y} + x + z)\,.$

    g. $x + yz + \overline{x}y + \overline{y}xz.$ **h.** $x + y\,(x + y)\,.$

    i. $(x + y) + \overline{x}\,\overline{y}.$ **j.** $(x + y)\,(\overline{y} + x)\,(\overline{x} + y)\,.$

    k. $xy + x + y.$ **l.** $(x + y)\,xy.$

6. For each part of Exercise 2, draw two logic circuits. One circuit should implement the expression on the left side of the equality. The other circuit should implement the expression on the right side of the equality. Each circuit should use the same number of gates as there are operations in the expression.

7. Write down the dual of each of the following Boolean expressions.

    a. $x + 1.$ **b.** $x\,(y + z)\,.$ **c.** $xy + xz.$

    d. $xy + z.$ **e.** $y + \overline{x}z.$ **f.** $z\overline{y}\,\overline{x} + xz\overline{x}.$

8. Show that, in a Boolean algebra, $1 + x = 1$ for every element x.

9. Show that, in a Boolean algebra, $\overline{xy} = \overline{x} + \overline{y}$ for all elements x and y.

Challenges

10. Let B be the carrier of a Boolean algebra. Suppose B is a finite set, and suppose $0 \neq 1$. Show that the cardinality of B is an even number.

11. A Boolean algebra can be made into a partially ordered set by letting $x \preceq y$ mean $x = xy$.

    a. Show that $\preceq$ is reflexive, antisymmetric, and transitive.

    b. Show that $x \preceq y$ if and only if $y = x + y$.

12. A Boolean algebra, when considered as a poset—as in Exercise 11—is also a lattice. Prove that $\mathrm{glb}(x, y) = xy$ and $\mathrm{lub}(x, y) = x + y$.

13. In Example 10.13 we considered the set B_n of positive divisors of n together with the operations of lcm, gcd, n/x, where n is one and 1 is zero. Prove that this algebra is not Boolean if a prime p occurs more than once as a factor of n. *Hint:* Consider the complement of p.

10.3   Abstract Data Types As Algebras

Programming problems involve data objects that are processed by a computer. To process data objects, we need operations to act on them. So algebra enters the programming picture. In computer science, an *abstract data type* consists of one or more sets of data objects together with one or more operations on the sets and some axioms to describe the operations. In other words, an abstract data type is an algebra. There is, however, a restriction on the carriers of abstract data types. A carrier must be able to be constructed in some way that will allow the data objects and the operations to be implemented on a computer.

Programming languages normally contain some built-in abstract data types. But it's not possible for a programming language to contain all possible ways to represent and operate on data objects. Therefore, programmers must often design and implement new abstract data types. The axioms of an abstract data type can be used by a programmer to check whether an implementation is correct. In other words, the implemented operations can be checked to see whether they satisfy the axioms.

An abstract data type allows us to program with its data objects and operations without having to worry about implementation details. For example, suppose we need to create an abstract data type for processing polynomials. We might agree to use the expression $\mathrm{add}(p, q)$ to represent the sum of two polynomials p and q. To implement the abstract data type, we might represent a polynomial as an array of coefficients and then implement the add operation by adding corresponding array components. Of course, there are other interesting and useful ways to represent polynomials and their addition. But no matter what implementation is used, the statement $\mathrm{add}(p, q)$ always means the same thing. So we've abstracted away the implementation details.

In this section we'll introduce some of the basic abstract data types of computer science.

10.3.1   Natural Numbers

In Chapter 3 we discussed the problem of trying to describe the natural numbers to a robot. Let's revisit the problem by trying to describe the natural numbers to ourselves from an algebraic point of view. We can start by trying out the following inductive definition:

1. $0 \in \mathbb{N}$.

2. There is a function $s : \mathbb{N} \to \mathbb{N}$ called "successor" with the following property: If $x \in \mathbb{N}$, then $s(x) \in \mathbb{N}$.

Does this inductive definition adequately describe the natural numbers? It depends on what we mean by "successor." For example, if $s(0) = 0$, then the set $\{0\}$ satisfies the definition. So the property $s(0) = 0$ must be ruled out. Let's try the additional axiom:

3. $s(x) \neq 0$ for all $x \in \mathbb{N}$.

Now $\{0\}$ doesn't satisfy the three axioms. But if we assume that $s(s(0)) = s(0)$, then the set $\{0, s(0)\}$ satisfies them. The problem here is that s sends two elements to the same place. We can eliminate this problem if we require s to be injective (one to one):

4. If $s(x) = s(y)$, then $x = y$.

This gives us the set of natural numbers in the form $0, s(0), s(s(0)), \ldots$, where we set $s(0) = 1$, $s(s(0)) = 2$, and so on.

Historically, the first description of the natural numbers using axioms 1–4 was given by Peano. He also included a fifth axiom to describe the principle of mathematical induction:

5. If $q(x)$ is a property of x such that: $q(0)$ is true, and $q(x)$ implies $q(s(x))$; then $q(x)$ is true for all $x \in \mathbb{N}$.

Let's stop for a minute to write down an algebraic description of the natural numbers in terms of the first four rules:

Carrier: $\mathbb{N}$.

Operations: $0 \in \mathbb{N}$,
$\quad\quad\quad\quad s : \mathbb{N} \to \mathbb{N}$.

Axioms: $s(x) \neq 0$,
$\quad\quad\quad$ If $s(x) = s(y)$, then $x = y$.

An algebra is useful as an abstract data type if we can define useful operations on the type in terms of its primitive operations. For example, can we define addition of natural numbers in this algebra? Sure. We can define the "plus" operation using only the successor operation as follows:

$$\text{plus}(0, y) = y,$$
$$\text{plus}(s(x), y) = s(\text{plus}(x, y)).$$

For example, plus(2, 1) is computed by first writing $2 = s(s(0))$ and $1 = s(0)$. Then we can apply the definition recursively as follows:

$$
\begin{aligned}
\text{plus}\,(2,1) &= \text{plus}\,(s\,(s\,(0)),\, s\,(0)) \\
&= s\,(\text{plus}\,(s\,(0),\, s\,(0))) \\
&= s\,(s\,(\text{plus}\,(0,\, s\,(0)))) \\
&= s\,(s\,(s\,(0))) \\
&= 3.
\end{aligned}
$$

An alternative definition for the plus operation can be given as

$$
\begin{aligned}
\text{plus}\,(0, y) &= y, \\
\text{plus}\,(s\,(x), y) &= \text{plus}\,(x, s\,(y)).
\end{aligned}
$$

For example, using this definition, we can evaluate plus(2, 2) as follows:

$$
\text{plus}\,(2,2) = \text{plus}\,(1,3) = \text{plus}\,(0,4) = 4.
$$

Now that we have the plus operation, we can use it to define the multiplication operation as follows:

$$
\begin{aligned}
\text{mult}\,(0, y) &= 0, \\
\text{mult}\,(s\,(x), y) &= \text{plus}\,(\text{mult}\,(x, y), y).
\end{aligned}
$$

For example, we'll evaluate mult(3, 4) as follows—assuming that plus does its job properly:

$$
\begin{aligned}
\text{mult}\,(3,4) &= \text{plus}\,(\text{mult}\,(2,4),4) \\
&= \text{plus}\,(\text{plus}\,(\text{mult}\,(1,4),4),4) \\
&= \text{plus}\,(\text{plus}\,(\text{plus}\,(\text{mult}\,(0,4),4),4),4) \\
&= \text{plus}\,(\text{plus}\,(\text{plus}\,(0,4),4),4) \\
&= \text{plus}\,(\text{plus}\,(4,4),4) \\
&= \text{plus}\,(8,4) = 12.
\end{aligned}
$$

Let's see whether we can write the definitions for plus and mult in if-then-else form. To do so, we need the idea of a predecessor. Letting $p(x)$ denote the "predecessor" of x, we can write the definition of plus in either of the following ways:

$$
\text{plus}\,(x, y) = \text{if } x = 0 \text{ then } y \text{ else } s\,(\text{plus}\,(p\,(x), y))
$$

or

$$
\text{plus}\,(x, y) = \text{if } x = 0 \text{ then } y \text{ else } \text{plus}\,(p\,(x), s\,(y)).
$$

We'll leave it as an exercise to prove that these two definitions are equivalent. We can write the definition of mult as follows:

$$\text{mult}(x, y) = \text{if } x = 0 \text{ then } 0 \text{ else plus}(\text{mult}(p(x), y), y).$$

We can define the predecessor operation in terms of successor using the equation $p(s(x)) = x$. Since we're dealing only with natural numbers, we should either make $p(0)$ undefined or else define it so that it won't cause trouble. The usual definition is to say that $p(0) = 0$. It's interesting to note that we can't write an if-then-else definition for the predecessor using only the successor operation. So in some sense, the predecessor is a primitive operation too. So we'll add the definition of p to our algebra. We also need a test for zero to handle the test "$x = 0$" that occurs in if-then-else definitions.

To describe the algebra that includes these notions, we'll need another carrier to contain the true and false results that are returned by the test for zero. Letting Boolean = {True, False} and replacing s and p by the more descriptive names "succ" and "pred," we obtain the following algebra to represent the abstract data type of natural numbers:

Abstract Data Type of Natural Numbers   (10.11)

Carriers: $\mathbb{N}$, Boolean.

Operations: $0 \in \mathbb{N}$,

   isZero : $\mathbb{N} \to$ Boolean,

   succ : $\mathbb{N} \to \mathbb{N}$,

   pred : $\mathbb{N} \to \mathbb{N}$.

Axioms: isZero(0) = True,

   isZero(succ(x)) = False,

   pred(0) = 0,

   pred(succ(x)) = x.

Notice that we've made some replacements. The old axiom succ(x) $\neq$ 0 has been replaced by the new axiom, isZero(succ(x)) = False, which expresses the same idea. Also, the old axiom "If succ(x) = succ(y), then $x = y$" has been replaced by the new axiom pred(succ(x)) = x. To see this, notice that succ(x) = succ(y) implies that pred(succ(x)) = pred(succ(y)). Therefore, we can conclude that $x = y$ because $x = $ pred(succ(x)) = pred(succ(y)) = y.

For example, we can rewrite the plus function in terms of the primitives of this algebra as

$$\text{plus}(x, y) = \text{if isZero}(x) \text{ then } y \text{ else succ}(\text{plus}(\text{pred}(x), y)).$$

We can also write the mult function in terms of the primitives of (10.11) together with the plus function as follows:

$$\text{mult}(x, y) = \text{if isZero}(x) \text{ then } 0 \text{ else plus}(\text{mult}(\text{pred}(x), y), y).$$

example 10.17 Less-Than

Let's define the "less" relation on natural numbers using only the primitives of the algebra (10.11). To get an idea of how we might proceed, consider the following evaluation of the expression less(2, 4):

$$\text{less}(2,\,4) = \text{less}(1,\,3) = \text{less}(0,\,2) = \text{True}.$$

We simply replace each argument by its predecessor until one of the arguments is zero. Therefore, less can be computed from a recursive definition such as the following:

$$\text{less}\,(0,0) = \text{False},$$
$$\text{less}\,(\text{succ}\,(x),0) = \text{False},$$
$$\text{less}\,(0,\text{succ}\,(y)) = \text{True},$$
$$\text{less}\,(\text{succ}\,(x),\text{succ}\,(y)) = \text{less}\,(x,y)\,.$$

Using the if-then-else form, we obtain the following definition:

$$\text{less}\,(x,y) = \text{if isZero}\,(y)\ \text{then False}$$
$$\text{else if isZero}\,(x)\ \text{then True}$$
$$\text{else less}\,(\text{pred}\,(x),\text{pred}\,(y))\,.$$

end example

The following paragraphs describe several fundamental algebras of computer science. As we have said, they are also called abstract data types. The need for abstraction can be seen by considering questions like the following: What do lists and stacks have in common? How can we be sure that a queue is implemented correctly? How can we be sure that any data structure is implemented correctly? The answers to these questions depend on how we define the structures that we are talking about, without regard to any particular implementation.

10.3.2   Lists and Strings

Lists

Recall that the set of lists over a set A can be defined inductively by using the empty list, $\langle\,\rangle$, and the cons operation (with infix form ::) as constructors. If we denote the set of all lists over A by lists(A), we have the following inductive definition:

Basis: $\langle\,\rangle \in$ lists(A).

Induction: If $x \in A$ and $L \in$ lists(A), then cons(x, L) $\in$ lists(A).

The algebra of lists can be defined by the constructors $\langle\ \rangle$ and cons together with the primitive operations isEmptyL, head, and tail. With these operations we can describe the *list abstract data type* as the following algebra of lists over A:

Abstract Data Type of Lists

Carriers: lists(A), A, Boolean.

Operations: $\langle\ \rangle \in$ lists(A),

   isEmptyL : lists(A)$\rightarrow$ Boolean,

   cons : $A \times$ lists(A) $\rightarrow$ lists(A),

   head : lists(A)$\rightarrow A$,

   tail : lists(A) $\rightarrow$ lists(A).

Axioms: isEmptyL($\langle\ \rangle$) = True,

   isEmptyL(cons(x, L)) = False,

   head(cons(x, L)) = x,

   tail(cons(x, L)) = L.

Can all desired list functions be written in terms of the "primitive" operations of this algebra? The answer probably depends on the definition of "desired." For example, we saw in Chapter 3 that the following functions can be written in terms of the operations of the list algebra.

length:	lists(A) $\rightarrow \mathbb{N}$	Finds length of a list.
member:	$A\times$ lists(A) $\rightarrow$ Boolean	Tests membership in a list.
last:	lists(A) $\rightarrow A$	Finds last element of a list.
concatenate:	lists(A) $\times$ lists(A) $\rightarrow$ lists(A)	
putLast:	$A\times$ lists(A) $\rightarrow$ lists(A)	Puts element at right end.

Let's look at a couple of these functions to see whether we can implement them. Assume that all the operations in the signature of the list algebra are implemented. Then a definition for "length" can be written as follows:

$$\text{length}\,(L) = \text{if isEmptyL}\,(L)\ \text{then } 0$$
$$\text{else } 1 + \text{length}(\text{tail}\,(L)\,).$$

In this case the algebra $\langle \mathbb{N};\ +,\ 0 \rangle$ must also be implemented for the length function to work properly.

Similarly, suppose we define "member" as follows:

$$\text{member}\,(a, L) = \text{if isEmptyL}\,(L)\ \text{then False}$$
$$\text{else if } a = \text{head}\,(L)\ \text{then True}$$
$$\text{else member}\,(a, \text{tail}\,(L)\,).$$

In this case the predicate "$a = \text{head}(L)$" must be computed. Thus an equality relation must be implemented for the carrier A.

As these two examples have shown, although we can define list functions in terms of the algebra of lists, we often need other algebras, such as $\langle \mathbb{N}; +, 0 \rangle$, or other relations, such as equality on A.

Strings

Strings may look different than lists, but these structures have a lot in common. For example, they both have length, and their constructions are similar. For example, the set of all strings over an alphabet A can be defined inductively from the empty string, Λ, and the append operation to attach a letter to a string (which we'll denote by $\cdot$). Letting A^* denote the set of all strings over A, we have the following inductive definition:

Basis: $\Lambda \in A^*$.

Induction: If $x \in A$ and $s \in A^*$, then $x \cdot s \in A^*$.

The algebra of strings can be defined by the constructors Λ and append together with the primitive operations isEmptyS, headS, and tailS. With these operations we can describe the *string abstract data type* as the following algebra of strings over A:

Abstract Data Type of Strings

     Carriers: A, A^*, Boolean.

Operations: $\Lambda \in A^*$,

           isEmptyS : $A^* \rightarrow$ Boolean,

           $\cdot$: $A \times A^* \rightarrow A^*$,

           headS : $A^* \rightarrow A$,

           tailS : $A^* \rightarrow A^*$.

    Axioms: isEmptyS(Λ) = True,

           isEmptyS($a \cdot s$) = False,

           headS($a \cdot s$) = a,

           tailS($a \cdot s$) = s.

When working with strings, we want to be able to combine strings, compare strings, and so on. We can define functions to accomplish these things using the string algebra. For example, let's write a definition for the "cat" function to combine two strings. For example, cat(cb, aba) = $cbaba$. Cat has type $A^* \times A^*$ $\to A^*$ and can be defined as follows:

$$\text{cat}\,(s, t) = \text{if isEmptyS}\,(s)\ \text{then}\ t$$
$$\text{else headS}\,(s) \cdot \text{cat}\,(\text{tailS}\,(s)\,, t)\,.$$

10.3.3   Stacks and Queues

Stacks

A *stack* is a structure satisfying the LIFO property of last in, first out. In other words, the last element input is the first element output. The main stack operations are *push*, which pushes a new element onto a stack; *pop*, which removes the top element from a stack; and *top*, which examines the top element of a stack. We also need an indication of when a stack is empty.

Let's describe the *stack abstract data type* as an algebra. For any set A, let Stks[A] denote the set of stacks whose elements are from A. We'll include error messages in our description for those cases in which the operators are not defined. Here's the algebra.

Abstract Data Type of Stacks

Carriers: A, Stks[A], Boolean, Errors.

Operations: emptyStk $\in$ Stks[A],

   isEmptyStk : Stks[A] $\to$ Boolean,

   push : $A \times$ Stks[A] $\to$ Stks[A],

   pop : Stks[A] $\to$ Stks[A] $\cup$ Errors,

   top : Stks[A] $\to$ $A \cup$ Errors.

Axioms: isEmptyStk(emptyStk) = True,

   isEmptyStk(push(a, s)) = False,

   pop(push(a, s)) = s,

   pop(emptyStk) = stackError,

   top(push(a, s)) = a,

   top(emptyStk) = valueError.

Notice the similarity between the stack algebra and the list algebra. In fact, we can implement the stack algebra as a list algebra by assigning the following meanings to the stack symbols:

$$\text{Stks}\,[A] = \text{lists}\,(A),$$
$$\text{emptyStk} = \langle\,\rangle,$$
$$\text{isEmptyStk} = \text{isEmptyL},$$
$$\text{push} = \text{cons},$$
$$\text{pop} = \text{tail},$$
$$\text{top} = \text{head}.$$

To prove that this implementation is correct, we need to show that the axioms of a stack are true for the above assignment. They are all trivial. For example, the proof of the third axiom is a one-liner:

$$\text{pop}(\text{push}(a,\,s)) = \text{tail}(\text{cons}(a,\,s)) = s. \text{ QED}.$$

example **10.18 Evaluating a Postfix Expression**

Let's look at the general approach to evaluate an arithmetic expression represented in postfix notation. For example, the postfix expression $abc+-$ can be evaluated by pushing a, b, and c onto a stack. Then c and b are popped, and the value $b + c$ is pushed onto the stack. Finally, $b + c$ and a are popped, and the value $a - (b + c)$ is pushed. We'll assume that all operators are binary and that there is a function "val," which takes an operator and two operands and returns the value of the operator applied to the two operands.

The general algorithm for evaluating a postfix expression can be given as follows, where the initial call has the form $\text{post}(L, \langle\,\rangle)$ and L is the list representation of the postfix expression:

$$\text{post}\,(\langle\,\rangle, \text{stk}) = \text{top}(\text{stk}),$$
$$\text{post}\,(x :: t, \text{stk}) = \text{if } x \text{ is an argument then}$$
$$\text{post}\,(t, \text{push}\,(x, \text{stk}))$$
$$\text{else } \{x \text{ is an operator}\}$$
$$\text{post}\,(t, \text{eval}\,(x, \text{stk})),$$

where eval is defined by the equation

$$\text{eval}(\text{op}, \text{push}(z, \text{push}(y, \text{stk}))) = \text{push}(\text{val}(y, \text{op}, z), \text{stk}).$$

For example, we'll evaluate the expression $\text{post}(\langle 2, 5, + \rangle, \langle \ \rangle)$.

$$
\begin{aligned}
\text{post}\left(\langle 2,5,+\rangle, \langle \ \rangle\right) &= \text{post}\left(\langle 5,+\rangle, \langle 2 \rangle\right) \\
&= \text{post}\left(\langle + \rangle, \langle 5,2 \rangle\right) \\
&= \text{post}\left(\langle \ \rangle, \text{eval}\left(+, \langle 5,2 \rangle\right)\right) \\
&= \text{top}\left(\text{eval}\left(+, \langle 5,2 \rangle\right)\right) \\
&= \text{top}\left(\text{push}\left(\text{val}\left(2,+,5\right), \langle \ \rangle\right)\right) \\
&= \text{val}\left(2,+,5\right) \\
&= 7.
\end{aligned}
$$

end example

Queues

A *queue* is a structure satisfying the FIFO property of first in, first out. In other words, the first element input is the first element output. So a queue is a fair waiting line. The main operations on a queue involve adding a new element, examining the front element, and deleting the front element.

To describe the *queue abstract data type* as an algebra, we'll let A be a set and $Q[A]$ be the set of queues over A. Here's the algebra.

Abstract Data Type of Queues

Carriers: A, $Q[A]$, Boolean.

Operations: $\text{emptyQ} \in Q[A]$,

  $\text{isEmptyQ} : Q[A] \rightarrow$ Boolean,

  $\text{addQ} : A \times Q[A] \rightarrow Q[A]$,

  $\text{frontQ} : Q[A] \rightarrow A$,

  $\text{delQ} : Q[A] \rightarrow Q[A]$.

Axioms: $\text{isEmptyQ}(\text{emptyQ}) = \text{True}$,

  $\text{isEmptyQ}(\text{addQ}(a, q)) = \text{False}$,

  $\text{frontQ}(\text{addQ}(a, q)) = \text{if isEmptyQ}(q) \text{ then } a$
    $\text{else frontQ}(q)$,

  $\text{delQ}(\text{addQ}(a, q)) = \text{if isEmptyQ}(q) \text{ then } q$
    $\text{else addQ}(a, \text{delQ}(q))$.

Although we haven't stated it in the axioms, an error will occur if either frontQ or delQ is applied to an empty queue.

Suppose we represent a queue as a list. For example, the list $\langle a, b \rangle$ represents a queue with a at the front and b at the rear. If we add a new item c to this queue, we obtain the queue $\langle a, b, c \rangle$. So $\mathrm{addQ}(c, \langle a, b \rangle) = \langle a, b, c \rangle$. Thus addQ can be implemented as the putLast function. The implementation of a queue algebra as a list algebra can be given as follows:

$$Q\,[A] = \mathrm{lists}\,(A)\,,$$
$$\mathrm{emptyQ} = \langle\,\rangle\,,$$
$$\mathrm{isEmptyQ} = \mathrm{isEmptyL},$$
$$\mathrm{frontQ} = \mathrm{head},$$
$$\mathrm{delQ} = \mathrm{tail},$$
$$\mathrm{addQ} = \mathrm{putLast}.$$

The proof of correctness of this implementation is more interesting (not trivial) because two queue axioms include conditionals, and putLast is written in terms of the list primitives. For example, we'll prove the correctness of the third axiom for the algebra of queues, leaving the proof of the fourth axiom as an exercise. Since the third axiom is an if-then-else statement, we'll consider two cases:

Case 1: Assume that $q = \mathrm{emptyQ}$. In this case the axiom becomes

$$\begin{aligned}
\mathrm{frontQ}(\mathrm{addQ}(a, \mathrm{emptyQ})) &= \mathrm{head}(\mathrm{putLast}(a, \mathrm{emptyQ})) \\
&= \mathrm{head}(a :: \mathrm{emptyQ}) \\
&= a.
\end{aligned}$$

Case 2: Assume that $q \neq \mathrm{emptyQ}$. In this case the axiom becomes

$$\begin{aligned}
\mathrm{frontQ}(\mathrm{addQ}(a, q)) &= \mathrm{head}\,(\mathrm{putLast}\,(a, q)) \\
&= \mathrm{head}\,(\mathrm{head}\,(q) :: \mathrm{putLast}\,(a, \mathrm{tail}\,(q))) \\
&= \mathrm{head}\,(q) \\
&= \mathrm{frontQ}\,(q)\,.
\end{aligned}$$

example **10.19 The Append Function**

Let's use the queue algebra to define the append function, apQ, which joins two queues together. It can be written in terms of the primitive operations of a queue algebra as follows:

$$\begin{aligned}
\mathrm{apQ}\,(x, y) = {}&\text{if } \mathrm{isEmptyQ}\,(y) \text{ then } x \\
&\text{else } \mathrm{apQ}\,(\mathrm{addQ}\,(\mathrm{frontQ}\,(y)\,, x)\,, \mathrm{delQ}\,(y))\,.
\end{aligned}$$

For example, suppose $x = \langle a, b \rangle$ and $y = \langle c, d \rangle$ are two queues, where a is the front of x and c is the front of y. We'll evaluate the expression apQ(x, y).

$$\begin{aligned} \text{apQ}\,(x, y) &= \text{apQ}\,(\langle a, b \rangle, \langle c, d \rangle) \\ &= \text{apQ}\,(\langle a, b, c \rangle, \langle d \rangle) \\ &= \text{apQ}\,(\langle a, b, c, d \rangle, \langle\ \rangle) \\ &= \langle a, b, c, d \rangle. \end{aligned}$$

end example

example   10.20   Decimal to Binary

Let's convert a natural number to a binary number and represent the output as a queue of binary digits. Let bin(n) represent the queue of binary digits representing n. For example, we should have bin$(4) = \langle 1, 0, 0 \rangle$, assuming that the front of the queue is the head of the list. Let's get to the definition.

If $n = 0$ or $n = 1$, we should return the queue $\langle n \rangle$, which is constructed by addQ$(n,$ emptyQ$)$. If n is not 0 or 1, then we should return the queue addQ$(n$ mod 2, bin(floor$(n/2)))$. In other words, we can define bin as follows:

$$\begin{aligned} \text{bin}\,(n) = \ &\text{if}\ n = 0\ \text{or}\ n = 1\ \text{then} \\ &\quad \text{addQ}\,(n, \text{emptyQ}) \\ &\text{else} \\ &\quad \text{addQ}\,(n \bmod 2, \text{bin}\,(\text{floor}\,(n/2))). \end{aligned}$$

We leave it as an exercise to check that bin works. For example, try to evaluate the expression bin(4) to see whether you get the list $\langle 1, 0, 0 \rangle$.

end example

10.3.4   Binary Trees and Priority Queues

Binary Trees

Let $B[A]$ denote the set of binary trees over a set A. The main operations on binary trees involve constructing a tree, picking the root, and picking the left and right subtrees. If $a \in A$ and $l, r \in B[A]$, let tree(l, a, r) denote the tree whose root is a, whose left subtree is l, and whose right subtree is r. We can describe the *binary tree abstract data type* as the following algebra of binary trees:

Abstract Data Type of Binary Trees

Carriers: A, $B[A]$, Boolean.

Operations: emptyTree $\in B[A]$,

isEmptyTree : $B[A] \to$ Boolean,

root : $B[A] \to A$,

tree : $B[A] \times A \times B[A] \to B[A]$,

left : $B[A] \to B[A]$,

right : $B[A] \to B[A]$.

Axioms: isEmptyTree(emptyTree) = True,

isEmptyTree(tree(l, a, r)) = False,

left(tree(l, a, r)) = l,

right(tree(l, a, r)) = r,

root(tree(l, a, r)) = a.

Although we haven't stated it in the axioms, an error will occur if the functions left, right, and root are applied to the empty tree. Next, we'll give a few examples to show how useful functions can be constructed from the basic tree operations.

example **10.21 Nodes and Depth**

We'll look at two typical functions, "count" and "depth." Count returns the number of nodes in a binary tree. Its type is $B[A] \to \mathbb{N}$, and its definition follows:

$$\text{count}(t) = \text{if isEmptyTree}(t) \text{ then } 0$$
$$\text{else } 1 + \text{count}(\text{left}(t)) + \text{count}(\text{right}(t)).$$

Depth returns the length of the longest path from the root to the leaves of a binary tree. Assume that an empty binary tree has depth –1. Its type is $B[A] \to \mathbb{N} \cup \{-1\}$, and its definition follows:

$$\text{depth}(t) = \text{if isEmptyTree}(t) \text{ then } -1$$
$$\text{else } 1 + \text{max}(\text{depth}(\text{left}(t)), \text{depth}(\text{right}(t))).$$

end example

example   ## 10.22   Inorder Traversal

Suppose we want to write a function "inorder" to perform an inorder traversal of a binary tree and place the nodes in a queue. So we want to define a function of type $B[A] \rightarrow Q[A]$. For example, we might use the following definition:

$$\text{inorder}\,(t) = \text{if isEmptyTree}\,(t)\ \text{then emptyQ}$$
$$\text{else apQ}\,(\text{addQ}\,(\text{root}\,(t)\,,\text{inorder}\,(\text{left}\,(t)))\,,\text{inorder}\,(\text{right}\,(t)))\,.$$

We'll leave the preorder and postorder traversals as exercises.

end example

Priority Queues

A *priority queue* is a structure satisfying the BIFO property: best in, first out. For example, a stack is a priority queue if we let Best = Last. Similarly, a queue is a priority queue if we let Best = First. The main operations of a priority queue involve adding a new element, accessing the best element, and deleting the best element.

Let $P[A]$ denote the set of priority queues over A. If $a \in A$ and $p \in P[A]$, then insert(a, p) denotes the priority queue obtained by adding a to p. We can describe the *priority queue abstract data type* as the following algebra:

Carriers: A, $P[A]$, Boolean.

Operations: emptyP $\in P[A]$,

   isEmptyP : $P[A] \rightarrow$ Boolean,

   better : $A \times A \rightarrow$ Boolean,

   best : $P[A] \rightarrow A$,

   insert : $A \times P[A] \rightarrow P[A]$,

   delBest : $P[A] \rightarrow P[A]$.

We'll note here that we are assuming that the function "better" is a binary relation on A. Now for the axioms:

Axioms: isEmptyP(emptyP) = True,

   isEmptyP(insert(a, p)) = False,

   best(insert(a, p)) = if isEmptyP(p) then a

$\quad\quad\quad\quad\quad\quad\quad$ else if better$(a, \text{best}(p))$ then a

$\quad\quad\quad\quad\quad\quad\quad$ else best(p),

   delBest(insert(a, p))= if isEmptyP(p) then emptyP

$\quad\quad\quad\quad\quad\quad\quad$ else if better$(a, \text{best}(p))$ then p

$\quad\quad\quad\quad\quad\quad\quad$ else insert$(a, \text{delBest}(p))$.

We should note that the operations best and delBest are defined only on nonempty priority queues. Priority queues can be implemented in many different ways, depending on the definitions of "better" and "best" for the set A.

To show the power of priority queues, we'll write a sorting function that sorts the elements of a priority queue into a sorted list. The initial call to sort the priority queue p is $\text{sort}(p, \langle\,\rangle)$. The definition can be written as follows:

$$\text{sort}\,(p, L) = \text{if isEmptyP}\,(p)\ \text{then}\ L$$
$$\text{else sort}\,(\text{delBest}\,(p)\,, \text{best}\,(p) :: L)\,.$$

 Exercises

Natural Numbers

1. The *monus* operation on natural numbers is like subtraction, except that it always gives a natural number as a result. An informal definition of monus can be written as follows:

$$\text{monus}(x,\ y) = \text{if}\ x \geq y\ \text{then}\ x - y\ \text{else}\ 0.$$

   Write down a recursive definition of monus that uses only the primitive operations isZero, succ, and pred.

2. The exponentiation function is defined by $\exp(a,\ b) = a^b$. Write down a recursive definition of exp that uses primitive operations or functions that are defined in terms of the primitive operations on the natural numbers. *Note:* Assume that $\exp(0, 0) = 0$.

Lists

3. Use the algebra of lists to write a definition of the function "reverse" to reverse the elements of a list. For example, $\text{reverse}(\langle x,\ y,\ z \rangle) = \langle z,\ y,\ x \rangle$.

4. Use lists to describe an implementation of the algebra of strings over some alphabet.

5. Write an algebraic specification for general lists over a set A (where the elements of a list may also be lists).

6. Use the algebra of lists to write a definition for the "flatten" function that takes a general list over a set A and returns the list of its elements from A. For example, $\text{flatten}(\langle\langle a,\ b \rangle,\ c,\ d \rangle) = \langle a,\ b,\ c,\ d \rangle$. *Hint:* Assume that there is a function isAtom to check whether its argument is an atom (not a list). Also assume that the other list operations work on general lists.

7. Evaluate the expression $\text{post}(\langle 4,\ 5,\ -,\ 2,\ + \rangle,\ \langle\,\rangle)$ by unfolding the definition in Example 10.18.

8. Evaluate the expression bin(4) by unfolding the definition in Example 10.20.

Binary Trees

9. Write down a definition for the function "preorder," which performs a pre-order traversal of a binary tree and places the node values in a queue.

10. Write down a definition for the function "postorder," which performs a post-order traversal of a binary tree and places the node values in a queue.

Stacks and Queues

11. Find a descriptive name for the "mystery" function f, which has the type $A \times \text{Stks}[A] \rightarrow \text{Stks}[A]$ and is defined by the following equations:

$$f\,(a, \text{emptyStk}) = \text{emptyStk},$$
$$f\,(a, \text{push}\,(a, s)) = f\,(a, s)\,,$$
$$f\,(a, \text{push}\,(b, s)) = \text{push}\,(b, f\,(a, s)) \quad \text{if } a \neq b.$$

12. Find a descriptive name for the "mystery" function f, which has type $Q[A] \rightarrow Q[A]$ and is defined as follows:

$$f\,(q) = \text{if isEmptyQ}\,(q) \text{ then } q$$
$$\text{else addQ}\,(\text{frontQ}\,(q)\,, f\,(\text{delQ}\,(q)))\,.$$

13. A *deque*, pronounced "deck," is a double-ended queue in which insertions and deletions can be made at either end of the deque. Write down an algebraic specification for deques over a set A.

14. For the list implementation of a queue, prove the correctness of the following axiom:

$$\text{delQ}\,(\text{addQ}\,(a, q)) = \text{if isEmptyQ}\,(q) \text{ then } q$$
$$\text{else addQ}\,(a, \text{delQ}\,(q))\,.$$

15. Implement a queue by using the operations of a deque. Prove the correctness of your implementation.

16. Suppose the "better" function used in a priority queue has the following type definition:

$$\text{better} : A \times A \rightarrow A.$$

How would the axioms change? Do we need any new operations?

Proofs and Challenges

17. Consider the following two definitions for adding natural numbers, where p and s denote the predecessor and successor operations.

$$\text{plus}\,(x, y) = \text{if } x = 0 \text{ then } y \text{ else } s\,(\text{plus}\,(p\,(x), y)),$$
$$\text{add}\,(x, y) = \text{if } x = 0 \text{ then } y \text{ else } \text{add}\,(p\,(x), s\,(y)).$$

   a. Use induction to prove that $\text{plus}(x,\ s(y)) = s(\text{plus}(x,\ y))$ for all x, $y \in \mathbb{N}$.
   b. Use induction to prove that $\text{plus}(x,\ y) = \text{add}(x,\ y)$ for all x, $y \in \mathbb{N}$. *Hint:* Part (a) can be useful.

18. Use induction to prove the following property over a queue algebra, where apQ is the append function defined in Example 10.19.

$$\text{apQ}(x,\ \text{addQ}(a,\ y)) = \text{addQ}(a,\ \text{apQ}(x,\ y)).$$

   Hint: To simplify notation, let $x{:}a$ denote $\text{addQ}(a,\ x)$. Then the equation becomes $\text{apQ}(x,\ y{:}a) = \text{apQ}(x,\ y){:}a$.

19. Use induction to prove the following property over a queue algebra, where apQ is the append function defined in Example 10.19.

$$\text{apQ}(x,\ \text{apQ}(y,\ z)) = \text{apQ}(\text{apQ}(x,\ y),\ z).$$

   Hint: Exercise 18 may be helpful.

10.4 Computational Algebras

In this section we present some important examples of algebras that are useful in the computation process. First we'll look at relational algebra as a tool for representing relational databases. Then we'll discuss functional algebra as a tool not only for programming, but for reasoning about programs.

10.4.1 Relational Algebras

Relations can be combined in various ways to build new relations that solve problems. An algebra is called a *relational algebra* if its carrier is a set of relations. We'll discuss three useful operations on relations: *select*, *project*, and *join*. Each of these operations builds a new relation by selecting certain tuples, by eliminating certain attributes, or by combining attributes of two relations. We'll motivate the definitions with some examples.

Rooms

Place	Seats	Boardtype	Computer
CH171	80	Chalk	No
HH101	250	No	Yes
SC211	35	White	Yes
CH301	90	Chalk	Yes
. . .	. . .	. . .	. . .

Figure 10.11   Classrooms in a school.

Let *Rooms* be the relation with attributes {Place, Seats, Boardtype, Computer} to describe classrooms in a college. For example, Figure 10.11 shows a few sample entries for Rooms.

Notice that Rooms can be represented as a relation consisting of a set of 4-tuples as follows:

$$\text{Rooms} = \{ \text{(CH171, 80, Chalk, No)},$$
$$\text{(HH101, 250, No, Yes)},$$
$$\text{(SC211, 35, White, Yes)},$$
$$\text{(CH301, 90, Chalk, Yes)}, \dots \}.$$

The Select Operation

The *select* operation on a relation forms a new relation that is a subset of the relation consisting of those tuples that have a common value in one of the attributes.

For example, suppose that we want to construct the relation A of tuples that represent all the rooms with chalk boards. In other words, we want to select from Rooms those tuples that have Boardtype equal to Chalk. We'll represent this by the notation

$$A = \text{select(Rooms, Boardtype, Chalk)}.$$

The value of this expression is

$$A = \{\text{(CH171, 80, Chalk, No)}, \text{(CH301, 90, Chalk, Yes)}, \dots \}.$$

example   **10.23   Selecting Tuples**

Suppose we want to construct the relation B of tuples that represent the rooms with chalk boards and computers. In this case we can select the tuples from the relation A.

$$B = \text{select}(A, \text{Computer, Yes})$$
$$= \text{select(select(Rooms, Boardtype, Chalk), Computer, Yes)}$$
$$= \{\text{(CH301, 90, Chalk, Yes)}, \dots \}.$$

end example

The Project Operation

The *project* operation on a relation forms a new relation consisting of tuples indexed by a subset of the attributes of the relation.

For example, suppose that we want to construct the relation *Size* of tuples with only the two attributes Place and Seats. In other words, we want to restrict ourselves to the first and second columns of the table for Rooms. We'll represent this by the notation

$$\text{Size} = \text{project}(\text{Rooms}, \{\text{Place, Seats}\}).$$

The value of this expression is

$$\text{Size} = \{(\text{CH171, 80}), (\text{HH101, 250}), (\text{SC211, 35}), (\text{CH301, 90}), \ldots\}.$$

example 10.24 Specific Properties

Here are a few more questions that ask for specific properties about the Rooms relation.

1. What rooms have chalk boards?

project(select(Rooms, Boardtype, Chalk), {Place})
$$= \{(\text{CH171}), (\text{CH301}), \ldots\}.$$

2. How large are the rooms with computers?

project(select(Rooms, Computer, Yes), {Place, Seats})
$$= \{(\text{HH101, 250}), (\text{SC211, 35}), (\text{CH301, 90}), \ldots\}.$$

3. What kind of board is in SC211?

project(select(Rooms, Place, SC211), {Boardtype})
$$= \{(\text{White})\}.$$

end example

The Join Operation

The *join* operation on two relations forms a new relation consisting of tuples that are indexed by the union of the attributes of the two relations. The new tuples in the join are constructed from pairs of tuples whose values agree on the common attributes of the two relations.

For example, let *Channel* and *Program* be two relations with attributes {Station, Satellite, Cable} and {Station, Type}, respectively, that describe information about television networks. Figure 10.12 shows a few sample entries for the two relations.

Suppose we want to join the two relations into a single relation called TV with attributes Station, Satellite, Cable, and Type. We'll represent this operation by the notation

$$\text{TV} = \text{join}(\text{Channel, Program}).$$

Channel

Station	Satellite	Cable
AMC	130	48
CNN	200	96
TCM	132	54
ESPN	140	32

Program

Station	Type
AMC	Movie
CNN	News
TCM	Movie
ESPN	Sports

Figure 10.12   TV channels and programs.

example   10.25   TV Questions

Now we can answer some TV questions. For example, what are the cable movie channels? One solution is to select the tuples in TV that have Movie as the Type attribute. Then project onto the Cable attribute:

$$\text{project}(\text{select}(\text{TV}, \text{Type}, \text{Movie}), \{\text{Cable}\}).$$

The expression evaluates to the channel numbers $\{(48), (54), \ldots\}$.

For another example, what type of programming is on satellite channel 140? One solution is to select the tuples in TV that have 140 as the Satellite attribute. Then project onto the Type attribute:

$$\text{project}(\text{select}(\text{TV}, \text{Satellite}, 140), \{\text{Type}\}).$$

The expression evaluates to $\{(\text{Sports})\}$.

end example

example   10.26   Class Schedules

Let *Schedule* be the class schedule with attributes {Dept, Course, Section, Credit, Time, Day, Place, Teacher}. Figure 10.13 shows a few sample entries.

Each entry of the table can be represented as a tuple. For example, the first row of Schedule can be represented as the tuple

$$(\text{CS}, 252, 1, 4, 1600\text{--}1750, \text{TTh}, \text{CH171}, \text{Hein}).$$

Schedule

Dept	Course	Section	Credit	Time	Day	Place	Teacher
CS	252	1	4	1600-1750	TTh	CH171	Hein
CS	252	2	4	1200-1350	MW	SC211	Jones
Mth	201	1	4	1000-1150	TTh	CH301	Appleby
Mth	256	1	4	1400-1550	MW	NH356	Ames
EE	300	1	4	0800-0950	TTh	SC211	Brand

Figure 10.13   A class schedule.

Here are some sample questions and answers.

1. What is the mathematics schedule?

   select(Schedule, Dept, Mth).

2. What mathematics classes meet TTh?

   select(select(Schedule, Dept, Mth), Day, TTh)

   $= \{(\text{Mth, 201, 1, 4, 1000–1150, TTh, NH325, Appleby}), \dots \}$.

3. What are the times and days that CS 252 is taught? If we want the set

   $\{(1600\text{–}1750, \text{TTh}), (1200\text{–}1350, \text{MW})\}$,

   then we can use the following expression:

   project(select(select(Schedule, Dept, CS), Course, 252), {Time, Day})

   $= \{(1600\text{–}1750, \text{TTh}), (1200\text{-}1350, \text{MW})\}$.

4. What classes are in rooms with computers?

   project(select(join(Rooms, Schedule), Computer, Yes), {Dept, Course, Section})

   $= \{(\text{CS, 252, 2}), (\text{Mth, 201, 1}), (\text{EE, 300, 1}), \dots \}$.

Formal Definitions of Select, Project, and Join

Let R be a relation, A an attribute of R, and a a possible value of A. The relation consisting of all tuples in R with attribute A having value a is denoted by select(R, A, a) and is defined as follows:

Select Operation

$$\text{select}(R, A, a) = \{t \mid t \in R \text{ and } t(A) = a\}.$$

For example, using the Rooms relation in Figure 10.11 we have

   select(Rooms, Boardtype, Chalk)

   $= \{t \mid t \in \text{Rooms and } t(\text{Boardtype}) = \text{Chalk}\}$

   $= \{(\text{CH171, 80, Chalk, No}), (\text{CH301, 90, Chalk, Yes}), \dots \}$.

If A and a are fixed, then select(R, A, a) is sometimes denoted by select$_{A=a}(R)$.

   If X is a subset of the set of attributes of the relation R, then the project operation of R on X is denoted project(R, X) and consists of all tuples indexed by X constructed from the tuples of R. In formal terms we have the following definition:

Project Operation

project$(R,\ X) = \{s \mid$ there exists $t \in R$ such that
$$s(A) = t(A) \text{ for all } A \in X\}.$$

For example, using the Rooms relation in Figure 10.11 we have

project(Rooms, {Place, Seats})

$= \{s \mid$ there exists $t \in$ Rooms such that
$s(\text{Place}) = t(\text{Place})$ and $s(\text{Seats}) = t(\text{Seats})\}$

$= \{(\text{CH171, 80}), (\text{HH101, 250}), (\text{SC211, 35}), (\text{CH301, 90}), \dots\}.$

If X is fixed, then project$(R,\ X)$ is sometimes denoted by project$_X(R)$.

Let R and S be two relations with attribute sets I and J, respectively. The *join* of R and S is the set of all tuples over the attribute set $I \cup J$ that are constructed from R and S by "joining" those tuples with equal values on the common attribute set $I \cap J$. We denote the join of R and S by join(R, S). Here's the formal definition:

Join Operation

join$(R, S) = \{t \mid$ there exist $r \in R$ and $s \in S$ such that
$$t(A) = r(A) \text{ for all } A \in I \text{ and}$$
$$t(B) = s(B) \text{ for all } B \in J\}.$$

There are two special cases. Let R and S be two relations with attribute sets I and J. If $I \cap J = \varnothing$, then join$(R,\ S)$ is obtained by concatenating all pairs of tuples in $R \times S$. For example, if we have tuples $(a,\ b) \in R$ and $(c,\ d,\ e) \in S$, then $(a,\ b,\ c,\ d,\ e) \in$ join$(R,\ S)$. If $I = J$, then join$(R,\ S) = R \cap S$.

A Relational Algebra

Now we have the ingredients of a relational algebra. The carrier is the set of all possible relations, and the three operations are select, project, and join. We should remark that join$(R,\ S)$ is often denoted by $R \bowtie S$. The properties of this relational algebra are too numerous to mention here. But we'll list a few properties that can be readily verified from the definitions.

$$\text{select}_{A=a}\left(\text{select}_{B=b}\left(R\right)\right)=\text{select}_{B=b}\left(\text{select}_{A=a}\left(R\right)\right),$$
$$R\bowtie R=R,$$
$$\left(R\bowtie S\right)\bowtie T=R\bowtie\left(S\bowtie T\right),$$
$$\text{project}_{X}\left(\text{select}_{A=a}\left(R\right)\right)=\text{select}_{A=a}\left(\text{project}_{X}\left(R\right)\right)\quad(\text{where }A\in X).$$

If R and S have the same set of attributes, then the select operation has some nice properties when combined with the set operations: $\cup$, $\cap$, and $-$. For example, we have the following properties:

$$\text{select}_{A=a}\left(R\cup S\right)=\text{select}_{A=a}\left(R\right)\cup\text{select}_{A=a}\left(S\right),$$
$$\text{select}_{A=a}\left(R\cap S\right)=\text{select}_{A=a}\left(R\right)\cap\text{select}_{A=a}\left(S\right),$$
$$\text{select}_{A=a}\left(R-S\right)=\text{select}_{A=a}\left(R\right)-\text{select}_{A=a}\left(S\right).$$

There are many other useful operations on relations, some of which can be defined in terms of the ones we have discussed. Relational algebra provides a set of tools for constructing, maintaining, and accessing databases.

10.4.2 Functional Algebras

From a programming point of view, a *functional algebra* consists of functions together with operations to combine functions in order to process data objects. Let's look at a particular functional algebra that is both a programming language and an algebra for reasoning about programs.

FP: A Functional Algebra

The correctness problem for programs can sometimes be solved by showing that the program under consideration is equivalent to another program that we "know" is correct. Methods for showing equivalence depend very much on the programming language. The FP language was introduced by Backus [1978]. FP stands for *functional programming*, and it is a fundamental example of a programming language that allows us to reason about programs in the programming language itself. To do this, we need a set of rules that allow us to do some reasoning. In this case the rules are axioms in the algebra of FP programs.

FP functions are defined on a set of objects that include atoms (numbers, and strings of characters), and lists. To apply an FP function f to an object x, we write down $f{:}x$ instead of the familiar $f(x)$. To compose two FP functions f and g, we write $f\ @\ g$ instead of the familiar $f\circ g$. Suppose we have the following definition for an if-then-else function:

$$f(x)=\text{if }a(x)\text{ then }b(x)\text{ else }c(x).$$

We would make f into an FP function by writing

$$f = a \rightarrow b;\ c.$$

We evaluate $f{:}x$ just like the evaluation of an if-then-else statement:

$$f{:}x = (a \rightarrow b;\ c){:}x = a{:}x \rightarrow b{:}x;\ c{:}x.$$

A function can be defined as a tuple of functions. For example, the following expression defines f as a 3-tuple of functions:

$$f = [g,\ h,\ k].$$

In this case $f{:}x = \langle g{:}x,\ h{:}x,\ k{:}x \rangle$. A constant function has the form $f = {\sim}\ c$, where $f{:}x = c$ for all objects x.

Our objective is to get the flavor of the language to see its algebraic nature. So we'll describe some operations and axioms of an algebra of FP programs. We'll limit the operations to those that will be useful in the examples and exercises. We'll also include a few axioms to show some useful relationships between composition, tupling, and if-then-else.

Operations to construct new functions:

@	composition (e.g., $f\ @\ g$),
$\rightarrow$	if-then-else (e.g., $p \rightarrow a; b$),
$[\ldots]$	tuple of functions (e.g., $[f, g, h]$),
$\sim$	constant (e.g., ~ 2).

Primitive operations:

id	the identity function,
hd, tl	head and tail,
apndl, apndr	cons and consR,
1, 2, ...	selectors (e.g., $2{:}\langle a, b, c \rangle = b$),
and, or, not	Boolean operations,
null	test for empty list,
atom	test for an atom,
$\langle\ \rangle$	empty list,
?	undefined symbol,
eq	test for equality of two atoms,
$<, >, +, -, *, /$	arithmetic relations and operations.

Axioms:

$$f\ @\ (a \rightarrow b\ ;\ c) = a \rightarrow f\ @\ b\ ;\ f\ @\ c,$$
$$(a \rightarrow b\ ;\ c)\ @\ d = a\ @\ d \rightarrow b\ @\ d\ ;\ c\ @\ d,$$
$$[f_1, \ldots, f_n]\ @\ g = [f_1\ @\ g,\ \ldots, f_n\ @\ g],$$
$$f\ @\ [\ldots, (a \rightarrow b\ ;\ c), \ldots] = a \rightarrow f\ @\ [\ldots, b, \ldots]\ ;\ f\ @\ [\ldots, c, \ldots].$$

Now we'll give some examples of FP program definitions. In the last example we'll use FP algebra to prove the equivalence of two FP programs.

example 10.27 Testing for Zero

Let eq0 be the function that tests its argument for zero. An FP definition for eq0 can be written as follows, where eq tests equality of atoms.

$$eq0 = eq \ @ \ [id, \sim 0].$$

For example, we'll evaluate the expression eq0 : 3 as follows:

$$eq0 : 3 = eq \ @ \ [id, \sim 0] : 3 = eq : \langle id : 3, \sim 0 : 3 \rangle = eq : \langle 3, 0 \rangle = False.$$

end example

example 10.28 Subtracting 1

Let sub1 be the function that subtracts 1 from its argument. An FP definition for sub1 can be written as follows, where – is subtraction.

$$sub1 = - \ @ \ [id, \sim 1].$$

For example, we'll evaluate the expression sub1 : 4 as follows:

$$sub1 : 4 = - \ @ \ [id, \sim 1] : 4 = - : \langle id : 4, \sim 1 : 4 \rangle = - : \langle 4, 1 \rangle = 3.$$

end example

example 10.29 Length of a List

Let "length" be the function that calculates the length of a list. An informal if-then-else definition of length can be written as follows.

$$length(x) = if \ x = \langle \ \rangle \ then \ 0 \ else \ 1 + length(tail(x)).$$

The corresponding FP definition can be written as follows, where null tests for the empty list and tl computes the tail of a list:

$$length = null \rightarrow \sim 0; \ + \ @ \ [\sim 1, length \ @ \ tl].$$

For example, we'll evaluate the expression length : $\langle a \rangle$.

$$
\begin{aligned}
\text{length} : \langle a \rangle &= (\text{null} \to\, \sim 0; +\, @[\sim 1, \text{length } @ \text{ tl}]) : \langle a \rangle \\
&= \text{null} : \langle a \rangle \to\, \sim 0 : \langle a \rangle; +\, @\,[\sim 1, \text{length}@\text{tl}] : \langle a \rangle \\
&= \text{False} \to 0; +\, : \langle 1, \text{ length:} \langle\ \rangle\rangle \\
&= +\, : \langle 1, \text{ length:} \langle\ \rangle\rangle \\
&= +\, : \langle 1, (\text{null} \to\, \sim 0; +@\,[\sim 1, \text{length } @ \text{ tl}]) : \langle\ \rangle\rangle \\
&= +\, : \langle 1, \text{ True} \to 0; +@\,[\sim 1, \text{ length } @ \text{ tl}] : \langle\ \rangle\rangle \\
&= +\, : \langle 1, 0 \rangle \\
&= 1.
\end{aligned}
$$

example   **10.30   An Equivalence Proof**

An FP program to compute $n!$ can be constructed directly from the following recursive definition.

$$\text{fact}(x) = \text{if } x = 0 \text{ then } 1 \text{ else } x * \text{fact}(x - 1).$$

The FP version of fact is

$$\text{fact} = \text{eq0} \to\, \sim 1; *\, @\,[\text{id}, \text{fact } @ \text{ sub1}].$$

An alternative FP program to compute $n!$ can be defined as follows.

$$\text{newfact} = g\, @\,[\sim 1, \text{id}],$$

where g is the FP program defined by

$$g = \text{eq0 } @\, 2 \to 1;\ g\, @\,[*, \text{sub1 } @\, 2].$$

Notice that g is iterative because it has a tail-recursive form (i.e., it has the form $g = a \to b;\ g\, @\, d$), which can be replaced by a loop. Therefore, newfact is also iterative. So newfact may be more efficient than fact. To prove the two programs are equivalent, we'll need the following relation involving g:

$$*\, @\,[a,\, g\, @\,[b,\, c]] = g\, @\,[*\, @\,[a,\, b],\, c], \tag{10.12}$$

where a, b, and c are functions that return natural numbers. We'll leave the proof of (10.12) as an exercise. Now we can prove that newfact = fact.

Proof: If the input to either function is 0, then eq0 is true, which gives us the base case fact:0 = newfact:0. Now we'll make the induction assumption that fact @ sub1 = newfact @ sub1 and show that newfact = fact. Starting with newfact, we have the following sequence of algebraic equations:

$$
\begin{aligned}
\text{newfact} \quad &= g\ @\ [\sim1,\ \text{id}] && \text{(definition)} \\
&= (\text{eq0}\ @\ 2 \to 1;\ g\ @\ [*,\ \text{sub1}\ @\ 2])\ @\ [\sim1,\ \text{id}] && \text{(definition)} \\
&= \text{eq0}\ @\ \text{id} \to \sim1;\ g\ @\ [*\ @\ [\sim1,\ \text{id}],\ \text{sub1}\ @\ \text{id}] && \text{(FP algebra)} \\
&= \text{eq0} \to \sim1;\ g\ @\ [*\ @\ [\sim1,\ \text{id}],\ \text{sub1}] && \text{(FP algebra)} \\
&= \text{eq0} \to \sim1;\ *\ @\ [\text{id},\ g\ @\ [\sim1,\ \text{sub1}]] && (10.12) \\
&= \text{eq0} \to \sim1;\ *\ @\ [\text{id},\ \text{newfact}\ @\ \text{sub1}] && \text{(FP algebra)} \\
&= \text{eq0} \to \sim1;\ *\ @\ [\text{id},\ \text{fact}\ @\ \text{sub1}] && \text{(induction)} \\
&= \text{fact} && \text{(definition). QED.}
\end{aligned}
$$

So newfact is correct if we assume that fact is correct. This is a plausible assumption because fact is just a translation of the definition of the factorial function.

Exercises

Relational Algebra

1. Given the following relations R and S with attribute sets $\{A,\ B,\ C,\ D\}$ and $\{B,\ C,\ D,\ E\}$, respectively.

R

A	B	C	D
1	a	#	M
2	a	*	N
1	b	#	M
3	a	%	N

S

B	C	D	E
a	#	M	x
b	*	N	y
a	#	M	z
b	%	M	w

Compute each of the following relations.

   a. select(R, B, a).
   b. project($R, \{B, D\}$).
   c. join(R, S).
   d. project($R, \{B, C, D\}$) $\cup$ project($S, \{B, C, D\}$).
   e. project($R, \{B, D\}$) $\cap$ project($S, \{B, D\}$).

2. Give an example of two relations R and S that have the same set of attributes $\{A,\ B\}$ and such that join($R,\ S$) $\neq \varnothing$ and join($R,\ S$) $\neq R \cup S$.

3. Find a relational algebra expression for each of the following questions about the example relations in Figures 10.11, 10.12, and 10.13.

    a. Construct a relation consisting of cable channel stations. The expression should evaluate to {(AMC, 48), (CNN, 96), ... }.

    b. What are the names of the movie channel stations? The expression should evaluate to {(AMC), (TCM), ... }.

    c. Which rooms with computers have white boards too? The expression should evaluate to {(SC211), ... }.

    d. How many seats are in CH301? The expression should evaluate to {(90)}.

    e. What information exists about ESPN? The expression should evaluate to {(ESPN, 140, 32, Sports)}.

    f. Is there a computer in the room where CS 252 Section 1 is taught? The expression should evaluate to {(Yes)}.

4. Use the definitions for the operators select and join to prove each of the following listed properties.

    a. $\text{select}_{A=a}\left(\text{select}_{B=b}\left(R\right)\right) = \text{select}_{B=b}\left(\text{select}_{A=a}\left(R\right)\right)$.

    b. $R \bowtie R = R$.

    c. $(R \bowtie S) \bowtie T = R \bowtie (S \bowtie T)$.

5. Let R be a relation, X a set of attributes of R, and A an attribute in X. Prove the following relationship between project and select.

$$\text{project}_X\left(\text{select}_{A=a}\left(R\right)\right) = \text{select}_{A=a}\left(\text{project}_X\left(R\right)\right).$$

Functional Algebra

6. Write an FP function to implement each of the following definitions.

    a. $f(n) = \langle(0, 0), (1, 1), \ldots, (n, n)\rangle$.

    b. $f((x_1, \ldots, x_n), (y_1, \ldots, y_n)) = \langle(x_1, y_1), \ldots, (x_n, y_n)\rangle$.

7. Prove each of the following FP equations.

    a. $+ \, @ \, [1, 2] = + \, @ \, [2, 1] = +$.

    b. $1 \, @ \sim \langle a, b\rangle = \sim a$ and $2 \, @ \sim \langle a, b\rangle = \sim b$.

8. Prove the following FP equation (10.12) from Example 10.30.

$$* \, @ \, [a, g \, @ \, [b, c]] = g \, @ \, [* \, @ \, [a, b], c],$$

where a, b, and c are any functions that return natural numbers and g has the following definition.

$$g = \text{eq0} \, @ \, 2 \to 1; \, g \, @ \, [*, \text{sub1} \, @ \, 2].$$

9. The following FP function is a translation of the recursive definition for the nth Fibonacci number, where sub2 is the FP function to subtract 2:

$$\text{slow} = \text{eq0} \rightarrow \sim 0;\ \text{eq1} \rightarrow \sim 1;\ +\ @\ [\text{slow} @ \text{sub1}, \text{slow} @ \text{sub2}].$$

The following FP function claims to compute the nth Fibonacci number by iteration:

$$\text{fast} = 1\ @\ g, \quad \text{where} \quad g = \text{eq0} \rightarrow \sim \langle 0,\ 1 \rangle;\ [2\ ,\ +]\ @\ g\ @\ \text{sub1}.$$

Prove that slow and fast are equivalent FP functions.

■

10.5 Other Algebraic Ideas

In this section we introduce some general algebraic tools that are used to solve some computational problems. We'll introduce congruence mod n as a computational tool and after studying some simple properties we'll see how they apply to computations in cryptology.

We'll also give a short description of some tools and techniques for constructing new algebras from existing algebras. We'll describe subalgebras, which have applications to defining new abstract data types from existing ones. Then we'll describe morphisms, which are used to transform one object into another without destroying certain properties.

10.5.1 Congruence

We're going to examine a useful property that sometimes occurs when an equivalence relation interacts with algebraic operations in a certain way. We'll use the familiar mod function that we discussed in Chapter 2.

Recall that if x is an integer and n is a positive integer, then x mod n denotes the remainder upon division of x by n. We can define an equivalence relation on the integers by relating two numbers x and y if the following equation holds:

$$x \bmod n = y \bmod n.$$

For example, if $n = 4$, then the integers are partitioned into the following four equivalence classes, where $[k] = \{x \mid x \bmod 4 = k \bmod 4\}$.

$$[0] = \{\ldots -8, -4, 0, 4, 8, \ldots\},$$
$$[1] = \{\ldots -7, -3, 1, 5, 9, \ldots\},$$
$$[2] = \{\ldots -6, -2, 2, 6, 10, \ldots\},$$
$$[3] = \{\ldots -5, -1, 3, 7, 11, \ldots\}.$$

We'll soon see that we can consider such classes to be elements of an algebra. Before we do this we'll introduce a notational convenience. Since the equation x mod $n = y$ mod n is used repeatedly when dealing with numbers, the following notation has been developed.

$$x \equiv y \ (\text{mod } n).$$

We say that x is *congruent* to y mod n. Remember, it still just means the same old thing that $x - y$ is divisible by n.

Now, we can notice two very interesting arithmetic properties of the mod function and how it interacts with addition and multiplication.

Two Properties of the Mod Function.   (10.13)

If $a \equiv b \ (\text{mod } n)$ and $c \equiv d \ (\text{mod } n)$, then

$$a + c \equiv b + d \ (\text{mod } n) \quad \text{and} \quad ac \equiv bd \ (\text{mod } n).$$

Proof: The hypotheses tell us that there are integers k and l such that

$$a = b + kn \quad \text{and} \quad c = d + ln.$$

Adding the two equations, we get

$$a + c = b + d + (k + l)n$$

so that $a + c \equiv b + d \ (\text{mod } n)$. Now multiply the two equations to get

$$ac = bd + (bl + kd + kln)n,$$

which gives $ac \equiv bd \ (\text{mod } n)$. QED.

The two properties (10.13) are an example of operations that interact with an equivalence relation in a special way, which we'll now describe. Suppose $\sim$ is an equivalence relation on a set A. An n-ary operation f on A is said to *preserve* $\sim$ if it satisfies the following property:

If $a_1 \sim b_1, \ldots, a_n \sim b_n$, then $f(a_1, \ldots, a_n) \sim f(b_1, \ldots, b_n)$.

For example, (10.13) says that the mod n relation is preserved by addition and multiplication.

When an equivalence relation on the carrier of an algebra is preserved by each operation of the algebra, then the relation is called a *congruence relation* on the algebra, and the expression $x \sim y$ is called a *congruence*. For example, the mod n relation is a congruence relation on the algebra $\langle \mathbb{Z}; +, \cdot \rangle$.

A Finite Algebra: The Integers Mod n

The two properties (10.13) allow us to think of the equivalence classes

$$[0], [1], \ldots, [n-1]$$

as the elements of an algebra where we can add and multiply them with the following definition:

$$[a] + [b] = [a + b] \quad \text{and} \quad [a] \cdot [b] = [a \cdot b].$$

We should note that these definitions make sense. In other words, if $[a] = [c]$ and $[b] = [d]$, then we must show that $[a + b] = [c + d]$ and $[a \cdot b] = [c \cdot d]$. Since $[a] = [c]$ and $[b] = [d]$, it follows that $a \equiv c \pmod{n}$ and $b \equiv d \pmod{n}$. So (10.13) tells us that $a + b \equiv c + d \pmod{n}$ and $a \cdot b \equiv c \cdot d \pmod{n}$. In other words, $[a + b] = [c + d]$ and $[a \cdot b] = [c \cdot d]$. So addition and multiplication of equivalence classes are indeed valid operations.

example **10.31 A Finite Algebra**

The mod 4 equivalence relation partitions the integers into the four classes $[0]$, $[1]$, $[2]$, and $[3]$. These four classes are elements of an algebra. Here are a few sample calculations.

$$[2] + [3] = [2 + 3] = [5] = [1],$$
$$[2] \cdot [3] = [2 \cdot 3] = [6] = [2],$$
$$[0] + [3] = [0 + 3] = [3],$$
$$[1] \cdot [2] = [1 \cdot 2] = [2].$$

We'll leave it as an exercise to write the addition and multiplication tables.

end example

Fermat's Little Theorem

We'll use the previous results to prove an old and quite useful result about numbers that is due to Fermat and is often called Fermat's little theorem.

> **Fermat's Little Theorem**
>
> If p is prime and a is not divisible by p, then $a^{p-1} \equiv 1 \pmod{p}$.

Proof: Let $[0], [1], \ldots, [p-1]$ denote the equivalence classes that partition the integers with respect to the mod p relation. Since a is not divisible by p, it follows that $[a] \neq [0]$ because $[0]$ is the set of all multiples of p (i.e., the set of all integers divisible by p). Now look at the sequence of classes

$$[1 \cdot a], [2 \cdot a], \ldots, [(p-1) \cdot a].$$

We can observe that each of these classes is not $[0]$. For example, if we had $[i \cdot a] = [0]$, this would imply that $i \cdot a$ is divisible by p. But p is relatively prime to a, so by (2.2d) it would have to divide i. But $i < p$, so i can't be divided by p. Therefore, $[i \cdot a] \neq [0]$. We can also observe that the classes are distinct. For example, if $[i \cdot a] = [j \cdot a]$, then we would have $i \cdot a \bmod p = j \cdot a \bmod p$. Since $\gcd(a, p) = 1$, it follows by (2.4d) that $i \bmod p = j \bmod p$, which tells us that $[i] = [j]$.

Since the classes $[1 \cdot a], [2 \cdot a], \ldots, [(p-1) \cdot a]$ are all distinct and not equal to $[0]$, they must be an arrangement of the sets $[1], [2], \ldots, [p-1]$. So the product of the classes $[1], [2], \ldots, [p-1]$ must be equal to the product of the classes $[1 \cdot a], [2 \cdot a], \ldots, [(p-1) \cdot a]$. In other words we have the following equality:

$$[1 \cdot a] \cdot [2 \cdot a] \cdot \cdots \cdot [(p-1) \cdot a] = [1] \cdot [2] \cdot \cdots \cdot [p-1].$$

Since $[x \cdot y] = [x] \cdot [y]$, we can rewrite this equation to obtain

$$[1 \cdot a \cdot 2 \cdot a \cdot \cdots \cdot (p-1) \cdot a] = [1 \cdot 2 \cdot \cdots \cdot (p-1)].$$

Put all the a's together to obtain the following equation.

$$[a^{p-1} \cdot 1 \cdot 2 \cdot \cdots \cdot (p-1)] = [1 \cdot 2 \cdot \cdots \cdot (p-1)].$$

This class equation means that

$$a^{p-1} \cdot 1 \cdot 2 \cdot \cdots \cdot (p-1) \bmod p = 1 \cdot 2 \cdot \cdots \cdot (p-1) \bmod p.$$

Since each of the numbers $1, 2, \ldots, p-1$ is relatively prime to p, it follows from (2.4d) that they can each be cancelled from the equation to obtain the following equation:

$$a^{p-1} \bmod p = 1 \bmod p = 1.$$

In other words, we have $a^{p-1} \equiv 1 \pmod{p}$, which is the desired result. QED.

10.5.2   Cryptology: The RSA Algorithm

In cryptology a *public-key cryptosystem* is a system for encrypting and decrypting messages in which the public is aware of the key that is needed to encrypt messages sent to the receiver. But the receiver is the only one who knows the private key needed to decrypt a message.

The first working system to accomplish this task is called the RSA algorithm, named after its founders Ronald Rivest, Adi Shamir, and Leonard Adleman [1978].

We'll describe the general idea of how the algorithm is used. Then we'll discuss the implementation and why it works. The algorithm works in the following way, where the message to be sent is a number. (This is no problem because any text can be transformed into a number in many ways.)

1. The receiver constructs a public encryption key, which is a pair of numbers (e, n), and makes it available to the public. The receiver also constructs a private decryption key d that no one else knows.

2. Any person with the public key (e, n) can send a message a to the receiver if $0 \leq a < n$. The sender encrypts a to a number c with the following calculation.

$$c = a^e \bmod n.$$

The sender then sends c to the receiver.

3. The receiver upon receiving c makes the following calculation to decrypt c, where d is the private key and n is taken from the public key (e, n).

$$c^d \bmod n.$$

This value is the desired message a.

The Details of the Keys

To construct the keys, choose two large distinct prime numbers p and q and let $n = pq$. Then choose a positive integer d that is relatively prime to the product $(p - 1)(q - 1)$. Then let e be a positive integer that satisfies the equation $ed \bmod ((p - 1)(q - 1)) = 1$.

If the keys are constructed in this way, then the system works. In other words, we have the following theorem.

RSA Theorem

If $0 \leq a < n$ and $c = a^e \bmod n$, then $c^d \bmod n = a$.

The proof depends on Fermat's little theorem. We'll give it in stages, beginning with two lemmas.

Lemma 1: $a^{ed} \bmod p = a \bmod p$ and $a^{ed} \bmod q = a \bmod q$.

Proof: The numbers e and d were chosen so that $ed \bmod (p - 1)(q - 1) = 1$. So we can write $ed = 1 + (p - 1)(q - 1)k$ for some integer k. We'll start with the following equations:

$$a^{ed} \bmod p = a^{1+(p-1)(q-1)k} \bmod p$$

$$= \left(a^1 \bmod p\right) \left(a^{(p-1)(q-1)k} \bmod p\right) \bmod p \qquad \text{(by 2.4c)}$$

$$= (a \bmod p) \left(a^{p-1} \bmod p\right)^{(q-1)k} \bmod p \qquad \text{(by 2.4c)}$$

Now we consider two cases. If $\gcd(a, p) = 1$, then we can apply Fermat's little theorem to obtain $a^{p-1} \bmod p = 1$. So the equations continue as follows:

$$= (a \bmod p)\,(1)^{(q-1)k} \bmod p$$
$$= a \bmod p.$$

If $\gcd(a, p) \neq 1$, then, since p is prime, it must be the case that $\gcd(a, p) = p$ so that p divides a. In this case $a \bmod p = 0$ and thus also $a^{ed} \bmod p = 0$. So in either case, we obtain the desired result $a^{ed} \bmod p = a \bmod p$. A similar argument shows that $a^{ed} \bmod q = a \bmod q$. QED.

Lemma 2: $a^{ed} \bmod n = a$ for $0 \leq a < n$.

Proof: By Lemma 1 we know that $a^{ed} \bmod p = a \bmod p$. So p divides $a^{ed} - a$. In other words, $a^{ed} - a = pk$ for some integer k. But Lemma 1 also tells us that $a^{ed} \bmod q = a \bmod q$. So q divides $a^{ed} - a = pk$. Since p and q are distinct primes, we have $\gcd(p, q) = 1$. So q divides k. It follows that $k = ql$ for some integer l. Thus $a^{ed} - a = pql$ and it follows that pq divides $a^{ed} - a$. Therefore, $a^{ed} \bmod pq = a \bmod pq = a$ because $a < n = pq$. QED.

Proof of the RSA Theorem:

The proof of the RSA Theorem is now a simple observation based on Lemma 2. Let $c = a^e \bmod n$. We must show that $c^d \bmod n = a$. The following sequence of equations does the job.

$$c^d \bmod n = (a^e \bmod n)^d \bmod n$$
$$= \left(a^{ed} \bmod n\right)$$
$$= a. \text{ QED.}$$

Practical Use of the RSA Algorithm

The practical use of the RSA algorithm is based on several pieces of mathematical knowledge. The security of the system is based on the fact that factoring large numbers is a very hard problem. If n is chosen to be the product of two very large prime numbers, then it will be very hard for someone to factor n to find the two prime numbers. So it will be very hard to construct the private decryption key d from the public encryption key (e, n).

The speed of the system is based on the fact that it is very easy to encrypt and decrypt numbers. This follows from some basic results about the mod function. For example, the RSA paper [1978] includes a simple algorithm to calculate $a^e \bmod n$ that requires at most $2(\log_2 e)$ multiplications and $2(\log_2 e)$ divisions. The calculation of $c^d \bmod n$ is similar. The paper also discusses fast methods to construct the keys in the first place: to find large prime numbers p and q; to construct the product $(p-1)(q-1)$; to choose d; and to compute e.

example 10.32 **Generating the Keys**

We'll construct keys for a simple example to see the construction method. Let $p = 13$ and $q = 17$. Then $n = pq = 221$ and $(p-1)(q-1) = 12 \cdot 16 = 192$. For the private key we'll choose $d = 23$, which is a prime number larger than either p or q, so we know it is relatively prime to $(p-1)(q-1) = 192$. To construct e, we must satisfy the equation $ed \bmod ((p-1)(q-1)) = 1$, which becomes $23e \bmod 192 = 1$. Since $\gcd(23, 192) = 1$, we can use Euclid's algorithm in reverse to find two numbers e and s such that $1 = 23e + 192s$. For example, to compute $\gcd(23, 192)$ we proceed as follows, where each equation has the form $a = b \cdot q + r$ with $0 \le r < |b|$:

$$192 = 23 \cdot 8 + 8$$
$$23 = 8 \cdot 2 + 7$$
$$8 = 7 \cdot 1 + 1.$$

The gcd is the last nonzero remainder, in this case 1. So we start with the remainder 1 in the third equation and work backward eliminating the remainders 7 and 8.

$$
\begin{aligned}
1 &= 8 - 7 \cdot 1 \\
&= 8 - (23 - 8 \cdot 2) \cdot 1 \\
&= 8 \cdot 3 - 23 \cdot 1 \\
&= (192 - 23 \cdot 8) \cdot 3 - 23 \cdot 1 \\
&= 23 \cdot (-25) + 192 \cdot 3.
\end{aligned}
$$

Therefore, $1 = 23 \cdot (-25) \bmod 192$. But we can't choose e to be -25 because e must be positive. This is no problem because we can add any multiple of 192 to -25. For example, let $e = 192 + (-25) = 167$. We'll verify that this value of e works.

$$
\begin{aligned}
de \bmod 192 &= 23 \cdot 167 \bmod 192 \\
&= 23 \cdot (192 - 25) \bmod 192 \\
&= ((23 \cdot 192 \bmod 192) + (23 \cdot (-25) \bmod 192) \bmod 192 \\
&= (0 + 1) \bmod 192 \\
&= 1.
\end{aligned}
$$

Therefore, the public key is $(e, n) = (167, 192)$ and the private key d is 23.

end example

example **10.33** Sending and Receiving a Message

We'll use the RSA algorithm to encrypt and decrypt a message. But first we need to agree on a way to represent the letters in the message. To keep things simple we'll identify the uppercase letters A, B, ..., Z with the integers 1, 2, ..., 26 and the blank space with 0. So each symbol can be represented by a 2-digit number. In other words, A = 01, B = 02, ..., Z = 26, and blank = 00.

To keep numbers a reasonable size, we'll break each message into a sequence of two-letter blocks. For example, to send the message HELLO WORLD we'll break it up into the following six 2-letter blocks.

HE = 0805, LL = 1212, O = 1500, WO = 2315, RL = 1812, D = 0400.

The largest number for any two letter block is ZZ = 2626. So we'll have to construct a public key (e, n), where $n > 2626$. We'll let $n = 2773 = 47 \cdot 59$, which is the smallest product of two primes that is greater then 2626. For this choice of n, an example in the RSA paper [1978] chose $d = 157$ and then computed $e = 17$. We'll use these values too.

For example, to encrypt a number x, we must calulate

$$x^{17} \bmod 2773.$$

For example, we'll use (2.4c) to encrypt the first block HE = 0805 as follows:

$$
\begin{aligned}
805^{17} \bmod 2773 \ &= [(805)((((805)^2)^2)^2)^2] \bmod 2773 \\
&= [(805)(((805^2 \bmod 2773)^2)^2)^2] \bmod 2773 \\
&= [(805)(((1916)^2)^2)^2] \bmod 2773 \\
&= [(805)((1916^2 \bmod 2773)^2)^2] \bmod 2773 \\
&= [(805)((2377)^2)^2] \bmod 2773 \\
&= [(805)(2377^2 \bmod 2773)^2] \bmod 2773 \\
&= [(805)(1528)^2] \bmod 2773 \\
&= [(805)(1528^2 \bmod 2773)] \bmod 2773 \\
&= [(805)(2691)] \bmod 2773 \\
&= 542.
\end{aligned}
$$

After encrypting all six blocks, we obtain the following six little messages to be sent out.

0542, 2345, 2417, 2639, 2056, 0017.

We'll leave it as an exercise to decrypt this sequence of numbers into the original message. For example, $0542^{157} \bmod 2773 = 0805$.

end example

10.5.3 Subalgebras

Programmers often need to create new data types to represent information. Sometimes a new data type can use the same operations of an existing type. For example, suppose we have an integer type available to us but we need to detect an error condition whenever a negative integer is encountered. One way to solve the problem is to define a new type for the natural numbers that uses some of the operations of the integer type.

For example, we can still use $+$, $\cdot$, mod, and div (integer divide) because $\mathbb{N}$ is "closed" with respect to these operations. In other words, these operations return values in $\mathbb{N}$ if their arguments are in $\mathbb{N}$. On the other hand, we can't use the subtraction operation because $\mathbb{N}$ isn't closed with respect to it (e.g., $3 - 4 \notin \mathbb{N}$). We say that our new type "inherits" the operations $+$, $\cdot$, mod, and div from the existing integer type. In algebraic terms we've created a new algebra

$$\langle \mathbb{N}; +, \cdot, \text{mod}, \text{div} \rangle,$$

which is a "subalgebra" of $\langle \mathbb{Z}; +, \cdot, \text{mod}, \text{div} \rangle$.

Let's describe the general idea of a subalgebra. Let A be the carrier of an algebra, and let B be a subset of A. We say that B is *closed* with respect to an operation if the operation returns a value in B whenever its arguments are from B. The diagram in Figure 10.14 gives a graphical picture showing B is closed with respect to the binary operation $\circ$.

Definition of Subalgebra

If A is the carrier of an algebra and B is a subset of A that is closed with respect to all the operations of A, then B is the carrier of an algebra called a *subalgebra* of the algebra of A. In other words, if $\langle A; \Omega \rangle$ is an algebra, where Ω is the set of operations, and if B is a subset of A that is closed with respect to all operations in Ω, then $\langle B; \Omega \rangle$ is an algebra, called a *subalgebra* of $\langle A; \Omega \rangle$.

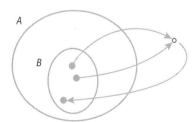

Figure 10.14 B is closed with respect to $\circ$.

example   **10.34   Some Sample Subalgebras**

Here are three examples of subalgebras.

1. $\langle \mathbb{N}; +, \cdot, \text{mod}, \text{div} \rangle$ is a subalgebra of $\langle \mathbb{Z}; +, \cdot, \text{mod}, \text{div} \rangle$.

2. Let $\Omega = \{+, -, \cdot, 0, 1\}$. Then we have a sequence of subalgebras, where each one is a subalgebra of the next: $\langle \mathbb{Z}; \Omega \rangle$, $\langle \mathbb{Q}; \Omega \rangle$, $\langle \mathbb{R}; \Omega \rangle$.

3. Consider the algebra $\langle \mathbb{N}_8; +_8, 0 \rangle$, where $+_8$ means addition mod 8. The set $\{0, 2, 4, 6\}$ forms the carrier of a subalgebra. But $\{0, 3, 6\}$ is not the carrier of a subalgebra because $3 +_8 6 = 1$ and $1 \notin \{0, 3, 6\}$.

end example

Combining Subalgebras

We can combine subalgebras by forming the intersection of the carriers. For example, consider the algebra $\langle \mathbb{N}_{12}; +_{12}, 0 \rangle$, where $+_{12}$ means addition mod 12. Two subalgebras of this algebra have carriers $\{0, 2, 4, 6, 8, 10\}$ and $\{0, 3, 6, 9\}$. The intersection of these two carriers is the set $\{0, 6\}$, which forms the carrier of another subalgebra of $\langle \mathbb{N}_{12}; +_{12}, 0 \rangle$. This is no fluke. It follows because the carrier of each subalgebra is closed with respect to the operations. So it follows that the intersection of carriers is also closed.

Generating a Subalgebra

One way to generate a new subalgebra is to take any subset you like—say, S—from the carrier of an algebra. If the operations of the algebra are closed with respect to S, then we have a new subalgebra. If not, then keep applying the operations of the algebra to elements of S. If an operation gives a result x and $x \notin S$, then enlarge S by adding x to form the bigger set $S \cup \{x\}$. Each time a bigger set is constructed, the process must start over again until the set is closed under the operations of the algebra. The resulting subalgebra has the smallest carrier that contains S.

example   **10.35   Smallest Subalgebra**

We'll start with the algebra $\langle \mathbb{N}_{12}; +_{12}, 0 \rangle$ and try to find the smallest subalgebra whose carrier contains the subset $\{4, 10\}$ of $\mathbb{N}_{12}$. Notice that this set is not closed under the operation $+_{12}$ because $4 +_{12} 4 = 8$, and $8 \notin \{4, 10\}$. So we'll add the number 8 to get the new subset $\{4, 8, 10\}$. Still there are problems because $4 +_{12} 8 = 0$, and $8 +_{12} 10 = 6$. So we'll add 0 and 6 to our set to obtain the subset $\{0, 4, 6, 8, 10\}$. We aren't done yet, because $6 +_{12} 8 = 2$. After adding 2, we obtain the set $\{0, 2, 4, 6, 8, 10\}$, which is closed under the operation of $+_{12}$ and contains the constant 0.

Therefore, the algebra $\langle \{0, 2, 4, 6, 8, 10\}; +_{12}, 0 \rangle$ is the smallest subalgebra of $\langle \mathbb{N}_{12}; +_{12}, 0 \rangle$ that contains the set $\{4, 10\}$.

end example

10.5.4 Morphisms

This little discussion is about some tools and techniques that can be used to compare two different entities for common properties. For example, if A is an alphabet, then we know that a string over A is different from a list over A. In other words, we know that A^* and lists(A) contain different kinds of objects. But we also know that A^* and lists(A) have a lot in common. For example, we know that the operations on A^* are similar to the operations on lists(A). We know that the algebra of strings and the algebra of lists both have an empty object and that they construct new objects in a similar way. In fact, we know that strings can be represented by lists.

On the other hand, we know that A^* is quite different from the set of binary trees over A. For example, the construction of a string is not at all like the construction of a binary tree.

The Transformation Problem

We would like to be able to decide whether two different entities are alike in some way. When two things are alike, we are often more familiar with one of the things. So we can apply our knowledge about the familiar one and learn something about the unfamiliar one. This is a bit vague. So let's start off with a general problem of computer science:

> *The Transformation Problem*
> Transform an object into another object with some particular property.

This is a very general statement. So let's look at a few interpretations. For example, we may want the transformed object to be "simpler" than the original object. This usually means that the new object has the same meaning as the given object but uses fewer symbols. For example, the expression $x + 1$ might be a simplification of $(x^2 + x)/x$, and the FP program f @ (true $\rightarrow$ c; d) can be simplified to f @ c.

We may want the transformed object to act as the meaning of the given object. For example, we usually think of the meaning of the expression $3 + 4$ as its value, which is 7. On the other hand, the meaning of the expression $x + 1$ is $x + 1$ if we don't know the value of x.

Whenever a light bulb goes on in our brain and we finally understand the meaning of some idea or object, we usually make statements like "Oh yes, I see it now" or "Yes, I understand." These statements usually mean that we have made a connection between the thing we're trying to understand and some other thing that is already familiar to us. So there is a transformation (i.e., a function) from the new idea to a familiar old idea.

Introductory Example: Semantics of Numerals

Suppose we want to describe the meaning of the base 10 numerals (i.e., nonempty strings of decimal digits) or the base 2 numerals (i.e., nonempty strings of binary digits). Let m_{ten} denote the meaning function for base 10 numerals, and let m_{two} denote the meaning function for base 2 numerals. If we can agree on anything, we most probably will agree that $m_{\text{ten}}(16) = m_{\text{two}}(10000)$ and $m_{\text{ten}}(14) = m_{\text{two}}(1110)$. Further, if we let m_{rom} denote the meaning function for Roman numerals, then we most probably also agree that $m_{\text{rom}}(\text{XII}) = m_{\text{ten}}(12) = m_{\text{two}}(1100)$.

For this example we'll use the set $\mathbb{N}$ of natural numbers to represent the meanings of the numerals. For base 10 and base 2 numerals there may be some confusion because, for example, the string 25 denotes a base 10 numeral and it also represents the natural number that we call 25. Given that this confusion exists, we have

$$m_{\text{ten}}(25) = m_{\text{two}}(11001) = m_{\text{rom}}(\text{XXV}) = 25.$$

So we can write down three functions from the three kinds of numerals (the syntax) to natural numbers (the semantics):

$$
\begin{aligned}
m_{\text{ten}} : & \quad \text{DecimalNumerals} \to \mathbb{N}, \\
m_{\text{two}} : & \quad \text{BinaryNumerals} \to \mathbb{N}, \\
m_{\text{rom}} : & \quad \text{RomanNumerals} \to \mathbb{N} - \{0\}.
\end{aligned}
$$

Can we give definitions of these functions? Sure. For example, a natural definition for m_{ten} can be given as follows: If $d_k d_{k-1} \ldots d_1 d_0$ is a base 10 numeral, then

$$m_{\text{ten}}(d_k d_{k-1} \ldots d_1 d_0) = 10^k d_k + 10^{k-1} d_{k-1} + \cdots + 10 d_1 + d_0.$$

Preserving Operations

What properties, if any, should a semantics function possess? Certain operations defined on numerals should be, in some sense, "preserved" by the semantics function. For example, suppose we let $+_{bi}$ denote the usual binary addition defined on binary numerals. We would like to say that the meaning of the binary sum of two binary numerals is the same as the result obtained by adding the two individual meanings in the algebra $\langle \mathbb{N}; + \rangle$. In other words, for any binary numerals x and y, the following equation holds:

$$m_{\text{two}}(x +_{bi} y) = m_{\text{two}}(x) + m_{\text{two}}(y).$$

The idea of a function preserving an operation can be defined in a general way. Let $f : A \to A'$ be a function between the carriers of two algebras. Suppose

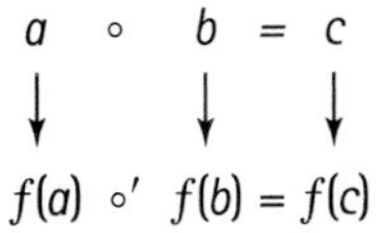

$$a \quad \circ \quad b \; = \; c$$
$$\downarrow \qquad \downarrow \qquad \downarrow$$
$$f(a) \; \circ' \; f(b) = f(c)$$

Figure 10.15   Preserving a binary operation.

ω is an n-ary operation on A. We say that f *preserves* the operation ω if there is a corresponding operation ω' on A' such that, for every $x_1, \ldots, x_n \in A$, the following equality holds:

$$f(\omega (x_1, \ldots, x_n)) = \omega'(f(x_1), \ldots, f(x_n)).$$

Of course, if ω is a binary operation, then we can write the above equation in its infix form as follows:

$$f(x \; \omega \; y) = f(x) \; \omega' \; f(y).$$

For example, the binary numeral meaning function m_{two} preserves $+_{bi}$. We can write the equation using the prefix form of $+_{bi}$ as follows:

$$m_{\text{two}}(+_{bi}(x, \; y)) = + \, (m_{\text{two}}(x), \; m_{\text{two}}(y)).$$

Here's the thing to remember about an operation that is preserved by a function $f : A \to A'$: You can apply the operation to arguments in A and then use f to map the result to A', or you can use f to map each argument from A to A' and then apply the corresponding operation on A' to these arguments. In either case you get the same result.

Figure 10.15 illustrates this property for two binary operators $\circ$ and $\circ'$. In other words, if $a \circ b = c$ in A, then $f(a \circ b) = f(c) = f(a) \circ' f(b)$ in A'.

Definition of Morphism

We say that $f : A \to A'$ is a *morphism* (also called a *homomorphism*) if every operation in the algebra of A is preserved by f. If a morphism is injective, then it's called a *monomorphism*. If a morphism is surjective, then it's called an *epimorphism*. If a morphism is bijective, then it's called an *isomorphism*. If there is an isomorphism between two algebras, we say that the algebras are *isomorphic*. Two isomorphic algebras are very much alike, and, hopefully, one of them is easier to understand.

For example, m_{two} is a morphism from $\langle \text{BinaryNumerals}; +_{bi} \rangle$ to $\langle \mathbb{N}; + \rangle$. In fact, we can say that m_{two} is an epimorphism because it's surjective. Notice that distinct binary numerals like 011 and 11 both represent the number 3.

Therefore, m_{two} is not injective, so it is not a monomorphism, and thus it is not an isomorphism.

example 10.36 A Morphism

Suppose we define $f : \mathbb{Z} \to \mathbb{Q}$ by $f(n) = 2^n$. Notice that

$$f(n + m) = 2^{n+m} = 2^n \cdot 2^m = f(n) \cdot f(m).$$

So f is a morphism from the algebra $\langle \mathbb{Z}; + \rangle$ to the algebra $\langle \mathbb{Q}; \cdot \rangle$. Notice that $f(0) = 2^0 = 1$. So f is a morphism from the algebra $\langle \mathbb{Z}; +, 0 \rangle$ to the algebra $\langle \mathbb{Q}; \cdot, 1 \rangle$. Notice that $f(-n) = 2^{-n} = (2^n)^{-1} = f(n)^{-1}$. Therefore, f is a morphism from the algebra $\langle \mathbb{Z}; +, -, 0 \rangle$ to the algebra $\langle \mathbb{Q}; \cdot, {}^{-1}, 1 \rangle$. It's easy to see that f is injective and that f is not surjective. Therefore, f is a monomorphism, but it is neither an epimorphism nor an isomorphism.

end example

example 10.37 The Mod Function

Let $m > 1$ be a natural number, and let the function $f : \mathbb{N} \to \mathbb{N}_m$ be defined by $f(x) = x \bmod m$. We'll show that f is a morphism from $\langle \mathbb{N}, +, \cdot, 0, 1 \rangle$ to the algebra $\langle \mathbb{N}_m, +_m, \cdot_m, 0, 1 \rangle$. For f to be a morphism we must have $f(0) = 0$, $f(1) = 1$, and for all $x, y \in \mathbb{N}$:

$$f(x + y) = f(x) +_m f(y) \text{ and } f(x \cdot y) = f(x) \cdot_m f(y).$$

It's clear that $f(0) = 0$ and $f(1) = 1$. The other equations are just restatements of the congruences (10.13).

end example

example 10.38 Strings and Lists

For any alphabet A we can define a function $f : A^* \to \text{lists}(A)$ by mapping any string to the list consisting of all letters in the string. For example, $f(\Lambda) = \langle \, \rangle$, $f(a) = \langle a \rangle$, and $f(aba) = \langle a, b, a \rangle$. We can give a formal definition of f as follows:

$$f(\Lambda) = \langle \, \rangle,$$
$$f(a \cdot t) = a :: f(t) \text{ for every } a \in A \text{ and } t \in A^*.$$

For example, if $a \in A$, then $f(a) = f(a \cdot \Lambda) = a :: f(\Lambda) = a :: \langle \, \rangle = \langle a \rangle$. It's easy to see that f is bijective because any two distinct strings get mapped to two distinct lists and any list is the image of some string.

We'll show that f preserves the concatenation of strings. Let "cat" denote both the concatenation of strings and the concatenation of lists. Then we must verify that $f(\text{cat}(s,\ t)) = \text{cat}(f(s),\ f(t))$ for any two strings s and t. We'll do it by induction on the length of s. If $s = \Lambda$, then we have

$$f(\text{cat}(\Lambda,\ t)) = f(t) = \text{cat}(\langle\ \rangle,\ f(t)) = \text{cat}(f(\Lambda),\ f(t)).$$

Now assume that s has length $n > 0$ and $f(\text{cat}(u,\ t)) = \text{cat}(f(u),\ f(t))$ for all strings u of length less than n. Since the length of s is greater than 0, we can write $s = a \cdot x$ for some $a \in A$ and $x \in A^*$. Then we have

$$
\begin{aligned}
f\left(\text{cat}\left(a \cdot x, t\right)\right) &= f\left(a \cdot \text{cat}\left(x, t\right)\right) & \text{(definition of string cat)} \\
&= a :: f\left(\text{cat}\left(x, t\right)\right) & \text{(definition of } f) \\
&= a :: \text{cat}\left(f\left(x\right), f\left(t\right)\right) & \text{(induction assumption)} \\
&= \text{cat}\left(a :: f\left(x\right), f\left(t\right)\right) & \text{(definition of list cat)} \\
&= \text{cat}\left(f\left(a \cdot x\right), f\left(t\right)\right) & \text{(definition of } f).
\end{aligned}
$$

Therefore, f preserves concatenation. Thus f is a morphism from the algebra $\langle A^*;\ \text{cat},\ \Lambda \rangle$ to the algebra $\langle \text{lists}(A);\ \text{cat},\ \langle\ \rangle \rangle$. Since f is also a bijection, it follows that the two algebras are isomorphic.

end example

Constructing Morphisms

Now let's consider the problem of constructing a morphism. We'll demonstrate the ideas with an example. Suppose we need a function $f : \mathbb{N}_8 \to \mathbb{N}_8$ with the property that $f(1) = 3$ and also f must be a morphism from the algebra $\langle \mathbb{N}_8;\ +_8,\ 0 \rangle$ to itself, where $+_8$ is the operation of addition mod 8. We'll finish the definition of f. For f to be a morphism it must preserve $+_8$ and 0. So we must set $f(0) = 0$. What value should we assign to $f(2)$? Notice that we can write $2 = 1 +_8 1$. Since $f(1) = 3$ and f must preserve the operation $+_8$, we can obtain the value $f(2)$ as follows:

$$f(2) = f(1 +_8 1) = f(1) +_8 f(1) = 3 +_8 3 = 6.$$

Now we can compute $f(3) = f(1 +_8 2) = f(1) +_8 f(2) = 3 +_8 6 = 1$. Continuing, we get the following values: $f(4) = 4$, $f(5) = 7$, $f(6) = 2$, and $f(7) = 5$. So the two facts $f(0) = 0$ and $f(1) = 3$ are sufficient to define f.

But does this definition of f result in a morphism? We must be sure that $f(x +_8 y) = f(x) +_8 f(y)$ for all $x,\ y \in \mathbb{N}_8$. For example, is $f(3 +_8 6) = f(3) +_8 f(6)$? We can check it out easily by computing the left- and right-hand sides of the equation:

$$f(3 +_8 6) = f(1) = 3 \quad \text{and} \quad f(3) +_8 f(6) = 1 +_8 2 = 3.$$

Do we have to check the function for all possible pairs $(x,\ y)$? No. Our method for defining f was to force the following equation to be true:

$$f(1 +_8 \cdots +_8 1) = f(1) +_8 \cdots +_8 f(1).$$

Since any number in $\mathbb{N}_8$ is a sum of 1's, we are assured that f is a morphism. Let's write this out for an example:

$$f\left(3 +_8 4\right) = f\left(1 +_8 1 +_8 1 +_8 1 +_8 1 +_8 1 +_8 1\right)$$
$$= f\left(1\right) +_8 f\left(1\right) +_8 f\left(1\right) +_8 f\left(1\right) +_8 f\left(1\right) +_8 f\left(1\right) +_8 f\left(1\right)$$
$$= \left[f\left(1\right) +_8 f\left(1\right) +_8 f\left(1\right)\right] +_8 \left[f\left(1\right) +_8 f\left(1\right) +_8 f\left(1\right) +_8 f\left(1\right)\right]$$
$$= f\left(1 +_8 1 +_8 1\right) +_8 f\left(1 +_8 1 +_8 1 +_8 1\right)$$
$$= f\left(3\right) +_8 f\left(4\right).$$

The above discussion might convince you that once we pick $f(1)$, then we know $f(x)$ for all x. But if the codomain is a different carrier, then things can break down. For example, suppose we want to define a morphism f from the algebra $\langle \mathbb{N}_3; +_3, 0 \rangle$ to the algebra $\langle \mathbb{N}_6; +_6, 0 \rangle$. Then we must have $f(0) = 0$. Now, suppose we try to set $f(1) = 3$. Then we must have $f(2) = 0$.

$$f(2) = f(1 +_3 1) = f(1) +_6 f(1) = 3 +_6 3 = 0.$$

Is this definition of f a morphism? The answer is No! Notice that $f(1 +_3 2) \neq f(1) +_6 f(2)$, because $f(1 +_3 2) = f(0) = 0$ and $f(1) +_6 f(2) = 3 +_6 0 = 3$. So morphisms are not as numerous as one might think.

example   **10.39 Language Morphisms**

If A and B are alphabets, then a function $f : A^* \to B^*$ is called a *language morphism* if $f(\Lambda) = \Lambda$ and $f(uv) = f(u)f(v)$ for any strings u, $v \in A^*$. In other words, a language morphism from A^* to B^* is a morphism from the algebra $\langle A^*; \mathrm{cat}, \Lambda \rangle$ to the algebra $\langle B^*; \mathrm{cat}, \Lambda \rangle$. Since concatenation must be preserved, a language morphism is completely determined by defining the values $f(a)$ for each $a \in A$.

For example, let $A = B = \{a, b\}$ and define $f : \{a, b\}^* \to \{a, b\}^*$ by setting $f(a) = b$ and $f(b) = ab$. Then we can make statements like

$$f(bab) = f(b)f(a)f(b) = abbab \quad \text{and} \quad f(b^2) = (ab)^2.$$

Language morphisms can be used to transform one language into another language with a similar grammar. For example, the grammar

$$S \to aSb \mid \Lambda$$

defines the language $\{a^n b^n \mid n \in \mathbb{N}\}$. Since $f(a^n b^n) = b^n (ab)^n$ for $n \in \mathbb{N}$, the set $\{a^n b^n \mid n \in \mathbb{N}\}$ is transformed by f into the set $\{b^n (ab)^n \mid n \in \mathbb{N}\}$. This language can be generated by the grammar $S \to f(a)Sf(b) \mid f(\Lambda)$, which becomes $S \to bSab \mid \Lambda$.

end example

example **10.40 Casting Out by Nines, Threes, etc.**

An old technique for finding some answers and checking errors in some arithmetic operations is called "casting out by nines." We want to study the technique and see why it works (so it's not magic). Is 44,820 divisible by 9? Is $43 \cdot 768 + 9579 = 41593$? We can use casting out by nines to answer yes to the first question and no to the second question. How does the idea work? It's a consequence of the following result:

Casting Out by Nines   **(10.14)**

If K is a natural number with decimal representation $d_n \ldots d_0$, then

$$K \bmod 9 = (d_n \bmod 9) +_9 \cdots +_9 (d_0 \bmod 9).$$

Proof: For the two algebras $\langle \mathbb{N}; +, \cdot, 0, 1 \rangle$ and $\langle \mathbb{N}_9; +_9, \cdot_9, 0, 1 \rangle$, the function $f \colon \mathbb{N} \to \mathbb{N}_9$ defined by $f(x) = x \bmod 9$ is a morphism. We can also observe that $f(10) = 1$, and in fact $f(10^n) = 1$ for any natural number n. Now, since $d_n \ldots d_0$ is the decimal representation of K, we can write

$$K = d_n \cdot 10^n + \cdots + d_1 \cdot 10 + d_0.$$

Now apply f to both sides of the equation to get the desired result.

$$\begin{aligned}
f(K) &= f(d_n \cdot 10^n + \cdots + d_1 \cdot 10 + d_0) \\
&= f(d_n) \cdot_9 f(10^n) +_9 \cdots +_9 f(d_1) \cdot_9 f(10) +_9 f(d_0) \\
&= f(d_n) \cdot_9 1 +_9 \cdots +_9 f(d_1) \cdot_9 1 +_9 f(d_0) \\
&= f(d_n) +_9 \cdots +_9 f(d_1) +_9 f(d_0). \quad \text{QED.}
\end{aligned}$$

Casting out by nines works because $10 \bmod 9 = 1$. Therefore, casting out by threes also works because $10 \bmod 3 = 1$. In general, for a base B number system, casting out by the predecessor of B works if we have the equation

$$B \bmod \mathrm{pred}(B) = 1.$$

For example, in octal, casting out by sevens works. (Do any other numbers work in octal?) But in binary, casting out by ones does not work because $2 \bmod 1 = 0$.

end example

 Exercises

Congruences

1. The equivalence relation $x \equiv y \pmod 4$ partitions the integers into the four equivalence classes $[0], [1], [2], [3]$. Construct the addition and multiplication tables for these classes.

2. (*Chinese Remainder Theorem*). Given n congruences

$$x \equiv a_1 \ (\text{mod } m_1), \ \ldots, \ x \equiv a_n \ (\text{mod } m_n),$$

where $\gcd(m_i, m_j) = 1$ for each $i \neq j$. The following algorithm finds a unique solution x such that $0 \leq x < m$, where $m = m_1 \cdots m_n$.

1. For each i find b_i such that $(m/m_i)b_i \equiv 1 \ (\text{mod } m_i)$.
2. Set $x = (m/m_1)b_1 a_1 + \cdots + (m/m_n)b_n a_n$.
3. If x is not in the proper range, then add or subtract a multiple of m.

Find the unique solution to each of the following sets of congruences.

 a. $x \equiv 8 \ (\text{mod } 13)$ b. $x \equiv 34 \ (\text{mod } 9)$ **c.** $x \equiv 17 \ (\text{mod } 6)$
   $x \equiv 3 \ (\text{mod } 8)$. $x \equiv 23 \ (\text{mod } 10)$. $x \equiv 15 \ (\text{mod } 11)$.
 d. $x \equiv 1 \ (\text{mod } 2)$ **e.** $x \equiv 3 \ (\text{mod } 2)$ f. $x \equiv 12 \ (\text{mod } 3)$
   $x \equiv 2 \ (\text{mod } 3)$. $x \equiv 1 \ (\text{mod } 5)$. $x \equiv 4 \ (\text{mod } 7)$.
   $x \equiv 1 \ (\text{mod } 5)$. $x \equiv 0 \ (\text{mod } 7)$. $x \equiv 15 \ (\text{mod } 11)$.

3. Prove the following statement about integers: If $x < 0$, then there is some $y > 0$ such that $y \equiv x \ (\text{mod } n)$.

4. (*A Pigeonhole Proof*). An interesting result about integers states that if $\gcd(a, n) = 1$, then there is an integer k in the range $1 \leq k \leq n$ such that $a^k \equiv 1 \ (\text{mod } n)$. A proof of this fact starts out as follows: Consider the set of $n + 1$ numbers

$$a, \ a^2, \ a^3, \ \ldots, \ a^{n+1}.$$

Calculate the set of remainders of these numbers upon division by n. In other words, we have the set of numbers

$$a \ \text{mod} \ n, \ a^2 \ \text{mod} \ n, \ a^3 \ \text{mod} \ n, \ \ldots, \ a^{n+1} \ \text{mod} \ n.$$

These $n + 1$ numbers are all in the range 0 to $n - 1$. By the pigeonhole principle two of the numbers must be identical. So for some $i < j$ we have $a^i \ \text{mod} \ n = a^j \ \text{mod} \ n$—in other words, $a^i \equiv a^j \ (\text{mod } n)$. This means that n divides $a^j - a^i = a^i(a^{j-i} - 1)$. Finish the proof by using properties of mod and gcd.

Cryptology

5. For each of the following cases, verify that n and d satisfy the requirements of the RSA algorithm and construct an encryption key e.

 a. $p = 5$, $q = 7$, $n = 35$, $d = 11$.
 b. $p = 7$, $q = 11$, $n = 77$, $d = 13$.
 c. $p = 47$, $q = 59$, $n = 2773$, $d = 101$.

6. Given the value $n = 47 \cdot 59 = 2773$ from Example 10.33.

   a. Verify that 17 is a valid choice for the encrypting key e.
   b. Verify that 157 is a valid choice for the decrypting key d.
   c. Verify the values of the encrypted numbers for HELLO WORLD.
   d. Decrypt the encrypted numbers to verify that they are the original six numbers for HELLO WORLD. *Hint:* The decryption key 157 can be written $157 = 1 + 4 + 8 + 16 + 128$. So for any number x we have $x^{157} \bmod 2773 = (x \cdot x^4 \cdot x^8 \cdot x^{16} \cdot x^{128}) \bmod 2773$.

Subalgebras

7. For each of the following sets, state whether the set is the carrier of a subalgebra of the algebra $\langle \mathbb{N}_9; +_9, 0 \rangle$.

   a. $\{0, 3, 6\}$.   b. $\{1, 4, 5\}$.   **c.** $\{0, 2, 4, 6, 8\}$.

8. Given the algebra $\langle \mathbb{N}_{12}; +_{12}, 0 \rangle$, find the carriers of the subalgebras generated by each of the following sets.

   a. $\{6\}$.   b. $\{3\}$.   **c.** $\{5\}$.

Morphisms

9. Find the three morphisms that exist from the algebra $\langle \mathbb{N}_3; +_3, 0 \rangle$ to the algebra $\langle \mathbb{N}_6; +_6, 0 \rangle$.

10. Let A be an alphabet and $f : A^* \to \mathbb{N}$ be defined by $f(x) = \text{length}(x)$. Show that f is a morphism from the algebra $\langle A^*; \text{cat}, \Lambda \rangle$ to $\langle \mathbb{N}; +, 0 \rangle$, where cat denotes the concatenation of strings.

11. Give an example to show that the absolute value function abs : $\mathbb{Z} \to \mathbb{N}$ defined by $\text{abs}(x) = |x|$ is not a morphism from the algebra $\langle \mathbb{Z}; + \rangle$ to the algebra $\langle \mathbb{N}; + \rangle$.

12. Let's assume that we know that the operation $+_n$ is associative over $\mathbb{N}_n$. Let $\circ$ be the binary operation over $\{a, b, c\}$ defined by the following table:

$\circ$	a	b	c
a	c	a	b
b	a	b	c
c	b	c	a

   Show that $\circ$ is associative by finding an isomorphism of the two algebras $\langle \{a, b, c\}; \circ \rangle$ and $\langle \mathbb{N}_3; +_3 \rangle$.

13. Given the language morphism $f : \{a, b\}^* \to \{a, b\}^*$ defined by $f(a) = b$ and $f(b) = ab$, compute the value of each of the following expressions.

    a. $f(\{b^n a \mid n \in \mathbb{N}\})$.
               b. $f(\{ba^n \mid n \in \mathbb{N}\})$.

    c. $f^{-1}(\{b^n a \mid n \in \mathbb{N}\})$.
         d. $f^{-1}(\{ba^n \mid n \in \mathbb{N}\})$.

    e. $f^{-1}(\{ab^{n+1} \mid n \in \mathbb{N}\})$.

10.6 Chapter Summary

An algebra consists of one or more sets, called carriers, together with operations on the sets. An algebra is useful for solving problems when we have a good knowledge of its operations. We can use the properties of the operations to transform algebraic expressions into equivalent simpler expressions. In high school algebra the carrier is the set of real numbers, and the operations are addition, multiplication, and so on.

An abstract algebra is described by giving a set of axioms to describe the properties of its operations. An abstract algebra is useful when it has lots of concrete examples. Two especially useful concrete examples of Boolean algebra are the algebra of sets and the algebra of propositions. Some important properties of Boolean algebra operations are the idempotent properties, the absorption laws, the involution law, and De Morgan's laws. Digital circuits are modeled by Boolean algebraic expressions. Thus Boolean algebra can be used to simplify a digital circuit by simplifying the corresponding algebraic expression.

The abstract data types of computer science can be described as algebras. When an abstract data type is described as an algebra, its operations can be implemented and then checked for correctness against the axioms. Some fundamental abstract data types are the natural numbers, lists, strings, stacks, queues, binary trees, and priority queues.

Two algebras that are useful as computational tools are relational algebras for databases and functional algebras for reasoning about functional programs.

Many other algebraic ideas are quite useful for computational problems. Congruences are useful for describing properties of numbers and there are direct applications to cryptology. Subalgebras can be used to define new abstract data types. Morphisms allow us to transform one algebra into another—often simpler—algebra and still preserve the meaning of the operations. Language morphisms can be used to generate new languages along with their grammars.

chapter 11

Regular Languages and Finite Automata

Unlearn'd, he knew no schoolman's subtle art,
No language, but the language of the heart.

— Alexander Pope (1688–1744)

Can a machine recognize a language? The answer is yes for some machines and some languages. In this chapter we'll study an elementary class of languages called regular languages and an elementary class of machines called finite automata. We'll see that regular languages can be represented by certain kinds of algebraic expressions, by finite automata, and by certain grammars.

To start things off, let's look at a familiar problem to programmers. The problem is to check for correctly formed input. Here's the statement.

The Recognition Problem
Write an algorithm to recognize input strings with a certain property.

For example, suppose that we need to compute with numbers that are represented in scientific notation. Can we write an algorithm to recognize strings of symbols represented in this way? Most of us will answer yes after we have the answers to a few more questions. For example, do we want to allow leading + signs? What is scientific notation? This example is an instance of a general class of problems that can be solved by some special techniques that we'll discuss in this chapter.

chapter guide

11.1 (Regular Languages) introduces the regular languages that we'll be discussing in the chapter. We'll see that these languages can be represented algebraically by "regular" expressions. We'll also discuss some techniques to simplify regular expressions.

11.2 (Finite Automata) introduces finite automata as machines that recognize regular languages. We'll present algorithms to transform between finite automata and regular expressions. We'll also introduce finite automata as output devices, and we'll present interpreters for finite automata.

11.3 (Constructing Efficient Finite Automata) introduces algorithms to help construct efficient finite automata. We'll see how to start with a regular expression and end up with a minimum-state deterministic finite automaton.

11.4 (Regular Language Topics) introduces grammars for regular languages. We'll see how to transform between regular grammars and nondeterministic finite automata. We'll introduce some properties of regular languages given by a pumping lemma, set operations, and morphisms. We'll also see some examples of languages that are not regular.

11.1 Regular Languages

Recall that a language over a finite alphabet A is a set of strings of letters from A. So a language over A is a subset of A^*. If we are given a language L and a string w, can we tell whether w is in L? The answer depends on our ability to describe the language L. Some languages are easy to describe, and others are not so easy to describe. In this section we'll introduce a class of languages that is easy to describe and for which algorithms can be found to solve the recognition problem.

The languages that we are talking about can be constructed from the letters of an alphabet by using the language operations of union, concatenation, and closure. These languages are called regular languages. Let's give a specific definition and then some examples. The collection of *regular languages* over A is defined inductively as follows:

The Regular Languages

1. $\varnothing$, $\{\Lambda\}$, and $\{a\}$ are regular languages for all $a \in A$.

2. If L and M are regular languages, then the following languages are also regular: $L \cup M$, ML, and L^*.

For example, the basis case of the definition gives us the following four regular languages over the alphabet $A = \{a, b\}$:

$$\varnothing, \quad \{\Lambda\}, \quad \{a\}, \quad \{b\}.$$

Now let's use the induction part of the definition to construct some more regular languages over $\{a, b\}$. Is the language $\{\Lambda, b\}$ regular? Yes. We can write it as the union of the two regular languages $\{\Lambda\}$ and $\{b\}$:

$$\{\Lambda, b\} = \{\Lambda\} \cup \{b\}.$$

Is the language $\{a, ab\}$ regular? Yes. We can write it as the product of the two regular languages $\{a\}$ and $\{\Lambda, b\}$:

$$\{a, ab\} = \{a\}\{\Lambda, b\}.$$

Is the language $\{\Lambda, b, bb, \ldots, b^n, \ldots\}$ regular? Yes. It's just the closure of the regular language $\{b\}$:

$$\{b\}^* = \{\Lambda, b, bb, \ldots, b^n, \ldots\}.$$

Here are two more regular languages over $\{a, b\}$, along with factorizations to show the reasons why:

$$\{a, ab, abb, \ldots, ab^n, \ldots\} = \{a\}\{\Lambda, b, bb, \ldots, b^n \ldots\} = \{a\}\{b\}^*,$$
$$\{\Lambda, a, b, aa, bb, \ldots, a^n, b^n, \ldots\} = \{a\}^* \cup \{b\}^*.$$

This little example demonstrates that there are many regular languages to consider. From a computational point of view, we want to find algorithms that can recognize whether a string belongs to a regular language. To accomplish this task, we'll introduce a convenient algebraic notation for regular languages.

11.1.1 Regular Expressions

A regular language is often described by means of an algebraic expression called a regular expression. We'll define the regular expressions and then relate them to regular languages. The set of *regular expressions* over an alphabet A is defined inductively as follows, where $+$ and $\cdot$ are binary operations and $*$ is a unary operation:

The Regular Expressions

1. Λ, $\varnothing$, and a are regular expressions for all $a \in A$.

2. If R and S are regular expressions, then the following expressions are also regular: (R), $R + S$, $R \cdot S$, and R^*.

For example, here are a few of the infinitely many regular expressions over the alphabet $A = \{a, b\}$:

$$\Lambda, \varnothing, a, b, \Lambda + b, b^*, a + (b \cdot a), (a + b) \cdot a, a \cdot b^*, a^* + b^*.$$

To avoid using too many parentheses, we assume that the operations have the following hierarchy:

$$* \quad \text{highest (do it first)},$$

$$\cdot$$

$$+ \quad \text{lowest (do it last)}.$$

For example, the regular expression

$$a + b \cdot a^*$$

can be written in fully parenthesized form as

$$(a + (b \cdot (a^*))).$$

We'll often use juxtaposition instead of $\cdot$ between two expressions. For example, we can write the preceding expression as

$$a + ba^*.$$

At this point in the discussion a regular expression is just a string of symbols with no specific meaning or purpose. For each regular expression E we'll associate a regular language $L(E)$ as follows, where A is an alphabet and R and S are any regular expressions:

$$L(\varnothing) = \varnothing,$$
$$L(\Lambda) = \{\Lambda\},$$
$$L(a) = \{a\} \qquad\qquad \text{for each } a \in A,$$
$$L(R + S) = L(R) \cup L(S),$$
$$L(R \cdot S) = L(R) L(S) \qquad \text{(language product)},$$
$$L(R^*) = L(R)^* \qquad \text{(language closure)}.$$

From this association it is clear that each regular expression represents a regular language and, conversely, each regular language is represented by a regular expression.

example 11.1 Finding a Language

Let's find the language of the regular expression $a + bc^*$. We can evaluate the expression $L(a + bc^*)$ as follows:

$$
\begin{aligned}
L(a + bc^*) &= L(a) \cup L(bc^*) \\
&= L(a) \cup (L(b) L(c^*)) \\
&= L(a) \cup \left(L(b) L(c)^*\right) \\
&= \{a\} \cup \left(\{b\} \{c\}^*\right) \\
&= \{a\} \cup \left(\{b\} \{\Lambda, c, c^2, \dots, c^n, \dots\}\right) \\
&= \{a\} \cup \{b, bc, bc^2, \dots, bc^n, \dots\} \\
&= \{a, b, bc, bc^2, \dots, bc^n, \dots\}.
\end{aligned}
$$

end example

Regular expressions give nice descriptive clues to the languages that they represent. For example, the regular expression $a + bc^*$ represents the language containing the single letter a or strings of the form b followed by zero or more occurrences of c.

Sometimes it's an easy matter to find a regular expression for a regular language. For example, the language

$$\{a,\ b,\ c\}$$

is represented by the regular expression $a + b + c$. In fact it's easy to find a regular expression for any finite language. If the language is empty, then its regular expression is $\varnothing$. Otherwise, form the regular expression consisting of the strings in the language separated by $+$ symbols.

Infinite languages are another story. An infinite language might not be regular. We'll see an example of a nonregular language later in this chapter. But many infinite languages are easily seen to be regular. For example, the language

$$\{a,\ aa,\ aaa,\ \dots,\ a^n,\ \dots\}$$

is regular because it can be written as the regular language $\{a\}\{a\}^*$, which is represented by the regular expression aa^*. The slightly more complicated language

$$\{\Lambda,\ a,\ b,\ ab,\ abb,\ abbb,\ \dots,\ ab^n,\ \dots\}$$

is also regular because it can be represented by the regular expression

$$\Lambda + b + ab^*.$$

Do distinct regular expressions always represent distinct languages? The answer is no. For example, the regular expressions $a + b$ and $b + a$ are different, but they both represent the same language,

$$L(a + b) = L(b + a) = \{a,\ b\}.$$

We want to equate those regular expressions that represent the same language. We say that regular expressions R and S are *equal* if $L(R) = L(S)$, and we denote this equality by writing the following familiar relation:

$$R = S.$$

For example, we know that $L(a + b) = \{a,\ b\} = \{b,\ a\} = L(b + a)$. Therefore, we can write $a + b = b + a$. We also have the equality

$$(a + b) + (a + b) = a + b.$$

On the other hand, we have $L(ab) = \{ab\}$ and $L(ba) = \{ba\}$. Therefore, $ab \neq ba$. Similarly, $a(b + c) \neq (b + c)a$. So although the expressions might make us think of high school algebra, we must remember that we are talking about regular expressions and languages, not numbers.

11.1.2 The Algebra of Regular Expressions

There are many general equalities for regular expressions. We'll list a few simple equalities together with some that are not so simple. All the properties can be verified by using properties of languages and sets. We'll assume that R, S, and T denote arbitrary regular expressions.

Properties of Regular Expressions   (11.1)

1. (+ properties)

$R + T = T + R$,

$R + \varnothing = \varnothing + R = R$,

$R + R = R$,

$(R + S) + T = R + (S + T)$.

2. ($\cdot$ properties)

$R\varnothing = \varnothing R = \varnothing$,

$R\Lambda = \Lambda R = R$,

$(RS)T = R(ST)$.

3. (Distributive properties)

$R(S + T) = RS + RT$,

$(S + T)R = SR + TR$.

(Closure properties)

4. $\varnothing^* = \Lambda^* = \Lambda$.

5. $R^* = R^*R^* = (R^*)^* = R + R^*$,

$R^* = \Lambda + R^* = (\Lambda + R)^* = (\Lambda + R)R^* = \Lambda + RR^*$,

$R^* = (R + \cdots + R^k)^*$ for any $k \geq 1$,

$R^* = \Lambda + R + \cdots + R^{k-1} + R^k R^*$ for any $k \geq 1$.

6. $R^*R = RR^*$.

7. $(R + S)^* = (R^* + S^*)^* = (R^*S^*)^* = (R^*S)^*R^* = R^*(SR^*)^*$.

8. $R(SR)^* = (RS)^*R$.

9. $(R^*S)^* = \Lambda + (R + S)^*S$,

$(RS^*)^* = \Lambda + R\,(R + S)^*$.

Proof: We'll prove three of the properties and leave the rest as exercises. First we'll prove $R + R = R$ by noticing the following equalities:

$$L(R + R) = L(R) \cup L(R) = L(R).$$

Next, let's prove the distributive property $R(S + T) = RS + RT$. The following series of equalities will do the job:

$$\begin{aligned} L\left(R\left(S+T\right)\right) &= L\left(R\right)L\left(S+T\right) \\ &= L\left(R\right)\left(L\left(S\right)\cup L\left(T\right)\right) \\ &= L\left(R\right)L\left(S\right)\cup L\left(R\right)L\left(T\right) \quad \text{(language property)} \\ &= L\left(RS\right)\cup L\left(RT\right) \\ &= L\left(RS+RT\right). \end{aligned}$$

Lastly, we'll prove that $R^* = R^*R^*$. Since $L(R^*) = L(R)^*$, we need to show that $L(R)^* = L(R)^*L(R)^*$. Let $x \in L(R)^*$. Then $x = x\Lambda \in L(R)^*L(R)^*$. Therefore, $L(R)^* \subseteq L(R)^*L(R)^*$. For the other way, suppose $x \in L(R)^*L(R)^*$. Then $x = yz$, where $y \in L(R)^*$ and $z \in L(R)^*$. Thus $y \in L(R)^k$ and $z \in L(R)^n$ for some k and n. Therefore, $yz \in L(R)^{k+n}$, which says that $x = yz \in L(R)^*$. So $L(R)^*L(R)^* \subseteq L(R)^*$. Thus $L(R)^* = L(R)^*L(R)^*$, so $R^* = R^*R^*$. QED.

The properties (11.1) can be used to simplify regular expressions and to prove the equality of two regular expressions. So we have an algebra of regular expressions. For example, suppose that we want to prove the following equality:

$$ba^*(baa^*)^* = b(a + ba)^*.$$

Since both expressions start with the letter b, we'll be done if we prove the simpler equality obtained by canceling b from both sides:

$$a^*(baa^*)^* = (a + ba)^*.$$

But this equality is an instance of property 7 of (11.1). To see why, we'll let $R = a$ and $S = ba$. Then we have

$$(a + ba)^* = (R + S)^* = R^*(SR^*)^* = a^*(baa^*)^*.$$

Therefore, the example equation is true. Let's look at a few more examples.

example 11.2 Equal Expressions

We'll show that $(\varnothing + a + b)^* = a^*(ba^*)^*$ by starting with the left side and using properties of regular expressions.

$$\begin{aligned} (\varnothing + a + b)^* &= (a + b)^* \quad \text{(property 1)} \\ &= a^*\left(ba^*\right)^* \quad \text{(property 7)}. \end{aligned}$$

end example

example 11.3 Equal Expressions

We'll show that the following two regular expressions are equal.

$$b^*(abb^* + aabb^* + aaabb^*)^* = (b + ab + aab + aaab)^*.$$

Start with the left side and use properties of regular expressions.

$$b^* \left(abb^* + aabb^* + aaabb^*\right)^* = b^* \left((ab + aab + aaab)\, b^*\right)^* \quad \text{(property 3)}$$
$$= (b + ab + aab + aaab)^* \quad \text{(property 7)}.$$

end example

 11.4 Equal Expressions

We'll show that $R + RS^*S = a^*bS^*$, where $R = b + aa^*b$ and S is any regular expression.

$$
\begin{aligned}
R + RS^*S &= R\Lambda + RS^*S & \text{(property 2)} \\
&= R\left(\Lambda + S^*S\right) & \text{(property 3)} \\
&= R\left(\Lambda + SS^*\right) & \text{(property 6)} \\
&= RS^* & \text{(property 5)} \\
&= (b + aa^*b)\, S^* & \text{(definition of } R) \\
&= (\Lambda + aa^*)\, bS^* & \text{(properties 2 and 3)} \\
&= a^*bS^*. & \text{(property 5)}.
\end{aligned}
$$

end example

Exercises

Regular Languages

1. Find a language (i.e., a set of strings) to describe each of the following regular expressions.

  a. $a + b$.   **b.** $a + bc$.   **c.** $a + b^*$.

  d. $ab^* + c$.   **e.** $ab^* + bc^*$.   **f.** $a^*bc^* + ac$.

Regular Expressions

2. Find a regular expression to describe each of the following languages.

  a. $\{a, b, c\}$.

  b. $\{aa, ab, ac\}$.

  c. $\{a, b, ab, ba, abb, baa, \ldots, ab^n, ba^n, \ldots\}$.

  d. $\{a, aaa, aaaaa, \ldots, a^{2n+1}, \ldots\}$.

  e. $\{\Lambda, a, abb, abbbb, \ldots, ab^{2n}, \ldots\}$.

  f. $\{\Lambda, a, b, c, aa, bb, cc, \ldots, a^n, b^n, c^n, \ldots\}$.

g. $\{\Lambda,\ a,\ b,\ ca,\ bc,\ cca,\ bcc,\ \ldots,\ c^{n}a,\ bc^{n},\ \ldots\}$.

h. $\{a^{2k}\ |\ k \in \mathbb{N}\} \cup \{b^{2k+1}\ |\ k \in \mathbb{N}\}$.

i. $\{a^{m}bc^{n}\ |\ m,\ n \in \mathbb{N}\}$.

3. Find a regular expression over the alphabet $\{0, 1\}$ to describe the set of all binary numerals without leading zeros (except 0 itself). So the language is the set $\{0, 1, 10, 11, 100, 101, 110, 111, \ldots\}$.

4. Find a regular expression for each of the following languages over the alphabet $\{a, b\}$.

   a. Strings with even length.

   b. Strings whose length is a multiple of 3.

   c. Strings containing the substring *aba*.

   d. Strings with an odd number of *a*'s.

Algebra of Regular Expressions

5. Simplify each of the following regular expressions.

   a. $\Lambda + ab + abab(ab)^{*}$.

   b. $aa(b^{*} + a) + a(ab^{*} + aa)$.

   c. $a(a + b)^{*} + aa(a + b)^{*} + aaa(a + b)^{*}$.

6. Prove each of the following equalities of regular expressions.

   a. $b + ab^{*} + aa^{*}b + aa^{*}ab^{*} = a^{*}(b + ab^{*})$.

   b. $a^{*}(b + ab^{*}) = b + aa^{*}b^{*}$.

   c. $ab^{*}a(a + bb^{*}a)^{*}b = a(b + aa^{*}b)^{*}aa^{*}b$.

7. Prove each of the following properties of regular expressions.

   a. $\varnothing^{*} = \Lambda^{*} = \Lambda$.

   b. $R^{*} = (R^{*})^{*} = R + R^{*}$.

   c. $R^{*} = \Lambda + R^{*} = (\Lambda + R)^{*} = (\Lambda + R)R^{*} = \Lambda + RR^{*}$.

   d. $R^{*} = (R + \cdots + R^{k})^{*}$ for any $k \geq 1$.

   e. $R^{*} = \Lambda + R + \cdots + R^{k-1} + R^{k}R^{*}$ for any $k \geq 1$.

   f. $(R + S)^{*} = (R^{*} + S^{*})^{*} = (R^{*}S^{*})^{*} = (R^{*}S)^{*}R^{*} = R^{*}(SR^{*})^{*}$.

   g. $R(SR)^{*} = (RS)^{*}R$.

   h. $(R^{*}S)^{*} = \Lambda + (R + S)^{*}S$.

Challenges

8. Answer each of the following questions for the algebra of regular expressions over an alphabet A.

   a. Is there an identity for the $+$ operation?

   b. Is there an identity for the $\cdot$ operation?

    c. Is there a zero for the · operation?

    d. Is there a zero for the + operation?

9. Find regular expressions for each of the following languages over the alphabet $\{a, b\}$.

    a. No string contains the substring *aa*.

    b. No string contains the substring *aaa*.

    c. No string contains the substring *aaaa*.

10. Find regular expressions to show that the following statement about cancellation is false: If $R \neq \varnothing$, $S \neq \Lambda$, $T \neq \Lambda$, and $RS = RT$, then $S = T$.

11. Let R and S be regular expressions, and let X represent a variable in the following equation:

$$X = RX + S.$$

    a. Show that $X = R^*S$ is a solution to the given equation.

    b. Find two distinct solutions to the equation $X = a^*X + ab$.

    c. Show that if $\Lambda \notin L(R)$, then the solution in Part (a) is unique.

11.2 Finite Automata

We've described regular languages in terms of regular expressions. Now let's see whether we can solve the recognition problem for regular languages. In other words, let's see whether we can find algorithms to recognize the strings from a regular language. To do this, we need to discuss some basic computing machines called finite automata.

11.2.1 Deterministic Finite Automata

Informally, a *deterministic finite automaton* over a finite alphabet A can be thought of as a finite directed graph with the property that each node emits one labeled edge for each distinct element of A. The nodes are called *states*. There is one special state called the *start* or *initial state*, and there is a—possibly empty—set of states called *final states*. We use the abbreviation DFA to stand for deterministic finite automaton.

For example, the labeled graph in Figure 11.1 represents a DFA over the alphabet $A = \{a, b\}$ with start state 0 and final state 3. We'll always indicate

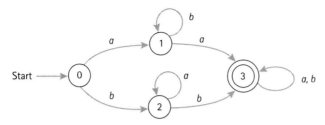

Figure 11.1   Sample DFA.

the start state by writing the word *Start* with an arrow pointing at it. Final states are indicated by a double circle.

The single arrow out of state 3 labeled with a, b is shorthand for two arrows from state 3 going to the same place, one labeled a and one labeled b. It's easy to check that this digraph represents a DFA over $\{a, b\}$ because there is a start state, and each state emits exactly two arrows, one labeled with a and one labeled with b.

The Execution of a DFA

We need to say what it means to execute a DFA over an alphabet A. For an input string $w \in A^*$, the execution of a DFA consists of scanning the letters of w from left to right and following the path that begins at the start state such that the letters on the edges of the path correspond to the letters of w that are being scanned. This means that the input string is equal to the concatenation of the letters on the edges of the path. The computation is deterministic because for each letter of the input string there is a unique edge emitted from each state labeled with the letter.

The movement along an edge from one state to another is called a *state transition*, and the input letter corresponding to the label on the edge is *consumed* by the state transition. So an execution consumes all the letters of an input string.

A DFA *accepts* a string w in A^* if the execution path ends in a final state. Otherwise, the DFA *rejects* w. The *language* of a DFA is the set of strings that it accepts.

For example, the DFA in Figure 11.1 has a path 0, 1, 1, 3 with edges labeled a, b, a. Since 0 is the start state and 3 is a final state, we conclude that the DFA accepts the string *aba*. The DFA also accepts the string *baaabab* by traveling along the path 0, 2, 2, 2, 2, 3, 3, 3. It's easy to see that the DFA accepts infinitely many strings because we can traverse the loop out of and into states 1, 2, or 3 any number of times. The DFA also rejects infinitely many strings. For example, any string of the form ab^n is rejected for any natural number n.

Now we're in position to state a remarkable result that is due to the mathematician and logician Stephen Kleene. Kleene [1956] showed that the languages recognized by DFAs are exactly the regular languages. We'll state the result for the record:

Theorem   **(11.2)**

The regular languages are exactly the same as the languages accepted by DFAs.

In fact, there is an algorithm to transform any regular expression into a DFA, and there is an algorithm to transform any DFA into a regular expression. We'll get to them soon enough. For now let's look at some examples.

example   11.5   DFAs for Regular Expressions

We'll give DFAs to recognize the regular languages represented by the regular expressions $(a + b)^*$ and $a(a + b)^*$ over the alphabet $A = \{a, b\}$. The language for the regular expression $(a + b)^*$ is the set $\{a, b\}^*$ of all strings over $\{a, b\}$, and it can be recognized by the following DFA:

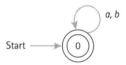

The language for the regular expression $a(a + b)^*$ is the set of all strings over A that begin with a, and it can be recognized by the following DFA:

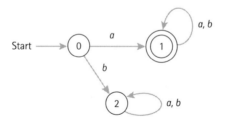

Notice that state 2 acts as an error state. It's necessary because of the requirement that each state of a DFA must emit one labeled arrow for each symbol of the alphabet.

Some programming languages define an *identifier* as a string beginning with a letter followed by any number of letters or digits. If we let a and b stand for letter and digit, respectively, then the set of all identifiers can be described by the regular expression $a(a + b)^*$ and thus be recognized by the preceding DFA.

end example

11.6 DFA for a Regular Expression

Suppose we want to build a DFA to recognize the regular language represented by the regular expression $(a + b)^*abb$ over the alphabet $A = \{a, b\}$. The language is the set of strings that begin with anything, but must end with the string abb. The diagram in Figure 11.2 shows a DFA to recognize this language. Try it out on a few strings. For example, does it recognize abb and $bbaabb$?

Example 11.6 brings up two questions: How was the DFA in Figure 11.2 created? How do we know that its language is represented by $(a + b)^*abb$? At this point it's a hit-or-miss operation to answer these questions, but we will soon see that there is an algorithm to construct a DFA, and there is also an algorithm to find the language of a DFA.

State Transitions

The graphical definition of a DFA is an important visual aid to help us understand finite automata. But we can also represent a DFA by a *state transition function*, which we'll denote by T, where any state transition of the form

is represented by

$$T(i, a) = j.$$

When a DFA is represented in this form, it can be easily programmed as a recognizer of strings. In fact, we'll soon give an interpreter for DFAs that are written in this form.

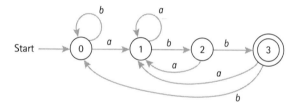

Figure 11.2   DFA for $(a + b)^*abb$.

11.2.2   Nondeterministic Finite Automata

A machine that is similar to a DFA, but less restrictive, is a *nondeterministic finite automaton* (NFA for short). An NFA over an alphabet A is a finite directed graph, where each state (i.e., node) emits zero or more edges, and where each edge is labeled either with a letter from A or with Λ. A state may emit more than one edge labeled with the same symbol. There is a single start state and there is a—possibly empty—set of final states.

The Execution of an NFA

The execution of an NFA is similar to that of a DFA, except that an edge labeled with the empty string Λ is traversed without consuming an input letter. An NFA *accepts* a string w if there is a path from the start state to some final state such that w is the concatenation of the letters on the edges of the path. Otherwise, the NFA *rejects* w. We should note that, since a state may emit an edge labeled with Λ or it may emit more than one edge labeled with the same symbol, there may be alternative paths to acceptance (i.e., nondeterminism). We should note that since a state does not have to emit an edge for every letter of the alphabet, an input string can be rejected because no path exists to consume the string. The *language* of an NFA is the set of strings that it accepts.

NFAs are usually simpler than DFAs because they don't need an edge out of each node for each letter of the alphabet. Now we're in position to state a remarkable result of Rabin and Scott [1959], which tells us that NFAs recognize a very special class of languages—the regular languages:

Theorem   **(11.3)**

The regular languages are exactly the same as the languages accepted by NFAs.

Combining (11.3) with Kleene's result (11.2), we can say that NFAs and DFAs recognize the same class of languages, the regular languages. In other words, we have the following statements about regular languages.

Regular expressions represent regular languages.

DFAs recognize regular languages.

NFAs recognize regular languages.

If we examine the definitions for DFA and NFA, it's easy to see that any DFA is an NFA. Later on, we'll give an algorithm that transforms any NFA into a DFA that recognizes the same regular language.

For now, let's get an idea of the utility of NFAs by considering the simple regular expression a^*a. An NFA for a^*a can be drawn as follows:

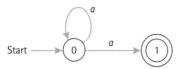

This NFA is certainly not a DFA because two edges are emitted from state 0, both labeled with the letter a. Thus there is nondeterminism. Another reason this NFA is not a DFA is that state 1 does not emit any edges.

We can also draw a DFA for a^*a. If we remember the equality $a^*a = aa^*$, then it's an easy matter to construct the following DFA:

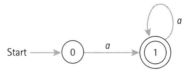

Sometimes it's much easier to find an NFA for a regular expression than it is to find a DFA for the expression. The next example should convince you.

example **11.7 NFAs for a Regular Expression**

We'll draw two NFAs to recognize the language of the regular expression $ab + a^*a$. The NFA in Figure 11.3 has a Λ edge, which allows us to travel to state 2 without consuming an input letter. Thus a^*a will be recognized on the path from state 0 to state 2 to state 3.

The NFA in Figure 11.4 also recognizes the same language. Perhaps it's easier to see this by considering the equality $ab + a^*a = ab + aa^*$.

Both NFAs are simple and easy to construct from the given regular expression. Now try to construct a DFA for the regular expression $ab + a^*a$.

end example

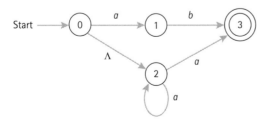

Figure 11.3   NFA with a Λ edge.

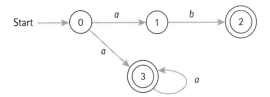

Figure 11.4   An equivalent NFA with no Λ edge.

As with DFAs, we can represent an NFA by a *state transition function*. Since there may be nondeterminism, we'll let the values of the function be sets of states. For example, if there are no edges from state k labeled with a, we'll write

$$T(k,\ a) = \varnothing.$$

If there are three edges from state k, all labeled with a, going to states i, j, and k, we'll write

$$T(k,\ a) = \{i,\ j,\ k\}.$$

We'll soon give an interpreter for NFAs that are written in this form.

11.2.3   Transforming Regular Expressions into Finite Automata

Now we're going to look at a simple algorithm that we can use to transform any regular expression into a finite automaton that accepts the regular language of the given regular expression.

Regular Expression to Finite Automaton (11.4)

Given a regular expression, we start the algorithm with a machine that has a start state, a single final state, and an edge labeled with the given regular expression as follows:

Continued ➡

Now transform this machine by applying the following rules until all edges are labeled with either a letter or Λ:

1. If an edge is labeled with $\varnothing$, then erase the edge.

2. Transform any diagram like

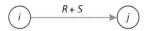

into the diagram

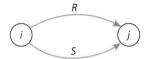

3. Transform any diagram like

into the diagram

4. Transform any diagram like

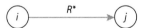

into the diagram

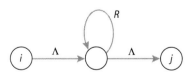

example 11.8 Constructing an NFA

To construct an NFA for $a^* + ab$, we'll start with the diagram

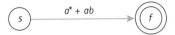

Next we apply rule 2 to obtain the following NFA:

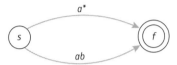

Next we'll apply rule 4 to a^* to obtain the following NFA:

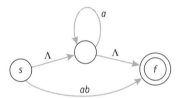

Finally, we apply rule 3 to ab to obtain the desired NFA for $a^* + ab$:

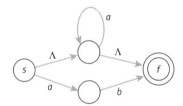

end example

Algorithm (11.4) has a shortcoming when it comes to implementation because rule 2 can cause many edges to be emitted from the same state. For example, if the algorithm is applied to an edge labeled with $(a + b) + c$, then two applications of rule 2 cause three new edges to be emitted from the same state. So it's easy to see that there is no bound on the number of edges that might be emitted from NFA states constructed by using (11.4). In Section 11.3 we'll give an alternative algorithm that's easy to implement because it limits the number of edges emitted from each node to at most 2.

11.2.4 Transforming Finite Automata into Regular Expressions

Now let's look at the opposite problem of transforming a finite automaton into a regular expression that represents the regular language accepted by the machine. Starting with either a DFA or an NFA, the algorithm performs a series of transformations into new machines, where these new machines have edges that may be labeled with regular expressions. The algorithm stops when a machine is obtained that has two states, a start state and a final state, and there is a regular expression associated with them that represents the language of the original automaton.

Finite Automaton to Regular Expression (11.5)

Assume that we have a DFA or an NFA. Perform the following steps:

1. Create a new start state s, and draw a new edge labeled with Λ from s to the original start state.

2. Create a new final state f, and draw new edges labeled with Λ from all the original final states to f.

3. For each pair of states i and j that have more than one edge from i to j, replace all the edges from i to j by a single edge labeled with the regular expression formed by the sum of the labels on each of the edges from i to j.

4. Construct a sequence of new machines by eliminating one state at a time until the only states remaining are s and f. As each state is eliminated, a new machine is constructed from the previous machine as follows:

Eliminate State k

Let $\text{old}(i, j)$ denote the label on edge (i, j) of the current machine. If there is no edge (i, j), then set $\text{old}(i, j) = \varnothing$. Now for each pair of edges (i, k) and (k, j), where $i \neq k$ and $j \neq k$, calculate a new edge label, $\text{new}(i, j)$, as follows:

$$\text{new}(i, j) = \text{old}(i, j) + \text{old}(i, k)\,\text{old}(k, k)^*\,\text{old}(k, j).$$

For all other edges (i, j) where $i \neq k$ and $j \neq k$, set

$$\text{new}(i, j) = \text{old}(i, j).$$

The states of the new machine are those of the current machine with state k eliminated. The edges of the new machine are the edges (i, j) for which label $\text{new}(i, j)$ has been calculated.

Now s and f are the two remaining states. If there is an edge (s, f), then the regular expression $\text{new}(s, f)$ represents the language of the original automaton. If there is no edge (s, f), then the language of the original automaton is empty, which is signified by the regular expression $\varnothing$.

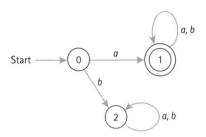

Figure 11.5   Sample DFA.

Let's walk through a simple example to demonstrate the algorithm. Suppose we start with the DFA in Figure 11.5.

The first three steps of (11.5) transform this machine into the following machine, where s is the start state and f is the final state:

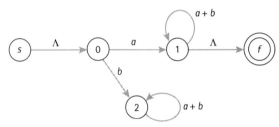

Now let's eliminate the states 0, 1, and 2. We can eliminate state 2 without any work because there are no paths passing through state 2 between states that are adjacent to state 2. In other words, $new(i, j) = old(i, j)$ for each edge (i, j), where $i \neq 2$ and $j \neq 2$. This gives us the machine

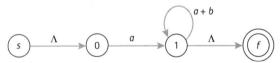

Now we'll eliminate state 0 from this machine by adding a new edge $(s, 1)$ that is labeled with the following regular expression:

$$new\,(s, 1) = old\,(s, 1) + old\,(s, 0)\; old\,(0, 0)^*\; old\,(0, 1)$$

$$= \varnothing + \Lambda\,\varnothing^* a$$

$$= a.$$

Therefore, we delete state 0 and add the new edge $(s, 1)$ labeled with a to obtain the following machine:

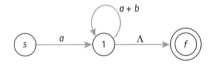

Next we eliminate state 1 in the same way by adding a new edge (s, f) labeled with the following regular expression:

$$\begin{aligned} \text{new} \, (s, f) &= \text{old} \, (s, f) + \text{old} \, (s, 1) \, \text{old} \, (1, 1)^* \, \text{old} \, (1, f) \\ &= \varnothing + a \, (a + b)^* \, \Lambda \\ &= a \, (a + b)^* . \end{aligned}$$

Therefore, we delete state 1 and label the edge (s, f) with $a(a + b)^*$ to obtain the following machine:

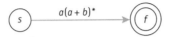

This machine terminates the algorithm. The label $a(a + b)^*$ on edge (s, f) is the regular expression representing the regular language of the original DFA given in Figure 11.5.

example **11.9 From a DFA to a Regular Expression**

We'll verify that the regular expression $(a + b)^* abb$ does indeed represent the regular language accepted by the DFA in Figure 11.2 from Example 11.6. We start the algorithm by attaching start state s and final state f to the DFA in Figure 11.2 to obtain the following machine:

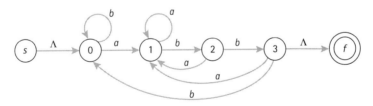

Now we need to eliminate the internal states. As we construct new edge labels, we'll simplify the regular expressions as we go. First we'll eliminate the state 0. To eliminate state 0, we construct the following new edges:

$$\text{new} \, (s, 1) = \varnothing + \Lambda b^* a = b^* a,$$

$$\text{new} \, (3, 1) = a + bb^* a = (\Lambda + bb^*) \, a = b^* a.$$

These new edges eliminate state 0 and we obtain the following machine:

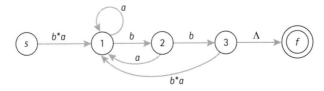

The states can be eliminated in any order. For example, we'll eliminate state 3 next, which forces us to create the following new edges:

$$\text{new}\,(2, f) = \varnothing + b\,\varnothing^* \Lambda = b,$$

$$\text{new}\,(2, 1) = a + b\,\varnothing^* b^* a = a + bb^* a = (\Lambda + bb^*)\,a = b^* a.$$

With state 3 eliminated, we obtain the following machine:

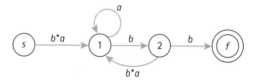

Now eliminate state 2, which forces us to create the following new edges:

$$\text{new}\,(1, f) = \varnothing + b\varnothing^* b = bb,$$

$$\text{new}\,(1, 1) = a + b\varnothing^* b^* a = a + bb^* a = (\Lambda + bb^*)\,a = b^* a.$$

With state 2 eliminated, we obtain the following machine:

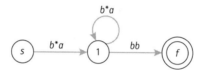

Finally, we remove state 1 by creating a new edge,

$$\text{new}\,(s, f) = \varnothing + b^* a\,(b^* a)^* bb$$

$$= b^*\,(ab^*)^*\,abb \qquad \text{(by 8 of (11.1))}$$

$$= (a + b)^*\,abb \qquad \text{(by 7 of (11.1))}.$$

With state 1 eliminated, we obtain the desired regular expression for the DFA in Figure 11.2.

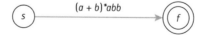

end example

The process of constructing a regular expression from a finite automaton can produce some complex regular expressions. If we remove the states in different orders, then we might obtain different regular expressions, some of which might be more complex than others. So the algebraic properties of regular expressions (11.1) are nice tools to simplify these complex regular expressions. As we've indicated in the example, it's better to simplify the regular expressions at each stage of the process to keep things manageable.

11.2.5 Finite Automata as Output Devices

The automata that we've discussed so far have only a limited output capability to indicate the acceptance or rejection of an input string. We want to introduce two classic models for finite automata that have output capability. We'll consider machines that transform input strings into output strings. These machines are like DFAs, except that we associate an output symbol with each state or with each state transition, and there are no final states because we are not interested in acceptance or rejection.

Mealy and Moore Machines

The first model, invented by Mealy [1955], is called a *Mealy machine*. It associates an output symbol with each transition. For example, if the output associated with the edge labeled with the letter a is x, we'll write a/x on that edge. A state transition for a Mealy machine can be pictured as follows:

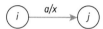

In a Mealy machine, an output always takes place during a transition of states.

The second model, invented by Moore [1956], is called a *Moore machine*. It associates an output symbol with each state. For example, if the output associated with state i is x, we'll always write i/x inside the state circle. A typical state transition for a Moore machine can be pictured as follows:

Each time a state is entered, an output takes place. So the first output always occurs as soon as the machine is started.

One way to remember the output structure of the two machines is to associate the word "mean" with "Mealy," where the output occurs at the "mean" or "middle" of two states. It turns out that Mealy and Moore machines are equivalent. In other words, any problem that is solvable by one type of machine can also be solved by the other type of machine.

Let's look at an example problem, which we'll solve with a Mealy machine and with a Moore machine. Suppose we want to compute the number of substrings of the form

$$bab$$

that occur in an arbitrary input string over the alphabet $\{a, b\}$. For example, there are three such substrings in the string

$$ababababaababb.$$

A Mealy machine to solve this problem is given by the following graph:

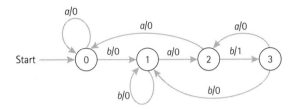

For example, the output of this Mealy machine for the sample string

$$ababababaababb$$

is

$$0\ 0\ 0\ 1\ 0\ 1\ 0\ 0\ 0\ 0\ 1\ 0.$$

The problem can also be solved by the following Moore machine:

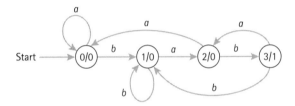

For example, the output of this Moore machine for the sample string

$$ababababaababb$$

is

$$0\ 0\ 0\ 0\ 1\ 0\ 1\ 0\ 0\ 0\ 0\ 1\ 0.$$

We can count the number of 1's in the output string to obtain the number of occurrences of the substring *bab*.

From a practical point of view we might wish to let some output symbol mean that no output takes place. For example, if we replaced the output symbol 0 with Λ in the preceding machines, then the output for the sample string would be the string 111. Let's look at a few more examples.

example 11.10 The Successor for Binary Numbers

Suppose we represent a natural number in the form of a binary string. To compute the successor, we need to add 1. Using the standard addition algorithm, which involves carrying, we can write down a Mealy machine to do the job. Since addition starts on the right end of the string, we'll assume that the input is the binary representation in reverse order.

We must consider the special case when a natural number has the form $2^k - 1$, which is represented by a string of k 1's. Thus when 1 is added to $2^k - 1$, we get 2^k, which is represented by a string of length $k + 1$. In this case, our machine will output a string of k 0's, which we will interpret as the number 2^k. With this assumption the Mealy machine looks like the following, where state 1 is the carry state and state 2 is the no-carry state:

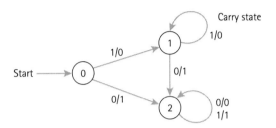

For example, let's find the successor of 13. The binary representation of 13 is 1101. So we take its reverse, which is 1011, as input. The sequence of states for this input is 1, 2, 2, 2, which gives the output string 0111. The reverse of this string is 1110, which is the binary representation of 14.

end example

example 11.11 A Simple Vending Machine

Suppose we have a simple vending machine that allows the user to pick from two 10-cent items A and B. To simplify things, the slot will accept only dimes. There are four inputs to the machine: d (dime), a (select item A), b (select item B), and r (return coins). The outputs will be n (do nothing), A (vend item A), B (vend item B), and d (dime). A Mealy machine to model this simple vending machine can be pictured as follows:

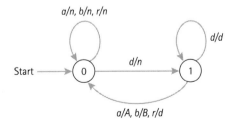

end example

example 11.12 A Simple Traffic Signal

Suppose we have a simple traffic intersection, where a north-south highway intersects with an east-west highway. We'll assume that the east-west highway always has a green light unless some north-south traffic is detected by sensors. When north-south traffic is detected, after a certain time delay the signals change and stay that way for a fixed period of time.

The input symbols will be 0 (no traffic detected) and 1 (traffic detected). Let G, Y, and R mean green, yellow, and red, respectively. The output symbols will be GR, YR, RG, and RY, where the first letter is the color of the east-west light and the second letter is the color of the north-south light. Here's a Moore machine for this simple traffic intersection.

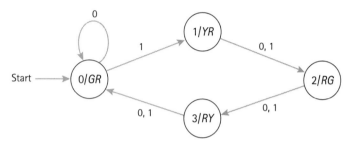

end example

Mealy machines appear to be more useful than Moore machines. But problems like traffic signal control have nice Moore machine solutions because each state is associated with a new output configuration. The exercises include some more problems.

11.2.6 Representing and Executing Finite Automata

We've seen that the graphical definition of finite automata is an important visual aid to help us understand and construct the machines. In the next few paragraphs we'll introduce some other ways to represent finite automata that will help us describe algorithms to execute DFAs and NFAs.

DFA Representation and Execution

We've seen that when a DFA has an edge from i to j labeled with a, we can denote it by writing $T(i, a) = j$. It will be convenient to extend this idea so that T takes an arbitrary string as a second argument rather than just a single letter. In other words, we want to define $T(i, w)$, where w is any string. A recursive definition of this extension of T can be written as follows:

$$T(i, \Lambda) = i,$$

$$T(i, a \cdot s) = T(T(i, a), s).$$

This gives us an easy way to see whether a string w is accepted by a DFA. Just evaluate $T(i, w)$, where i is the start state of the DFA. If the resulting state is final, then w is accepted by the DFA. Otherwise, w is rejected.

Transition Tables

To construct a program to execute a DFA, it's useful to represent the transition function in a tabular form called a *transition table*. The rows are labeled with the states and the columns are labeled with the elements of the alphabet. The start state and the final states are also labeled.

For example, suppose we have the following DFA, which is the same as the DFA in Figure 11.2:

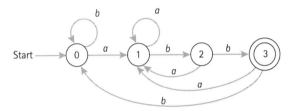

We've seen that this DFA recognizes the language of $(a + b)^*abb$ over the alphabet $A = \{a, b\}$. Figure 11.6 shows the transition table for this DFA.

This table is a complete representation for the DFA. For example, the value $T(0, b) = 0$ represents the transition from state 0 to state 0 along the edge labeled with b. Of course, we can also use the table to compute values of the extended transition function. For example, we can compute $T(0, aab)$ as follows:

$$T(0, aab) = T(T(0, a), ab) = T(1, ab) = T(T(1, a), b) = T(1, b) = 2.$$

Since state 2 is not a final state, it follows that aab is rejected by the DFA.

Formal Representation of a DFA

When we discuss DFAs in general terms, it is sometimes convenient to represent a DFA by listing five things: the set of states, the alphabet, the transition function

	T	a	b
start	0	1	0
	1	1	2
	2	1	3
final	3	1	0

Figure 11.6   Transition table.

or graph, the start state, and the set of final states. Traditionally, these five items are listed as a 5-tuple as follows:

(states, alphabet, transition function, start state, final states).

For example, someone may say, "Let (S, A, T, s, F) be a DFA." We can then assume that S is the set of states, A is the alphabet, T is the transition function, s is the start state, and F is the set of final states. We can also use the 5-tuple notation to represent a particular DFA. For example, we can represent the DFA described in Figure 11.6 by the 5-tuple

$$(\{0,\ 1,\ 2,\ 3\},\ \{a,\ b\},\ T,\ 0,\ \{3\}),$$

where T represents the transition function.

We can also represent any DFA as an algebraic structure. This kind of representation can help us discover a general algorithm for executing any DFA. Suppose we have a DFA (S, A, T, s, F). We can represent this DFA as an algebra as follows:

Carriers: A, A^*, S, Boolean.   (11.6)

Operations: $s \in S$,

$F \subseteq S$,

$T : S \times A \to S$,

accept $: A^* \to$ Boolean,

path $: S \times A^* \to$ Boolean.

Axioms: accept$(w) =$ path$(s,\ w)$,

path$(k, \Lambda\ ) =$ **if** k is a final state **then** True **else** False,

path$(k,\ a{\cdot}t) =$ path$(T(k,\ a),\ t)$.

The axioms are the important part because they form the basis for an algorithm to recognize the language of any DFA. In other words, for any DFA and any string w, the value accept(w) is true if w is accepted by the DFA and false if w is rejected. Let's look at a particularly simple implementation of the algorithm as a logic program.

example 11.13 A Logic Program Interpreter for DFAs

We'll write a logic program to compute any DFA. The axioms in (11.6) return Boolean values, which makes it easy to write them as logic program clauses. We'll represent an arbitrary DFA by facts of the following form:

t(state, letter, nextstate).

start(state).

final(state).

For example, the DFA in Figure 11.6 becomes the following list of facts:

$$\text{start}\,(0)\,.$$

$$t\,(0, a, 1)\,.$$

$$t\,(0, b, 0)\,.$$

$$t\,(1, a, 1)\,.$$

$$t\,(1, b, 2)\,.$$

$$t\,(2, a, 1)\,.$$

$$t\,(2, b, 3)\,.$$

$$t\,(3, a, 1)\,.$$

$$t\,(3, b, 0)\,.$$

$$\text{final}\,(3)\,.$$

To keep things simple, we'll represent each string as a list. The interpreter consists of the following three logic program clauses, where any argument beginning with an uppercase letter is a variable.

$$\text{accept}\,(S) \leftarrow \text{start}\,(I)\,,\ \text{path}\,(I, S)\,.$$

$$\text{path}\,(K, \langle\,\rangle) \leftarrow \text{final}\,(K)\,.$$

$$\text{path}(K,\ \text{Head} :: \text{Tail}) \leftarrow t(K,\ \text{Head},\ M),\ \text{path}(M,\ \text{Tail})\,.$$

For example, to test whether the string *aaab* is accepted by the DFA, we would write the following goal:

$$\leftarrow \text{accept}(\langle a,\ a,\ a,\ b\rangle)\,.$$

This interpreter can be easily translated into the Prolog language with three simple notational changes: use brackets [and] to represent lists; replace $\leftarrow$ with ":–"; and replace "$X :: Y$" with $[X\,|\,Y]$.

end example

Representations for NFAs

Just as we did with DFAs, we can associate a transition table with any NFA. Recall that the transition function for an NFA returns a set of states because of the nondeterminism. For example, $T(k, b) = \varnothing$ if there are no edges from state k labeled with b. The notation $T(k, a) = \{i, j, k\}$ means that three edges are emitted from state k, all labeled with a, pointing at states i, j, and k.

Now we can construct a transition table for any NFA. For example, suppose we have the following NFA:

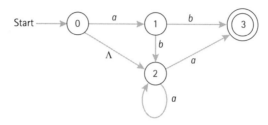

The transition table for this NFA is pictured as follows:

	T	a	b	Λ
start	0	$\{1\}$	$\varnothing$	$\{2\}$
	1	$\varnothing$	$\{2,3\}$	$\varnothing$
	2	$\{2,3\}$	$\varnothing$	$\varnothing$
final	3	$\varnothing$	$\varnothing$	$\varnothing$

For example, we have $T(2, a) = \{2, 3\}$ because state 2 emits two edges labeled with a, one going to state 2 and one going to state 3.

We can represent an NFA as a 5-tuple having the same form as a DFA:

(states, alphabet, transition function, start state, final states).

The only difference is that the transition function for an NFA returns values that are sets of states because there may be nondeterminism.

We can also represent any NFA as an algebraic structure. This kind of representation can help us discover a general algorithm for executing any NFA. Suppose we have an NFA (S, A, T, s, F). We can represent this NFA as an algebra as follows:

Carriers: A, A^*, S, power(S), Boolean.

Operations: $s \in S$,

$F \subseteq S$,

$T : S \times A \cup \{\Lambda\} \rightarrow$ power(S),

accept : $A^* \rightarrow$ Boolean,

path : $S \times A^* \rightarrow$ Boolean.

Axioms: accept$(w) =$ path(s, w),

path$(k, \Lambda) =$ **if** k is final **or** a final state is reachable from k by following Λ edges
then True **else** False,

path$(k, a{\cdot}t) =$ **if** there is $m \in T(k, a)$ such that path$(m, t) =$ True
or there is $m \in T(k, \Lambda)$ such that path$(m, a{\cdot}t) =$ True
then True **else** False.

The rather complicated-looking equation for path(k, $a \cdot t$) is necessary to allow for the nondeterminism that may take place either by having more than one edge labeled with the same letter emitted from state k or by traversing a Λ edge from state k without consuming an input letter.

The next example shows how an NFA interpreter can be implemented with a very simple logic program.

example 11.14 A Logic Program Interpreter for NFAs

It's almost as easy to program an NFA in logic as a DFA. We'll represent an arbitrary NFA by facts of the following form.

$$t(\text{state, symbol, nextstate}).$$

$$\text{start(state)}.$$

$$\text{final(state)}.$$

To keep things simple we'll represent strings as lists. So the empty string Λ will be represented by the empty list $\langle \, \rangle$. Consider the following NFA table.

	T	a	b	Λ
start	0	{1,2}	$\varnothing$	{1}
	1	$\varnothing$	{2}	$\varnothing$
final	2	{2}	$\varnothing$	$\varnothing$

The facts to represent this NFA are listed as follows, where there are two facts to take care of the table value $T(0, a) = \{1, 2\}$.

$$\text{start}(0).$$

$$t(0, a, 1).$$

$$t(0, a, 2).$$

$$t(0, \langle \, \rangle, 1).$$

$$t(1, b, 2).$$

$$t(2, a, 2).$$

$$\text{final}(2).$$

The NFA interpreter consists of the following four clauses, where any argument beginning with an uppercase letter is a variable.

$$\text{accept}(S) \leftarrow \text{start}(I), \text{path}(I, S).$$

$$\text{path}(K, \langle \, \rangle) \leftarrow \text{final}(K).$$

$$\text{path}(K, \text{Head :: Tail}) \leftarrow t(K, \text{Head}, M), \text{path}(M, \text{Tail}).$$

$$\text{path}(K, X) \leftarrow t(K, \langle \, \rangle, M), \text{path}(M, X).$$

For example, to test whether the string *abaa* is accepted by the given NFA, we would write the following goal:

$$\leftarrow \text{accept}(\langle a,\ b,\ a,\ a \rangle).$$

The fourth clause of the interpreter is used to follow Λ edges without consuming any input. For example, the letter b is accepted by the sample NFA by traveling along a Λ edge. So the fourth clause does the trick.

This interpreter can be easily translated into the Prolog language with three simple notational changes: use brackets [and] to represent lists; replace $\leftarrow$ with ":–"; and replace "$X :: Y$" with $[X|Y]$.

end example

 Exercises

Deterministic Finite Automata

1. Write down the transition function for the following DFA.

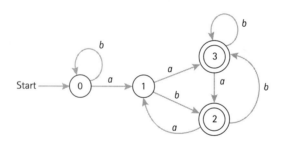

2. Use your wits to construct a DFA for each of the following regular expressions.

   a. $a + b$.   **b.** $a + bc$.   **c.** $a + b^*$.
   d. $ab^* + c$.   **e.** $ab^* + bc^*$.   **f.** $a^*bc^* + ac$.

3. Suppose we need a DFA to recognize decimal representations of rational numbers with no repeating decimal patterns. We can represent the strings by the following regular expression, where d represents a decimal digit and the vertical line "|" denotes the usual + for regular expressions, since + is now used as the arithmetic plus sign:

$$(- \mid \Lambda \mid +)dd^*.dd^*.$$

Find a DFA for this regular expression.

Nondeterministic Finite Automata

4. Write down the transition function for the following NFA:

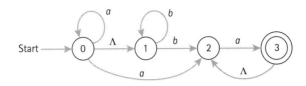

5. Use your wits to construct an NFA for each of the following regular expressions.

    a. $a^*bc^* + ac.$ **b.** $(a + b)^*a.$ **c.** $a^* + ab.$

6. For each of the following regular expressions, use (11.4) to construct an NFA.

    a. $(ab)^*.$ **b.** $a^*b^*.$ **c.** $(a + b)^*.$ **d.** $a^* + b^*.$

7. Show why Step 4 of (11.4) can't be simplified by transforming any diagram like

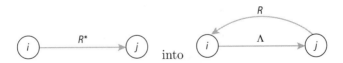

    Hint: Look at the NFA obtained for the regular expression a^*b^*.

8. Given the following NFA:

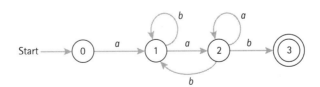

    Use algorithm (11.5) to find two regular expressions for the language accepted by the NFA as follows.

    a. Delete state 1 before deleting state 2.
    b. Delete state 2 before deleting state 1.
    c. Prove that the regular expressions obtained in Parts (a) and (b) are equal.

9. Use algorithm (11.5) to find a regular expression for the language accepted by the following NFA:

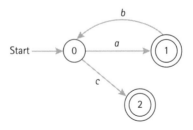

Challenges

10. Suppose we have a vending machine with two kinds of soda pop, A and B, costing 20 cents each. (It's an old fashioned machine from the 1970s.) Use the following input alphabet: n (nickel), d (dime), q (quarter), a (select A), b (select B), and r (return coins). For the output, use strings over the alphabet $\{c, n, d, q, A, B, \Lambda\}$, where c denotes the coins inserted so far and Λ denotes no output. Assume also that the message "correct change only" is off. For example, if the input is qa, then the output should be a string like An, which represents a can of A soda and five cents change. Construct a Mealy machine to model the behavior of this machine.

11. Suppose there are traffic signals at the intersection of two highways, an east-west highway and a north-south highway. The east-west highway is the major highway, with signals for through traffic and left-turn lanes with left-turn signals. The north-south highway has signals only for through traffic. There are two kinds of sensors to indicate left-turn traffic on the east-west highway and to indicate regular traffic on the north-south highway. Once a signal light turns green in response to a sensor, it stays green for only a finite period of time. Priority is given to the left-turn lanes when there is left-turn traffic and north-south traffic. Construct a Moore machine to model the behavior of the traffic lights.

11.3   Constructing Efficient Finite Automata

In this section we'll see how any regular expression can be automatically transformed into an efficient DFA that recognizes the regular language of the given expression. The construction will be in three parts. First, we transform a regular expression into an NFA. Next, we transform the NFA into a DFA. Finally, we transform the DFA into an efficient DFA having the minimum number of states. So let's begin.

11.3.1 Another Regular Expression to NFA Algorithm

In the preceding section we gave an algorithm to transform a regular expression into an NFA. The algorithm was easy to understand but not so easy to implement efficiently. Here we'll give a more mechanical algorithm that has an efficient implementation.

The algorithm will always construct an NFA with the following two properties:

1. There is exactly one final state with no edges emitted.

2. Every nonfinal state emits at most two edges.

These properties allow the algorithm to be efficiently implemented because each entry of the transition table for an NFA constructed by the algorithm is a set consisting of at most two states. Let's get to the algorithm, which is due to Thompson [1968]:

Regular Expression to NFA   **(11.7)**

Apply the following rules inductively to any regular expression. The letters s and f represent the start state and the final state.

1. Construct an NFA of the following form for each occurrence of the symbol ∅ in the regular expression:

The final state can never be reached. Thus the empty language is accepted by this NFA.

2. Construct an NFA of the following form for each occurrence of the symbol Λ in the regular expression:

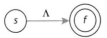

3. Construct an NFA of the following form for each occurrence of a letter x in the regular expression:

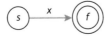

Continued ➡

---

4. Let M and N be NFAs constructed by this algorithm for the regular expressions R and S, respectively. The next three rules show how to construct the NFAs for the regular expressions $R + S$, RS, and R^*.

   a. The NFA for $R + S$ has the following form, where the dots indicate the previous start and final states of M and N:

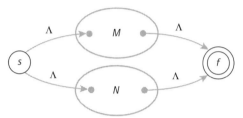

   b. The NFA for RS has the following form, where the dot combines the final state of M and the start state of N into a single state, s is the start state of M, and f is the final state of N:

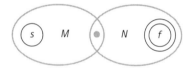

   c. The NFA for R^* has the following form, where the dots are the start and final states of M:

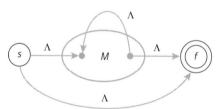

The key to applying algorithm (11.7) is to construct the little NFAs first and work in a bottom-up fashion to the bigger ones. Here's an example.

example **11.15 From Regular Expression to NFA**

We'll use (11.7) to construct an NFA for the regular expression $a^* + ab$. Since the letter a occurs twice and b occurs once in the expression, we begin by constructing the following three little NFAs.

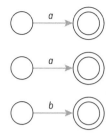

Now construct the NFA for the subexpression ab from the latter two NFAs.

Now construct the NFA for the subexpression a^* from the first little NFA.

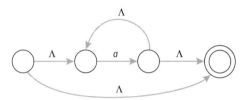

Now construct the NFA for the expression $a^* + ab$ from the preceding two NFAs to get the desired result.

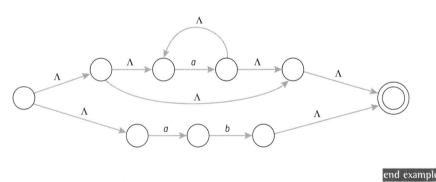

end example

11.3.2 Transforming an NFA into a DFA

We now have an automatic process to transform any regular expression into an NFA. Since NFAs have fewer restrictions than DFAs, it's also easier to create an NFA by using our wits. But no matter how we construct an NFA, it may be inefficient to execute because of nondeterminism. So it's nice to know that we can automatically transform any NFA into a DFA. Let's describe the process.

The key idea is that each state of the new DFA will actually be a subset of the NFA states. That's why it is often called "subset construction." We'll use two kinds of building blocks to construct the new subsets:

1. The sets of states that occur in the NFA transition table.

2. Certain sets of NFA states that are reachable by traversing Λ edges.

Let's describe the meaning of "reachable by traversing Λ edges."

Definition of Lambda Closure

If s is an NFA state, then the *lambda closure of s*, denoted $\lambda(s)$, is the set of states that can be reached from s by traversing zero or more Λ edges. We can define $\lambda(s)$ inductively as follows for any state s in an NFA.

1. $s \in \lambda(s)$.

2. If $p \in \lambda(s)$ and there is a Λ edge from p to q, then $q \in \lambda(s)$.

---

example **11.16 Constructing Lambda Closures**

Lambda closures are easy to construct. For example, suppose we have the following NFA, which we've described as a graph as well as a table.

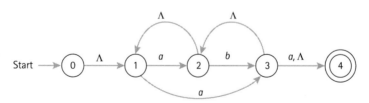

T_N		a	b	Λ
start	0	$\varnothing$	$\varnothing$	$\{1\}$
	1	$\{2,3\}$	$\varnothing$	$\varnothing$
	2	$\varnothing$	$\{3\}$	$\{1\}$
	3	$\{4\}$	$\varnothing$	$\{2,4\}$
final	4	$\varnothing$	$\varnothing$	$\varnothing$

The lambda closures for the five states of the NFA are as follows:

$$\lambda(0) = \{0, 1\},$$
$$\lambda(1) = \{1\},$$
$$\lambda(2) = \{1, 2\},$$
$$\lambda(3) = \{1, 2, 3, 4\},$$
$$\lambda(4) = \{4\}.$$

end example

Let's extend the definition of lambda closure to a set of states. If S is a set of states, then the *lambda closure of S*, denoted $\lambda(S)$, is the set of states that can be reached from states in S by traversing zero or more Λ edges. A useful property of lambda closure involves taking unions. If C and D are any sets of states, then we have

$$\lambda(C \cup D) = \lambda(C) \cup \lambda(D).$$

This property extends easily to the union of two or more sets of states. So the lambda closure of a union of sets is the union of the lambda closures of the sets. Therefore, we can compute the lambda closure of a set by calculating the union of the lambda closures of the individual elements in the set. In other words, if $S = \{s_1, ..., s_n\}$, then

$$\lambda(S) = \lambda(\{s_1, \ldots, s_n\}) = \lambda(s_1) \cup \cdots \cup \lambda(s_n).$$

For example, the lambda closure of the set $\{0, 2, 4\}$ for the NFA in Example 11.16 can be computed as follows:

$$\lambda(\{0, 2, 4\}) = \lambda(0) \cup \lambda(2) \cup \lambda(4) = \{0, 1\} \cup \{1, 2\} \cup \{4\} = \{0, 1, 2, 4\}.$$

Now that we have the tools, let's describe the algorithm for transforming any NFA into a DFA that recognizes the same language. Here's the algorithm in all its glory:

NFA to DFA Algorithm   (11.8)

Input: An NFA over alphabet A with transition function T_N.

Output: A DFA over A with transition function T_D that accepts the same language as the NFA. The states of the DFA are represented as certain subsets of NFA states.

Continued ➡

➡ ➡

---

1. The DFA start state is $\lambda(s)$, where s is the NFA start state. Now perform Step 2 for this DFA start state.

2. If $\{s_1, \ldots, s_n\}$ is a DFA state and $a \in A$, then construct the following DFA state as a DFA table entry in either of two ways:

$$T_D(\{s_1, \ldots, s_n\}, a)$$
$$= \lambda(T_N(s_1, a) \cup \cdots \cup T_N(s_n, a)) \qquad \text{(closure of union)}$$
$$= \lambda(T_N(s_1, a)) \cup \cdots \cup \lambda(T_N(s_n, a)) \quad \text{(union of closure)}.$$

Repeat Step 2 for each new DFA state constructed in this way.

3. A DFA state is final if one of its elements is an NFA final state.

Let's work through an example to show how to use the algorithm. We'll transform the NFA of Example 11.16 into a DFA by constructing the transition table for the DFA. The first step of (11.8) computes the DFA's start state: $\lambda(0) = \{0, 1\}$. So we can begin constructing the transition table T_D for the DFA as follows:

T_D	a	b
start $\{0, 1\}$		

Our next step is to fill in the table entries for the first row. In other words, we want to compute $T_D(\{0, 1\}, a)$ and $T_D(\{0, 1\}, b)$. Using the closure of union formula from Step 2 of (11.8), we calculate $T_D(\{0, 1\}, a)$ as follows:

$$T_D(\{0, 1\}, a) = \lambda(T_N(0, a) \cup T_N(1, a))$$
$$= \lambda(\varnothing \cup \{2, 3\})$$
$$= \lambda(\{2, 3\})$$
$$= \lambda(2) \cup \lambda(3)$$
$$= \{1, 2\} \cup \{1, 2, 3, 4\}$$
$$= \{1, 2, 3, 4\}.$$

We can describe this process in terms of the NFA table as follows: Using the state $\{0, 1\}$, we enter the NFA table at rows 0 and 1. To construct the entry in column a of T_D, we take the union of the sets in column a rows 0 and 1 to get $\varnothing \cup \{2, 3\} = \{2, 3\}$. Then we compute the lambda closure of $\{2, 3\}$, obtaining $\{1, 2, 3, 4\}$. Similarly, we obtain $T_D(\{0, 1\}, b) = \varnothing$. In terms of the table, we take the union of the sets in column b rows 0 and 1 to get $\varnothing \cup \varnothing = \varnothing$. The lambda closure of $\varnothing$ is $\varnothing$. So the table for T_D now looks like the following:

T_D	a	b
start $\{0, 1\}$	$\{1, 2, 3, 4\}$	$\varnothing$

Next, we check to see whether any new states have been added to the new table. Both $\{1, 2, 3, 4\}$ and $\varnothing$ are new states. So we choose one of them and make it a new row label:

T_D		a	b
start	$\{0, 1\}$	$\{1, 2, 3, 4\}$	$\varnothing$
	$\{1, 2, 3, 4\}$		

Now, we apply the same process to fill in the entries for this new row. In this case we get $T_D(\{1, 2, 3, 4\}, a) = \{1, 2, 3, 4\}$ and also $T_D(\{1, 2, 3, 4\}, b) = \{1, 2, 3, 4\}$. Therefore, the table for T_D looks like the following:

T_D		a	b
start	$\{0, 1\}$	$\{1, 2, 3, 4\}$	$\varnothing$
	$\{1, 2, 3, 4\}$	$\{1, 2, 3, 4\}$	$\{1, 2, 3, 4\}$

The state $\varnothing$ is in the table and is not yet a row label. So we add $\varnothing$ as a new row label to create the following table:

T_D		a	b
start	$\{0, 1\}$	$\{1, 2, 3, 4\}$	$\varnothing$
	$\{1, 2, 3, 4\}$	$\{1, 2, 3, 4\}$	$\{1, 2, 3, 4\}$
	$\varnothing$		

Now we continue the process by filling in the row labeled with $\varnothing$. We'll use the convention that an empty union is empty (there are no states in $\varnothing$). So we get $T_D(\varnothing, a) = \varnothing$ and $T_D(\varnothing, b) = \varnothing$. Therefore, the table looks like the following:

T_D		a	b
start	$\{0, 1\}$	$\{1, 2, 3, 4\}$	$\varnothing$
	$\{1, 2, 3, 4\}$	$\{1, 2, 3, 4\}$	$\{1, 2, 3, 4\}$
	$\varnothing$	$\varnothing$	$\varnothing$

Now every entry in the table is also a row label. Therefore, we've completed the definition of T_D. To finish things off, we note that state $\{1, 2, 3, 4\}$ contains a final state of the NFA. Therefore, we mark it as final in the DFA and obtain the following table for T_D:

T_D		a	b
start	$\{0, 1\}$	$\{1, 2, 3, 4\}$	$\varnothing$
final	$\{1, 2, 3, 4\}$	$\{1, 2, 3, 4\}$	$\{1, 2, 3, 4\}$
	$\varnothing$	$\varnothing$	$\varnothing$

Since the states of the constructed DFA are sets, it's awkward to draw a picture. So we'll make the following changes to simplify the process:

Replace $\{0, 1\}$ by the symbol 0.

Replace $\{1, 2, 3, 4\}$ by the symbol 1.

Replace $\varnothing$ by the symbol 2.

With these replacements, the table for T_D can be written in the following form:

T_D		a	b
start	0	1	2
final	1	1	1
	2	2	2

The graph version of the DFA for table T_D can be drawn as follows:

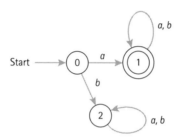

It's easy to see that the regular expression for this DFA is $a(a + b)^*$. Therefore, $a(a + b)^*$ is also the regular expression for the NFA of Example 11.16.

If an NFA does not have any Λ edges, then the lambda closure of a set is just the set itself. In this case, (11.8) simplifies to the following:

From NFA (without Λ edges) to DFA

1. The DFA start state is $\{s\}$, where s is the NFA start state. Now perform Step 2 for this DFA start state.

2. If $\{s_1, \ldots, s_n\}$ is a DFA state and $a \in A$, then construct the following DFA state as a DFA table entry:

$$T_D(\{s_1, \ldots, s_n\}, a) = T_N(s_1, a) \cup \cdots \cup T_N(s_n, a).$$

Repeat Step 2 for each new DFA state constructed in this way.

3. A DFA state is final if one of its elements is an NFA final state.

Let's summarize the situation as it now stands. We can start with a regular expression and use (11.4) or (11.7) to construct an NFA for the expression. Next we can use (11.8) to transform the NFA into a DFA. Since the NFAs constructed by (11.4) or (11.7) might have a large number of states, it follows that the DFA constructed by (11.8) might have a large number of states. In the next few paragraphs we'll see that any DFA can be automatically transformed into a DFA with the minimum number of states. Then, putting everything together, we'll have an automatic process for constructing the most efficient DFA for any regular expression.

11.3.3 Minimum-State DFAs

One way to try and simplify the DFA for some regular expression is to algebraically transform the regular expression into a simpler one before starting construction of the DFA. For example, from the properties (11.1) we have

$$\Lambda + a + aaa^* = a^*.$$

If we use our wits, most of us can construct a simpler DFA for a^* than for $\Lambda + a + aaa^*$. If we use the algorithms, we can also obtain a simpler DFA for a^* than for $\Lambda + a + aaa^*$. But we still might not have obtained the simplest DFA.

It's nice to know that no matter what DFA we come up with, we can always transform it into a DFA with the minimum number of states that recognizes the same language. The basic result is given by the following theorem, which is named after Myhill [1957] and Nerode [1958]:

Theorem   (11.9)
Every regular expression has a unique minimum-state DFA.

The word "unique" in (11.9) means that the only difference that can occur between any two minimum-state DFAs for a regular expression is not in the number of states, but rather in the names given to the states. So we could rename the states of one DFA so that it becomes the same as the other DFA.

We already know how to transform a regular expression into an NFA and then into a DFA. Now let's see how to transform a DFA into a minimum-state DFA. The key idea is to define two states s and t to be *equivalent* if for every string w, the transitions

$$T(s, w) \text{ and } T(t, w) \text{ are either both final or both nonfinal.}$$

In other words, to say that s and t are equivalent means that whenever the execution of the DFA reaches either s or t with the same string w left to consume, then the DFA will consume w and, in either case, enter the same type of state—either both reject or both accept.

It's easy to see that equivalence is an equivalence relation. Stop and check it out. So once we know the equivalent pairs of states, we can partition the states of the DFA into a collection of subsets, where each subset contains states that are equivalent to each other. These subsets become the states of the new minimum-state DFA.

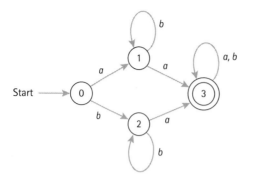

	T	a	b
start	0	1	2
	1	3	1
	2	3	2
final	3	3	3

Figure 11.7   DFA as graph and table.

Before we present the algorithm, let's try to make the idea more precise with a very simple example. Suppose we're given the four-state DFA in Figure 11.7, which is represented in graphical form and as a transition table.

It's pretty easy to see that states 1 and 2 are equivalent. For example, if the DFA is in either state 1 or 2 with any string starting with a, then the DFA will consume a and enter state 3. From this point the DFA stays in state 3. On the other hand, if the DFA is in either state 1 or 2 with any string starting with b, then the DFA will consume b and it will stay in state 1 or 2 as long as b is present at the beginning of the resulting string. So for any string w, both $T(1, w)$ and $T(2, w)$ are either both final or both nonfinal.

It's also pretty easy to see that no other distinct pairs of states are equivalent. For example, states 0 and 1 are not equivalent because $T(0, a) = 1$, which is a reject state, and $T(1, a) = 3$, which is an accept state. Since the only distinct equivalent states are 1 and 2, we can partition the states of the DFA into the subsets

$$\{0\}, \{1, 2\}, \text{ and } \{3\}.$$

These three subsets form the states of the minimum-state DFA. This minimum-state DFA can be represented by either one of the two forms shown in Figure 11.8.

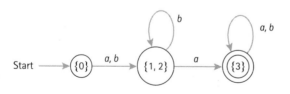

	T_{min}	a	b
start	{0}	{1, 2}	{1, 2}
	{1, 2}	{3}	{1, 2}
final	{3}	{3}	{3}

Figure 11.8   Minimum-state DFA.

Partitioning the States

There are several methods to compute the equivalence relation and its corresponding partition. The method that we'll present is easy to understand, and it can be programmed. We start the process by forming the set

$$E_0$$

of distinct pairs of the form $\{s, t\}$, where s and t are either both final or both nonfinal. This collection contains the possible equivalent pairs.

Next we construct a new collection E_1 from E_0 by throwing away any pair $\{s, t\}$ if there is some letter a such that $\{T(s, a),\ T(t, a)\}$ is a distinct pair that does not occur in E_0. This means that the pair $\{T(s, a),\ T(t, a)\}$ contains two states of different types. So we must throw $\{s, t\}$ away.

The process continues by constructing a new collection E_2 from E_1 by throwing away $\{s, t\}$ if there is some letter a such that $\{T(s, a),\ T(t, a)\}$ is a distinct pair that does not occur in E_1. This means that there is a string of length 2 such that the DFA, if started from either s or t, consumes the string and enters two different types of states. So we must throw $\{s, t\}$ out of E_1.

We continue the process by constructing a descending sequence

$$E_0 \supseteq E_1 \supseteq E_2 \supseteq \cdots \supseteq E_n \supseteq \cdots.$$

Each set E_n in the sequence has been constructed to have the property that for each pair $\{s, t\}$ in E_n and for any string of length less than or equal to n, the DFA, if started from either s or t, will consume the string and enter the same type of states—either both reject or both accept. Since E_0 is a finite set, the sequence of sets must eventually stop with some set E_k such that

$$E_{k+1} = E_k.$$

This means E_k is the desired set of equivalent pairs of states, because for any pair $\{s, t\}$ in E_k and any length string, the DFA, if started from either s or t, will consume the string and enter the same type of states—either both reject or both accept.

For example, from the DFA in Figure 11.7 we start with

$$E_0 = \{\{0, 1\}, \{0, 2\}, \{1, 2\}\}.$$

To construct E_1 from E_0, we throw away $\{0, 1\}$ because

$$\{T(0, a),\ T(1, a)\} = \{1, 3\},$$

which is not in E_0. We must also throw away $\{0, 2\}$ because

$$\{T(0, a),\ T(2, a)\} = \{1, 3\},$$

which is not in E_0. This leaves us with the set

$$E_1 = \{\{1, 2\}\}.$$

We can't throw away any pairs from E_1. Therefore, $E_2 = E_1$, which says that the desired set of equivalent pairs is $E_1 = \{\{1, 2\}\}$. Notice that $\{1, 2\}$ is a state in the minimum-state DFA shown in Figure 11.8.

Now we're ready to present the actual algorithm to transform a DFA into a minimum-state DFA.

Algorithm to Construct a Minimum–State DFA   (11.10)

Given: A DFA with set of states S and transition table T. Assume that all states that cannot be reached from the start state have already been thrown away.

Output: A minimum-state DFA recognizing the same regular language as the input DFA.

1. Construct the equivalent pairs of states by calculating the descending sequence of sets of pairs $E_0 \supseteq E_1 \supseteq \cdots$ defined as follows:

   $E_0 = \{\{s, t\} \mid s$ and t are distinct and either both states are final or both states are nonfinal$\}$.

   $E_{i+1} = \{\{s, t\} \mid \{s, t\} \in E_i$ and for every $a \in A$ either $T(s, a) = T(t, a)$ or $\{T(s, a), T(t, a)\} \in E_i\}$.

   The computation stops when $E_k = E_{k+1}$ for some index k. E_k is the desired set of equivalent pairs.

2. Use the equivalence relation generated by the pairs in E_k to partition S into a set of equivalence classes. These equivalence classes are the states of the new DFA.

3. The *start state* is the equivalence class containing the start state of the input DFA.

4. A *final state* is any equivalence class containing a final state of the input DFA.

5. The transition table T_{min} for the minimum-state DFA is defined as follows, where $[s]$ denotes the equivalence class containing s and a is any letter: $T_{min}([s], a) = [T(s, a)]$.

---

example **11.17 A Minimum–State DFA Construction**

We'll compute the minimum-state DFA for the following DFA.

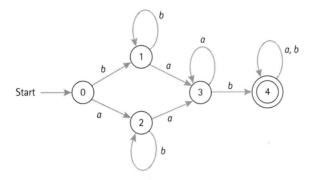

The set of states is $S = \{0, 1, 2, 3, 4\}$. For Step 1 we'll start by calculating E_0 as the set of pairs $\{s, t\}$, where s and t are both final or both nonfinal:

$$E_0 = \{\{0, 1\}, \{0, 2\}, \{0, 3\}, \{1, 2\}, \{1, 3\}, \{2, 3\}\}.$$

To calculate E_1 we throw away $\{0, 3\}$ because $\{T(0, b), T(3, b)\} = \{1, 4\}$, which is not in E_0. We also throw away $\{1, 3\}$ and $\{2, 3\}$. That leaves us with

$$E_1 = \{\{0, 1\}, \{0, 2\}, \{1, 2\}\}.$$

To calculate E_2 we throw away $\{0, 2\}$ because $\{T(0, a), T(2, a)\} = \{2, 3\}$, which is not in E_1. That leaves us with

$$E_2 = \{\{1, 2\}\}.$$

To calculate E_3 we don't throw anything away from E_2. So we stop with

$$E_3 = E_2 = \{\{1, 2\}\}.$$

So the only distinct equivalence pair is $\{1, 2\}$. Therefore, the set S of states is partitioned into the following four equivalence classes:

$$\{0\}, \{1, 2\}, \{3\}, \{4\}.$$

These are the states for the new DFA. The start state is $\{0\}$, and the final state is $\{4\}$. Using equivalence class notation we have

$$[0] = \{0\}, \quad [1] = [2] = \{1, 2\}, \quad [3] = \{3\}, \quad \text{and} \quad [4] = \{4\}.$$

Thus we can apply Step 5 to construct the table for $T_{\min}$. For example, we'll compute $T_{\min}(\{0\}, a)$ and $T_{\min}(\{1, 2\}, b)$ as follows:

$$T_{\min}(\{0\}, a) = T_{\min}([0], a) = [T(0, a)] = [2] = \{1, 2\},$$
$$T_{\min}(\{1, 2\}, b) = T_{\min}([1], b) = [T(1, b)] = [1] = \{1, 2\}.$$

Similar computations yield the table for $T_{\min}$, which is listed as follows:

$T_{\min}$		a	b
start	$\{0\}$	$\{1,2\}$	$\{1,2\}$
	$\{1,2\}$	$\{3\}$	$\{1,2\}$
	$\{3\}$	$\{3\}$	$\{4\}$
final	$\{4\}$	$\{4\}$	$\{4\}$

We can simplify the table by assigning a single number to each state. For example, assign 0, 1, 2, and 3 to the states $\{0\}$, $\{1, 2\}$, $\{3\}$, and $\{4\}$. Then the preceding table can be written in the following familiar form.

$T_{\min}$		a	b
start	0	1	1
	1	2	1
	2	2	3
final	3	3	3

Be sure to draw a picture of this minimum-state DFA.

end example

example **11.18 A Minimum-State DFA Construction**

We'll compute the minimum-state DFA for the following DFA.

T		a	b
start	0	1	2
	1	4	1
	2	4	3
	3	4	3
final	4	4	5
final	5	5	5

The set of states is $S = \{0, 1, 2, 3, 4, 5\}$. For Step 1 we get the following sequence of relations:

$$E_0 = \{\{0,1\}, \{0,2\}, \{0,3\}, \{1,2\}, \{1,3\}, \{2,3\}, \{4,5\}\},$$
$$E_1 = \{\{1,2\}, \{1,3\}, \{2,3\}, \{4,5\}\},$$
$$E_2 = E_1.$$

Therefore, the equivalence relation is generated by the four equivalent pairs $\{1, 2\}$, $\{1, 3\}$, $\{2, 3\}$, and $\{4, 5\}$. Thus we obtain a partition of S into the following three equivalence classes:

$$\{0\}, \{1, 2, 3\}, \{4, 5\}.$$

These are the states for the new DFA. The start state is $\{0\}$, and the final state is $\{4, 5\}$. Using the standard notation for equivalence classes, we have

$$[0] = \{0\}, \quad [1] = [2] = [3] = \{1, 2, 3\}, \quad \text{and} \quad [4] = [5] = \{4, 5\}.$$

So we can apply Step 5 to construct the table for $T_{\min}$. For example, we can compute $T_{\min}(\{1, 2, 3\}, b)$ and $T_{\min}(\{4, 5\}, a)$ as follows:

$$T_{\min}(\{1, 2, 3\}, b) = T_{\min}([1], b) = [T(1, b)] = [1] = \{1, 2, 3\},$$
$$T_{\min}(\{4, 5\}, a) = T_{\min}([4], a) = [T(4, a)] = [4] = \{4, 5\}.$$

Similar computations will yield the table for $T_{\min}$. We'll leave these calculations as an exercise.

end example

Exercises

Construction Algorithms

1. Use (11.7) to build an NFA for each of the following regular expressions.

   a. a^*b^*.   **b.** $(a + b)^*$.   **c.** $a^* + b^*$.

2. Construct a DFA table for the following NFA in two ways using (11.8):

		a	b	Λ
start	0	$\varnothing$	$\{1, 2\}$	$\{1\}$
	1	$\{2\}$	$\varnothing$	$\varnothing$
final	2	$\varnothing$	$\{2\}$	$\{1\}$

   a. Take unions of lambda closures of the NFA entries.

   b. Take lambda closures of unions of the NFA entries.

3. Suppose we are given the following NFA over the alphabet $\{a, b\}$:

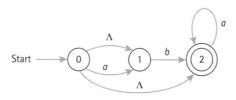

   a. Find a regular expression for the language accepted by the NFA.

   b. Write down the transition table for the NFA.

   c. Use (11.8) to transform the NFA into a DFA.

   d. Draw a picture of the resulting DFA.

4. Transform each of the following NFAs into a DFA.

   a. $T(0, a) = \{0, 1\}$, where 0 is start and 1 is final.
   b. $T(0, a) = \{1, 2\}$, $T(1, b) = \{1, 2\}$, where 0 is start and 2 is final.

5. Transform each of the following regular expressions into a DFA by using (11.7) and (11.8). Note that the NFAs obtained by (11.7) are the answers to Exercise 1.

   a. $a*b*$.   **b.** $(a + b)*$.   **c.** $a* + b*$.

Minimum-State DFAs

6. Let the set of states for a DFA be $S = \{0, 1, 2, 3, 4, 5\}$, where the start state is 0 and the final states are 2 and 5. Let the equivalence relation on S for a minimum-state DFA be generated by the following set of equivalent pairs of states:

$$\{\{0, 1\}, \{0, 4\}, \{1, 4\}, \{2, 5\}\}.$$

Write down the states of the minimum-state DFA.

7. Finish Example 11.18 by calculating the transition table for the new minimum-state DFA.

8. Given the following DFA table, use algorithm (11.10) to compute the minimum-state DFA. Answer Parts (a), (b), and (c) as you proceed through the algorithm.

		a	b
start	0	1	2
	1	1	4
final	2	2	3
	3	3	4
final	4	4	2

   a. Write down the set of equivalent pairs.
   b. Write down the states of the minimum-state DFA.
   c. Write down the transition table for the minimum-state DFA.

9. For each of the following DFAs, use (11.10) to find the minimum-state DFA.

a.

		a	b
start	0	1	2
final	1	1	2
	2	3	2
final	3	4	2
final	4	1	2

b.

		a	b
start	0	1	2
final	1	4	4
final	2	3	4
final	3	3	4
	4	4	4

10. For each of the following regular expressions, start by writing down the NFA obtained by algorithm (11.7). Then use (11.8) to transform the NFA into a DFA. Then use (11.10) to find the minimum-state DFA.

    a. $a^* + a^*$ (don't simplify).   **b.** $(a + b)^*a$.

    c. a^*b^*.   **d.** $(a + b)^*$.

11. Suppose we're given the following NFA table:

		a	b	Λ
start	0	$\{1\}$	$\{1\}$	$\{2\}$
	1	$\{1, 2\}$	$\varnothing$	$\varnothing$
final	2	$\varnothing$	$\{0\}$	$\{1\}$

Find a simple regular expression for the regular language recognized by this NFA. *Hint:* Transform the NFA into a DFA, and then find the minimum-state DFA.

Challenge

12. What can you say about the regular language accepted by a DFA in which all states are final?

11.4   Regular Language Topics

We've already seen characterizations of regular languages by regular expressions, languages accepted by DFAs, and languages accepted by NFAs. In this section we'll introduce still another characterization of regular languages in terms of certain restricted grammars. We'll discuss some properties of regular languages that can be used to find languages that are not regular.

11.4.1 Regular Grammars

A regular language can be described by a special kind of grammar in which the productions take a certain form. A grammar is called a *regular grammar* if each production takes one of the following forms, where the uppercase letters are nonterminals and w is a nonempty string of terminals:

$$S \to \Lambda,$$
$$S \to w,$$
$$S \to T,$$
$$S \to wT.$$

The thing to keep in mind here is that only one nonterminal can appear on the right side of a production, and it must appear at the right end of the right side. For example, the productions $A \to aBc$ and $S \to TU$ are not part of a regular grammar. But the production $A \to abcA$ is OK.

The most important aspect of grammar writing is knowledge of the language under discussion. We should also remember that grammars are not unique. So we shouldn't be surprised when two people come up with two different grammars for the same language.

To start things off, we'll look at a few regular grammars for some simple regular languages. Each line of the following list describes a regular language in terms of a regular expression and a regular grammar. As you look through the following list, cover up the grammar column with your hand and try to discover your own version of a regular grammar for the regular language of each regular expression.

Regular Expression	Regular Grammar
a^*	$S \to \Lambda \mid aS$
$(a+b)^*$	$S \to \Lambda \mid aS \mid bS$
$a^* + b^*$	$S \to \Lambda \mid A \mid B$ $A \to a \mid aA$ $B \to b \mid bB$
a^*b	$S \to b \mid aS$
ba^*	$S \to bA$ $A \to \Lambda \mid aA$
$(ab)^*$	$S \to \Lambda \mid abS$

The last three examples in the preceding list involve products of languages. Most problems occur in trying to construct a regular grammar for a language that is the product of languages. Let's look at an example to see whether we can get some insight into constructing such grammars.

Suppose we want to construct a regular grammar for the language of the regular expression a^*bc^*. First we observe that the strings of a^*bc^* start with either the letter a or the letter b. We can represent this property by writing down the following two productions, where S is the start symbol:

$$S \rightarrow a\ S \mid b\ C.$$

These productions allow us to derive strings of the form bC, abC, $aabC$, and so on. Now all we need is a definition for C to derive the language of c^*. The following two productions do the job:

$$C \rightarrow \Lambda \mid c\ C.$$

Therefore, a regular grammar for a^*bc^* can be written as follows:

$$S \rightarrow aS \mid bC$$
$$C \rightarrow \Lambda \mid cC.$$

example 11.19 Sample Regular Grammars

We'll consider some regular languages, all of which consist of strings of a's followed by strings of b's. The most general language of this form is the language $\{a^m b^n \mid m,\ n \in \mathbb{N}\}$, which is represented by the regular expression a^*b^*. A regular grammar for this language can be written as follows:

$$S \rightarrow \Lambda \mid aS \mid B$$
$$B \rightarrow b \mid bB.$$

Let's look at four sublanguages of $\{a^m b^n \mid m,\ n \in \mathbb{N}\}$ that are defined by whether each string contains occurrences of a or b. The following list shows each language together with a regular expression and a regular grammar.

Language	Expression	Regular Grammar
$\{a^m b^n \mid m \geq 0 \text{ and } n > 0\}$	a^*bb^*	$S \rightarrow aS \mid B$
		$B \rightarrow b \mid bB.$
$\{a^m b^n \mid m > 0 \text{ and } n \geq 0\}$	aa^*b^*	$S \rightarrow aA$
		$A \rightarrow aA \mid B$
		$B \rightarrow \Lambda \mid bB.$

Language	Expression	Regular Grammar
$\{a^m b^n \mid m > 0 \text{ and } n > 0\}$	$aa^* bb^*$	$S \rightarrow aA$
		$A \rightarrow aA \mid B$
		$B \rightarrow b \mid bB.$
$\{a^m b^n \mid m > 0 \text{ or } n > 0\}$	$aa^* b^* + a^* bb^*$	$S \rightarrow aA \mid bB$
		$A \rightarrow \Lambda \mid aA \mid B$
		$B \rightarrow \Lambda \mid bB.$

end example

Any regular language has a regular grammar; conversely, any regular grammar generates a regular language. To see this, we'll give two algorithms: one to transform an NFA to a regular grammar and the other to transform a regular grammar to an NFA, where the language accepted by the NFA is identical to the language generated by the regular grammar.

NFA to Regular Grammar (11.11)

Perform the following steps to construct a regular grammar that generates the language of a given NFA:

1. Rename the states to a set of uppercase letters.

2. The start symbol is the NFA's start state.

3. For each state transition from I to J labeled with a, create the production $I \rightarrow aJ$.

4. For each state transition from I to J labeled with Λ, create the production $I \rightarrow J$.

5. For each final state K, create the production $K \rightarrow \Lambda$.

It's easy to see that the language of the NFA and the language of the constructed grammar are the same. Just notice that each state transition in the NFA corresponds exactly with a production in the grammar so that the acceptance path in the NFA for some string corresponds to a derivation by the grammar for the same string. Let's do an example.

example 11.20 From NFA to Regular Grammar

Let's see how (11.11) transforms the following NFA into a regular grammar:

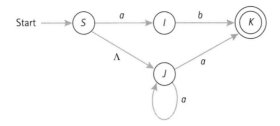

The algorithm takes this NFA and constructs the following regular grammar with start symbol S:

$$S \to aI \mid J$$
$$I \to bK$$
$$J \to aJ \mid aK$$
$$K \to \Lambda.$$

For example, to accept the string aa, the NFA follows the path S, J, J, K with edges labeled Λ, a, a, respectively. The grammar derives this string with the following sequence of productions:

$$S \to J,\ J \to aJ,\ J \to aK,\ K \to \Lambda.$$

Now let's look at the converse problem of constructing an NFA from a regular grammar. For the opposite transformation we'll first take a regular grammar and rewrite it so that all the productions have one of two forms $S \to x$ or $S \to xT$, where x is either Λ or a single letter. Let's see how to do this so that we don't lose any generality. For example, if we have a production like

$$A \to bcB,$$

we can replace it by the following two productions, where C is a new nonterminal:

$$A \to bC \quad \text{and} \quad C \to cB.$$

Now let's look at an algorithm that does the job of transforming a regular grammar into an NFA.

Regular Grammar to NFA   (11.12)

Perform the following steps to construct an NFA that accepts the language of a given regular grammar:

1. If necessary, transform the grammar so that all productions have the form $A \to x$ or $A \to xB$, where x is either a single letter or Λ.

2. The start state of the NFA is the grammar's start symbol.

3. For each production $I \to aJ$, construct a state transition from I to J labeled with the letter a.

4. For each production $I \to J$, construct a state transition from I to J labeled with Λ.

5. If there are productions of the form $I \to a$ for some letter a, then create a single new state symbol F. For each production $I \to a$, construct a state transition from I to F labeled with a.

6. The final states of the NFA are F together with all I for which there is a production $I \to \Lambda$.

It's easy to see that the language of the NFA is the same as the language of the given regular grammar because the productions used in the derivation of any string correspond exactly with the state transitions on the path of acceptance for the string. Here's an example.

example **11.21 From Regular Grammar to NFA**

Let's use (11.12) to transform the following regular grammar into an NFA:

$$S \to aS \,|\, bI$$
$$I \to a \,|\, aI.$$

Since there is a production $I \to a$, we need to introduce a new state F, which then gives us the following NFA:

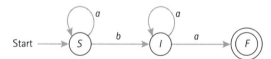

end example

11.4.2 Properties of Regular Languages

We need to face the fact that not all languages are regular. To see this, let's look at a classic example. Suppose we want to find a DFA or NFA to recognize the following language.

$$\{a^n b^n \mid n \geq 0\}.$$

After a few attempts at trying to find a DFA or an NFA or a regular expression or a regular grammar, we might get the idea that it can't be done. But how can we be sure that a language is not regular? We can try to prove it. A proof usually proceeds by assuming that the language is regular and then trying to find a contradiction of some kind. For example, we might be able to find some property of regular languages that the given language doesn't satisfy. So let's look at a few properties of regular languages.

Pumping Lemma

One useful property of regular languages comes from the observation that any DFA for an infinite regular language must contain a loop to recognize infinitely many strings. For example, suppose a DFA with four states accepts the 4-letter string $abcd$. To accept $abcd$ the DFA must enter five states. For example, if the states of the DFA are numbered 0, 1, 2, and 3, where 0 is the start state and 3 is the final state, then there must be a path through the DFA starting at 0 and ending at 3 with edges labeled a, b, c, and d. For example, if the path is 0, 1, 2, 1, 3, then the following graph represents a portion of the DFA that contains the path to accept $abcd$.

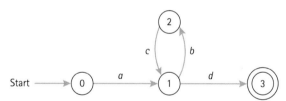

Of course, the loop 1, 2, 1 can be traveled any number of times. For example, the path 0, 1, 2, 1, 2, 1, 3 accepts the string $abcbcd$. So the DFA will accept the strings, ad, $abcd$, $abcbcd$, ..., $a(bc)^n d$, ... This is the property that we want to describe.

We'll generalize the idea illustrated in our little example. Suppose a DFA with m states recognizes an infinite regular language. If s is a string accepted by the DFA, then there must be a path from the start state to a final state that traverses $|s| + 1$ states. If $|s| \geq m$, then $|s| + 1 > m$, which tells us that some state must be traversed twice or more. So the DFA must have at least one loop that is traversed at least once on the path to accept s. Let x be the string of letters along the path from the start state to the state that begins the first traversal of a loop. Let y be the string of letters along one traversal of the loop and let z be the string of letters along the rest of the path of acceptance to the final state. So we can write $s = xyz$. Note that z may include more traversals of the loop or any subsequent loops. To illustrate from our little example, if $s = abcd$, then $x = a$, $y = bc$, and $z = d$. If $s = abcbcd$, then $x = a$, $y = bc$, and $z = bcd$. If $s = abcbcbcbcd$, then $x = a$, $y = bc$, and $z = bcbcbcd$.

The following graph symbolizes the path to accept s, where the arrows labeled x and y represent paths along distinct states of the DFA while the arrow labeled z represents the rest of the path to the final state.

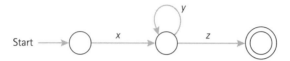

Since $|s| \geq m$, the path must traverse the loop at least once. So $y \neq \Lambda$. Since the paths for x and y consist of distinct states (remember that y is the string on just one traversal of the loop), it follows that $|xy| \leq m$. Finally, since the path through the loop may be traversed any number of times, it follows that the DFA must accept all strings of the form $xy^k z$ for all $k \geq 0$.

The property that we've been discussing is called the *pumping property* because the string y can be pumped up to y^k by traveling through the same loop k times. Our discussion serves as an informal proof of the following pumping lemma.

Pumping Lemma for Regular Languages   (11.13)

Let L be an infinite regular language. Then there is an integer $m > 0$ (m is the number of states in a DFA to recognize L) such that for any string $s \in L$ with $|s| \geq m$, there exist strings x, y, and z such that $s = xyz$, $y \neq \Lambda$, $|xy| \leq m$, and $xy^k z \in L$ for all $k \geq 0$. The last property tells us that $\{xz,\ xyz,\ xy^2z\ ,\ \dots,\ xy^kz\ ,\ \dots\ \} \subseteq L$.

If an infinite language does not satisfy the conclusion of (11.13), then it can't be regular. We can sometimes use this fact to prove that an infinite language is not regular by assuming that it is regular, applying the conclusion of (11.13), and then finding a contradiction. Here's an example.

example 11.22 Using the Pumping Lemma

Let's show that the language $L = \{a^n b^n \mid n \geq 0\}$ is not regular. We'll assume, by way of contradiction, that L is regular. This allows us to use the pumping lemma (11.13). Since m exists but is unknown, it must remain a symbol and not be given any specific value. With this in mind, we'll choose a string $s \in L$ such that $|s| \geq m$ and try to contradict some property of the pumping lemma. Let $s = a^m b^m$. The pumping lemma tells us that s can be written as

$$s = a^m b^m = xyz,$$

where $y \neq \Lambda$ and $|xy| \leq m$. So x and y consist only of a's. Therefore, we can write $y = a^n$ for some $n > 0$. Now the pumping lemma also tells us $xy^k z \in L$ for all $k \geq 0$. If we look at the case $k = 2$ we have $xy^2 z = a^{m+n} b^m$, which means that $xy^2 z$ has more a's than b's. So $xy^2 z \notin L$ and this contradicts the pumping lemma. So L cannot be regular. *Note:* We can also find a contradiction with $k = 0$ by observing that $xz = a^{m-n} b^m$, which has fewer a's than b's.

end example

 11.23 Using the Pumping Lemma

We'll show that the language P of palindromes over the alphabet $\{a, b\}$ is not regular. Assume, by way of contradiction, that P is regular. Then P can be recognized by some DFA. Let m be the number of states in the DFA. We'll choose s to be the following palindrome:

$$s = a^m b a^m.$$

Since $|s| \geq m$, the pumping lemma (11.13) asserts the existence of strings x, y, z such that

$$a^m b a^m = xyz,$$

where $y \neq \Lambda$ and $|xy| \leq m$. It follows that x and y are both strings of a's and we can write $y = a^n$ for some $n > 0$. If we pump up y to y^2 we obtain the form

$$xy^2 z = a^{m+n} b a^m,$$

which is not a palindrome. This contradicts the fact that it must be a palindrome according to (11.13). Therefore, P is not regular.

end example

Additional Properties

If we're trying to find out whether a language is regular, it may help to know how regular languages can be combined or transformed to form new regular languages. We know by definition that regular languages can be combined by union, language product, and closure to form new regular languages. We'll list these three properties along with two others.

Properties of Regular Languages   (11.14)

1. The union of two regular languages is regular.

2. The language product of two regular languages is regular.

3. The closure of a regular language is regular.

4. The complement of a regular language is regular.

5. The intersection of two regular languages is regular.

Proof: We'll prove statement 4 and leave statement 5 as an exercise. If D is a DFA for the regular language L, then construct a new DFA, say D', from D by making all the final states nonfinal and by making all the nonfinal states final. If we let A be the alphabet for L, it follows that D' recognizes the complement $A^* - L$. Thus the complement of L is regular. QED.

example 11.24 **An Intersection Argument**

Suppose L is the language over alphabet $\{a, b\}$ consisting of all strings with an equal number of a's and b's. For example, the strings *abba*, *ab*, and *babbaa* are all in L. Is L a regular language? We'll show that the answer is no. Let M be the language of the regular expression a^*b^*. It follows that

$$L \cap M = \{a^n b^n \mid n \geq 0\}.$$

Now we're in position to use (11.14). Suppose on the contrary that L is regular. We know that M is regular because it's the language of the regular expression a^*b^*. Therefore, $L \cap M$ must be regular by (11.14). In other words, the language $\{a^n b^n \mid n \geq 0\}$ must be regular. But we know that the language $\{a^n b^n \mid n \geq 0\}$ is *not* regular. Therefore, our assumption that L is regular leads to a contradiction. Thus L is not regular.

end example

We'll finish by listing two interesting properties of regular languages that can also be used to determine nonregularity.

Regular Language Morphisms   (11.15)

Let $f : A^* \to A^*$ be a language morphism. In other words, $f(\Lambda) = \Lambda$ and $f(uv) = f(u)f(v)$ for all strings u and v. Let L be a language over A.

1. If L is regular, then $f(L)$ is regular.

2. If L is regular, then $f^{-1}(L)$ is regular.

Proof: We'll prove statement 1 and leave statement 2 as an exercise. Since L is regular, it has a regular grammar. We'll create a regular grammar for $f(L)$ as follows: Transform productions like $S \to w$ and $S \to wT$ into new productions of the form $S \to f(w)$ and $S \to f(w)T$. The new grammar is regular, and any string in $f(L)$ is derived by this new grammar. QED.

example 11.25 **A Morphism Argument**

We'll use (11.15) to show that $L = \{a^n b c^n \mid n \in \mathbb{N}\}$ is not regular. Define the morphism $f : \{a, b, c\}^* \to \{a, b, c\}^*$ by $f(a) = a$, $f(b) = \Lambda$, and $f(c) = b$. Then $f(L) = \{a^n b^n \mid n \geq 0\}$. If L is regular, then we must also conclude by (11.15) that $f(L)$ is regular. But we know that $f(L)$ is not regular. Therefore, L is not regular.

end example

There are some very simple nonregular languages. For example, we know that the language $\{a^n b^n \mid n \geq 0\}$ is not regular. So finite automata are not powerful enough to recognize the language. Therefore, finite automata can't recognize some simple programming constructs, such as nested begin-end pairs, nested open and closed parentheses, brackets, and braces. We'll see in the next chapter that there are more powerful machines to recognize such constructs.

 Exercises

Regular Grammars

1. Find a regular grammar for each of the following regular expressions.

   a. $a + b$.
   b. $a + bc$.
   c. $a + b^*$.
   d. $ab^* + c$.
   e. $ab^* + bc^*$.
   f. $a^* bc^* + ac$.
   g. $(aa + bb)^*$.
   h. $(aa + bb)(aa + bb)^*$.
   i. $(ab)^* c(a + b)^*$.

2. Find a regular grammar to describe each of the following languages.

   a. $\{a, b, c\}$.
   b. $\{aa, ab, ac\}$.
   c. $\{a, b, ab, ba, abb, baa, \ldots, ab^n, ba^n, \ldots\}$.
   d. $\{a, aaa, aaaaa, \ldots, a^{2n+1}, \ldots\}$.
   e. $\{\Lambda, a, abb, abbbb, \ldots, ab^{2n}, \ldots\}$.
   f. $\{\Lambda, a, b, c, aa, bb, cc, \ldots, a^n, b^n, c^n, \ldots\}$.
   g. $\{\Lambda, a, b, ca, bc, cca, bcc, \ldots, c^n a, bc^n, \ldots\}$.
   h. $\{a^{2k} \mid k \in \mathbb{N}\} \cup \{b^{2k+1} \mid k \in \mathbb{N}\}$.
   i. $\{a^m bc^n \mid m, n \in \mathbb{N}\}$.

3. Find a regular grammar for each of the following languages over the alphabet $\{a, b\}$.

   a. All strings have even length.
   b. All strings have length that is a multiple of 3.
   c. All strings contain the substring aba.
   d. All strings have an odd number of a's.

4. Any regular language can also be defined by a grammar with productions of the following form, where w is a nonempty string of terminals:

$$S \to \Lambda, \quad S \to w, \quad S \to T, \quad \text{or} \quad S \to Tw.$$

Find a grammar of this form for the language of each of the following regular expressions.

a. $a(ab)^*$. **b.** $(ab)^*a$. **c.** $(ab)^*c(a + b)^*$.

5. It's easy to see that the regular expression $(a + b)^* abb$ represents the language recognized by the following NFA:

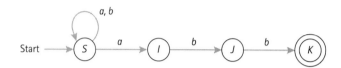

Use (11.11) to find a grammar for the language represented by the NFA.

6. Use (11.12) to construct an NFA to recognize the language of the following regular grammar.

$$S \to aI \,|\, bJ$$
$$I \to bI \,|\, \Lambda$$
$$J \to aJ \,|\, \Lambda.$$

Pumping Lemma

7. Show that each of the following languages is not regular by using the pumping lemma (11.13).

a. $\{a^n ba^n \mid n \in \mathbb{N}\}$.
b. $\{w \mid w \in \{a, b\}^*$ and w is a palindrome of even length$\}$.
c. $\{a^n b^k \mid n, k \in \mathbb{N}$ and $n \le k\}$.
d. $\{a^n b^k \mid n, k \in \mathbb{N}$ and $n \ge k\}$.
e. $\{w \mid w \in \{a, b\}^*$ and w has an equal number of a's and b's$\}$.
f. $\{a^p \mid p$ is a prime number$\}$.

Challenges

8. Prove that the intersection of two regular languages is regular.

9. Let $f : A^* \to A^*$ be a language morphism (i.e., $f(\Lambda) = \Lambda$ and $f(uv) = f(u)f(v)$ for all strings u and v), and let L be a regular language over A. Prove that $f^{-1}(L)$ is regular. *Hint:* Construct a DFA for $f^{-1}(L)$ from a DFA for L.

10. Show that the language $\{a^n ba^n \mid n \in \mathbb{N}\}$ is not regular by performing the following tasks.

a. Given the morphism $f : \{a, b, c\}^* \to \{a, b, c\}^*$ defined by

$$f(a) = a, \quad f(b) = b, \quad \text{and} \quad f(c) = a,$$

describe $f^{-1}(\{a^n b a^n \mid n \in \mathbb{N}\})$.

b. Show that

$$f^{-1}\left(\{a^n b a^n \mid n \in \mathbb{N}\}\right) \cap \{a^m b c^n \mid m, n \in \mathbb{N}\} = \{a^n b c^n \mid n \in \mathbb{N}\}.$$

c. Define a morphism $g : \{a, b, c\}^* \to \{a, b, c\}^*$ such that

$$g\left(\{a^n b c^n \mid n \in \mathbb{N}\}\right) = \{a^n b^n \mid n \in \mathbb{N}\}.$$

d. Argue that $\{a^n b a^n \mid n \in \mathbb{N}\}$ is not regular by using Parts (a), (b), and (c) together with (11.14) and (11.15).

11.5 Chapter Summary

This chapter introduced several formulations for regular languages. Regular expressions are algebraic representations of regular languages. Finite automata— DFAs and NFAs—are machines that recognize regular languages. Regular grammars derive regular languages. The most important point is that there are algorithms to transform from any one of these formulations into any other one. In other words, each formulation is equivalent to any of the other formulations.

There is also an algorithm to transform any DFA into a minimum-state DFA. Therefore, we can start with a regular language as either a regular expression or a regular grammar and automatically transform it into a minimum-state DFA to recognize the regular language.

We also observed some other things. Finite automata can be used as output devices—Mealy and Moore machines. There are some very simple interpreters for DFAs and NFAs. There are some basic properties of regular languages given by the pumping lemma, set operations, and morphisms. Many simple languages are not regular. For example, the fact that the language $\{a^n b^n \mid n \geq 0\}$ is not regular tells us that finite automata can't recognize some simple programming constructs such as nested begin-end pairs and nested open and closed parentheses.

chapter 12

Context–Free Languages and Pushdown Automata

If it keeps up, man will atrophy all his limbs but
the pushbutton finger.

—Frank Lloyd Wright (1869–1959)

Can a machine recognize a language that is not regular? The answer is yes for many languages that we encounter. In this chapter we'll study context-free languages together with the machines, called pushdown automata, that recognize them. We'll also look at some classical parsing methods and some general properties of context-free languages.

chapter guide

12.1 Context-Free Languages

In Chapter 11 we studied the class of regular languages and their representations via regular expressions, regular grammars, and finite automata. We also noticed that not all languages are regular. So it's time again to consider the recognition problem and find out whether we can solve it for a larger class of languages.

We know that there are nonregular languages. In Example 11.22 we showed that the following language is not regular:

$$\{a^n b^n \mid n \geq 0\}. \tag{12.1}$$

Therefore, we can't describe this language by any of the four representations of regular languages: regular expressions, DFAs, NFAs, and regular grammars.

Language (12.1) can be easily described by the nonregular grammar:

$$S \to \Lambda \mid aSb. \tag{12.2}$$

This grammar is an example of a more general kind of grammar, which we'll now define.

Definition of Context-Free

A *context-free grammar* is a grammar whose productions are of the form

$$S \to w,$$

where S is a nonterminal and w is any string over the alphabet of terminals and nonterminals.

For example, the grammar (12.2) is context-free. Also, any regular grammar is context-free. A language is *context-free* if it is generated by a context-free grammar. So language (12.1) is a context-free language. Regular languages are context-free. On the other hand, we know language (12.1) is context-free but not regular. Therefore, the set of all regular languages is a proper subset of the set of all context-free languages.

The term "context-free" comes from the requirement that all productions contain a single nonterminal on the left. When this is the case, any production $S \to w$ can be used in a derivation without regard to the "context" in which S appears. For example, we can use this rule to make the following derivation step:

$$aS \Rightarrow aw.$$

A grammar that is not context-free must contain a production whose left side is a string of two or more symbols. For example, the production $Sc \to w$ is not part of any context-free grammar. A derivation that uses this production can replace the nonterminal S only in a "context" that has c on the right. For example, we can use this rule to make the following derivation step:

$$aSc \Rightarrow aw.$$

We'll see some examples of languages that are not context-free later.

Most programming languages are context-free. For example, a grammar for some typical statements in an imperative language might look like the following, where the words in boldface are considered to be single terminals:

$$S \rightarrow \textbf{while } E \textbf{ do } S \,|\, \textbf{if } E \textbf{ then } S \textbf{ else } S \,|\, \{SL\} \,|\, I := E$$
$$L \rightarrow \, ; SL \,|\, \Lambda$$
$$E \rightarrow \ldots \quad \text{(description of an expression)}$$
$$I \rightarrow \ldots \quad \text{(description of an identifier)}.$$

We can combine context-free languages by union, language product, and closure to form new context-free languages. This follows from (3.16), which we'll reproduce here in terms of context-free languages.

Combining Context-Free Languages   (12.3)

Suppose M and N are context-free languages whose grammars have disjoint sets of nonterminals (rename them if necessary). Suppose also that the start symbols for the grammars of M and N are A and B, respectively. Then we have the following new languages and grammars:

1. The language $M \cup N$ is context-free, and its grammar starts with the two productions

$$S \rightarrow A \,|\, B.$$

2. The language MN is context-free, and its grammar starts with the production

$$S \rightarrow AB.$$

3. The language M^* is context-free, and its grammar starts with the two productions

$$S \rightarrow \Lambda \,|\, AS.$$

Now let's get back to our main topic of discussion. Since there are context-free languages that aren't regular and thus can't be recognized by DFAs and NFAs, we have a natural question to ask: Are there other kinds of automata that will recognize context-free languages? The answer is yes! We'll discuss them in the next section.

◣ Exercises

1. Find a context-free grammar for each of the following languages over the alphabet $\{a, b\}$.

   a. $\{a^n b^{2n} \mid n \geq 0\}$.

   b. $\{a^n b^{n+2} \mid n \geq 0\}$.

   c. The palindromes of even length.

   d. The palindromes of odd length.

   e. All palindromes.

   f. All strings with the same number of a's and b's.

2. Find a context-free grammar for each of the following languages.

   a. $\{a^n b^n \mid n \geq 0\} \cup \{a^n b^{2n} \mid n \geq 0\}$.

   b. $\{a^n b^n \mid n \geq 0\}\{a^n b^{2n} \mid n \geq 0\}$.

   c. $\{a^n b^n \mid n \geq 0\}^*$.

   d. $\{a^n b^m \mid n \geq m \geq 0\}$.

12.2 Pushdown Automata

From an informal point of view, a *pushdown automaton* is a finite automaton with a stack. A *stack* is a structure with the LIFO property of last in, first out. In other words, the last element put into a stack is the first element taken out. There is one start state and there is a—possibly empty—set of final states. We can imagine a pushdown automaton as a machine with the ability to read the letters of an input string, perform stack operations, and make state changes. We'll let PDA stand for pushdown automaton.

The Execution of a PDA

The execution of a PDA always begins with one symbol on the stack. So we must observe the following:

     Always specify the initial symbol on the stack.

We could eliminate this specification by simply assuming that a PDA always begins execution with a particular symbol on the stack, but we'll designate whatever symbol we please as the starting stack symbol. A PDA will use three stack operations as follows:

   The *pop* operation reads the top symbol and removes it from the stack.

   The *push* operation writes a designated symbol onto the top of the stack. For example, push(X) means put X on top of the stack.

   The *nop* operation does nothing to the stack.

We can represent a pushdown automaton as a finite directed graph in which each state (i.e., node) emits zero or more labeled edges. Each edge from state i to state j is labeled with three items as shown in the following diagram, where L is either a letter of an alphabet or Λ, S is a stack symbol, and O is the stack operation to be performed:

$$
\begin{array}{ccc}
& \dfrac{L,\,S}{O} & \\
(i) & \xrightarrow{\hspace{3cm}} & (j)
\end{array}
$$

Since it takes five pieces of information to describe a labeled edge, we'll also represent it by the following 5-tuple, which is called a PDA *instruction:*

$$(i, L, S, O, j).$$

An instruction of this form is executed as follows, where w is an input string whose letters are scanned from left to right:

> If the PDA is in state i, and either L is the current letter of w being scanned or $L = \Lambda$, and the symbol on top of the stack is S, then perform the following actions: (1) execute the stack operation O; (2) move to state j; and (3) if $L \neq \Lambda$, then scan right to the next letter of w (i.e., consume the current letter of w).

A string is *accepted* by a PDA if there is some path (i.e., sequence of instructions) from the start state to a final state that consumes all the letters of the string. Otherwise, the string is *rejected* by the PDA. The *language* of a PDA is the set of strings that it accepts.

Nondeterminism

A PDA is *deterministic* if there is at most one move possible from each state. Otherwise, the PDA is *nondeterministic*. There are two types of nondeterminism that may occur. One kind of nondeterminism occurs when a state emits two or more edges labeled with the same input symbol and the same stack symbol. In other words, there are two 5-tuples with the same first three components. For example, the following two 5-tuples represent nondeterminism:

$$(i, b, C, \text{ pop}, j),$$

$$(i, b, C, \text{ push}(D), k).$$

The second kind of nondeterminism occurs when a state emits two edges labeled with the same stack symbol, where one input symbol is Λ and the other input symbol is not. For example, the following two 5-tuples represent nondeterminism because the machine has the option of consuming the input letter b or leaving it alone:

$$(i, \Lambda, C, \text{ pop}, j),$$

$$(i, b, C, \text{ push}(D), k).$$

We will always use the designation PDA to mean a pushdown automaton that may be either deterministic or nondeterministic.

Representing a Computation

Before we do an example, let's discuss a way to represent the computation of a PDA. We'll represent a computation as a sequence of 3-tuples of the following form:

$$\text{(current state, unconsumed input, stack contents)}.$$

Such a 3-tuple is called an *instantaneous description*, or ID for short. For example, the ID

$$(i,\ abc,\ XYZW)$$

means that the PDA is in state i, reading the letter a, where X is at the top of the stack. Let's do an example.

example **12.1 A Sample PDA**

The language $\{a^n b^n \mid n \geq 0\}$ can be accepted by a PDA. We'll keep track of the number of a's in an input string by pushing the symbol Y onto the stack for each a. A second state will be used to pop the stack for each b encountered. The following PDA will do the job, where X is the initial symbol on the stack.

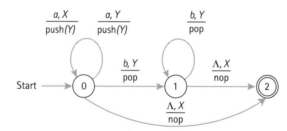

This PDA can be represented by the following six instructions:

$$(0, \Lambda, X,\ \text{nop}, 2),$$

$$(0, a, X,\ \text{push}\,(Y), 0),$$

$$(0, a, Y,\ \text{push}\,(Y), 0),$$

$$(0, b, Y,\ \text{pop}, 1),$$

$$(1, b, Y,\ \text{pop}, 1),$$

$$(1, \Lambda, X,\ \text{nop}, 2).$$

This PDA is nondeterministic because either of the first two instructions in the list can be executed if the first input letter is a and X is on top of the stack.

Let's see how a computation proceeds. For example, a computation sequence for the input string *aabb* can be written as follows:

$(0, aabb, X)$	Start in state 0 with X on the stack.
$(0, abb, YX)$	Consume a and push Y.
$(0, bb, YYX)$	Consume a and push Y.
$(1, b, YX)$	Consume b and pop.
$(1, \Lambda, X)$	Consume b and pop.
$(2, \Lambda, X)$	Move to final state.

12.2.1 Equivalent Forms of Acceptance

We defined acceptance of a string by a PDA in terms of final-state acceptance. That is, a string is accepted if it has been consumed and the PDA is in a final state. But there is an alternative definition of acceptance called *empty-stack acceptance*, which requires the input string to be consumed and the stack to be empty, with no requirement that the machine be in any particular state. These definitions of acceptance are equivalent. In other words, the class of languages accepted by PDAs that use empty-stack acceptance is the same class of languages accepted by PDAs that use final-state acceptance.

example 12.2 An Empty–Stack PDA

Let's consider the language $\{a^n b^n \mid n \geq 0\}$. The PDA that follows will accept this language by empty stack, where X is the initial symbol on the stack:

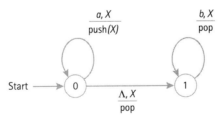

This PDA can be represented by the following three instructions:

$$(0, a, X, \ \text{push}(X), 0),$$

$$(0, \Lambda, X, \ \text{pop}, 1),$$

$$(1, b, X, \ \text{pop}, 1).$$

This PDA is nondeterministic. Can you see why? Let's see how a computation proceeds. For example, a computation sequence for the input string *aabb* can be written as follows:

$(0, aabb, X)$	Start in state 0 with X on the stack.
$(0, abb, XX)$	Consume a and push X.
$(0, bb, XXX)$	Consume a and push X.
$(1, bb, XX)$	Pop.
$(1, b, X)$	Consume b and pop.
$(1, \Lambda, \Lambda)$	Consume b and pop (stack is empty).

Equivalence of Acceptance by Final State and Empty Stack

Acceptance by final state is more common than acceptance by empty stack. But we need to consider empty-stack acceptance when we discuss why the context-free languages are exactly the class of languages accepted by PDAs. So let's convince ourselves that we get the same class of languages with either type of acceptance. We'll give two algorithms. One algorithm transforms a final-state acceptance PDA into an empty-stack acceptance PDA, and the second algorithm does the reverse, where both PDAs accept the same language.

From Final State to Empty Stack

We'll start with an algorithm to transform a PDA that accepts by final state into an empty-stack-accepting PDA. The idea is to create a new empty-stack state that can be entered from any final state of the given PDA without consuming any input. Then the computation simply empties the stack. We can do this by creating a new start state with a new stack symbol Y. Then add a Λ edge from the new start state to the old start state that pushes the old start stack symbol X onto the stack. Here is the algorithm to construct an empty-stack PDA from a final-state PDA.

Transforming a Final–State PDA into an Empty–Stack PDA   (12.4)

1. Create a new start state s, a new "empty stack" state e, and a new stack symbol Y that is at the top of the stack when the new PDA starts its execution.

2. Connect the new start state to the old start state by an edge labeled with the following expression, where X is the starting stack symbol for the given PDA:

$$\frac{\Lambda, Y}{\text{push}\,(X)}.$$

Continued ➡

3. Connect each final state to the new "empty stack" state e with one edge for each stack symbol. Label the edges with the expressions of the following form, where Z denotes any stack symbol, including Y:

$$\frac{\Lambda, Z}{\text{pop}}.$$

4. Add new edges from e to e labeled with the same expressions that are described in Step 3.

We can observe from the algorithm that *if the final-state PDA is deterministic, then the empty-stack PDA might be nondeterministic.*

example **12.3 From Final State to Empty Stack**

A deterministic PDA to accept the little language $\{\Lambda, a\}$ by final state is given as follows, where X is the initial stack symbol:

After applying algorithm (12.4) to this PDA, we obtain the following PDA, which accepts $\{\Lambda, a\}$ by empty stack, where Y is the initial stack symbol:

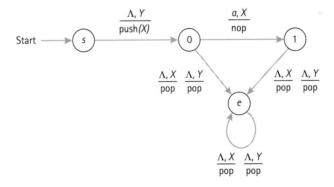

We should observe that this PDA is nondeterministic even though the given PDA is deterministic.

end example

As the example shows, we don't always get pretty-looking results. Sometimes we can come up with simpler results by using our wits. For example, the following

PDA also accepts—by empty stack—the language $\{\Lambda, a\}$, where X is the initial stack symbol:

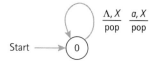

This PDA is also nondeterministic because either of the two instructions can be executed when a is the input letter. In fact, all PDAs that accept $\{\Lambda, a\}$ by empty stack must be nondeterministic. To see this, remember that a PDA must start with an initial symbol on the stack. If the letter a is the input symbol, then there must be a transition that eventually causes the stack to become empty. But if there is no input, then another transition from the same state must also cause the stack to become empty. Thus there is a nondeterministic choice at that state. This argument holds whenever the language under consideration contains Λ and at least one other string.

From Empty Stack to Final State

Now we'll discuss the other direction of our goal, which is to transform a PDA that accepts by empty stack into a final-state-accepting PDA. The idea is to create a new final state that can be entered when an empty stack occurs during the execution of the given PDA. We can do this by creating a new start state with a new stack symbol Y. Then add a Λ edge from the new start state to the old start state that pushes the old start stack symbol X onto the stack. Now an empty stack of the given PDA will be detected whenever Y appears at the top of the stack. Here is the algorithm to construct a final-state PDA from an empty-stack PDA.

Transforming an Empty-Stack PDA into a Final-State PDA   (12.5)

1. Create a new start state s, a new final state f, and a new stack symbol Y that is on top of the stack when the new PDA starts executing.

2. Connect the new start state to the old start state by an edge labeled with the following expression, where X is the starting stack symbol for the given PDA:

$$\frac{\Lambda, Y}{\text{push}\,(X)}.$$

3. Connect each state of the given PDA to the new final state f, and label each of these new edges with the expression

$$\frac{\Lambda, Y}{\text{nop}}.$$

We can also observe from this algorithm that *if the empty-stack PDA is deterministic, then the final-state PDA is deterministic*. This is easy to see because the new edges created by the algorithm are all labeled with the new stack symbol Y, which doesn't occur in the original PDA. Let's do a simple example.

example 12.4 Empty Stack to Final State

The following PDA accepts the little language $\{\Lambda\}$ by empty stack, where X is the initial stack symbol:

The algorithm creates the following PDA that accepts $\{\Lambda\}$ by final state:

end example

As the example shows, algorithm (12.5) doesn't always give the simplest results. For example, a simpler PDA to accept $\{\Lambda\}$ by final state can be written as follows:

12.2.2 Context-Free Grammars and Pushdown Automata

Now we're in the proper position to state the main result that connects context-free languages to pushdown automata.

Theorem $\hspace{8cm}$ **(12.6)**
The context-free languages are exactly the languages accepted by PDAs.

The proof of (12.6) consists of two algorithms. One algorithm transforms a context-free grammar into a PDA, and the other algorithm transforms a PDA into a context-free grammar. In each case the grammar generates the same language that the PDA accepts. Let's look at the algorithms.

Transforming a Context-Free Grammar into a PDA

We'll give here an algorithm to transform any context-free grammar into a PDA such that the PDA recognizes the same language as the grammar. For convenience we'll allow the operation field of a PDA instruction to hold a list of stack instructions. For example, the 5-tuple

$$(i, \ a, \ C, \ \langle \text{pop}, \text{push}(X), \text{push}(Y)\rangle, \ j)$$

is executed by performing the three operations

$$\text{pop}, \text{push}(X), \text{push}(Y).$$

We can implement these actions in a "normal" PDA by placing enough new symbols on the stack at the start of the computation to make sure that any sequence of pop operations will not empty the stack if it is followed by a push operation. For example, we can execute the example instruction by the following sequence of normal instructions, where k and l are new states:

$$(i, a, C, \text{pop}, k)$$
$$(k, \Lambda, ?, \text{push}(X), l) \qquad (? \text{ represents some stack symbol})$$
$$(l, \Lambda, X, \text{push}(Y), j).$$

Here's the algorithm to transform any context-free grammar into a PDA that accepts by empty stack.

Context-Free Grammar to PDA (Empty-Stack Acceptance)   (12.7)

The PDA will have a single state 0. The stack symbols will be the set of terminals and nonterminals. The initial symbol on the stack will be the grammar's start symbol. Construct the PDA instructions as follows:

1. For each terminal symbol a, create the instruction $(0, \ a, \ a, \text{pop}, 0)$.

2. For each production $A \rightarrow B_1 B_2 \ldots B_n$, where each B_i represents either a terminal or a nonterminal, create the instruction

$$(0, \ \Lambda, \ A, \ \langle \text{pop}, \text{push}(B_n), \text{push}(B_{n-1}), \ldots, \text{push}(B_1)\rangle, \ 0).$$

3. For each production $A \rightarrow \Lambda$, create the instruction $(0, \ \Lambda, \ A, \text{pop}, 0)$.

The PDA built by the algorithm accepts the language of the grammar because each state transition of the PDA corresponds exactly to one derivation step in a derivation. Let's do an example to get the idea.

example   **12.5 Context-Free Grammar to PDA**

Let's consider the following context-free grammar for $\{a^n b^n \mid n \geq 0\}$:

$$S \rightarrow aSb \mid \Lambda.$$

We can apply algorithm (12.7) to this grammar to construct a PDA. From the terminals a and b, we'll use rule 1 to create the two instructions:

$$(0, a, a, \text{ pop, } 0),$$

$$(0, b, b, \text{ pop, } 0).$$

From the production $S \to \Lambda$ we'll use rule 3 to create the instruction

$$(0, \Lambda, S, \text{pop, } 0).$$

From the production $S \to aSb$, we'll use rule 2 to create the instruction

$$(0, \Lambda, S, \langle \text{pop, push}(b), \text{push}(S), \text{push}(a)\rangle, 0).$$

We'll write down the PDA computation sequence for the input string $aabb$:

ID	PDA Instruction to Obtain ID
$(0, aabb, S)$	Initial ID
$(0, aabb, aSb)$	$(0, \Lambda, S, \langle \text{pop, push}(b), \text{push}(S), \text{push}(a)\rangle, 0)$
$(0, abb, Sb)$	$(0, a, a, \text{ pop, } 0)$
$(0, abb, aSbb)$	$(0, \Lambda, S, \langle \text{pop, push}(b), \text{push}(S), \text{push}(a)\rangle, 0)$
$(0, bb, Sbb)$	$(0, a, a, \text{ pop, } 0)$
$(0, bb, bb)$	$(0, \Lambda, S, \text{ pop, } 0)$
$(0, b, b)$	$(0, b, b, \text{ pop, } 0)$
$(0, \Lambda, \Lambda)$	$(0, b, b, \text{ pop, } 0)$

See whether you can tell which steps of this computation correspond to the steps in the following derivation of $aabb$:

$$S \Rightarrow aSb \Rightarrow aaSbb \Rightarrow aabb.$$

end example

Transforming a PDA into a Context-Free Grammar

Now let's go in the other direction and transform any PDA into a context-free grammar that accepts the same language. We will assume that the PDA accepts strings by empty stack. The idea is to construct a grammar so that a leftmost derivation for a string w corresponds to a computation sequence of the PDA that accepts w. To do this, we'll define the nonterminals of the grammar in terms of the stack symbols of the PDA. Here's the algorithm:

PDA (Empty-Stack Acceptance) to Context-Free Grammar   **(12.8)**

For each stack symbol B and each pair of states i and j of the PDA, we construct a nonterminal of the grammar and denote it by B_{ij}. We can think of B_{ij} as deriving all strings that cause the PDA to move, in one or more steps, from state i to state j in such a way that the stack at state j is obtained

---

Continued ➡

from the stack at state i by popping B. We create one additional nonterminal S to denote the start symbol for the grammar.

1. For each state j of the PDA, construct a production of the following form, where s is the start state and E is the starting stack symbol:

$$S \to E_{sj}.$$

2. For each instruction of the form (p, a, B, pop, q), construct a production of the following form:

$$B_{pq} \to a.$$

3. For each instruction of the form (p, a, B, nop, q), construct a production of the following form for each state j:

$$B_{pj} \to aB_{qj}.$$

4. For each instruction of the form $(p, a, B, \text{push}(C), q)$ construct productions of the following form for all states i and j:

$$B_{pj} \to aC_{qi}B_{ij}.$$

Note: This algorithm normally produces many useless productions that can't derive terminal strings. For example, if a nonterminal occurs on the right side of a production but not on the left side of any production, then the production can't derive a string of terminals. Similarly, if a recursive production doesn't have a basis case, then it can't derive a terminal string. We can safely discard these productions. Let's do an example to get the idea.

example 12.6 PDA to Context-Free Grammar

The following PDA accepts the language $\{a^n b^{n+2} \mid n \geq 1\}$ by empty stack, where X is the starting stack symbol:

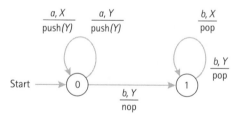

We can apply algorithm (12.8) to this PDA to construct the following context-free grammar; we've omitted the productions that can't derive terminal strings:

$$S \to X_{01}$$
$$X_{01} \to aY_{01}X_{11}$$
$$Y_{01} \to aY_{01}Y_{11} \mid bY_{11}$$
$$X_{11} \to b$$
$$Y_{11} \to b.$$

We'll do a sample leftmost derivation of the string $aabbbb$ as follows:

$$S \Rightarrow X_{01} \Rightarrow aY_{01}X_{11} \Rightarrow aaY_{01}Y_{11}X_{11} \Rightarrow aabY_{11}Y_{11}X_{11}$$
$$\Rightarrow aabbY_{11}X_{11} \Rightarrow aabbbX_{11} \Rightarrow aabbbb.$$

We should note from this example that the algorithm doesn't always construct the nicest-looking grammar. For example, the constructed grammar can be transformed into the following grammar for $\{a^n b^{n+2} \mid n \geq 1\}$:

$$S \to aBb$$
$$B \to aBb \mid bb.$$

end example

Nondeterministic PDAs Are More Powerful

Recall that DFAs accept the same class of languages as NFAs, namely, the regular languages. We know by (12.6) that the context-free languages coincide with the languages accepted by PDAs. But we haven't said anything about determinism versus nondeterminism with respect to PDAs. In fact, there are some context-free languages that can't be recognized by any deterministic PDA. We'll state the result for the record.

Theorem   **(12.9)**

There are some context-free languages that are accepted only by nondeterministic PDAs.

Although we won't prove (12.9), we'll give an indication of the kind of property that requires nondeterminism. For example, the language of even palindromes over $\{a, b\}$ can be generated by the following context-free grammar:

$$S \to aSa \mid bSb \mid \Lambda.$$

So by (12.6) the even palindromes can be recognized by a PDA. But this language can't be recognized by any deterministic PDA. In other words, every PDA to accept the even palindromes over $\{a, b\}$ must be nondeterministic. To see why this is the case, notice that we can describe the even palindromes over $\{a, b\}$ as follows, where w^R is the reverse of w:

$$\{ww^R \mid w \in \{a, b\}^*\}.$$

We can recognize a string of the form ww^R by stacking the letters of w and then unstacking the letters of w^R. But a deterministic PDA cannot tell when the middle of such an arbitrary string has been reached because it looks at one letter at a time from one end of the string. This is the crux of showing why no deterministic PDA can recognize the even palindromes over $\{a, b\}$.

Since we've been discussing the fact that even palindromes over $\{a, b\}$ can be recognized only by nondeterministic PDAs, let's give an example.

example **12.7 A PDA for Even Palindromes**

We'll find a PDA—necessarily nondeterministic—for the even palindromes over $\{a, b\}$. The start state is 0, the final state is 2, and X is the initial stack symbol. To simplify things, we'll let "?" stand for any stack symbol. With these assumptions we can write the following instructions for the PDA:

$$(0, a, ?, \text{push}(a), 0), \quad \text{(push string } w \text{ on the stack)}$$
$$(0, b, ?, \text{push}(b), 0),$$
$$(0, \Lambda, ?, \text{nop}, 1),$$
$$(1, a, a, \text{pop}, 1), \qquad \text{(pop string } w^R \text{ off the stack)}$$
$$(1, b, b, \text{pop}, 1),$$
$$(1, \Lambda, X, \text{nop}, 2).$$

How many instructions does the PDA have if we don't allow question marks? For practice, draw the graphical version of the PDA.

end example

12.2.3 Representing and Executing Pushdown Automata

When we talk about PDAs, it's sometimes convenient to represent them by listing seven things: the set of states, the alphabet, the stack alphabet, the instructions, the starting stack symbol, the start state, and the set of final states. Traditionally, these seven items are listed as a 7-tuple. For example, if someone says, "Let (S, A, B, I, E, s, F) be a PDA," then we can assume that S is the set of states, A is the alphabet, B is the set of stack symbols, I is the instruction set, E is the starting stack symbol, s is the start state, and F is the set of final states.

We can also represent any PDA as an algebraic structure, which will help us discover a general algorithm for a PDA interpreter. To start things off, suppose we have a PDA represented as the 7-tuple (S, A, B, I, E, s, F). Assume that we also have an algebra of stacks, where "Stacks" is the set of stacks whose elements are from the set of stack symbols B and "StkCalls" is the set of stack operations. We can describe the algebra for a PDA as follows:

Carriers: S, A, A^*, B, Stacks, StkCalls, Boolean.

Operations: $s \in S$,

$F \subseteq S$,

$I \subseteq S \times A \cup \{\Lambda\} \times B \times$ StkCalls $\times S$,

path: $S \times A^* \times$ Stacks $\to$ Boolean,

accept: $A^* \to$ Boolean.

Axioms: To simplify the presentation of the axioms, we'll represent a stack as a list of the form $X :: Y$, where X is the top of the stack. If we have an instruction of the form $(?, ?, X, O, ?)$, then newStack denotes the stack obtained by applying operation O to the stack $X :: Y$. With these assumptions we can write the axioms as follows, where w is an input string and E is the starting stack symbol:

$\text{accept}(w) = \text{path}(s, w, \langle E \rangle)$.

$\text{path}(k, \Lambda, X :: Y) = $ **if** k is a final state **then** True
else if there is an instruction
(k, Λ, X, O, n)
then path$(n, \Lambda,$ new Stack$)$ **else** False.

$\text{path}(k, a \cdot t, X :: Y) = $ **if** there is an instruction (k, a, X, O, n)
and path$(n, t,$ newStack$) =$ True
or there is an instruction (k, Λ, X, O, n)
and path$(n, a \cdot t,$ newStack$) =$ True
then True **else** False.

A Logic Program Interpreter for PDAs

The axioms give us the basic information we need to build an interpreter that executes PDAs. We'll construct a logic program version. To start things off we need to agree on the representation of a PDA. We'll write a PDA as a set of logic program facts taking one of the following forms:

t(state, letter, top, operation, nextState).

state(state).

final(state).

To write a simple interpreter for PDAs, we'll make a few assumptions. The stack will be a list that is initialized with the value $\langle e \rangle$, which means that e is always the starting stack symbol. We'll reserve the letters p and n for the operations pop and nop, and we'll agree to let the push instruction be represented by the symbol that is to be pushed. For example, in the instruction

$$t(0,\ a,\ e,\ b,\ 1)$$

the letter b means push b.

To keep the interpreter simple, we'll represent strings as lists. So the empty string Λ will be represented by the empty list $\langle\ \rangle$.

example **12.8   A PDA as Logic Program Data**

For example, a PDA to recognize the language $\{a^n b^n \mid n \geq 0\}$ can be written as the following set of facts:

$$\text{start}\,(0)\,.$$
$$t\,(0,a,e,a,0)\,.$$
$$t\,(0,a,a,a,0)\,.$$
$$t\,(0,b,a,p,1)\,.$$
$$t\,(0,\langle\ \rangle,e,n,2)\,.$$
$$t\,(1,b,a,p,1)\,.$$
$$t\,(1,\langle\ \rangle,e,n,2)\,.$$
$$\text{final}\,(2)\,.$$

end example

Since we're representing strings as lists, to test whether a string such as $aabb$ is accepted, we'll write the goal

$$\leftarrow\ \text{accept}(\langle a,\ a,\ b,\ b \rangle).$$

This goal starts the execution of the PDA interpreter, so we better describe what it means.

The interpreter executes a computation sequence, where the "path" predicate represents an ID containing the current state, the current input string, and the current stack. If the input is empty and the current state is final, then the computation ends successfully. Otherwise, if the input is not empty, the computation continues by looking up an appropriate instruction.

The predicate oper(Stack, O, NewStack) means "perform stack operation O on Stack, resulting in NewStack." Recall that $H :: T$ denotes the list whose head is H and whose tail is T. Now we can write the interpreter, where S is the input string, and all variables start with uppercase letters.

Logic Program Interpreter for PDAs

$\text{accept}(S) \leftarrow \text{start}(I), \text{path}(I, S, \langle e \rangle).$

$\text{path}(K, \langle \rangle, \text{Stack}) \leftarrow \text{final}(K).$

$\text{path}(K, A :: B, H :: T) \leftarrow t(K, A, H, O, N),$

$\qquad\qquad\qquad\qquad \text{oper}(H :: T, O, \text{NewStack}),$

$\qquad\qquad\qquad\qquad \text{path}(N, B, \text{NewStack}).$

$\text{path}(K, \text{String}, H :: T) \leftarrow t(K, \langle \rangle, H, O, N),$

$\qquad\qquad\qquad\qquad \text{oper}(H :: T, O, \text{NewStack}),$

$\qquad\qquad\qquad\qquad \text{path}(N, \text{String}, \text{NewStack}).$

$\text{oper}(H :: T, p, T).$

$\text{oper}(\text{Stack}, n, \text{Stack}).$

$\text{oper}(\text{Stack}, A, A :: \text{Stack}).$

This interpreter can be easily translated into the Prolog language with three simple notational changes: Use brackets [and] to represent lists; replace $\leftarrow$ with ":–"; and replace "$X :: Y$" with $[X | Y]$.

 Exercises

Pushdown Automata

1. Find a pushdown automaton for each of the following languages.
   - **a.** $\{ab^n cd^n \mid n \geq 0\}$.
   - b. All strings over $\{a, b\}$ with the same number of a's and b's.
   - **c.** $\{wcw^R \mid w \in \{a, b\}^*\}$.
   - d. The palindromes of odd length over $\{a, b\}$.
   - **e.** $\{a^n b^{n+2} \mid n \geq 0\}$.

2. Find a single state PDA to recognize the language $\{a^n b^m \mid n, m \in \mathbb{N}\}$.

3. For each of the following languages, find a deterministic PDA that accepts by final state.
   - **a.** $\{a^n b^n \mid n \geq 0\}$.
   - b. $\{a^n b^{2n} \mid n \geq 0\}$.

4. If we allow each PDA instruction to contain any finite sequence of stack operations, then we can reduce the number of states required for any PDA. Let $L = \{a^n b^n \mid n \in \mathbb{N}\}$. Find PDAs that accept L by final state with the given restrictions.

   a. A two-state PDA that contains one or more Λ instructions.
   b. A two-state PDA that does not contain any Λ instructions.

Construction Algorithms

5. Use (12.4) to transform the final-state PDA from Example 12.1 into an empty-stack PDA.

6. Use (12.5) to transform the empty-stack PDA from Example 12.2 into a final-state PDA.

7. In each of the following cases, use (12.7) to construct a PDA that accepts the language of the given grammar.

   a. $S \rightarrow c \mid aSb$.
   b. $S \rightarrow \Lambda \mid aSb \mid aaS$.

8. Use (12.8) to construct a grammar for the language of the following PDA that accepts by empty stack, where 0 is the start state and X is the initial stack symbol: $(0, a, X, \text{push}(X), 0)$, $(0, \Lambda, X, \text{pop}, 1)$, $(1, b, X, \text{pop}, 1)$.

9. Suppose we're given the following PDA that accepts by empty stack, where X is the initial stack symbol:

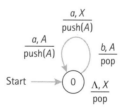

   a. Use your wits to describe the language recognized by the PDA.
   b. Use (12.8) to construct a grammar for the language of the PDA.
   c. Do your answers to Parts (a) and (b) describe the same language?

Challenge

10. Give an argument to show that the following context-free language is not accepted by any deterministic PDA: $\{a^n b^n \mid n \geq 0\} \cup \{a^n b^{2n} \mid n \geq 0\}$.

12.3 Context-Free Parsing

An important part of compiler construction for programming languages is the study of techniques to construct parsers. Virtually all programming language constructs can be represented by context-free grammars. Since the context-free languages are exactly those that can be recognized by PDAs, it follows that parsers can be constructed for these languages (i.e., a parser is a PDA). Context-free languages that are not regular don't have algebraic representations like the regular expressions that represent regular languages. Therefore, the grammar is the important factor in trying to construct parsers for programming languages.

One goal of the compiler writer is to build a parser that is efficient. If a parser is nondeterministic, then time can be wasted by backtracking to find a proper derivation path. It's nice to know that most programming language constructs have deterministic parsers. In other words, the languages are recognized by deterministic PDAs. If a context-free language can be recognized by a deterministic PDA by final state, then the language is said to be a *deterministic context-free language*. For example, the language $\{a^n b^n \mid n \in \mathbb{N}\}$ is deterministic context-free because it has a deterministic PDA that accepts by final state (Exercise 3a of Section 12.2). We'll confine our remarks to deterministic context-free languages.

When a parse tree for a string is constructed by starting at the root and proceeding downward toward the leaves, the construction is called *top-down parsing*. A top-down parser constructs a derivation by starting with the grammar's start symbol and working toward the string. Another type of parsing is *bottom-up parsing*, in which the parse tree for a string is constructed by starting with the leaves and working up to the root of the tree. A bottom-up parser constructs a derivation by starting with the string and working backward to the start symbol.

12.3.1 Top-Down Parsing and LL(k) Grammars

Many deterministic context-free languages can be parsed top-down if they can be described by a special kind of grammar called an LL(k) grammar. An LL(k) *grammar* has the property that a parser can be constructed that scans an input string from left to right and builds a leftmost derivation of the string by examining the next k symbols of the input string. In other words, the next k input symbols of a string are enough to determine the unique production to be used at each step of the derivation. The next k symbols of the input string are often called *lookahead symbols*. LL(k) grammars were introduced by Lewis and Stearns [1968]. The first letter L stands for the left-to-right scan of input, and the second letter L stands for the leftmost derivation.

It's often quite easy to inspect a grammar and determine whether it's an LL(k) grammar for some k. So we'll spend a little time discussing LL(k) grammars. Let's get our feet wet with some examples.

example **12.9**   An LL(1) Grammar

Let's consider the following language:

$$\{a^n b c^n \mid n \in \mathbb{N}\}.$$

A grammar for this language can be written as follows:

$$S \to aSc \mid b.$$

This grammar is LL(1) because the right sides of the two S productions begin with distinct letters a and b. Therefore, each step of a leftmost derivation is uniquely determined by examining the current input symbol (i.e., one lookahead symbol). In other words, if the lookahead symbol is a, then the production $S \to aSc$ is used; if the lookahead symbol is b, then the production $S \to b$ is used. For example, the derivation of the string

$$aabcc$$

can be constructed as follows, where we've written a reason for each step:

$$S \Rightarrow aSc \qquad \text{Use } S \to aSc \text{ because } aabcc \text{ begins with } a.$$
$$\Rightarrow aaScc \qquad \text{Use } S \to aSc \text{ because } abcc \text{ begins with } a.$$
$$\Rightarrow aabcc \qquad \text{Use } S \to b \text{ because } bcc \text{ begins with } b.$$

This derivation is a leftmost derivation by default because there is only one nonterminal to replace in each *sentential form* (i.e., string of terminals and/or nonterminals).

end example

We can use the same reasoning as in the example to show that the following grammar for $\{a^n b^n \mid n \in \mathbb{N}\}$ is LL(1):

$$S \to aSb \mid \Lambda.$$

Just notice that a string in the language either begins with the letter a or is the empty string. Thus if the lookahead symbol is a, then the production $S \to aSb$ is used; if the lookahead symbol signals the end of a string, then the production $S \to \Lambda$ is used.

example **12.10**   An LL(2) Grammar

Let's consider a grammar for the following language:

$$\{a^m b^n c \mid m \geq 1 \text{ and } n \geq 0\}.$$

A grammar for this language can be written as follows:

$$S \to AB \qquad\qquad\qquad\qquad (12.10)$$
$$A \to aA \mid a$$
$$B \to bB \mid c.$$

This grammar is not LL(1) because the right sides of the two A productions begin with the same letter a. For example, the first letter of the string abc is not enough information to choose the correct A production to continue the following leftmost derivation:

$$S \Rightarrow AB \qquad \text{No other choice.}$$
$$\Rightarrow ? \qquad \text{Don't know which } A \text{ production to choose.}$$

After some thought we can see that (12.10) is LL(2) because a string starting with aa causes the production $A \to aA$ to be chosen and a string starting with either ab or ac forces the production $A \to a$ to be chosen. For example, we'll construct a leftmost derivation of the string $aabbc$:

$$S \Rightarrow AB \qquad \text{No other choice.}$$
$$\Rightarrow aAB \qquad \text{Use } A \to aA \text{ because } aabbc \text{ begins with } aa.$$
$$\Rightarrow aaB \qquad \text{Use } A \to a \text{ because } abbc \text{ begins with } ab.$$
$$\Rightarrow aabB \qquad \text{Use } B \to bB \text{ because } bbc \text{ begins with } b.$$
$$\Rightarrow aabbB \qquad \text{Use } B \to bB \text{ because } bc \text{ begins with } b.$$
$$\Rightarrow aabbc \qquad \text{Use } B \to c \text{ because } c \text{ begins with } c.$$

end example

Suppose we have a grammar that contains two productions—where one or both right sides begin with nonterminals—like the two S productions in the following grammar:

$$S \to A \mid B$$
$$A \to aA \mid \Lambda$$
$$B \to bB \mid c.$$

Is this an LL(k) grammar? The answer is yes. In fact it's an LL(1) grammar. The A and B productions are clearly LL(1). The only problem is to figure out which S production should be chosen to start a derivation. If the first letter of the input string is a or if the input string is empty, we use production $S \to A$. Otherwise, if the first letter is b or c, then we use the production $S \to B$. In either case, all we need is one lookahead symbol. So we might have to chase through a few productions to check the LL(k) property. Most programming constructs can be described by LL(1) grammars that are easy to check.

Grammar Transformations

Just because we write down an LL(k) grammar for some language doesn't mean we've found an LL grammar with the smallest such k. For example, grammar (12.10) is an LL(2) grammar for $\{a^m b^n c \mid m \geq 1 \text{ and } n \geq 0\}$. But it's easy to see that the following grammar is an LL(1) grammar for this language:

$$S \rightarrow aAB$$
$$A \rightarrow aA \mid \Lambda$$
$$B \rightarrow bB \mid c.$$

Left-Factoring

Sometimes it's possible to transform an LL(k) grammar into an LL(n) grammar for some $n < k$ by a process called *left-factoring*. An example should suffice to describe the process. Suppose we're given the following LL(3) grammar fragment:

$$S \rightarrow abcC \mid abdD.$$

Since the two right sides have the common prefix ab, we can "factor out" the string ab to obtain the following equivalent productions, where B is a new non-terminal:

$$S \rightarrow abB$$
$$B \rightarrow cC \mid dD.$$

This grammar fragment is LL(1). So an LL(3) grammar fragment has been transformed into an LL(1) grammar fragment by left-factoring.

Removing Left Recursion

Sometimes we can transform a grammar that is not LL(k) into an LL(k) grammar for the same language. A grammar is *left-recursive* if, for some nonterminal A, there is a derivation of the form $A \Rightarrow \cdots \Rightarrow Aw$ for some nonempty string w. An LL(k) grammar can't be left-recursive because there is no way to tell how many times a left-recursive derivation may need to be repeated before an alternative production is chosen to stop the recursion. Here's an example of a grammar that is not LL(k) for any k.

example   **12.11   A Non-LL(k) Grammar**

The language $\{ba^n \mid n \in \mathbb{N}\}$ has the following left-recursive grammar:

$$A \rightarrow Aa \mid b.$$

We can see that this grammar is not LL(1) because if the string ba is input, then the first letter b is not enough to determine which production to use to start the leftmost derivation of ba.

The grammar is not LL(2) because if the input string is baa, then the first two-letter string ba is enough to start the derivation of baa with the production $A \rightarrow Aa$. But the letter b of the input string can't be consumed because it doesn't occur at the left of Aa. Thus the same two-letter string ba must determine the next step of the derivation, causing $A \to Aa$ to be chosen. This goes on forever, obtaining an infinite derivation. The same idea can be used to show that the grammar is not LL(k) for any k.

end example

Sometimes we can remove the left recursion from a grammar and the resulting grammar is an LL(k) grammar for the same language. A simple form of left recursion that occurs frequently is called *immediate* left recursion. This type of recursion occurs when the grammar contains a production of the form $A \to Aw$. In this case there must be at least one other A production to stop the recursion. Thus the simplest form of immediate left recursion takes the following form, where w and y are nonempty strings and y does not begin with A:

$$A \to Aw \,|\, y.$$

Notice that any string derived from A starts with y and is followed by any number of w's. We can use this observation to remove the left recursion by replacing the two A productions with the following productions, where B is a new nonterminal:

$$A \to yB$$
$$B \to wB \,|\, \Lambda.$$

But there may be more than one A production that is left-recursive. Here is a general method for removing immediate left recursion. Suppose that we have the following left-recursive A productions, where x_i and w_j denote arbitrary nonempty strings and no x_i begins with A:

$$A \to Aw_1 \,|\, \ldots \,|\, Aw_n \,|\, x_1 \,|\, \ldots \,|\, x_m.$$

It's easy to remove this immediate left recursion. Notice that any string derived from A must start with x_i for some i and is followed by any number and combination of w_j's. So we can replace the A productions by the following productions, where B is a new nonterminal:

$$A \to x_1 B \,|\, \ldots \,|\, x_m B$$
$$B \to w_1 B \,|\, \ldots \,|\, w_n B \,|\, \Lambda.$$

This grammar may or may not be LL(k). It depends on the value of the strings x_i and w_j. For example, if they all are single distinct terminals, then the grammar is LL(1). Here are two examples.

12.12 Removing Left Recursion

Let's look again at the language $\{ba^n \mid n \in \mathbb{N}\}$ and the following left-recursive grammar:

$$A \to Aa \mid b.$$

We saw in Example 12.11 that this grammar is not $LL(k)$ for any k. But we can remove the immediate left recursion in this grammar to obtain the following $LL(1)$ grammar for the same language:

$$A \to bB$$
$$B \to aB \mid \Lambda.$$

end example

example 12.13 Removing Left Recursion

Let's look at an example that occurs in programming languages that process arithmetic expressions. Suppose we want to parse the set of all arithmetic expressions described by the following grammar:

$$E \to E + T \mid T$$
$$T \to T * F \mid F$$
$$F \to (E) \mid a.$$

This grammar is not $LL(k)$ for any k because it's left-recursive. For example, the expression $a*a*a+a$ requires a scan of the first six symbols to determine that the first production in a derivation is $E \to E+T$.

Let's remove the immediate left recursion for the nonterminals E and T. The result is the following $LL(1)$ grammar for the same language of expressions:

$$E \to TR$$
$$R \to +TR \mid \Lambda$$
$$T \to FV$$
$$V \to *FV \mid \Lambda$$
$$F \to (E) \mid a.$$

For example, we'll construct a leftmost derivation of $(a+a)*a$. Check the $LL(1)$ property by verifying that each step of the derivation is uniquely determined by the single current input symbol:

$$E \Rightarrow TR \Rightarrow FVR \Rightarrow (E)\,VR \Rightarrow (TR)\,VR \Rightarrow (FVR)\,VR$$
$$\Rightarrow (aVR)\,VR \Rightarrow (aR)\,VR \Rightarrow (a+TR)\,VR$$
$$\Rightarrow (a+FVR)\,VR \Rightarrow (a+aVR)\,VR$$
$$\Rightarrow (a+aR)\,VR \Rightarrow (a+a)\,VR \Rightarrow (a+a) * FVR$$
$$\Rightarrow (a+a) * aVR \Rightarrow (a+a) * aR \Rightarrow (a+a) * a.$$

end example

Removing Indirect Left Recursion

The other kind of left recursion that can occur in a grammar is called *indirect* left recursion. This type of recursion occurs when at least two nonterminals are involved in the recursion. For example, the following grammar is left-recursive because it has indirect left recursion:

$$S \to Bb$$
$$B \to Sa \,|\, a.$$

To see the left recursion in this grammar, notice the following derivation:

$$S \Rightarrow Bb \Rightarrow Sab.$$

We can remove indirect left recursion from this grammar in two steps. First, replace B in the S production by the right side of the B production to obtain the following grammar:

$$S \to Sab \,|\, ab.$$

Now remove the immediate left recursion in the usual manner to obtain the following LL(1) grammar:

$$S \to abT$$
$$T \to abT \,|\, \Lambda.$$

For another example, suppose in the following grammar that we want to remove the indirect left recursion that begins with the nonterminal A:

$$A \to Bb \,|\, e$$
$$B \to Cc \,|\, f$$
$$C \to Ad \,|\, g.$$

First, replace each occurrence of B (just one in this example) in the A productions by the right sides of the B productions to obtain the following A productions:

$$A \to Ccb \,|\, fb \,|\, e.$$

Next, replace each occurrence of C (just one in this example) in these A productions by the right sides of the C productions to obtain the following A productions:

$$A \to Adcb \,|\, gcb \,|\, fb \,|\, e.$$

Lastly, remove the immediate left recursion from these A productions to obtain the following grammar:

$$A \to gcbD \,|\, fbD \,|\, eD$$
$$D \to dcbD \,|\, \Lambda.$$

This idea can be generalized to remove all left recursion in context-free grammars.

Top-Down Parsing by Recursive Descent

LL(k) grammars have top-down parsing algorithms because a leftmost derivation can be constructed by starting with the start symbol and proceeding through sentential forms until the desired string is obtained. We'll illustrate the ideas of top-down parsing with examples. One method of top-down parsing, called *recursive descent*, can be accomplished by associating a procedure with each nonterminal. The parse begins by calling the procedure associated with the start symbol.

For example, suppose we have the following LL(1) grammar fragment for two statements in a programming language:

$$S \to \text{id} = E \mid \textbf{while } E \textbf{ do } S.$$

We'll assume that any program statement can be broken down into "tokens," which are numbers that represent the syntactic objects. For example, the statement

$$\textbf{while } x < y \textbf{ do } x = x + 1$$

might be represented by the following string of tokens, which we've represented by capitalized words:

WHILE ID LESS ID DO ID EQ ID PLUS CONSTANT

To parse a program statement, we'll assume that there is a variable "lookahead" that holds the current input token. Initially, lookahead is given the value of the first token, which in our case is the WHILE token. We'll also assume that there is a procedure "match," where match(x) checks to see whether x matches the lookahead value. If a match occurs, then lookahead is given the next token value in the input string. Otherwise, an error message is produced. The match procedure can be described as follows:

> **procedure** match (x)
>   **if** lookahead $= x$ **then**
>   lookahead := next input token
>   **else**
>   error
>   **fi**

For example, if lookahead $=$ WHILE and we call match(WHILE), then a match occurs, and lookahead is given the new value ID. We'll assume that the procedure for the nonterminal E, to recognize expressions, is already written.

Now the procedure for the nonterminal S can be written as follows:

procedure S
    if lookahead = ID **then**
        match(ID);
        match(EQ);
        E
    else if lookahead = WHILE **then**
        match(WHILE);
        E;
        match(DO);
        S
    else
        error
fi

This parser is a deterministic PDA in disguise: The "match" procedure consumes an item of input; the state transitions are statements in the then and else clauses; and the stack is hidden because the procedures are recursive. In an actual implementation, procedure S would also contain output statements that could be used to construct the machine code to be generated by the compiler. Thus an actual parser is a PDA with output. A PDA with output is often called a *pushdown transducer*. So we can say that parsers are pushdown transducers.

Top-Down Parsing with a Parse Table

Another top-down parsing method uses a parse table and an explicit stack instead of the recursive descent procedures. The details of how to construct such parse tables can be found in any compiler construction text. We'll be satisfied with an example of a parse table and its use. Suppose we have the following LL(1) grammar.

$$S \rightarrow ASb \,|\, C$$
$$A \rightarrow a$$
$$C \rightarrow cC \,|\, \Lambda.$$

A parse table for this grammar is represented as follows, where a blank entry in the table indicates an error message.

	a	b	c	$
S	$S \rightarrow ASb$	$S \rightarrow C$	$S \rightarrow C$	$S \rightarrow C$
A	$A \rightarrow a$			
C		$C \rightarrow \Lambda$	$C \rightarrow cC$	$C \rightarrow \Lambda$

The symbol $ is placed on the bottom of the stack and at the right end of the input string. The parsing algorithm begins with the grammar's start symbol on top of the stack. The algorithm repeatedly examines entries in the parse table by accessing two symbols—the symbol on top of the stack and the current input symbol. A production in the table indicates that the stack should be popped and the grammar symbols on the right side of the production should be pushed onto the stack in reverse order. An input string is accepted when the top of the stack is $ and the current input symbol is $. The algorithm consists of a loop that stops when either the input string is accepted or an error is detected.

Let's do a parse of the string *aaccbb* using this table. We'll represent each step of the parse by a line containing the stack contents and the unconsumed input, where the top of the stack is at the right end of the stack string and the current input symbol is at the left end of the input string. The third column of each line contains the actions to perform to obtain the next line, where consume means get the next input symbol.

Stack	Input	Actions to Perform
$S	aaccbb$	Pop, push b, push S, push A
$bSA	aaccbb$	Pop, push a
$bSa	aaccbb$	Pop, consume
$bS	accbb$	Pop, push b, push S, push A
$bbSA	accbb$	Pop, push a
$bbSa	accbb$	Pop, consume
$bbS	ccbb$	Pop, push C
$bbC	ccbb$	Pop, push C, push c
$bbCc	ccbb$	Pop, consume
$bbC	cbb$	Pop, push C, push c
$bbCc	cbb$	Pop, consume
$bbC	bb$	Pop
$bb	bb$	Pop, consume
$b	b$	Pop, consume
$	$	Accept

LL(*k*) Facts

An important result about LL(k) grammars is that they describe a proper hierarchy of languages. In other words, for any $k \in \mathbb{N}$ there is a proper containment of languages.

Theorem

LL(k) languages $\subset$ LL($k + 1$) languages.

This result is due to Kurki-Suonio [1969]. In particular, Kurki-Suonio showed that for each $k > 1$ the following grammar is an LL(k) grammar whose language has no LL($k - 1$) grammar:

$$S \to aSA \mid \Lambda$$
$$A \to a^{k-1}bS \mid c.$$

We should emphasize that the LL(k) grammars can't describe every deterministic context-free language. This result was shown by Rosenkrantz and Stearns [1970]. For example, consider the following language:

$$\{a^n \mid n \geq 0\} \cup \{a^n b^n \mid n \geq 0\}.$$

This is a classic example of a deterministic context-free language that is not LL(k) for any k. It is deterministic context-free because it can be accepted by the following deterministic PDA, where X is the starting stack symbol:

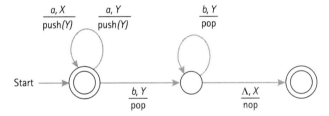

Let's see why the language is not LL(k) for any k. Any grammar for the language needs two productions, such as $S \to A \mid C$, where A generates strings of the form a^n and C generates strings of the form $a^n b^n$. Notice that the two strings a^{k+1} and $a^{k+1} b^{k+1}$ both start with $k + 1$ a's. Therefore, k symbols of lookahead are not sufficient to decide whether to use $S \to A$ or $S \to C$.

12.3.2 Bottom–Up Parsing and LR(k) Grammars

A powerful class of grammars whose languages can be parsed bottom-up is the set of LR(k) grammars, which were introduced by Knuth [1965]. These grammars allow a string to be parsed in a bottom-up fashion by constructing a rightmost derivation in reverse.

We'll start with a little example to introduce the idea of bottom-up parsing with the following grammar:

$$S \to aSB \mid d$$
$$B \to b.$$

We'll consider the following rightmost derivation of the string *aadbb*:

$$S \Rightarrow aSB \Rightarrow aSb \Rightarrow aaSBb \Rightarrow aaSbb \Rightarrow aadbb.$$

We want to give an informal description of how such a derivation can be constructed in reverse, starting on the right with the string *aadbb*. The derivation steps are found by using a table-driven "shift-reduce" parser. We'll give a rough idea of how things work.

The action of the parser will be represented by the following table with three columns labeled Stack, Input, and Action. The top of the stack is the rightmost symbol of Stack. The current input symbol is the leftmost symbol of Input. Each line of the table is obtained by performing the action on the previous line. The Shift action means to move the current input symbol to the top of the stack. The Reduce action causes the right side of a production represented by the symbols nearest to the top of the stack to be replaced by the left side of the production. The $ is a marker for an empty stack and an empty input string. The reduce actions correspond to the reductions used in a rightmost derivation. The Stack is initially empty and the Input contains the string *aadbb*.

Stack	Input	Action to Perform
$	*aadbb*$	Shift
$*a*	*adbb*$	Shift
$*aa*	*dbb*$	Shift
$*aad*	*bb*$	Reduce by $S \rightarrow d$
$*aaS*	*bb*$	Shift
$*aaSb*	*b*$	Reduce by $B \rightarrow b$
$*aaSB*	*b*$	Reduce by $S \rightarrow aSB$
$*aS*	*b*$	Shift
$*aSb*	$	Reduce by $B \rightarrow b$
$*aSB*	$	Reduce by $S \rightarrow aSB$
$*S*	$	Accept

To describe LR(k) grammars and the parsing technique for their languages, we need to define an object called a *handle*. We'll introduce the idea with an example and then give the formal definition. Suppose we have a grammar with the production $A \rightarrow aCb$ that is used to make the following derivation step in some rightmost derivation, where we've underlined the occurrence of the production's right side in the derived sentential form:

$$BaAbc \Rightarrow Ba\underline{aCb}bc.$$

The problem of bottom-up parsing is to perform this derivation step in reverse. In other words, we must discover this derivation step from our knowledge of the grammar and by scanning the sentential form

$$BaaCbbc.$$

By scanning $BaaCbbc$ we must find two things: the production $A \rightarrow aCb$ and the occurrence of its right side in $BaaCbbc$. With these two pieces of information we can reduce $Ba\underline{aCb}bc$ to $BaAbc$.

We'll denote the occurrence of a substring within a string by the position of the substring's rightmost character. For example, the position of aCb in $BaaCbbc$ is 5. This allows us to represent the production $A \rightarrow aCb$ and the occurrence of its right side in $BaaCbbc$ by an ordered pair of the form

$$(A \rightarrow aCb, 5),$$

which we call a *handle* of $BaaCbbc$.

Now let's formalize the definition of a handle. A *handle* of a sentential form w is a pair

$$(A \rightarrow y, p),$$

where w can be written in the form $w = xyz$, p is the length of the string xy, and z is a string of terminals. This allows us to say that there is a rightmost derivation step

$$xAz \Rightarrow xyz = w.$$

example **12.14 A Bottom-Up Parse**

Suppose we have the following grammar for arithmetic expressions, where a stands for an identifier:

$$E \rightarrow E + T \mid T$$
$$T \rightarrow T * F \mid F$$
$$F \rightarrow (E) \mid a.$$

We'll perform a bottom-up parse of the string $a+a*a$ by constructing a rightmost derivation in reverse. Each row of the following table represents a step of the reverse derivation by listing a sentential form together with the handle used to make the reduction for the next step.

Sentential Form	Handle
$a + a * a$	$(F \rightarrow a, 1)$
$F + a * a$	$(T \rightarrow F, 1)$
$T + a * a$	$(E \rightarrow T, 1)$
$E + a * a$	$(F \rightarrow a, 3)$
$E + F * a$	$(T \rightarrow F, 3)$
$E + T * a$	$(F \rightarrow a, 5)$
$E + T * F$	$(T \rightarrow T * F, 5)$
$E + T$	$(E \rightarrow E + T, 3)$
E	

So, we've constructed the following rightmost derivation in reverse:

$$E \Rightarrow E + T \Rightarrow E + T * F \Rightarrow E + T * a \Rightarrow E + F * a \Rightarrow E + a * a$$
$$\Rightarrow T + a * a \Rightarrow F + a * a \Rightarrow a + a * a.$$

end example

LR(k) Grammars

Now let's get down to business and discuss LR(k) grammars. An LR(k) *grammar* has the property that every string has a unique rightmost derivation that can be constructed in reverse order, where the handle of each sentential form is found by scanning the form's symbols from left to right, including up to k symbols past the handle. By "past the handle" we mean: If $(A \rightarrow y, p)$ is the handle of xyz, then we can determine it by a left-to-right scan of xyz, including up to k symbols in z. We should also say that the L in LR(k) means a left-to-right scan of the input and the R means construct a rightmost derivation in reverse.

For example, in an LR(0) grammar we can't look at any symbols beyond the handle to find it. Knuth showed that the number k doesn't affect the collection of languages defined by such grammars for $k \geq 1$. He also showed that the deterministic context-free languages are exactly the languages that can be described by LR(1) grammars. Thus we have the following relationships for all $k \geq 1$.

Theorem (Knuth)

      LR(k) languages = LR(1) languages

                 = deterministic context-free languages.

---

example   **12.15   An LR(0) Grammar**

Let's convince ourselves that the following grammar is LR(0):

$$S \rightarrow aAc$$
$$A \rightarrow Abb \,|\, b.$$

This grammar generates the language $\{ab^{2n+1}c \mid n \geq 0\}$. We need to see that the handle of any sentential form can be found without scanning past it. There are only three kinds of sentential forms, other than the start symbol, that occur in any derivation:

$$ab^{2n+1}c, \quad aAb^{2n}c, \quad \text{and} \quad aAc.$$

For example, the string $abbbbbc$ is derived as follows:

$$S \Rightarrow aAc \Rightarrow aAbbc \Rightarrow aAbbbbc \Rightarrow abbbbbc.$$

Scanning the prefix ab in $ab^{2n+1}c$ is sufficient to conclude that the handle is $(A \rightarrow b, 2)$. So we don't need to scan beyond the handle to discover it. Similarly, scanning the prefix $aAbb$ of $aAb^{2n}c$ is enough to conclude that its handle is $(A \rightarrow Abb, 4)$. Here, too, we don't need to scan beyond the handle to find it. Lastly, scanning all of aAc tells us that its handle is $(S \rightarrow aAc, 3)$ and we don't need to scan beyond the c.

Since we can determine the handle of any sentential form in a rightmost derivation without looking at any symbols beyond the handle, it follows that the grammar is LR(0).

end example

To get some more practice with LR(k) grammars, let's look at a grammar for the language of Example 12.15 that is not LR(k) for any k.

example 12.16 A Non–LR(k) Grammar

In the preceding example we gave an LR(0) grammar for the language $\{ab^{2n+1}c \mid n \geq 0\}$. Here's an example of a grammar for the same language that is not LR(k) for any k:

$$S \rightarrow aAc$$
$$A \rightarrow bAb \mid b.$$

For example, the handle of $abbbc$ is $(A \rightarrow b, 3)$, but we can discover this fact only by examining the entire string, which includes two symbols beyond the handle. Similarly, the handle of $abbbbbc$ is $(A \rightarrow b, 4)$, but we can discover this fact only by examining the entire string, which in this case includes three symbols beyond the handle.

In general, the handle for any string $ab^{2n+1}c$ with $n > 0$ is $(A \rightarrow b, n + 2)$, and we can discover it only by examining all symbols of the string, including $n + 1$ symbols beyond the handle. Since n can be any positive integer, we can't constrain the number of symbols that need to be examined past a handle to find it. Therefore, the grammar is not LR(k) for any natural number k.

end example

example 12.17 An LR(1) Grammar

Let's show that the following grammar is LR(1):

$$S \rightarrow aCd \mid bCD$$
$$C \rightarrow cC \mid c$$
$$D \rightarrow d.$$

To see that this grammar is LR(1), we'll examine the possible kinds of sentential forms that can occur in a rightmost derivation. The following two rightmost derivations are typical:

$$S \Rightarrow aCd \Rightarrow acCd \Rightarrow accCd \Rightarrow acccd.$$
$$S \Rightarrow bCD \Rightarrow bCD \Rightarrow bcCd \Rightarrow bccCd \Rightarrow bcccd.$$

So we can say with some confidence that any sentential form in a rightmost derivation looks like one of the following forms, where $n \geq 0$:

$$aCd,\ ac^{n+1}Cd,\ ac^{n+1}d,\ bCD,\ bCd,\ bc^{n+1}Cd,\ bc^{n+1}d.$$

It's easy to check that for each of these forms the handle is determined by at most one symbol to its right. In fact, for most of these forms the handles are determined with no lookahead. In the following table we've listed each sentential form together with its handle and the number of lookahead symbols to its right that are necessary to determine it.

Sentential Form	Handle	Lookahead
aCd	$(S \to aCd, 3)$	0
$ac^{n+1}Cd$	$(C \to cC, n+3)$	0
$ac^{n+1}d$	$(C \to c, n+2)$	1
bCD	$(S \to bCD, 3)$	0
bCd	$(D \to d, 3)$	0
$bc^{n+1}Cd$	$(C \to cC, n+3)$	0
$bc^{n+1}d$	$(C \to c, n+2)$	1

So each handle can be determined by observing at most one character to its right. The only situation in which we need to look beyond the handle is when the substring cd occurs in a sentential form. In this case we must examine the d to conclude that the handle's production is $C \to c$. Therefore, the grammar is LR(1).

end example

LR(1) Parsing

Now let's discuss LR(1) parsing, which is sufficient for all deterministic context-free languages. The goal of an LR(1) parser is to build a rightmost derivation in reverse by using one symbol of lookahead to find handles. To make sure that there is always one symbol of lookahead available to be scanned, we'll attach an end-of-string symbol $ to the right end of the input string. For example, to parse the string abc, we input the string abc$.

An LR(1) parser is a table-driven algorithm that uses an explicit stack that always contains the part of a sentential form to the left of the currently scanned symbol. We'll describe the parsing process with an example. The grammar of Example 12.17 is simple enough for us to easily describe the possible sentential

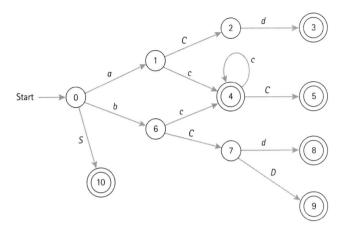

Figure 12.1   DFA for eight forms.

forms with handles at the right end. There are eight possible forms, where we've also included S itself:

$$S,\ aCd,\ ac^{n+1}C,\ ac^{n+1},\ bCD,\ bCd,\ bc^{n+1}C,\ bc^{n+1}.$$

The next task is to construct a DFA that accepts all strings having these eight forms. The diagram in Figure 12.1 represents such a DFA, where any missing edges go to an error state that we've also omitted.

Here's the connection between the DFA and the parser: Any path that is traversed by the DFA is represented as a string of symbols on the stack. For example, the path whose labeled edges concatenate to bC is represented by the stack

$$0\ b\ 6\ C\ 7.$$

The path whose labeled edges concatenate to $accc$ is represented by the stack

$$0\ a\ 1\ c\ 4\ c\ 4\ c\ 4.$$

The state on top of the stack always represents the current state of the DFA. The parsing process starts with 0 on the stack. Each step of the parse performs an action found in the parse table based on top of the stack and the current input symbol. The parse continues in this way until an error occurs or until an accept entry is found in the table.

The two main actions of the parser are shifting input symbols onto the stack and reducing handles, which is why this method of parsing is often called *shift-reduce* parsing. The best thing about the parser is that when a handle has been found, its symbols are sitting on the topmost portion of the stack. So a reduction can be performed by popping these symbols off the stack and pushing a nonterminal onto the stack. Here's a description of the shift and reduce actions:

"Shift j" means to shift the current input symbol and state j onto the stack.

"Reduce $A \to w$" means to reduce the handle by popping the symbols of w from the stack, which leaves some state k on top. Then push A onto the stack and lastly push the next state in the transition from state k with edge labeled A in the DFA.

For example, the following table shows each step in the parse of $bccd$:

Stack	Input	Action to Perform
0	$bccd\$$	Shift 6
0 b 6	$ccd\$$	Shift 4
0 b 6 c 4	$cd\$$	Shift 4
0 b 6 c 4 c 4	$d\$$	Reduce by $C \to c$
0 b 6 c 4 C 5	$d\$$	Reduce by $C \to cC$
0 b 6 C 7	$d\$$	Shift 8
0 b 6 C 7 d 8	$\$$	Reduce by $D \to d$
0 b 6 C 7 D 9	$\$$	Reduce by $S \to bCD$
0 S 10	$\$$	Accept

There are two main points about LR(1) grammars. They describe the class of deterministic context-free languages, and they produce efficient parsing algorithms to perform rightmost derivations in reverse. It's nice to know that there are algorithms to automatically construct LR(1) parse tables because for even the simplest grammars it can be a tedious process. You'll find all the details about algorithms to construct LR(1) parsers in any book about compilers.

 Exercises

LL(k) Grammars

1. Find an LL(1) grammar for each of the following languages.

    a. $\{a, ba, bba\}$.
    b. $\{a^n b \mid n \in \mathbb{N}\}$.
    c. $\{a^{n+1} bc^n \mid n \in \mathbb{N}\}$.
    d. $\{a^m b^n c^{m+n} \mid m, n \in \mathbb{N}\}$.

2. Find an LL(k) grammar for the language $\{aa^n \mid n \in \mathbb{N}\} \cup \{aab^n \mid n \in \mathbb{N}\}$. What is k for your grammar?

3. For each of the following grammars, perform the left-factoring process, where possible, to find an equivalent LL(k) grammar where k is as small as possible.

    a. $S \to abS \mid a$.
    b. $S \to abA \mid abcA$
       $A \to aA \mid \Lambda$.

4. For each of the following grammars, find an equivalent grammar with no left recursion. Are the resulting grammars LL(k)?

    a. $S \rightarrow Sa \mid Sb \mid c.$

    b. $S \rightarrow SaaS \mid ab.$

5. Write down the recursive descent procedures to parse strings in the language of expressions defined by the following grammar:

$$E \rightarrow TR$$
$$R \rightarrow +TR \mid \Lambda$$
$$T \rightarrow FV$$
$$V \rightarrow *FV \mid \Lambda$$
$$F \rightarrow (E) \mid a.$$

LR(k) Grammars

6. Show that each of the following grammars is LR(0).

    a. $S \rightarrow aA \mid bB$ b. $S \rightarrow aAc \mid b$

         $A \rightarrow cA \mid d$ $A \rightarrow aSc \mid b.$

         $B \rightarrow cB \mid d.$

7. Show that the following grammar is LR(1): $S \rightarrow aSb \mid \Lambda$.

8. The language $\{a^n b^n \mid n \geq 0\} \cup \{a^n b^{2n} \mid n \geq 0\}$ is not deterministic context-free. So it doesn't have any LR(k) grammars for any k. For example, give an argument to show that the following grammar for this language is not LR(k) for any k:

$$S \rightarrow A \mid B \mid \Lambda$$
$$A \rightarrow aAb \mid ab$$
$$B \rightarrow aBbb \mid abb.$$

12.4   Context–Free Language Topics

In this section we'll look at a few properties of context-free grammars and languages. We'll start by discussing some restricted grammars that still generate all the context-free languages. Then we'll discuss a tool that can be used to show that some languages are not context-free.

12.4.1 Transforming Grammars

Context-free grammars appear to be very general because the right side of a production can be any string of any length. It's interesting and useful to know that we can put more restrictions on the productions and still generate the same context-free languages. We'll see that for languages that don't contain Λ, we can modify their grammars so that the productions don't contain Λ. Then we'll introduce two classic special grammars that have many applications.

Removing Λ-Productions

A context-free language that does not contain Λ can be written with a grammar that does not contain Λ on the right side of any production. For example, suppose we have the following grammar:

$$S \to aDaE$$
$$D \to bD \mid E$$
$$E \to cE \mid \Lambda.$$

Although Λ appears in this grammar, it's clear that Λ does not occur in the language generated by the grammar. After some thought, we can see that this grammar generates all strings of the form $ab^k c^m ac^n$, where k, m, and n are nonnegative integers. Since the language does not contain Λ, we can write a grammar whose productions don't contain Λ. Try it on your own, and then look at the following three-step algorithm:

Algorithm to Remove Lambda Productions   (12.11)

1. Find the set of all nonterminals N such that N derives Λ.

2. For each production of the form $A \to w$, create all possible productions of the form $A \to w'$, where w' is obtained from w by removing one or more occurrences of the nonterminals found in Step 1.

3. The desired grammar consists of the original productions together with the productions constructed in Step 2, minus any productions of the form $A \to \Lambda$.

---

example   **12.18 Remove Lambda Productions**

Let's try this algorithm on our example grammar. Step 1 gives us two nonterminals D and E because they both derive Λ as follows:

$$E \Rightarrow \Lambda \quad \text{and} \quad D \Rightarrow E \Rightarrow \Lambda.$$

For Step 2 we'll list each original production together with all new productions that it creates:

Original Productions	New Productions
$S \to aDaE$	$S \to aaE \mid aDa \mid aa$
$D \to bD$	$D \to b$
$D \to E$	$D \to \Lambda$
$E \to cE$	$E \to c$
$E \to \Lambda$	None

For Step 3, we take the originals together with the new productions and throw away those containing Λ to obtain the following grammar:

$$S \to aDaE \mid aaE \mid aDa \mid aa$$
$$D \to bD \mid b \mid E$$
$$E \to cE \mid c.$$

end example

Chomsky Normal Form

Any context-free grammar can be written in a special form called *Chomsky normal form*, which appears in Chomsky [1959]. Each production has one of the following forms:

$$A \to BC,$$
$$A \to a,$$
$$S \to \Lambda,$$

where B and C are nonterminals, a is a terminal, and S is the start symbol. Also, if the production $S \to \Lambda$ occurs, then S does not appear on the right side of any production.

The Chomsky normal form has several uses, both practical and theoretical. For example, any string of length $n > 0$ can be derived in $2n - 1$ steps. Also, the derivation trees are binary trees. Here's an algorithm that will construct a Chomsky normal form with no occurrence of the start symbol on the right side of any production.

Transforming to Chomsky Normal Form   (12.12)

1. If the start symbol S of the given grammar occurs on the right side of some production, then create a new start symbol S' and a new production $S' \to S$.

Continued ➡

➡ ➡

2. If there is a production $A \to \Lambda$, where A is not the start symbol, then use (12.11) to remove all productions that contain Λ. If this process removes a Λ production from the start symbol, then add it back.

3. This step removes all *unit* productions $A \to B$, where A and B are nonterminals. For each pair of nonterminals A and B, if $A \to B$ is a unit production or if there is a derivation $A \Rightarrow^+ B$, then add all productions of the form $A \to w$, where $B \to w$ is not a unit production. Now remove all the unit productions.

4. For each production whose right side has two or more symbols, replace all occurrences of each terminal a with a new nonterminal A, and also add the new production $A \to a$.

5. Replace each production of the form $B \to C_1 C_2 ... C_n$, where $n > 2$, with the following two productions, where D is a new nonterminal:

$$B \to C_1 D \quad \text{and} \quad D \to C_2 \dots C_n.$$

Continue this step until all right sides have two nonterminals.

example 12.19 Finding a Chomsky Normal Form

We'll transform the following grammar into Chomsky normal form:

$$S \to aSb \,|\, T$$
$$T \to cT \,|\, \Lambda.$$

Since S occurs on the right side of a production, we'll apply step 1 to create a new start symbol S' and obtain the grammar

$$S' \to S$$
$$S \to aSb \,|\, T$$
$$T \to cT \,|\, \Lambda.$$

This grammar contains the production $T \to \Lambda$, where T is not the start symbol. So we'll apply Step 2, which tells us to apply (12.11) to remove all lambda productions. The algorithm produces the production $S' \to \Lambda$, so we'll add it back to obtain the grammar

$$S' \to S \,|\, \Lambda$$
$$S \to aSb \,|\, ab \,|\, T$$
$$T \to cT \,|\, c.$$

Now we'll apply Step 3 to remove the unit productions. From the unit productions $S' \to S$ and $S \to T$ we add the new productions $S' \to aSb \mid ab$ and $S \to cT \mid c$. From the derivation $S' \Rightarrow^+ T$ we add the new productions $S' \to cT \mid c$. Now remove the unit productions to obtain the grammar

$$S' \to aSb \mid ab \mid cT \mid c \mid \Lambda$$
$$S \to aSb \mid ab \mid cT \mid c$$
$$T \to cT \mid c.$$

Now we can do Step 4. Replace the letters a, b, and c by A, B, and C, respectively and add the new productions $A \to a$, $B \to b$, and $C \to c$. This gives us the grammar

$$S' \to ASB \mid AB \mid CT \mid c \mid \Lambda$$
$$S \to ASB \mid AB \mid CT \mid c$$
$$T \to CT \mid c$$
$$A \to a$$
$$B \to b$$
$$C \to c.$$

Next is Step 5. Replace each occurrence of ASB with AD, where $D \to SB$. This gives us the following Chomsky normal form:

$$S' \to AD \mid AB \mid CT \mid c \mid \Lambda$$
$$S \to AD \mid AB \mid CT \mid c$$
$$T \to CT \mid c$$
$$D \to SB$$
$$A \to a$$
$$B \to b$$
$$C \to c.$$

end example

Greibach Normal Form

Any context-free grammar can be written in a special form called *Greibach normal form*, which appears in Greibach [1965]. Each production has one of the following forms:

$$A \to aB_1 B_2 \ldots B_n,$$
$$A \to a,$$
$$\overline{ S \to \Lambda, }$$

where S is the start symbol, a is a terminal, and each B_k is a nonterminal not equal to S.

Notice that there can be no left recursion in this grammar. Also, Λ can occur only with the production $S \to \Lambda$. So let's look at a method to remove immediate left recursion without introducing Λ. For example, suppose we have the following A productions, where w and y are not Λ:

$$A \to Aw \,|\, y.$$

Any string derived from A begins with y and is followed one or more occurrences of w. The following A productions derive the same strings, where T is a new nonterminal:

$$A \to yT \,|\, y$$
$$T \to wT \,|\, w.$$

For the general case, suppose we have the following set of A productions:

$$A \to Aw_1 \,|\, \ldots \,|\, Aw_n \,|\, x_1 \,|\, \ldots \,|\, x_m.$$

We can remove the left recursion by replacing these productions with the following productions, where T is a new nonterminal:

$$A \to x_1 T \,|\, \ldots \,|\, x_m T \,|\, x_1 \,|\, \ldots \,|\, x_m.$$
$$T \to w_1 T \,|\, \ldots \,|\, w_n T \,|\, w_1 \,|\, \ldots \,|\, w_n.$$

The Greibach normal form has several uses, both practical and theoretical. For example, any string of length $n > 0$ can be derived in n steps, which makes parsing quite efficient. Here's an algorithm that will construct a Greibach normal form.

1. Perform Steps 1, 2, and 3 of (12.12).

2. Remove all left recursion, including indirect, without adding Λ.

3. Make substitutions to put the grammar into the proper form.

example   **12.20 Finding a Greibach Normal Form**

Let's do a "simple" example to get the general idea. Suppose we have the following grammar:

$$S \to aAB \,|\, \Lambda$$
$$A \to BA \,|\, a$$
$$B \to AB \,|\, b.$$

Notice that start symbol S does not appear on the right side of any production and $S \to \Lambda$ is the only lambda production. Further, there are no unit clauses. So we can skip steps 1, 2, and 3 of (12.12). The grammar does have indirect left recursion. So we'll have to make a substitution to remove the indirection. We'll replace A on the third line by the right hand sides of the two A productions on the second line to obtain the grammar

$$S \to aAB \mid \Lambda$$
$$A \to BA \mid a$$
$$B \to BAB \mid aB \mid b.$$

This gives us a left-recursive production $B \to BAB$. We'll remove the left recursion without introducing Λ by replacing $B \to BAB \mid aB \mid b$ with the productions

$$B \to aBT \mid bT \mid aB \mid b$$
$$T \to ABT \mid AB.$$

With these replacements our grammar takes the form

$$S \to aAB \mid \Lambda$$
$$A \to BA \mid a$$
$$B \to aBT \mid bT \mid aB \mid b$$
$$T \to ABT \mid AB.$$

There is no longer any left recursion. So we can start making substitutions to obtain the desired form. We'll start by using the third line to replace B on the second line. This gives us the grammar

$$S \to aAB \mid \Lambda$$
$$A \to aBTA \mid bTA \mid aBA \mid bA \mid a$$
$$B \to aBT \mid bT \mid aB \mid b$$
$$T \to ABT \mid AB.$$

Now we can work on the T productions. We'll use the second line to replace the two occurrences of A on the fourth line. This gives us the grammar in Greibach normal form:

$$S \to aAB \mid \Lambda$$
$$A \to aBTA \mid bTA \mid aBA \mid bA \mid a$$
$$B \to aBT \mid bT \mid aB \mid b$$
$$T \to aBTABT \mid bTABT \mid aBABT \mid bABT \mid aBT \mid$$
$$aBTAB \mid bTAB \mid aBAB \mid bAB \mid aB.$$

end example

12.4.2 Properties of Context–Free Languages

Although most languages we encounter are context-free languages, we need to face the fact that not all languages are context-free. For example, suppose we want to find a PDA or a context-free grammar for the language $\{a^n b^n c^n \mid n \geq 0\}$. After a few attempts we might get the idea that the language is not context-free. How can we be sure? In some cases we can use a pumping argument similar to the one used to show that a language is not regular. So let's discuss a pumping lemma for context-free languages.

Pumping Lemma

If a context-free language has an infinite number of strings, then any grammar for the language must be recursive. In other words, there must be a production that is recursive or indirectly recursive. For example, a grammar for an infinite context-free language will contain a fragment similar to the following:

$$S \rightarrow uNy$$
$$N \rightarrow vNx \mid w.$$

Notice that either v or x must be nonempty. Otherwise, the language derived is finite, consisting of the single string uwy. The grammar allows us to derive infinitely many strings having a certain pattern. For example, the derivation to recognize the string $uv^3 wx^3 y$ can be written as follows:

$$S \Rightarrow uNy \Rightarrow uvNxy \Rightarrow uvvNxxy \Rightarrow uvvvNxxxy \Rightarrow uv^3 wx^3 y.$$

This derivation can be shortened or lengthened to obtain the set of all strings of the form $uv^k wx^k y$ for all $k \geq 0$. This example illustrates the main result of the pumping lemma for context-free languages, which we'll state in all its detail as follows:

Pumping Lemma for Context-Free Languages   (12.13)

Let L be an infinite context-free language. Then there is a positive integer m such that for all strings $z \in L$ with $|z| \geq m$, z can be written in the form $z = uvwxy$, where the following properties hold:

$$|vx| \geq 1,$$
$$|vwx| \leq m,$$
$$uv^k wx^k y \in L \text{ for all } k \geq 0.$$

The positive integer m in (12.13) depends on the grammar for the language L. Without going into the proof, suffice it to say that m is large enough to ensure a recursive derivation of any string of length m or more. Let's use the lemma to show that a particularly simple language is not context-free.

12.21 Using the Pumping Lemma

Let $L = \{a^n b^n c^n \mid n \geq 0\}$. We'll show that L is not context-free by assuming that it is context-free and then trying to find a contradiction.

Proof: If L is context-free, then by (12.13) we can pick a string $z = a^m b^m c^m$ in L, where m is the positive integer mentioned in the lemma. Since $|z| \geq m$, we can write it in the form $z = uvwxy$, such that $|vx| \geq 1$, $|vwx| \leq m$, and such that $uv^k wx^k y \in L$ for all $k \geq 0$.

Now we need to come up with a contradiction. One thing to observe is that the pumped variable v can't contain two distinct letters. For example, if the substring ab occurs in v, then the substring $ab \ldots ab$ occurs in v^2, which means that the pumped string $uv^2 wx^2 y$ can't be in L, contrary to the pumping lemma conclusion. Therefore, v is a string of a's, or v is a string of b's, or v is a string of c's. A similar argument shows that x can't contain two distinct letters.

Since $|vx| \geq 1$, we know that at least one of v and x is a nonempty string of the form a^i, or b^i, or c^i for some $i > 0$. Therefore, the pumped string $uv^2 wx^2 y$ can't contain the same number of a's, b's, and c's because one of the three letters a, b, and c does not get pumped up. For example, if $v = a^i$ for some $i > 0$, and $x = \Lambda$, then $uv^2 wx^2 y = a^{m+i} b^m c^m$, which is not in L. The other cases for v and x are handled in a similar way. Thus $uv^2 wx^2 y$ can't be in L, which contradicts the pumping lemma (12.13). So it follows that the language L is not context-free. QED.

Additional Properites

In (12.3) we saw that the operations of union, product, and closure can be used to construct new context-free languages from other context-free languages. Now that we have an example of a language that is not context-free, we're in position to show that the operations of intersection and complement can't always be used in this way. Here's the first statement:

$$\text{Context-free languages are not closed under intersection.} \qquad (12.14)$$

For example, we know from Example 12.21 that the language $L = \{a^n b^n c^n \mid n \geq 0\}$ is not context-free. It's easy to see that L is the intersection of the two languages

$$L_1 = \{a^n b^n c^k \mid n, k \in \mathbb{N}\} \quad \text{and} \quad L_2 = \{a^k b^n c^n \mid n, k \in \mathbb{N}\}.$$

It's also easy to see that these two languages are context-free. Just find a context-free grammar for each language. Thus we have an example of two context-free languages whose intersection is not context-free.

Now we're in position to prove the following result about complements:

$$\text{Context-free languages are not closed under complement.} \qquad (12.15)$$

Proof: Suppose, by way of contradiction, that complements of context-free languages are context-free. Then we can take the two languages L_1 and L_2 from the proof of (12.14) and make the following sequence of statements: Since L_1 and L_2 are context-free, it follows that the complements L_1' and L_2' are context-free. We can take the union of these two complements to obtain another context-free language. Further, we can take the complement of this union to obtain the following context-free language:

$$(L_1' \cup L_2')'.$$

Now let's describe a contradiction. Using De Morgan's laws, we have the following statement:

$$(L_1' \cup L_2')' = L_1 \cap L_2.$$

So we're forced to conclude that $L_1 \cap L_2$ is context-free. But we know that

$$L_1 \cap L_2 = \{a^n b^n c^n \mid n \geq 0\},$$

and we've shown that this language is not context-free. This contradiction proves (12.15). QED.

Although (12.14) says that we can't expect the intersection of context-free languages to be context-free, we can say that the intersection of a regular language with a context-free language is context-free. We won't prove this, but we'll include it with the closure properties that we do know about. Here is a listing of them:

Properties of Context-Free Languages   (12.16)

1. The union of two context-free languages is context-free.

2. The language product of two context-free languages is context-free.

3. The closure of a context-free language is context-free.

4. The intersection of a regular language with a context-free language is context-free.

We'll finish with two more properties of context-free languages that can be quite useful in showing that a language is not context-free:

> **Context-Free Language Morphisms**   **(12.17)**
>
> Let $f : A^* \to A^*$ be a language morphism. In other words, $f(\Lambda) = \Lambda$ and $f(uv) = f(u)f(v)$ for all strings u and v.
>
> Let L be a language over A.
>
> **1.** If L is context-free, then $f(L)$ is context-free.
>
> **2.** If L is context-free, then $f^{-1}(L)$ is context-free.

Proof: We'll prove statement 1 (statement 2 is a bit complicated). Since L is context-free, it has a context-free grammar. We'll create a context-free grammar for $f(L)$ as follows: Transform each production $A \to w$ into a new production of the form $A \to w'$, where w' is obtained from w by replacing each terminal a in w by $f(a)$. The new grammar is context-free, and any string in $f(L)$ is derived by this new grammar. QED.

example 12.22 Using a Morphism

Let's use (12.17) to show that $L = \{a^n bc^n de^n \mid n \geq 0\}$ is not context-free. We can define a morphism $f : \{a, b, c, d, e\}^* \to \{a, b, c, d, e\}^*$ by

$$f(a) = a, \quad f(b) = \Lambda, \quad f(c) = b, \quad f(d) = \Lambda, \quad f(e) = c.$$

Then $f(L) = \{a^n b^n c^n \mid n \geq 0\}$. If L is context-free, then we must also conclude by (12.17) that $f(L)$ is context-free. But we know that $f(L)$ is not context-free. Therefore, L is not context-free.

end example

It might occur to you that the language $\{a^n b^n c^n \mid n \geq 0\}$ could be recognized by a pushdown automaton with two stacks available rather than just one stack. For example, we could push the a's onto one stack. Then we pop the a's as we push the b's onto the second stack. Finally, we pop the b's from the second stack as we read the c's.

So it might make sense to take the next step and study pushdown automata with two stacks. Instead, we're going to switch gears and discuss another type of device, called a Turing machine, which is closer to the idea of a computer. The interesting thing is that Turing machines are equivalent in power to pushdown automata with two stacks. In fact, Turing machines are equivalent to pushdown automata with n stacks for any $n \geq 2$. We'll discuss them in the next chapter.

◢ Exercises

Grammar Transformations

1. For each of the following grammars, find a grammar without Λ productions that generates the same language.

   a. $S \rightarrow aA \mid aBb$
      $A \rightarrow aA \mid \Lambda$
      $B \rightarrow aBb \mid \Lambda.$

   b. $S \rightarrow aAB$
      $A \rightarrow aAb \mid \Lambda$
      $B \rightarrow bB \mid \Lambda.$

2. Find a Chomsky normal form for each of the following grammars.

   a. $S \rightarrow abT \mid Rab$
      $T \rightarrow aT \mid b$
      $R \rightarrow aRb \mid \Lambda.$

   b. $S \rightarrow TR$
      $T \rightarrow aTb \mid \Lambda$
      $R \rightarrow bR \mid c.$

   c. $S \rightarrow aSb \mid T$
      $T \rightarrow Tb \mid \Lambda.$

3. Find a Greibach normal form for each of the following grammars.

   a. $S \rightarrow AB \mid bA$
      $A \rightarrow BA \mid c$
      $B \rightarrow Bb \mid a.$

   b. $S \rightarrow bAB$
      $A \rightarrow BAa \mid a$
      $B \rightarrow bB \mid \Lambda.$

Pumping Lemma

4. Use the pumping lemma (12.13) to show that each of the following languages is not context-free.

   a. $\{a^n b^n a^n \mid n \geq 0\}$. *Hint:* Look at Example 12.21.

   b. $\{a^i b^j c^k \mid 0 < i < j < k\}$. *Hint:* Let $z = a^m b^{m+1} c^{m+2} = uvwxy$, and consider the following two cases: (1) There is at least one a in either v or x. (2) Neither v nor x contains any a's.

   c. $\{a^p \mid p \text{ is a prime number}\}$. *Hint:* Let $z = a^p = uvwxy$, where p is prime and $p > m + 1$. Let $k = |uwy|$. Show $|uv^k wx^k y|$ is not prime.

Challenges

5. Show that the language $\{a^n b^n a^n \mid n \in \mathbb{N}\}$ is not context-free by performing the following tasks:

   a. Given the morphism $f : \{a, b, c\}^* \rightarrow \{a, b, c\}^*$ defined by $f(a) = a$, $f(b) = b$, and $f(c) = a$, describe $f^{-1}(\{a^n b^n a^n \mid n \in \mathbb{N}\})$.

   b. Show that

$$f^{-1}(\{a^n b^n a^n \mid n \in \mathbb{N}\}) \cap \{a^k b^m c^n \mid k, m, n \in \mathbb{N}\} =$$
$$\{a^n b^n c^n \mid n \in \mathbb{N}\}.$$

   c. Argue that $\{a^n b^n a^n \mid n \in \mathbb{N}\}$ is not context-free by using Parts (a) and (b) together with (12.16) and (12.17).

12.5 Chapter Summary

This chapter introduced context-free languages as languages derived by context-free grammars, which are grammars in which every production contains exactly one nonterminal on its left side. Context-free grammars can be easily constructed for unions, products, and closures of context-free languages.

A pushdown automaton (PDA) is like a finite automaton with a stack attached. PDAs can accept strings by final state or by empty stack. Either form of acceptance is equivalent to the other. PDAs are equivalent to context-free grammars. The important point is that there are algorithms to transform back and forth between PDAs and context-free grammars. Therefore, we can start with a context-free grammar and automatically transform it into a PDA that recognizes the context-free language. Deterministic PDAs are less powerful than nondeterministic PDAs, which means that deterministic PDAs recognize a proper subcollection of the context-free languages. A simple interpreter can be constructed for PDAs.

A deterministic context-free language is a context-free language that is recognized by a deterministic PDA using final-state acceptance. Most programming language constructs can be recognized by efficient deterministic parsers. $LL(k)$ parsing is an efficient top-down technique for parsing many—but not all—deterministic context-free languages. $LR(k)$ parsing is an efficient bottom-up technique for parsing any deterministic context-free language.

We observed some other things too. Although context-free grammars have few restrictions, any context-free grammar can be transformed into either of two special restricted forms—Chomsky normal form and Greibach normal form. There are some basic properties of context-free languages given by a pumping lemma, set operations, and morphisms. Many simple languages, including $\{a^n b^n c^n \mid n \geq 0\}$, are not context-free.

chapter 13

Turing Machines and Equivalent Models

*Machines are worshipped because they are beautiful
and valued because they confer power; they are hated
because they are hideous and loathed because they
impose slavery.*

—Bertrand Russell (1872–1970)

Is there a computing device more powerful than any other computing device?
What does "powerful" mean? Can we easily compare machines to see whether
they have the same power? In this chapter we try to answer these questions by
studying Turing machines and the Church-Turing thesis, which claims that there
are no models of computation more powerful than Turing machines.

13.1 Turing Machines

It's time to discuss a simple yet powerful computing device that was invented
by the mathematician and logician Alan Turing (1912–1954). The machine is
described in the paper by Turing [1936]. It models the actions of a person doing

a primitive calculation on a long strip of paper divided up into contiguous individual cells, each of which contains a symbol from a fixed alphabet. The person uses a pencil with an eraser. Starting at some cell, the person observes the symbol in the cell and decides either to leave it alone or to erase it and write a new symbol in its place. The person can then perform the same action on one of the adjacent cells. The computation continues in this manner, moving from one cell to the next along the paper in either direction. We assume that there is always enough paper to continue the computation in either direction as far as we want. The computation can stop at some point or continue indefinitely.

13.1.1   Definition of a Turing Machine

Let's give a more precise description of this machine, which is named after its creator. A *Turing machine* consists of two major components, a tape and a control unit. The *tape* is a sequence of cells that extends to infinity in both directions. Each cell contains a symbol from a finite alphabet. There is a tape head that reads from a cell and writes into the same cell. The *control unit* contains a finite set of instructions, which are executed as follows: Each instruction causes the tape head to read the symbol from a cell, to write a symbol into the same cell, and either to move the tape head to an adjacent cell or to leave it at the same cell. Here is a picture of a Turing machine.

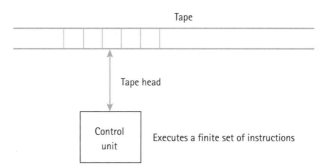

Each Turing machine instruction contains the following five parts:

The current machine state.
A tape symbol read from the current tape cell.
A tape symbol to write into the current tape cell.
A direction for the tape head to move.
The next machine state.

We'll agree to let the letters L, S, and R mean "move left one cell," "stay at the current cell," and "move right one cell," respectively. We can represent an instruction as a 5-tuple or in graphical form. For example, the 5-tuple

$$(i, \ a, \ b, \ L, \ j)$$

is interpreted as follows:

> If the current state of the machine is i, and if the symbol in the current tape cell is a, then write b into the current tape cell, move left one cell, and go to state j.

We can also write the instruction in graphical form as follows:

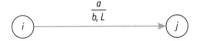

If a Turing machine has at least two instructions with the same state and input letter, then the machine is *nondeterministic*. Otherwise, it's *deterministic*. For example, the following two instructions are nondeterministic:

$$(i, a, b, L, j),$$
$$(i, a, a, R, j).$$

Turing Machine Computations

The tape is used much like the memory in a modern computer, to store the input, to store data needed during execution, and to store the output. To describe a Turing machine computation, we need to make a few more assumptions.

Turing Machine Assumptions

1. An input string is represented on the tape by placing the letters of the string in contiguous tape cells. All other cells of the tape contain the blank symbol, which we'll denote by Λ.

2. The tape head is positioned at the leftmost cell of the input string unless specified otherwise.

3. There is one *start state*.

4. There is one *halt state*, which we denote by "Halt."

The execution of a Turing machine stops when it enters the Halt state or when it enters a state for which there is no valid move. For example, if a Turing machine enters state i and reads a in the current cell, but there is no instruction of the form $(i, a, \dots)$, then the machine stops in state i.

The Language of a Turing Machine

We say that an input string is *accepted* by a Turing machine if the machine enters the Halt state. Otherwise, the input string is *rejected*. There are two ways to reject an input string: Either the machine stops by entering a state other than the Halt state from which there is no move, or the machine runs forever. The *language of a Turing machine* is the set of all input strings accepted by the machine.

It's easy to see that Turing machines can solve all the problems that PDAs can solve because a stack can be maintained on some portion of the tape. In fact a Turing machine can maintain any number of stacks on the tape by allocating some space on the tape for each stack.

Let's do a few examples to see how Turing machines are constructed. Some things to keep in mind when constructing Turing machines to solve problems are: find a strategy, let each state have a purpose, document the instructions, and test the machine. In other words, use good programming practice.

example   13.1   A Sample Turing Machine

Suppose we want to write a Turing machine to recognize the language $\{a^n b^m \mid m, n \in \mathbb{N}\}$. Of course, this is a regular language, represented by the regular expression a^*b^*. So there is a DFA to recognize it. Of course, there is also a PDA to recognize it. So there had better be a Turing machine to recognize it.

The machine will scan the tape to the right, looking for the empty symbol and making sure that no a's are scanned after any occurrence of b. Here are the instructions, where the start state is 0:

$$
\begin{array}{ll}
(0,\ \Lambda,\ \Lambda,\ S,\ \text{Halt}) & \text{Accept } \Lambda \text{ or only } a\text{'s.} \\
(0,\ a,\ a,\ R,\ 0) & \text{Scan } a\text{'s.} \\
(0,\ b,\ b,\ R,\ 1) & \\
(1,\ b,\ b,\ R,\ 1) & \text{Scan } b\text{'s.} \\
(1,\ \Lambda,\ \Lambda,\ S,\ \text{Halt}) &
\end{array}
$$

For example, to accept the string abb, the machine enters the following sequence of states: 0, 0, 1, 1, Halt. This Turing machine also has the following graphical definition, where H stands for the Halt state:

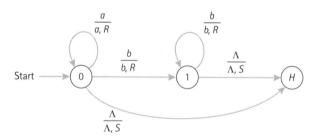

example 13.2 **An Example of Power**

To show the power of Turing machines, we'll construct a Turing machine to recognize the following language.

$$\{a^n b^n c^n \mid n \geq 0\}.$$

We've already shown that this language cannot be recognized by a PDA. A Turing machine to recognize the language can be written from the following informal algorithm:

> If the current cell is empty, then halt with success. Otherwise, if the current cell contains a, then write an X in the cell and scan right, looking for a corresponding b to the right of any a's, and replace it by Y. Then continue scanning to the right, looking for a corresponding c to the right of any b's, and replace it by Z. Now scan left to the X and see whether there is an a to its right. If so, then start the process again. If there are no a's, then scan right to make sure there are no b's or c's.

Now let's write a Turing machine to implement this algorithm. The state 0 will be the initial state. The instructions for each state are preceded by a prose description. In addition, each line contains a short comment.

If Λ is found, then halt. If a is found, then write X and scan right. If Y is found, then scan over Y's and Z's to find the right end of the string.

$(0, a, X, R, 1)$	Replace a by X and scan right.
$(0, Y, Y, R, 0)$	Scan right.
$(0, Z, Z, R, 4)$	Go make the final check.
$(0, \Lambda, \Lambda, S, \text{Halt})$	Success.

Scan right, looking for b. If found, replace it by Y.

$(1, a, a, R, 1)$	Scan right.
$(1, b, Y, R, 2)$	Replace b by Y and scan right.
$(1, Y, Y, R, 1)$	Scan right.

Scan right, looking for c. If found, replace it by Z.

$(2, c, Z, L, 3)$	Replace c by Z and scan left.
$(2, b, b, R, 2)$	Scan right.
$(2, Z, Z, R, 2)$	Scan right.

Scan left, looking for X. Then move right and repeat the process.

$(3, a, a, L, 3)$	Scan left.
$(3, b, b, L, 3)$	Scan left.
$(3, X, X, R, 0)$	Found X. Move right one cell.
$(3, Y, Y, L, 3)$	Scan left.
$(3, Z, Z, L, 3)$	Scan left.

Scan right, looking for Λ. Then halt.

$$(4,\ Z,\ Z,\ R,\ 4) \qquad \text{Scan right.}$$
$$(4,\ \Lambda,\ \Lambda,\ S,\ \text{Halt}) \qquad \text{Success.}$$

end example

13.1.2   Turing Machines with Output

Turing machines can also be used to compute functions. As usual, the input is placed on the tape in contiguous cells. We usually specify the form of the output along with the final position of the tape head when the machine halts. Here are a few examples.

example   **13.3   Adding 2 to a Natural Number**

Let a natural number be represented in unary form. For example, the number 4 is represented by the string 1111. We'll agree to represent 0 by the empty string Λ. Now it's easy to construct a Turing machine to add 2 to a natural number. The initial state is 0. When the machine halts, the tape head will point at the left end of the string. There are just three instructions. Comments are written to the right of each instruction:

$$(0,\ 1,\ 1,\ L,\ 0) \qquad \text{Move left to blank cell.}$$
$$(0,\ \Lambda,\ 1,\ L,\ 1) \qquad \text{Add 1 and move left.}$$
$$(1,\ \Lambda,\ 1,\ S,\ \text{Halt}) \qquad \text{Add 1 and halt.}$$

The following diagram is a graphical picture of this Turing machine:

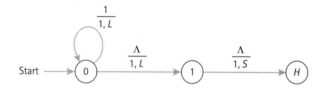

end example

13.4 Adding 1 to a Binary Natural Number

Here we'll represent natural numbers as binary strings. For example, the number 5 will be represented as the string 101 placed in three tape cells. The algorithm can be described as follows:

> Move to right end of string;
>
> **repeat**
>
>> If current cell contains 1, write 0 and move left
>
> **until** current cell contains 0 or Λ;
>
> Write 1;
>
> Move to left end of string and halt.

Here's a Turing machine to implement the algorithm.

$(0, 0, 0, R, 0)$ Scan right.
$(0, 1, 1, R, 0)$ Scan right.
$(0, \Lambda, \Lambda, L, 1)$ Found right end of string.

$(1, 0, 1, L, 2)$ Write 1, done adding.
$(1, 1, 0, L, 1)$ Write 0 and move left with carry bit.
$(1, \Lambda, 1, S, \text{Halt})$ Write 1, done and in proper position.

$(2, 0, 0, L, 2)$ Move to left end and halt.
$(2, 1, 1, L, 2)$
$(2, \Lambda, \Lambda, R, \text{Halt})$

end example

example 13.5 An Equality Test

Let's write a Turing machine to test the equality of two natural numbers, representing the numbers as unary strings separated by #. We'll assume that the number 0 is denoted by a blank cell. For example, the string Λ #11 represents the two numbers 0 and 2. The two numbers 3 and 4 are represented by the string 111#1111. The idea is to repeatedly cancel leftmost and rightmost 1's until none remain. The machine will halt with a 1 in the current cell if the numbers are not equal and Λ if they are equal. A Turing machine program to accomplish this follows:

$(0, 1, \Lambda, R, 1)$ Cancel leftmost 1.
$(0, \Lambda, \Lambda, R, 4)$ Left number is zero.
$(0, \#, \#, R, 4)$ Finished with left number.

$$(1, \ 1, \ 1, \ R, \ 1) \qquad \text{Scan right.}$$
$$(1, \ \Lambda, \ \Lambda, \ L, \ 2) \qquad \text{Found the right end.}$$
$$(1, \ \#, \ \#, \ R, \ 1) \qquad \text{Scan right.}$$

$$(2, \ 1, \ \Lambda, \ L, \ 3) \qquad \text{Cancel rightmost 1.}$$
$$(2, \ \#, \ 1, \ S, \ \text{Halt}) \quad \text{Not equal, first } > \text{ second.}$$

$$(3, \ 1, \ 1, \ L, \ 3) \qquad \text{Scan left.}$$
$$(3, \ \Lambda, \ \Lambda, \ R, \ 0) \qquad \text{Found left end.}$$
$$(3, \ \#, \ \#, \ L, \ 3) \qquad \text{Scan left.}$$

$$(4, \ 1, \ 1, \ S, \ \text{Halt}) \qquad \text{Not equal, first } < \text{ second.}$$
$$(4, \ \Lambda, \ \Lambda, \ S, \ \text{Halt}) \qquad \text{Equal.}$$
$$(4, \ \#, \ \#, \ R, \ 4) \qquad \text{Scan right.}$$

If the two numbers are not equal, then it's easy to modify the Turing machine so that it can detect the inequality relationship. For example, the second instruction of state 2 could be modified to write the letter G to mean that the first number is greater than the second number. Similarly, the first instruction of state 4 could write the letter L to mean that the first number is less than the second number.

end example

Hard–Working Turing Machines (Busy Beavers)

What does it mean for a Turing machine to work hard? To discuss the question we'll consider Turing machines that are deterministic, can write either Λ or 1 on a tape cell, and must shift left or right after each move. Such a Turing machine is called a *busy beaver* if it accepts the empty string (i.e., it starts with a blank tape) and, after it halts, the number of tape cells containing 1 is the maximum of all such Turing machines with the same number of states that accept the empty string. So a busy beaver is a hard-working Turing machine.

Let $b(n)$ denote the number of 1's that can be written by a busy beaver with n states, not including the halt state. It's pretty easy to see that $b(1) = 1$. For example, the following 1-state Turing machine writes a 1 and then halts:

$$(0, \ \Lambda, \ 1, \ R, \ \text{Halt}).$$

If we try to build a 1-state machine that writes two or more 1's, then our machine would need an instruction like

$$(0, \ \Lambda, \ 1, \ R, \ 0) \quad \text{or} \quad (0, \ \Lambda, \ 1, \ L, \ 0).$$

But this causes the machine to loop forever without halting. So $b(1) = 1$.

Busy beavers and the problem of finding the values of $b(n)$ were introduced by Rado [1962], where he observed that $b(1) = 1$ and $b(2) = 4$. Rado proved that the function b cannot be computed by any algorithm. We'll discuss this further in the next chapter. So there will never be a formula to compute b. Yet, some progress has been made. Lin and Rado [1965] proved that $b(3) = 6$. Brady [1983] proved that $b(4) = 13$. For $n \geq 5$, things get out of hand quickly. Marxen and Buntrock [1990] constructed a 5-state Turing machine that writes 4,098 1's before it halts. Therefore, we can say that $b(5) \geq 4,098$.

example 13.6 A Busy Beaver

The following Turing machine is a busy beaver that satisfies $b(2) = 4$. In other words, the machine has two states (not including the halt state) and it writes four 1's before halting.

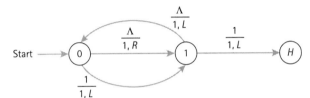

end example

13.1.3 Alternative Definitions

We should point out that there are many different definitions of Turing machines. Our definition is similar to the machine originally defined by Turing. Some definitions allow the tape to be infinite in one direction only. In other words, the tape has a definite left end and extends infinitely to the right.

A *multihead* Turing machine has two or more tape heads positioned on the tape. A *multitape* Turing machine has two or more tapes with corresponding tape heads. It's important to note that all these Turing machines are *equivalent* in power. In other words, any problem solved by one type of Turing machine can also be solved by any other type of Turing machine.

Simulating a Multitape Turing Machine

Let's give an informal description of how a multitape Turing machine can be simulated by a single-tape Turing machine. For our description we'll assume that we have a Turing machine T that has two tapes, each with a single tape head. We'll describe a new single-tape, single-head machine M that will start with its tape containing the two nonblank portions taken from the tapes of T, separated by a new tape symbol

@.

Whenever T executes an instruction (which is actually a pair of instructions, one for each tape), M simulates the action by performing two corresponding instructions, one instruction for the left side of @ and the other instruction for the right side of @.

Since M has only one tape head, it must chase back and forth across @ to execute instructions. So it needs to keep track of the positions of the two tape heads that it is simulating. One way to do this is to place a position marker $\cdot$ in every other tape cell. To indicate a current cell, we'll write the symbol

$$\char`^$$

in place of $\cdot$ in the adjacent cell to the right of the current cell for the left tape and to the adjacent cell to the left of the current cell for the right tape. For example, if the two tapes of T contain the strings abc and $xyzw$, with tape heads pointing at b and z, then the tape for M has the following form, where the symbol # marks the left end and the right end of the relevant portions of the tape:

$$\ldots \Lambda \cdot \# \cdot a \cdot b \char`^ c \cdot @ \cdot x \cdot y \char`^ z \cdot w \cdot \# \cdot \Lambda \ldots$$

Suppose now that T writes a into the current cell of its abc tape and then moves right and that it writes w into the current cell of its $xyzw$ tape and then moves left. These actions would be simulated by M to produce the following tape:

$$\ldots \Lambda \cdot \# \cdot a \cdot a \cdot c \char`^ @ \cdot x \char`^ y \cdot w \cdot w \cdot \# \cdot \Lambda \ldots$$

A problem can occur if the movement of one of T's tape heads causes M's tape head to bump into either @ or #. In either case, we need to make room for a new cell. If M's tape head bumps into @, then the entire representation on that side of @ must be moved to make room for a new tape cell next to @. This is where the # is needed to signal the end of the relevant portion of the tape. If M's tape head bumps into #, then # must be moved farther out to make room for a new tape cell.

Constructing Multitape Turing Machines

Multitape Turing machines are usually easier to construct because distinct data sets can be stored on distinct tapes. This eliminates the tedious scanning back and forth required to maintain different data sets.

Instruction Format

An instruction of a multitape Turing machine is still a 5-tuple. But now the elements in positions 2, 3, and 4 are tuples. For example, a typical instruction for a 3-tape Turing machine has the following form:

$$(i, (a, b, c), (x, y, z), (R, L, S), j).$$

This instruction is interpreted as follows:

> If the machine state is i and if the current three tape cells contain the symbols a, b, and c, respectively, then overwrite the cells with x, y, and z. Then move the first tape head right, move the second tape head left, and keep the third tape head stationary. Then go to state j.

The same instruction format can also be used for multihead Turing machines. In the next example, we'll use a multitape Turing machine to multiply two natural numbers.

example 13.7 Multiplying Natural Numbers

Suppose we want to construct a Turing machine to multiply two natural numbers, each represented as a unary string of ones. We'll use a three-tape Turing machine, where the first two tapes hold the input numbers and the third tape will hold the answer. If either number is zero, the product is zero. Otherwise, the machine will use the first number as a counter to repeatedly add the second number to itself, by repeatedly copying it onto the third tape.

For example, the diagrams in Figure 13.1 show the contents of the three tapes before the computation of 3·4 and before the start of the second of the three additions:

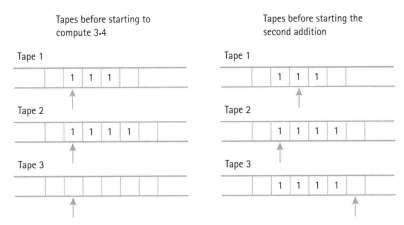

Figure 13.1   Multiplying two numbers.

The following three-tape Turing machine will perform the multiplication of two natural numbers by repeated addition, where 0 is the start state:

Start by checking to see whether either number is zero:

$$(0, (\Lambda, \Lambda, \Lambda), (\Lambda, \Lambda, \Lambda), (S, S, S), \text{Halt}) \quad \text{Both are zero.}$$
$$(0, (\Lambda, 1, \Lambda), (\Lambda, 1, \Lambda), (S, S, S), \text{Halt}) \quad \text{First is zero.}$$
$$(0, (1, \Lambda, \Lambda), (1, \Lambda, \Lambda), (S, S, S), \text{Halt}) \quad \text{Second is zero.}$$
$$(0, (1, 1, \Lambda), (1, 1, \Lambda), (S, S, S), 1) \quad \text{Both are nonzero.}$$

Add the number on the second tape to the third tape:

$(1, (1, 1, \Lambda), (1, 1, 1), (S, R, R), 1)$ Copy.

$(1, (1, \Lambda, \Lambda), (1, \Lambda, \Lambda), (S, L, S), 2)$ Done copying.

Move the tape head of the second tape back to the left end of the number, and also move the tape head of the first number one cell to the right:

$(2, (1, 1, \Lambda), (1, 1, \Lambda), (S, L, S), 2)$ Move to the left end.

$(2, (1, \Lambda, \Lambda), (1, \Lambda, \Lambda), (R, R, S), 3)$ Move both tape heads to the right one cell.

Check the first tape head to see if all the additions have been performed:

$(3, (\Lambda, 1, \Lambda), (\Lambda, 1, \Lambda), (S, S, L), \text{Halt})$ Done.

$(3, (1, 1, \Lambda), (1, 1, \Lambda), (S, S, S), 1)$ Do another add.

Nondeterministic Turing Machines

We haven't yet classified Turing machines as deterministic or nondeterministic. Let's do so now. All our preceding examples are deterministic Turing machines. It's natural to wonder whether nondeterministic Turing machines are more powerful than deterministic Turing machines. We've seen that nondeterminism is more powerful than determinism for pushdown automata. But we've also seen that the two ideas are equal in power for finite automata.

For Turing machines we don't get any more power by allowing nondeterminism. In other words, we have the following result.

Theorem   **(13.1)**

If a nondeterministic Turing machine accepts a language L, then there is a deterministic Turing machine that also accepts L.

We'll give an informal idea of the proof. Suppose N is a nondeterministic Turing machine that accepts language L. We define a deterministic Turing machine D that simulates the execution of N by exhaustively executing all possible paths caused by N's nondeterminism. But since N might very well have an input that leads to an infinite computation, we must be careful in the way D simulates the actions of N. First D simulates all possible computations of N that take one step. Next D simulates all possible computations of N that take two steps. This process continues, with three steps, four steps, and so on. If, during this process, D simulates the action of N entering the Halt state, then D enters its Halt state. So N halts on an input string if and only if D halts on the same input string.

The problem is to write D so that it does all these wonderful things in a deterministic manner. The usual approach is to define D as a three-tape machine. One tape holds a permanent copy of the input string for N. The second tape keeps track of the next computation sequence and the number of steps of N that must be simulated by D. The third tape is used repeatedly by D to simulate the computation sequences for N that are specified on the second tape.

The computation sequences on the second tape are the interesting part of D. Since N is nondeterministic, there may be more than one instruction of the form (state, input, ?, ?, ?). Let m be the maximum number of instructions for any (state, input) pair. For purposes of illustration, suppose $m = 3$. Then for any (state, input) pair there are no more than three instructions of the form (state, input, ?, ?, ?), and we can number them 1, 2, and 3. If some (state, input) pair doesn't have three nondeterministic instructions, then we'll simply write down extra copies of one instruction to make the total three. This gives us exactly three choices for each (state, input) pair. For convenience we'll use the letters a, b, and c rather than 1, 2, and 3.

Each simulation by D will be guided by a string over $\{a, b, c\}$ that is sitting on the second tape. For example, the string *ccab* tells D to simulate four steps of N because length(*ccab*) = 4. For the first simulation step we pick the third of the possible instructions because *ccab* starts with c. For the second simulation step we also pick the third of the possible instructions because the second letter of *ccab* is c. The third letter of *ccab* is a, which says that the third simulation step should choose the first of the possible instructions. And the fourth letter of *ccab* is b, which says that the fourth simulation step should choose the second of the possible instructions.

To make sure D simulates all possible computation sequences, it needs to generate all the strings over $\{a, b, c\}$. One way to do this is to generate the nonempty strings in standard order, where $a \prec b \prec c$. Recall that this means that strings are ordered by length, and strings of equal length are ordered lexicographically. For example, here are the first few strings in the standard ordering, not including Λ:

$$a, b, c, aa, ab, ac, ba, bb, bc, ca, cb, cc, aaa, aab, \ldots .$$

So D needs to generate a new string in this ordering before it starts a new simulation. We've left the job of finding a Turing machine to compute the successor as an exercise.

13.1.4 A Universal Turing Machine

In the examples that we've seen up to this point, each problem required us to build a special-purpose Turing machine to solve only that problem. Is there a more general Turing machine that acts like a general-purpose computer? The answer is yes. We'll see that a Turing machine can be built to interpret any other Turing machine. In other words, there is a Turing machine that can take as input an arbitrary Turing machine M together with an arbitrary input for M and then perform the execution of M on its input. Such a machine is called a *universal Turing machine*. A universal Turing machine acts like a general purpose computer that stores a program and its data in memory and then executes the program.

We'll give a description of a universal Turing machine U. Since U can have only a finite number of instructions and a finite alphabet of tape cell symbols, we have to discuss the representation of any Turing machine in terms of the fixed symbols of U. We begin by selecting a fixed infinite set of states, say $\mathbb{N}$, and a fixed infinite set of tape cell symbols, say $L = \{a_i \mid i \in \mathbb{N}\}$. Now we require that every Turing machine must use states from the set $\mathbb{N}$ and tape cell symbols from L. This is easy to do by simply renaming the symbols used in any Turing machine.

Now we select a fixed finite alphabet A for the machine U and find a way to encode any Turing machine (i.e., the instructions for any Turing machine) into a string over A. Similarly, we encode any input string for a Turing machine into a string over A.

Now that we have the two strings over A, one for the Turing machine and one for its input, we can get down to business and describe the action of machine U. We'll describe U as a three-tape Turing machine. We use three tapes because it's easier to describe the machine's actions. Recall that any k-tape machine can be simulated by a one-tape machine.

Before U starts its execution, we place the two strings over A on tapes 1 and 2, where tape 1 holds the representation for a Turing machine and tape 2 holds the representation of an input string. We also place the start state on tape 3. Now U repeatedly performs the following actions: If the state on tape 3 is the halt state, then halt. Otherwise, get the current state from tape 3 and the current input symbol from tape 2. With this information, find the proper instruction on tape 1. Write the next state at the beginning of tape 3, and then perform the indicated write and move operations on tape 2.

A Logic Program Interpreter for Turing Machines

It's much easier to build an interpreter for Turing machines by using a programming language than to build a universal Turing machine. So as an example, we'll write a logic program version of an interpreter for Turing machines. Our representation of the input tape in the logic program should give you an indication of the difficult problem a universal Turing machine has in maintaining the input portion of its tape.

The interpreter will execute any Turing machine with a single two-way infinite tape. We'll make the following assumptions about any Turing machine that is to be executed by the interpreter:

The read/write head is at the left end of any nonempty input string.
The letters l, s, and r represent the moves of the read/write head.
The symbol # denotes a blank tape cell.
Instructions are represented as facts of the following form:

$$t(\text{state, letterToRead, letterToWrite, move, nextState}).$$

The start state is represented by a fact of the following form:

$$\text{start(state)}.$$

example **13.8 A Turing Machine as Data**

A Turing machine to add 1 to a binary number can be written as the following set of facts.

$$\text{start}(0).$$
$$t(0,\ 0,\ 0,\ r,\ 0).$$
$$t(0,\ 1,\ 1,\ r,\ 0).$$
$$t(0,\ \#,\ \#,\ l,\ 1).$$
$$t(1,\ 0,\ 1,\ s,\ \text{halt}).$$
$$t(1,\ 1,\ 0,\ l,\ 1).$$
$$t(1,\ \#,\ 1,\ s,\ \text{halt}).$$

end example

To keep the interpreter simple, we'll represent strings as lists. For example, to find the result of adding 1 to the binary number 1011, we'll type the goal

$$\leftarrow \text{compute}(\langle 1,\ 0,\ 1,\ 1\rangle,\ \text{Out}).$$

This goal starts the execution of the Turing machine interpreter, so we better describe what it means. The interpreter starts by placing the list $\langle 1,\ 0,\ 1,\ 1\rangle$ onto the tape. Then the instructions of the given Turing machine are executed. When a "halt" instruction is executed the variable Out is returned as a list that represents the tape.

The tape is represented by the following three objects.

1. "Cell" is a variable that holds the tape symbol in the current cell pointed at by the read/write head.

2. "Left" is a list that holds the information on the tape to the left of the current cell. The head of the list holds the symbol immediately to the left of the current cell.

3. "Right" is a list that holds the information on the tape to the right of the current cell. The head of the list holds the symbol immediately to the right of the current cell.

For example, the input list $\langle 1,\ 0,\ 1,\ 1\rangle$ is represented on the tape as follows:

$$\text{Left} = \langle\ \rangle,\quad \text{Cell} = 1,\quad \text{Right} = \langle 0,\ 1,\ 1\rangle.$$

Description of the Predicates

The "find" predicate tries to find and execute an instruction. Its first argument is the state. The next three arguments are the representation of the tape, and the last argument holds the variable for the output tape. The "move" predicate makes a move and returns a new representation of the tape in variables A, B, and C. The "continue" predicate checks for the halt state. If it's found, the output tape is constructed and placed in the last variable. Otherwise the execution continues by calling the "find" predicate to execute another instruction. Here are the clauses for the interpreter.

compute($\langle\,\rangle$, OutTape)   (blank input tape)
    ← start(I), find(I, $\langle\,\rangle$, #, $\langle\,\rangle$, OutTape).

compute(Head :: Tail, OutTape)   (nonblank input tape)
    ← start(I), find(I, $\langle\,\rangle$, Head, Tail, OutTape).

find(State, Left, Cell, Right, OutTape)
    ← t(State, Cell, Write, Move, Next),
        move(Move, Left, Write, Right, A, B, C),
        continue(Next, A, B, C, OutTape).

continue(halt, Left, Cell, Right, OutTape)   (build output tape)
    ← reverse(Left, R),
        cat(R, Cell :: Right, OutTape).

continue(State, Left, Cell, Right, OutTape)
    ← find(State, Left, Cell, Right, OutTape).

move(l, $\langle\,\rangle$, Cell, Right, $\langle\,\rangle$, #, Cell :: Right).
move(l, Head :: Tail, Cell, Right, Tail, Head, Cell :: Right).
move(s, Left, Cell, Right, Left, Cell, Right).
move(r, Left, Cell, $\langle\,\rangle$, Cell :: Left, #, $\langle\,\rangle$).
move(r, Left, Cell, Head :: Tail, Cell :: Left, Head, Tail).

When the halt state is reached, the tape is reconstructed as a single list with entries in the proper order. To do this we need to reverse the "Left" part of the tape before concatenating it to the list Cell::Right. The predicate "reverse(X, Y)" sets Y to the reverse of list X. The output tape consists of all cells that were used during the computation. The predicate "cat(X, Y, Z)" sets Z to the concatenation of the two lists X and Y.

In our example, to add 1 to the binary number 1011, we write the goal

$$\leftarrow \text{compute}(\langle 1,\, 0,\, 1,\, 1\rangle,\, \text{Out}).$$

The Turing machine interpreter returns

$$\text{Out} = \langle 1,\, 1,\, 0,\, 0,\, \# \rangle,$$

which represents the binary number 1100. Notice in this example that the symbol # is included in the output tape. This is because the computation used that extra cell when it scanned to the right looking for a blank.

This interpreter can be easily translated into the Prolog language with three simple notational changes: use brackets [and] to represent lists; replace ← with ":–"; and replace "$X :: Y$" with $[X\,|\,Y]$.

 Exercises

Constructing Turing Machines

1. Construct a Turing machine to recognize the language of all palindromes over $\{a,\, b\}$.

2. Construct a Turing machine that starts with the symbol # in one cell, where all other tape cells are blank. The beginning position of the tape head is not known. The machine should halt with the tape head pointing at the cell containing #, with all other tape cells being blank.

3. Construct a Turing machine to move an input string over $\{a,\, b\}$ to the right one cell position. Assume that the tape head is at the left end of the input string if the string is nonempty. The rest of the tape cells are blank. The machine moves the entire string to the right one cell position, leaving all remaining tape cells blank.

4. Construct a Turing machine to implement each function. The inputs are pairs of natural numbers represented as unary strings and separated by the symbol #. Where necessary, represent zero by the tape symbol Λ.

   a. Add two natural numbers, neither of which is zero.

   b. Add two natural numbers, either of which may be zero.

5. Construct a Turing machine to perform each task.

   a. Complement the binary representation of a natural number, and then add 1 to the result.

   b. Add 2 to a natural number represented as a binary string.

   c. Add 3 to a natural number represented as a binary string.

6. Construct a Turing machine to test for equality of two strings over the alphabet $\{a,\, b\}$, where the strings are separated by a cell containing #. Output a 0 if the strings are not equal and a 1 if they are equal.

7. Construct a three-tape Turing machine to add two binary numbers, where the first two tapes hold the input strings and the tape heads are positioned at the right end of each string. The third tape will hold the output.

Challenges

8. Construct a single-tape Turing machine that inputs any string over the alphabet $\{a, b, c\}$ and outputs its successor in the standard ordering, where we assume that $a \prec b \prec c$. Recall that in the standard ordering, strings are ordered by length, strings of the same length being ordered lexicographically.

9. For busy beaver Turing machines, it is known that $b(3) = 6$, which means that 3-state busy beavers write six 1's before halting. Try to construct a 3-state busy beaver.

13.2   The Church-Turing Thesis

The word "computable" is meaningful to most of us because we have a certain intuition about it, and we actually feel quite comfortable with it. We might even say something like, "A thing is computable if it can be computed." Or we might say, "A thing is computable if there is some computation that computes it." Of course, we might also say, "A thing is computable if it can be described by an algorithm."

The Meaning of Computability

So the word "computable" is defined by using words like "computation" and "algorithm." We can relate these two words by saying that a computation is the execution of an algorithm. So we can say that "computable" has something to do with a formal process (execution) and a formal description (algorithm). Let's list some examples of formal processes and formal descriptions that have something to do with our intuitive notion of computable:

The derivation process associated with grammars

The evaluation process associated with functions

The state transition process associated with machines

The execution process associated with programs and programming languages

For example, we can talk about the strings derived by regular grammars or the strings derived by context-free grammars. We can talk about the evaluation of a certain class of functions. We can discuss the state transitions of Turing machines, or pushdown automata, or real computers. We can also discuss the execution of programs written in our favorite programming language.

So when we think of computability, we most often try to formalize it in some way. For our purposes, a *model* is a formalization of an idea. So we'll use the word "model" instead of "formalization." Since there are many ways to model the idea of computability, the following questions need to be answered.

Questions about Computational Models

Is one model more powerful than another? That is, does one model solve all the problems of another model and also solve some problem that is not solvable by the other model?

Is there a most powerful model?

We've already seen some answers to the first question. For example, we know that Turing machines are more powerful than pushdown automata and that pushdown automata are more powerful than deterministic finite automata. We also know that nondeterministic finite automata have the same power as deterministic finite automata.

What about the second question? One of the goals of these paragraphs is to convince you that the answer is yes. In fact there are many equivalent most powerful models. In particular, we would like to convince you that the following statement is true:

Church-Turing Thesis

Anything that is intuitively computable can be computed by a Turing machine.

The Church-Turing thesis says that any problem that is intuitively solvable in some way can also be solved by a Turing machine. Now let's discuss why the word "thesis" is used instead of the word "theorem." Each of us has some idea of what it means to be computable, even though the idea is informal. On the other hand, the idea of a Turing machine is formal and precise. So there is no possibility of ever proving that everyone's idea of computability is equivalent to the formal idea of a Turing machine. That's why the statement is a thesis rather than a theorem.

The Church-Turing thesis is important because no one has ever invented a computational model more powerful than a Turing machine! The name Church in the Church-Turing thesis belongs to the mathematician and logician Alonzo Church. He proposed an alternative formalization for the notion of algorithm in Church [1936].

13.2.1 Equivalence of Computational Models

In these paragraphs we'll discuss some computational models, each of which is equal in power to the Turing machine model. To say that two computational models are *equivalent* (i.e., *equal in power*) means that they both solve the same class of problems. Once we know that some computational model, say *M*, is equivalent to the Turing machine model, then we have an alternative form of the Church-Turing thesis:

Church–Turing Thesis for M

Anything that is intuitively computable can be computed by the M computational model.

For example, a normal task of any programming language designer is to make sure that the new language being developed—call it X—has the same power as a Turing machine. Is this hard to do? Maybe yes and maybe no. If we already know that some other language, say Y, is equivalent in power to a Turing machine, then we don't need to concern ourselves with Turing machines. All we need to do is show that languages X and Y are equal in power.

We'll see that a programming language does not need to be sophisticated to be powerful. In fact, we'll see that there are just a few properties that need to be present in any language. At first glance it may be hard to accept the Church-Turing thesis. Most of the results of this chapter should help convince the skeptic. If we accept the Church-Turing thesis, then we can associate "computable" with the phrase "computed by a Turing machine." If we don't want to make the leap of assuming the Church-Turing thesis, then we can refer to a thing being *Turing-computable* if it can be computed by a Turing machine.

In the following paragraphs we'll survey a variety of computational models, all of which are equivalent in power to the Turing machine model. Although some models process different kinds of data, they can still be compared because any piece of data can be represented by a string of symbols, which in turn can be represented by a natural number. So the common denominator for comparing models is the ability to process natural numbers.

13.2.2 A Simple Programming Language

Let's look at a little imperative language that contains a small set of commands and a few other minimal features. The language that we present is a slight variation of a formalism called an unbounded register machine (URM), which was introduced by Shepherdson and Sturgis [1963]. An informal description of the language, which we'll call the *simple language*, is given in the following definition:

1. There are variables that take values in the set $\mathbb{N}$ of natural numbers.

2. There is a while statement of the form

$$\text{while } X \neq 0 \text{ do statement od.}$$

3. There is an assignment statement taking one of the three forms:

$$X := 0, \quad X := \text{succ}(Y), \quad \text{and} \quad X := \text{pred}(Y).$$

4. A statement is either a while statement or an assignment statement, or a sequence of two or more statements separated by semicolons.

5. A simple program is a statement.

We should note that the statements $X := \text{succ}(Y)$ and $X := \text{pred}(Y)$ compute the successor and predecessor, respectively. Also note that $\text{pred}(0) = 0$ to keep all values in $\mathbb{N}$. The language doesn't have any input or output statements. We'll take care of this problem by assuming that all the variables in a program have been given initial values. Similarly, we'll assume that the output consists of the collection of values of the variables at program termination.

example **13.9 Some Simple Macros**

Let's see whether this language can do anything. In Figure 13.2 we have listed some "macro statements" together with the code for each macro in the simple language. With the aid of macros, we can construct some familiar-looking programs. We'll leave some more problems as exercises.

end example

The simple language has the same power as a Turing machine. In other words, any problem that can be solved by a Turing machine can be solved with a simple program; conversely, any problem that can be solved by a simple program can be solved by a Turing machine. The details can be found in many books, so we'll gladly omit them.

Macro Statement	Simple Code for Macro
$X := Y$	$X := \text{succ}(Y); X := \text{pred}(X)$
$X := 3$	$X := 0; X := \text{succ}(X);$ $X := \text{succ}(X); X := \text{succ}(X)$
$X := X + Y$	$I := Y;$ while $I \neq 0$ do $X := \text{succ}(X); I := \text{pred}(I)$ od
Loop Forever	$X := 0; X := \text{succ}(X);$ while $X \neq 0$ do $Y := 0$ od
repeat S until $X = 0$	$S;$ while $X \neq 0$ do S od

Figure 13.2   Some simple macros.

13.2.3   Partial Recursive Functions

Now we'll look at a collection of functions whose arguments and values are natural numbers. If we believe anything, we most likely believe that the following three functions are computable.

$$f(x) = 0, \quad g(x) = x + 1, \quad \text{and} \quad h(x, y, z) = x.$$

Interestingly, functions like these together with some simple combining rules are all we need to construct all possible computable functions. What follows is a description of the functions and combining rules used to construct the collection of functions, which are called *partial recursive functions*.

Initial Functions   (13.2)

$\text{zero}(x) = 0$ (the zero function)

$\text{succ}(x) = x + 1$ (the successor function)

$\text{project}_i\,(x_1, \dots, x_n) = x_i$ (the projection functions).

Composition Rule   (13.3)

Define a new function by composition.

Primitive Recursion Rule   (13.4)

Define a new function f in terms of functions h and g as follows (where x represents any number of arguments).

$$f(x, 0) = h(x),$$
$$f(x, \text{succ}(y)) = g(x, y, f(x, y)).$$

Unbounded Search Rule   (13.5)

Define a new function f in terms of a total function g as follows (where x represents any number of arguments). The value $f(x)$ is determined by searching the following sequence for the smallest y such that $g(x, y) = 0$.

$$g(x, 0), \; g(x, 1), \; g(x, 2), \dots.$$

If such a y exists, then define $f(x) = y$. Otherwise, $f(x)$ is undefined. We'll represent this definition of f by the notation

$$f(x) = \min(y, g(x, y) = 0).$$

Informal Definition	Definition using (13.2-13.5)
identity(x) = x	identity(x) = project$_1$(x)
two(x) = 2	two(x) = succ(succ(zero(x)))
one(x, y) = 1	one(x, y) = project$_2$(x, succ(zero(y)))
add(x, y) = x + y	add(x, 0) = x add(x, succ(y)) = succ(add(x, y))
pred(x) = if x = 0 then 0 else x − 1	pred(0) = 0 pred(succ(x)) = x
monus(x, y) = if x ≥ y then x − y else 0.	monus(x, 0) = x monus(x, succ(y)) = pred(monus(x, y))

Figure 13.3   Some elementary functions.

example 13.10 Some Elementary Functions

Figure 13.3 shows how some elementary functions can be defined by combining rules (13.2–13.5). See if you can find definitions for other functions, such as the product of two numbers.

end example

example 13.11 Equality and Inequality

Assume that False and True are represented by 0 and 1, respectively. Then we can define equality and inequality predicates. We'll start with the "less" relation, which has the following informal description:

$$\text{less}(x,\ y) = \text{if } x < y \text{ then 1 else 0.}$$

Notice that $x < y$ if and only if monus$(y,\ x) > 0$. So we can write the following informal description of less:

$$\text{less}(x,\ y) = \text{if monus}(y,\ x) > 0 \text{ then 1 else 0.}$$

Now all we need is a test to decide whether a number is positive or not. Let "sign" be defined by

$$\text{sign}(x) = \text{if } x > 0 \text{ then 1 else 0.}$$

The sign function can be defined by primitive recursion. So the less function can be defined by composition. Figure 13.4 includes definitions for sign and less together with a few other common relations. See if you can find definitions for the "less than or equal to" and "greater than or equal to" relations.

end example

Informal Definition	Definition using (13.2-13.5)
sign(x) = if x = 0 then 0 else 1	sign(0) = 0 sign(succ(x)) = 1
less(x, y) = "x < y"	less(x, y) = sign(monus(y, x))
greater(x, y) = "x > y"	greater(x, y) = sign(monus(x, y))
eq(x, y) = "x = y"	eq(x, y) = monus(1, less(x, y) + less(y, x))
notEq(x, y) = "x ≠ y"	notEq(x, y) = monus(1, eq(x, y))

Figure 13.4   Relational functions.

example **13.12 Unbounded Search**

We'll examine the unbounded search rule with three examples. To begin, we should notice that the condition $g(x, y) = 0$ can be replaced by any condition that has an equivalent form. For example, we can use the condition $x = y + 1$ because it can be expressed as $\text{notEq}(x, y + 1) = 0$.

1. (*An Infinite Loop*) Let $f(x) = \min(y, x + y + 1 = 0)$. Notice that the equation $x + y + 1 = 0$ has no solutions over the natural numbers. So $f(x)$ is undefined for all $x \in \mathbb{N}$.

2. Let $f(x) = \min(y, x + y = 2)$. In this case $f(0) = 2$, $f(1) = 1$, $f(2) = 0$, and $f(x)$ is undefined for $x \geq 3$.

3. Let $f(x, y) = \min(z, x = y + z)$. For example, $f(5, 2) = 3$ and $f(2, 5)$ is undefined. After a few more test cases, it's easy to see that if $x \geq y$, then $f(x, y) = x - y$. Otherwise, $f(x, y)$ is undefined.

end example

The fact that the collection of partial recursive functions is equivalent to the collection of functions computed by Turing machines was proven by Kleene [1936]. So now we have the following equivalent models of computation: Turing machines, simple programs, and partial recursive functions.

Functions constructed without using unbounded search are called *primitive recursive functions*. It's easy to see that the primitive recursive functions are total functions (i.e, they are defined for all values in $\mathbb{N}$). The unbounded search rule is used to construct partial functions that are not total as well as some total functions that are not primitive recursive.

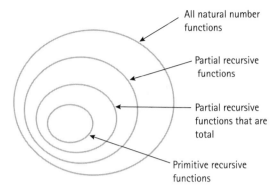

Figure 13.5   A hierarchy of functions.

Ackermann's Function

A famous example of a partial recursive function that is total but not primitive recursive is *Ackermann's function*:

$$A(x, y) = \text{if } x = 0 \text{ then } y + 1$$
$$\text{else if } y = 0 \text{ then } A(x - 1, 1)$$
$$\text{else } A(x - 1, A(x, y - 1)).$$

Since we can implement A in most computer languages, it follows by the Church-Turing thesis that A is a partial recursive function. But it can't be defined using only initial functions, composition, and primitive recursion. The proof follows from the fact (which we won't prove) that for any primitive recursive function f there is a natural number n such that $f(x) < A(n, x)$ for all $x \in \mathbb{N}$. So if A was primitive recursive, then there would be a natural number n such that $A(x, x) < A(n, x)$ for all $x \in \mathbb{N}$. Letting $x = n$, we get the contradiction $A(n, n) < A(n, n)$.

We know that there are an uncountable number of natural number functions, and it's easy to see that the collection of partial recursive functions is countable. So the collections of functions that we have been discussing can be pictured as proper subsets in the Venn diagram shown in Figure 13.5.

13.2.4 Machines That Transform Strings

We'll turn our attention now to some powerful models that process strings rather than numbers.

Markov Algorithms

A *Markov algorithm* over an alphabet A is a finite ordered sequence of productions $x \to y$, where $x, y \in A^*$. Some productions may be labeled with the word "halt," although this is not a requirement. A Markov algorithm transforms an input string into an output string. In other words, a Markov algorithm computes a function from A^* to A^*. Here's how the execution proceeds.

Execution of a Markov Algorithm

Given an input string w, the productions are scanned, starting at the beginning of the ordered sequence. If there is a production $x \to y$ such that x occurs as a substring of w, then the leftmost occurrence of x in w is replaced by y to obtain a transformed string. If $x \to y$ is a halt production, then the process halts with the transformed string as output.

Otherwise, the process starts all over again with the transformed string, where again the scan starts at the beginning of the ordered sequence of productions. If a scan of the instructions occurs without any new replacements of the current string, then the process halts with the current string as output.

If a production has the form $\Lambda \to y$, then it transforms any string w into the string yw. For example, suppose we wish to transform any string of the form a^i into the string a^{i+1}. The following single production Markov algorithm will do the job:

$$\Lambda \to a \text{ (halt)}.$$

This production causes the letter a to be appended to the left of any input string, after which the process halts.

Let's look at another example to get a better idea of the execution process. Suppose M is the Markov algorithm over $\{a, b\}$ consisting of the following sequence of three productions:

1. $aba \to b$

2. $ba \to b$

3. $b \to \Lambda$.

We'll trace the execution of M for the input string $w = aabaaa$. Each arrow indicates the transformation of a string caused by the execution of the Markov instruction whose number is in parentheses.

$$
\begin{aligned}
w = aabaaa \quad &\to \quad abaa \quad (1) \\
&\to \quad ba \quad (1) \\
&\to \quad b \quad (2) \\
&\to \quad \Lambda \quad (3).
\end{aligned}
$$

It's easy to see that M computes Λ for all strings of the form $a^i ba^j$, where $i \leq j$. It's also easy to see that M computes a^{i-j} for all input strings of the form $a^i ba^j$, where $i \geq j$. So if we think of a^i as a representation of the natural number i, then M transforms $a^i ba^j$ into a representation of the monus operation applied to i and j. Let's look at another example.

example **13.13 Reversing a String**

We'll write a Markov algorithm to reverse any string over $\{a, b, c\}$. For example, if the input string is abc, then the output string is cba. The algorithm will move the leftmost letter of a string to the right end of the string by swapping adjacent letters. Then it will move the leftmost letter of the transformed string to its proper position, and so on.

We'll use the symbol X to keep track of the swapping process, and we'll use the symbol Y to help remove the X's after all swapping has been completed. The algorithm follows:

1. $XX \rightarrow Y$ (clean up instructions)

2. $Ya \rightarrow aY$

3. $Yb \rightarrow bY$

4. $Yc \rightarrow cY$

5. $YX \rightarrow Y$

6. $Y \rightarrow \Lambda$ (halt)

7. $Xaa \rightarrow aXa$ (swapping instructions)

8. $Xab \rightarrow bXa$

9. $Xac \rightarrow cXa$

10. $Xba \rightarrow aXb$

11. $Xbb \rightarrow bXb$

12. $Xbc \rightarrow cXb$

13. $Xca \rightarrow aXc$

14. $Xcb \rightarrow bXc$

15. $Xcc \rightarrow cXc$

16. $\Lambda \rightarrow X$ (introduce X at left).

The following transformations trace the execution of the algorithm for the input string abc. Be sure to fill in the production used for each step.

$$abc \rightarrow Xabc \rightarrow bXac \rightarrow bcXa \rightarrow XbcXa \rightarrow cXbXa \rightarrow XcXbXa \rightarrow XXcXbXa$$
$$\rightarrow YcXbXa \rightarrow cYXbXa \rightarrow cYbXa \rightarrow cbYXa \rightarrow cbYa \rightarrow cbaY \rightarrow cba.$$

end example

Markov algorithms are described in Markov [1954]. Are Markov algorithms as powerful as Turing machines? The answer is yes, they are equivalent in power. So now we have the following equivalent models of computation: Turing machines, simple programs, partial recursive functions, and Markov algorithms.

Post Algorithms

Let's look at another string processing model—due to the mathematician Emil Post (1897–1954)—which appears in Post [1943]. A *Post algorithm* over an alphabet A is a finite set of productions that are used to transform strings. The productions have the form $s \to t$, where s and t are strings made up of symbols from A and possibly some variables. If a variable X occurs in t, then X must also occur in s. There is no particular ordering of the productions in a Post algorithm, unlike the ordering of productions in Markov algorithms. Some productions may be labeled with the word "halt," although this is not required.

Execution of a Post Algorithm

The computation of a Post algorithm proceeds by string pattern matching. If the input string matches the left side of some production, then we construct a new string to match the right side of the same production. If the production is a halt production, then the computation halts, and the new string is output.

Otherwise, the process continues by trying to match the new string with the left side of some production. If no matches can be found, then the process halts, and the output is the current string.

A Post algorithm can be either deterministic or nondeterministic. Nondeterminism occurs if some computation has a string that matches the left side of more than one production or matches the left side of a production in more than one way.

Let's start off with a simple example. The following single production makes up a Post algorithm over the alphabet $\{a, b\}$:

$$aXb \to X.$$

We'll execute the algorithm for the string *aab*. Execution starts by matching *aab* with *aXb*, where $X = a$. Thus *aab* is transformed into the string *a*. Since *a* doesn't match the left side of the production, the computation halts. Notice that X can match the empty string too. For example, if the input string is *ab*, then it matches *aXb*, where $X = \Lambda$. So *ab* gets transformed to Λ. This Post algorithm does many things. For example, it transforms any string of the form $a^i b$ to a^{i-1} for $i > 0$.

example 13.14 Replacing All Occurrences

Suppose that we want to replace all occurrences of *a* by *b* in any string over $\{a, b, c\}$. We'll construct two Post algorithms to solve the problem. The first is nondeterministic, consisting of the single production

$$XaY \to XbY.$$

For example, the string $acab$ is transformed to $bcbb$ in two different ways as follows, depending on which a is matched:

$$acab \rightarrow bcab \rightarrow bcbb,$$
$$acab \rightarrow acbb \rightarrow bcbb.$$

Now we'll write a deterministic Post algorithm. We'll use the symbol $\#$ to mark the current position in a left to right scan of a string, and we'll use the symbol $@$ to mark the left end of the string so exactly one instruction can be used at each step. Here's the algorithm:

$$aX \rightarrow @b\#X$$
$$bX \rightarrow @b\#X$$
$$cX \rightarrow @c\#X$$
$$@X\#aY \rightarrow @Xb\#Y$$
$$@X\#bY \rightarrow @Xb\#Y$$
$$@X\#cY \rightarrow @Xc\#Y$$
$$@X\# \rightarrow X \text{ (halt).}$$

The following trace shows the unique sequence of transformations made by the algorithm for the input string $acab$.

$$acab \rightarrow @b\#cab \rightarrow @bc\#ab \rightarrow @bcb\#b \rightarrow @bcbb\# \rightarrow bcbb.$$

end example

Are Post algorithms as powerful as Turing machines? The answer is yes, they have equivalent power. So now we have the following equivalent models of computation: Turing machines, simple programs, partial recursive functions, Markov algorithms, and Post algorithms.

Post Systems

Now let's look at systems that generate sets of strings. A *Post system* over the alphabet A (actually it's called a Post canonical system) is a finite set of inference rules and a finite set of strings in A^* that act as axioms to start the process. The inference rules are like the productions of a Post algorithm, except that an inference rule may contain more than one string on the left side. For example, a rule might have the form

$$s_1, s_2, s_3 \rightarrow t.$$

Execution in a Post System

Computation proceeds by finding axioms to match the patterns on the left side of some inference rule and then constructing a new string from the right side of the same rule. The new string can then be used as an axiom.

In this way, each Post system generates a set of strings over A consisting of the axioms and all strings inferred by the axioms.

example 13.15 Balanced Parentheses

Let's write a Post system to generate the set of all balanced parentheses. The alphabet will be the two symbols (and). The axioms and inference rules have the following form:

Axiom: Λ.

Inference rules: $X \to (X)$

$\qquad\qquad\qquad X, Y \to XY.$

We'll do a few sample computations. The axiom Λ matches the left side of the first rule. Thus we can infer the string ( ). Now the string ( ) matches the left side of the same rule, which allows us to infer the string (( )). If we let $X = ( )$ and $Y = (( ))$, the second rule allows us to infer the string ( )(( )). So the set of strings generated by this Post system is

$$\{\Lambda, ( ), (( )), ( )(( )), \dots\}.$$

end example

example 13.16 Generating Palindromes

We can generate the set of all palindromes over the alphabet $\{a, b\}$ by the Post system with the following axioms and inference rules:

Axioms: $\Lambda, a, b.$

Inference rules: $X \to aXa$

$\qquad\qquad\qquad X \to bXb.$

end example

Post Computable Functions

If A is an alphabet and $f : A^* \to A^*$ is a function, then f is said to be *Post-computable* if there is a Post system that computes the set of pairs

$$\{(x, f(x)) \mid x \in A^*\}.$$

To simplify things, we'll agree to represent the pair $(x, f(x))$ as the string $x \# f(x)$, where $\#$ is a new symbol not in A.

For example, the function $f(x) = xa$ over the alphabet $\{a\}$ is Post-computable by the Post system consisting of the axiom $\#a$ together with the following rule:

$$X\#Xa \rightarrow Xa\#Xaa.$$

This Post system computes f as the set $\{\#a,\ a\#aa,\ aa\#aaa,\dots\}$.

Are Post systems as powerful as Turing machines? The answer is yes, they have equivalent power. So now we have the following equivalent models of computation: Turing machines, simple programs, partial recursive functions, Markov algorithms, Post algorithms, and Post systems.

Other Models of Computation

There are many other computational models that have the same power as Turing machines. For example, most modern programming languages are as powerful as Turing machines if we assume that they can handle numbers of arbitrary size and if we assume that enough memory is always available.

Some models of computation may seem to be more powerful than others. For example, a parallel computation may speed up the solution to a problem. But a parallel language is no more powerful than a nonparallel language. That is, if some problem can be solved by using n processors running in parallel, then the same problem can be solved by using one processor, by simulating the use of n processors.

We'll end this section by reasserting the fact that every computational model invented so far has no more power than that of the Turing machine model. So it's easy to see why most people believe the Church-Turing thesis.

 Exercises

Simple Programs

1. Write *simple* programs to perform the actions indicated by each of the following macros. *Hint:* Each problem can be solved with the aid of a macro in a previous problem or a macro defined in the text.

   a. $Z := X + Y$.
   b. $Z := X * Y$.
   c. **if** $X \neq 0$ **then** S **fi**.
   d. **if** $X \neq 0$ **then** S **else** T **fi**.
   e. $Z := X$ monus Y, where X monus $Y = $ if $X \geq Y$ then $X - Y$ else 0.
   f. **while** $X < Y$ **do** S **od**.
   g. **while** $X \leq Y$ **do** S **od**.
   h. $Z := $ absoluteDiff$(X,\ Y)$, which is the absolute value of the difference between X and Y.

    i.   while $X \neq Y$ do S od.

    j.   while $X \neq 0$ and $Y \neq 0$ do S od.

    k.   while $X \neq 0$ or $Y \neq 0$ do S od.

Partial Recursive Functions

2. Find a definition as a partial recursive function for each of the following functions, where LE means "less than or equal to" and GE means "greater or equal to."

    a. $\mathrm{LE}(x, y) =$ if $x \leq y$ then 1 else 0.

    b. $\mathrm{GE}(x, y) =$ if $x \geq y$ then 1 else 0.

3. Give an informal description of each of the following functions.

    a. $f(x) = \min(y, x + 1 = y)$.   **b.** $f(x) = \min(y, x = y + 1)$.

    c. $f(x, y) = \min(z, x + z = y)$.   **d.** $f(x, y) = \min(z, x = y*z)$.

4. Let $f(x) = \min(y, g(x, y) = 0)$. Find a *simple* program to compute f under the assumption that the simple macro statement $Z := g(X, Y)$ computes a value of the total function g.

Markov Algorithms

5. For each of the following Markov algorithms over $\{a, b\}$, describe the form of the output for an input string of the form $a^i ba^j$.

    a. $ba \to b$ **b.** $b \to \Lambda$ **c.** $ba \to b$

        $aba \to b$ $ba \to b$ $b \to \Lambda$

        $b \to \Lambda$. $aba \to b$. $aba \to b$.

6. Write Markov algorithms to accomplish each of the following actions.

    a.   An infinite loop occurs when the input is the letter a.

    b.   Delete the leftmost occurrence of b in any string over $\{a, b, c\}$.

    c.   For inputs of the form $a^i * a^j$ the output is a^{i+j}.

    d.   Transform strings of the form $a^n bc^n$ to $c^n b^2 a^n$ for any $n \in \mathbb{N}$.

    e.   Interchange all a's and b's in any string over $\{a, b\}$.

Post Algorithms

7. Write Post algorithms to accomplish each of the following actions.

    a.   An infinite loop occurs when the input is the letter a.

    b.   Add 1 to a unary string of 1's.

    c.   For inputs of the form $a^i * a^j$ the output is a^{i+j}.

    d.   Transform strings of the form $a^n bc^n$ to $c^n b^2 a^n$ for any $n \in \mathbb{N}$.

    e.   Interchange all a's and b's in any string over $\{a, b\}$.

    f.   Delete the leftmost occurrence of b in any string over $\{a, b, c\}$.

Post Systems

8. Write a Post system to generate each of the following sets of strings.

    a. The even palindromes over the alphabet $\{a,\ b\}$.

    b. The odd palindromes over the alphabet $\{a,\ b,\ c\}$.

    c. $\{a^n * b^n \# c^n \mid n \in \mathbb{N}\}$.

    d. The binary strings that represent the natural numbers, where each string except 0 begins with 1 on the left end.

13.3 Chapter Summary

This chapter discussed Turing machines and other equivalent computational models. Turing machines may be defined with multiple tapes, multiple heads, and having either two-way infinite tapes or one-way infinite tapes. All formulations are equivalent. Turing machines can recognize languages, and they can generate output. They are general-purpose computation devices that are more powerful than pushdown automata. Deterministic Turing machines have the same power as nondeterministic Turing machines. A simple interpreter can be constructed for Turing machines.

The Church-Turing thesis claims that anything that is intuitively computable can be computed by a Turing machine. We can compare computational models because there are ways to transform between the different types of data that they process. We discussed a variety of computational models that are equal in power to the Turing machine model: simple programs, partial recursive functions, Markov algorithms, Post algorithms, and Post systems. All known computational models are no more powerful than the Turing machine model. So there is much support for the Church-Turing thesis.

chapter 14

Computational Notions

Give us the tools, and we will finish the job.
> —Winston Churchill (1874–1965)

Some problems can't be solved by any machine and some problems that can be solved might be impractical because they could take too much time or space. In this chapter we'll look at the limits of computation by discussing some classic problems that are not solvable by any machine and some classic problems that are solvable. We'll also introduce complexity classes as a way to partition problems with respect to whether solutions take a reasonable amount of time and space.

chapter guide

14.1 (Computability) discusses some general problems that can't be solved by any machine. We'll also see examples of unsolvable problems that are partially solvable, and we'll see examples of unsolvable problems that can be modified to create solvable problems.

14.2 (A Hierarchy of Languages) presents a hierarchy of languages, some of which can be recognized by machines and others that can't be recognized by any machine.

14.3 (Complexity Classes) returns to the positive side of the discussion where we'll see how decision problems are classified by whether there exist solutions whose execution times are "reasonable." After we clarify the meaning of "reasonable," we'll discuss the complexity classes *P*, *NP*, and *PSPACE* and what it means for a problem to be *NP*-complete. Lastly, we'll describe complexity from a formal point of view.

14.1 Computability

It's fun when we solve a problem, and it's not fun when we can't solve a problem. If we can't solve a problem, we can sometimes alter it in some way and then solve the modified problem. If no one can solve a problem, then it could mean that the problem can't be solved, but it could also mean that the problem is hard and a solution will eventually be found. We're going to introduce some classic problems that cannot be solved by any machine. But first we need to discuss a few preliminaries.

If something is computable, the Church-Turing thesis tells us that it can be computed by a Turing machine. A Turing machine or any other equivalent model of computation can be described by a finite number of symbols. So each Turing machine can be considered as a finite string. If we let S be a countable set of symbols, then any Turing machine can be coded as a finite string of symbols over S. Since there are a countable number of strings over S, it follows that there are a countable number of Turing machines.

We can make the same statement for any computational model. That is, there are a countable number of instances of the model. The word "countable" as we've used it means "countably infinite." So there are a countably infinite number of problems that can be solved by computational models. This is nice to know because we won't ever run out of work trying to find algorithms to solve problems. That's good. But we should also be aware that some problems are too general to be solvable by any machine.

Since the inputs and outputs of any computation can be represented by natural numbers, we'll restrict our discussion to functions that have natural numbers as arguments and as values. It follows from Example 2.28 that this set of functions is uncountable. If we assume the Church-Turing thesis, which we do, then there are only countably many of these functions that are "computable," which means that they can be computed by Turing machines or other equivalent computational models.

14.1.1   Effective Enumerations

To discuss computability, we need to define the idea of an effective enumeration of a set. An *effective enumeration* of a set is a listing of its elements by an algorithm. There is no requirement that an effective enumeration should list the elements in any particular order, and it's OK for an effective enumeration to output repeated values.

Let's get to the important part of the discussion. We want to be able to effectively enumerate all instances of any particular computational model. For example, we want to be able to effectively enumerate the set of all Turing machines, or the set of all simple programs, or the set of all programs in your favorite programming language, and so on. Since any instance of a computational model can be thought of as a string of symbols, we'll associate each natural number with an appropriate string of symbols.

One way to accomplish this is to let $b(n)$ denote the binary representation of a natural number n. For example,

$$b(7018) = 1101101101010.$$

Next we'll partition $b(n)$ into seven-bit blocks by starting at the right end of the string. If necessary we can add leading zeros to the left end of the string to make sure that all blocks contain seven bits. For example, $b(7018)$ gets partitioned into the two seven-bit blocks

$$0110110 \quad \text{and} \quad 1101010.$$

Recall that each character in the ASCII character set is represented by a block of seven bits. For example, the block 0110110 represents the character 6 and the block 1101010 represents the character j. Let $a(b(n))$ denote the string of ASCII characters represented by the partitioning of $b(n)$ into seven-bit blocks. For example, we have

$$a(b(7018)) = 6j.$$

Now we're in position to effectively enumerate all of the instances of any computational model. Here's the general idea: If the string $a(b(n))$ represents a syntactically correct definition for an instance of the model, then we'll use it as the nth instance of the model. If $a(b(n))$ doesn't make any sense, we set the nth instance of the model to be some specifically chosen instance. We observed in our little example that $a(b(7018)) = 6j$. So n will have to be a very large number before $a(b(n))$ has a chance of being a syntactically correct instance of the model. But that's OK. All we're interested in is effectively enumerating all instances of a computational model. Since we have forever, we'll eventually get them all.

example 14.1 Some Sample Enumerations

Here are three sample enumerations to clarify the discussion.

Turing machines: For each natural number n, let T_n denote the Turing machine defined as follows: If $a(b(n))$ represents a string of valid Turing machine instructions, then let $T_n = a(b(n))$. Otherwise, let T_n be the simple machine $T_n =$ "(0, a, a, S, Halt)." So we can effectively enumerate all the Turing machines T_0, T_1, $T_2, \ldots$.

Simple programs: For each natural number n, let S_n denote the simple program defined as follows: If $a(b(n))$ represents a string of valid simple program instructions, then let $S_n = a(b(n))$. Otherwise, let $S_n =$ "$X := 0$." So we can effectively enumerate all the simple programs S_0, S_1, $S_2, \ldots$.

Partial recursive functions: For each natural number n, let P_n denote the partial recursive function defined as follows: If $a(b(n))$ represents a string defining a partial recursive function, then let $P_n = a(b(n))$. Otherwise, let $P_n =$ "zero(x) $= 0$." So we can effectively enumerate all the partial recursive functions P_0, P_1, $P_2, \ldots$.

end example

Enumerating Computable Functions

We can effectively enumerate all possible instances of any computational model in a similar way. Therefore, by the Church-Turing thesis, we can effectively enumerate all possible computable functions. For our discussion, we'll assume that we have the following effective enumeration of all the computable functions:

$$f_0, f_1, f_2, \ldots, f_n, \ldots . \tag{14.1}$$

If we like Turing machines, we can think of f_n as the function computed by T_n. If we like partial recursive functions, we can think of f_n as P_n. If we like algorithms expressed in English mixed with other symbols, then we can think of f_n in this way too. The important point is that we can effectively enumerate all the computable functions. Thus the list (14.1) contains all the usual functions that we think of as computable, such as successor, addition, multiplication, and others like

$$p(k) = \text{the } k\text{th prime number.}$$

The list (14.1) also contains many functions that we might not even think about. For example, suppose we define the following function:

$$g(x) = \text{if the fifth digit of } \pi \text{ is 7 then 1 else 0.}$$

Since we know $\pi = 3.14159...$, it follows that g is the constant function $g(x) = 0$. So g is clearly computable. Now let's define the following function:

$$f(x) = \text{if the googolth digit of } \pi \text{ is 7 then 1 else 0.}$$

This function is computable because we know that the condition "the googolth digit of π is 7," is either true or false. Therefore, either f is the constant function 1 or f is the constant function 0. We just don't know the value of f because no one has computed π to a googol places.

The list (14.1) also contains functions that are partially defined, such as the following samples:

$$f(n) = \text{if } n \text{ is odd then } n+1 \text{ else error,}$$
$$h(k) = \text{if } k \text{ is even then 1 else loop forever.}$$

At times we'll want to talk about all the computable functions that take a single argument. Can they be effectively enumerated? Sure. For example, suppose we are enumerating partial recursive functions. If $a(b(n))$ represents a valid string for a partial recursive function of a single variable, then set $P_n = a(b(n))$. Otherwise, set $P_n = \text{"zero}(x) = 0.\text{"}$

Decision Problems

Now we're prepared to study a few classic problems. For convenience of expression, we'll only consider decision problems. A *decision problem* is a problem that asks a question that has a yes or no answer.

A decision problem is *decidable* if there is an algorithm that can input any arbitrary instance of the problem and halt with the correct answer. If no such algorithm exists, then the problem is *undecidable*.

A decision problem is *partially decidable* if there is an algorithm that halts with the answer yes for those instances of the problem that have yes answers, but that may run forever for those instances of the problem whose answers are no.

14.1.2 The Halting Problem

Is there a way to examine a computer program to see whether it halts on an arbitrary input? Depending on the program, we might be able to do it. For example, suppose we're given the function $f : \mathbb{N} \to \mathbb{N}$ defined by $f(x) = 2x + 1$, and we're given the input value 17. We can certainly say that f halts on input 17. In fact we can see that f halts for all natural numbers x. For another example, suppose we're given the function h defined by

$$h(x) = \text{if } x = 7 \text{ then 2 else loop forever.}$$

In this case we can see that $h(7)$ halts. We can also see that $h(x)$ does not halt for all $x \neq 7$.

In these two examples we were able to tell whether the programs halted on arbitrary inputs. Now let's consider a more general question. The *halting problem* asks the following question about programs:

The Halting Problem   (14.2)

Is there an algorithm that can decide whether the execution of an arbitrary program halts on an arbitrary input?

The answer to the question is no. In other words, the halting problem is undecidable. The halting problem was introduced and proved undecidable by Turing [1936]. He considered the problem in terms of Turing machines rather than programs. Of course, we can replace "Turing machine" by any equivalent computational model. If we assume the Church-Turing thesis, then we can replace "algorithm" by any computational model that is equivalent to the Turing machine model. For example, we can state the halting problem as follows in terms of computable functions.

The Halting Problem   (14.3)

Is there a computable function that can decide whether the execution of an arbitrary computable function halts on an arbitrary input?

The answer is no. The halting problem is undecidable.

Proof: Suppose, by way of contradiction, that the answer is yes. Then we must conclude that a computable function exists to compute the condition, "f_x halts on input x," where we assume that the functions f_x in (14.1) take single arguments. Now we need to find some kind of contradiction. A classic way to do this is to define the following function that uses the computable condition about halting:

$$g(x) = \text{if } f_x \text{ halts on input } x \text{ then loop forever else } 0.$$

Notice that g is computable because "f_x halts on input x" is computable. Therefore, g must occur somewhere in the list of computable functions (14.1). So there is a natural number n such that $g = f_n$. In other words,

$$g(x) = f_n(x) \text{ for all natural numbers } x.$$

We can obtain a contradiction by studying the situations that can occur when g is applied to n. If $g(n)$ loops forever (i.e., does not halt), then the condition "f_n halts on input n" is true, which means that $f_n(n)$ halts. But $g(n) = f_n(n)$. So we have the following non sequitur:

$$\text{If } g(n) \text{ does not halt, then } g(n) \text{ halts.}$$

Now look at the other case. If $g(n)$ halts with the value 0, then the condition "f_n halts on input n" is false, which means that $f_n(n)$ does not halt. But $g(n) = f_n(n)$. So we have the following non sequitur:

$$\text{If } g(n) \text{ halts, then } g(n) \text{ does not halt.}$$

So there is no computable function to tell whether an arbitrary computable function halts on its own index. Thus there certainly is no computable function that can tell whether an arbitrary computable function halts on an arbitrary input. Therefore, the halting problem is undecidable. QED.

A restricted form of the halting problem, which is decidable, asks the question: Is there a computable function that, when given f_n, m, and k, can tell whether f_n halts on input m in k units of time? Can you see why this function is computable? We'll leave it as an exercise.

14.1.3   The Total Problem

Is there a way to examine a computer program to see whether it halts on all inputs? In terms of computable functions, is there a way to tell whether an arbitrary computable function is total? For example, it's easy to see that the function f with domain $\mathbb{N}$ defined by $f(x) = x + 1$ is total. But is there an algorithm to decide the question for all computable functions? That's the question asked by the *total problem*.

> ### The Total Problem   (14.4)
> Is there an algorithm to tell whether an arbitrary computable function is total?

The answer is no. The total problem is undecidable.

To prove (14.4), we need the following intermediate result about listing total computable functions.

> ### Theorem   (14.5)
> There is no effective enumeration of all the total computable functions.

Proof: We'll prove (14.5) for the case of natural number functions having a single variable. Suppose, by way of contradiction, that we have an effective enumeration of all the total computable functions:

$$h_0, h_1, h_2, \ldots, h_n, \ldots .$$

Now define a new function H by diagonalization as follows:

$$H(n) = h_n(n) + 1.$$

Since each h_n is total, it follows that H is total. Since the listing is an effective enumeration, there is an algorithm that, when given n, produces h_n. Therefore, $h_n(n) + 1$ can be computed. Thus H is computable. Therefore, H is a total computable function that is not in the listing because it differs from each function in the list at the diagonal entries $h_n(n)$. QED.

Now we'll prove that the total problem (14.4) is undecidable.

Proof: Suppose, by way of contradiction, that the answer is yes. Then we must conclude that a computable function exists to compute the condition, "f_x is a total function," where we are considering the functions f_x in (14.1). We will obtain a contradiction by exhibiting an effective enumeration of all total computable functions, which we know can't happen by (14.5). One way to accomplish this is to define the function g as follows:

$g(0)$ is the smallest index k such that f_k is a total function.

$g(n+1)$ is the smallest index $k > g(n)$ such that f_k is a total function.

Since the condition "f_k is a total function" is computable, it follows that g is computable. Therefore, we have the following effective enumeration of all the total computable functions.

$$f_{g(0)}, f_{g(1)}, f_{g(2)}, \ldots, f_{g(n)}, \ldots .$$

But this contradicts (14.5). So the total problem is undecidable. QED.

14.1.4 Other Problems

We'll conclude this section with a few more examples of undecidable problems. Then we'll have a short discussion about partially decidable problems and decidable problems.

The Equivalence Problem

Can we tell by examining two programs whether they produce the same output when given the same input? Depending on the programs, we might be able to do it. For example, we can certainly tell that the two functions f and g are equal, where $f(x) = x + x$ and $g(x) = 2x$. The *equivalence problem* asks a more general question:

The Equivalence Problem   (14.6)
Does there exist an algorithm that can decide whether two arbitrary computable functions produce the same output?

The answer to this question is no. So the equivalence problem is undecidable. If we restrict the problem to deciding whether an arbitrary computable function is the identity function, the answer is still no. Most proofs of the equivalence problem show that this restricted version is undecidable.

Post's Correspondence Problem

The problem known as *Post's correspondence problem* was introduced by Post [1946]. An *instance* of the problem can be stated as follows: Given a finite sequence of pairs of strings

$$(s_1, t_1), \ldots, (s_n, t_n),$$

is there a sequence of indexes $i_1, \ldots, i_k$, with repetitions allowed, such that

$$s_{i_1} \ldots s_{i_k} = t_{i_1} \ldots t_{i_k}?$$

For example, let's consider the instance of the problem consisting of the following sequence of pairs, where we'll number the pairs sequentially as 1, 2, 3, and 4:

$$(ab, a), (b, bb), (aa, b), (b, aab).$$

After a little fiddling we can find that the sequence 1, 2, 1, 3, 4 will produce the following equality:

$$abbabaab = abbabaab.$$

For another example, let's consider the instance of the problem described by the following sequence of pairs, which we'll refer to as 1 and 2:

$$(ab, a), \; (b, ab) \, .$$

This instance of the problem has no solution. To see this, notice that any solution sequence would have to contain an equal number of 1's and 2's to make sure that the two strings have equal length. But this implies that the left side would have twice as many b's than a's and the right side would have twice as many a's than b's. So the strings could not be equal. Here's the problem.

Post's Correspondence Problem   (14.7)

Is there an algorithm that can decide whether an arbitrary instance of the problem has a solution?

Post's correspondence problem is undecidable. At first glance it might seem that the problem is solvable by an algorithm that exhaustively checks sequences of pairs for a desired equality of strings. But if there is no solution for some instance of the problem, then the algorithm will go on checking ever larger sequences for an equality that doesn't exist. So it won't be able to halt and tell us that there is no solution sequence.

Hilbert's Tenth Problem

In 1900, Hilbert stated 23 problems that—at the time—were not solved. Here's the 10th problem that he gave.

Hilbert's Tenth Problem   (14.8)

Does a polynomial equation $p(x_1, \ldots, x_n) = 0$ with integer coefficients have a solution consisting of integers?

Of course, we can solve specific instances of the problem. For example, it's easy for us to find integer solutions to the equation

$$2x + 3y + 1 = 0.$$

It's also easy for us to see that there are no integer solutions to the equation

$$x^2 - 2 = 0.$$

In 1970, Matiyasevich proved that Hilbert's tenth problem is undecidable. So there is no algorithm to decide whether an arbitrary polynomial equation with integer coefficients has a solution consisting of integers.

Turing Machine Problems

There are many undecidable problems that deal with halting in one way or another and they are often expressed in terms of Turing machines. For example, each of the following problems is undecidable, where M is an arbitrary Turing machine:

>   Does M halt when started on the empty tape?

>   Is there an input string for which M halts?

>   Does M halt on every input string?

Of course, with a little effort, these problems can also be stated in terms of other computational models. For example, in terms of partial recursive functions, the three problems take the following form, where f is an arbitrary partial recursive function:

>   Is $f(0)$ defined?

>   Is there a natural number n for which $f(n)$ is defined?

>   Is f a total function?

The Busy Beaver Problem

Let's spend a little time discussing a problem from the last chapter about hard-working Turing machines. In Section 13.1 we defined a busy beaver as a deterministic Turing machine that writes either Λ or 1 on a tape cell such that it accepts the empty string and, after it halts, the number of tape cells containing 1 is the maximum of all such Turing machines with the same number of states that accept the empty string.

   The *busy beaver function* is the function b with domain the positive integers such that $b(n)$ is the number of 1's that can be written by a busy beaver with n states, not including the halt state. We noted that $b(1) = 1$, $b(2) = 4$, $b(3) = 6$, $b(4) = 13$, and $b(5) \geq 4{,}098$. We also noted that Rado [1962] proved that the busy beaver function is not computable. In other words, the following problem is undecidable:

The Busy Beaver Problem

For an arbitrary value of n, can $b(n)$ be computed?

We'll prove that b is not computable by proving four statements, the last of which is the desired result.

1. $b(n + 1) > b(n)$ for all $n \geq 1$.

Proof: If T is a busy beaver with n states, then we can modify T as follows: Replace the Halt state by a state that looks to the right for an empty cell and when it is found, it writes a 1 and then halts. This new machine has $n + 1$ states and writes one more 1 than T. Therefore, a busy beaver with $n + 1$ states writes more 1's than T. Therefore, $b(n + 1) > b(n)$. QED.

2. If f is computable by a Turing machine with k states, then $b(n + k) \geq f(n)$ for all $n \geq 1$.

Proof: Let T be the k-state Turing machine that computes f by starting with a tape that contains a string of n 1's and halting with a tape that contains a string of $f(n)$ 1's. For any given n, construct a new machine that starts with an empty tape and uses n states to write n 1's onto the tape. Then it transfers to the start state of T. The new machine has $n + k$ states and when it is started on the empty tape, it halts with $f(n)$ 1's on the tape. Therefore, a busy beaver with $n + k$ states must write at least $f(n)$ 1's on the tape. Therefore, we have the desired result $b(n + k) \geq f(n)$. QED.

3. The composition of two computable functions is computable.

Proof: We'll leave this general result as an exercise. QED.

4. The busy beaver function b is not computable.

Proof: Suppose, by way of contradiction, that b is computable. Pick the simple computable function g defined by $g(n) = 2n$. Let f be the composition of b with g. That is, let $f(n) = b(g(n)) = b(2n)$. The function f is computable by Statement 3. So we can apply Statement 2 to obtain the result $b(n + k) \geq f(n)$ for all $n \geq 1$. But $f(n) = b(2n)$. So Statement 2 becomes $b(n + k) \geq b(2n)$ for all $n \geq 1$. If we let $n = k + 1$, then we obtain the inequality

$$b(k + 1 + k) \geq b(2(k + 1)),$$

which can be written as $b(2k + 1) \geq b(2k + 2)$. But this inequality contradicts Statement 1. Therefore, b is not computable. QED.

We should note that busy beavers and the busy beaver function can be described with other computational models. For example, Morales-Bueno [1995] uses the simple programming language to describe busy beavers and to prove that b is not computable.

Partially Decidable Problems

Recall that a decision problem is partially decidable if there is an algorithm that halts with the answer yes for those instances of the problem with yes answers, but that may run forever for those instances whose answers are no. Many undecidable problems are in fact partially undecidable. Whenever we can search for a yes answer to a decision problem and be sure that it takes a finite amount of time, then we know that the problem is partially decidable.

For example, the halting problem is partially decidable because, for any computable function f_n and any input x, we can evaluate the expression $f_n(x)$. If the evaluation halts with a value, then we output yes. We don't care what happens if $f_n(x)$ is undefined or its evaluation never halts.

Another partially decidable problem is Post's correspondence problem. In this case, we can check for a solution by systematically looking at all sequences of length 1, then length 2, and so on. If there is a sequence that gives two matching strings, we'll eventually find it and output yes. Otherwise, we don't care what happens.

The total problem is not even partially decidable. Although we won't prove this fact, we can at least observe that an arbitrary computable function f_n can't be proven total by testing it on every input because there are infinitely many inputs.

Decidable Problems

The things that we really want to deal with are decidable problems. In fact, this book is devoted to presenting ideas and techniques for solving problems. For example, in Chapter 11 we studied several techniques for solving the recognition problem for regular languages. In Chapter 12 we saw that the recognition problem for context-free languages is decidable.

It's also nice to know that most undecidable problems have specific instances that are decidable. For example, the following problems are decidable.

1. Given the function $f(x) = x + 1$, does f halt on input $x = 35$? Does f halt on any input?

2. Is Ackermann's function a total function?

3. Are the following two functions f and g equivalent?

$$f(x) = \text{if } x \text{ is odd then } x \text{ else } x + 1,$$
$$g(x) = \text{if } x \text{ is even then } 2x - x + 1 \text{ else } x.$$

 Exercises

Computable Functions

1. Show that the composition of two computable functions is computable. In other words, show that if $h(x) = f(g(x))$, where f and g are computable and the range of g is a subset of the domain of f, then h is computable.

2. Show that the following function is computable.

$$h(x) = \text{if } f_x \text{ halts on input } x \text{ then } 1 \text{ else loop forever.}$$

3. Suppose we have the following effective enumeration of all the computable functions that take a single argument:

$$f_0, f_1, f_2, \ldots, f_n, \ldots.$$

For each of the following functions g, explain what is *wrong* with the following diagonalization argument claiming to show that g is a computable function that isn't in the list. "Since the enumeration is effective, there is an algorithm to transform each n into the function f_n. Since each f_n is computable, it follows that g is computable. It is easy to see that g is not in the list. Therefore, g is a computable function that isn't in the list."

   a. $g(n) = f_n(n) + 1$.
   b. $g(n) = $ if $f_n(n) = 4$ then 3 else 4.
   c. $g(n) = $ if $f_n(n)$ halts and $f_n(n) = 4$ then 3 else 4.
   d. $g(n) = $ if $f_n(n)$ halts and $f_n(n) = 4$ then 3 else loop forever.

Decidability

4. Show that the following problem is decidable: Is there a computable function that, when given f_n, m, and k, can tell whether f_n halts on input m in k units of time?

5. Show that the problem of deciding whether two DFAs over the same alphabet are equivalent is decidable.

6. For each of the following instances of Post's correspondence problem, find a solution or state that no solution exists.

   a. $\{(a, abbbbb), (bb, b)\}$.
   b. $\{(ab, a), (ba, b), (a, ba), (b, ab)\}$.
   c. $\{(10, 100), (0, 01), (0, 00)\}$.
   d. $\{(1, 111), (0111, 0), (10, 0)\}$.
   e. $\{(ab, aba), (ba, abb), (b, ab), (abb, b), (a, bab)\}$.

14.2 A Hierarchy of Languages

We now have enough tools to help us describe a hierarchy of languages. In addition to meeting some old friends, we'll also meet some new kids on the block.

14.2.1 The Languages

Starting with the smallest class of languages—the regular languages—we'll work our way up to the largest class of languages—the languages without grammars.

Regular Languages

Regular languages are described by regular expressions and they are the languages that are accepted by NFAs and DFAs. Regular languages are also defined by grammars with productions having the form

$$A \to \Lambda, \quad A \to w, \quad A \to B, \quad \text{or} \quad A \to wB,$$

where A and B are nonterminals and w is a nonempty string of terminals. A typical regular language is $\{a^m b^n \mid m,\ n \in \mathbb{N}\}$.

Deterministic Context-Free Languages

Deterministic context-free languages are recognized by deterministic PDAs that accept by final state, and they are described by LR(1) grammars. The language $\{a^n b^n \mid n \in \mathbb{N}\}$ is the standard example of a deterministic context-free language that is not regular.

Context-Free Languages

Context-free languages are recognized by PDAs that may be deterministic or nondeterministic. Context-free languages are also defined by grammars with productions having the form $A \to w$, where A is a nonterminal and w is a string of grammar symbols. The language of palindromes over $\{a,\ b\}$ is a classic example of a context-free language that is not deterministic context-free.

Context-Sensitive Languages

A *context-sensitive grammar* has productions of the form $xAz \to xyz$, where A is a nonterminal and x, y, and z are strings of grammar symbols with $y \neq \Lambda$. The production $S \to \Lambda$ is also allowed if S is the start symbol and it does not appear on the right side of any production. The *context-sensitive languages* are those that have context-sensitive grammars. The term context-sensitive is well chosen because a production $xAz \to xyz$ can be used to replace A by y only within the surrounding context x and z.

We saw in Section 12.4 that any context-free grammar can be transformed into Chomsky normal form with productions having the form

$$A \to BC, \ A \to a, \ \text{or} \ S \to \Lambda,$$

where B and C are nonterminals, a is a terminal, and S is the start symbol. Also, if the production $S \to \Lambda$ occurs, then S does not appear on the right side of any production. The productions $A \to BC$ and $A \to a$ satisfy the context-sensitive property, where $x = z = \Lambda$. Therefore, any context-free language is context-sensitive.

The language $\{a^n b^n c^n \mid n \geq 0\}$ is the standard example of a context-sensitive language that is not context-free. (We showed that it is not context-free in Example 12.21.) Here's a context-sensitive grammar for the language:

$$S \to \Lambda \mid abc \mid aTBc$$
$$T \to abC \mid aTBC$$
$$CB \to CX \to BX \to BC$$
$$bB \to bb$$
$$Cc \to cc.$$

Notice that the three productions $CB \to CX \to BX \to BC$ are used to change CB into BC. We can't accomplish this change with the single production $CB \to BC$ because this production is not context-sensitive. Let's look now at a less restrictive kind of grammar that allows such productions but still generates the same collection of context-sensitive languages.

A *monotonic* grammar has productions of the form $u \to v$, where $|u| \leq |v|$ and u contains at least one nonterminal. The production $S \to \Lambda$ is also allowed if S is the start symbol and it does not appear on the right side of any production. Notice that with this grammar we don't have to worry about maintaining the context. For example, the production $CB \to BC$ is allowed.

Notice that any context-sensitive grammar is monotonic because any production of the form $xAz \to xyz$, where $y \neq \Lambda$, satisfies $|xAz| \leq |xyz|$. It is somewhat surprising that even though monotonic grammars have fewer restrictions, they still generate only the context-sensitive languages. We'll sketch the proof. First we observe that for either type of grammar, we can obtain an equivalent grammar where each production consists of only nonterminals or has the form $A \to a$ when a is a terminal. For example, the production

$$aSb \to acSb$$

would be put in the following form, where A, B, and C are new nonterminals:

$$ASB \to ACSB$$
$$A \to a$$
$$B \to b$$
$$C \to c.$$

With grammars in this new form, it is easy to show that the language of a monotonic grammar is context-sensitive. For example, the monotonic production $CB \to BC$ can be replaced by the three context-sensitive productions

$$CB \to CX \to BX \to BC,$$

where X is a new nonterminal that is used only in these productions. This idea can be extended to any monotonic production. For another example, suppose we have the monotonic production $STU \to ABCD$. We can replace this production by the context-sensitive productions

$$STU \to XTU \to XYU \to XYZ \to XYCD \to XBCD \to ABCD,$$

where X, Y, and Z are new nonterminals that are used only in these productions. So we have the following theorem.

> **Theorem**
>
> The context-sensitive languages are those with monotonic grammars. In other words, monotonic grammars are equivalent to context-sensitive grammars.

This allows some simplification in the writing of a grammar. For example, here's a monotonic grammar for the standard example $\{a^n b^n c^n\} \,|\, n \geq 0\}$.

$$S \to \Lambda \,|\, abc \,|\, aTbc$$
$$T \to abC \,|\, aTbC$$
$$Cb \to bC$$
$$Cc \to cc.$$

If we restrict a Turing machine to a finite tape consisting of the input tape cells together with two boundary cells that may not be changed, and if we allow nondeterminism, then the resulting machine is called a *linear bounded automaton* (LBA). Here's the connection between LBAs and context-sensitive languages:

> **Theorem**
>
> The context-sensitive languages coincide with the languages that are accepted by LBAs.

For Example, in Example 13.2 we constructed a Turing machine to recognize the language $\{a^n b^n c^n \,|\, n \geq 0\}$. In fact the Turing machine that we constructed is an LBA because it uses only the tape cells of the input string.

An interesting fact about context-sensitive languages is that they can be recognized by LBAs that always halt. In other words,

> **Theorem**
>
> The recognition problem for context-sensitive languages is decidable.

The idea for the proof is to show that for any context-sensitive grammar and any natural number n, it is possible to construct all the derivations from the start symbol that produces sentential forms of length n or less. So if w is a string of length n, then we simply check to see whether w coincides with any of these sentential forms.

Recursively Enumerable Languages

The most general kind of grammars are the *unrestricted grammars* with productions of the form $v \to w$, where v is any nonempty string and w is any string. So the general definition of a grammar (3.12) is that of an unrestricted grammar. Unrestricted grammars are also called *phrase-structure grammars*. The most important thing about unrestricted grammars is that the class of languages that they generate is exactly the class of languages that are accepted by Turing machines. Although we won't prove this statement, we should note that a proof consists of transforming any unrestricted grammar into a Turing machine and transforming any Turing machine into an unrestricted grammar. The resulting algorithms have the same flavor as the transformation algorithms (12.7) and (12.8) for context-free grammars and PDAs.

A language is called *recursively enumerable* if there is a Turing machine that outputs (i.e., enumerates) all the strings of the language. If we assume the Church-Turing thesis, we can replace "Turing machine" with "algorithm." For example, let's show that the language $\{a^n \mid f_n(n) \text{ halts}\}$ is recursively enumerable. We are assuming that the functions f_n are the computable functions that take only single arguments and are enumerated as in (14.1). Here's an algorithm to enumerate the language $\{a^n \mid f_n(n) \text{ halts}\}$:

1. Set $k = 0$.

2. For each pair (m, n) such that $m + n = k$, do the following:

   if $f_n(n)$ halts in m steps, then output a^n.

3. Increment k, and go to Step 2.

An important fact that we won't prove is that the class of recursively enumerable languages is exactly the same as the class of languages that are accepted by Turing machines. Therefore, we have three different ways to say the same thing.

Theorem

The languages generated by unrestricted grammars are the languages accepted by Turing machines, which are the recursively enumerable languages.

The language $\{a^n \mid f_n(n) \text{ halts}\}$ is a classic example of a recursively enumerable language that is not context-sensitive. The preceding algorithm shows that the language is recursively enumerable. Now if the language were context-sensitive, then it could be recognized by an algorithm that always halts. This means that we would have a solution to the halting problem, which we know to be undecidable. Therefore, $\{a^n \mid f_n(n) \text{ halts}\}$ is not context-sensitive.

Nongrammatical Languages

There are many languages that are not definable by any grammar. In other words, they are not recursively enumerable, which means that they can't be recognized by Turing machines. The reason for this is that there are an uncountable number of languages and only a countable number of Turing machines (we enumerated them all in the last section). Even for the little alphabet $\{a\}$ we know that power($\{a\}^*$) is uncountable. In other words, there are an uncountable number of languages over $\{a\}$.

The language $\{a^n \mid f_n \text{ is total}\}$ is a standard example of a language that is not recursively enumerable. This is easy to see. Again we're assuming that the functions f_n are the computable functions listed in (14.1). Suppose, by way of contradiction, that the language is recursively enumerable. This means that we can effectively enumerate all the total computable functions. But this contradicts (14.5), which says that the total computable functions can't be effectively enumerated. Therefore, the language $\{a^n \mid f_n \text{ is total}\}$ is not recursively enumerable.

14.2.2 Summary

Let's summarize the hierarchy that we've been discussing. Figure 14.1 shows a table where each line represents a particular class of grammars and/or languages. Each line also contains an example language together with the type of machine used to recognize languages of the class. Each line represents a more general class than the next lower line of the table.

The example language given on each line is a classic example of a language that belongs on that line but not on a lower line of the table. The symbols DPDA, LBA, TM, and NTM mean deterministic pushdown automaton, linear bounded automaton, Turing machine, and nondeterministic Turing machine.

The four grammars—unrestricted, context-sensitive, context-free, and regular—were originally introduced by Chomsky [1956, 1959]. He called them type 0, type 1, type 2, and type 3, respectively.

Grammar or Language	Classic Example	Machine
Arbitrary (grammatical or nongrammatical)	$\{a^n \mid f_n \text{ is total}\}$	None
Unrestricted (recursively enumerable)	$\{a^n \mid f_n(n) \text{ halts}\}$	TM or NTM
Context-sensitive	$\{a^n b^n c^n \mid n \in \mathbb{N}\}$	LBA
Context-free	Palindromes over $\{a, b\}$	PDA
LR(1) (deterministic context-free)	$\{a^n b^n \mid n \in \mathbb{N}\}$	DPDA
Regular	$\{a^m b^n \mid m, n \in \mathbb{N}\}$	DFA or NFA

Figure 14.1   A language hierarchy table.

◣ Exercises

1. Construct a two-tape Turing machine to enumerate all strings in the language $\{a^n b^n c^n \mid n \geq 0\}$. Use the first tape to keep track of n, the number of a's, b's, and c's to write for each string. Use the second tape to write the strings, each separated by the symbol #.

2. Write a monotonic grammar for each of the following context-sensitive languages.

   a. The language of all strings over $\{a, b, c\}$ that contain the same number of a's, b's, and c's.
   b. $\{a^n b^n a^n \mid n > 0\}$.

3. Write a context-sensitive grammar for each of the following context-sensitive languages.

   a. The language of all strings over $\{a, b, c\}$ that contain the same number of a's, b's, and c's.
   b. $\{a^n b^n a^n \mid n > 0\}$.

◥

14.3   Complexity Classes

Some computational problems are impractical because the known algorithms to solve them take too much time or space. In this section we'll make the idea of "too much" more precise by introducing some fundamental complexity classes of problems. The main results of the theory are stated in terms of *decision problems* that ask a question whose answer is either yes or no. So we'll consider only decision problems.

   We should note that a computational problem can often be rephrased as a decision problem that reflects the original nature of the problem. For example, the problem of finding the prime factors of a natural number greater than 1 can be rephrased as a decision problem that asks whether a natural number greater than 1 is composite (not prime).

   Let's look at a famous problem. The traveling salesman problem is to find the shortest tour of a set of cities that starts and ends at the same city. This problem can be modified to form a decision problem by giving a number B and asking whether there is a tour of the cities that starts and ends at the same city and the length of the tour is less than or equal to B. We'll make a more precise statement of the decision version of the traveling salesman problem:

> **Traveling Salesman Problem**
>
> Given a set of cities $\{c_1, \ldots, c_m\}$, a set of distances $d(c_i, c_j) > 0$ for each $i \neq j$, and a bound $B > 0$, does there exist a tour $(v_1, \ldots, v_m)$ of the m cities (a permutation of the cities) such that
>
> $$d(v_1, v_2) + \cdots + d(v_{m-1}, v_m) + d(v_m, v_1) \leq B?$$

An *instance* of a decision problem is a specific example of the given part of the problem. For example, an instance I of the traveling salesman problem can be represented as follows:

$$I = \{\{c_1, c_2, c_3, c_4\}, B = 27, d(c_1, c_2) = 10, d(c_1, c_3) = 5, d(c_1, c_4) = 9,$$
$$d(c_2, c_3) = 6, d(c_2, c_4) = 9, d(c_3, c_4) = 3\}.$$

The *length* of an instance is an indication of the space required to represent the instance. For example, the length of the preceding instance I might be the number of characters that occur between the two braces { and }. Or the length might be some other measure like the number of bits required to represent the instance as an ASCII string. We often approximate the length of an instance. For example, an instance I with m cities contains $m(m-1)/2$ distances and one bounding relation. We can assume that each of these entities takes no more than some constant amount of space. If c is this constant, then the length of I is no more than

$$c[m + 1 + m(m-1)/2].$$

So we can assume that the length of I is $O(m^2)$.

Sometimes we want more than just a yes or no answer to a decision problem. A *solution* for an instance of a decision problem is a structure that yields a yes answer to the problem. If an instance has a solution, then the instance is called a *yes-instance*. Otherwise, the instance is a *no-instance*. For example, the tour (c_1, c_2, c_4, c_3) is a solution for the instance I given previously for the traveling salesman problem because its total distance is 27. So I is a yes-instance of the traveling salesman problem.

14.3.1 The Class *P*

Informally, the class P consists of all problems that can be solved in polynomial time. Let's clarify this statement a bit. For our purposes, a *deterministic algorithm* is an algorithm that never makes an arbitrary choice of what to do next during a computation. In other words, each step of the algorithm is uniquely determined. We say that a deterministic algorithm *solves* a decision problem if for each instance of the problem the algorithm halts with the correct answer, yes or no.

Now we can be a bit more precise and say that the class P consists of those decision problems that can be solved by deterministic algorithms that have worst-case running times of polynomial order. In other words, a decision problem is in the class P if there is a deterministic algorithm A that solves the problem and there is a polynomial p such that for each instance I of the problem we have $W_A(n) \leq p(n)$, where n is the size of I. In short, the class P consists of those decision problems that can be solved by deterministic algorithms of order $O(p(n))$ for some polynomial p.

There are many familiar problems in the class P. For example, consider the problem determining whether an item can be found in an n-element list. A simple search that compares the item to each element of the list takes at most n comparisons. If we assume that the size of the input is n, then the algorithm solves the problem in time $O(n)$. Thus the problem is in P.

A problem is said to be *tractable* if it is in P and *intractable* if it is not in P. In other words, a problem is intractable if it has a lower bound worst-case complexity greater than any polynomial.

14.3.2   The Class *NP*

Informally, the class NP consists of all problems for which a solution can be checked in polynomial time. Problems in NP can have algorithms that search in a nondeterministic manner for a solution. The stipulation is that a solution path must take no more than polynomial time. Let's clarify things a bit. For our purposes, a *nondeterministic algorithm* for an instance I of a decision problem has two distinct stages as follows:

Guessing Stage

A guess is made at a possible solution S for instance I.

Checking Stage

A deterministic algorithm starts up to supposedly check whether the guess S from the guessing stage is a solution to instance I. This checking algorithm will halt with the answer yes if and only if S is a solution of I. But it may or may not halt if S is not a solution of I.

In theory, the guess at a possible solution S could be made out of thin air. Also in theory, S could be a structure of infinite length so that the guessing stage would never halt. And also in theory, the checking stage may not even consider S.

We say that a nondeterministic algorithm solves a decision problem in polynomial time if there is a polynomial p such that for each yes-instance I there is a solution S that when guessed in the guessing stage will lead the checking stage to halt with yes, and the time for the checking stage is less than or equal to $p(n)$, where $n = \text{length}(I)$.

Now we can be a bit more precise and say that the class NP consists of those decision problems that can be solved by nondeterministic algorithms in polynomial time.

Let's consider some relationships between P and NP. The first result is that P is a subset of NP. To see this, notice that any problem π in P has a deterministic algorithm A that solves any instance of π in polynomial time. We can construct a nondeterministic polynomial time algorithm to solve π as follows: For any instance I of π, let the guessing stage make a guess S. Let the checking stage ignore S and run algorithm A on I. This stage will halt with yes or no depending on whether I is a yes-instance or not. Therefore, π is in NP. So we've proven the following simple relationship:

$$P \subseteq NP. \tag{14.9}$$

So we have a lot of problems—all those in P—that are in NP. It is not known whether P and NP are equal. In other words, no one has been able to find an NP problem that is not in P, which would prove that $P \neq NP$, and no one has been able to prove that all NP problems are in P, which would prove that $P = NP$. This problem is one of the foremost open questions of mathematics and computer science.

example   **14.2   The Traveling Salesman Problem Is in NP**

The traveling salesman problem is an NP problem. Let's see why. A guessing stage can guess a tour $(v_1, \ldots, v_m)$ of the m cities. Then the checking stage can check whether

$$d(v_1, v_2) + \cdots + d(v_{m-1}, v_m) + d(v_m, v_1) \leq B.$$

This check takes $m - 1$ addition operations and one comparison operation. We can assume that each of these operations takes a constant amount of time. Therefore, a guess can be checked in time $O(m)$. Since the length of an instance of the traveling salesman problem is $O(m^2)$, it follows that the checking stage takes time $O(\sqrt{n})$, where n is the length of an instance of the problem. So the checking stage can be done in polynomial time—actually better than linear time. Therefore, the traveling salesman problem is in NP.

end example

It is not known whether the traveling salesman problem is in P. It appears that any deterministic algorithm to solve the problem might have to check all possible tours of m cities. Since each tour begins and ends at the same city, there are $(m - 1)!$ possible tours to check. But $(m - 1)!$ has higher order than any polynomial in m, and thus it also has higher order than any polynomial in m^2. Since the length of an input instance is $O(m^2)$, it follows that the worst-case running time of such an algorithm is greater than any polynomial with respect to the length of an input instance.

14.3.3   The Class *PSPACE*

Now let's look at a class of decision problems that are characterized by the space required by algorithms to solve them. The class *PSPACE* is the set of decision problems that can be solved by deterministic algorithms that use no more memory cells than a polynomial in the length of an instance. In other words, a problem is in *PSPACE* if there is a deterministic algorithm to solve it and there is a polynomial p such that the algorithm uses no more than $p(n)$ memory cells, where n is the length of an instance.

It's interesting to observe that $NP \subseteq PSPACE$. Here's why. For any problem π in *NP* there is a nondeterministic algorithm A and a polynomial p such that A takes at most $p(n)$ steps to check a solution for a yes-instance I of length n. Any step of A can access at most a fixed number k of memory cells. So A uses at most $kp(n)$ memory cells to check a solution for I. Since p is a polynomial, kp is also a polynomial. So the checking stage uses polynomial space. If S is a solution for I, then the part of S used by the checking stage fits within $kp(n)$ memory cells. So we can assume that S is a string of length at most $kp(n)$ over a finite alphabet of symbols—one symbol per memory cell.

Now let's define a deterministic algorithm B to solve π. For an instance of length n, B generates and checks—one at a time—all possible strings of length at most $kp(n)$. The checking is done by the checking stage of A modified to stop after $p(n)$ steps if it hasn't stopped yet. If a solution is found, then B stops with a yes-answer. Otherwise, B stops with a no-answer after generating and checking all possible strings of length at most $kp(n)$.

The generating stage uses polynomial space because it generates a string of length at most $kp(n)$. The checking stage uses polynomial space because it is the checking stage of A modified by adding a clock. The finite alphabet and other local variables use a constant amount of space. So B uses polynomial space. Therefore, $NP \subseteq PSPACE$. We can put this together with (14.9) to obtain the following relationships:

$$P \subseteq NP \subseteq PSPACE. \tag{14.10}$$

Just as it is not known whether P and NP are equal, it is also not known whether NP and $PSPACE$ are equal.

Quantified Boolean Formulas

Let's look at a classic example of a *PSPACE* problem. That is, a problem that is in *PSPACE* but for which it is not known whether it is in *NP*. Before we can state the problem, we need a definition. A *quantified Boolean formula* is a logical expression of the form

$$Q_1 x_1 Q_2 x_2 \ldots Q_n x_n E$$

where $n \geq 1$, each Q_i is either $\forall$ or $\exists$, each x_i is distinct, and E is a wff of the propositional calculus that is restricted to using the variables $x_1, \ldots, x_n$, the operations $\neg$, $\wedge$, and $\vee$, and parentheses. For example, the following wffs are quantified Boolean formulas:

$$\exists x \ x,$$
$$\forall x \ \exists y \ (\neg x \vee y),$$
$$\forall x \ \exists y \ \forall z \ ((x \vee \neg y) \wedge z).$$

A quantified Boolean formula is true if its value is True over the domain {True, False}. Otherwise, it is false. For example, the preceding three formulas are true, true, and false, respectively.

The problem we want to consider is whether a given quantified Boolean formula is true. Here's the statement of the problem.

Quantified Boolean Formula Problem (*QBF*)

Given a quantified Boolean formula, is it true?

We'll show that *QBF* is in *PSPACE*. Let "val" compute the truth value of a formula. The following equations give a recursive definition of val.

val(True) = True	val($A \wedge B$) = val(A) **and** val(B)
val(False) = False	val($A \vee B$) = val(A) **or** val(B)
val((A)) = val(A)	val($\forall x E$) = val($E(x/\text{True})$) **and** val($E(x/\text{False})$)
val($\neg A$) = **not** val(A)	val($\exists x E$) = val($E(x/\text{True})$) **or** val($E(x/\text{False})$)

The number of operations and distinct variables in a formula is at most the length k of the formula. So the time required by the algorithm is $O(2^k)$. But the depth of recursion is $O(k)$ and the space required for each recursive call is $O(k)$. So the space used by the algorithm is $O(k^2)$. Therefore, *QBF* is in *PSPACE*. It is not known whether *QBF* is in *NP*.

14.3.4 Intractable Problems

We haven't yet given an example of an intractable problem (i.e., a problem that is not in *P*). Of course, any undecidable problem is intractable because there is no algorithm to solve it. So there is no polynomial time algorithm to solve it. But this isn't very satisfying. So we'll give some real live examples of problems that are intractable.

Presburger Arithmetic

The first intractable problem involves arithmetic formulas. The problem is to decide the truth of statements in a simple theory about addition of natural numbers. The statements of the theory are expressed as closed wffs of a first-order predicate calculus that uses just + and =. For example, the following formulas are wffs of the theory:

$$\forall x \, \forall y \, (x + y = y + x) \, ,$$
$$\exists y \, \forall x \, (x + y = x) \, ,$$
$$\forall x \, \forall y \, \exists z \, (\neg \, (x = y) \rightarrow x + z = y) \, ,$$
$$\forall x \, \forall y \, \forall z \, (x + (y + z) = (x + y) + z) \, ,$$
$$\forall x \, \forall y \, (x + x = x \rightarrow x + y = y) \, .$$

Each wff of the theory is either true or false when interpreted over the natural numbers. You might notice that one of the preceding example wffs is false and the other four are true. In 1930, Presburger showed that this theory is decidable. In other words, there is an algorithm that can decide whether any wff of the theory is true. The theory is called *Presburger arithmetic*.

Fischer and Rabin [1974] proved that any algorithm to solve the decision problem for Presburger arithmetic must have an exponential lower bound for the number of computational steps. Here's the result.

Theorem

There is a constant $c > 0$ such that for every nondeterministic or deterministic algorithm A that solves the decision problem for Presburger arithmetic, there is a natural number k such that for every natural number $n > k$, there is a wff of length n for which A requires more than $2^{2^{cn}}$ computational steps to decide whether the wff is true.

This statement tells us that the decision problem for Presburger arithmetic is not in *NP*. Therefore, it is not in *P*. So it must be intractable.

Now let's look at another problem about Presburger arithmetic that has exponential space complexity. The problem concerns the length of formal proofs in Presburger arithmetic. Fischer and Rabin [1974] proved that any proof system for wffs of Presburger arithmetic contains proofs of exponential length. Here's the result.

Theorem

There is a positive constant $c > 0$ such that for every formal proof system for Presburger arithmetic, there is a natural number k such that for every natural number $n > k$ there is a true wff of length n for which its shortest formal proof requires more than $2^{2^{cn}}$ symbols.

There is no decision problem mentioned in this theorem. So we don't have a problem to classify. But we can at least conclude that any formal proof system that contains the simple Presburger arithmetic will use a lot of space for some formal proofs.

Generalized Regular Expressions

The second intractable problem involves regular expressions. Recall that the set of *regular expressions* over an alphabet A is defined inductively as follows, where $+$ and $\cdot$ are binary operations and * is a unary operation:

> *Basis*: Λ, $\varnothing$, and a are regular expressions for all $a \in A$.

> *Induction*: If R and S are regular expressions, then the following expressions are also regular: (R), $R + S$, $R \cdot S$, and R^*.

For example, here are a few of the infinitely many regular expressions over the alphabet $A = \{a, b\}$: Λ, $\varnothing$, a, b, $\Lambda + b$, b^*, $a + (b \cdot a)$, $(a + b) \cdot a$, $a \cdot b^*$, $a^* + b^*$.

Each regular expression represents a regular language. For example, Λ represents the language $\{\Lambda\}$; $\varnothing$ represents the empty language $\varnothing$; $a \cdot b^*$ represents the language of all strings that begin with a and are followed by zero or more occurrences of b; and $(a + b)^*$ represents the language $\{a, b\}^*$.

Suppose we extend the definition of regular expressions to include the additional notation $(R)^2$ as an abbreviation for $R \cdot R$. For example, we have

$$a \cdot a \cdot a \cdot a \cdot a = a \cdot \left(\left(a \right)^2 \right)^2 = a \cdot \left(a^2 \right) \cdot \left(a^2 \right).$$

A *generalized regular expression* is a regular expression that may use this additional notation. Now we're in position to state an intractable problem.

Inequivalence of Generalized Regular Expressions

Given a generalized regular expression R over a finite alphabet A, does the language of R differ from A^*?

Here are some examples to help us get the idea:

> The language of $(a + b)^*$ is the same as $\{a, b\}^*$.

> The language of $\Lambda + (a \cdot b)^2 + a^* + b^*$ differs from $\{a, b\}^*$.

> The language of $\Lambda + (a \cdot b)^2 + (a + b)^*$ is the same as $\{a, b\}^*$.

Meyer and Stockmeyer [1972] showed that the problem of inequivalence of generalized regular expressions is intractable. They showed it by proving that any algorithm to solve the problem requires exponential space. So the problem is not in *PSPACE*. Therefore, it is not in *NP* and not in *P*. So it is intractable. We should note that the intractability comes about because we allow abbreviations of the form $(R)^2$. That is, the *inequivalence of regular expressions*, where the abbreviation is not allowed, is in *PSPACE*.

14.3.5 Completeness

Whenever we have a class of things, there is the possibility that some object in the class is a good representative of the class. From the point of view of complexity, we want to find representatives of *NP* that are connected to the other problems in *NP* in such a way that if the representative can be solved efficiently, then so can every other problem in *NP*. This is a bit vague. So let's get down to brass tacks and discuss a mechanism for transforming one problem into another so that an efficient solution for one will automatically give an efficient solution for the other.

Polynomially Reducible

A problem A is *polynomially reducible* to a problem B if there is a polynomial time computable function f that maps instances of A to instances of B such that

$$I \text{ is a yes-instance of } A \text{ iff } f(I) \text{ is a yes-instance of } B.$$

This property of f says that yes-instances of A get mapped to yes-instances of B and that no-instances of A get mapped to no-instances of B.

Let's see how we can use this idea to find an efficient algorithm for A from an efficient algorithm for B. Suppose we have found a polynomial time algorithm, say M, to solve B. We'll use M and f to find a polynomial time algorithm that solves A. The algorithm can be described as follows:

Algorithm for Finding an Efficient Algorithm

Take an arbitrary instance I of problem A. To find out whether I is a yes-instance of A we first construct the instance $f(I)$ of problem B. (There is an efficient algorithm to construct $f(I)$ because f is polynomial time computable.) Then we run algorithm M on the instance $f(I)$.

If M finds that $f(I)$ is a yes-instance of B, then it follows from the property of f that I is a yes-instance of A. If M finds that $f(I)$ is a no-instance of B, then it follows from the property of f that I is a no-instance of A. So the efficient algorithm to solve problem A is just the composition of the algorithm M and the algorithm to compute f.

This algorithm solves A and it does it in polynomial time because both the algorithm for B and the algorithm for f are polynomial time algorithms. We'll give an example of polynomial reducibility shortly.

NP-Completeness

The importance of polynomial reducibility comes into play in discussions about the class *NP*. A decision problem in *NP* is said to be *NP-complete* if every other decision problem in *NP* can be polynomially transformed to it. For example, suppose we could find some problem B in *NP* such that every other problem in

NP could be polynomially reduced to *B*. Then it might make sense to concentrate on trying to find an efficient algorithm for *B*. If we found a deterministic polynomial time algorithm for *B*, then every problem in *NP* would have a deterministic polynomial time algorithm. In addition to providing us with efficient solutions to many well-known problems, we would also be able to say that *NP* and *P* are equal.

But we can't start thinking about such things until we know whether there are any *NP*-complete problems.

CNF–Satisfiability

The first example of an *NP*-complete problem is due to Cook [1971]. The problem asks whether a propositional wff in conjunctive normal form is satisfiable. In other words, is there an assignment of truth values to the letters of the formula such that the value of the formula is true? We'll state the problem as follows:

CNF–Satisfiability Problem

Given a propositional wff in conjunctive normal form, is it satisfiable?

For example, $(x \lor y \lor \neg z) \land (x \lor z)$ is satisfiable by letting $x =$ True, $y =$ False, and $z =$ False. But $(x \lor y) \land (x \lor \neg y) \land \neg x$ is not satisfiable because it is false for any assignments of truth values to x and y.

Cook proved that the CNF-satisfiability problem is *NP*-complete. We'll state the result for the record:

Cook's Theorem

The CNF-satisfiability problem is *NP*-complete.

It's easy to see that CNF-satisfiability is in *NP*. Let the length of a wff be the total number of literals that appear in it. If n is the length of a wff, then the number of distinct variables in the wff is at most n. For example, the length of

$$(x \lor y \lor \neg z) \land (x \lor z)$$

is five and the wff has three variables. The guessing stage of a nondeterministic algorithm can produce some assignment of truth values for the variables of the wff. Then the checking stage must check to see whether each fundamental disjunction of the wff is true for the given assignment. Once a literal in some fundamental disjunction is found to be true, then the fundamental disjunction is true. So the checking process can proceed to the next fundamental disjunction. Since there are n literals in the wff, there are at most n literals to check. So the checking stage can be done in $O(n)$ time. Therefore, CNF-satisfiability is in *NP*.

Cook proved that CNF-satisfiability is *NP*-complete by showing that any *NP* problem can be polynomially reduced to CNF-satisfiability. The proof is complicated and lengthy, so we won't discuss it here. Cook's theorem opened the flood gates for finding other *NP*-complete problems. The reason is that once we have an *NP*-complete problem (e.g., CNF-satisfiability), then to show that some other problem *A* is *NP*-complete, all we have to do is show that *A* is in *NP* and then show that some known *NP*-complete problem (e.g., CNF-satisfiability) can be polynomially reduced to *A*. It follows that any *NP* problem can be polynomially reduced to *A* by polynomially reducing it to the known *NP*-complete problem that can be polynomially reduced to *A*. Let's state this result for the record.

Algorithm to Find an *NP*-Complete Problem

If *A* is an *NP* problem and *B* is an *NP*-complete problem that is polynomially reducible to *A*, then *A* is *NP*-complete.

Restricting CNF–Satisfiability

Cook also proved that a restricted form of CNF-satisfiability is *NP*-complete. The problem is called the *3-satisfiability problem* because the wffs contain at most three literals in each fundamental disjunction. For example, fundamental disjunctions like

$$(x), \quad (x \vee y), \quad \text{and} \quad (x \vee y \vee z)$$

are allowed. But $(x \vee y \vee z \vee w)$ is not allowed. There are no other restrictions. So a CNF can still have any number of variables and fundamental disjunctions. For example, the following expression, which contains five fundamental disjunctions and uses four variables, is OK because each fundamental disjunction has at most three literals:

$$(x \vee \neg y \vee z) \wedge (x \vee \neg y) \wedge (\neg x \vee y \vee w) \wedge (w \vee \neg y) \wedge (x \vee \neg y).$$

Cook proved that the 3-satisfiability problem is *NP*-complete by showing that CNF-satisfiability can be polynomially reduced to it. Let's state and prove this transformation result.

Theorem

The 3-satisfiability problem is *NP*-complete.

Proof: The 3-satisfiability problem is in *NP* because it is just a restricted form of the CNF-satisfiability problem, which we know is in *NP*. To show that 3-satisfiability is *NP*-complete, we'll show that CNF-satisfiability can be polynomially reduced to it. The basic idea is to transform each fundamental disjunction that has four or more literals into a conjunctive normal form where each fundamental disjunction has three literals with the property that, for some assignment of truth values to variables, the original fundamental disjunction is true if and only if the replacement conjunctive normal form is true. Here's how the transformation is accomplished. Suppose we have the following fundamental disjunction that contains k literals, where $k \geq 4$:

$$(l_1 \vee l_2 \vee \cdots \vee l_k).$$

We transform this fundamental disjunction into the following conjunctive normal form, where x_1, x_2, ..., x_{k-3} are new variables :

$$(l_1 \vee l_2 \vee x_1) \wedge (l_3 \vee \neg\, x_1 \vee x_2) \wedge (l_4 \vee \neg\, x_2 \vee x_3) \wedge \cdots$$
$$\wedge\, (l_{k-2} \vee \neg\, x_{k-4} \vee x_{k-3}) \wedge (l_{k-1} \vee l_k \vee \neg\, x_{k-3}).$$

This transformation can be applied to each fundamental disjunction (containing four or more literals) of a conjunctive normal form, resulting in a conjunctive normal form where each fundamental disjunction has three or fewer literals. For example, the fundamental disjunction

$$(u \vee \neg\, w \vee x \vee \neg\, y \vee z)$$

is transformed into

$$(u \vee \neg\, w \vee x_1) \wedge (x \vee \neg\, x_1 \vee x_2) \wedge (\neg\, y \vee z \vee \neg\, x_2)$$

where x_1 and x_2 are new variables.

The point about the transformation is that there is some assignment to the new variables such that the original fundamental disjunction is true (i.e., one of its literals is true) if and only if the new expression is true. For example, if $l_1 =$ True or $l_2 =$ True, then we can set all the variables x_i to False to make the new expression true. If $l_3 =$ True, then we can set x_1 to True and all the other variables x_i to False to make the new expression true. In general, if $l_i =$ True, then set x_j to True for $j \leq i - 2$ and set x_j to False for $j > i - 2$. This will make the new expression true.

Conversely, suppose there is some truth assignment to the variables x_j that makes the new expression true. Then some literal in the original fundamental disjunction must be true. For example, if $x_1 =$ False, then either l_1 or l_2 must be true. Similarly, if $x_{k-3} =$ True, then either l_{k-1} or l_k must be true. Now assume that $x_1 =$ True and $x_{k-3} =$ False. In this case, there must be an index i in the range $1 \leq i < k - 3$ such that $x_i =$ True and $x_{i+1} =$ False. It follows that l_{i+2} must be true. Therefore, some literal must be true, which makes the original fundamental disjunction true.

We need to show that the transformation can be done in polynomial time. A straightforward algorithm to accomplish the transformation applies the definition to each fundamental disjunction that contains four or more literals. If an input wff has length n (i.e., n literals), then there are at most $n/4$ fundamental disjunctions of length four or more. Each of these fundamental disjunctions is transformed into a conjunctive normal form containing at most $3(n - 2)$ literals. Therefore, the algorithm constructs at most $3n(n - 2)/4$ literals. Since each new literal can be constructed in a constant amount of time, the algorithm will run in time $O(n^2)$. Therefore, the CNF-satisfiability problem can be polynomially reduced to the 3-satisfiability problem. QED.

Now we have another NP-complete problem. Many NP-complete problems have been obtained by exhibiting a polynomial reduction from the 3-satisfiability problem. No one knows whether the 3-satisfiablity problem is in P. If it were, then we could tell the world that P and NP are the same.

Further Restricting CNF–Satisfiability

It's natural to wonder about what would happen if we restricted things further and considered the 2-*satisfiability problem*, where each fundamental disjunction of a conjunctive normal form has at most two literals. It turns out that the 2-satisfiability problem is in P. In other words, there is a deterministic polynomial time algorithm that solves the 2-satisfiability problem.

> **Question**
>
> Why hasn't anyone been able to say whether the 3-satisfiability problem is in P?

We don't know. But we'll look at a deterministic polynomial time algorithm for the 2-satisfiability problem and discuss why it becomes exponential when extended to the 3-satisfiability problem.

example **14.3** **The 2-Satisfiability Problem Is in P**

We'll describe a deterministic polynomial time algorithm to solve the 2-satisfiability problem. The algorithm will use the resolution rule for propositions (9.3). Suppose that the input wff has the form

$$C_1 \wedge C_2 \wedge \cdots \wedge C_n,$$

where each C_i is a clause (i.e., fundamental disjunction) consisting of either a single literal or the disjunction of two literals.

Now we can describe the algorithm. We list the clauses $C_1, C_2, \ldots, C_n$ as premises. Then we apply the resolution rule (9.3) to all possible pairs of existing or new clauses until we obtain the empty clause or until no new clauses are obtainable. If we obtain the empty clause, then the original wff is unsatisfiable and we output the answer no. Otherwise, the wff is satisfiable and we output the answer yes.

The algorithm is clearly deterministic. Let's discuss why it halts and why it runs in polynomial time. Since each clause has at most two literals, the resolvant of any two clauses contains at most two literals. Two examples:

The resolvant of $x \vee y$ and $\neg\, x \vee z$ is $y \vee z$.

The resolvant of $x \vee x$ and $\neg\, x \vee y$ is y.

Since there are n clauses in the wff, it follows that it has at most $2n$ distinct literals. Since we have the equivalence $x \vee y \equiv y \vee x$, we'll assume that two clauses are identical (not distinct) if they contain the same literals without regard to order. It follows from (5.32) that there are at most $n(2n + 1)$ distinct clauses that contain two literals. There are at most $2n$ distinct clauses consisting of one literal, and there is one empty clause. So the algorithm generates at most $2n^2 + 3n + 1$ distinct clauses. Therefore, the algorithm will always halt.

We'll leave it as an exercise to show that the number of times that the resolution rule is performed by the algorithm is $O(n^4)$. The algorithm must also spend some time on overhead, like checking to see whether each resolvant is distinct from those that already exist. We'll also leave it as an exercise to show that this checking takes $O(n^6)$ comparisons. So the algorithm runs in polynomial time. Therefore, the 2-satisfiability problem is in P.

end example

The reason that the resolution algorithm for 2-satisfiability runs in polynomial time is that each resolvant has at most two literals. If we were to apply the algorithm to the 3-satisfiability problem, then the number of possible clauses would explode because resolvants might contain up to as many literals as there are in the original wff—for example, the resolvant of

$$(x \vee y \vee z) \text{ and } (\neg\, x \vee v \vee w) \text{ is } (y \vee z \vee v \vee w).$$

So resolvants can get bigger and bigger. If a wff contains n clauses and each clause contains at most three literals, then there are at most $3n$ literals in the wff. If we agree to remove repeated literals from each resolvant, then the maximum number of literals in any resolvant is $3n$. Now we'll count the number of distinct clauses that consist of at most $3n$ literals. Since we're not concerned about the order of occurrence of literals in a clause, it follows that the number of distinct clauses consisting of at most $3n$ literals is equal to the number of subsets of a set with $3n$ elements, which is 2^{3n}. So in the worst case, the algorithm takes exponential time.

On the surface, the 2- and 3-satisfiability problems appear to be similar in difficulty. But after a little analysis, we can see that they appear to be worlds apart. Perhaps someone will eventually use these two problems to help explain whether P and NP are the same or are distinct.

We should mention the notion of NP-hard problems. A problem is NP-hard if all NP problems can be polynomially reduced to it. So the difference between NP-complete and NP-hard is that an NP-complete problem must be in NP. An NP-hard problem need not be in NP.

14.3.6 Formal Complexity Theory

Many results in complexity theory are very hard to state and very hard to prove. So practitioners normally try to simplify things as much as possible and formalize things so that statements can be clear and concise, and so that there can be a common means of communicating results. This is done by discussing complexity theory in terms of languages and Turing machines.

We can still discuss any decision problem that we like. But we'll think of each instance of a decision problem as a string over an alphabet of symbols. The set of all yes-instances for the problem forms a language over this alphabet. An algorithm to solve the decision problem will be a Turing machine that recognizes the language of yes-instances.

For example, the problem of deciding whether a natural number $n \geq 2$ is prime can easily be stated as a language recognition problem. We can pick some letter, say a, and define the language

$$L = \{a^n \mid n \text{ is a prime number}\}.$$

Now we can decide whether a natural number $n \geq 2$ is prime by deciding whether a string of two or more a's is in the language L.

Time Complexity

Let's get on with things and describe some formal complexity theory. A Turing machine has *time complexity* $t(n)$ if for every input string of length n, it executes at most $t(n)$ instructions before stopping. A language has *time complexity* $t(n)$ if it is accepted by a Turing machine of time complexity $t(n)$. The class $TIME(t(n))$ is the set of all languages of time complexity $t(n)$.

If a Turing machine M has time complexity $O(n^k)$ for some natural number k, we say that M has a *polynomial time complexity*. We'll let $DPTIME$ be the set of languages that are accepted by deterministic Turing machines in polynomial time. Similarly, we'll let $NPTIME$ be the set of languages that are accepted by nondeterministic Turing machines in polynomial time. Since any language accepted by a deterministic Turing machine can also be accepted by a nondeterministic Turing machine—just add a nondeterministic instruction to the deterministic machine—it follows that

$$DPTIME \subseteq NPTIME. \tag{14.11}$$

Space Complexity

Let's see if we can bring space complexity into the discussion. A Turing machine has *space complexity* $s(n)$ if for every input string of length n, it uses at most $s(n)$ tape cells before stopping. A language has *space complexity* $s(n)$ if it is accepted by a Turing machine of space complexity $s(n)$. The class $SPACE(s(n))$ is the set of all languages of space complexity $s(n)$. If a Turing machine M has space complexity $O(n^k)$ for some natural number k, we say that M has a *polynomial space complexity*.

Let $DSPACE(s(n))$ be the set of languages accepted by deterministic Turing machines of space complexity $s(n)$. Let $NSPACE(s(n))$ be the set languages accepted by nondeterministic Turing machines of space complexity $s(n)$. Let $DPSPACE$ be the set languages accepted by deterministic Turing machines in polynomial space. Let $NPSPACE$ be the set of languages accepted by nondeterministic Turing machines in polynomial space. The nice thing about these classes of languages is the following equality, which we'll discuss in an exercise:

$$DPSPACE = NPSPACE.$$

Because of this equality, we'll use $PSPACE$ to denote the common value of these two equal sets. In other words, $PSPACE$ is the set of languages that are accepted by either deterministic or nondeterministic Turing machines in polynomial space.

Now let's discuss the connection between the time and space classes. If a Turing machine has time complexity $t(n)$, then it uses n tape cells to store the input string and, since it can access only one tape cell per instruction, it uses at most $t(n)$ more cells during its computation. So the space complexity of the Turing machine is

$$t(n) + n.$$

If $t(n)$ is a polynomial, then $t(n) + n$ is also a polynomial. So if a language is accepted by a nondeterministic Turing machine in polynomial time, then it is also accepted by the same Turing machine in polynomial space. In other words, we can write

$$NPTIME \subseteq PSPACE. \tag{14.12}$$

We can put (14.11) and (14.12) together to obtain the following fundamental relationships between classes:

$$DPTIME \subseteq NPTIME \subseteq PSPACE. \tag{14.13}$$

The class $DPTIME$ is usually denoted by P and the class $NPTIME$ is usually denoted by NP. This allows us to write (14.13) in the following more popular form, which is the same as (14.10):

$$P \subseteq NP \subseteq PSPACE.$$

It is widely believed that these containments are proper. But no one has been able to prove or disprove whether either containment is proper. In other words, it is not known whether there is a language in NP that is not in P and it is not known whether there is a language in $PSPACE$ that is not in NP. These are among the most important unresolved questions in present-day computer science.

Polynomial Reducibility

Now let's discuss the idea of polynomial reducibility in terms of languages and Turing machines. First, we'll say that a function is polynomial time computable if it can be computed in polynomial time by some Turing machine. Now, we say that a language K over alphabet A is *polynomially reducible* to a language L over alphabet B if there is a polynomial time computable function $f : A^* \to B^*$ such that

$$x \in K \quad \text{if and only if} \quad f(x) \in L.$$

This property of f says that the language K gets mapped by f into the language L and its complement gets mapped by f into the complement of L. In other words, the property can be written as follows:

$$f(K) \subseteq L \quad \text{and} \quad f(A^* - K) \subseteq (B^* - L).$$

Let's see how we can use this idea. If we happen to find an efficient Turing machine, say M, that accepts language L, then we can also find an efficient Turing machine that accepts language K. Here's how. Take an arbitrary string $x \in A^*$. To find out whether x is in K we first construct $f(x)$. (There is an efficient Turing machine to construct f because it is polynomial time computable.) Then we run Turing machine M with input $f(x)$. If M accepts $f(x)$, then $f(x) \in L$, which, by the property of f, implies that $x \in K$. If M rejects $f(x)$, then $f(x) \notin L$, which, by the property of f, implies that $x \notin K$. So the efficient Turing machine to accept language K is just the composition of the Turing machine M and the Turing machine to compute f.

The purpose of this section has been to give a very brief introduction to complexity classes. The book by Garey and Johnson [1979] is a good choice to begin further study. Among other good things, it contains a very large list of *NP*-complete problems.

◢ Exercises

1. Transform $(u \vee v \vee w \vee x \vee y \vee z)$ into an equivalent conjunctive normal form where each fundamental disjunction has three literals.

2. The 1-*satisfiability problem* is to determine whether a conjunction of literals is satisfiable. Prove that the 1-satisfiability problem is in P by finding a deterministic polynomial time algorithm to solve it.

3. Show that the worst-case complexity for the number of resolution steps required in the resolution algorithm for the 2-satisfiability problem in Example 14.3 is $O(n^4)$. Then show that the worst-case complexity for the number of comparisons that must be made to see whether a resolvant is distinct from those already in existence is $O(n^6)$.

4. For each of the following questions, assume that A is an NP problem.

   a. Suppose you prove a theorem showing that a lower bound for the running time of any algorithm to solve A is $\Theta(2^n)$. What would you tell the world?

   b. Suppose you find a deterministic algorithm that solves A in polynomial time. What would you tell the world?

   c. Suppose A is NP-complete and you find a deterministic algorithm that solves A in polynomial time. What would you tell the world?

   d. Suppose you prove that the 3-satisfiablility problem is polynomially reducible to A. What would you tell the world?

5. Given the following three statements: (1) $P = NP$; (2) $P \neq NP$; (3) $P \neq NP$ and all NP-complete problems are intractable. For each of the following statements, select one of the preceding statements to fill in the blank.

   a. If problem p is in NP and p is intractable, then_____.
   b. If problem p is NP-complete and p is intractable, then_____.
   c. If problem p is NP-complete and p is in P, then_____.

6. Let $DSPACE(s(n))$ be the languages accepted by deterministic Turing machines of space complexity $s(n)$. Let $NSPACE(s(n))$ be the set of languages accepted by nondeterministic Turing machines of space complexity $s(n)$. Let $DPSPACE$ be the set of languages accepted by deterministic Turing machines in polynomial space. Let $NPSPACE$ be the set of languages accepted by nondeterministic Turing machines in polynomial space. There is a theorem stating that $NSPACE(n^i) \subseteq DSPACE(n^{2i})$ for $i \geq 1$. Use this result to prove that $DPSPACE = NPSPACE$.

14.4   Chapter Summary

Some general problems—such as the halting problem, the total problem, the equivalence problem, and the Post correspondence problem—can't be solved by any machine. The halting problem and Post's correspondence problem are partially decidable. It's often the case that specific instances of undecidable problems are decidable. For example, we can certainly tell whether f halts on any input x if f is defined by $f(x) = 2x$.

There is a hierarchy of languages extending from the class of regular languages up to the class of languages that have no restrictions on their grammars and finally to the class of languages that may or may not have grammars. All the grammatical languages can be recognized by certain kinds of machines. But the class of languages without grammars can't be recognized by any machine.

The class P consists of the decision problems that can be solved by deterministic polynomial time algorithms. The class NP consists of the decision problems that can be solved by nondeterministic polynomial time algorithms. The class *PSPACE* consists of the decision problems that can be solved by polynomial space algorithms. It follows that

$$P \subseteq NP \subseteq PSPACE.$$

The question of whether these containments are proper is an important open problem.

This is not the end. It is not even the beginning of the end.
But it is, perhaps, the end of the beginning.

—Winston Churchill (1874–1965)

Answers to Selected Exercises

Chapter 1

Section 1.1

1. Consider the four statements "if $1 = 1$ then $2 = 2$," "if $1 = 1$ then $2 = 3$," "if $1 = 0$ then $2 = 2$," and "if $1 = 0$ then $2 = 3$." The second statement is the only one that is false.

3. a. 47 is a prime between 45 and 54. **c.** The statement is true. **e.** The statement is false. For example, the numbers 2 and 5 have the desired form: $2 = 3(0) + 2$ and $5 = 3(1) + 2$. But the product $2(5) = 10$ can't be written as $10 = 3k + 2$ because the equation does not have an integer solution for k.

4. a. Let x and y be any two even integers. Then they can be written in the form $x = 2m$ and $y = 2n$ for some integers m and n. Therefore, the sum $x + y$ can be written as $x + y = 2m + 2n = 2(m + n)$, which is an even integer. **c.** Let x and y be any odd integers. Then they can be written $x = 2m + 1$ and $y = 2n + 1$ for some integers m and n. Therefore, we have $x - y = 2m + 1 - 2n - 1 = 2(m - n)$, which is an even integer.

6. a. Let $x = 3m + 4$, and let $y = 3n + 4$ for some integers m and n. Then $xy = (3m + 4)(3n + 4) = 9mn + 12m + 12n + 16 = 3(3mn + 4m + 4n + 4) + 4$, which has the desired form. **c.** Let $x = 7m + 8$, and let $y = 7n + 8$ for some integers m and n. Then $xy = (7m + 8)(7n + 8) = 49mn + 56m + 56n + 64 = 7(7mn + 8m + 8n + 8) + 8$, which has the desired form.

7. a. Let $d \mid (da + b)$. Then $da + b = dk$ for some integer k. Solving the equation for b gives $b = d(k - a)$, which says that $d \mid b$. **c.** Let $d \mid a$ and $a \mid b$. It follows from the definition of divisibility that there are integers m and n such that $a = dm$ and $b = an$. Now substitute for a in the second equation to obtain $b = an = (dm)n = d(mn)$. This equation says that $d \mid b$.

e. Let $5 \mid 2n$. Then $2n = 5k$ for some integer k. We can write $2n = 5k = 4k + k$. So $k = 2n - 4k = 2(n - 2k)$. Now substitute for k to obtain $2n = 5k = 5(2(n - 2k)) = 10(n - 2k)$. Divide both ends of the equation by 2 to get $n = 5(n - 2k)$. So $5 \mid n$.

8. a. First we'll prove the statement "If x is even then x^2 is even." If x is even, then $x = 2n$ for some integer n. Therefore, $x^2 = (2n)(2n) = 2(2n^2)$, which is even. Next we'll prove the statement "If x^2 is even then x is even," by proving the contrapositive "If x is odd, then x^2 is odd." So if x is odd, then $x = 2n + 1$ for some integer n. Thus $x^2 = (2n + 1)(2n + 1) = 4n^2 + 4n + 1 = 2(2n^2 + 2n) + 1$, which is an odd integer. So the iff statement is proven.
c. $x^2 + 6x + 9$ is even iff $(x + 3)^2$ is even iff $(x + 3)$ is even iff x is odd.

9. Let x be a positive integer. If x is odd, then we can write $x = 1 \cdot x = 2^0 \cdot x$, where x is odd. If x is even, let 2^k be the highest power of 2 that divides x. Then $x = 2^k n$ for some integer n. We claim that n is odd. For if 2 divided n, then 2^{k+1} would divide x, contrary to 2^k being the highest power of 2 that divides x. Therefore, n is odd.

Section 1.2

1. a. $\{1, 2, 3, 4, 5, 6, 7\}$. **c.** $\{3, 5, 7, 11, 13, 17, 19\}$. **e.** $\{M, I, S, P, R, V, E\}$.
2. a. $\{x \mid x \in \mathbb{N} \text{ and } 1 \le x \le 31\}$.
c. $\{x \mid x = n^2 \text{ and } n \in \mathbb{N} \text{ and } 1 \le n \le 8\}$ or $\{x^2 \mid x \in \mathbb{N} \text{ and } 1 \le x \le 8\}$.
3. a. True. **c.** False. **e.** True. **g.** True.
5. For example, let $A = \{x\}$ and $B = \{x, \{x\}\}$.
6. a. $\{\varnothing, \{x\}, \{y\}, \{z\}, \{w\}, \{x, y\}, \{x, z\}, \{x, w\}, \{y, z\}, \{y, w\}, \{z, w\}, \{x, y, z\}, \{x, y, w\}, \{x, z, w\}, \{y, z, w\}, \{x, y, z, w\}\}$. **c.** $\{\varnothing\}$. **e.** $\{\varnothing, \{\{a\}\}, \{\varnothing\}, \{\{a\}, \varnothing\}\}$.
7. a. $\{a, b, c\}$. **c.** $\{a, \{a\}\}$.
8. a. Let $x \in A$. Then $x = 6k + 5$ for some $k \in \mathbb{N}$. We can write $x = 6k + 5 = 3(2k + 1) + 2$, where $2k + 1 \in \mathbb{N}$ because $k \in \mathbb{N}$. So $x \in B$. Therefore, $A \subseteq B$. **b.** Notice that 2 is an element of B that does not occur in A. Therefore, $A \ne B$.
9. a. $x = 4$. **c.** $x = 5$.
10. $A = \{1, 2, 3, 4\}$.
11. $A = \{2, 3, 5, 6, 7\}$.
12. $A = \{1, 2, 4, 5\}$.
13. a. $\{4, 5, 6, 7, 8, 9\}$. **c.** $\{6, 7, 8, 9\}$. **e.** $\{0, 1, 4, 5, 6, 7, 8, 9\}$.
14. a. $A \cap B - C$. **c.** $B \oplus C$.
15. a. $D_0 = \mathbb{N}$, $D_6 = \{1, 2, 3, 6\}$, $D_{12} = \{1, 2, 3, 4, 6, 12\}$, and $D_{18} = \{1, 2, 3, 6, 9, 18\}$.
c. Assume that $m \mid n$. Let $d \in D_m$. Then $d \mid m$ and it follows from (1.1a) that $d \mid n$. Therefore, $d \in D_n$. So $D_m \subseteq D_n$.

16. a. $M_0 = \{0\}$, $M_1 = \mathbb{N}$, $M_3 = \{3, 6, 9, \ldots , \}$, $M_4 = \{4, 8, 12, \ldots , \}$, and $M_{12} = \{12, 24, 36, \ldots , \}$.
c. The assumption that $m \mid n$ tells us that we can write $n = qm$ for some natural number q. Let $x \in M_n$. Then $x = kn$ for some natural number k. We can substitute for n in this equation to obtain $x = kn = kqm$. So $x \in M_m$. Therefore, $M_n \subseteq M_m$.

17. $\mid A \mid + \mid B \mid + \mid C \mid + \mid D \mid - \mid A \cap B \mid - \mid A \cap C \mid - \mid A \cap D \mid - \mid B \cap C \mid$
$- \mid B \cap D \mid - \mid C \cap D \mid + \mid A \cap B \cap C \mid + \mid A \cap B \cap D \mid + \mid A \cap C \cap D \mid$
$+ \mid B \cap C \cap D \mid - \mid A \cap B \cap C \cap D \mid$.

19. a. 82. **c.** 23.

20. At most 20 drivers were smoking, talking, and tuning the radio.

22. The answer is 15. Let A, B, and C be sets of senators who voted for the three bills, with cardinalities 70, 65, and 80, respectively. Since the total number voting is 100, it follows that $\mid A \cap B \mid \geq 35$. So we can consider the set $(A \cap B) \cup C$, which we can use to find the minimum value for $\mid A \cap B \cap C \mid$ as follows: $100 \geq \mid (A \cap B) \cup C \mid = \mid A \cap B \mid + \mid C \mid - \mid A \cap B \cap C \mid \geq 35 + 80 - \mid A \cap B \cap C \mid$. Solving for $\mid A \cap B \cap C \mid$ we obtain $\mid A \cap B \cap C \mid \geq 15$. For an alternative solution use complements and De Morgan's laws.

24. a. $[x, y, z]$, $[x, y]$. **c.** $[a, a, a, b, b, c]$, $[a, a, b]$.
e. $[x, x, a, a, [a, a], [a, a]]$, $[x, x]$.

26. Let A and B be bags, and let m and n be the number of times x occurs in A and B, respectively. If $m \geq n$, then put $m - n$ occurrences of x in $A - B$, and if $m < n$, then do not put any occurrences of x in $A - B$.

27. a. $x \in A \cup \varnothing$ iff $x \in A$ or $x \in \varnothing$ iff $x \in A$. Therefore, $A \cup \varnothing = A$.
c. $x \in A \cup A$ iff $x \in A$ or $x \in A$ iff $x \in A$. Therefore, $A \cup A = A$.

28. a. Since there are no elements in $\varnothing$, there can be no elements in both A and $\varnothing$. Therefore, $A \cap \varnothing = \varnothing$.
c. $x \in A \cap (B \cap C)$ iff $x \in A$ and $x \in B \cap C$ iff $x \in A$ and $x \in B$ and $x \in C$ iff $x \in A \cap B$ and $x \in C$ iff $x \in (A \cap B) \cap C$. Therefore, $A \cap (B \cap C) = (A \cap B) \cap C$. **e.** First prove that $A \subseteq B$ implies $A \cap B = A$. Assume that $A \subseteq B$. If $x \in A \cap B$, then $x \in A$ and $x \in B$, and it follows that $x \in A$. Thus $A \cap B \subseteq A$. If $x \in A$, then $x \in B$ (by assumption), and it follows that $x \in A \cap B$. So $A \subseteq A \cap B$. Therefore, $A \cap B = A$. Next prove that $A \cap B = A$ implies $A \subseteq B$. Assume that $A \cap B = A$. Let $x \in A$. Then $x \in A \cap B$, which says $x \in A$ and $x \in B$. So $x \in B$. Therefore, we have $A \subseteq B$. So the iff statement has been proven.

29. Let $S \in \text{power}(A \cap B)$. Then $S \subseteq A \cap B$, which says that $S \subseteq A$ and $S \subseteq B$. Therefore, $S \in \text{power}(A)$ and $S \in \text{power}(B)$, which says that $S \in \text{power}(A) \cap \text{power}(B)$. This proves that $\text{power}(A \cap B) \subseteq \text{power}(A) \cap \text{power}(B)$. The other containment is similar.

31. a. Let $x \in A \cap (B \cup A)$. Then $x \in A$, so we have $A \cap (B \cup A) \subseteq A$. For the other containment, let $x \in A$. Then $x \in B \cup A$. Therefore, $x \in A \cap (B \cup A)$, which says that $A \subseteq A \cap (B \cup A)$. This proves the equality by set containment.

We can also prove the equality by using a property of intersection (1.6e) applied to the two sets A and $B \cup A$. Thus (1.6e) becomes $A \subseteq B \cup A$ if and only if $A \cap (B \cup A) = A$. Since we know that $A \subseteq B \cup A$ is always true, the equality follows.

32. Assume that $(A \cap B) \cup C = A \cap (B \cup C)$. If $x \in C$, then $x \in (A \cap B) \cup C = A \cap (B \cup C)$, which says that $x \in A$. Thus $C \subseteq A$. Assume that $C \subseteq A$. If $x \in (A \cap B) \cup C$, then $x \in A \cap B$ or $x \in C$. In either case it follows that $x \in A \cap (B \cup C)$ because $C \subseteq A$. Thus $(A \cap B) \cup C \subseteq A \cap (B \cup C)$. The other containment is similar. Therefore, $(A \cap B) \cup C = A \cap (B \cup C)$.

33. a. Counterexample: $A = \{a\}$, $B = \{b\}$.
c. Counterexample: $A = \{a\}$, $B = \{b\}$, $C = \{b\}$.

34. a. $x \in (A')'$ iff $x \in U$ and $x \notin A'$ iff $x \notin U - A$ iff $x \in A$.
c. $x \in A \cap A'$ means that $x \in A$ and $x \in U - A$. This says that $x \in A$ and $x \notin A$, which can't happen. Therefore, $A \cap A' = \varnothing$. To see that $A \cup A' = U$, observe that any element of U must be either in A or not in A. **e.** $x \in (A \cap B)'$ iff $x \notin A \cap B$ iff $x \notin A$ or $x \notin B$ iff $x \in A'$ or $x \in B'$ iff $x \in A' \cup B'$. Therefore, $(A \cap B)' = A' \cup B'$. **g.** $A \cup (A' \cap B) = (A \cup A') \cap (A \cup B) = U \cap (A \cup B) = A \cup B$.

Section 1.3

1. $(x, x, x), (x, x, y), (x, y, x), (y, x, x), (x, y, y), (y, x, y), (y, y, x), (y, y, y)$.
2. a. $\{(a, a), (a, b), (b, a), (b, b), (c, a), (c, b)\}$. **c.** $\{(\ )\}$.
e. $\{(a, a), (a, b), (a, c), (b, a), (b, b), (b, c), (c, a), (c, b), (c, c)\}$.
3. $\langle\ \rangle, \langle a \rangle, \langle b \rangle, \langle a, a \rangle, \langle a, b \rangle, \langle b, a \rangle, \langle b, b \rangle$.
4. a. $\text{head}(\langle a \rangle) = a$ and $\text{tail}(\langle a \rangle) = \langle\ \rangle$.
c. $\text{head}(\langle\langle a, b \rangle, c \rangle) = \langle a, b \rangle$ and $\text{tail}(\langle\langle a, b \rangle, c \rangle) = \langle c \rangle$.
5. a. $\langle 24, 60, \langle 2, 3, 4, 6, 12 \rangle \rangle$. **c.** $\langle 14, 15, \langle\ \rangle \rangle$.
8. a. $LM = \{bba, ab, a, abbbba, abbab, abba, bbba, bab, ba\}$.
c. $L^0 = \{\Lambda\}$. **e.** $L^2 = LL = \{\Lambda, abb, b, abbabb, abbb, babb, bb\}$.
9. a. $L = \{b, ba\}$. **c.** $L = \{a, b\}$. **e.** $L = \{\Lambda, ba\}$.
10. a. $x = uvw$, where $u, w \in L$ and $v \in M$.
c. $x = \Lambda$ or $x = u_1 \ldots u_n$, where $u_k \in L \cup M$. **e.** $x = \Lambda$ or $s_1 \ldots s_n$ or $u_1 \ldots u_m$ or $(s_1 \ldots s_n)(u_1 \ldots u_m)$, where $s_k \in L$ and $u_k \in M$.
11. a. $\{a, b\}^* \cap \{b, c\}^* = \{b\}^*$.
c. $\{a, b, c\}^* - \{a\}^*$ is the set of strings over $\{a, b, c\}$ that contain at least one b or at least one c.
12. a. $\{(1, 12), (2, 12), (3, 12), (4, 12), (6, 12), (12, 12)\}$.
c. $\{(2, 1, 1), (3, 1, 2), (3, 2, 1)\}$.
e. $U = \{(a, 1), (a, 2), (b, 1), (b, 2)\}$.
13. a. $\{z \mid (\text{Michigan}, y, z) \in \text{Borders for some } y\}$.
c. $\{x \mid (x, y, \text{None}) \in \text{Borders for some } y\}$.

e. $\{(x, z) \mid (x, \text{Canada}, z) \in \text{Borders}\}$.

14. a. $2(5^3) - 2(4^3) = 122$. **c.** $1(2)(4^4) = 512$.

15. Let U be the set of n-tuples over A. In other words, $U = A^n$. Let S be the subset of U whose n-tuples do not contain any occurrences of letters from the set B. So $S = (A - B)^n$. Then $U - S$ is the set of n-tuples over A that contain at least one occurrence of an element from B. We have $\mid U - S \mid = \mid U \mid - \mid S \mid = \mid A^n \mid - \mid (A - B)^n \mid = \mid A \mid^n - \mid A - B \mid^n = \mid A \mid^n - (\mid A \mid - \mid B \mid)^n$.

16. a. $(x, y) \in (A \cup B) \times C$ iff $x \in A \cup B$ and $y \in C$ iff $(x \in A$ or $x \in B)$ and $y \in C$ iff $(x, y) \in A \times C$ or $(x, y) \in B \times C$ iff $(x, y) \in (A \times C) \cup (B \times C)$. Therefore, $(A \cup B) \times C = (A \times C) \cup (B \times C)$.
c. Prove that $(A \cap B) \times C = (A \times C) \cap (B \times C)$. $(x, y) \in (A \cap B) \times C$ iff $x \in A \cap B$ and $y \in C$ iff $(x \in A$ and $x \in B)$ and $y \in C$ iff $(x, y) \in (A \times C) \cap (B \times C)$, which proves the statement.

17. a. The statement is true because $s\Lambda = \Lambda s = s$ for any string s.
c. $x \in L(M \cup N)$
    iff $x = yz$, where $y \in L$ and $z \in M \cup N$
    iff $x = yz$, where $y \in L$ and $(z \in M$ or $z \in N)$
    iff $x = yz$, where $yz \in LM$ or $yz \in LN$
    iff $x \in LM$ or $x \in LN$
    iff $x \in LM \cup LN$.
Therefore, $L(M \cup N) = LM \cup LN$. The second equality is proved the same way.

18. a. The statement is true because $A^0 = \{\Lambda\}$ for any language A.
c. First we'll prove $L^* = L^*L^*$. Since $L^* = L^*\{\Lambda\}$, we have $L^* = L^*\{\Lambda\} \subseteq L^*L^*$. So $L^* \subseteq L^*L^*$. If $x \in L^*L^*$, then $x = yz$, where $y, z \in L^*$. So $y \in L^m$ and $z \in L^n$ for some numbers m and n. Therefore, $x = yz \in L^{m+n} \subseteq L^*$. Thus $L^*L^* \subseteq L^*$. So we have the equality $L^* = L^*L^*$.

    Next we'll prove the equality $L^* = (L^*)^*$. L^* is a subset of $(L^*)^*$ by definition. If $x \in (L^*)^*$, then there is a number n such that $x \in (L^*)^n$. So x is a concatenation of n strings, each one from L^*. So x is a concatenation of n strings, each from some power of L. Therefore, $x \in L^*$. Thus $(L^*)^* \subseteq L^*$. So we have the equality $L^* = (L^*)^*$.

19. a. Notice that $(3, 7) = \{\{3\}, \{3, 7\}\}$ and $(7, 3) = \{\{7\}, \{7, 3\}\}$ and that the two sets cannot be equal. **c.** $(\{a\}, b) = \{\{a\}, \{b\}\} = \{\{b\}, \{a\}\} = (\{b\}, a)$.

20. a. From Exercise 19 we have $(x, y) = S = \{\{x\}, \{x, y\}\}$. Therefore,

$$(x, y, z) = \{\{S\}, \{S, z\}\} = \{\{\{\{x\}, \{x, y\}\}\}, \{\{\{x\}, \{x, y\}\}, z\}\}.$$

c. $(a, b, a) = \{\{a\}, \{a, b\}, \{a, b, a\}\} = \{\{a\}, \{a, b\}, \{a, b\}\} = \{\{a\}, \{a, b\}\} = \{\{a\}, \{a, a\}, \{a, a, b\}\} = (a, a, b)$.

21. a. If a is a 3 by 4 matrix, then the address polynomial for the column-major location of $a[i, j]$ is $B + 3M(j - 1) + M(i - 1)$.
c. If a is a 3-dimensional array of size l by m by n stored as an l-tuple of m by n matrices, each in row-major form, then the address polynomial for the location of $a[i, j, k]$ is $B + mnM(i - 1) + nM(j - 1) + M(k - 1)$.

Section 1.4

2.

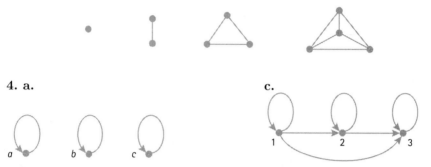

4. a.   **c.**

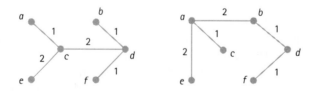

5. a. The four possible strings are $f\,d\,g\,e\,b\,a\,c$, $f\,d\,g\,b\,e\,a\,c$, $f\,g\,d\,e\,b\,a\,c$, and $f\,g\,d\,b\,e\,a\,c$.

b. The three possible strings are $f\,g\,e\,d\,b\,a\,c$, $f\,d\,e\,g\,b\,a\,c$, and $f\,d\,b\,a\,c\,e\,g$.

7. a. One answer is $a\,b\,c\,d\,e\,f$.

b. One answer is $a\,b\,c\,e\,d\,f$.

8. Pick a vertex and color each of the five edges connected to it with a different color. At least three of these edges must be one color. Look at the three vertices at the other end of these three edges and analyze the possible colors of the edges between them.

10.

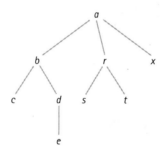

12. Here are two answers:

14. a. One possible answer: **b.** One possible answer:

16. a. A height zero binary tree has one node, the root, which is also the leaf. Each node has a maximum of two children. So a height one binary tree has a maximum of two leaves, a height two binary tree has a maximum of four leaves, a height three binary tree has a maximum of 8 leaves, and so on.

b. There is one node at the root and a maximum of two children for each node. So the maximum number of nodes for a binary tree of height 0 is 1, which we can write as $2^{0+1} - 1$. For height one the maximum number of nodes is $1 + 2 = 3$, which we can write as $2^{1+1} - 1$. For height two the maximum number of nodes is $1 + 2 + 4 = 7$, which we can write as $2^{2+1} - 1$. For height three the maximum number of nodes is $1 + 2 + 4 + 8 = 15$, which we can write as $2^{3+1} - 1$. In general, the maximum number of nodes for height n is $1 + 2 + 2^2 + \cdots + 2^n = 2^{n+1} - 1$. To see this, just multiply the left side by 1 in the form $(2 - 1)$.

18. The graph is connected and either none or two vertices have odd degree.

Chapter 2

Section 2.1

1. a. There is one function of type $\{a, b\} \rightarrow \{1\}$; it maps both a and b to 1.
c. There are four functions of type $\{a, b\} \rightarrow \{1, 2\}$: one maps both a and b to 1; one maps both a and b to 2; one maps a to 1 and b to 2; and one maps a to 2 and b to 1.

2. a. O. **c.** $\{x \mid x = 4k + 3 \text{ where } k \in \mathbb{N}\}$. **e.** $\varnothing$.

3. a. –5. **c.** 4.

5. For example, $x = y = \,^1/_2$.

6. a. 3. **c.** 1.

7. $\gcd(296, 872) = 8 = (-53) \cdot 296 + 18 \cdot 872$.

8. a. 3. **c.** 3.

9. a. $f(\varnothing) = \varnothing$. **c.** $f(\{2, 5\}) = \{4\}$. **e.** $f(\{1, 2, 3\}) = \{0, 2, 4\}$.

10. a. $\text{floor}(x) = $ if $x \geq 0$ then $\text{trunc}(x)$ else if $x = \text{trunc}(x)$ then x else $\text{trunc}(x - 1)$.

11. When x is negative, $f(x, y)$ can be different than $x \bmod y$. For example, $f(-16, 3) = -1$ and $-16 \bmod 3 = 2$.

13. a. 4. **c.** –3. **e.** $2(1/5)^{\log_5 2} = 2 \cdot 2^{\log_5(1/5)} = 2 \cdot 2^{\log_5 1 - \log_5 5} = 2 \cdot 2^{0-1} = 1$.
14. a. $(a/b)^{\log_b c} = c^{\log_b(a/b)} = c^{\log_b a - \log_b b} = c^{\log_b a - 1} = (1/c)c^{\log_b a} = (1/c)a^{\log_b c}$.

c. $a^k(n/b^{2k})^{\log_b a} = a^k a^{\log_b(n/b^{2k})} = a^k a^{\log_b n - \log_b b^{2k}} = a^k a^{\log_b n - 2k \log_b b}$
$= a^k a^{\log_b n - 2k} = a^{\log_b n - k} = a^{\log_b n - \log_b b^k} = a^{\log_b(n/b^k)} = (n/b^k)^{\log_b a}$.

15. a. If $x \in A \cup B$ then $x \in A$ or $x \in B$. So $\chi_{A \cup B}(x) = 1$ and either $\chi_A(x) = 1$ or $\chi_B(x) = 1$. If $\chi_A(x) = \chi_B(x) = 1$, the equation becomes $1 = 1 + 1 - 1(1)$, which is true. If $\chi_A(x) = 1$ and $\chi_B(x) = 0$, the equation becomes $1 = 1 + 0 - 1(0)$, which is true. If $x \notin A \cup B$, then $x \notin A$ and $x \notin B$. So $\chi_{A \cup B}(x) = \chi_A(x) = \chi_B(x) = 0$ and the equation becomes $0 = 0 + 0 - 0(0)$, which is true.
c. $\chi_{A-B}(x) = \chi_A(x)(1 - \chi_B(x))$.

16. a. A. **c.** $\{0\}$.

17. For any real number x there is an integer n such that $n \leq x < n + 1$. It follows that $n = \lfloor x \rfloor$ and $n + 1 = \lceil x \rceil$. Multiply the inequality by –1 to obtain $-n - 1 < -x < -n$. So $\lceil -x \rceil = -n = -\lfloor x \rfloor$.

18. a. For any real number x there is an integer m such that $k \leq x < k + 1$ and $k = \lfloor x \rfloor$. Now add an integer n to each term of the inequality to obtain $k + n \leq x + n < k + n + 1$. Since $k + n$ is an integer, the inequality tells us that $\lfloor x + n \rfloor = k + n$. Since $k = \lfloor x \rfloor$, we obtain the desired result that $\lfloor x + n \rfloor = \lfloor x \rfloor + n$.

19. If n is even, then $n = 2k$ for some integer k. Therefore, $\lceil n/2 \rceil = \lceil 2k/2 \rceil = \lceil k \rceil = k$, and $\lfloor (n + 1)/2 \rfloor = \lfloor (2k + 1)/2 \rfloor = \lfloor 2k/2 + 1/2 \rfloor = \lfloor k + 1/2 \rfloor$. Now apply (2.1a) to obtain $\lfloor k + 1/2 \rfloor = k + \lfloor 1/2 \rfloor = k + 0 = k$. So the equation holds when n is even. If n is odd, then $n = 2k + 1$ for some integer k and the reasoning is similar to the above.

20. a. We always have the inequality $\lfloor x \rfloor \leq x < \lfloor x \rfloor + 1$ and there are no integers between $\lfloor x \rfloor$ and $\lfloor x \rfloor + 1$. So if $\lfloor x \rfloor < n$, it follows that $\lfloor x \rfloor + 1 \leq n$. So $x < n$. Conversely, if $x < n$, then $\lfloor x \rfloor \leq x < n$. Therefore, $\lfloor x \rfloor < n$.
c. We always have the inequality $\lfloor x \rfloor \leq x$. So if $n \leq \lfloor x \rfloor$, it follows that $n \leq x$. Conversely, assume that $n \leq x$. Since there are no integers between $\lfloor x \rfloor$ and $\lfloor x \rfloor + 1$, it follows that $n \leq \lfloor x \rfloor$.

21. a. We know that $\lfloor x \rfloor \leq x$ and $\lfloor y \rfloor \leq y$. So $\lfloor x \rfloor + \lfloor y \rfloor \leq x + y$. Now apply property (2.1c) to obtain $\lfloor x \rfloor + \lfloor y \rfloor \leq \lfloor x + y \rfloor \leq x + y$.

22. a. The equation holds if and only if $(x - 1)/2 \leq n \leq x/2$ for some integer n. Multiply the inequality by 2 to obtain $x - 1 \leq 2n \leq x$. Add 1 to each term in the inequality to obtain the inequality $x \leq 2n + 1 \leq x + 1$. Therefore, $2n \leq x \leq 2n + 1$. It follows that for any integer n, if $2n \leq x \leq 2n + 1$, then the equation holds.

23. a. $p(x) = 1 + 2x + 3x^2$. **c.** $p(x) = 5x^2$.

24. a. $\log_b(b^x) = x$ means $b^x = b^x$, which is true.
c. Let $r = \log_b(x^y)$ and $s = \log_b x$, and proceed as in Part (b) to show that $r = ys$. **e.** Let $r = \log_a x$, $s = \log_a b$, and $t = \log_b x$. Proceed as in (b) to show that $r = st$. **g.** Apply $\log_b$ to both sides and then use (2.5c) to see that the resulting expressions are equal.

25. a. The equalities follow because $\gcd(a, b)$ is the largest common divisor of a and b.

c. Since $\gcd(d, a) = 1$, it follows from (2.2c) that there are integers m and n such that $1 = md + na$. Multiply the equation by b to obtain $b = bmd + bna = bmd + abn$. Since d divides both terms on the right side, d also divides the left side. Therefore, $d \mid b$.

27. a. We'll prove both containments with iff statements: $x \in f(E \cup F)$ iff $x = f(y)$, where $y \in E \cup F$ iff $x = f(y)$, where $y \in E$ or $y \in F$ iff $x \in f(E)$ or $x \in f(F)$ iff $x \in f(E) \cup f(F)$.

c. Consider the function $f : \{a, b, c\} \rightarrow \{1, 2, 3\}$, defined by $f(a) = f(b) = 1$ and $f(c) = 2$. Then $\{a\} \cap \{b, c\} = \varnothing$, which gives $f(\{a\} \cap \{b, c\}) = f(\varnothing) = \varnothing$. But we have $f(\{a\}) \cap f(\{b, c\}) = \{1\}$. So $f(\{a\} \cap \{b, c\}) \neq f(\{a\}) \cap f(\{b, c\})$.

28. a. We'll prove both containments at once: $x \in f^{-1}(G \cup H)$ iff $f(x) \in G \cup H$ iff $f(x) \in G$ or $f(x) \in H$ iff $x \in f^{-1}(G)$ or $x \in f^{-1}(H)$ iff $x \in f^{-1}(G) \cup f^{-1}(H)$.

c. If $x \in E$, then $f(x) \in f(E)$, which says that $x \in f^{-1}(f(E))$. This proves the containment.

e. Consider the function $f : \{a, b, c\} \rightarrow \{1, 2, 3\}$, defined by $f(a) = f(b) = 1$ and $f(c) = 2$. Let $E = \{a\}$. Then $f^{-1}(f(E)) = f^{-1}(f(\{a\})) = f^{-1}(\{1\}) = \{a, b\}$. So E is a proper subset of $f^{-1}(f(E))$. For the other example, let $G = \{2, 3\}$. Then $f(f^{-1}(G)) = f(\{c\}) = \{2\}$. So $f(f^{-1}(G))$ is a proper subset of G.

29. a. By (2.4a) it suffices to show that n divides $(x + y) - ((x \bmod n) + (y \bmod n))$. By the definition of mod we have $x \bmod n = x - nq_1$ and $y \bmod n = y - nq_2$. So $(x + y) - ((x \bmod n) + (y \bmod n)) = (x + y) - ((x - nq_1) + (x - nq_2)) = n(q_1 - q_2)$. So n divides $(x + y) - ((x \bmod n) + (y \bmod n))$.

c. Since $\gcd(a, n) = 1$, it follows from (2.2c) that there are integers x and y such that $1 = ax + ny$. Now we have the sequence of equations

$$
\begin{aligned}
1 \bmod n &= (ax + ny) \bmod n \\
&= ((ax \bmod n) + (ny \bmod n)) \bmod n && \text{(by (a))} \\
&= ((ax \bmod n) + ((n \bmod n)(y \bmod n) \bmod n)) \bmod n && \text{(by (b))} \\
&= ((ax \bmod n) + 0) \bmod n \\
&= ax \bmod n.
\end{aligned}
$$

30. Let $a = dq + r$, where $0 \leq r < d$. Solve for r to obtain $r = a - dq$. Now use the fact that $d = ax + by$ to substitute for d to obtain

$$
\begin{aligned}
r &= a - dq \\
&= a - (ax + by)q \\
&= a(1 - xq) + b(-yq)
\end{aligned}
$$

which, if $r > 0$, has the form of a number in S. But d is the smallest number in S and $0 \leq r < d$. So if $r > 0$, then it would be a number in S smaller than the smallest number, a contradiction. Therefore, $r = 0$ and consequently $d \mid a$. Similarly, we have $d \mid b$. QED.

31. Let $m, n \in \mathbb{N}$ and let $d = \gcd(m, n)$. Then $d \mid m$ and $d \mid n$. So by (1.1a) any divisor of d divides m and n. Therefore, $D_d \subseteq D_m \cap D_n$. For the other direction, let c be a common divisor of m and n. Since d is the greatest common divisor of m and n, it follows (from Exercise 23) that c divides d. Therefore, $D_m \cap D_n \subseteq D_k$. So the two sets are equal.

33. Assume that $m, n \in \mathbb{N}$ and let k be the least common multiple of m and n. So $k = ma = nb$ for some natural numbers a and b. Let $x \in M_k$. Then $x = qk$ for some natural number q. We can substitute for k in the equation to obtain $x = qk = qma = qnb$, which tells us that $x \in M_m \cap M_n$. So $M_k \subseteq M_m \cap M_n$. Now let $x \in M_m \cap M_n$. Then x is a common multiple of m and n. Therefore, since k is the least common multiple of m and n, it follows (from Exercise 25) that $k \mid x$. In other words, x is a multiple of k, which tells us that $x \in M_k$. So $M_m \cap M_n \subseteq M_k$. So the two sets are equal.

35. a. Let $n = \lfloor x \rfloor$. Then $n \le x < n + 1$. Split the interval up so that either $n \le x < n + 1/2$ or $n + 1/2 \le x < n + 1$. Consider each case. For the first case, multiply the inequality by 2 to obtain $2n \le 2x < 2n + 1$, which gives $\lfloor 2x \rfloor = 2n$. By adding $1/2$ to the first case inequality we get $n + 1/2 \le x + 1/2 < n + 1$, which gives $\lfloor x + 1/2 \rfloor = n$. Therefore, $\lfloor x \rfloor + \lfloor x + 1/2 \rfloor = n + n = 2n = \lfloor 2x \rfloor$. Now consider the second case. Multiply the inequality by 2 to obtain $2n + 1 \le 2x < 2n + 2$. So we get $\lfloor 2x \rfloor = 2n + 1$. By adding $1/2$ to the second case inequality we get $n + 1 \le x + 1/2 < n + 3/2$, which gives $\lfloor x + 1/2 \rfloor = n + 1$. Therefore, $\lfloor x \rfloor + \lfloor x + 1/2 \rfloor = n + (n + 1) = 2n + 1 = \lfloor 2x \rfloor$. Therefore, the equation holds in either case.

Section 2.2

1. a. 4. **c.** 2. **e.** $\langle (4, 0), (4, 1), (4, 2), (4, 3) \rangle$.
g. $\langle (+, (0, 0)), (+, (1, 1)), (+, (2, 2)) \rangle$.

2. a. $f(g(x)) = \text{ceiling}(x)$, $g(f(x)) = (2)\text{ceiling}(x/2)$, $f(g(1)) = 1$, and $g(f(1)) = 0$.
c. $f(g(x)) = \gcd(x \bmod 5, 10)$, $g(f(x)) = \gcd(x, 10) \bmod 5$, $f(g(5)) = 10$, and $g(f(5)) = 0$.

3. a. $f(g(x, y))$. **c.** $f(g(x, g(y, z)))$.

4. a. $2^7 \le x < 2^8$.

5. One solution is $\text{max4}(w, x, y, z) = \max(\max(\max(w, x), y), z)$.

6. $\text{floor}(\log_2(x)) + 1$.

7. a. $\langle 0, 1, 1, 2, 2, 2, 2, 3, 3, 3, 3, 3, 3, 3, 3, 4 \rangle$.

8. a. $f(n) = \text{map}(+, \text{pairs}(\text{seq}(n), \text{seq}(n)))$.

9. a. $f(n, k) = \text{map}(+, \text{dist}(n, \text{seq}(k)))$.
c. $f(n, m) = \text{map}(+, \text{dist}(n, \text{seq}(m - n)))$.

e. $f(n) = \text{pairs}(\text{seq}(n), g(n))$, where $g(n)$ is the solution to Part (d).
g. $f(g, n) = \text{pairs}(\text{seq}(n), \text{map}(g, \text{seq}(n)))$.
i. $f(g, h, \langle x_1, \ldots, x_n \rangle) = \text{pairs}(\text{map}(g, \langle x_1, \ldots, x_n \rangle), \text{map}(h, \langle x_1, \ldots, x_n \rangle))$.
10. $f(n) = \text{map}(+, \text{dist}(1, \text{seq}(n - 1)))$.
12. If $x \geq 1$, then $2^n \leq x < 2^{n+1}$ for some integer n. Therefore, $2^n \leq \lfloor x \rfloor < 2^{n+1}$.
Take the log of both inequalities to get $n \leq \log_2(x) < n + 1$ and $n \leq \log_2(\lfloor x \rfloor)$
$< n + 1$. Therefore, $\lfloor \log_2(x) \rfloor = n = \lfloor \log_2(\lfloor x \rfloor) \rfloor$.

Section 2.3

1. The fatherOf function is not injective because some fathers have more than one child. For example, if John and Mary have the same father, then fatherOf(John) = fatherOf(Mary). The fatherOf function is not surjective because there are people who are not fathers. For example, Mary is not a father. So fatherOf(x) $\neq$ Mary for all people x.
2. a. $f : C \to B$, where $f(1) = x$, $f(2) = y$.
c. $f : A \to B$, where $f(a) = x$, $f(b) = y$, $f(c) = z$.

3. a. Eight functions; no injections, six surjections, no bijections, and two with none of the properties.
c. 27 functions; six satisfy the three properties (injective, surjective, and bijective), 21 with none of the properties.

4. a. If $f(x) = f(y)$, then $2x = 2y$, which upon dividing by 2 yields $x = y$. So f is injective. f is not surjective because the range of f is the set of even natural numbers, which is not equal to the codomain $\mathbb{N}$. E.g., no x maps to 1.
c. If $y \in \mathbb{N}$, then $f(2y) = \text{floor}(2y/2) = \text{floor}(y) = y$. So f is surjective. f is not injective because, for example, $f(0) = f(1) = 0$.
e. Let $f(x) = f(y)$. If x is odd and y is even, then $x - 1 = y + 1$, which implies $x = y + 2$. This tells us that x and y are either both even or both odd, a contradiction. We get a similar contradiction if x is even and y is odd. If x and y are both even, then $x + 1 = y + 1$, which implies $x = y$. If x and y are both odd, then $x - 1 = y - 1$, which implies that $x = y$. So f is injective. Let $y \in \mathbb{N}$. If y is odd, then $y - 1$ is even and $f(y - 1) = y - 1 + 1 = y$. If y is even, then $y + 1$ is odd and $f(y + 1) = y + 1 - 1 = y$. So f is surjective. Therefore, f is bijective.

5. a. Surjective. **c.** Surjective. **e.** Injective. **g.** Surjective. **i.** Surjective.

6. a. Let $f(x) = f(y)$. Then $(b - a)x + a = (b - a)y + a$. Subtract a from both sides and divide the resulting equation by $(b - a)$ to obtain $x = y$. So f is injective. To show f is surjective, let $y \in (a, b)$ and solve the equation $f(x) = y$ for x to obtain $(b - a)x + a = y$, which gives $x = (y - a)/(b - a)$. It follows that $f((y - a)/(b - a)) = y$ and since $a < y < b$, we have $0 < (y - a)/(b - a) < 1$. Thus f is surjective. Therefore, f is a bijection.
c. Let $f(x) = f(y)$. Then $1/(2x - 1) - 1 = 1/(2y - 1) - 1$, which by elementary algebra implies that $x = y$. Therefore, f is an injection. To show f is surjective, let $y > 0$ and then find some x such that $f(x) = y$. Solve the equation $f(x) = y$

to get $x = (2 + y)/(2y + 2)$. It follows that $f((2 + y)/(2y + 2)) = y$, and since $y > 0$, it follows that $1/2 < (2 + y)/(2y + 2) < 1$. To see this, we can obtain a contradiction if $(2 + y)/(2y + 2) < 1/2$ or if $(2 + y)/(2y + 2) > 1$. So f is surjective. Therefore, f is a bijection.

e. f is a bijection because it is defined in terms of the bijections given in Parts (c) and (d). To see this, let $f(x) = f(y)$. If this value is positive, then $x = y$ by part (c). If the value is negative, then $x = y$ by Part (d). If $f(x) = f(y) = 0$, then $x = y = 1/2$ by the definition of f. So f is injective. To show f is surjective, let $y \in \mathbb{R}$ and find some $x \in (0, 1)$ such that $f(x) = y$. Again, take the three cases. If $y > 0$, then part (c) gives an element $x \in (1/2, 1)$ such that $f(x) = y$. If $y < 0$, then part (d) gives an element $x \in (0, 1/2)$ such that $f(x) = y$. If $y = 0$, then $f(1/2) = y$. So f is surjective. Therefore, f is a bijection.

7. a. 15.

c. Any nonempty string over $\{a, b, c\}$ has one of nine possible patterns of beginning and ending letters. So any set of 10 such strings will contain two strings with the same beginning and ending letters.

8. a. Since there are 10 decimal digits, we can be assured that any set of 11 numbers will contain two numbers that use a common digit in their representations.

c. Of the ten numbers listed, at least nine are in the range from 1 to 8. (It could happen that some $x_k = 8$, so that $x_k + 1 = 9$ is not in the set.) So the pigeonhole principle tells us that two of the nine numbers are equal. Since the two lists are each distinct, it follows that $x_i = x_j + 1$ for some i and j.

9. a. Not bijective because $\gcd(2, 6) \neq 1$. The fixed point is 0.

c. Bijective and $f^{-1} = f$. The fixed points are 0 and 3.

e. Bijective and $f^{-1}(x) = (4x + 2) \bmod 7$. The fixed point is 4.

g. Bijective and $f^{-1}(x) = (9x + 1) \bmod 16$. There are no fixed points.

10. a. $\{a \mid \gcd(a, 26) = 1\} = \{1, 3, 5, 7, 9, 11, 15, 17, 19, 21, 23, 25\}$.

11. a. one, two, six, four, five, nine, three, seven, eight.

c. Only one, two, and three can be placed in the table: one, blank, blank, two, blank, blank, three, blank, blank.

12. a. Wednesday, Monday, Friday, Tuesday, Sunday, Thursday, Saturday.

c. Wednesday, Monday, Sunday, Tuesday, Friday, Thursday, Saturday.

13. a. March, April, January, February, July, August, May, June.

c. March, May, January, February, August, July, April, June.

15. a. Let f and g be injective, and assume that $g \circ f(x) = g \circ f(y)$ for some $y \in A$. Since g is injective, it follows that $f(x) = f(y)$, and it follows that $x = y$ because f is injective. Therefore, $g \circ f$ is injective.

c. If f and g are bijective, then they are both injective and surjective. So by Parts (a) and (b) the composition $g \circ f$ is injective and surjective, hence bijective.

16. Let $x \in A$ and let $f(g(x)) = y$. Apply g to both sides to obtain $g(f(g(x))) = g(y)$. Since $g(f(x)) = x$ for all x, it follows that $g(f(g(x))) = g(x)$. So $g(x) = g(y)$. Since g is bijective, hence injective, it follows that $x = y$. Therefore, $f(g(x)) = x$.

17. a. If $g \circ f$ is surjective, then for each element $z \in C$ there exists an element $x \in A$ such that $z = (g \circ f)(x) = g(f(x))$. So it follows that $f(x)$ is an element of B such that $z = g(f(x))$. Therefore, g is surjective if $g \circ f$ is surjective.

18. a. Let f be surjective, and let $b \in B$ and $c \in C$. Then there exists an element $a \in A$ such that $f(a) = (b, c)$. But $f(a) = (g(a), h(a))$. Therefore, $b = g(a)$ and $c = h(a)$. So g and h are surjective. Now let $A = \{1, 2, 3\}$, $B = \{4, 5\}$, and $C = \{6, 7\}$. The set $B \times C$ has four elements, and A has three elements. So there can be no surjection from A to $B \times C$.

19. Let $g = \gcd(a, n)$. Let x be an integer such that $ax \bmod n = b \bmod n$. Then n divides $(ax - b)$. So there is an integer q such that $ax - b = nq$, or $b = a - nq$. Since g divides a and g divides n, it follows from (1.1b) that g divides b. For the converse, suppose that g divides b. Then we can write $b = gk$ for some integer k. Since $g = \gcd(a, n)$ it follows from (2.2c) that $g = as + nt$ for some integers s and t. Multiply this equation by k to obtain $b = gk = ask + ntk$. Apply mod n to both sides to obtain $b \bmod n = ask \bmod n$. Therefore, $x = sk$ is a solution to the equation $ax \bmod n = b \bmod n$.

21. If we show that g is a bijection and $f(g(x)) = g(f(x)) = x$ for all $x \in \mathbb{N}_n$, then it follows that $g = f^{-1}$. Since $1 = ak + nm$, it follows that $\gcd(k, n) = 1$. So g is a bijection by the first part of (2.6). We have the following sequence of equations:

$$\begin{aligned}
f(g(x)) &= (ag(x) + b) \bmod n \\
&= (a((kx + c) \bmod n) + b) \bmod n \\
&= (a(kx + c + nq) + b) \bmod n && \text{(for some integer } q\text{)} \\
&= (akx + ac + b) \bmod n && \text{(by (2.4))} \\
&= (akx + f(c)) \bmod n \\
&= (akx + 0) \bmod n && (f(c) = 0) \\
&= (akx) \bmod n \\
&= ((1 - nm)x) \bmod n && (1 = ak + nm) \\
&= (x - nmx) \bmod n \\
&= x \bmod n \\
&= x.
\end{aligned}$$

Note also that if we let $g(f(x)) = y$, then we can apply f to both sides to get $f(g(f(x))) = f(y)$. But $f(g(f(x))) = f(x)$ because we just showed that $f(g(x)) = x$ for all $x \in \mathbb{N}_n$. So $f(x) = f(y)$, from which we conclude that $x = y$ because f is injective. Therefore, $g(f(x)) = x$ for all $x \in \mathbb{N}_n$. So $g = f^{-1}$. QED.

Section 2.4

1. a. Let A be the set. The smallest number in A is $2(0) + 5 = 5$ and the largest number in A is $2(46) + 5 = 97$. So the function $f : \{0, 1, \ldots, 46\} \to A$ defined by mapping $f(x) = 2x + 5$ is a bijection. Therefore, $|A| = 47$.
c. The function $f : \{0, 1, \ldots, 15\} \to \{2, 5, 8, 11, 14, 17, \ldots, 44, 47\}$ defined by $f(k) = 2 + 3k$ is a bijection. So the cardinality of the set is 16.

2. a. Let Even be the set of even natural numbers. For example, the function $f : \mathbb{N} \to$ Even defined by $f(k) = 2k$ is a bijection. So Even is countable.
c. Let S be the set of strings over $\{a\}$. So $S = \{a\}^* = \{\Lambda, a, aa, aaa, \ldots \}$. The mapping from $\mathbb{N}$ to S that maps each n to the string of length n is a bijection. So S is countable.
e. For example, the function $f : \mathbb{Z} \to \mathbb{N}$ defined $f(x) = 2x - 1$ when $x > 0$ and $f(x) = -2x$ when $x \leq 0$ is a bijection. So $\mathbb{Z}$ is countable.
g. Let E be the set of even integers. For example, the function $f : \mathbb{N} \to E$ defined $f(x) = x - 1$ when x is odd and $f(x) = -(x + 2)$ when x is even is a bijection. So E is countable.

3. a. Let S be the set of strings over $\{a, b\}$ that have odd length. For each number n let S_n be the set of all strings over $\{a, b\}$ that have length $2n + 1$. For example, $S_0 = \{a, b\}$ and $S_1 = \{aaa, aab, aba, baa, bbb, bba, bab, abb\}$. It follows that $S = S_0 \cup S_1 \cup \ldots \cup S_n \cup \ldots$. Since each set S_n is finite, hence countable, it follows from (2.11) that the union is countable.
c. Let B be the set of all binary trees over $\{a, b\}$. For each natural number n let S_n be the set of all binary trees over $\{a, b\}$ that have n nodes. It follows that $B = S_0 \cup S_1 \cup \ldots \cup S_n \cup \ldots$. Since each set S_n is finite, hence countable, it follows from (2.11) that the union is countable.

4. a. Let $g(n) = $ hello if $f_n(n) = $ world, and let $g(n) = $ world if $f_n(n) = $ hello. Then the sequence $(g(0), g(1), \ldots, g(n), \ldots)$ is not in the given set.
c. Let $g(n) = 2$ if $a_{nn} = 4$, and let $g(n) = 6$ if $a_{nn} \neq 4$. Then the sequence $(g(0), g(1), \ldots, g(n), \ldots)$ is not in the given set.

5. We can represent each subset S of $\mathbb{N}$ as a sequence of 1's and 0's where 1 in the kth position means that $k \in S$ and 0 means $k \notin S$. For example, $\mathbb{N}$ is represented by $(1, 1, 1, \ldots)$ and the empty set by $(0, 0, 0, \ldots)$. So each set S_n can be represented by an infinite sequence of 0's and 1's. But now (2.13) applies to say that there is some sequence of 1's and 0's that is not listed. This contradicts the statement that all subsets of $\mathbb{N}$ are listed.

7. a. Let S be a subset of the countable set A. Then the mapping from S to A that sends every element to itself is an injection. So $|S| \leq |A|$. Since A is countable, there is an injection from A to $\mathbb{N}$ by countable property (b). Since a composition of injections is an injection, we have an injection from S to $\mathbb{N}$. Therefore, $|S| \leq |\mathbb{N}|$.

8. a. For each natural number k, let S_k be the set of strings of length n over the alphabet $\{a_0, a_1, \ldots, a_k\}$. It follows that $A_n = S_0 \cup S_1 \cup \ldots \cup S_n \cup \ldots$. Since each set S_n is finite, hence countable, it follows from (2.11) that the union is countable.

9. For each n let F_n be the collection of subsets of $\{0, 1, \ldots, n\}$. In other words, $F_n = \text{power}(\{0, 1, \ldots, n\})$. Since any finite subset S of $\mathbb{N}$ has a largest element n, it follows that S is in the collection F_n. So $\text{Finite}(\mathbb{N}) = F_0 \cup F_1 \cup \ldots \cup F_n \cup \ldots$. Each set F_n is countable because it is finite. So (2.11) tells us that the union is countable.

Chapter 3

Section 3.1

1. a. 3, 5, 9, 17, 33, 65, 129, 257, 513, 1025.

2. a. Basis: $1 \in S$; Induction: If $x \in S$, then $x + 2 \in S$.
c. Basis: $-3 \in S$; Induction: If $x \in S$, then $x + 2 \in S$.
e. Basis: $1 \in S$; Induction: If $x \in S$, then $(\sqrt{x} + 1)^2 \in S$.

3. a. Basis: 4, 3 $\in S$. Induction: If $x \in S$, then $x + 3 \in S$.

4. a. Basis: 0, 1 $\in S$; Induction: If $x \in S$, then $x + 4 \in S$.
c. Basis: $2 \in S$; Induction: If $x \in S$, then $x + 5 \in S$.

5. $4 = 3 \cup \{3\} = 2 \cup \{2\} \cup \{3\} = 1 \cup \{1\} \cup \{2\} \cup \{3\} = 0 \cup \{0\} \cup \{1\} \cup \{2\}$ $\cup \{3\} = \varnothing \cup \{0\} \cup \{1\} \cup \{2\} \cup \{3\} = \{0, 1, 2, 3\}$.

6. a. Basis: $b \in S$. Induction: If $x \in S$, then $axc \in S$.
c. Basis: $a \in S$. Induction: If $x \in S$, then $aax \in S$.
e. Basis: $b \in S$. Induction: If $x \in S$, then $ax, xc \in S$.
g. Basis: $b \in S$. Induction: If $x \in S$, then $ax, xb \in S$.
i. Basis: $a, b \in S$. Induction: If $x \in S$, then $ax, xb \in S$.
k. Basis: $\Lambda \in S$. Induction: If $x \in S$, then $abx, bax, axb, bxa \in S$.

7. a. Basis: $\Lambda \in S$. Induction: If $x \in S$, then $axa, bxb \in S$.
c. Basis: $\Lambda, a, b \in S$. Induction: If $x \in S$, then $axa, bxb \in S$.

8. Basis: $a, b, c, x, y, z \in T$. Induction: If $t \in T$, then $f(t), g(t) \in T$.

9. a. $\langle a \rangle, \langle b, a \rangle, \langle b, b, a \rangle, \langle b, b, b, a \rangle, \langle b, b, b, b, a \rangle$.

10. a. Basis: $\langle a \rangle \in S$. Induction: If $L \in S$, then $a :: L \in S$.
c. Basis: $\langle a, b \rangle, \langle b, a \rangle \in S$. Induction: If $L \in S$, then $\text{head}(L) :: L \in S$.
e. Basis: $\langle \, \rangle \in S$. Induction: If $L \in S$ and $a, b \in \{0, 1, 2\}$, then $a :: b :: L \in S$.
g. Basis: $\langle a \rangle \in S$. Induction: If $L \in S$, then $a :: a :: L \in S$.
i. Basis: $\langle a \rangle \in S$ for all $a \in A$. Induction: If $L \in S$ and $a, b \in A$, then $a :: b :: L \in S$.

11. a. Basis: $\langle a \rangle \in S$. Induction: If $L \in S$, then $\text{consR}(L, b) \in S$.
c. Basis: $\langle \, \rangle \in S$. Induction: If $L \in S$, then put the following four lists in S: $\text{cons}(a, \text{cons}(b, L)), \text{cons}(b, \text{cons}(a, L)), \text{cons}(a, \text{consR}(L, b))$, and $\text{cons}(b, \text{consR}(L, a))$.

13. Each nonleaf node has a leaf as the left child and a tree with the same property as the right child.

15. Basis: tree($\langle\,\rangle$, a, $\langle\,\rangle$) $\in B$. Induction: If $T \in B$, then tree(T, a, tree($\langle\,\rangle$, a, $\langle\,\rangle$)), tree(tree($\langle\,\rangle$, a, $\langle\,\rangle$), a, T) $\in B$.

16. a. $B = \{(x, y) \mid x, y \in \mathbb{N} \text{ and } x \geq y\}$.

17. a. Basis: $(0, 0) \in S$.
Induction: If $(x, y) \in S$ and $x = y$, then $(x, y + 1)$, $(x + 1, y + 1) \in S$.

18. a. Basis: $(\langle\,\rangle, \langle\,\rangle) \in S$. Induction: If $(x, y) \in S$ and $a \in A$, then $(a :: x, y)$, $(x, a :: y) \in S$.
c. Basis: $(0, \langle\,\rangle) \in S$. Induction: If $(x, L) \in S$ and $m \in \mathbb{N}$, then $(x, m :: L)$, $(x + 1, L) \in S$.

19. Basis: $\langle\,\rangle \in E$ and $\langle a\rangle \in O$ for all $a \in A$. Induction:
If S, $T \in E$ and $a \in A$, then tree(S, a, T) $\in O$.
If S, $T \in O$ and $a \in A$, then tree(S, a, T) $\in O$.
If $S \in E$ and $T \in O$ and $a \in A$, then tree(S, a, T), tree(T, a, S) in E.

20. Basis: $(a, g(a)) \in A$. Induction:
If $(x, y) \in A$ and $y < f(x)$, then $(x, y + 1) \in A$.
If $(x, y) \in A$ and $x < b$, then $(x + 1, g(x + 1)) \in A$.

Section 3.2

1. We'll evaluate the leftmost term in each expression.
fib(4) = fib(3) + fib(2) = fib(2) + fib(1) + fib(2) = fib(1) + fib(0) + fib(1) + fib(2) = 1 + fib(0) + fib(1) + fib(2) = 1 + 0 + fib(1) + fib(2) = 1 + 0 + 1 + fib(2) = 1 + 0 + 1 + fib(1) + fib(0) = 1 + 0 + 1 + 1 + fib(0) = 1 + 0 + 1 + 1 + 0 = 3.

3. For (3.9): makeTree($\langle\,\rangle$, $\langle 3, 2, 4\rangle$) = makeTree(insert(3, $\langle\,\rangle$), $\langle 2, 4\rangle$)
$\qquad\qquad$ = makeTree(insert(2, insert(3, $\langle\,\rangle$)), $\langle 4\rangle$)
$\qquad\qquad$ = makeTree(insert(4, insert(2, insert(3, $\langle\,\rangle$))), $\langle\,\rangle$)
$\qquad\qquad$ = insert(4, insert(2, insert(3, $\langle\,\rangle$))).
For (3.10): makeTree($\langle\,\rangle$, $\langle 3, 2, 4\rangle$ = insert(3, makeTree($\langle\,\rangle$, $\langle 2, 4\rangle$))
$\qquad\qquad$ = insert(3, insert(2, makeTree($\langle\,\rangle$, $\langle 4\rangle$)))
$\qquad\qquad$ = insert(3, insert(2, insert(4, makeTree($\langle\,\rangle$, $\langle\,\rangle$))))
$\qquad\qquad$ = insert(3, insert(2, insert(4, $\langle\,\rangle$))).

4. a. $f(0) = 0$ and $f(n) = f(n - 1) + 2n$.
c. $f(1, n) = gcd(1, n)$ and $f(k, n) = f(k - 1, n) + gcd(k, n)$.
e. $f(0, k) = 0$ and $f(n, k) = f(n - 1, k) + nk$.

5. a. $f(\Lambda) = \Lambda$ and $f(ax) = f(x)a$ and $f(bx) = f(x)b$.
c. $f(x, y) = $ if $x = \Lambda$ then True
$\qquad\qquad$ else if $x = as$ and $y = at$ or $x = bs$ and $y = bt$ then $f(s, t)$
$\qquad\qquad$ else False.
e. $f(x) = $ if $x = \Lambda$ or $x = a$ or $x = b$ then True
$\qquad\quad$ else if $x = asa$ or $x = bsb$ then $f(s)$
$\qquad\quad$ else False.

6. a. $f(0) = \langle 0 \rangle$ and $f(n) = 2n :: f(n-1)$.
c. $\max(\langle x \rangle) = x$, and $\max(h :: t) = $ if $h > \max(t)$ then h else $\max(t)$.
e. $f(\langle \ \rangle) = \langle \ \rangle$, and $f(h :: t) = $ if $P(h)$ then $h :: f(t)$ else $f(t)$.
g. $f(a, \langle \ \rangle) = \langle \ \rangle$, and $f(a, (x, y) :: t) = (x + a, y) :: f(a, t)$.
i. $f(0) = \langle (0, 0) \rangle$ and $f(n) = (0, n) :: g(1, f(n-1))$, where g adds 1 to the first component of each ordered pair in a list of ordered pairs. [See Part (g).]

7. a. $f(0) = \langle 0 \rangle$ and $f(n+1) = \mathrm{cat}(f(n), \langle n+1 \rangle)$.
c. $f(0) = \langle 1 \rangle$ and $f(n) = \mathrm{cat}(f(n-1), \langle 2n+1 \rangle)$.
e. $f(0, k) = \langle 0 \rangle$ and $f(n, k) = \mathrm{cat}(f(n-1, k), \langle nk \rangle)$.
g. $f(n, n) = \langle n \rangle$ and $f(n, m) = \mathrm{cat}(f(n, m-1), \langle m \rangle)$.

8. $\mathrm{insert}(f, \langle a, b \rangle) = f(a, b)$,
  $\mathrm{insert}(f, \mathrm{cons}(h, t)) = f(a, \mathrm{insert}(f, t))$.

10. a. $\mathrm{last}(\langle x \rangle) = x$,
    $\mathrm{last}(\mathrm{cons}(h, t)) = \mathrm{last}(t)$.

12. Let $\mathrm{rem}(L)$ denote the list obtained from L by removing repetitions of elements and keeping the rightmost occurrence of each element.

$$\mathrm{rem}\,(L) = \text{if } L = \langle \ \rangle \text{ then } \langle \ \rangle$$
$$\text{else } \mathrm{cat}(\mathrm{rem}(\mathrm{removeAll}(\mathrm{last}\,(L), \mathrm{front}\,(L))), \mathrm{last}\,(L) :: \langle \ \rangle).$$

14. a. $\mathrm{In}(T)$: **if** $T \neq \langle \ \rangle$ **then** $\mathrm{In}(\mathrm{left}(T))$; $\mathrm{print}(\mathrm{root}(T))$; $\mathrm{In}(\mathrm{right}(T))$ **fi**.

15. a. Equational form: $\mathrm{leaves}(\langle \ \rangle) = 0$,
        $\mathrm{leaves}(\mathrm{tree}(\langle \ \rangle, a, \langle \ \rangle)) = 1$,
        $\mathrm{leaves}(\mathrm{tree}(l, a, r)) = \mathrm{leaves}(l) + \mathrm{leaves}(r)$.
If-then form: $\mathrm{leaves}(t) = $ if $t = \langle \ \rangle$ then 0
        else if $\mathrm{left}(t) = \mathrm{right}(t) = \langle \ \rangle$ then 1
        else $\mathrm{leaves}(\mathrm{left}(t)) + \mathrm{leaves}(\mathrm{right}(t))$.

c. Equational form:
   $\mathrm{postOrd}(\langle \ \rangle) = \langle \ \rangle$
   $\mathrm{postOrd}(\mathrm{tree}(L, r, R)) = \mathrm{cat}(\mathrm{postOrd}(L), \mathrm{cat}(\mathrm{postOrd}(R), \langle r \rangle))$.
If-then form:
   $\mathrm{postOrd}(T) = $ if $T = \langle \ \rangle$ then $\langle \ \rangle$
        else $\mathrm{cat}(\mathrm{postOrd}(\mathrm{left}(T)), \mathrm{cat}(\mathrm{postOrd}(\mathrm{right}(T)), \langle \mathrm{root}(T) \rangle))$.

16. a. $f(\langle \ \rangle) = \langle \ \rangle$ and $f(\langle L, r, R \rangle) = r + f(L) + \mathrm{f}(R)$.
c. $f(\langle \ \rangle) = \langle \ \rangle$ and $f(\langle L, r, R \rangle) = $ if $p(r)$ then $r :: \mathrm{cat}(f(L), f(R))$
else $\mathrm{cat}(f(L), f(R))$.

18. a. $f(k, \langle \ \rangle) = \langle \ \rangle$ and $f(k, x :: t) = x_k :: f(k, t)$.

19. a. $\mathrm{isMember}(x, L) = $ if $L = \langle \ \rangle$ then False
          else if $x = \mathrm{head}(L)$ then True
          else $\mathrm{isMember}(x, \mathrm{tail}(L))$.
c. $\mathrm{areEqual}(K, L) = $ if $\mathrm{isSubset}(K, L)$ then $\mathrm{isSubset}(L, K)$ else False.

e. intersect$(K, L) =$ if $K = \langle\,\rangle$ then $\langle\,\rangle$
    else if isMember(head(K), L) then
        head(K) :: intersect(tail(K), L)
    else intersect(tail(K), L).

20. 1, 1, 2, 2, 3, 4, 4, 4, 5, 6, 7, 7, 8, 8, 8, 8, 9.

22. Assume that the product of the empty list $\langle\,\rangle$ with any list is $\langle\,\rangle$. Then define product as follows:

$$\text{product}\,(A, B) = \text{if } A = \langle\,\rangle \text{ or } B = \langle\,\rangle \text{ then } \langle\,\rangle$$

else concatenate the four lists

$$\langle(\text{head}\,(A)\,,\,\text{head}\,(B))\rangle\,,$$
$$\text{product}(\,\langle\text{head}\,(A)\rangle\,,\,\text{tail}\,(B)\,),$$
$$\text{product}(\text{tail}\,(A)\,,\,\langle\text{head}\,(B)\rangle\,)\text{, and}$$
$$\text{product}(\text{tail}\,(A)\,,\,\text{tail}\,(B)\,).$$

23. a. 1, 2.5, 2.05. **c.** 3, 2.166..., 2.0064.... **e.** 1, 5, 3.4.

24. a. Square$(x :: s) = x^2 ::$ Square(s).
c. Prod$(n, s) =$ if $n = 0$ then 1 else head(s)Prod$(n - 1,$ tail$(s))$.
e. Skip$(x, k) = x ::$ Skip$(x + k, k)$.
g. ListOf$(n, s) =$ if $n = 0$ then $\langle\,\rangle$ else head$(s) ::$ ListOf$(n - 1,$ tail$(s))$.

25. a. head(Primes) $=$ head(sieve(inst(2))) $=$ head(sieve(2 :: ints(3))) $=$ head(2 :: sieve(remove(2, ints(3)))) $= 2$.
c. remove(2, ints(0))) $=$ remove(2, 0 :: ints(1)) $=$ remove(2, ints(1))
$=$ remove(2, 1 :: ints(2)) $= 1 ::$ remove(2, ints(2)) $= 1 ::$ remove(2, 2 :: ints(3))
$= 1 ::$ remove(2, ints(3)) $= 1 ::$ remove(2, 3 :: ints(4))
$= 1 :: 3 ::$ remove(2, ints(4)).

26. $f(x) =$ if $x \leq 10$ then 1 else $x - 10$.

Section 3.3

1. a. $S \to DS$, $D \to 7$, $S \to DS$, $S \to DS$, $D \to 8$, $D \to 0$, $S \to D$, $D \to 1$.
c. $S \Rightarrow DS \Rightarrow DDS \Rightarrow DDDS \Rightarrow DDDD \Rightarrow DDD1 \Rightarrow DD01 \Rightarrow D801 \Rightarrow 7801$.

2. a. Leftmost: $S \Rightarrow S[S] \Rightarrow [S] \Rightarrow [\,]$. Rightmost: $S \Rightarrow S[S] \Rightarrow S[\,] \Rightarrow [\,]$.
c. Leftmost: $S \Rightarrow S[S] \Rightarrow S[S]\,[S] \Rightarrow [S]\,[S] \Rightarrow [\,]\,[S] \Rightarrow [\,]\,[\,]$.
Rightmost: $S \Rightarrow S[S] \Rightarrow S[\,] \Rightarrow S[S]\,[\,] \Rightarrow S[\,]\,[\,] \Rightarrow [\,]\,[\,]$.

3. a. $S \to bb \,|\, bbS$.
c. $S \to \Lambda \,|\, abS$.
e. $S \to ab \,|\, abS$.
g. $S \to b \,|\, bbS$.
i. $S \to aBc$ and $B \to \Lambda \,|\, bB$.

4. a. $S \to AB$ and $A \to \Lambda \,|\, aA$ and $B \to \Lambda \,|\, bB$.
c. $S \to AB$ and $A \to a \,|\, aA$ and $B \to \Lambda \,|\, bB$.
e. $S \to AB$ and $A \to a \,|\, aA$ and $B \to b \,|\, bB$.

5. a. $S \rightarrow \Lambda \mid aSa \mid bSb \mid cSc$.
c. $S \rightarrow A \mid B$ and $A \rightarrow \Lambda \mid aaA$ and $B \rightarrow b \mid bbB$.
e. $S \rightarrow aAB \mid ABb$ and $A \rightarrow \Lambda \mid aA$ and $B \rightarrow \Lambda \mid bB$.

6. a. $O \rightarrow B1$ and $B \rightarrow \Lambda \mid B0 \mid B1$.
c. $O \rightarrow D1 \mid D3 \mid D5 \mid D7 \mid D9$ and $D \rightarrow \Lambda \mid D0 \mid D1 \mid D2 \mid D3 \mid D4 \mid D5 \mid D6 \mid D7 \mid D8 \mid D9$.

7. a. $S \rightarrow D \mid S + S \mid (S)$, and D denotes a decimal numeral.

8. a. $S \rightarrow a \mid b \mid c \mid x \mid y \mid z \mid f(S) \mid g(S)$.

9. a. $S \rightarrow a \mid b \mid c \mid x \mid y \mid z \mid f(S) \mid g(S, S)$.

11. a. The string $ababa$ has two parse trees.
c. The string aa has two parse trees.
e. The string $[\,]\,[\,]$ has two parse trees.

13. a. $S \rightarrow a \mid abS$.
c. $S \rightarrow a \mid aS$.
e. $S \rightarrow S[S] \mid \Lambda$.

14. a. $S \rightarrow A \mid AB$ and $A \rightarrow Aa \mid a$ and $B \rightarrow Bb \mid b$.

15. a. Basis: $\Lambda \in L(G)$. Induction: If $w \in L(G)$, then put $aaw \in L(G)$.

Chapter 4

Section 4.1

1. a. Reflexive, symmetric, transitive. **c.** Reflexive, symmetric, transitive.
e. Reflexive, symmetric, transitive. **g.** Irreflexive, transitive. **i.** Reflexive.

2. a. Symmetric. **c.** Reflexive, antisymmetric, and transitive.
e. Symmetric.

3. a. The irreflexive property follows because $(x, x) \notin \varnothing$ for any x. The symmetric, antisymmetric, and transitive properties are conditional statements that are always true because their hypotheses are false.

4. a. $\{(a, a), (b, b), (c, c), (a, b), (b, c)\}$.
c. $\{(a, b)\}$.
e. $\{(a, a), (b, b), (c, c), (a, b)\}$.
g. $\{(a, a), (b, b), (c, c)\}$.

5. a. isGrandchildOf. **c.** isNephewOf.

6. isFatherOf $\circ$ isBrotherOf.

7. a. Let $R = \{(a, b), (b, a)\}$. Then R is irreflexive, and $R^2 = \{(a, a), (b, b)\}$, which is not irreflexive.

8. a. $\{(x, y) \mid x < y - 1\}$.

9. a. $\mathbb{N} \times \mathbb{N}$. **c.** $\{(x, y) \mid y \neq 0\} - \{(0, 1)\}$.

11. $r(\varnothing) = \{(a, a) \mid a \in A\}$, which is basic equality over A.

12. a. $\varnothing$. **c.** $\{(a, b), (b, a), (b, c), (c, b)\}$.

13. a. $\varnothing$. **c.** $\{(a, b), (b, a), (a, a), (b, b)\}$.

15. a. isAncestorOf. **c.** greaterThan.

16. a.

	1	2	3	4
1	0	20	∞	5
2	∞	0	10	∞
3	∞	∞	0	10
4	∞	10	5	0

b.

	1	2	3	4
1	0	15	10	5
2	∞	0	10	20
3	∞	20	0	10
4	∞	10	5	0

	1	2	3	4
1	0	4	4	0
2	0	0	0	3
3	0	4	0	0
4	0	0	0	0

17. Let "path" be the function to compute the list of edges on a shortest path from i to j. We'll use the "cat" function to concatenate two lists:

path(i, j) = if $P_{ij} = 0$ then $\langle (i, j) \rangle$ else cat(path(i, P_{ij}), path(P_{ij}, j)).

19. Let M be the adjacency matrix for R. **a.** Check to see if $M_{ii} = 1$ for all i.
c. Check to see that "$M_{ij} = M_{jk} = 1$ implies $M_{ik} = 1$" for all i, j, and k.
e. For all i and j check M_{ij}. If $M_{ij} = 1$, then set $M_{ji} = 1$.

20. a. Let R be reflexive. Then $a \, R \, a$ and $a \, R \, a$ for all a, which implies that $a \, R^2 \, a$ for all a. Therefore, R^2 is reflexive.
c. Let R be transitive, and let $a \, R^2 \, b$ and $b \, R^2 \, c$. Then $a \, R \, x$ and $x \, R \, b$, and $b \, R \, y$ and $y \, R \, c$ for some x and y. Since R is transitive, it follows that $a \, R \, b$ and $b \, R \, c$. Therefore, $a \, R^2 \, c$. Thus R^2 is transitive.

21. Since less is transitive we have $t(\text{less}) = \text{less}$. It follows that $st(\text{less}) = s(\text{less}) = \{(m, n) \mid m \neq n\}$. But $ts(\text{less}) = t(\{(m, n) \mid m \neq n\}) = \mathbb{N} \times \mathbb{N}$.

22. a. $(x, y) \in R \circ (S \circ T)$ iff $(x, w) \in R$ and $(w, y) \in S \circ T$ for some w iff $(x, w) \in R$ and $(w, z) \in S$ and $(z, y) \in T$ for some w and z iff $(x, z) \in R \circ S$ and $(z, y) \in T$ for some z iff $(x, y) \in (R \circ S) \circ T$.
c. If $(x, y) \in R \circ (S \cap T)$, then $(x, w) \in R$ and $(w, y) \in S \cap T$ for some w. Thus $(x, y) \in R \circ S$ and $(x, y) \in R \circ T$, which gives $(x, y) \in R \circ S \cap R \circ T$.

24. a. If R is reflexive, then it contains the set $\{(a, a) \mid a \in A\}$. Since $s(R)$ and $t(R)$ contain R as a subset, it follows that they each contain $\{(a, a) \mid a \in A\}$.
c. Suppose R is transitive. Let (a, b), $(b, c) \in r(R)$. If $a = b$ or $b = c$, then certainly $(a, c) \in r(R)$. So suppose $a \neq b$ and $b \neq c$. Then (a, b), $(b, c) \in R$. Since R is transitive, it follows that $(a, c) \in R$, which of course also says that $(a, c) \in r(R)$. Therefore, $r(R)$ is transitive.

25. a. A proof by containment goes as follows: If $(a, b) \in rt(R)$, then either $a = b$ or there is a sequence of elements $a = x_1, x_2, \ldots, x_n = b$ such that $(x_i, x_{i+1}) \in R$ for $1 \leq i < n$. Since $R \subseteq r(R)$, we also have $(x_i, x_{i+1}) \in r(R)$ for $1 \leq i < n$, which says that $(a, b) \in tr(R)$. For the other containment, let

$(a, b) \in tr(R)$. If $a = b$, then $(a, b) \in rt(R)$. If $a \neq b$, then there is a sequence of elements $a = x_1, x_2, \ldots, x_n = b$ such that $(x_i, x_{i+1}) \in r(R)$ for $1 \leq i < n$. If $x_i = x_{i+1}$, then we can remove x_i from the sequence. So we can assume that $x_i \neq x_{i+1}$ for $1 \leq i < n$. Therefore, $(x_i, x_{i+1}) \in R$ for $1 \leq i < n$, which says that $(a, b) \in t(R)$, and thus also $(a, b) \in rt(R)$.

c. If $(a, b) \in st(R)$, then either $(a, b) \in t(R)$ or $(b, a) \in t(R)$. Without loss of generality we can assume that $(a, b) \in t(R)$. Then there is a sequence of elements $a = x_1, x_2, \ldots, x_n = b$ such that $(x_i, x_{i+1}) \in R$ for $1 \leq i < n$. Since $R \subseteq s(R)$, we also have $(x_i, x_{i+1}) \in s(R)$ for $1 \leq i < n$, which says that $(a, b) \in ts(R)$. The symmetry also puts $(b, a) \in ts(R)$.

26. a. Let R be asymmetric. Then ($x \mathrel{R} y$ and $y \mathrel{R} x$) is false for all $x, y \in A$ and it follows that $x \mathrel{R} x$ is false for all $x \in A$, by letting $x = y$. So R is irreflexive.

c. Parts (a) and (b) show that the asymmetric property implies both the irreflexive and the antisymmetric properties. For the other direction, let R be irreflexive and antisymmetric. And suppose, by way of contradiction, that R is not asymmetric. This means there must be $x, y \in A$ such that $x \mathrel{R} y$ and $y \mathrel{R} x$. But now the antisymmetric property implies that $x = y$. So we have $x \mathrel{R} x$, which contradicts the irreflexive property. Therefore, R is asymmetric.

Section 4.2

1. a. Any point x is the same distance from the point as x. So $\sim$ is reflexive. If x and y are equidistant from the point, then y and x are too. So $\sim$ is symmetric. If x and y are equidistant from the point and y and z are equidistant from the point, then x, y, and z are equidistant from the point. Thus x and z are equidistant from the point. So $\sim$ is transitive.

c. $x + x$ is even for all natural numbers x. So $\sim$ is reflexive. If $x + y$ is even, then $y + x$ is even. So $\sim$ is symmetric. Let $x + y$ be even and let $y + z$ be even. Then $x + y = 2m$ and $y + z = 2n$ for some integers m and n. Solve for x and z to obtain $x = 2m - y$ and $z = 2n - y$. Add the two equations to obtain the equation $x + z = 2(m + n - y)$. Therefore, $x + z$ is even. So $\sim$ is transitive.

e. $xx > 0$ for all nonzero x. So $\sim$ is reflexive. If $xy > 0$, then $yx > 0$. So $\sim$ is symmetric. Let $xy > 0$ and $yz > 0$. Then x and y are either both positive or both negative. The same is true for y and z. So if y is positive, then x and z must be positive and if y is negative, then x and z must be negative. So in either case, we have $xz > 0$. So $\sim$ is transitive.

2. a. The relation is not reflexive because $a + a$ is always even. It is not transitive because, for example, $3 + 4$ is odd and $4 + 5$ is odd, but $3 + 5$ is not odd.

c. Not transitive. For example, $\mid 2 - 7 \mid \leq 5$ and $\mid 7 - 12 \mid \leq 5$, but $\mid 2 - 12 \mid > 5$.

e. Not reflexive: $(10, 10) \notin R$. Not symmetric: $(11, 10) \in R$, but $(10, 11) \notin R$.

3. a. $[0] = \mathbb{N}$.

c. $[2n] = \{2n, 2n + 1\}$ for each $n \in \mathbb{N}$.

e. $[4n] = \{4n, 4n + 1, 4n + 2, 4n + 3\}$ for each $n \in \mathbb{N}$.

g. $[0] = \{0, 1, 2, 3, 4, 5, 6, 7, 8, 9, 10, 11\}$ and $[x] = \{x\}$ for $x \geq 12$.

4. a. $[x] = \{x, -x\}$ for $x \in \mathbb{Z}$.

5. a. Six classes $[0]$, $[1]$, $[2]$, $[3]$, $[4]$, $[5]$, where $[n] = \{6k + n \mid k \in \mathbb{N}\}$.
c. Twelve classes $[0]$, $[1]$, $[2]$, ..., $[11]$, where $[n] = \{12k + n \mid k \in \mathbb{N}\}$.

6. a. Two classes $\{rot,\ roto,\ root\}$ and $\{tot,\ toot,\ toto,\ too,\ to,\ otto\}$.

7. a. The weight is 7. One answer is $\{\{a,\ c\}, \{e,\ c\}, \{c,\ d\}, \{b,\ d\}, \{d,\ f\}\}$.

9. Let $x \in A$. Since E and F are reflexive, we have $(x,\ x) \in E$ and $(x,\ x) \in F$, so it follows that $(x,\ x) \in E \cap F$. Thus $E \cap F$ is reflexive. Let $(x,\ y) \in E \cap F$. Then $(x,\ y) \in E$ and $(x,\ y) \in F$. Since E and F are symmetric, we have $(y,\ x) \in E$ and $(y,\ x) \in F$. So $(y,\ x) \in E \cap F$. Thus $E \cap F$ is symmetric. Let $(x,\ y)$, $(y,\ z) \in E \cap F$. Then $(x,\ y)$, $(y,\ z) \in E$ and $(x,\ y)$, $(y,\ z) \in F$. Since E and F are transitive it follows that $(x,\ z) \in E$ and $(x,\ z) \in F$. So $(x,\ z) \in E \cap F$. Thus $E \cap F$ is transitive, and hence is an equivalence relation on A.

11. $tsr(R) = trs(R) = rts(R)$ and $str(R) = srt(R) = rst(R)$.

13. The function $s : A \to P$ defined by $s(a) = [a]$ is a surjection because every element in P has the form $[a]$ for some $a \in A$. The function $i : P \to B$ defined by $i([a]) = f(a)$ is an injection because if $i([a]) = i([b])$, then $f(a) = f(b)$. But this says that $[a] = [b]$, which implies that i is injective. To see that $f = i \circ s$, notice that $(i \circ s)(a) = i(s(a)) = i([a]) = f(a)$.

Section 4.3

1. a. False. **c.** True.

2. a. No. **c.** Yes. **e.** No.

3. a.   **b.**   **c.**

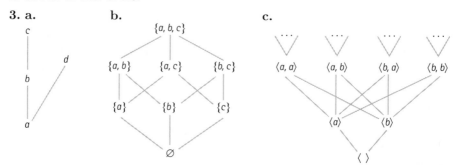

5. The glb of two elements is their greatest common divisor, and the lub is their least common multiple.

7. a. No tree has fewer than zero nodes. Therefore, every descending chain of trees is finite if the order is by the number of nodes.
c. No list has length less than zero. Therefore, every descending chain of lists is finite if the order is by the length of the list.

9. Yes.

11. An element i is a source if the ith column is full of 0's. When a source i is output, set the elements in the ith row to 0.

13. Suppose A is well-founded and S is a nonempty subset of A. If S does not have a minimal element, then there is an infinite descending chain of elements in S, which contradicts the assumption that A is well-founded. For the converse, suppose that every nonempty subset of A has a minimal element. So any descending chain of elements from A is a nonempty subset of A that must have a minimal element. Thus the descending chain must be finite. Therefore, A is well-founded.

14. a. Yes. **c.** No. For example, $-2 < 1$, but $f(-2) > f(1)$. **e.** Yes. **g.** No.

Section 4.4

1. 2900.

2. a. The equation is true if $n = 1$. So assume that the equation is true for n, and prove that it's true for $n + 1$. Starting on the left-hand side, we get

$$1 + 3 + \cdots + (2n - 1) + (2(n + 1) - 1)$$
$$= (1 + 3 + \cdots + (2n - 1)) + (2(n + 1) - 1)$$
$$= n^2 + (2(n + 1) - 1)$$
$$= n^2 + 2n + 1 = (n + 1)^2.$$

c. The equation is true for $n = 1$. So assume that the equation is true for n, and prove that it's true for $n + 1$. Starting on the left-hand side, we get

$$3 + 7 + 11 + \cdots + (4n - 1) + [4(n + 1) - 1]$$
$$= (3 + 7 + 11 + \cdots + (4n - 1)) + [4(n + 1) - 1]$$
$$= n(2n + 1) + 4(n + 1) - 1$$
$$= 2n^2 + 5n + 3$$
$$= (n + 1)(2n + 3)$$
$$= (n + 1)(2(n + 1) + 1).$$

e. The equation is true for $n = 0$. So assume that the equation is true for n, and prove that it's true for $n + 1$. Starting on the left-hand side, we get

$$1 + 5 + 9 + \cdots + (4n + 1) + [4(n + 1) + 1]$$
$$= (1 + 5 + 9 + \cdots + (4n + 1)) + [4(n + 1) + 1]$$
$$= (n + 1)(2n + 1) + 4(n + 1) + 1$$
$$= 2n^2 + 7n + 6$$
$$= (2n + 3)(n + 2)$$
$$= (2(n + 1) + 1)((n + 1) + 1).$$

g. The equation is true for $n = 1$. So assume that the equation is true for n, and prove that it's true for $n + 1$. Starting with the left side of the equation for $n + 1$, we get

$$1^2 + 2^2 + \cdots + (n+1)^2 = \left(1^2 + 2^2 + \cdots + n^2\right) + (n+1)^2$$
$$= \frac{n(n+1)(2n+1)}{6} + (n+1)^2$$
$$= \frac{(n+1)((n+1)+1)(2(n+1)+1)}{6}.$$

i. The equation is true for $n = 2$. So assume that the equation is true for n, and prove that it's true for $n + 1$. Starting on the left-hand side, we get

$$2 + 6 + 12 + \cdots + \left[(n+1)^2 - (n+1)\right]$$
$$= 2 + 6 + 12 + \cdots + \left(n^2 - n\right) + \left[(n+1)^2 - (n+1)\right]$$
$$= \frac{n(n^2 - 1)}{3} + \left[(n+1)^2 - (n+1)\right]$$
$$= \frac{n(n+1)(n-1) + 3\left[(n+1)^2 - (n+1)\right]}{3}$$
$$= \frac{(n+1)[n(n-1) + 3(n+1) - 3]}{3}$$
$$= \frac{(n+1)[n^2 + 2n]}{3}$$
$$= \frac{(n+1)[(n^2 + 2n + 1) - 1]}{3}$$
$$= \frac{(n+1)[(n+1)^2 - 1]}{3}.$$

3. a. For $n = 0$ the equation becomes $0 = 1 - 1$. Assume that the equation is true for n. Then the case for $n + 1$ goes as follows:

$$F_0 + F_1 + \cdots + F_n + F_{n+1} = (F_0 + F_1 + \cdots + F_n) + F_{n+1}$$
$$= F_{n+2} - 1 + F_{n+1} = F_{n+3} - 1 = F_{(n+1)+2} - 1.$$

c. Since the equation contains the term F_{m-1}, we must have $m \geq 1$. For $n = 0$ or $n = 1$ the equation holds for any $m \geq 1$. Let $n \geq 2$ and assume that the equation is true for all $(i, j) \prec (m, n)$. Then the case for (m, n) goes as follows:

$$F_{m+n} = F_{m+n-1} + F_{m+n-2}$$
$$= (F_{m-1}F_{n-1} + F_m F_n) + (F_{m-1}F_{n-2} + F_m F_{n-1})$$
$$= F_{m-1}(F_{n-1} + F_{n-2}) + F_m(F_n + F_{n-1})$$
$$= F_{m-1}F_n + F_m F_{n+1}.$$

4. a. For $n = 0$ the equation becomes $2 = 3 - 1$. Assume that the equation is true for n. Then the case for $n + 1$ goes as follows:

$$L_0 + L_1 + \cdots + L_n + L_{n+1} = (L_0 + L_1 + \cdots + L_n) + L_{n+1}$$
$$= L_{n+2} - 1 + L_{n+1} = L_{n+3} - 1.$$

5. Let $P(m, n)$ denote the equation. Use induction on the variable n. For any m we have $\text{sum}(m + 0) = \text{sum}(m) = \text{sum}(m) + \text{sum}(0) + m0$. So $P(m, 0)$ is true for arbitrary m. Now assume that $P(m, n)$ is true, and prove that $P(m, n + 1)$ is true. Starting on the left-hand side we get

$$\text{sum}\,(m + (n + 1)) = \text{sum}\,((m + n) + 1)$$
$$= \text{sum}\,(m + n) + m + n + 1$$
$$= \text{sum}\,(m) + \text{sum}\,(n) + mn + m + n + 1$$
$$= \text{sum}\,(m) + \text{sum}\,(n + 1) + m\,(n + 1).$$

Therefore, $P(m, n + 1)$ is true. Therefore, $P(m, n)$ is true for all m and n.

7. Since L is transitive and $R \subseteq L$, it follows that $t(R) \subseteq L$. For the other direction, let $(x, y) \in L$. In other words, $x < y$. Therefore, there is some natural number $k \geq 1$ such that $y = x + k$. We'll use induction on k. If $k = 1$, then we have $(x, y) = (x, x + 1) \in R$. So $(x, y) \in t(R)$. Now assume that $(x, x + k) \in t(R)$ and prove $(x, x + k + 1) \in t(R)$. Since $(x, x + k) \in t(R)$ and $(x + k, x + k + 1) \in R \subseteq t(R)$, it follows by the transitivity of $t(R)$ that $(x, x + k + 1) \in t(R)$. Therefore, $(x, y) \in t(R)$. Therefore, $L \subseteq t(R)$. So we have $t(R) = L$.

9. For lists K and L, let $K \prec L$ mean the length of K is less than the length of L. This forms a well-founded ordering on lists. Let $P(L)$ denote the statement "$f(L)$ is the length of L." Notice that $f(\langle \, \rangle) = 0$. Therefore, $P(\langle \, \rangle)$ is true. Now let $L \neq \langle \, \rangle$ and assume $P(K)$ is true for all lists $K \prec L$. In other words, we are assuming "$f(K)$ is the length of K" for all $K \prec L$. We must show $P(L)$ is true. Since $L \neq \langle \, \rangle$, we have $f(L) = 1 + f(\text{tail}(L))$. Since $\text{tail}(L) \prec L$, our induction assumption applies and we have $P(\text{tail}(L))$ is true. In other words, $f(\text{tail}(L))$ is the length of $\text{tail}(L)$. Thus $f(L)$ is 1 plus the length of $\text{tail}(L)$, which of course is the length of L.

10. a. Let T be a binary tree. We know that an empty tree has no nodes. Since $g(\langle \, \rangle) = 0$, we know that the function is correct when $T = \langle \, \rangle$. For the induction part we need a well-founded ordering on binary trees. For example, let $t \prec s$ mean that t is a subtree of s. Now assume that T is a nonempty binary tree, and also assume that the function is correct for all subtrees of T. Since T is nonempty, it has the form $T = \text{tree}(L, x, R)$. We know that the number of nodes in T is equal to the number of nodes in L plus those in R plus 1. The function g, when given argument T, returns $1 + g(L) + g(R)$. Since L and R are subtrees of T, it follows by assumption that $g(L)$ and $g(R)$ represent the number of nodes in L and R, respectively. Thus $g(T)$ is the number of nodes in T.

11. a. If $L = \langle x \rangle$, then $\text{forward}(L) = \{\text{print}(\text{head}(L)); \text{forward}(\text{tail}(L))\} = \{\text{print}(x); \text{forward}(\langle \, \rangle)\} = \{\text{print}(x)\}$. We'll use the well-founded ordering based on the length of lists. Let L be a list with n elements, where $n > 1$, and assume that forward is correct for all lists with fewer than n elements. Then forward(L) = {print(head(L)); forward(tail(L))}. Since tail(L) has fewer than n elements, forward(tail(L)) correctly prints out the elements of tail(L) in the order listed. Since print(head(L)) is executed before forward(tail(L)), it follows that forward(L) is correct.

12. a. We can use well-founded induction, where $L \prec M$ if length(L) < length(M). Since an empty list is sorted and sort($\langle\,\rangle$) = $\langle\,\rangle$, it follows that the function is correct for the basis case $\langle\,\rangle$. For the induction case, assume that sort(L) is sorted for all lists L of length n, and show that sort($x :: L$) is sorted. By definition, we have sort($x :: L$) = insert(x, sort(L)). The induction assumption implies that sort(L) is sorted. Therefore, insert(x, sort(L)) is sorted by the assumption in the problem. Thus sort($x :: L$)) is sorted.

13. Let's define the following order on pairs of positive integers: $(a,\ b) \prec (c,\ d)$ iff $a < c$ and $b \leq d$, or $a \leq c$ and $b < d$. This is a well-founded ordering with least element $(1,\ 1)$. For the base case we have $g(1,\ 1) = 1$. So g is correct in this case because $\gcd(1,\ 1) = 1$. For the induction case, assume $(x,\ y)$ is a pair of positive integers and assume $g(x',\ y') = \gcd(x',\ y')$ for all $(x',\ y') \prec (x,\ y)$. If $x = y$ then of course $g(x,\ y) = x = \gcd(x,\ y)$. So assume $x < y$. Then $g(x, y) = g(x,\ y - x)$. Since $(x,\ y - x) \prec (x,\ y)$ the induction assumption says that $g(x,\ y - x) = \gcd(x,\ y - x)$. Since $\gcd(x, y) = \gcd(y, x) = \gcd(x, y - x)$ (by 2.2a and 2.2b), it follows that $g(x,\ y) = \gcd(x,\ y)$. The argument is similar if $y < x$.

15. We'll use induction on the list variable. So we need a well-founded ordering on lists. For lists L and M, let $L \prec M$ mean length(L) < length(M). Let $P(x,\ L)$ be the statement, "delete(x, L) returns L with the first occurrence of x deleted." We need to show that $P(x,\ L)$ is true for all lists L. The single minimal element is $\langle\,\rangle$. The definition of delete gives delete(x, $\langle\,\rangle$) = $\langle\,\rangle$. This makes sense because $\langle\,\rangle$ is the result of deleting x from $\langle\,\rangle$. Therefore, the base case $P(x, \langle\,\rangle)$ is true. For the induction case, let K be a nonempty list and assume $P(x,\ L)$ is true for all $L \prec K$. We need to show $P(x,\ K)$ is true. There are two cases. The first case is x = head(K). Then delete(x, K) = tail(K), which is clearly the result of removing the first occurrence of x from K. Therefore, $P(x,\ K)$ is true. Now assume $x \neq$ head(K). Then the definition of delete gives

$$\text{delete}(x,\ K) = \text{head(K)} :: \text{delete}(x, \text{tail}(K)).$$

Since tail(K) $\prec K$, the induction assumption says that $P(x, \text{tail}(K))$ is true. Therefore, $P(x,\ K)$ is true because the first element of delete(x, K) is not equal to x. Therefore, (4.29) applies to say delete(x, L) is true for all lists L.

17. If $a = b$, then the equation holds. So assume $a \neq b$. The equation holds for $L = \langle\,\rangle$. Assume $L = x :: M$ and assume the equation holds for all lists having length less than that of L. Then the left side of the equation becomes $r(a, r(b, x :: M))$. If $x = b$, then the expression becomes $r(a, r(b, b :: M))$. But $r(b, b :: M) = r(b, M)$. Therefore, the left side becomes $r(a, r(b, M))$. The induction assumption then allows us to write this expression as $r(b, r(a, M))$. Now look at the right side of the equation. We have $r(b, r(a, x :: M))$. Still assuming $x = b$, we write $r(b, r(a, b :: M)) = r(b, b :: r(a, M)) = r(b, r(a, M))$. Thus the equation holds if $x = b$. A similar argument tells us that the equation holds if $x = a$. Lastly, assume $x \neq a$ and $x \neq b$. Then we can write $r(a, r(b, x :: M)) = r(a, x :: r(b, M)) = x :: r(a, r(b, M))$. Now apply the induction assumption to the last expression to get $x :: r(b, r(a, M))$. But we can reach this expression if we start on the right side: $r(b, r(a, x :: M)) = r(b, x :: r(a, M)) = x :: r(b, r(a, M))$. Thus the equation is true for any list.

19. a. If $L = \langle\ \rangle$, then isMember$(a, L) =$ False, which is correct. Now assume that L has length n and that isMember(a, K) is correct for all lists K of length less than n. If $a =$ head(L), then isMember$(a, L) =$ True, which is correct. So assume that $a \neq$ head(L). It follows that $a \in L$ iff $a \in$ tail(L). Since $a \neq$ head(L), it follows that isMember$(a, L) =$ isMember$(a,$ tail$(L))$. Since tail(L) has fewer than n elements, the induction assumption says that isMember$(a,$ tail$(L))$ is correct. Therefore, isMember(a, L) is correct for any list L.

20. If $x = \langle\ \rangle$, then the definition of cat implies that cat$(\langle\ \rangle,$ cat$(y, z)) =$ cat$(y, z) =$ cat$($cat$(\langle\ \rangle, y), z)$. Now assume that the statement is true for x, and prove the statement for $a :: x$:

$$
\begin{aligned}
\text{cat}(a :: x, \text{cat}(y, z)) &= a :: \text{cat}(x, \text{cat}(y, z)) && \text{(definition)} \\
&= a :: \text{cat}(\text{cat}(x, y), z) && \text{(induction)} \\
&= \text{cat}(a :: \text{cat}(x, y), z) && \text{(definition)} \\
&= \text{cat}(\text{cat}(a :: x, y), z) && \text{(definition)}.
\end{aligned}
$$

So the statement is true for $a :: x$ under the assumption that it is true for x. It follows by induction that the statement is true for all x, y, and z.

22. Let W be a well-founded set, and let S be a nonempty subset of W. We'll assume condition 2 of (4.28): Whenever an element x in W has the property that all its predecessors are elements in S, then x also is an element in S. We want to prove condition 1 of (4.28): S contains all the minimal elements of W. Suppose, by way of contradiction, that there is some minimal element $x \in W$ such that $x \notin S$. Then all predecessors of x are in S because there aren't any predecessors of x. Condition 2 of (4.28) now forces us to conclude that $x \in S$, a contradiction. Therefore, condition 1 of (4.28) follows from condition 2 of (4.28).

24. a. If we can show that $f(n, 0, 1) = f(k, F_{n-k}, F_{n-k+1})$ for all $0 \leq k \leq n$, then for $k = 0$ we have $f(n, 0, 1) = f(0, F_n, F_{n+1}) = F_n$, by the definition of f. To prove that $f(n, 0, 1) = f(k, F_{n-k}, F_{n-k+1})$ for all $0 \leq k \leq n$, we'll fix n and use induction on the variable k as it ranges from n down to 0. So the basis case is $k = n$. In this case we have

$$
f(n, 0, 1) = f(n, F_0, F_1) = f(k, F_{n-k}, F_{n-k+1}).
$$

For the induction case, assume that $f(n, 0, 1) = f(k, F_{n-k}, F_{n-k+1})$ for some k such that $0 < k \leq n$, and prove that

$$
f(n, 0, 1) = f(k - 1, F_{n-k+1}, F_{n-k+2}).
$$

We have the following equations:

$$f(n, 0, 1) = f(k, F_{n-k}, F_{n-k+1}) \qquad \text{(induction assumption)}$$
$$= f(k - 1, F_{n-k+1}, F_{n-k} + F_{n-k+1}) \quad \text{(definition of } f)$$
$$= f(k - 1, F_{n-k+1}, F_{n-k+2}). \qquad \text{(definition of } F_{n-k+2}).$$

Therefore, $f(n, 0, 1) = f(k, F_{n-k}, F_{n-k+1})$ for all $0 \le k \le n$.

25. Both formulas give $d(2) = 1$. For the induction part let $n > 2$ and assume that $d(k) = kd(k - 1) + (-1)^k$ for $k < n$. Show that $d(n) = nd(n - 1) + (-1)^n$. Using the original formula we have

$$d(n) = (n - 1)(d(n - 1) + d(n - 2))$$
$$= nd(n - 1) - d(n - 1) + nd(n - 2) - d(n - 2).$$

Now use the hypothesis to replace the second occurrence of $d(n - 1)$ to obtain

$$= nd(n - 1) - \left[(n - 1)d(n - 2) + (-1)^{n-1}\right] + nd(n - 2) - d(n - 2)$$
$$= nd(n - 1) - nd(n - 2) + d(n - 2) - (-1)^{n-1} + nd(n - 2) - d(n - 2)$$
$$= nd(n - 1) - (-1)^{n-1} = nd(n - 1) + (-1)^n.$$

Chapter 5

Section 5.1

1. In the following tree, move to the left child of (a, b) whenever $a > b$.

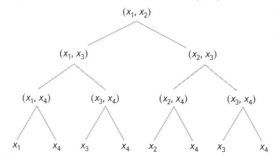

2. a. 7. **c.** 4.

3. There are 61 possible outcomes. So there must be at least 61 leaves on any tree that solves the problem. If h is the height of a ternary decision tree, then the tree has at most 3^h leaves. Therefore, $61 \le 3^h$. Take the log to conclude that $h \ge \text{ceiling}(\log_3(61)) = 4$. So 4 is a reasonable lower bound.

Section 5.2

1. a. $2(1) + 3 + 2(2) + 3 + 2(3) + 3 + 2(4) + 3 + 2(5) + 3$.
c. $5(3^0) + 4(3^1) + 3(3^2) + 2(3^3) + 1(3^4)$.

2. a. $\sum_{k=0}^{n-1} g(k)a_{k+1}x^{k+2}$. **c.** $\sum_{k=-1}^{n-2} g(k+1)a_{k+2}x^{k+3}$. **e.** $\sum_{k=-2}^{n-3} g(k+2)a_{k+3}x^{k+4}$.

3. a. $\sum_{k=1}^{n} 3k = 3 \sum_{k=1}^{n} k = \frac{3n(n+1)}{2}$. **c.** $\sum_{k=0}^{n} 3(2^k) = 3 \sum_{k=0}^{n} 2^k = 3(2^{n+1} - 1)$.

4. a. $n^2 + 3n$. **c.** $n^2 + 4n$. **e.** $2n^2$. **g.** $\frac{n(n+1)(n+2)}{3}$.

8. a. $\sum_{i=0}^{k-1} 2^i(1/5^i)^{\log_5 2} = \sum_{i=0}^{k-1} 2^i 5^{-i\log_5 2} = \sum_{i=0}^{k-1} 2^i 5^{\log_5 2^{-i}} = \sum_{i=0}^{k-1} 2^i 2^{-i} = \sum_{i=0}^{k-1} 1 = k$.

c. $\sum_{i=0}^{k-1} (1/5^i)^{\log_5 2} = \sum_{i=0}^{k-1} 5^{-i\log_5 2} = \sum_{i=0}^{k-1} 5^{\log_5 2^{-i}} = \sum_{i=0}^{k-1} (1/2)^i = \frac{(1/2)^k - 1}{(1/2) - 1} = $

$2 - (1/2)^{k-1}$.

9. $\sum_{i=0}^{k-1} \frac{1}{\log_2(n/2^i)} = \sum_{i=0}^{k-1} \frac{1}{\log_2 n - \log_2 2^i} = \sum_{i=0}^{k-1} \frac{1}{\log_2 n - i} = \sum_{i=0}^{k-1} \frac{1}{k-i} = \sum_{i=1}^{k} \frac{1}{i} = H_k$.

10. a. $H_{2n+1} - (1/2)H_n$.

11. a. $\sum_{k=1}^{n} \frac{k}{k+1} = \sum_{k=1}^{n} \left(1 - \frac{1}{k+1}\right) = n - (H_{n+1} - 1) = n + 1 - H_{n+1}$.

12.

$$\sum_{k=1}^{n} kH_k = \sum_{k=1}^{n} (1/k)(k + \cdots + n)$$

$$= \sum_{k=1}^{n} (1/k) \left(\sum_{i=1}^{n} i - \sum_{i=1}^{k-1} i\right)$$

$$= \sum_{k=1}^{n} (1/k) \left((1/2)n(n+1) - (1/2)(k-1)k\right)$$

$$= (1/2)n(n+1) \sum_{k=1}^{n} (1/k) - (1/2) \sum_{k=1}^{n} (k-1)$$

$$= \frac{n(n+1)}{2} H_n - \frac{n(n-1)}{4}.$$

13. a. $n^2 + 3n$.

14. a. $2 + 3 + \cdots + (n+1) = (1/2)(n+1)(n+2) - 1$.

15. a. $3 + 5 + \cdots + (2k+3)$, where $k = \lceil n/2 \rceil - 1$. This sum has the value $(k+1)(k+3) = (k+2)^2 - 1 = (\lceil n/2 \rceil + 1)^2 - 1$.

16. $(1/4)n^2(n+1)^2$.

17. $\frac{a}{(a-1)^3} \left(n^2 a^{n+2} + (1 - 2n - 2n^2)a^{n+1} + (n+1)^2 a^n - a - 1\right)$.

18. a. $\frac{n^2}{2} \leq \frac{n(n+1)}{2} \leq \frac{(n+1)^2}{2} - \frac{1}{2} = \frac{n(n+2)}{2}$.

19. $\ln n + 1 - \ln\left(\dfrac{n+1}{2}\right) - 1 = \ln n - \ln\left(\dfrac{n+1}{2}\right) = \ln\left(\dfrac{2n}{n+1}\right)$

$< \ln\left(\dfrac{2(n+1)}{n+1}\right) = \ln 2.$

20. a. Starting with $\ln n$, we obtain

$$\ln n = \int_1^n (1/x)dx = \sum_{k=1}^{n-1}\left(\int_k^{k+1}(1/x)dx\right) < \sum_{k=1}^{n-1}\left((1/2)\left(\frac{1}{k}+\frac{1}{k+1}\right)\right)$$

$$= (1/2)\sum_{k=1}^{n-1}(1/k) + (1/2)\sum_{k=1}^{n-1}(1/(k+1)) = (1/2)(H_{n-1}+H_n-1)$$

$$= (1/2)(H_n - (1/n) + H_n - 1) = H_n - (1/2n) - (1/2).$$

Therefore, $\ln n + (1/2n) + (1/2) < H_n$.

Section 5.3

1. a. 720. **c.** 30. **e.** 10.

2. a. $abc,\ acb,\ bac,\ bca,\ cab,\ cba.$ **c.** $\{a,\ b,\ c\}.$
e. $[a,\ a],\ [a,\ b],\ [a,\ c],\ [b,\ b],\ [b,\ c],\ [c,\ c].$

3. a. $P(3,\ 3)$. **c.** $C(3,\ 3)$. **e.** $C(3+2-1,\ 2)$.

4. The number of bag permutations of B is $4!/(2!2!) = 6$. They can be listed as follows: $aabb,\ abab,\ abba,\ bbaa,\ baba,\ baab.$

5. a. $8! = 40{,}320$. **c.** $6!/(2!2!1!1!) = 180$. **e.** $9!/(4!2!2!1!) = 3780$.

6. a. There are none. **c.** $BCA,\ CAB.$

7. $n = 7$, $k = 3$, and $m = 4$.

9. a. Either floor$(n/2)$ or ceiling$(n/2)$.

10. a. 5-card hands: $C(52,\ 5) = 2{,}598{,}960$.
b. A straight flush can begin with any of the ten cards Ace, 2, ..., 10 and all 5 cards must be the same suit. There are 4 suits, so the total number of straight flushes is $10 \cdot 4 = 40$.
c. There are 13 ranks for 4-of-a-kind and 48 choices for the fifth card. So the number of 4-of-a-kind hands is $13 \cdot C(4,4) \cdot C(48,1) = 13 \cdot 1 \cdot 48 = 624$.
d. There are $C(4,\ 3)$ ways to choose three cards of the same rank and $C(4,\ 2)$ ways to choose two cards of the same rank. There are 13 possible ranks for one set and 12 possible ranks for the other set. So the number of full houses is $13 \cdot C(4,3) \cdot 12 \cdot C(4,2) = 3744$.
e. There are $C(13,\ 5)$ flushes of one suit. So the number of flushes that are not straight flushes is $4 \cdot C(13,5) - 40 = 5{,}108$.
f. A straight can begin with any of the ten cards Ace, 2, ..., 10. So the number of straights that are not straight flushes is $10 \cdot 4^5 - 40 = 10{,}200$.

g. There are $C(4, 3)$ ways to choose three cards of the same rank and $C(48, 2)$ ways to choose the remaining two cards that are not of that rank. There are 13 ranks, so the number of 3-of-a-kind hands that are not full houses is $13 \cdot C(4,3) \cdot C(48,2) - 3744 = 54{,}912$.

h. There are $C(13, 2)$ ways to choose two different ranks and $C(4, 2)$ ways to choose two cards of the same rank. There are then 44 ways to choose the fifth card of a different rank. So the number of 2-pair hands of different ranks with a different fifth card is $C(13,2) \cdot C(4,2) \cdot C(4,2) \cdot 44 = 123{,}552$.

i. There are $C(4,2)$ ways to choose two cards of the same rank. So there are $13 \cdot C(4,2)$ ways to choose one pair of cards. There are 12 ranks other than the rank of the pair. So there are $C(12,3)$ ways to choose the remaining three card ranks. But each of the three cards can be any of the four suits. So the number of 1-pair hands with the other three cards having a rank different from the pair is $13 \cdot C(4,2) \cdot C(12,3) \cdot 4^3 = 1{,}098{,}240$.

j. There are $C(13, 5)$ ways to choose five cards of different ranks. But the 10 straights must be excluded. There are 4^5 possible suits for the five cards. But the 4 possible flushes must be excluded. So the number of high-card hands with no straights and no flushes is $(C(13,5) - 10) \cdot (4^5 - 4) = 1{,}302{,}540$.

12. There are nk actions to schedule. Since the k actions of each process must be done in order, we can represent each process as a bag consisting of k identical elements. Assume that the bags are disjoint from each other. Then the union B of the n bags contains nk elements, and each bag permuation of B is one schedule. Therefore there are as many schedules as there are bag permutations of B. That number is $(nk)!/(k!)^n$.

14. Let $|S| = n$. If $n = 1$, then there is just one bijection, the identity mapping on S. Let $n > 1$ and assume there are $k!$ bijections between any two sets with k elements when $k < n$. Pick some element $x \in S$. Any bijection of S must map x to one of its n elements y. The remaining elements in $S - \{x\}$ must be mapped to $S - \{y\}$. These sets each have $n - 1$ elements. The induction assumption tells us there are $(n - 1)!$ bijections from $S - \{x\}$ to $S - \{y\}$. Therefore there are $n(n - 1)! = n!$ bijections from S to S.

Section 5.4

1. There are eight possible outcomes when three coins are tossed.
a. 0.375. **c.** 0.875.

2. There are 36 possible outcomes. **a.** 0.1666.... **c.** 0.222....

3. a. 80/243. **c.** 4/9.

4. a. 0.21.

6. 1/3, 2/15, 8/15.

7. a. 1/2. **c.** no.

9. a. $1/C(49,6)$, since order is not important.

c. $[C(6, 5)C(43, 1) + C(6, 4)C(43, 2)]/C(49, 6)$. To obtain the answer, notice that there are $C(49, 6)$ 6-element subsets of a 49-element set. Now suppose that $\{a, b, c, d, e, f\}$ is the winning set of numbers. First, we'll count all the 6-element sets that contain exactly five winners. To do this, notice that there are $C(6, 5)$ 5-element subsets of $\{a, b, c, d, e, f\}$. To each of these 5-element subsets we add a non-winner from the set $\{1, ..., 49\} - \{a, b, c, d, e, f\}$. Since there are $43 (= C(43, 1))$ non-winners, it follows that there are $C(6, 5)C(43, 1)$ sets of 6 numbers that contain exactly five winners. Next, we'll count the 6-element sets that contain exactly four winners. To do this, notice that there are $C(6, 4)$ 4-element subsets of $\{a, b, c, d, e, f\}$. To each of these 4-element subsets we add two non-winners from the set $\{1, ..., 49\} - \{a, b, c, d, e, f\}$. Since there are $C(43, 2)$ possible pairs of non-winners, it follows that there are $C(6, 4)C(43, 2)$ sets of six numbers that contain exactly four winners. So the probability of choosing either four or five of the six winning numbers is given by $[C(6, 5)C(43, 1) + C(6, 4)C(43, 2)]/C(49, 6)$.

10. a. 3.5.

11. a. $(1/x^2)(x^2 - \pi(x^2/4))$. **b.** $(45/115)x^2$.

12. The following answers are calculated to five decimal places.

a. $P^4 = \begin{pmatrix} 0.4375 & 0.5625 \\ 0.3750 & 0.6250 \end{pmatrix}$, $P^8 = \begin{pmatrix} 0.40234 & 0.59766 \\ 0.39844 & 0.60156 \end{pmatrix}$,

$P^{16} = \begin{pmatrix} 0.40001 & 0.59999 \\ 0.39999 & 0.60001 \end{pmatrix}$.

b. $X_4 = (0.40625, 0.59375)$, $X_8 = (0.40039, 0.59961)$, $X_{16} = (0.40000, 0.60000)$.
c. $X_4 = (0.38542, 0.61458)$, $X_8 = (0.39909, 0.60091)$, $X_{16} = (0.40000, 0.60000)$.

13. a. $P^2 = \begin{pmatrix} 0.40 & 0.10 & 0.50 \\ 0.19 & 0.29 & 0.52 \\ 0.21 & 0.15 & 0.64 \end{pmatrix}$.

b. To five decimal places, $X = (0.25532, 0.15957, 0.58511)$.

c. $P^4 = \begin{pmatrix} 0.2840 & 0.1440 & 0.5720 \\ 0.2403 & 0.1811 & 0.5786 \\ 0.2469 & 0.1605 & 0.5926 \end{pmatrix}$, $P^8 = \begin{pmatrix} 0.25649 & 0.15878 & 0.58473 \\ 0.25462 & 0.16027 & 0.58511 \\ 0.25520 & 0.15973 & 0.58527 \end{pmatrix}$.

d. To five decimal places, $X_1 = (0.43000, 0.21000, 0.36000)$, $X_2 = (0.21300, 0.25700, 0.53000)$, $X_4 = (0.24533, 0.17533, 0.57934)$, and $X_8 = (0.25484, 0.16006, 0.58509)$.

e. To five decimal places, $X_1 = (0.23000, 0.17000, 0.60000)$, $X_2 = (0.26500, 0.14900, 0.58600)$, $X_4 = (0.25737, 0.15761, 0.58502)$, and $X_8 = (0.25541, 0.15950, 0.58509)$.

f. The company should manufacture 306, 192, and 702 units of A, B, and C, respectively.

14. Since A is a subset S we have $A \cap S = A$. So $P(A \cap S) = P(A) = P(A)P(S)$. Thus S and A are independent. If $A \cap B = \varnothing$, then the only way for A and B to be independent is if $A = B = \varnothing$.

16. a. 109/30 (or about 3.63).
b. $(p/n)[(n + 1) \log_2(n + 1) - n] + (1 - p)\log_2(n + 1)$.

17. a. Solve the inequality $0.99 \leq P(\text{at least one head in } n \text{ flips}) = 1 - P(\text{no heads in } n \text{ flips}) = 1 - C(n, 0)(1/2)^0(1/2)^n = 1 - (1/2)^n$. Rewriting, we obtain $(1/2)^n \leq 0.01$, or $2^n \geq 100$. Therefore, $n = 7$.

Section 5.5

1. a. $4(n - 1)$. **c.** $2^{n+2} - 3$.

2. a. Let a_n be the number of cons operations when L has length n. Then $a_0 = 0$ and $a_n = 1 + a_{n-1}$, which has solution $a_n = n$. **c.** Let a_n be the number of cons operations when L has length n. Then $a_0 = 1$, and $a_n = a_{n-1} + 5 \cdot 2^{n-1}$, which has solution $a_n = 5 \cdot 2^n - 4$.

3. The recurrence is given by $H_0 = 0$ and $H_n = 2H_{n-1} + 1$. The solution is $H_n = 2^n - 1$.

5. Let r_n be the number of regions created by n lines. Then $r_0 = 1$ since a plane with no lines is one region. It's easy to see that $r_1 = 2$, $r_2 = 4$, $r_3 = 7$, and $r_4 = 11$. After some thought, we see that when there are $n - 1$ lines in the plane and we add one more line, it intersects each of the existing $n - 1$ lines and splits up n existing regions. So $r_n = r_{n-1} + n$. This recurrence can be solved by substitution or cancellation to get $r_n = (1/2)(n^2 + n + 2)$.

6. a. $a_n = -1/(2^{n+1}) - 2(-3)^n$. **c.** $a_n = (-1/2)(3/2)^n - n - 1$.

7. a. $a_n = 3^n + (-1)^{n+1}$. **c.** $a_n = (1/3)(2^n - (-1)^n)$.

8. a. $A(x) = 4/(1 - x)^2 - 8/(1 - x) + 4$, which yields $a_n = 4(n - 1)$.
c. $A(x) = -3/(1 - x) + 4/(1 - 2x)$, which yields $a_n = 2^{n+2} - 3$.

9. a. If $n = 1$, then the equation evaluates to $2 = 2$. Assume that the equation holds for n, and show that it holds for $n + 1$. Starting with the left side for the $n + 1$ case we have

$$\frac{(1)\,(1)\,(3)\,(5)\cdots(2n - 3)\,(2n - 1)}{(n + 1)!} 2^{n+1}$$

$$= \frac{(1)\,(1)\,(3)\,(5)\cdots(2n - 3)}{n!} 2^n \frac{2n - 1}{n + 1} 2$$

$$= \frac{2}{n}\binom{2n - 2}{n - 1}\frac{2n - 1}{n + 1}2 = \frac{2}{n + 1}\binom{2(n + 1) - 2}{n + 1 - 1},$$

which is the right side for the $n + 1$ case.

10. Letting $F(x)$ be the generating function for F_n, we get

$$F(x) = x/(1 - x - x^2).$$

The denominator factors into $1 - x - x^2 = (1 - \alpha\, x)(1 - \beta\, x)$, where

$$\alpha = \frac{1}{2}\left(1 + \sqrt{5}\right) \quad \text{and} \quad \beta = \frac{1}{2}\left(1 - \sqrt{5}\right).$$

Now use partial fractions to obtain

$$F(x) = \frac{1}{\sqrt{5}}\left(\frac{1}{1 - \alpha x} - \frac{1}{1 - \beta x}\right).$$

This yields the closed formula $F_n = \frac{1}{\sqrt{5}}\left(\alpha^n - \beta^n\right)$.

Section 5.6

1. a. From the hypothesis it follows that $|g(n)| \le |f(n)|$ for all n. So by letting $c = 1$ and $m = 1$ we obtain $g(n) = O(f(n))$.

2. a. $(1)(1) \le 1 + 1/n \le 2(1)$ for $n \ge 1$.

3. If $r \ge 0$, then $n^r \le (n+1)^r \le (n+n)^r = (2n)^r = 2^r n^r$. If $r < 0$, then $n^r \ge (n+1)^r \ge (n+n)^r = (2n)^r = 2^r n^r$. So in either case, we have $(n+1)^r = \Theta(n^r)$.

4. a. The quotient $\log(kn)/\log n$ approaches 1 as n approaches infinity.

5. a. If $\epsilon > 0$ and $f(n) = O(n^{k-\epsilon})$, then $|f(n)| \le c|\, n^{k-\epsilon}|$. So $|f(n)/n^k| \le c|\, n^{-\epsilon}|$ which goes to 0 as n goes to infinity. Therefore, $f(n) = o(n^k)$.

6. $1 \prec \log \log n \prec \log n$.

8. Take limits.

9. In each case, replace $n!$ by its Stirling's approximation (5.61). Then take limits.

10. a. $5! = 120$; Stirling ≈ 118.02; diff $= 1.98$.

11. a. $f(n) = \Theta(n \log n)$. Notice that $f(n) = \log(1 \ldots n) = \log(n!)$. Now use (5.61) to approximate $n!$ and take the log of Stirling's formula to obtain $\Theta(n \log n)$.

12. a. $T(n) = \Theta(n^{\log_2 3})$. **b.** $T(n) = \Theta(n \log n)$.

13. a. $T(n) = \Theta(n)$. **b.** $T(n) = \Theta\left(\sqrt{n \log n}\right)$. **c.** $T(n) = \Theta\left(\sqrt{n} \log \log n\right)$.
d. $T(n) = \Theta\left(\sqrt{n}\right)$. **e.** $T(n) = \Theta\left(\sqrt{n}\right)$. **f.** $T(n) = \Theta\left(n^2 \log n\right)$. **g.** $T(n) = O(n^2)$.

14. a. $|f(n)| \le |f(n)|$ for all n. So $f(n) = O(f(n))$ by letting $c = 1$ and $m = 1$.
c. If $f(n) = O(g(n))$, then there are positive constants c and m such that $|f(n)| \le c|g(n)|$ for $n \ge m$. If $a = 0$, then $af(n) = 0$ for all n. So $af(n) = O(g(n))$. If $a \ne 0$ then multiply both sides of the inequality by $|a|$ to obtain $|af(n)| \le (c|a|)|g(n)|$ for $n \ge m$. Thus $af(n) = O(g(n))$.
e. Since $f(n) = O(g(n))$, there are positive numbers c and m such that $|f(n)| \le c|g(n)|$ for $n \ge m$. Let $n \ge bm$. Then $n/b \ge m$. So $|f(n/b)| \le c|g(n/b)|$ and thus $f(n/b) = O(g(n/b))$.
g. The hypotheses tell us that $0 \le f_1(n) \le c_1 g_1(n)$ for $n \ge m_1$ and $0 \le f_2(n) \le c_2 g_2(n)$ for $n \ge m_2$. Let c be the larger of c_1 and c_2 and let m be the larger of

m_1 and m_2. Then we have $0 \le f_1(n) + f_2(n) \le c_1 g_1(n) + c_2 g_2(n) \le c(g_1(n) + g_2(n))$ for $n \ge m$. Therefore, $f_1(n) + f_2(n) = O(g_1(n)) + g_2(n))$.

15. a. To say that $g(n) = O(O(f(n)))$ means that $g(n) = O(h(n))$, where $h(n) = O(f(n))$. It follows from Exercise 14b that $g(n) = O(f(n))$. Therefore, $O(O(f(n))) = O(f(n))$.

c. Let $h(n) = O(g(n))$. Then there are positive constants c and m such that $|h(n)| \le c|g(n)|$ for $n \ge m$. Multiply the inequality by $|f(n)|$ to obtain $|f(n)h(n)| \le c|f(n)g(n)|$. This says that $f(n)h(n) = O(f(n)g(n))$. Therefore, $f(n)O(g(n)) = O(f(n)g(n))$.

e. Let $h_1(n) = O(f(n))$ and $h_2(n) = O(g(n))$. Then there are positive c_1, m_1, c_2, and m_2 such that $|h_1(n)| \le c_1 f(n)$ for $n \ge m_1$ and $|h_2(n)| \le c_2 g(n)$ for $n \ge m_2$. Let c be the larger of c_1 and c_2 and let m be the larger of m_1 and m_2. Then $|h_1(n) + h_2(n)| \le |h_1(n)| + |h_2(n)| \le c(f(n) + g(n))$ for $n \ge m$. Therefore, $h_1(n) + h_2(n) = O(f(n) + g(n))$. So $O(f(n)) + O(g(n)) = O(f(n) + g(n))$.

16. a. $|f(n)| \ge |f(n)|$ for all n. So by letting $c = 1$ and $m = 1$ we obtain $f(n) = \Omega(f(n))$.

c. If $0 \le f(n) \le g(n)$ for $n \ge m$, then let $c = 1$ to obtain $g(n) = \Omega(f(n))$.

e. Since f_1 and f_2 are nonnegative, there are positive constants such that $f_1(n) \ge c_1|g(n)|$ for $n \ge m_1$ and $f_2(n) \ge c_2|g(n)|$ for $n \ge m_2$. Let $c = c_1 + c_2$ and let m be the larger of m_1 and m_2. Then $f_1(n) + f_2(n) \ge c_1|g(n)| + c_2|g(n)| = c|g(n)|$ for $n \ge m$. So $f_1(n) + f_2(n) = \Omega(g(n))$.

17. a. Since $|f(n)| \le |f(n)| \le |f(n)|$ for all n, we can let $c = d = 1$ and $m = 1$ to obtain $f(n) = \Theta(f(n))$.

c. The hypothesis tells us there are positive constants c_1, d_1, m_1, c_2, d_2, and m_2 such that $c_1|g(n)| \le |f(n)| \le d_1|g(n)|$ for $n \ge m_1$ and $c_2|h(n)| \le |g(n)| \le d_2|h(n)|$ for $n \ge m_2$. Let m be the larger of m_1 and m_2. It follows that $c_1 c_2|h(n)| \le c_1|g(n)| \le |f(n)| \le d_1|g(n)| \le d_1 d_2|h(n)|$ for $n \ge m$. Therefore, $f(n) = \Theta(h(n))$.

e. Since $f(n) = \Theta(g(n))$, there are positive constants c, d, and m such that $c|g(n)| \le |f(n)| \le d|g(n)|$ for $n \ge m$. Let $n \ge bm$. Then $n/b \ge m$ and it follows that $c|g(n/b)| \le |f(n/b)| \le d|g(n/b)|$ for $n/b \ge m$. Therefore, $f(n/b) = \Theta(g(n/b))$.

18. a. To say that $g(n) = \Theta(\Theta(f(n)))$ means that $g(n) = \Theta(h(n))$, where $h(n) = \Theta(f(n))$. It follows from Exercise 17c that $g(n) = \Theta(f(n))$. Therefore, $\Theta(\Theta(f(n))) = \Theta(f(n))$.

c. The hypothesis tells us that there are positive constants c_1, d_1, m_1, c_2, d_2, and m_2 such that $c_1 g_1(n) \le f_1(n) \le d_1 g_1(n)$ for $n \ge m_1$ and $c_2 g_2(n) \le f_2(n) \le d_2 g_2(n)$ for $n \ge m_2$. Let c be the smaller of c_1 and c_2, let d be the larger of d_1 and d_2, and let m be the larger of m_1 and m_2. Then $c(g_1(n) + g_2(n)) \le c_1 g_1(n) + c_2 g_2(n) \le f_1(n) + f_2(n) \le d_1 g_1(n) + d_2 g_2(n) \le d(g_1(n) + g_2(n))$ for $n \ge m$. Therefore, $f_1(n) + f_2(n) = \Theta(g_1(n) + g_2(n))$.

19. a. Let $f(n) = o(g(n))$ and $g(n) = o(h(n))$. Let $\epsilon > 0$. Then there are positive constants m_1 and m_2 such that $|f(n)| \le (\epsilon/2)|g(n)|$ for $n \ge m_1$ and

$|g(n)| \leq (\epsilon/2)|h(n)|$ for $n \geq m_2$. Let m be the larger of m_1 and m_2. It follows that $|f(n)| \leq (\epsilon/2)|g(n)| \leq (\epsilon/2)(\epsilon/2)|h(n)| \leq \epsilon|h(n)|$ for $n \geq m$. Therefore, $f(n) = o(h(n))$.

c. Let $h_1(n) = o(f(n))$ and $h_2(n) = o(g(n))$. Let $\epsilon > 0$. Then there are positive constants m_1 and m_2 such that $|h_1(n)| \leq (\epsilon/2)|f(n)|$ for $n \geq m_1$ and $|h_2(n)| \leq (\epsilon/2)|g(n)|$ for $n \geq m_2$. Let m be the larger of m_1 and m_2. It follows that $|h_1(n)h_2(n)| \leq (\epsilon/2)|f(n)|(\epsilon/2)|g(n)| \leq \epsilon|f(n)g(n)|$ for $n \geq m$. So $h_1(n)h_2(n) = o(f(n)g(n))$. Therefore, $o(f(n))o(g(n)) = o(f(n)g(n))$.

20. a. Let $h(n) = \Theta(f(n))$ and $g(n) = O(f(n))$. Then there are positive constants c, d, m, c_1, and m_1 such that $c|f(n)| \leq |h(n)| \leq d|f(n)|$ for $n \geq m_1$ and $|g(n)| \leq c_1|f(n)|$ for $n \geq m_2$. Let m be the larger of m_1 and m_2. It follows that $|h(n) + g(n)| \leq |h(n)| + |g(n)| \leq d|f(n)| + c_1|f(n)| = (d + c_1)|f(n)|$ for $n \geq m$. Thus $h(n) + g(n) = O(f(n))$. Therefore, $\Theta(f(n)) + O(f(n)) = O(f(n))$.

c. Let $f(n) = O(g(n))$ and $h(n) = \Theta(f(n))$. Then there are positive constants c_1, m_1, c_2, d_2, and m_2 such that $|f(n)| \leq c_1|g(n)|$ for $n \geq m_1$ and $c_2|f(n)| \leq |h(n)| \leq d_2|f(n)|$ for $n \geq m_2$. Let m be the larger of m_1 and m_2. It follows that $|h(n)| \leq d_2|f(n)| \leq d_2c_1|g(n)|$ for $n \geq m$. Thus $h(n) = O(g(n))$. Therefore, $\Theta(f(n)) = O(g(n))$.

e. Let $f(n) = o(g(n)$ and $h(n) = O(f(n))$. Then there are positive constants c and m_1 such that $|h(n)| \leq c|f(n)|$ for $n \geq m_1$. Let $\epsilon > 0$. Since $f(n) = o(g(n)$ we will pick ϵ/c for which there is a positive constant m_2 such that $|f(n)| \leq (\epsilon/c)|g(n)|$ for $n \geq m_2$. Let m be the larger of m_1 and m_2. It follows that $|h(n)| \leq c|f(n)| \leq c(\epsilon/c)|g(n)| = \epsilon|g(n)|$ for $n \geq m$. Thus $h(n) = o(g(n))$. Therefore, $O(f(n)) = o(g(n))$.

g. Let $f(n) = o(g(n))$. Part (e) tells us that $O(f(n)) = o(g(n))$. Since $\Theta(f(n)) = O(f(n))$, it follows that $\Theta(f(n)) = o(g(n))$.

Chapter 6

Section 6.2

1. a. $((\neg P) \wedge Q) \rightarrow (P \vee R)$.

c. $(A \rightarrow (B \vee (((\neg C) \wedge D) \wedge E))) \rightarrow F$.

2. a. $(P \vee Q \rightarrow \neg R) \vee \neg Q \wedge R \wedge P$.

3. $(A \rightarrow B) \wedge (\neg A \rightarrow C)$ or $(A \wedge B) \vee (\neg A \wedge C)$.

5. $A \wedge \neg B \rightarrow \text{False} \equiv \neg (A \wedge \neg B) \vee \text{False} \equiv \neg (A \wedge \neg B) \equiv \neg A \vee \neg \neg B \equiv \neg A \vee B \equiv A \rightarrow B$.

7. a. If $B = \text{True}$, then the wff is true. If $B = \text{False}$ and $A = \text{True}$, then the wff is false.

c. If $A = \text{True}$, then the wff is true. If $A = \text{False}$ and $C = \text{True}$, then the wff is false.

e. If $B = \text{True}$, then the wff is true. If $B = \text{False}$ and $A = C = \text{True}$, then the wff is false.

8. a. If $C = \text{True}$, $A \rightarrow C$ is true, so the wff is trivially true too. If $C = \text{False}$, then the wff becomes $(A \rightarrow B) \wedge (B \rightarrow \text{False}) \rightarrow (A \rightarrow \text{False})$, which is equivalent to $(A \rightarrow B) \wedge \neg B \rightarrow \neg A$. If $A =$ False, then the wff is trivially true. If $A =$ True, the wff becomes

$$(\text{True} \rightarrow B) \wedge \neg B \rightarrow \text{False} \equiv B \wedge \neg B \rightarrow \text{False} \equiv \text{False} \rightarrow \text{False} \equiv \text{True}.$$

c. If $A =$ False or $B =$ False, then the consequent is true, so the statement is trivially true. If $A = B =$ True, then the wff becomes

$$(\text{True} \rightarrow C) \wedge (\text{True} \rightarrow D) \wedge (\neg C \vee \neg D) \rightarrow \text{False}$$
$$\equiv C \wedge D \wedge (\neg C \vee \neg D) \rightarrow \text{False}.$$

If $C =$ True, then the wff becomes

$$\text{True} \wedge D \wedge (\text{False} \vee \neg D) \rightarrow \text{False} \equiv D \wedge \neg D \rightarrow \text{False}$$
$$\equiv \text{False} \rightarrow \text{False} \equiv \text{True}.$$

If $C =$ False, then the wff becomes

$$\text{False} \wedge D \wedge (\text{True} \vee \neg D) \rightarrow \text{False} \equiv \text{False} \rightarrow \text{False} \equiv \text{True}.$$

e. If $B =$ True, then the wff is trivially true. If $B =$ False, then the wff becomes

$$(\text{True} \rightarrow \neg A) \rightarrow ((\text{True} \rightarrow A) \rightarrow \text{False}) \equiv \neg A \rightarrow (A \rightarrow \text{False})$$
$$\equiv \neg A \rightarrow \neg A \equiv \text{True}.$$

g. If $C =$ True, then the wff is trivially true. If $C =$ False, then the wff becomes

$$(A \rightarrow \text{False}) \rightarrow ((B \rightarrow \text{False}) \rightarrow (A \vee B \rightarrow \text{False}))$$
$$\equiv \neg A \rightarrow (\neg B \rightarrow \neg (A \vee B)).$$

If $A =$ True, then the wff is vacuously true. If $A =$ False, then the wff becomes

$$\text{True} \rightarrow (\neg B \rightarrow \neg (\text{False} \vee B)) \equiv \neg B \rightarrow \neg (\text{False} \vee B) \equiv \neg B \rightarrow \neg B \equiv \text{True}.$$

9. a. $(A \rightarrow B) \wedge (A \vee B) \equiv (\neg A \vee B) \wedge (A \vee B) \equiv (\neg A \wedge A) \vee B$
$\equiv \text{False} \vee B \equiv B.$
c. $A \wedge B \rightarrow C \equiv \neg (A \wedge B) \vee C \equiv (\neg A \vee \neg B) \vee C \equiv \neg A \vee (\neg B \vee C)$
$\equiv \neg A \vee (B \rightarrow C) \equiv A \rightarrow (B \rightarrow C).$
e. $A \rightarrow B \wedge C \equiv \neg A \vee (B \wedge C) \equiv (\neg A \vee B) \wedge (\neg A \vee C)$
$\equiv (A \rightarrow B) \wedge (A \rightarrow C).$

10. a. $A \rightarrow A \vee B \equiv \neg A \vee A \vee B \equiv \text{True} \vee B \equiv \text{True}.$
c. $(A \vee B) \wedge \neg A \rightarrow B \equiv \neg ((A \vee B) \wedge \neg A) \vee B \equiv \neg (A \vee B) \vee A \vee B \equiv$
$(\neg A \wedge \neg B) \vee (A \vee B) \equiv (\neg A \vee A \vee B) \wedge (\neg B \vee A \vee B) \equiv$
$(\text{True} \vee \neg B) \wedge (\text{True} \vee A) \equiv \text{True} \wedge \text{True} \equiv \text{True}.$
e. $(A \rightarrow B) \wedge \neg B \rightarrow \neg A \equiv (\neg A \vee B) \wedge \neg B \rightarrow \neg A \equiv \neg ((\neg A \vee B) \wedge \neg B)$
$\vee \neg A \equiv (A \wedge \neg B) \vee (B \vee \neg A) \equiv (A \vee B \vee \neg A) \wedge (\neg B \vee B \vee \neg A)$
$\equiv (\text{True} \vee B) \wedge (\text{True} \vee \neg A) \equiv \text{True} \wedge \text{True} \equiv \text{True}.$
g. $A \rightarrow (B \rightarrow (A \wedge B)) \equiv \neg A \vee (\neg B \vee (A \wedge B)) \equiv (\neg A \vee \neg B) \vee (A \wedge B)$
$\equiv \neg (A \wedge B) \vee (A \wedge B) \equiv \text{True}.$

11. a. $(P \wedge \neg Q) \vee P$ or $P.$ **c.** $\neg Q \vee P.$ **e.** $\neg P \vee (Q \wedge R).$

12. a. $P \wedge (\neg Q \vee P)$ or $P.$ **c.** $\neg Q \vee P.$ **e.** $(\neg P \vee Q) \wedge (\neg P \vee R).$
g. $(A \vee C \vee \neg E \vee F) \wedge (B \vee C \vee \neg E \vee F) \wedge (A \vee D \vee \neg E \vee F)$
$\wedge (B \vee D \vee \neg E \vee F).$

13. a. Full DNF: $(P \wedge Q) \vee (P \wedge \neg Q)$. Full CNF: $(P \vee \neg Q) \wedge (P \vee Q)$.

14. a. $(P \wedge Q) \vee (P \wedge \neg Q)$.

c. $(P \wedge Q) \vee (P \wedge \neg Q) \vee (\neg P \wedge Q) \vee (\neg P \wedge \neg Q)$.

e. $(P \wedge Q \wedge R) \vee (\neg P \wedge Q \wedge R) \vee (\neg P \wedge \neg Q \wedge R)$
$\vee (\neg P \wedge Q \wedge \neg R) \vee (\neg P \wedge \neg Q \wedge \neg R)$.

15. a. $(P \vee Q) \wedge (P \vee \neg Q)$. **c.** $\neg Q \vee P$. **e.** $(P \vee Q \vee R) \wedge (P \vee Q \vee \neg R)$
$\wedge (P \vee \neg Q \vee R) \wedge (\neg P \vee Q \vee R) \wedge (\neg P \vee \neg Q \vee R)$.

16. a. $A \vee B \equiv \neg (\neg A \wedge \neg B)$. So $\{\neg, \wedge\}$ is adequate because $\{\neg, \vee\}$ is adequate.

c. $\neg A \equiv A \rightarrow$ False. So $\{$False, $\rightarrow\}$ is an adequate set because $\{\neg, \rightarrow\}$ is adequate.

e. $\neg A \equiv$ NOR(A, A), and $A \vee B \equiv \neg$ NOR$(A, B) =$ NOR(NOR(A, B), NOR(A, B)). Therefore, NOR is adequate because $\{\neg, \vee\}$ is an adequate set of connectives.

Section 6.3

1. The proof uses an extra premise A on line 2. So the result of CP is not W, but instead it is the tautology $(A \vee B \rightarrow C) \wedge A \rightarrow (B \vee C)$.

2. Line 6 is not correct because it uses line 3, which is in a previous subproof. Only lines 1 and 5 can be used to infer something on line 6.

3. a. Three premises: A is the premise for the proof of the conditional, whose conclusion is $B \rightarrow (C \rightarrow D)$. B is the premise for the conditional proof whose conclusion is $C \rightarrow D$. Finally, C is the premise for the proof of $C \rightarrow D$.

4. Let D mean "I am dancing," H mean "I am happy," and M mean "There is a mouse in the house." Then a proof can be written as follows:

1.	$D \rightarrow H$	P
2.	$M \vee H$	P
3.	$\neg H$	P
4.	M	2, 3, DS
5.	$\neg D$	1, 3, MT
6.	$M \wedge \neg D$	4, 5, Conj
	QED	1–6, CP.

5. a.

1.	A	P
2.	B	P [for $B \rightarrow A \wedge B$]
3.	$A \wedge B$	1, 2, Conj
4.	$B \rightarrow A \wedge B$	2, 3, CP
	QED	1, 4, CP.

c. 1. $A \vee B \rightarrow C$ P
    2. A P
    3. $A \vee B$ 2, Add
    4. C 1, 3, MP
        QED   1–4, CP.

e. 1. $A \vee B \rightarrow C \wedge D$ P
    2. B P [for $B \rightarrow D$]
    3. $A \vee B$ 2, Add
    4. $C \wedge D$ 1, 3, MP
    5. D 4, Simp
    6. $B \rightarrow D$ 2–5, CP
        QED   1, 6, CP.

g. 1. $\neg A \vee \neg B$ P
    2. $B \vee C$ P
    3. $C \rightarrow D$ P
    4. A P [for $A \rightarrow D$]
    5. $\neg \neg A$ 4, DN
    6. $\neg B$ 1, 5, DS
    7. C 2, 6, DS
    8. D 3, 7, MP
    9. $A \rightarrow D$ 4–8, CP
        QED   1–3, 9, CP.

i. 1. $A \rightarrow C$ P
    2. $A \wedge B$ P [for $A \wedge B \rightarrow C$]
    3. A 2, Simp
    4. C 1, 3, MP
    5. $A \wedge B \rightarrow C$ 2–4, CP
        QED   1, 5, CP.

k. 1. $A \rightarrow B$ P
    2. $C \vee A$ P [for $C \vee A \rightarrow C \vee B$]
    3. C P [for $C \rightarrow C \vee B$]
    4. $C \vee B$ 3, Add
    5. $C \rightarrow C \vee B$ 3, 4, CP
    6. A P [for $A \rightarrow C \vee B$]

7.		B	1, 6, MP
8.		$C \lor B$	7, Add
9.		$A \to C \lor B$	6–8, CP
10.		$C \lor B$	2, 5, 9, Cases
11.	$C \lor A \to C \lor B$		2, 5, 9, 10, CP
	QED		1, 11, CP.

6. a.

1.	A			P
2.		B		P [for $B \to A$]
3.			$\neg A$	P [for A]
4.			False	1, 3, Contr
5.		A		3, 4, IP
6.	$B \to A$			2, 5, CP
	QED			1, 6, CP.

c.

1.	$\neg B$			P
2.		B		P [for $B \to C$]
3.			$\neg C$	P [for C]
4.			False	1, 2, Contr
5.		C		3, 4, IP
6.	$B \to C$			2, 5, CP
	QED			1, 6, CP.

e.

1.	$A \to B$			P
2.		$A \to \neg B$		P [for $(A \to \neg B) \to \neg A$]
3.			$\neg \neg A$	P [for $\neg A$]
4.			A	3, DN
5.			B	1, 4, MP
6.			$\neg B$	2, 4, MP
7.			False	5, 6, Contr
8.		$\neg A$		3–7, IP
9.	$(A \to \neg B) \to \neg A$			2, 8, CP
	QED			1, 9, CP.

g.
  1. $A \rightarrow B$ P

  2. $C \lor A$ P [for $C \lor A \rightarrow C \lor B$]

  3. $\neg (C \lor B)$ P [for $C \lor B$]

  4. $\neg C$ P [for C]

  5. A 2, 4, DS

  6. B 1, 5, MP

  7. $C \lor B$ 6, Add

  8.   False   3, 7, Contr

  9. C 4–8, IP

 10. $C \lor B$ 9, Add

 11.   False   3, 10, Contr

 12. $C \lor B$ 3, 9–11, IP

 13. $C \lor A \rightarrow C \lor B$ 2, 12, CP

    QED   1, 13, CP.

7. a.
  1. $A \lor B \rightarrow C$ P

  2. A P

  3. $\neg C$ P[for C]

  4. $\neg (A \lor B)$ 1, 3, MT

  5. $A \lor B$ 2, Add

  6.   False   4, 5, Contr

  7. C 3–6, IP

    QED   1, 2, 7, CP.

c.
  1. $A \lor B \rightarrow C \land D$ P

  2. $\neg (B \rightarrow D)$ P[for $B \rightarrow D$]

  3. B P[for $B \rightarrow D$]

  4. $A \lor B$ 3, Add

  5. $C \land D$ 1, 4, MP

  6. D 5, Simp

  7. $B \rightarrow D$ 3–6, CP

  8.   False   2, 7, Contr

  9. $B \rightarrow D$ 2, 7, 8, IP

    QED   1, 9, CP.

e. 1. $\neg (A \wedge B)$ P
  2. $B \vee C$ P
  3. $C \rightarrow D$ P
  4. A P [for $A \rightarrow D$]
  5. $\neg D$ P [for D]
  6. $\neg C$ 3, 5, MT
  7. B 2, 6, DS
  8. $A \wedge B$ 4, 7, Conj
  9.   False   1, 8, Contr
  10. D 5–9, IP
  11. $A \rightarrow D$ 4, 10, CP
       QED   1– 3, 11, CP.

g. 1. $A \rightarrow (B \rightarrow C)$ P
  2. B P [for $B \rightarrow (A \rightarrow C)$]
  3. A P [for $A \rightarrow C$]
  4. $\neg C$ P [for C]
  5. $B \rightarrow C$ 1, 3, MP
  6. C 2, 5, MP
  7.   False   4, 6, Contr
  8. C 4–7, IP
  9. $A \rightarrow C$ 3, 8, CP
  10. $B \rightarrow (A \rightarrow C)$ 2, 9, CP
       QED   1, 10, CP.

8. 1. $\neg C \vee \neg D$ P
  2. $A \rightarrow C$ P
  3. $B \rightarrow D$ P
  4. $\neg C$ P [for $\neg C \rightarrow \neg A \vee \neg B$]
  5. $\neg A$ 2, 4, MT
  6. $\neg A \vee \neg B$ 5, Add
  7. $\neg C \rightarrow \neg A \vee \neg B$ 4–6, CP
  8. $\neg D$ P [for $\neg D \rightarrow \neg A \vee \neg B$]
  9. $\neg B$ 3, 8, MT
  10. $\neg A \vee \neg B$ 5, Add
  11. $\neg D \rightarrow \neg A \vee \neg B$ 8–10, CP
  12. $\neg A \vee \neg B$ 1, 7, 11, Cases
       QED   1–3, 7, 11, 12, CP.

9. a.

1.	$(A \land B) \lor (\neg A \land C)$		P
2.		A	P [for $A \to B$]
3.		$\neg \neg (\neg A \land C)$	P [for $\neg (\neg A \land C)$]
4.		$\neg A \land C$	3, DN
5.		$\neg A$	4, Simp
6.		False	2, 5, Contr
7.		$\neg (\neg A \land C)]$	3–6, IP
8.		$A \land B$	1, 7, DS
9.		B	8, Simp
10.	$A \to B$		2, 7–9, CP
11.		$\neg A$	P [for $\neg A \to C$]
12.		$\neg \neg (A \land B)$	P [for $\neg (A \land B)$]
13.		$A \land B$	12, DN
14.		A	13, Simp
15.		False	11, 14, Contr
16.		$\neg (A \land B)$	12–15, IP
17.		$\neg A \land C$	1, 16, DS
18.		C	17, Simp
19.	$\neg A \to C$		11, 16–18, CP
20.	$(A \to B) \land (\neg A \to C)$		10, 19, Conj
	QED		1, 10, 19, 20, CP.

10. a. Proof of $A \land \neg$ False $\to A$:

1. $A \land \neg$ False P
2. A 1, Simp
   QED   1, 2, MP.

Proof of $A \to A \land \neg$ False:

1. A P
2. $\neg \neg$ False P [for $\neg$ False]
3.   False   2, DN
4. $\neg$ False   2, 3, IP
5. $A \land \neg$ False   1, 4, Conj
   QED   1, 4, 5, CP.

c. Proof of $A \land A \to A$:

1. $A \land A$ P
2. A 1, Simp
   QED   1, 2, CP.

Proof of $A \rightarrow A \land A$:

1. A P
2. $A \lor A$ 1, Add
3. $A \land (A \lor A)$ 1, 2, Conj
4. A 3, Simp
5. $A \land A$ 1, 4, Conj
    QED   1–5, CP.

e. Proof of $A \land B \rightarrow B \land A$:

1. $A \land B$ P
2. A 1, Simp
3. B 1, Simp
4. $B \land A$ 2, 3, Conj
    QED   1–4, CP.

The proof of $B \land A \rightarrow A \land B$ is similar to proof of $A \land B \rightarrow B \land A$.

11. a. Proof of $A \lor \neg$ False $\rightarrow \neg$ False:

1. $A \lor \neg$ False P
2. $\neg \neg$ False P [for $\neg$ False]
3.   False   2, DN
4. $\neg$ False   2, 3, IP
    QED   1, 4, CP.

Proof of $\neg$ False $\rightarrow A \lor \neg$ False:

1. $\neg$ False P
2. $A \lor \neg$ False   1, Add
    QED   1, 2, CP.

c. Proof of $A \lor A \rightarrow A$:

1. $A \lor A$ P
2. $\neg A$ P [for A]
3. A 1, 2, DS
4.   False   2, 3, Contr
5. A 2–4, IP
    QED   1, 5, CP.

Proof of $A \rightarrow A \lor A$:

1. A P
2. $A \lor A$ 1, Add
    QED   1, 2, CP.

12. a. Proof of $A \rightarrow \neg$ False:

1. A P
2. $\neg \neg$ False P [for $\neg$ False]
3.   False   2, DN
4. $\neg$ False   2, 3, IP
     QED   1, 4, CP.

c. Proof of $(\neg$ False $\rightarrow A) \rightarrow A$:

1. $\neg$ False $\rightarrow A$ P
2. $\neg A$ P [for A]
3. $\neg \neg$ False   1, 2, DS
4.   False   3, DN
5. A 2–4, IP
     QED   1, 5, CP.

Proof of $A \rightarrow (\neg$ False $\rightarrow A)$:

1. $A \rightarrow (\neg$ False $\rightarrow A)$ T [Example 6.13]
     QED.

13. a. Proof of $(A \rightarrow B) \rightarrow (\neg B \rightarrow \neg A)$:

1. $A \rightarrow B$ P
2. $\neg B$ P [for $\neg B \rightarrow \neg A$]
3. $\neg A$ 1, 2, MT
4. $\neg B \rightarrow \neg A$ 2, 3, CP
     QED   1, 4, CP.

Proof of $(\neg B \rightarrow \neg A) \rightarrow (A \rightarrow B)$:

1. $\neg B \rightarrow \neg A$ P
2. A P [for $A \rightarrow B$]
3. $\neg \neg A$ 2, DN
4. $\neg \neg B$ 1, 3, MT
5. B 4, DN
6. $A \rightarrow B$ 2–5, CP
     QED   1, 6, CP.

c. Proof of $\neg (A \wedge B) \rightarrow (\neg A \vee \neg B)$:

1.	$\neg (A \wedge B)$	P
2.	$\neg (\neg A \vee \neg B)$	P [for $\neg A \vee \neg B$]
3.	$\neg \neg A \wedge \neg \neg B$	2, T [Example 6.30]
4.	$\neg \neg A$	3, Simp
5.	A	4, DN
6.	$\neg \neg B$	3, Simp
7.	B	6, DN
8.	$A \wedge B$	5, 7, Conj
9.	False	1, 8, Contr
10.	$\neg A \vee \neg B$	2–9, IP
	QED	1, 10, CP.

Proof of $(\neg A \vee \neg B) \rightarrow \neg (A \wedge B)$:

1.	$\neg A \vee \neg B$	P
2.	$\neg \neg (A \wedge B)$	P[for $\neg (A \wedge B)$]
3.	$A \wedge B$	2, DN
4.	A	3, Simp
5.	$\neg \neg A$	4, DN
6.	$\neg B$	1, 5, DS
7.	B	3, Simp
8.	False	6, 7, Contr
9.	$\neg (A \wedge B)$	2–8, IP
	QED	1, 9, CP.

e. Proof of $A \vee (B \wedge C) \rightarrow (A \vee B) \wedge (A \vee C)$:

1.	$A \vee (B \wedge C)$	P
2.	A	P [for $A \rightarrow (A \vee B) \wedge (A \vee C)$]
3.	$A \vee B$	2, Add
4.	$A \vee C$	2, Add
5.	$(A \vee B) \wedge (A \vee C)$	3, 4, Conj
6.	$A \rightarrow (A \vee B) \wedge (A \vee C)$	2–5, CP
7.	$B \wedge C$	P [for $B \wedge C \rightarrow (A \vee B) \wedge (A \vee C)$]
8.	B	7, Simp
9.	$A \vee B$	8, Add
10.	C	7, Simp
11.	$A \vee C$	10, Add

12. $\quad (A \vee B) \wedge (A \vee C)$ $\qquad$ 9, 11, Conj
13. $\quad B \wedge C \rightarrow (A \vee B) \wedge (A \vee C)$ $\quad$ 7–12, CP
14. $\quad (A \vee B) \wedge (A \vee C)$ $\qquad$ 1, 6, 13, Cases
$\qquad$ QED $\qquad\qquad\qquad\qquad$ 1, 6, 13, 14, CP.

The proof of $(A \vee B) \wedge (A \vee C) \rightarrow A \vee (B \wedge C)$:

1. $\quad (A \vee B) \wedge (A \vee C)$ $\quad$ P
2. $\qquad A \vee B$ $\qquad\qquad$ 1, Simp
3. $\qquad A \vee C$ $\qquad\qquad$ 1, Simp
4. $\qquad\quad \neg A$ $\qquad\qquad$ P [for $\neg A \rightarrow B \wedge C$]
5. $\qquad\quad B$ $\qquad\qquad\quad$ 2, 4, DS
6. $\qquad\quad C$ $\qquad\qquad\quad$ 3, 4, DS
7. $\qquad\quad B \wedge C$ $\qquad\qquad$ 5, 6, Conj
8. $\qquad \neg A \rightarrow B \wedge C$ $\qquad$ 4–7, CP
9. $\qquad A \rightarrow A$ $\qquad\qquad$ T [Example 6.11]
10. $\qquad A \vee \neg A$ $\qquad\qquad$ T [Example 6.17]
11. $\qquad A \vee (B \wedge C)$ $\qquad$ 8, 9, 10, CD
$\qquad$ QED $\qquad\qquad\qquad$ 1–3, 8–11, CP.

14. a. Proof of $A \wedge (A \vee B) \rightarrow A$:

1. $\quad A \wedge (A \vee B)$ $\quad$ P
2. $\qquad A$ $\qquad\qquad\quad$ 1, Simp
$\qquad$ QED $\qquad\qquad$ 1, 2, CP.

Proof of $A \rightarrow A \wedge (A \vee B)$:

1. $\quad A$ $\qquad\qquad\qquad$ P
2. $\quad A \vee B$ $\qquad\qquad$ 1, Add
3. $\quad A \wedge (A \vee B)$ $\quad$ 1, 2, Conj
$\qquad$ QED $\qquad\qquad\quad$ 1–3, CP.

c. Proof of $A \wedge (\neg A \vee B) \rightarrow A \wedge B$:

1. $\quad A \wedge (\neg A \vee B)$ $\quad$ P
2. $\quad A$ $\qquad\qquad\qquad$ 1, Simp
3. $\quad \neg A \vee B$ $\qquad\qquad$ 1, Simp
4. $\quad \neg \neg A$ $\qquad\qquad$ 2, DN
5. $\quad B$ $\qquad\qquad\qquad$ 3, 4, DS
6. $\quad A \wedge B$ $\qquad\qquad$ 2, 5, Conj
$\qquad$ QED $\qquad\qquad\quad$ 1–6, CP.

Proof of $A \wedge B \to A \wedge (\neg A \vee B)$:

1.	$A \wedge B$	P
2.	A	1, Simp
3.	B	1, Simp
4.	$\neg A \vee B$	3, Add
5.	$A \wedge (\neg A \vee B)$	2, 4, Conj
	QED	1–5, CP.

Section 6.4

1.

1.	$\neg A \to B$	P
2.	$\neg A \to \neg B$	P
3.	$(\neg A \to \neg B) \to (B \to A)$	Axiom 3
4.	$B \to A$	2, 3, MP
5.	$(B \to A) \to ((\neg B \to A) \to A)$	T [by (6.11f)]
6.	$(\neg B \to A) \to A)$	4, 5, MP
7.	$(\neg A \to B) \to (\neg B \to \neg \neg A)$	T [by (6.11d)]
8.	$\neg B \to \neg \neg A$	1, 7, MP
9.	$\neg \neg A \to A$	T [by (6.11b)]
10.	$\neg B \to A$	8, 9, HS
11.	A	6, 10, MP
	QED	1–11, CP.

2.

1.	$A \to (B \to C)$	P
2.	$(A \to (B \to C)) \to ((A \to B) \to (A \to C))$	Axiom 2
3.	$(A \to B) \to (A \to C)$	1, 2, MP
4.	$\qquad\qquad B$	P [for $B \to (A \to C)$]
5.	$\qquad\qquad B \to (A \to B)$	Axiom 1
6.	$\qquad\qquad A \to B$	4, 5, MP
7.	$\qquad\qquad A \to C$	3, 6, MP
8.	$B \to (A \to C)$	4–7, CP
	QED	1–3, 8, CP.

4. a.

1.	$(A \to B) \to ((\neg C \vee A) \to (\neg C \vee B))$	Axiom 4
2.	$(A \to B) \to ((C \to A) \to (C \to B))$	1, Definition of $\to$
	QED.	

c.

1.	$A \to A \vee A$	Axiom 2
2.	$A \vee A \to A$	Axiom 1
3.	$A \to A$	1, 2, HS (i.e., Part (b))
	QED.	

e. 1. $\neg\, A \vee \neg\,\neg\, A$ Part (d)

   2. $A \to \neg\,\neg\, A$ 1, Definition of $\to$

      QED.

g. 1. $\neg\, A \to (\neg\, A \vee B)$ Axiom 2

   2. $\neg\, A \to (A \to B)$ 1, definition of $\to$

      QED.

i. 1. $\neg\,\neg\, B \to B$ Part (e)

   2. $(\neg\,\neg\, B \to B)$

          $\to ((\neg\, A \vee \neg\,\neg\, B) \to (\neg\, A \vee B))$ Axiom 4

   3. $(\neg\, A \vee \neg\,\neg\, B) \to (\neg\, A \vee B)$ 1, 2, MP

   4. $(\neg\,\neg\, B \vee \neg\, A) \to (\neg\, A \vee \neg\,\neg\, B)$ Axiom 3

   5. $(\neg\,\neg\, B \vee \neg\, A) \to (\neg\, A \vee B)$ 3, 4, HS (i.e., Part (b))

   6. $(\neg\, B \to \neg\, A) \to (A \to B)$ 5, Definition of $\to$

      QED.

6. We give a couple of hints to aid the reasoning process. *Hint:* Let each name, like A, mean that A won a position. Then transform each statement into a wff of the propositional calculus. Create a wff to describe the problem, and find an assignment of truth values to make the wff true. *Hint:* Make a table of possibilities with rows A, B, C, and D, and columns E, F, G, and H. Place a check in an entry if the row name and column name were not elected to the board.

Chapter 7

Section 7.1

1. a. $[p(0, 0) \wedge p(0, 1)] \vee [p(1, 0) \wedge p(1, 1)]$.

2. a. $\forall x\ q(x)$, where $x \in \{0, 1\}$.

c. $\forall\ y\ p(x, y)$, where $y \in \{0, 1\}$.

e. $\exists x\ p(x)$, where x is an odd natural number.

3. a. x is a term. Therefore, $p(x)$ is a wff, and it follows that $\exists x\ p(x)$ and $\forall x$ $p(x)$ are wffs. Thus $\exists x\ p(x) \to \forall x\ p(x)$ is a wff.

4. It is illegal to have an atom, $p(x)$ in this case, as an argument to a predicate.

5. a. The three occurrences of x, left to right, are free, bound, and bound. The four occurrences of y, left to right, are free, bound, bound, and free.

c. The three occurrences of x, left to right, are free, bound, and bound. Both occurrences of y are free.

6. $\forall x\ p(x, y, z) \to \exists z\ q(z)$.

7. a. False. **c.** False. **e.** True. **g.** False.

8. a. True. **c.** False.

9. a. With this interpretation, W can be written in more familiar notation as follows: $\forall x\ ((2x \bmod 3 = x) \to (x = 0))$. A bit of checking will convince us that W is true with respect to the interpretation.

10. a. Every bird eats every worm.

11. a. There is someone who eats every chocolate bar.

12. a. $\forall x\ (\neg\ e(x,\ a) \rightarrow \exists y\ p(y,\ x))$, where $a = 0$.

13. a. One interpretation has $p(a) =$ True, in which case both $\forall x\ p(x)$ and $\exists x$ $p(x)$ are true. Therefore, W is true. The other interpretation has $p(a) =$ False, in which case both $\forall x\ p(x)$ and $\exists x\ p(x)$ are false. Therefore, W is true.

14. a. Let the domain be the set $\{a,\ b\}$, and assign $p(a) =$ True and $p(b) =$ False. Finally, assign the constant $c = a$.
c and **d.** Let $p(x,\ y) =$ False for all elements x and y in any domain. Then the antecedent is false for both Parts (c) and (d). Therefore, both wffs are true for this interpretation. **e.** Let $D = \{a\}$, $f(a) = a$, $y = a$, and let p denote equality.

15. a. Let the domain be $\{a\}$, and let $p(a) =$ True and $c = a$.
c. Let $D = \mathbb{N}$, let $p(x)$ mean "x is odd," and let $q(x)$ mean "x is even." Then the antecedent is true, but the consequent is false.
e. Let $D = \mathbb{N}$, and let $p(x,\ y)$ mean "$y = x + 1$." Then the antecedent $\forall x\ \exists y$ $p(x,\ y)$ is true and the consequent $\exists y\ \forall x\ p(x,\ y)$ is false for this interpretation.
g. Let $D = \{a,\ b\}$, $p(a) =$ True, $p(b) =$ False, $q(a) =$ False, and $q(b) =$ True. Then $\forall x\ p(x)$ is false, so the antecedent is true. But $p(a) \rightarrow q(a)$ is false, so the consequent is false.

16. a. If the domain is $\{a\}$, then either $p(a) =$ True or $p(a) =$ False. In either case, W is true.

17. a. Let $\{a\}$ be the domain of the interpretation. If $p(a,\ a) =$ False, then W is true, since the antecedent is false. If $p(a,\ a) =$ True, then W is true, since the consequent is true.
c. Let $\{a,\ b,\ c\}$ be the domain. Let $p(a,\ a) = p(b,\ b) = p(c,\ c) =$ True and $p(a,\ b) = p(a,\ c) = p(b,\ c) =$ False. This assignment makes W false. Therefore, W is invalid.

18. $\forall x\ p(x,\ x) \rightarrow$
$\forall x\ \forall y\ \forall z\ \forall w\ (p(x,\ y) \lor p(x,\ z) \lor p(x,\ w) \lor p(y,\ z) \lor p(y,\ w) \lor p(z,\ w))$.

19. a. For any domain D and any element $d \in D$, $p(d) \rightarrow p(d)$ is true. Therefore, any interpretation is a model.
c. If the wff is invalid, then there is some interpretation making the wff false. This says that $\forall x\ p(x)$ is true and $\exists x\ p(x)$ is false. This is a contradiction because we can't have $p(x)$ true for all x in a domain while at the same time having $p(x)$ false for some x in the domain.
e. If the wff is not valid, then there is an interpretation with domain D for which the antecedent is true and the consequent is false. So $A(d)$ and $B(d)$ are false for some element $d \in D$. Therefore, $\forall x\ A(x)$ and $\forall x\ B(x)$ are false, contrary to assumption.
g and **h.** If the antecedent is true for a domain D, then $A(d) \rightarrow B(d)$ is true for all $d \in D$. If $A(d)$ is true for all $d \in D$, then $B(d)$ is also true for all $d \in D$ by MP. Thus the consequent is true for D.

20. a. Suppose the wff is satisfiable. Then there is an interpretation that assigns c a value in its domain such that $p(c) \land \neg\, p(c) =$ True. Of course, this is impossible. Therefore, the wff is unsatisfiable.

c. Suppose the wff is satisfiable. Then there is an interpretation making $\exists x\ \forall y$ $(p(x, y) \land \neg\, p(x, y))$ true. This says that there is an element d in the domain such that $\forall y\ (p(d, y) \land \neg\, p(d, y))$ is true. This says that $p(d, y) \land \neg\, p(d, y)$ is true for all y in the domain, which is impossible.

22. Assume that $A \to B$ is valid and A is also valid. Let I be an interpretation for B with domain D. Extend I to an interpretation J for A by using D to interpret all predicates, functions, free variables, and constants that occur in A but not in B. So J is an interpretation for $A \to B$, A, and B. Since we are assuming that $A \to B$ and A are valid, it follows that $A \to B$ and A are true with respect to J. Therefore, B is true with respect to J. But J and I are the same interpretation on B. So B is true with respect to I. Therefore, I is a model for B. Since I was arbitrary, it follows that B is valid. Now we go in the other direction. Assume that if A is valid, then B is valid. Let I be an interpretation for $A \to B$. Then I is also an interpretation for A and for B. Since A and B are valid, it follows that A and B are true with respect to I. Therefore, $A \to B$ is true with respect to I. Therefore, I is a model for $A \to B$. Since I was arbitrary, it follows that $A \to B$ is valid.

Section 7.2

1. For each part we'll assume that I is an interpretation with domain D. **a.** The left side is true for I iff $A\,(d) \land B(d)$ is true for all $d \in D$ iff $A(d)$ and $B(d)$ are both true for all $d \in D$ iff the right side is true for I.

c. Assume that the left side is true for I. Then $A(d) \to B(d)$ is true for some $d \in D$. If $A(d)$ is true, then $B(d)$ is true by MP. So $\exists x\ B(x)$ is true for I. If $A(d)$ is false, then $\forall x\ A(x)$ is false. So in either case the right side is true for I. Now assume the right side is true for I. If $\forall x\ A(x)$ is true, then $\exists x\ B(x)$ is also true. This means that $A(d)$ is true for all $d \in D$ and $B(d)$ is true for some $d \in D$. Thus $A(d) \to B(d)$ is true for some $d \in D$, which says that the left side is true for I. If $\forall x\ A(x)$ is false, then $A(d)$ is false for some $d \in D$. So $A(d) \to B(d)$ is true. Thus the left side is true for I.

e. $\exists x\ \exists y\ W(x, y)$ is true for I iff $W(d, e)$ is true for some elements $d, e \in D$ iff $\exists y\ \exists x\ W(x, y)$ is true for I.

2. The assumption that x is not free in C means that any substitution x/t does not change C. In other words, $C(x/t) = C$ for all possible terms t. We'll assume that I is an interpretation with domain D.

a. If I is a model for $\forall x\ C$, then $C(x/d)$ is true for I for all d in D. Since $C(x/d) = C$, it follows that C is true for I. Therefore, I is a model for C. If I is a model for C, then C is true for I. Since $C = C(x/d)$ for all d in D, it follows that $C(x/d)$ is true for I for all d in D. Therefore, I is a model for $\forall x\ C$.

c. If I is a model for $\exists x\ (C \lor A(x))$, then $(C \lor A(x))(x/d)$ is true for I for some d in D. But we have $(C \lor A(x))(x/d) = C(x/d) \lor A(x)(x/d) = C \lor A(x)(x/d)$ because x is not free in C. So $C \lor A(x)(x/d)$ is true for I. Since C

is not affected by any substitution for x, it follows that either C is true for I or $A(x)(x/d)$ is true for I. So either I is a model for C or I is a model for $\exists x\ A(x)$. Therefore, I is a model for $C \vee \exists x\ A(x)$.

Conversely, if I is a model for $C \vee \exists x\ A(x)$, then $C \vee \exists x\ A(x)$ is true for I. So either C is true for I or $\exists x\ A(x)$ is true for I. Suppose C is true for I. Since $C = C(x/d)$ for any d in D, it is true for some d in D. So $C(x/d) \vee A(x)(x/d)$ is true for I for some d in D. Substitution gives $C(x/d) \vee A(x)(x/d) = (C \vee A(x))(x/d)$. So $(C \vee A(x))(x/d)$ is true for I for some d in D. Thus I is a model for $\exists x\ (C \vee A(x))$. Suppose that $\exists x\ A(x)$ is true for I, then $A(x)(x/d)$ is true for I for some d in D. So $C(x/d) \vee A(x)(x/d)$ is true for I and thus $(C \vee A(x))(x/d)$ is true for I. So I is a model for $\exists x\ (C \vee A(x))$.

3. a. $\forall x\ (C \to A(x)) \equiv \forall x\ (\neg\ C \vee A(x)) \equiv \neg\ C \vee \forall\ x\ A(x) \equiv C \to \forall x\ A(x)$.
c. $\exists x\ (A(x) \to C) \equiv \exists x\ (\neg\ A(x) \vee C) \equiv \exists x\ \neg\ A(x) \vee C \equiv \neg\ \forall x\ A(x) \vee C$
$\equiv \forall x\ A(x) \to C$.

4. a. $\exists x\ \forall y\ \forall z\ ((\neg\ p(x) \vee p(y) \vee q(z)) \wedge (\neg\ q(x) \vee p(y) \vee q(z)))$.
c. $\exists x\ \forall y\ \exists z\ \forall w\ (\neg\ p(x,\ y) \vee p(w,\ z))$.
e. $\forall x\ \forall y\ \exists z\ ((\neg\ p(x,\ y) \vee p(x,\ z)) \wedge (\neg\ p(x,\ y) \vee p(y,\ z)))$.

5. a. $\exists x\ \forall y\ \forall z\ ((\neg\ p(x) \wedge \neg\ q(x)) \vee p(y) \vee q(z))$.
c. $\exists x\ \forall y\ \exists z\ \forall w\ (\neg\ p(x,\ y) \vee p(w,\ z))$.
e. $\forall x\ \forall y\ \exists z\ (\neg\ p(x,\ y) \vee (p(x,\ z) \wedge p(y,\ z)))$.

6. a. Let D be the domain $\{a,\ b\}$. Assume that C is false, $W(a)$ is true, and $W(b)$ is false. Then $(\forall x\ W(x) \equiv C)$ is true, but $\forall x\ (W(x) \equiv C)$ is false. Therefore, the statement is false.

7. a. $\forall x\ (C(x) \to R(x) \wedge F(x))$.
c. $\forall x\ (G(x) \to S(x))$.
e. $\exists x\ (G(x) \wedge \neg\ S(x))$.

8. a. $\forall x\ (B(x) \to \exists y\ (W(y) \wedge E(x,\ y)))$.
c. $\forall x\ \forall y\ (W(x) \wedge E(y,\ x) \to B(y))$.
e. $\forall x\ \forall y\ (B(x) \wedge E(x,\ y) \to W(y))$.
g. $\exists x\ (\neg\ B(x) \wedge \exists y\ (W(y) \wedge E(x,\ y)))$.

9. a. $\forall x\ (F(x) \to S(x)) \wedge \neg\ S(\text{John}) \to \neg\ F(\text{John})$.

10. a. $\forall x\ (P(x) \wedge \neg\ B(x) \to \forall y\ (C(y) \to K(x,\ y)))$.
c. $\forall x\ (B(x) \to \neg\ K(x,\ x))$.
e. $\forall x\ (P(x) \to \exists y\ (P(y) \wedge N(x,\ y) \wedge G(y)))$.
g. $\forall x\ \forall y\ (P(x) \wedge A(y) \wedge \neg\ K(x,\ y) \to \neg\ G(x))$.

Section 7.3

1. a. Line 2 is wrong because x is free in line 1, which is a premise. So line 1 can't be used with the UG rule to generalize x.
c. Line 2 is wrong because $f(y)$ is not free to replace x. That is, the substitution of $f(y)$ for x yields a new bound occurrence of y. Therefore, EG can't generalize to x from $f(y)$.
e. Line 4 is wrong because c already occurs in the proof on line 3.

2. Lines 3 and 4 are errors; they apply UG to a subexpression of a larger wff.

3. a. Let I be the interpretation with domain $D = \{0, 1\}$, where $P(0) = $ True, $P(1) = $ False, and $Q(0) = Q(1) = $ False. Then the antecedent is true for I and the consequent is false for I. So the wff is false for I. Therefore, I is a countermodel for the wff.

5. This reasoning is wrong because the premise on line 1 does not infer the wff on line 3. Instead, the wff on line 3 is the result of the CP rule applied to the little derivation from the premise on line 1 to the wff on line 2. So UG can be applied to line 3.

6. a.
1. $\forall x\, p(x)$ P
2. $p(x)$ 1, UI
3. $\exists x\, p(x)$ 2, EG
    QED   1–3, CP.

c.
1. $\exists x\, (p(x) \wedge q(x))$ P
2. $p(c) \wedge q(c)$ 1, EI
3. $p(c)$ 2, Simp
4. $\exists x\, p(x)$ 3, EG
5. $q(c)$ 2, Simp
6. $\exists x\, q(x)$ 5, EG
7. $\exists x\, p(x) \wedge \exists x\, q(x)$ 4, 6, Conj
    QED   1–7, CP.

e.
1. $\forall x\, (p(x) \rightarrow q(x))$ P
2. $\forall x\, p(x)$ P[for $\forall x\, p(x) \rightarrow \exists x\, q(x)$]
3. $p(x)$ 2, UI
4. $p(x) \rightarrow q(x)$ 1, UI
5. $q(x)$ 3, 4, MP
6. $\exists x\, q(x)$ 5, EG
7. $\forall x\, p(x) \rightarrow \exists x\, q(x)$ 2–6, CP
    QED   1, 7, CP.

g.
1. $\exists y\, \forall x\, p(x, y)$ P
2. $\forall x\, p(x, c)$ 1, EI
3. $p(x, c)$ 2, UI
4. $\exists y\, p(x, y)$ 3, EG
5. $\forall x\, \exists y\, p(x, y)$ 4, UG
    QED   1–5, CP.

7. a.
1. $\forall x\, p(x)$ P
2. $\neg\, \exists x\, p(x)$ P [for $\exists x\, p(x)$]
3. $p(x)$ 1, UI
4. $\exists x\, p(x)$ 3, EI
5.   False   2, 4, Contr
6. $\exists x\, p(x)$ 2–5, IP
    QED   1, 7, CP.

c. 1. $\exists y \; \forall x \; p(x, y)$ P

2. $\neg \; \forall x \; \exists y \; p(x, y)$ P [for $\forall x \; \exists y \; p(x, y)$]

3. $\exists x \; \forall y \; \neg \; p(x, y)$ 2, T [by (7.4)]

4. $\forall x \; p(x, c)$ 1, EI

5. $\forall y \; \neg \; p(d, y)$ 3, EI

6. $p(d, c)$ 4, UI

7. $\neg \; p(d, c)$ 5, UI

8.   False   6, 7, Contr

9. $\forall x \; \exists y \; p(x, y)$ 2–8, IP

    QED   1, 9, CP.

e. 1. $\forall x \; p(x) \vee \forall x \; q(x)$ P

2. $\neg \; \forall x \; (p(x) \vee q(x))$ P [for $\forall x \; (p(x) \vee q(x))$]

3. $\exists x \; (\neg \; p(x) \wedge \neg \; q(x))$ 2, T [by (7.4)]

4. $\neg \; p(c) \wedge \neg \; q(c)$ 3, EI

5. $\neg \; p(c)$ 4, Simp

6. $\exists x \; \neg \; p(x)$ 5, EG

7. $\neg \; \forall x \; p(x)$ 6, T [by (7.4)]

8. $\forall x \; q(x)$ 1, 7, DS

9. $q(c)$ 8, UI

10. $\neg \; q(c)$ 4, Simp

11.   False   9, 10, Contr

12. $\forall x \; (p(x) \vee q(x))$ 2–11, IP

    QED   1, 12, CP.

8. a. Let $D(x)$ mean that x is a dog, $L(x)$ mean that x likes people, $H(x)$ mean that x hates cats, and $a =$ Rover. Then the argument can be formalized as

$$\forall x \; (D(x) \rightarrow L(x) \vee H(x)) \wedge D(a) \wedge \neg \; H(a) \rightarrow \exists x \; (D(x) \wedge L(x)).$$

Proof: 1. $\forall x \; (D(x) \rightarrow L(x) \vee H(x))$ P

2. $D(a)$ P

3. $\neg \; H(a)$ P

4. $D(a) \rightarrow L(a) \vee H(a)$ 1, UI

5. $L(a) \vee H(a)$ 2, 4, MP

6. $L(a)$ 3, 5, DS

7. $D(a) \wedge L(a)$ 2, 6, Conj

8. $\exists x \; (D(x) \wedge L(x))$ 7, EG

    QED   1–8, CP.

c. Let $H(x)$ mean that x is a human being, $Q(x)$ mean that x is a quadruped, and $M(x)$ mean that x is a man. Then the argument can be formalized as

$$\forall x\ (H(x) \rightarrow \neg\ Q(x)) \wedge \forall x\ (M(x) \rightarrow H(x)) \rightarrow \forall x\ (M(x) \rightarrow \neg\ Q(x)).$$

Proof: 1. $\forall x\ (H(x) \rightarrow \neg\ Q(x))$ P
      2. $\forall x\ (M(x) \rightarrow H(x))$ P
      3. $H(x) \rightarrow \neg\ Q(x)$ 1, UI
      4. $M(x) \rightarrow H(x)$ 2, UI
      5. $M(x) \rightarrow \neg\ Q(x)$ 3, 4, HS
      6. $\forall x\ (M(x) \rightarrow \neg\ Q(x))$ 5, UG
      QED   1–6, CP.

e. Let $F(x)$ mean that x is a freshman, $S(x)$ mean that x is a sophomore, $J(x)$ mean that x is a junior, and $L(x, y)$ mean that x likes y. Then the argument can be formalized as $A \rightarrow B$, where

$$A = \exists x\ (F\ (x) \wedge \forall y\ (S\ (y) \rightarrow L\ (x, y))) \wedge \forall x\ (F\ (x) \rightarrow \forall y\ (J\ (y) \rightarrow \neg\ L\ (x, y)))$$
$$B = \forall x\ (S\ (x) \rightarrow \neg\ J\ (x))\,.$$

Proof: 1. $\exists x\ (F(x) \wedge \forall y\ (S(y) \rightarrow L(x,\ y)))$ P
      2. $\forall x\ (F(x) \rightarrow \forall y\ (J(y) \rightarrow \neg\ L(x,\ y)))$ P
      3. $F(c) \wedge \forall y\ (S(y) \rightarrow L(c,\ y))$ 1, EI
      4. $\forall y\ (S(y) \rightarrow L(c,\ y))$ 3, Simp
      5. $S(x) \rightarrow L(c,\ x)$ 4, UI
      6. $S(x)$ P [for $S(x) \rightarrow \neg J(x)$]
      7. $L(c,\ x)$ 5, 6, MP
      8. $F(c) \rightarrow \forall y\ (J(y) \rightarrow \neg\ L(c,\ y))$ 2, UI
      9. $F(c)$ 3, Simp
    10. $\forall y\ (J(y) \rightarrow \neg\ L(c,\ y))$ 8, 9, MP
    11. $J(x) \rightarrow \neg\ L(c,\ x)$ 10, UI
    12. $\neg\ J(x)$ 7, 11, MT
    13. $S(x) \rightarrow \neg\ J(x)$ 6–12, CP
    14. $\forall x\ (S(x) \rightarrow \neg\ J(x))$ 13, UG
      QED   1–5, 13, 14, CP.

9. First prove that the left side implies the right side, then the converse.
a.   1. $\exists x\ \exists y\ W(x,\ y)$ P
      2. $\exists y\ W(c,\ y)$ 1, EI
      3. $W(c,\ d)$ 2, EI
      4. $\exists x\ W(x,\ d)$ 3, EG
      5. $\exists y\ \exists x\ W(x,\ y)$ 4, EG
      QED   1–5, CP.

1. $\exists y\ \exists x\ W(x,\ y)$ P
2. $\exists x\ W(x,\ d)$ 1, EI
3. $W(c,\ d)$ 2, EI
4. $\exists y\ W(c,\ y)$ 3, EG
5. $\exists x\ \exists y\ W(x,\ y)$ 4, EG
    QED   1–5, CP.

c. 1. $\exists x\ (A(x) \lor B(x))$ P
  2. $\neg\ (\exists x\ A(\mathrm{x}) \lor \exists x\ B(x))$ P [for $\exists x\ A(\mathrm{x}) \lor \exists x\ B(x)$]
  3. $\forall x\ \neg\ A(x) \land \forall x\ \neg\ B(x)$ 2, T [by (7.4)]
  4. $\forall x\ \neg\ A(x)$ 3, Simp
  5. $A(c) \lor B(c)$ 1, EI
  6. $\neg\ A(c)$ 4, UI
  7. $B(c)$ 5, 6, DS
  8. $\forall x\ \neg\ B(x)$ 3, Simp
  9. $\neg\ B(c)$ 8, UI
  10.   False   7, 9, Contr
  11. $\exists x\ A(\mathrm{x}) \lor \exists x\ B(x)$ 2–10, IP
      QED   1, 11, CP.

1. $\exists x\ A(x) \lor \exists x\ B(x)$ P
2. $\neg\ \exists x\ (A(x) \lor B(x))$ P [for $\exists x\ (A(x) \lor B(x))$]
3. $\forall x\ (\neg\ A(x) \land \neg\ B(x))$ 2, T [by (7.4)]
4. $\forall x\ \neg\ A(x) \land \forall x\ \neg\ B(x)$ 3, T [Part (b)]
5. $\forall x\ \neg\ A(x)$ 4, Simp
6. $\neg\ \exists x\ A(x)$ 5, T [by (7.4)]
7. $\exists x\ B(x)$ 1, 6, DS
8. $\forall x\ \neg\ B(x)$ 4, Simp
9. $\neg\ \exists x\ B(\mathrm{x})$ 5, T [by (7.4)]
10.   False   7, 9, Contr
11. $\exists x\ (A(x) \lor B(x))$ 2–10, IP
    QED   1, 11, CP.

10. 1. $\forall x\ (\exists y\ (q(x, y) \land s(y)) \to \exists y\ (p(y) \land r(x, y)))$ P

   2. $\neg\ \exists x\ p(x)$ P [for $(\neg\ \exists x\ p(x) \to \forall x\ \forall y\ (q(x, y) \to \neg\ s(y)))$]

   3. $\neg\ \forall x\ \forall y\ (q(x, y) \to \neg\ s(y))$ P [for $\forall x\ \forall y\ (q(x, y) \to \neg\ s(y))$]

   4. $\exists x\ \exists y\ (q(x, y) \land s(y))$ 3, T

   5. $\exists y\ (q(c, y) \land s(y))$ 4, EI

   6. $\exists y\ (q(c, y) \land s(y)) \to \exists y\ (p(y) \land r(c, y))$ 1, UI

   7. $\exists y\ (p(y) \land r(c, y))$ 5, 6, MP

   8. $p(d) \land r(c, d)$ 7, EI

   9. $p(d)$ 8, Simp

  10. $\exists x\ p(x)$ 9, EG

  11.   False   2, 10, Contr

  12. $\forall x\ \forall y\ (q(x, y) \to \neg\ s(y))$ 3–11, IP

  13. $\neg\ \exists x\ p(x) \to \forall x\ \forall y\ (q(x, y) \to \neg\ s(y))$ 2, 12, CP

       QED   1, 13, CP.

12. a. 1. $\exists x\ A(x)$ P

   2. $A(x)$ 1, EI (wrong to use a variable)

   3. $\forall x\ A(x)$ 2, UG

       NOT QED   1–3, CP.

c. 1. $\forall x\ (A(x) \lor B(x))$ P

   2. $\neg\ (\forall x\ A(x) \lor \forall x\ B(x))$ P [for $\forall x\ A(x) \lor \forall x\ B(x)$]

   3. $\exists x\ \neg\ A(x) \land \exists x\ \neg\ B(x)$ 2, T

   4. $\exists x\ \neg\ A(x)$ 3, Simp

   5. $\exists x\ \neg\ B(x)$ 4, Simp

   6. $\neg\ A(c)$ 4, EI

   7. $\neg\ B(c)$ 5, EI (wrong to use an existing constant)

   8. $A(c) \lor B(c)$ 1, UI

   9. $B(c)$ 6, 8, DS

  10.   False   7, 9, Contr

  11. $\forall x\ A(x) \lor \forall x\ B(x)$ 2–10, IP

       NOT QED   1, 11, CP.

e. 1. $\forall x\ \exists y\ W(x, y)$ P

   2. $\exists y\ W(x, y)$ 1, UI

   3. $W(x, c)$ 2, EI

   4. $\forall x\ W(x, c)$ 3, UG (wrong because x is free in use of EI)

   5. $\exists y\ \forall x\ W(x, y)$ 4, EG (may be wrong if $W(x, c)$ contains bound y)

       NOT QED   1–5, CP.

13. a. Similar to proof of Exercise 9b. **c.** Use IP in both directions.
e. Similar to Part (c). **g.** Similar to Part (c).

14. a. Let $\neg B$ and $A \rightarrow B$ be valid wffs. Let I be an arbitrary interpretation of these two wffs. Then $\neg B$ and $A \rightarrow B$ are true for I. Thus we can apply MT to conclude that $\neg A$ is true for I. Since the interpretation was arbitrary, it follows that $\neg A$ is valid.

15. a. The variable x is not free within the scope of a quantifier in $W(x)$.

17.

1.	$\forall x \, \neg \, p(x, x)$	P
2.	$\forall x \, \forall y \, \forall z \, (p(x, y) \wedge p(y, z) \rightarrow p(x, z))$	P
3.	$\neg \, \forall x \forall y \, (p(x, y) \rightarrow \neg \, p(y, x))$	
		$P[\text{for } \forall x \, \forall y \, (p(x, y) \rightarrow \neg \, p(y, x))]$
4.	$\exists x \, \exists y \, (p(x, y) \wedge p(y, x))$	$3, T$
5.	$\exists y \, (p(a, y) \wedge p(y, a))$	$4, EI$
6.	$p(a, b) \wedge p(b, a))$	$5, EI$
7.	$\forall y \, \forall z \, (p(a, y) \wedge p(y, z) \rightarrow p(a, z))$	$2, UI$
8.	$\forall z \, (p(a, b) \wedge p(b, z) \rightarrow p(a, z))$	$7, UI$
9.	$p(a, b) \wedge p(b, a) \rightarrow p(a, a))$	$8, UI$
10.	$p(a, a)$	$6, 9, MP$
11.	$\neg \, p(a, a)$	$1, UI$
12.	False	$10, 11, Contr$
13.	$\forall x \, \forall y \, (p(x, y) \rightarrow \neg \, p(y, x))$	$3–12, IP$
	QED	$1, 2, 13, CP.$

Chapter 8

Section 8.1

1.

1.	$s = v$	P
2.	$t = w$	P
3.	$p(s, t)$	P
4.	$p(v, t)$	$1, 3, EE$
5.	$p(v, w)$	$2, 4, EE$
	QED	$1–5, CP.$

3.

1.	$c = a^i$	P
2.	$i \leq b$	P
3.	$\neg \, (i < b)$	P
4.	$(i < b) \vee (i = b)$	$2, T$ [definition]
5.	$i = b$	$3, 4, DS$
6.	$c = a^b$	$1, 5, EE$
	QED	$1–6, CP.$

4. a.
1. $t = u$ P
2. $\neg\, p(\dots t \dots)$ P
3. $p(\dots u \dots)$ P [for $\neg p(\dots u \dots)$]
4. $u = t$ 1, Symmetric
5. $p(\dots t\dots)$ 3, 4, EE
6.   False   2, 5, Contr
7. $\neg\, p(\dots u \dots)$ 3–6, IP
    QED   1, 2, 7, CP.

c.
1. $t = u$ P
2. $p(\dots t \dots) \vee q(\dots t \dots)$ P
3. $\neg\, (p(\dots u \dots) \vee q(\dots u\dots))$ P [for $p(\dots u \dots) \vee q(\dots u\dots)$]
4. $\neg\, p(\dots u \dots) \wedge \neg\, q(\dots u\dots)$ 3, T
5. $\neg\, p(\dots u \dots)$ 4, Simp
6. $\neg\, q(\dots u\dots)$ 4, Simp
7. $u = t$ 1, Symmetric
8. $\neg\, p(\dots t \dots)$ 5, 7, EE [from Part (a)]
9. $\neg\, q(\dots t\dots)$ 6, 7, EE [from Part (a)]
10. $\neg\, p(\dots t \dots) \wedge \neg\, q(\dots t \dots)$ 8, 9, Conj
11. $\neg\, (p(\dots t \dots) \vee q(\dots t \dots))$ 10, T
12.   False   2, 11, Contr
13. $p(\dots u \dots) \vee q(\dots u\dots)$ 3–12, IP
    QED   1, 2, 13, CP.

e.
1. $x = y$ P
2. $\forall z\, p(\dots x\dots)$ P
3. $p(\dots x\dots)$ 2, UI
4. $p(\dots y\dots)$ 1, 3, EE
5. $\forall z\, p(\dots y\dots)$ 4, UG
    QED   1–5, CP.

5.
1. $\neg\, \forall x\, \exists y\, (x = y)$ P [for $\forall x\, \exists y\, (x = y)$]
2. $\exists x\, \forall y\, (x \neq y)$ 1, T
3. $\forall y\, (c \neq y)$ 2, EI
4. $c \neq c$ 3, UI
5. $c = c$ EA
6. False   4, 5, Contr
    QED   1–6, IP.

6. a.   Proof of $p(x) \rightarrow \exists y\ ((x = y) \wedge p(y))$:

1.	$p(x)$	P
2.	$\neg\ \exists y\ ((x = y) \wedge p(y))$	P [for $\exists y\ ((x = y) \wedge p(y))$]
3.	$\forall y\ ((x \neq y) \vee \neg\ p(y))$	2, T
4.	$(x \neq x) \vee \neg\ p(x)$	3, UI
5.	$x \neq x$	1, 4, DS
6.	$x = x$	EA
7.	False	5, 6, Contr
8.	$\exists y\ ((x = y) \wedge p(y))$	2–7, IP
	QED	1, 8, CP.

Proof of $\exists y\ ((x = y) \wedge p(y)) \rightarrow p(x)$:

1.	$\exists y\ ((x = y) \wedge p(y))$	P
2.	$(x = c) \wedge p(c)$	1, EI
3.	$p(x)$	2, EE
	QED	1–3, CP.

7. a. $\mathrm{odd}(x) = \exists z\ (x = 2z + 1)$.
c. $\mathrm{div}(a,\ b) = (a \neq 0) \wedge \exists x\ (b = ax)$.
e. $\mathrm{div}(d,\ a) \wedge \mathrm{div}(d,\ b) \wedge \forall z\ (\mathrm{div}(z,\ a) \wedge \mathrm{div}(z,\ b) \rightarrow (z \leq d))$.

8. a. Possible answers include either of the following two equivalent wffs.

$$\forall x\ \forall y\ (A(x) \wedge A(y) \rightarrow (x = y)),$$
$$\neg\ \exists x\ A(x) \vee \exists x\ (A(x) \wedge \forall y\ (A(y) \rightarrow (x = y))).$$

c. One possible answer is $\forall x\ \forall y\ \forall z\ (A(x) \wedge A(y) \wedge A(z) \rightarrow (x = y) \vee (x = z) \vee (y = z))$. Another answer has the form: None $\vee$ Exactly One $\vee$ Exactly Two.

9. a.   Proof that (a) implies (b).

1.	$\exists x\ (A(x) \wedge \forall y\ (A(y) \rightarrow (x = y)))$	P
2.	$\neg\ (\exists x\ A(x) \wedge \forall x\ \forall y\ (A(x) \wedge A(y) \rightarrow (x = y)))$	P [for second wff]
3.	$\forall x\ \neg\ A(x) \vee \exists x\ \exists y\ (A(x) \wedge A(y) \wedge (x \neq y))$	2, T
4.	$A(c) \wedge \forall y\ (A(y) \rightarrow (c = y))$	1, EI
5.	$A(c)$	4, Simp
6.	$\exists x\ A(x)$	5, EG
7.	$\neg\ \forall x\ \neg\ A(x)$	6, T
8.	$\exists x\ \exists y\ (A(x) \wedge A(y) \wedge (x \neq y))$	3, 7, DS
9.	$A(a) \wedge A(b) \wedge (a \neq b)$	8, EI, EI
10.	$\forall y\ (A(y) \rightarrow (c = y))$	4, Simp
11.	$A(a) \rightarrow (c = a)$	10, UI
12.	$A(a)$	9, Simp

13.	$c = a$	11, 12, MP
14.	$A(b) \rightarrow (c = b)$	10, UI
15.	$A(b)$	9, Simp
16.	$c = b$	14, 15, MP
17.	$a = b$	13, Symmetry, 16, Transitive
18.	$a \neq b$	9, Simp
19.	False	17, 18, Contr
20.	$\exists x\, A(x) \wedge \forall x\, \forall y\, (A(x) \wedge A(y) \rightarrow (x = y))$	2–19, IP
	QED	1, 20, CP.

Section 8.2

1.
1. $\{odd(x + 1)\}\ y := x + 1\ \{odd(y)\}$ AA
2.   True $\wedge$ even(x) P [for CP]
3.   even(x) 2, Simp
4.   odd$(x + 1)$ 3, T
5.   True $\wedge$ even$(x) \rightarrow$ odd$(x + 1)$ 2–4, CP
6. $\{$True $\wedge$ even$(x)\}\ y := x + 1\ \{odd(y)\}$ 1, 5, Consequence
        QED.

2. a.
1. $\{x + b > 0\}\ y := b\ \{x + y > 0\}$ AA
2. $\{a + b > 0\}\ x := a\ \{x + b > 0\}$ AA
3. $(a > 0) \wedge (b > 0) \rightarrow (a + b > 0)$ T
4. $\{(a > 0) \wedge (b > 0)\}\ x := a\ \{x + b > 0\}$ 2, 3, Consequence
        QED   1, 4, Composition.

3. Use the composition rule (8.12) applied to a sequence of three statements.

a.
1. $\{temp < x\}\ y := temp\ \{y < x\ \}$ AA
2. $\{temp < y\}\ x := y\ \{temp < x\}$ AA
3. $\{x < y\}\ temp := x\ \{temp < y\}$ AA
        QED   3, 2, 1, Composition.

4. a. First, prove $\{(x < 10) \wedge (x \geq 5)\}\ x := 4\ \{x < 5\}$:

1. $\{4 < 5\}\ x := 4\ \{x < 5\}$ AA
2. $(x < 10) \wedge (x \geq 5) \rightarrow (4 < 5)$ T
3. $\{(x < 10) \wedge (x \geq 5)\}\ x := 4\ \{x < 5\}$ 1, 2, Consequence
        QED.

Second, prove that $(x < 10) \wedge \neg\, (x \geq 5) \rightarrow (x < 5)$. This is a valid wff because of the equivalence $\neg\, (x \geq 5) \equiv x < 5$. Thus the original wff is correct, by the if-then rule.

c. First, prove $\{$True $\wedge\ (x < y)\}\ x := y\ \{x \geq y\}$:

1. $\{y \geq y\}\ x := y\ \{x \geq y\}$ AA
2.   True $\wedge\ (x < y) \rightarrow (y \geq y)$ T
3. $\{$True $\wedge\ (x < y)\}\ x := y\ \{x \geq y\}$ 1, 2, Consequence
    QED.

Second, prove that True $\wedge\ \neg\ (x < y) \rightarrow (x \geq y)$. This is a valid wff because of the equivalence True $\wedge\ \neg\ (x < y) \equiv \neg\ (x < y) \equiv (x \geq y)$. Thus the original wff is correct, by the if-then rule.

5. a. Use the if-then-else rule. Thus we must prove the two statements

$$\{\text{True} \wedge (x < y)\}\ \max := y\ \{(\max \geq x) \wedge (\max \geq y)\}$$
$$\{\text{True} \wedge (x \geq y)\}\ \max := x\ \{(\max \geq x) \wedge (\max \geq y)\}.$$

For example, the first statement can be proved as follows:

1. $\{(y \geq x) \wedge (y \geq y)\}\ \max := y\ \{(\max \geq x) \wedge (\max \geq y)\}$ AA
2.   True $\wedge\ (x < y)$ P [for CP]
3. $x < y$ 2, Simp
4. $x \leq y$ 3, Add
5. $y \geq y$ T
6. $(y \geq x) \wedge (y \geq y)$ 4, 5, Conj
7.   True $\wedge\ (x < y) \rightarrow (y \geq x) \wedge (y \geq y)$ 2–6, CP
8. $\{$True $\wedge\ (x < y)\}\ \max := y$
        $\{(\max \geq x) \wedge (\max \geq y)\}$ 1, 7, Consequence
    QED.

6. a. The wff is incorrect if $x = 1$.

7. Since the wff fits the form of the while rule, we need to prove the following statement:

$$\{(x \geq y) \wedge \text{even}(x - y) \wedge (x \neq y)\}x := x - 1; y := y + 1\{(x \geq y) \wedge \text{even}(x - y)\}.$$

Proof:

1. $\{(x \geq y + 1) \wedge \text{even}\ (x - y - 1)\}$
        $y := y + 1\ \{(x \geq y) \wedge \text{even}\ (x - y)\}$ AA
2. $\{(x - 1 \geq y + 1) \wedge \text{even}\ (x - 1 - y - 1)\}$
        $x := x - 1\ \{(x \geq y + 1) \wedge \text{even}\ (x - y - 1)\}$ AA
3. $(x \geq y) \wedge \text{even}(x - y) \wedge (x \neq y)$ P [for CP]
4. $x \geq y + 2$ 3, T
5. $x - 1 \geq y + 1$ 4, T

6.	$\text{even}(x - 1 - y - 1)$	3, T
7.	$(x - 1 \geq y + 1) \wedge \text{even}(x - 1 - y - 1)$	5, 6, Conj
8.	$(x \geq y) \wedge \text{even}(x - y) \wedge (x \neq y)$	

$$\quad\quad\quad \rightarrow (x - 1 \geq y + 1) \wedge \text{even}(x - 1 - y - 1) \quad\quad \text{3–7, CP}$$

$$\quad\quad \text{QED} \quad\quad\quad\quad\quad\quad\quad\quad\quad\quad\quad\quad\quad\quad\quad\quad\quad 1, 2, 8,$$

Consequence,

Composition.

Now the result follows from the while rule.

8. a. The postcondition $i = \text{floor}(x)$ is equivalent to $(i \leq x) \wedge (x < i + 1)$. This statement has the form $Q \wedge \neg C$, where C is the condition of the while loop and Q is the suggested loop invariant. To show that the while loop is correct with respect to Q, show that $\{Q \wedge C\}\ i := i + 1\ \{Q\}$ is correct. Once this is done, show that $\{x \geq 0\}\ i := 0\ \{Q\}$ is correct.

c. The given wff fits the form of the if-then-else rule. Therefore, we need to prove the following two wffs:

$$\{\text{True} \wedge (x \geq 0)\}\ S_1\ \{i = \text{floor}(x)\} \text{ and } \{\text{True} \wedge (x < 0)\}\ S_2\ \{i = \text{floor}(x)\}.$$

These two wffs are equivalent to the two wffs of Parts (a) and (b). Therefore, the given wff is correct.

9. Let Q be the suggested loop invariant. Then the postcondition is implied by $Q \wedge \neg C$, where C is the while loop condition. Therefore, the program can be proven correct by proving the validity of the following two wffs:

$$\{Q \wedge C\}\ i := i + 1;\ s := s + i\ \{Q\} \text{ and } \{n \geq 0\}\ i := 0;\ s := 0;\ \{Q\}.$$

11. Letting Q denote the loop invariant, the while loop can be proved correct with respect to Q by proving the following wff:

$$\{Q \wedge (x \neq y)\}\ \textbf{if}\ x > y\ \textbf{then}\ x := x - y\ \textbf{else}\ y := y - x\ \{Q\}.$$

The parts of the program before and after the while loop can be proved correct by proving the following two wffs:

$$\{(a > 0) \wedge (b > 0)\}\ x := a;\ y := b\ \{Q\},$$
$$\{Q \wedge \neg (x \neq y)\}\ \text{great} := x\ \{\text{gcd}(a, b) = \text{great}\}.$$

13. a. $\{(\textbf{if}\ j = i - 1\ \textbf{then}\ 24\ \textbf{else}\ a[j]) = 24\}.$

c. We obtain the precondition

$$\{((\textbf{if}\ i = j - 1\ \textbf{then}\ 12\ \textbf{else}\ (\textbf{if}\ i = i + 1\ \textbf{then}\ 25\ \textbf{else}\ a[i])) = 12)$$
$$\wedge\ ((\textbf{if}\ j = j - 1\ \textbf{then}\ 12\ \textbf{else}\ (\textbf{if}\ j = i + 1\ \textbf{then}\ 25\ \textbf{else}\ a[j])) = 25)\}.$$

Since it is impossible to have $i = i + 1$ and $j = j - 1$, the precondition can be simplified to

$$\{((\textbf{if}\ i = j - 1\ \textbf{then}\ 12\ \textbf{else}\ a[i]) = 12) \wedge ((\textbf{if}\ j = i + 1\ \textbf{then}\ 25\ \textbf{else}\ a[j]) = 25)\}.$$

14. a. 1. $\{(\text{if } j = i - 1 \text{ then } 24 \text{ else a}[j]) = 24\}$
$\qquad$ $a[i - 1] := 24 \; \{a[j] = 24\}$ $\qquad\qquad$ AAA

$\qquad$ 2. $\qquad$ $(i = j + 1) \wedge (a[j] = 39)$ $\qquad\qquad$ P [for PC]

$\qquad$ 3. $\qquad$ $i = j + 1$ $\qquad\qquad\qquad\qquad$ 2, Simp

$\qquad$ 4. $\qquad$ $24 = 24$ $\qquad\qquad\qquad\qquad$ T

$\qquad$ 5. $\qquad$ $(i = j + 1) \rightarrow (24 = 24)$ $\qquad$ T [trivial]

$\qquad$ 6. $\qquad$ $(i \neq j + 1) \rightarrow (a[j] = 24)$ $\qquad$ $3, T$ [vacuous]

$\qquad$ 7. $\qquad$ $(\text{if } j = i - 1 \text{ then } 24 \text{ else a}[j]) = 24$ $\quad$ 5, 6, Conj, T

$\qquad$ 8. $\;$ $(i = j + 1) \wedge (a[j] = 39) \rightarrow$
$\qquad\qquad$ $((\text{if } j = i - 1 \text{ then } 24 \text{ else a}[j]) = 24)$ $\quad$ 2–7, CP

$\qquad$ QED $\qquad\qquad\qquad\qquad\qquad\qquad$ 1, 8, Consequence.

c. 1. $\{((\text{if } i = j - 1 \text{ then } 12 \text{ else } a[i]) = 12) \wedge$
$\qquad$ $((\text{if } j = j - 1 \text{ then } 12 \text{ else } a[j]) = 25)\}$
$\qquad$ $a[j - 1] := 12$
$\qquad$ $\{(a[i] = 12) \wedge (a[j] = 25)\}$ $\qquad\qquad\qquad$ AAA

$\qquad$ 2. $\{((\text{if } i = j - 1 \text{ then } 12 \text{ else } a[i]) = 12) \wedge (a[j] = 25)\}$
$\qquad$ $a[j - 1] := 12$
$\qquad$ $\{(a[i] = 12) \wedge (a[j] = 25)\}$ $\qquad\qquad\qquad$ 1, T

$\qquad$ 3. $\{((\text{if } i = j - 1 \text{ then } 12 \text{ else }$
$\qquad\qquad$ $(\text{if } i = i + 1 \text{ then } 25 \text{ else } a[i])) = 12)$
$\qquad$ $\wedge ((\text{if } j = i + 1 \text{ then } 25 \text{ else } a[j]) = 25)\}$
$\qquad$ $a[i + 1] := 25$
$\qquad$ $\{((\text{if } i = j - 1 \text{ then } 12 \text{ else } a[i]) = 12) \wedge (a[j]) = 25)\}$ $\quad$ AAA

$\qquad$ 4. $\{((\text{if } i = j - 1 \text{ then } 12 \text{ else } a[i]) = 12)$
$\qquad\qquad$ $\wedge ((\text{if } j = i + 1 \text{ then } 25 \text{ else } a[j]) = 25)\}$
$\qquad$ $a[i + 1] := 25$
$\qquad$ $\{((\text{if } i = j - 1 \text{ then } 12 \text{ else } a[i]) = 12) \wedge (a[j]) = 25)\}$ $\quad$ 3, T

$\qquad$ 5. $\qquad$ $(i = j - 1) \wedge (a[i] = 25) \wedge (a[j] = 12)$ $\qquad$ P [for CP]

$\qquad$ 6. $\qquad$ $i = j - 1$ $\qquad\qquad\qquad\qquad\qquad\qquad$ 5, Simp

$\qquad$ 7. $\qquad$ $((\text{if } i = j - 1 \text{ then } 12 \text{ else } a[i]) = 12)$
$\qquad\qquad$ $\wedge ((\text{if } j = i + 1 \text{ then } 25 \text{ else } a[j]) = 25)$ $\qquad$ 6, T

$\qquad$ 8. $(i = j - 1) \wedge (a[i] = 25) \wedge (a[j] = 12)$
$\qquad$ $\rightarrow ((\text{if } i = j - 1 \text{ then } 12 \text{ else } a[i]) = 12)$
$\qquad$ $\wedge ((\text{if } j = i + 1 \text{ then } 25 \text{ else } a[j]) = 25)$ $\qquad$ 5–7, CP

$\qquad$ QED $\qquad\qquad\qquad\qquad\qquad\qquad\qquad$ 2, 4, 8
$\qquad\qquad\qquad\qquad\qquad\qquad\qquad\qquad\qquad$ Consequence,
$\qquad\qquad\qquad\qquad\qquad\qquad\qquad\qquad\qquad$ Composition.

15. a. After applying AAA to the postcondition and assignment, we obtain the condition even($a[i] + 1$). It is clear that the precondition even($a[i]$) does not imply even($a[i] + 1$). **c.** After applying AAA twice to the postcondition and two assignments, we obtain the condition

$$\forall j \ ((1 \le j \le 5) \to (\text{if } j = 3 \text{ then } 355 \text{ else } a[j]) = 23).$$

This wff is the conjunction of five propositions, one for each j, where $1 \le j \le 5$. For $j = 3$ we obtain the proposition

$$((1 \le 3 \le 5) \to (\text{if } 3 = 3 \text{ then } 355 \text{ else } a[3]) = 23),$$

which is equivalent to the false statement $(1 \le 3 \le 5) \to (355 = 23)$. Therefore, the given precondition cannot imply the obtained condition.

16. a. Define $f(i, x) = x - i$. If $s = (i, x)$, then after the execution of the loop body the state will be $t = (i + 1, x)$. Thus $f(s) = x - i$ and $f(t) = x - i - 1$. To prove termination, assume P and C are true and prove that $f(s)$, $f(t) \in \mathbb{N}$ and $f(s) > f(t)$. So assume int(i) $\land$ (int(x)) $\land$ $i \le x$ and $i < x$. It follows that i and x are integers and $i < x$. So $x - i$ is a positive integer and $x - i - 1$ is a nonnegative integer. In other words, both $x - i$ and $x - i - 1$ are natural numbers, which tells us that $f(s)$, $f(t) \in \mathbb{N}$. Since subtraction by 1 yields a smaller number we have $x - i > x - i - 1$, so that $f(s) > f(t)$. Therefore, the loop terminates.

c. Define $f(x) = |\,x\,|$. If $s = x$, then after the execution of the loop body the state will be $t = x/2$. So $f(s) = |\,x\,|$ and $f(t) = |\,x/2\,|$. To prove termination, assume P and C are true and prove that $f(s)$, $f(t) \in \mathbb{N}$ and $f(s) > f(t)$. So assume int(x) and even(x) $\land$ $x \ne 0$. It follows that x is a nonzero even integer. Since x is even, it is divisible by 2. So $x/2$ is still an integer. Thus $|\,x\,|$ and $|\,x/2\,|$ are both natural numbers, so we have $f(s)$, $f(t) \in \mathbb{N}$. Since x is nonzero it follows that $|\,x\,| > |\,x/2\,|$, so that $f(s) > f(t)$. Therefore, the loop terminates.

17. a. We are given that $f(x, y) = x + y$ and $W = \mathbb{N}$. To prove termination, assume P and C are true and prove that $f(s)$, $f(t) \in \mathbb{N}$ and $f(s) > f(t)$. So assume pos(x) $\land$ pos(y) and $x \ne y$. If $s = (x, y)$, then the state after the execution of the loop body will depend on whether $x < y$. If $x < y$, then $t = (x, y - x)$, which gives $f(t) = y$. Otherwise, if $x > y$, then $t = (x - y, y)$, which gives $f(t) = x$. Since x and y are positive integers, it follows that both $x + y$ and x are natural numbers and $x + y > x$ and $x + y > y$. So in either case (i.e., $x < y$ or $x > y$) we have $f(s)$, $f(t) \in \mathbb{N}$ and $f(s) > f(t)$. Therefore, the loop terminates.

c. We are given that $f(x, y) = (x, y)$ and $W = \mathbb{N} \times \mathbb{N}$ with the lexicographic ordering. To prove termination, assume P and C are true and prove that $f(s)$, $f(t) \in \mathbb{N}$ and $f(s) > f(t)$. So assume pos(x) $\wedge$ pos(y) and $x \neq y$. If $s = (x, y)$, then the state t after the execution of the loop body has two possible values. If $x < y$, then $t = (x, y - x)$, so it follows that $f(s)$, $f(t) \in W$ and we also have

$$f(s) = f(x, y) = (x, y) \succ (x, y - x) = f(x, y - x) = f(t).$$

If $x > y$, then $t = (x - y, y)$, so it follows that $f(s)$, $f(t) \in W$ and we also have

$$f(s) = f(x, y) = (x, y) \succ (x - y, y) = f(x - y, y) = f(t).$$

Therefore, the loop terminates.

18. a. The definition $f(x, y) = \mid x - y \mid$ cannot be used because there are state values s and t such that $f(s) \leq f(t)$, which is contrary to the need in (8.20) for $f(s) > f(t)$. For example, if $s = (x, y) = (10, 13)$, then $f(s) = 3$. But after the body of the loop executes, we have $t = (x, y - x) = (10, 3)$, which gives $f(t) = 7$.

19. Let P be the loop invariant, $P = \exists x \ (a = xb + r) \wedge (0 \leq r)$. The post-condition $r = a \bmod b$ means that r is the remainder obtained on division of a by b, where $0 \leq r \leq b$. This is exactly the condition $P \wedge \neg (r \geq b)$ which is needed for the end of the while loop. So the proof of partial correctness follows by composition from the correctness of the following two statements:

1. $\{(a \geq 0) \wedge (b > 0)\} \ r := a \ \{P\}$.

2. $\{P\}$ **while** $r \geq b$ **do** $r := r - b$ **od** $\{P \wedge \neg (r \geq b)\}$.

Proof of Statement 1:

1.	$\{ \exists x \ (a = xb + a) \wedge (0 \leq a)\} \ r := a \ \{P\}$	AA
2.	$(a \geq 0) \wedge (b > 0)$	P [for CP]
3.	$a = (0)b + a$	T
4.	$\exists x \ (a = xb + a)$	3, EG
5.	$a \geq 0$	2, Simp
6.	$\exists x \ (a = xb + a) \wedge (0 \leq a)$	4, 5, Conj
7.	$(a \geq 0) \wedge (b > 0) \rightarrow \exists x \ (a = xb + a) \wedge (0 \leq a)$	2–6, CP
8.	$\{(a \geq 0) \wedge (b > 0)\} \ r := a \ \{P\}$	1, 7, Consequence
	QED.	

Proof of Statement 2:

1.	$\{\exists x \ (a = xb + r - b) \wedge (0 \le r - b)\} \ r := r - b \ \{P\}$	AA
2.	$\exists x \ (a = xb + r) \wedge (0 \le a) \wedge (r \ge b)$	P [for CP]
3.	$\exists x \ (a = xb + r)$	2, Simp
4.	$a = qb + r$	3, EI
5.	$a = (q + 1)b + r - b$	4, T
6.	$\exists x \ (a = xb + r - b)$	5, EG
7.	$r \ge b$	2, Simp
8.	$0 \le r - b$	7, T
9.	$\exists x \ (a = xb + r - b) \wedge (0 \le r - b)$	6, 8, Conj
10.	$\exists x \ (a = xb + r) \wedge (0 \le a) \wedge (r \ge b)$	
	$\quad\quad \to \exists x \ (a = xb + r - b) \wedge (0 \le r - b)$	2–9, CP
11.	$\{P \wedge (r \ge b)\} \ r := r - b \ \{P \ \}$	1, 10,
		Consequence
12.	$\{P\}$ **while** $r \ge b$ **do** $r := r - b$ **od** $\{P \wedge \neg \ (\ r \ge b)\}$	11, While-rule
	QED.	

Proof of Termination:

Let $W = \mathbb{N}$ with the usual ordering and let $f(a, b, r) = r$. If $s = (a, b, r)$, then the state t after the execution of the loop body is $t = (a, b, r - b)$. To prove termination, assume P and C are true and prove that $f(s), f(t) \in \mathbb{N}$ and $f(s) > f(t)$. So assume $\exists x \ (a = xb + r) \wedge (0 \le r)$ and $(r \ge b)$. It follows that $r \ge 0$ and also $r - b \ge 0$, so we have $f(s), f(t) \in \mathbb{N}$. Since $b > 0$, it follows that $f(s) > f(t)$. Therefore, the loop terminates.

Section 8.3

1. a. Second. **c.** Fifth. **e.** Third. **g.** Third. **i.** Fourth.

2. a. $\exists A \ \exists B \ \forall x \ \neg \ (A(x) \wedge B(x))$.

3. a. Let S be state and C be city. Then $\forall S \ \exists C \ (S(C) \wedge (C = \text{Springfield}))$. The wff is second order.
c. Let H, R, S, B, and A mean house, room, shelf, book, and author. Then $\exists H$ $\exists R \ \exists S \ \exists B \ (H(R) \wedge R(S) \wedge S(B) \wedge A(B, \text{Thoreau}))$. The wff is fourth order.
e. The statement can be expressed as follows:
$\exists S \ \exists A \ \exists B \ (\forall x \ (A(x) \vee B(x) \ \to \ S(x)) \wedge \forall x \ (S(x) \ \to \ A(x) \vee B(x)) \wedge$
$\forall x \ \neg \ (A(x) \wedge B(x)))$. The wff is second order.

5. $\forall R \ (B(R) \ \to \ (\forall x \ \neg \ R(x, x) \wedge \forall x \ \forall y \ \forall z \ (R(x, y) \wedge R(y, z) \ \to \ R(x, z))$
$\to \ \forall x \ \forall y \ (R(x, y) \ \to \ \neg \ R(y, x))))$.

7. Think of $S(x)$ as $x \in S$. **a.** For any domain D the antecedent is false because S can be the empty set. Thus the wff is true for all domains.
c. For any domain D the consequent is true because S can be chosen as D. Thus the wff is true for all domains.

8. Informal proof: Let I be an interpretation with domain D. Then the wff has the following meaning with respect to I. For every subset P of D there is a subset Q of D such that $x \in Q$ implies $x \in P$. This statement is true because we can choose Q to be P. So I is a model for the wff. Since I was an arbitrary interpretation, it follows that the wff is valid. QED.

In the formal proof, we'll represent instantiations of the variables P and Q with p and q, respectively.

1.	$\neg \forall P \, \exists Q \, \forall x \, (Q(x) \rightarrow P(x))$	P [for $\forall P \, \exists Q \, \forall x (Q(x) \rightarrow P(x))$]
2.	$\exists P \, \forall Q \, \exists x \, (Q(x) \wedge \neg P(x))$	1, T
3.	$\forall Q \, \exists x \, (Q(x) \wedge \neg p(x))$	2, EI
4.	$\exists x \, (p(x) \wedge \neg p(x))$	3, UI
5.	$p(c) \wedge \neg p(c)$	4, EI
6.	False	5, Contr
	QED	1–6, IP.

9. a. Assume that the statement is false. Then there is some line L containing every point. Now Axiom 4 says that there are three distinct points not on the same line. This is a contradiction. Thus the statement is true.

c. Let w be a point. By Axiom 4 there is another point x such that $x \neq w$. By Axiom 1 there is a line L on x and w. By Part (a) there is a point z not on L. By Axiom 1 there is a line M on w and z. Since z is on M and z is not on L, it follows that $L \neq M$.

10. Here are some sample formalizations.

a. $\forall L \, \exists x \, \neg \, L(x)$.

Proof:			
	1.	$\neg \forall L \, \exists \, x \, \neg \, L(x)$	P [for $\forall L \, \exists x \, \neg \, L(x)$]
	2.	$\exists L \, \forall x \, L(x)$	1, T
	3.	$\forall x \, l(x)$	2, EI
	4.	Axiom 4	
	5.	$l(a) \wedge l(b) \rightarrow \neg \, l(c)$	4, EI, EI, EI, Simp, UI
	6.	$l(a) \wedge l(b)$	3, UI, UI, Conj
	7.	$\neg \, l(c)$	5, 6, MP
	8.	$l(c)$	3, UI
	9.	False	7, 8, Contr
		QED	1–9, IP.

c. $\forall x \, \exists L \, \exists M \, (L\,(x) \wedge M\,(x) \wedge \exists y \, (\neg \, L\,(y) \wedge M\,(y)))$.

Proof:

1.	$\neg (\forall x \ \exists L \ \exists M \ (L(x) \wedge M(x) \wedge \exists y \ (\neg L(y) \wedge M(y))))$	P [for IP]
2.	$\exists x \ \forall L \ \forall M \ (\neg (L(x) \wedge M(x)) \vee \forall y \ (L(y) \vee \neg M(y)))$	$1, T$
3.	$\forall L \ \forall M \ (\neg (L(a) \wedge M(a)) \vee \forall y \ (L(y) \vee \neg M(y)))$	$2, EI$
4.	$(b \neq c) \wedge (b \neq d) \wedge (c \neq d)$	
	$\quad \wedge \ \forall L \ (L(b) \wedge L(c) \rightarrow \neg L(d))$	Axiom 4, EI, EI, EI
5.	$\qquad a = b$	P [for $a \neq b$]
6.	$\qquad b \neq c$	$4, Simp$
7.	$\qquad a \neq c$	$5, 6, EE$
8.	$\qquad (a \neq c) \rightarrow \exists L \ (L(a) \wedge L(c))$	Axiom 1, UI, UI
9.	$\qquad \exists L \ (L(a) \wedge L(c))$	$7, 8, MP$
10.	$\qquad l(a) \wedge l(c)$	$9, EI$
11.	$\qquad b \neq d$	$4, Simp$
12.	$\qquad a \neq d$	$5, 11, EE$
13.	$\qquad (a \neq d) \rightarrow \exists L \ (L(a) \wedge L(d))$	Axiom 1, UI, UI
14.	$\qquad \exists L \ (L(a) \wedge L(d))$	$12, 13, MP$
15.	$\qquad m(a) \wedge m(d)$	$14, EI$
16.	$\qquad \neg (l(a) \wedge m(a)) \vee \forall y \ (l(y) \vee \neg m(y))$	$3, UI, UI$
17.	$\qquad l(a) \wedge m(a)$	$10, 15, Simp, Conj$
18.	$\qquad \forall y \ (l(y) \vee \neg m(y))$	$16, 17, DS$
19.	$\qquad l(d) \vee \neg m(d)$	$18, UI$
20.	$\qquad l(d)$	$15, Simp, 19, DS$
21.	$\qquad \forall L \ (L(b) \wedge L(c) \rightarrow \neg L(d))$	$4, Simp$
22.	$\qquad l(b) \wedge l(c) \rightarrow \neg l(d)$	$21, UI$
23.	$\qquad l(b) \wedge l(c)$	$5, 10, EE$
24.	$\qquad \neg l(d)$	$22, 23, MP$
25.	$\qquad$ False	$20, 24, Contr$
26.	$a \neq b$	$5\text{--}25, IP$
27.	$a \neq c$	T [like $a \neq b$]
28.	$a \neq d$	T [like $a \neq b$]
29.	$(a \neq b) \rightarrow \exists L \ (L(a) \wedge L(b))$	Axiom 1, UI, UI

30.	$\exists L\,(L\,(a) \wedge L\,(b))$	26, 29, MP
31.	$l\,(a) \wedge l\,(b)$	30, EI
32.	$(a \neq c) \rightarrow \exists L\,(L\,(a) \wedge L\,(c))$	Axiom 1, UI, UI
33.	$\exists L\,(L\,(a) \wedge L\,(c))$	27, 32, MP
34.	$m\,(a) \wedge m\,(c)$	33, EI
35.	$(a \neq d) \rightarrow \exists L\,(L\,(a) \wedge L\,(d))$	Axiom 1, UI, UI
36.	$\exists L\,(L\,(a) \wedge L\,(d))$	28, 35, MP
37.	$n\,(a) \wedge n\,(d)$	36, EI
38.	$l\,(a) \wedge m\,(a)$	31, Simp, 34, Simp, Conj
39.	$\neg\,(l\,(a) \wedge m\,(a)) \vee \forall y\,(l\,(y) \vee \neg\, m\,(y))$	3, UI, UI
40.	$\forall y\,(l\,(y) \vee \neg\, m\,(y))$	38, 39, DS
41.	$l\,(c) \vee \neg\, m\,(c)$	40, UI
42.	$l\,(c)$	34, Simp, 41, DS
43.	$l\,(a) \wedge n\,(a)$	31, Simp, 37, Simp, Conj
44.	$\neg\,(l\,(a) \wedge n\,(a)) \vee \forall y\,(l\,(y) \vee \neg\, n\,(y))$	3, UI, UI
45.	$\forall y\,(l\,(y) \vee \neg\, n\,(y))$	43, 44, DS
46.	$l\,(d) \vee \neg\, n\,(d)$	45, UI
47.	$l\,(d)$	37, Simp, 46, DS
48.	$l\,(b) \wedge l\,(c)$	31, Simp, 42, Conj
49.	$\forall L\,(L\,(b) \wedge L\,(c) \rightarrow \neg\, L\,(d))$	4, Simp
50.	$l\,(b) \wedge l\,(c) \rightarrow \neg\, l\,(d)$	49, UI
51.	$\neg\, l\,(d)$	48, 50, MP
52.	False	47, 51, Contr
	QED	1–4, 26–52, IP.

Chapter 9

Section 9.1

1. a. $(A \vee C \vee D) \wedge (B \vee C \vee D)$.
c. $\forall x\,(\neg\, p(x, c) \vee q(x))$.
e. $\forall x\,\forall y\,(p(x, y) \vee q(x, y, f(x, y)))$.

2. $p \vee \neg\, p$ and $p \vee \neg\, p \vee q \vee q$.

3. a.

1.	$A \vee B$	P
2.	$\neg\, A$	P
3.	$\neg\, B \vee C$	P
4.	$\neg\, C$	P
5.	B	1, 2, R
6.	$\neg\, B$	3, 4, R
7.	$\square$	5, 6, R. QED.

c.
	1.	$A \lor B$	P
	2.	$A \lor \neg C$	P
	3.	$\neg A \lor C$	P
	4.	$\neg A \lor \neg B$	P
	5.	$C \lor \neg B$	P
	6.	$\neg C \lor B$	P
	7.	$B \lor C$	1, 3, R
	8.	$B \lor B$	6, 7, R
	9.	$\neg A$	4, 8, R
	10.	$\neg C$	2, 9, R
	11.	$\neg B$	5, 10, R
	12.	A	1, 11, R
	13.	$\square$	9, 12, R. QED.

4. a. $\{y/x\}$. **c.** $\{y/a\}$. **e.** $\{x/f(a),\ y/f(b),\ z/b\}$.

5. a. $\{x/f(a,\ b),\ v/f(y,\ a),\ z/y\}$ or $\{x/f(a,\ b),\ v/f(z,\ a),\ y/z\}$.
c. $\{x/g(a),\ z/g(b),\ y/b\}$.

6. a. $\{x/f(a,\ b),\ v/f(z,\ a),\ y/z\}$. **c.** $\{x/g(a),\ z/g(b),\ y/b\}$.

7. Make sure the clauses to be resolved have distinct sets of variables. The answers are $p(x) \lor \neg\, p(f(a))$ and $p(x) \lor \neg\, p(f(a)) \lor q(x) \lor q(f(a))$.

8. a.
	1.	$p(x)$	P
	2.	$q(y,\ a) \lor \neg\, p(a)$	P
	3.	$\neg\, q(a,\ a)$	P
	4.	$\neg\, p(a)$	2, 3, R, $\{y/a\}$
	5.	$\square$	1, 4, R, $\{x/a\}$. QED.

c.
	1.	$p(a) \lor p(x)$	P
	2.	$\neg\, p(a) \lor \neg\, p(y)$	P
	3.	$\square$	1, 2, R, $\{x/a,\ y/a\}$. QED.

e. Number the clauses 1, 2, and 3. Resolve 2 with 3 by unifying all four of the p atoms to obtain the clause $\neg\, q(a) \lor \neg\, q(a)$. Resolve this clause with 1 to obtain the empty clause.

9. a. After negating the statement and putting the result in clausal form, we obtain the following proof:

1.	$A \lor B$	P
2.	$\neg A$	P
3.	$\neg B$	P
4.	B	1, 2, R
5.	$\square$	3, 4, R. QED.

c. After negating the statement and putting the result in clausal form, we obtain the following proof:

1.	$p \lor q$	P
2.	$\neg q \lor r$	P
3.	$\neg r \lor s$	P
4.	$\neg p$	P
5.	$\neg s$	P
6.	$\neg r$	3, 5, R
7.	$\neg q$	2, 6, R
8.	q	1, 4, R
9.	$\square$	7, 8, R. QED.

10. a. After negating the statement and putting the result in clausal form, we obtain the following proof:

1.	$p(x)$	P
2.	$\neg p(y)$	P
3.	$\square$	1, 2, R, $\{x/y\}$. QED.

c. After negating the statement and putting the result in clausal form, we obtain the following proof:

1.	$p(x, a)$	P
2.	$\neg p(b, y)$	P
3.	$\square$	1, 2, R, $\{x/b, y/a\}$. QED.

e. After negating the statement and putting the result in clausal form, we obtain the following proof:

1.	$p(x) \lor q(y)$	P
2.	$\neg p(a)$	P
3.	$\neg q(a)$	P
4.	$q(y)$	1, 2, R, $\{x/a\}$
5.	$\square$	3, 4, R, $\{y/a\}$. QED.

11. a. We need to show that $x(\theta\sigma) = (x\theta)\sigma$ for each variable x in E. First, suppose $x/t \in \theta$ for some term t. If $x = t\sigma$, then $x(\theta\sigma) = x$ because the binding $x/t\sigma$ has been removed from $\theta\sigma$. But since $x/t \in \theta$, it follows that $x\theta = t$. Now apply σ to both sides to obtain $(x\theta)\sigma = t\sigma = x$. Therefore, $x(\theta\sigma) = x = (x\theta)\sigma$. If $x \neq t\sigma$, then $x(\theta\sigma) = t\sigma = (x\theta)\sigma$. Second, suppose that $x/t \in \sigma$ and x does not occur as a numerator of θ. Then $x(\theta\sigma) = t = x\sigma = (x\theta)\sigma$. Lastly, if x does not occur as a numerator of either σ or θ, then the substitutions have no effect on x. Thus $x(\theta\sigma) = x = (x\theta)\sigma$.

c. If $x/t \in \theta$, then $x/t = x/t\epsilon$, so it follows from the definition of composition that $\theta = \theta\epsilon$. For any variable x we have $x(\epsilon\theta) = (x\epsilon)\theta = x\theta$. Therefore, $\theta\epsilon = \epsilon\theta = \theta$.

e. $(A \cup B)\theta = \{E\theta \mid E \in A \cup B\} = \{E\theta \mid E \in A\} \cup \{E\theta \mid E \in B\} = A\theta \cup B\theta$.

12. a. In first-order predicate calculus the argument can be written as

$$\forall x \ (C(x) \rightarrow P(x)) \land \exists x \ (C(x) \land L(x)) \rightarrow \exists \ x \ (P(x) \land L(x)),$$

where $C(x)$ means that x is a computer science major, $P(x)$ means that x is a person, and $L(x)$ means that x is a logical thinker. After negating the wff and transforming the result into clausal form, we obtain the proof:

1. $\neg \ C(x) \lor P(x)$ P
2. $C(a)$ P
3. $L(a)$ P
4. $\neg \ P(z) \lor \neg \ L(z)$ P
5. $\neg \ P(a)$ 3, 4, R, $\{z/a\}$
6. $\neg \ C(a)$ 1, 5, R, $\{x/a\}$
7. $\square$ 2, 6, R, $\{ \ \}$. QED.

13. a. Let $D(x)$ mean that x is a dog, $L(x)$ mean that x likes people, $H(x)$ mean that x hates cats, and $a =$ Rover. Then the argument can be formalized as follows:

$$\forall x \ (D(x) \rightarrow L(x) \lor H(x)) \land D(a) \land \neg \ H(a) \rightarrow \exists x \ (D(x) \land L(x)).$$

After negating the wff and transforming the result into clausal form, we obtain the proof:

1. $\neg \ D(x) \lor L(x) \lor H(x)$ P
2. $D(a)$ P
3. $\neg \ H(a)$ P
4. $\neg \ D(y) \lor \neg \ L(y)$ P
5. $L(a) \lor H(a)$ 1, 2, R, $\{x/a\}$
6. $L(a)$ 3, 5, R, $\{ \ \}$
7. $\neg \ D(a)$ 4, 6, R, $\{y/a\}$
8. $\square$ 2, 7, R, $\{ \ \}$. QED.

c. Let $H(x)$ mean that x is a human being, $Q(x)$ mean that x is a quadruped, and $M(x)$ mean that x is a man. Then the argument can be formalized as

$$\forall x \ (H(x) \rightarrow \neg \ Q(x)) \land \forall x \ (M(x) \rightarrow H(x)) \rightarrow \forall x \ (M(x) \rightarrow \neg \ Q(x)).$$

After negating the wff and transforming the result into clausal form, we obtain the proof:

1. $\neg H(x) \lor \neg Q(x)$ P
2. $\neg M(y) \lor H(y)$ P
3. $M(a)$ P
4. $Q(a)$ P
5. $H(a)$ 2, 3, R, $\{y/a\}$
6. $\neg Q(a)$ 1, 5, R, $\{x/a\}$
7. $\square$ 4, 6, R, $\{\ \}$. QED.

e. Let $F(x)$ mean that x is a freshman, $S(x)$ mean that x is a sophomore, $J(x)$ mean that x is a junior, and $L(x, y)$ mean that x likes y. Then the argument can be formalized as $A \rightarrow B$, where

$$A = \exists x \ (F(x) \land \forall y \ (S(y) \rightarrow L(x, y))) \land \forall x \ (F(x) \rightarrow \forall y \ (J(y) \rightarrow \neg L(x, y)))$$

and $B = \forall x \ (S(x) \rightarrow \neg J(x))$. After negating the wff and transforming the result into clausal form, we obtain the proof:

1. $F(a)$ P
2. $\neg S(x) \lor L(a, x)$ P
3. $\neg F(y) \lor \neg J(z) \lor \neg L(y, z)$ P
4. $S(b)$ P
5. $J(b)$ P
6. $\neg J(z) \lor \neg L(a, z)$ 1, 3, R, $\{y/a\}$
7. $\neg L(a, b)$ 5, 6, R, $\{z/b\}$
8. $\neg S(b)$ 2, 7, R, $\{x/b\}$
9. $\square$ 4, 8, R, $\{\ \}$. QED.

14. a. Here is an indirect proof that W is valid.

1. $\neg \forall x \ \exists y \ (p(x, y) \lor \neg p(y, y))$ P [for IP]
2. $\exists x \ \forall y \ (\neg p(x, y) \land p(y, y))$ 1, T
3. $\forall y \ (\neg p(c, y) \land p(y, y))$ 2, EI
4. $\neg p(c, c) \land p(c, c)$ 3, UI
5.   False   4, Contr
    QED   1–5, IP.

15. a. Let I be an interpretation for W. If C is true for I, then $\exists y \ (p(y) \land \neg C)$ is false, so W is false. If C is false for I, then W becomes $(\exists x \ p(x) \rightarrow \text{False}) \land \exists y \ (p(y) \land \neg \text{False}) \equiv \neg \exists x \ p(x) \land \exists y \ p(y)$, which is false. Therefore, W is false for I. Since I was arbitrary, W is unsatisfiable.

c. After eliminating $\rightarrow$ from W, we apply Skolem's rule to obtain the wff $(\neg\, p(a) \vee C) \wedge (p(b) \wedge \neg\, C)$. Define an interpretation for this wff by letting $C = $ False, $p(a) = $ False, and $p(b) = $ True. This interpretation makes the wff true. So it is satisfiable.

Section 9.2

1. a. isChildOf$(x,\ y) \leftarrow$ isParentOf$(y,\ x)$.
c. isGreatGrandParentOf$(x,\ y)$
    $\leftarrow$ isParentOf$(x,\ w)$, isParentOf$(w,\ z)$, isParentOf$(z,\ y)$.

2. a. The following definition will work if $x \neq y$:

$$\text{isSiblingOf}(x,\ y) \leftarrow \text{isParentOf}(z,\ x),\ \text{isParentOf}(z,\ y).$$

c. Let s denote isSecondCousinOf. One possible definition is

$$s(x,\ y) \leftarrow \text{isParentOf}(z,\ x),\ \text{isParentOf}(w,\ y),\ \text{isCousinOf}(z,\ w).$$

3. a.

1.	$p(a,\ b)$	P
2.	$p(a,\ c)$	P
3.	$p(b,\ d)$	P
4.	$p(c,\ e)$	P
5.	$g(x,\ y) \leftarrow p(x,\ z),\ p(z,\ y)$	P
6.	$\leftarrow g(a,\ w)$	P [initial goal]
7.	$\leftarrow p(a,\ z),\ p(z,\ y)$	5, 6, R, $\theta_1 = \{x/a,\ w/y\}$.
8.	$\leftarrow p(b,\ y)$	1, 7, R, $\theta_2 = \{z/b\}$
9.	$\square$	3, 8, R, $\theta_3 = \{y/d\}$. QED.

b.

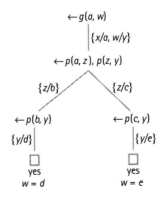

4. a.

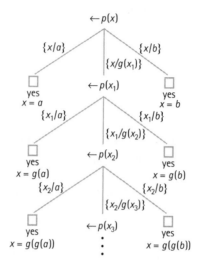

c. $\{g^n(a) \mid n \in \mathbb{N}\}$.

5. a. The program returns the answer yes.

6. a. The symmetric closure s can be defined by the following two classes:

$$s(x, y) \leftarrow r(x, y).$$
$$s(x, y) \leftarrow r(y, x).$$

7. a. fib(0, 0).

fib(1, 1).

fib(x, $y + z$) $\leftarrow$ fib($x - 1$, y), fib($x - 2$, z).

c. pnodes($\langle \, \rangle$, 0).

pnodes($\langle L, a, R \rangle$, $1 + x + y$) $\leftarrow$ pnodes(L, x), pnodes(R, y).

8. a. equalLists($\langle \, \rangle$, $\langle \, \rangle$).

equalLists($x :: t$, $x :: s$) $\leftarrow$ equalLists(t, s).

c. all(x, $\langle \, \rangle$, $\langle \, \rangle$).

all(x, $x :: t$, u) $\leftarrow$ all(x, t, u)

all(x, $y :: t$, $y :: u$) $\leftarrow$ all(x, t, u).

e. subset($\langle \, \rangle$, y).

subset($x :: t$, y) $\leftarrow$ member(x, y), subset(t, y).

g. Using the "remove" predicate from Example 9.26, which removes one occurrence of an element from a list, the program to test for a subbag can be written as follows:

subBag($\langle \, \rangle$, y).

subBag($x :: t$, y) $\leftarrow$ member(x, y), remove(x, y, w), subBag(t, w).

9. Let the predicate schedule(L, S) mean that S is a schedule for the list of classes L. For example, if $L = \langle$english102, math200$\rangle$, then S is a list of 4-tuples of the form (name, section, time, place). For the example, S might look like the following:

$$\langle(\text{english102, 2, 3pm, ivy238}), (\text{math200, 1, 10am, briar315})\rangle.$$

Assume that the available classes are listed as facts of the following form:

$$\text{class(name, section, time, place)}.$$

The following solution will yield one schedule of classes that might contain time conflicts. All schedules can be found by backtracking. If a class cannot be found, a note is made to that effect.

> schedule($\langle\ \rangle, \langle\ \rangle$).
> schedule($x :: y,\ S$) $\leftarrow$ class(x, Sect, Time, Place),
>   schedule($y,\ T$),
>   cat($\langle(x$, Sect, Time, Place)$\rangle,\ T,\ S$).
> schedule($x :: y,\ \langle\text{unfillable}\rangle$).

10. Let letters($A,\ L$) mean that L is the list of propositional letters that occur in the wff A. Let replace(p, true, $A,\ B$) mean $B = A(p/\text{True})$. Then we can start the process for a wff A with the goal $\leftarrow$ tautology(A, Answer), where A is a tautology if Answer = true. The initial definitions might go like the following, where uppercase letters denote variables:

> tautology(A, Answer) $\leftarrow$ letters($A,\ L$), evaluate($A,\ L$, Answer).
> evaluate($A,\ \langle\ \rangle$, Answer) $\leftarrow$ value(A, Answer).
> evaluate($A,\ H :: T$, Answer) $\leftarrow$ replace(H, true, $A,\ B$),
>   replace(H, false, $A,\ C$),
>   evaluate($B \wedge C$, Answer).

When "value" is called, A is a proposition containing only true and false terms. The definition for the "replace" predicate might include some clauses like the following:

> replace(X, true, X, true).
> replace(X, true, $\neg\ X$, false).
> replace(X, true, $\neg\ A,\ \neg\ B$) $\leftarrow$ replace(X, true, $A,\ B$).
> replace(X, true, $A \wedge X,\ B$) $\leftarrow$ replace(X, true, $A,\ B$).
> replace(X, true, $X \wedge A,\ B$) $\leftarrow$ replace(X, true, $A,\ B$).
> replace(X, true, $A \wedge B,\ C \wedge D$) $\leftarrow$ replace(X, true, $A,\ C$),
>   replace(X, true, $B,\ D$).

Continue by writing the clauses for the false case and for the other operators $\vee$ and $\rightarrow$. The first few clauses for the "value" predicate might include some clauses like the following:

value(true, true).
value(false, false).
value(¬ true, false).
value(¬ false, true).
value(¬ X, Y) ← value(X, A), value (¬ A, Y).
value(false ∧ X, false).
value(X ∧ false, false).
value(true ∧ X, Y) ← value(X, Y).
value(X ∧ true, Y) ← value(X, Y).
value(X ∧ Y, Z) ← value(X, U), value(Y, V), value(U ∧ V, Z).

Continue by writing the clauses to find the value of expressions containing the operators ∨ and →. The predicate to construct the list of propositional letters in a wff might start off something like the following:

letters(X, $\langle X \rangle$) ← atom(X).
letters(X ∧ Y, Z) ← letters(X, U), letters(Y, V), cat(U, V, Z).

Continue by writing the clauses for the other operations.

Chapter 10

Section 10.1

1. The zero is m because $\min(x, m) = \min(m, x) = m$ for all $x \in A$. The identity is n because $\min(x, n) = \min(n, x) = x$ for all $x \in A$. If $x, y \in A$ and $\min(x, y) = n$, then x and y are inverses of each other. Since n is the largest element of A, it follows that n is the only element with an inverse.

2. a. No; no; no. **c.** True; False; False is its own inverse.

3. $S = \{a, f(a), f^2(a), f^3(a), f^4(a)\}$.

4. a. An element z is a zero if both row z and column z contain only the element z. **c.** If x is an identity, then an element y has a right and left inverse w if x occurs in row y column w and also in row w column y of the table.

5. a.

∘	a	b	c	d
a	a	b	c	d
b	b	c	d	d
c	c	d	b	b
d	d	a	b	c

Notice that $d \circ b = a$, but $b \circ d \neq a$. So b and d have one-sided inverses but not inverses (two-sided).

c.

∘	a	b	c	d
a	a	a	a	a
b	a	c	d	b
c	a	d	a	b
d	a	a	b	c

Notice that $(b \circ b) \circ c = c \circ c = a$ and $b \circ (b \circ c) = b \circ d = b$. Therefore, $\circ$ is not associative.

e.

∘	a	b	c	d
a	a	b	c	d
b	b	a	a	a
c	c	a	a	a
d	d	a	a	a

Notice that $(b \circ b) \circ c = a \circ c = c$ and $b \circ (b \circ c) = b \circ a = b$. Therefore, $\circ$ is not associative.

6. a.

∘	a	b
a	a	b
b	b	a

c.

∘	a	b
a	b	b
b	b	b

7. Suppose the elements of table T are numbers $1, \ldots, n$. Check the equation $T(i, T(j, k)) = T(T(i, j), k)$ for all values of i, j, and k between 1 and n.

8. a. $f(g(x)) = f(f(f(x))) = g(f(x))$.

c. For example, a, $f(a)$, $g(a)$, $f(g(a))$.

9. a. $y = y \circ e = y \circ (x \circ x^{-1}) = (y \circ x) \circ x^{-1} = (z \circ x) \circ x^{-1} = z \circ (x \circ x^{-1}) = z \circ e = z$.

Section 10.2

1. No. Notice that $A \cdot B \neq B \cdot A$ because $A - B \neq B - A$. Similarly, 0 is not an identity for $+$, and 1 is not an identity for $\cdot$.

2. a. $x + x\,y = x(1 + y) = x\,1 = x$.
c. $x + \overline{x}y = (x + \overline{x})(x + y) = 1\,(x + y) = x + y$.

3. $\overline{e} = \overline{\overline{yz} + y\overline{z}} = \overline{\overline{yz}}\,\overline{y\overline{z}} = (y + \overline{z})(\overline{y} + z) = \overline{y}\,\overline{z} + yz$.

4. a. $\overline{x} + \overline{y} + xyz = \overline{xy} + (xy)z = \overline{xy} + z = \overline{x} + \overline{y} + z$.

5. a. $x + y$. **c.** $\overline{y}(x + z)$. **e.** $x\,y + z$. **g.** $x + y$. **i.** 1. **k.** $x + y$.

6. a.

c.

7. a. $x0$. **c.** $(x + y)(x + z)$. **e.** $y(\bar{x} + z)$.

9. Show that $\bar{x} + \bar{y}$ acts like the complement of xy. In other words, show that

$$(xy) + (\bar{x} + \bar{y}) = 1 \text{ and } (xy)(\bar{x} + \bar{y}) = 0.$$

The result then follows from (10.8). For the first equation we have

$$(xy) + (\bar{x} + \bar{y}) = (x + \bar{x} + \bar{y})(y + \bar{x} + \bar{y}) = (1 + \bar{y})(1 + \bar{x}) = 1 \cdot 1 = 1.$$

For the second equation we have

$$(xy)(\bar{x} + \bar{y}) = xy\bar{x} + xy\bar{y} = 0 + 0 = 0.$$

11. a. Since $x = xx$, we have $x \preceq x$. So $\preceq$ is reflexive. If $x \preceq y$ and $y \preceq x$, then $x = xy$ and $y = yx$. Therefore, $x = xy = yx = y$. Thus $\preceq$ is antisymmetric. If $x \preceq y$ and $y \preceq z$, then $x = xy$ and $y = yz$. Therefore, $x = xy = x(yz) = (xy)z = xz$. So $x \preceq z$. Thus $\preceq$ is transitive.

13. Since p occurs more than once in the factorization of n, it follows that n/p still contains at least one factor of p. For example, if $n = p^2q$, then $n/p = pq$. So $\text{lcm}(p, n/p) = n/p$, which is not equal to n (the unit of the algebra). Similarly, $\gcd(p, n/p) = p$, which is not 1 (the zero of the algebra). So properties of Part 3 of the definition of a Boolean algebra fail to hold.

Section 10.3

1. $\text{monus}(x, 0) = x$, $\text{monus}(0, y) = 0$, $\text{monus}(s(x), s(y)) = \text{monus}(x, y)$, where $s(x)$ denotes the successor of x.

3. $\text{reverse}(L) = \text{if isEmptyL}(L) \text{ then } L$
        $\text{else cat}(\text{reverse}(\text{tail}(L)), \langle\text{head}(L)\rangle)$.

5. Let $\text{genlists}(A)$ denote the set of general lists over A. The operations for general lists are similar to those for lists. The main difference is that the cons function and the head function have the following types to reflect the general nature of elements in a list.

$$\text{cons: } A \cup \text{genlists}(A) \times \text{genlists}(A) \rightarrow \text{genlists}(A),$$
$$\text{head: genlists}(A) \rightarrow \text{genlists}(A) \cup A.$$

The axioms are identical to those for lists.

7. post($\langle 4, 5, -, 2, + \rangle$, $\langle \; \rangle$) = post($\langle 5, -, 2, + \rangle$, $\langle 4 \rangle$) = post($\langle -, 2, + \rangle$, $\langle 5, 4 \rangle$) = post($\langle 2, + \rangle$, eval($-$, $\langle 5, 4 \rangle$)) = post($\langle 2, + \rangle$, $\langle -1 \rangle$) = post($\langle + \rangle$, $\langle 2, -1 \rangle$) = post($\langle \; \rangle$, eval($+$, $\langle 2, -1 \rangle$)) = post($\langle \; \rangle$, $\langle 1 \rangle$) = 1.

9. In equational form we have preorder(emptyTree) = emptyQ, and preorder(tree(L, x, R)) = apQ(addQ(x, emptyQ), apQ(preorder(L), preorder(R))).

11. Remove (all occurrences of an element from a stack).

13. Let D be the set of deques over the set A. Then the carriers should be A, D, and Boolean. The operators can be defined as

$$\text{emptyD} \in D,$$
$$\text{isEmptyD: } D \to \text{Boolean},$$
$$\text{addLeft: } A \times D \to D,$$
$$\text{addRight: } D \times A \to D,$$
$$\text{left: } D \to A,$$
$$\text{right: } D \to A,$$
$$\text{deleLeft: } D \to D,$$
$$\text{deleRight: } D \to D.$$

With axioms:

$$\text{isEmptyD(emptyD)} = \text{True},$$
$$\text{isEmptyD(addLeft}(a, d)) = \text{isEmptyD(addRight}(d, a)) = \text{False},$$
$$\text{left(addLeft}(a, d)) = \text{right(addRight}(d, a)) = a,$$
$$\text{left(addRight}(d, a)) = \text{if isEmptyD}(d) \text{ then } a$$
$$\qquad\qquad\qquad \text{else left}(d),$$
$$\text{right(addLeft}(a, d)) = \text{if isEmptyD}(d) \text{ then } a$$
$$\qquad\qquad\qquad \text{else right}(d),$$
$$\text{deleLeft(addLeft}(a, d)) = \text{deleRight(addRight}(d, a)) = d,$$
$$\text{deleLeft(addRight}(d, a)) = \text{if isEmptyD}(d) \text{ then emptyD}$$
$$\qquad\qquad\qquad \text{else addRight(deleLeft}(d), a),$$
$$\text{deleRight(addLeft}(a, d)) = \text{if isEmptyD}(d) \text{ then emptyD}$$
$$\qquad\qquad\qquad \text{else addLeft}(a, \text{deleRight}(d)).$$

15. Let $\quad Q[A] = D[A]$,

$$\text{emptyQ} = \text{emptyD},$$
$$\text{isEmptyQ} = \text{isEmptyD},$$
$$\text{frontQ} = \text{left},$$
$$\text{deleQ} = \text{deleLeft},$$
$$\text{addQ}(a, q) = \text{addRight}(q, a).$$

Then the axioms are proved as follows:

$$\text{isEmptyQ(emptyQ)} = \text{isEmptyD(emptyD)} = \text{True.}$$
$$\text{isEmptyQ(addQ}(a,\ q)) = \text{isEmptyD(addRight}(q,\ a)) = \text{False.}$$
$$\text{frontQ(addQ}(a,\ q)) = \text{left(addRight}(q,\ a))$$
$$= \text{if isEmptyD}(q) \text{ then } a \text{ else left}(q)$$
$$= \text{if isEmptyQ}(q) \text{ then } a \text{ else frontQ}(q).$$
$$\text{delQ(addQ}(a,\ q)) = \text{deleLeft(addRight}(q,\ a))$$
$$= \text{if isEmptyD}(q) \text{ then emptyD}$$
$$\text{else addRight(deleLeft}(q),\ a)$$
$$= \text{if isEmptyQ}(q) \text{ then emptyQ}$$
$$\text{else addQ}(a,\ \text{deleQ}(q)).$$

17. a. Let $P(x)$ denote the statement "plus$(x,\ s(y)) = s(\text{plus}(x,\ y))$ for all $y \in \mathbb{N}$." Certainly $P(0)$ is true because plus$(0,\ s(y)) = s(y) = s(\text{plus}(0,\ y))$. So assume that $P(x)$ is true, and prove that $P(s(x))$ is true. We can evaluate each expression in the statement of $P(s(x))$ as follows:

$$\text{plus}(s(x),\ s(y)) = s(\text{plus}(p(s(x)),\ s(y))) \quad \text{(by definition of plus)}$$
$$= s(\text{plus}(x,\ s(y))) \quad \text{(since } p(s(x)) = x)$$
$$= s(s(\text{plus}(x,\ y))) \quad \text{(by induction),}$$

and

$$s(\text{plus}(s(x),\ y)) = s(s(\text{plus}(p(s(x)),\ y))) \quad \text{(by definition of plus)}$$
$$= s(s(\text{plus}(x,\ y))) \quad \text{(since } p(s(x)) = x).$$

Both expressions are equal. So $P(s(x))$ is true.

18. Induction will be with respect to the length of y. We'll use the notation $y : a$ for addQ$(a,\ y)$. For the basis case we have the following equations, where $y = \text{emptyQ}$:

apQ$(x, \text{emptyQ} : a)$
$$= \text{apQ}(x : \text{front(emptyQ} : a),\ \text{delQ(emptyQ} : a)) \text{ (def. of apQ)}$$
$$= \text{apQ}(x : a,\ \text{emptyQ}) \qquad\qquad\qquad \text{(simplify)}$$
$$= x : a \qquad\qquad\qquad\qquad\qquad\qquad \text{(simplify)}$$
$$= \text{apQ}(x : a,\ \text{emptyQ})) \qquad\qquad\quad \text{(def. of apQ).}$$

For the induction case, assume that the equation is true for all queues y having length n, and show that the equation is true for the queue $y : b$, having length

$n + 1$. Starting with the left side of the equation, we have

$\text{apQ}(x, y:b:a)$

$\qquad = \text{apQ}(x:\text{front}(y:b:a), \text{delQ}(y:b:a)) \qquad$ (def of apQ)

$\qquad = \text{apQ}(x:\text{front}(y:b), \text{delQ}(y:b:a)) \qquad (\text{front}(y:b:a) = \text{front}(y:b))$

$\qquad = \text{apQ}(x:\text{front}(y:b), \text{delQ}(y:b):a) \qquad (\text{delQ}(y:b:a) = \text{delQ}(y:b):a)$

$\qquad = \text{apQ}(x:\text{front}(y:b), \text{delQ}(y:b)):a \qquad$ (induction)

$\qquad = \text{apQ}(x, y:b):a \qquad$ (def of apQ).

Section 10.4

1. a. $\{(1, a, \#, M), (2, a, *, N), (3, a, \%, N)\}$.

c. $\{(1, a, \#, M, x), (1, a, \#, M, z)\}$.

e. $\{(a, M), (b, M)\}$.

2. For example, if we let $R = \{(1, a), (2, b)\}$ and $S = \{(1, a)\}$, then join(R, S) $= \{(1, a)\}$ and $R \cup S = \{(1, a), (2, b)\}$.

3. a. project(Channel, {Station, Cable}).

c. project(select(select(Rooms, Computer, Yes), BoardType, White), {Place}).

e. select(join(Channel, Program), Station, ESPN).

4. a. $t \in \text{select}_{A=a}(\text{select}_{B=b}(R))$ iff $t \in \text{select}_{B=b}(R)$ and $t(A) = a$ iff $t \in R$ and $t(B) = b$ and $t(A) = a$ iff $t \in \text{select}_{A=a}(R)$ and $t(B) = b$ iff $t \in \text{select}_{B=b}(\text{select}_{A=a}(R))$.

c. Let I, J, and K be the attribute sets for R, S, and T, respectively. Use the definition of join to show that $u \in (R \bowtie S) \bowtie T$ iff there exist $r \in R$ and $s \in S$ and $t \in T$ such that $u(a) = r(a)$ for all $a \in I$ and $u(a) = s(a)$ for all $a \in J$ and $u(a) = t(a)$ for all $a \in K$ iff $u \in R \bowtie (S \bowtie T)$.

5. $s \in \text{project}_X(\text{select}_{A=a}(R))$ iff there exists $t \in \text{select}_{A=a}(R)$ such that $s(B) = t(B)$ for all $B \in X$ iff there exists $t \in R$ such that $t(A) = a$ and $s(B) = t(B)$ for all $B \in X$ iff $s \in \text{project}_X(R)$ and $s(A) = a$ iff $s \in \text{select}_{A=a}(\text{project}_X(R))$.

6. a. Let $f = \text{seqPairs}$, where

seqPairs = eq0 $\rightarrow \sim ((0, 0))$; apndr @ [seqPairs @ sub1, [id, id]].

7. a. For any pair of numbers (m, n), all three expressions compute the value of the expression $m + n$.

8. If c returns 0, then $* @ [a, g @ [b, c]] = * @ [a, b] = g @ [* @ [a, b], c]$, which proves the basis case. Now assume that c returns a positive number and (10.12) holds for sub1 @ c. We'll prove that (10.12) holds for c as follows, starting with the left side:

$* @ [a, g @ [b, c]] = * @ [a, (\text{eq0} @ 2 \rightarrow 1; g @ [*, sub1 @ 2]) @ [b, c]] \qquad$ (def of g)

$\qquad\qquad = * @ [a, g @ [* @ [b, c], sub1 @ c]] \qquad\qquad (\text{eq0} @ c = \text{False})$

$\qquad\qquad = g @ [* @ [a, * @ [b, c]], sub1 @ c] \qquad\qquad$ (induction).

Now look at the right side:

$$g @ [* @ [a, b], c] = g @ [*, \text{sub1} @ 2] @ [* @ [a, b], c] \qquad \text{(def of } g)$$
$$= g @ [* @ [* @ [a, b], c], \text{sub1} @ c].$$

It follows that the two sides are equal because multiplication is associative:

$$* @ [a, * @ [b, c]] = * @ [* @ [a, b], c].$$

Section 10.5

1.

+	[0]	[1]	[2]	[3]
[0]	[0]	[1]	[2]	[3]
[1]	[1]	[2]	[3]	[0]
[2]	[2]	[3]	[0]	[1]
[3]	[3]	[0]	[1]	[2]

·	[0]	[1]	[2]	[3]
[0]	[0]	[0]	[0]	[0]
[1]	[0]	[1]	[2]	[3]
[2]	[0]	[2]	[0]	[2]
[3]	[0]	[3]	[2]	[1]

2. a. $x = 99$. **c.** $x = 59$. **e.** 21.

3. Add multiples of n to x until the sum is positive. In other words, there is some k such that $x + kn > 0$. Set $y = x + kn$. It follows that $y \equiv x \pmod{n}$.

5. a. e = 11 works. **c.** e = 317 works.

7. a. Yes.

c. No. $4 +_9 6 = 1 \notin \{0, 2, 4, 6, 8\}$.

8. a. $\{0, 6\}$.

c. $\mathbb{N}_{12}$.

9. The three morphisms are f, g, and h, where: f is the zero function; $g(0) = 0$, $g(1) = 2$, $g(2) = 4$; $h(0) = 0$, $h(1) = 4$, $h(2) = 2$.

11. Notice that $\text{abs}(1 + (-1)) = \text{abs}(0) = 0$, but that $\text{abs}(1) + \text{abs}(-1) = 1 + 1 = 2$. So in general $\text{abs}(x + y) \neq \text{abs}(x) + \text{abs}(y)$.

13. a. $\{(ab)^n b \mid n \in \mathbb{N}\}$. **c.** $\varnothing$. **e.** $\{ba^n \mid n \in \mathbb{N}\}$.

Chapter 11

Section 11.1

1. a. $\{a, b\}$.

c. $\{a, \Lambda, b, bb, \ldots, b^n, \ldots\}$.

e. $\{a, b, ab, bc, abb, bcc, \ldots, ab^n, bc^n, \ldots\}$.

2. a. $a + b + c$. **c.** $ab^* + ba^*$. **e.** $\Lambda + a(bb)^*$. **g.** $\Lambda + c^*a + bc^*$. **i.** a^*bc^*.

3. $0 + 1(0 + 1)^*$.

4. a. $(aa + ab + ba + bb)^*$. **c.** $(a + b)^*aba(a + b)^*$.

5. a. $(ab)^*$. **c.** $a(a + b)^*$.

6. a.
$$b + ab^* + aa^*b + aa^*ab^* = b + ab^* + aa^*(b + ab^*)$$
$$= (\Lambda + aa^*)(b + ab^*)$$
$$= a^*(b + ab^*) \qquad \text{(by (11.1), property 5)}.$$

c. By using property 7 of (11.1) the subexpression $(a + bb^*a)^*$ of the left side can be written $(a^*bb^*a)^*a^*$. So the left expression has the following form:

$$ab^*a(a + bb^*a)^*b = ab^*a(a^*bb^*a)^*a^*b.$$

Similarly, the subexpression $(b + aa^*b)^*$ of the right side of the original equation can be written as $b^*(aa^*bb^*)^*$. So the right expression has the following form:

$$a(b + aa^*b)^*aa^*b = ab^*(aa^*bb^*)^*aa^*b.$$

So we'll be done if we can show that $ab^*a(a^*bb^*a)^*a^*b = ab^*(aa^*bb^*)^*aa^*b$. Since both expressions have ab^* on the left end and a^*b on the right end, it suffices to show that

$$a(a^*bb^*a)^* = (aa^*bb^*)^*a.$$

But this equation is just an instance of property 8 of (11.1).

7. The proofs follow from corresponding properties of languages given in Chapter 1. See, for example, properties (1.16) and Exercises 17 and 18 of Section 1.3.

8. a. $\varnothing$. **c.** $\varnothing$.

9. a. $(b + ab)^*(\Lambda + a)$. **c.** $(b + ab + aab + aaab)^*(\Lambda + a + aa + aaa)$.

11. a. Let $X = R^*S$. Then we can use properties 2, 3, and 5 of (11.1) to write the right side of the equation $X = RX + S$ as follows:

$$RX + S = R(R^*S) + S = RR^*S + \Lambda S = (RR^* + \Lambda)S = R^*S = X.$$

c. *Hint*: Assume that A and B are two solutions to the equation so that we have $A = RA + S$ and $B = RB + S$. Try a proof by contradiction by assuming that $A \neq B$. Then there is some string in one of $L(A)$ and $L(B)$ that is not in the other. Say w is the shortest string in $L(A) - L(B)$. Then $w \notin L(S)$ because if it were, then it would also be in $L(B)$. Thus $w \in L(RA)$. Now argue toward a contradiction.

Section 11.2

1. $T(0, a) = T(2, a) = 1$, $T(0, b) = 0$, $T(1, a) = T(2, b) = T(3, b) = 3$, $T(1, b) = T(3, a) = 2$, where 0 is the start and both 2 and 3 are final states.

2. a. States 0 (start), 1(final), and 2. $T(0, a) = T(0, b) = 1$ and all other transitions go to state 2.

c. States 0, 1, 2, 3, with start state 0 and final states 0, 1, and 2. $T(0, a) = 1$, $T(0, b) = 2$, $T(2, b) = 2$, and all other transitions go to state 3.

e. States 0 (start), 1 (final), 2 (final), and 3. $T(0, a) = T(1, b) = 1$, $T(0, b) = T(2, c) = 2$, and all other transitions go to state 3.

3. States 0 (start), 1, 2, 3, 4 (final), and 5. $T(0, -) = T(0, +) = 1$, $T(0, d)$ $= T(1, d) = T(2, d) = 2$, $T(2, .) = 3$, $T(3, d) = T(4, d) = 4$, and all other transitions go to state 5.

5. a. States 0 (start), 1, 2 (final), 3, and 4 (final). $T(0, a) = \{1\}$, $T(1, c) = \{2\}$, $T(0, \Lambda) = T(3, a) = \{3\}$, $T(3, b) = T(4, c) = \{4\}$, and all other transitions go to $\varnothing$.
c. States 0 (start), 1, 2 (final), and 3. $T(0, a) = \{1\}$, $T(1, b) = \{2\}$, $T(0, \Lambda) = T(3, a) = \{3\}$, $T(3, \Lambda) = \{2\}$, and all other transitions go to $\varnothing$.

6. a.   **c.**

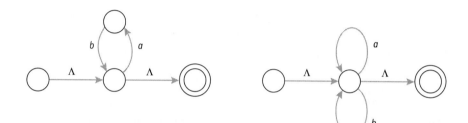

7. The NFA obtained is for $(a + b)^*$.

8. Without any simplification we obtain **a.** $ab^*a(a + bb^*a)^*b$. **c.** These two expressions are shown to be equal in the exercises of Section 11.1.

9. If we apply the algorithm by eliminating states 2, 1, and 0 in that order, then we obtain the expression $(ab)^*(c + a)$.

11. Let the digits 0 and 1 denote the output of the sensors, where 1 means that traffic is present. The strings 00, 01, 10, and 11 represent the four possible pairs of inputs, where the left digit is the left-turn sensor and the right digit is the north-south traffic sensor. Each state outputs a string of length 3 over the letters

$$G \text{ (green)}, R \text{ (red)}, \text{ and } Y \text{ (yellow)}.$$

In a string of length 3, the left letter denotes the left-turn light, the middle letter denotes the east-west light, and the right letter denotes the north-south light. The start state is 0 with output RGR, which gives priority to traffic on the main east-west highway. A Moore machine to model the behavior of the signals can be written as follows:

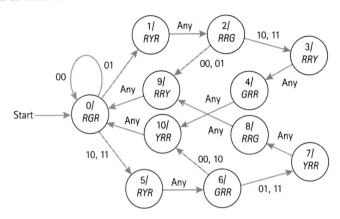

Section 11.3

1. a. The NFA has seven states: 0 (start), 1, 2, 3, 4, 5, and 6 (final). $T(0, \Lambda) = \{1, 3\}$, $T(1, a) = \{2\}$, $T(2, \Lambda) = \{1, 3\}$, $T(3, \Lambda) = \{4, 6\}$, $T(4, b) = \{5\}$, $T(5, \Lambda) = \{4, 6\}$, and all other transitions map to $\varnothing$.
c. The NFA has ten states: 0 (start), 1, 2, 3, 4, 5, 6, 7, 8, and 9 (final). $T(0, \Lambda) = \{1, 5\}$, $T(1, \Lambda) = \{2, 4\}$, $T(2, a) = \{3\}$, $T(3, \Lambda) = \{2, 4\}$, $T(4, \Lambda) = \{9\}$, $T(5, \Lambda) = \{6, 8\}$, $T(6, b) = \{7\}$, $T(7, \Lambda) = \{6, 8\}$, $T(8, \Lambda) = 9$, and all other transitions map to $\varnothing$.

2. The states are $\{0, 1\}$ (start) and $\{1, 2\}$ (final), and all transitions go to state $\{1, 2\}$.

3. a. $ba^* + aba^* + a^*$ over $A = \{a, b\}$.
c. States $\{0, 1, 2\}$ (start and final), $\{1, 2\}$ (final), $\{2\}$ (final), and $\varnothing$. $T_D(\{0, 1, 2\}, a) = \{1, 2\}$, $T_D(\{0, 1, 2\}, b) = \{2\}$, $T_D(\{1, 2\}, a) = T_D(\{1, 2\}, b) = T_D(\{2\}, a) = \{2\}$, where the other transitions go to $\varnothing$.

4. a. States $\{0\}$ (start) and $\{0, 1\}$ (final), where $T_D(\{0\}, a) = T_D(\{0, 1\}, a) = \{0, 1\}$.

5. a. The NFA has seven states: 0 (start), 1, 2, 3, 4, 5, and 6 (final). $T(0, \Lambda) = \{1, 3\}$, $T(1, a) = \{2\}$, $T(2, \Lambda) = \{1, 3\}$, $T(3, \Lambda) = \{4, 6\}$, $T(4, b) = \{5\}$, $T(5, \Lambda) = \{4, 6\}$, and all other transitions map to $\varnothing$. The DFA has four states: 0 (start, final), 1 (final), 2 (final), and 3. $T_D(0, a) = T_D(1, a) = 1$, $T_D(0, b) = T_D(1, b) = T_D(2, b) = 2$, $T_D(2, a) = T_D(3, a) = T_D(3, b) = 3$.
c. The NFA has ten states: 0 (start), 1, 2, 3, 4, 5, 6, 7, 8, and 9 (final). $T(0, \Lambda) = \{1, 5\}$, $T(1, \Lambda) = \{2, 4\}$, $T(2, a) = \{3\}$, $T(3, \Lambda) = \{2, 4\}$, $T(4, \Lambda) = \{9\}$, $T(5, \Lambda) = \{6, 8\}$, $T(6, b) = \{7\}$, $T(7, \Lambda) = \{6, 8\}$, $T(8, \Lambda) = 9$, and all other transitions map to $\varnothing$. The DFA has four states: 0 (start, final), 1 (final),

2 (final), and 3. $T_D(0, a) = T_D(1, a) = 1$, $T_D(0, b) = T_D(2, b) = 2$, $T_D(1, b)$ $= T_D(2, a) = T_D(3, a) = T_D(3, b) = 3$.

7. $T_{\min}([0], a) = T_{\min}([0], b) = T_{\min}([1], b) = [1]$, and $T_{\min}([1], a) = T_{\min}([4], a) = T_{\min}([4], b) = [4]$, where $[0] = \{0\}$, $[1] = \{1, 2, 3\}$, and $[4] = \{4, 5\}$.

8. a. $\{1, 3\}$.

c. $T_{\min}([0], a) = T_{\min}([1], a) = T_{\min}([2], b) = [1]$, $T_{\min}([0], b) = T_{\min}([2], a)$ $= T_{\min}([4], b) = [2]$, $T_{\min}([1], b) = T_{\min}([4], a) = [4]$.

9. a. The equivalence pairs are $\{0, 2\}$, $\{1, 3\}$, $\{1, 4\}$, $\{3, 4\}$. Therefore, the states are $\{0, 2\}$ and $\{1, 3, 4\}$, where $\{0, 2\}$ is the start state and $\{1, 3, 4\}$ is the final state. $T_{\min}([0], a) = T_{\min}([1], a) = [1]$, and $T_{\min}([0], b) = T_{\min}([1], b) = [0]$.

10. a. The NFA has ten states. It transforms into a DFA with two states, both of which are final. The minimum state DFA has the single state 0 (start, final), and $T(0, a) = 0$.

c. See the answer to Exercise 5a for the seven-state NFA and the four-state DFA. The minimum state DFA has three states, 0 (start, final), 1 (final), and 2. $T_{\min}(0, a) = 0$, $T_{\min}(0, b) = T_{\min}(1, b) = 1$, $T_{\min}(1, a) = T_{\min}(2, a) = T_{\min}(2, b) = 2$.

11. $(a + b)^*$.

Section 11.4

1. a. $S \rightarrow a \mid b$.
c. $S \rightarrow a \mid B$, $B \rightarrow \Lambda \mid bB$.
e. $S \rightarrow aB \mid bC$, $B \rightarrow \Lambda \mid bB$, $C \rightarrow \Lambda \mid cC$.
g. $S \rightarrow \Lambda \mid aaS \mid bbS$.
i. $S \rightarrow abS \mid cT$, $T \rightarrow aT \mid bT \mid \Lambda$.

2. a. $S \rightarrow a \mid b \mid c$.
c. $S \rightarrow aB \mid bC$, $B \rightarrow \Lambda \mid bB$, $C \rightarrow \Lambda \mid aC$.
e. $S \rightarrow \Lambda \mid aB$, $B \rightarrow \Lambda \mid bbB$.
g. $S \rightarrow \Lambda \mid a \mid cA \mid bC$, $A \rightarrow a \mid cA$, $C \rightarrow \Lambda \mid cC$.
i. $S \rightarrow aS \mid bC$, $C \rightarrow \Lambda \mid cC$.

3. a. $S \rightarrow \Lambda \mid aaS \mid abS \mid baS \mid bbS$.
c. $S \rightarrow aS \mid bS \mid abaT$, $T \rightarrow \Lambda \mid aT \mid bT$.

4. a. $S \rightarrow a \mid Sab$. **c.** $S \rightarrow Tc \mid Sa \mid Sb$, $T \rightarrow \Lambda \mid Tab$.

5. $S \rightarrow aS \mid bS \mid aI$, $I \rightarrow bJ$, $J \rightarrow bK$, $K \rightarrow \Lambda$.

7. a. Let $L = \{a^n ba^n \mid n \in \mathbb{N}\}$, and suppose that L is regular. Using the pumping lemma (11.13), we'll choose $s = a^m ba^m$. Then s can be written as $s = a^m ba^m = xyz$, where $y \neq \Lambda$ and $\mid xy \mid \leq m$. It follows that $y = a^i$ for some $i > 0$. Then $xy^2 z = a^{m+i} ba^m$, which is not in L. This contradicts the pumping lemma result that $xy^k z \in L$ for all $k \geq 0$. Thus L is not regular.

c. Let $L = \{a^n b^k \mid n, k \in \mathbb{N}$ and $n \leq k\}$ and suppose that L is regular. Using the pumping lemma (11.13) we'll choose $s = a^m b^m$. Then s can be written as $s = a^m b^m = xyz$, where $y \neq \Lambda$ and $\mid xy \mid \leq m$. It follows that $y = a^i$ for some

$i > 0$. Then $xy^2z = a^{m+i}b^m$, which is not in L. This contradicts the pumping lemma result that $xy^kz \in L$ for all $k \geq 0$. Thus L is not regular.

e. Let $L = \{w \mid w \in \{a, b\}^*$ and w has an equal number of a's and b's$\}$ and suppose that L is regular. Using the pumping lemma (11.13) we'll choose $s = a^m b^m$. Then s can be written as $s = a^m b^m = xyz$, where $y \neq \Lambda$ and $\mid xy \mid \leq m$. It follows that $y = a^i$ for some $i > 0$. Then $xy^2z = a^{m+i}b^m$, which is not in L. This contradicts the pumping lemma result that $xy^kz \in L$ for all $k \geq 0$. Thus L is not regular. *Note:* We could also compute xz to obtain a string that is not in L.

8. Let L and M be two regular languages. Let A be the union of the alphabets for L and M. By (11.14)(4) we know that the complements $L' = A^* - L$ and $M' = A^* - M$ are regular languages. By (11.14)(1) the union $L' \cup M'$ is regular. Since $L' \cup M' = (L \cap M)'$, one more application of (11.14)(4) tells us that $L \cap M$ is regular.

9. For each state i and letter $a \in A$, create an edge (i, j) labeled with a in the new DFA if there is a path from i to j in the DFA for L whose labels concatenate to $f(a)$. This new DFA accepts a string w exactly when the original DFA accepts $f(w)$. In other words, the new DFA accepts $f^{-1}(L)$.

10. a. $\{xby \mid x$ and y are strings of length n over $\{a, c\}^*\}$.
c. Let $g(a) = a$, $g(b) = \Lambda$, and $g(c) = b$.

Chapter 12

Section 12.1

1. a. $S \to aSbb \mid \Lambda$.
c. $S \to aSa \mid bSb \mid \Lambda$.
e. $S \to aSa \mid bSb \mid a \mid b \mid \Lambda$.
2. a. $S \to A \mid B$, $A \to aAb \mid \Lambda$, $B \to aBbb \mid \Lambda$.
c. $S \to AS \mid \Lambda$, $A \to aAb \mid \Lambda$.

Section 12.2

1. a. A PDA for $\{ab^n cd^n \mid n \geq 0\}$ has start state 0 and final state 3 with $\perp$ as the starting stack symbol:

$$(0, a, \perp , \text{nop}, 1),$$
$$(1, b, \perp , \text{push}(b), 1),$$
$$(1, b, b, \text{push}(b), 1),$$
$$(1, c, \perp , \text{nop}, 2),$$
$$(1, c, b, \text{nop}, 2),$$
$$(2, d, b, \text{pop}, 2),$$
$$(2, \Lambda, \perp, \text{nop}, 3).$$

c. A PDA for $\{wcw^R \mid w \in \{a, b\}^*\}$ has start state 0 and final state 2 with $\perp$ as the starting stack symbol:

$$(0, a, ?, \text{push}(a), 0),$$
$$(0, b, ?, \text{push}(b), 0),$$
$$(0, c, ?, \text{nop}, 1),$$
$$(1, a, a, \text{pop}, 1),$$
$$(1, b, b, \text{pop}, 1),$$
$$(1, \Lambda, \perp, \text{nop}, 2).$$

e. A PDA for $\{a^n b^{n+2} \mid n \geq 0\}$ has start state 0 and final state 2 with $\perp$ as the starting stack symbol:

$$(0, a, \perp, \text{push}(a), 0),$$
$$(0, a, a, \text{push}(a), 0),$$
$$(0, b, \perp, \text{nop}, 1),$$
$$(0, b, a, \text{nop}, 1),$$
$$(1, b, \perp, \text{nop}, 2),$$
$$(1, b, a, \text{pop}, 1).$$

2. Let 0 be the start and final state and $\perp$ be the starting stack symbol:

$$(0, a, \perp, \text{push}(a), 0),$$
$$(0, a, a, \text{nop}, 0),$$
$$(0, b, \perp, \text{push}(b), 0),$$
$$(0, b, a, \text{push}(b), 0),$$
$$(0, b, b, \text{nop}, 0).$$

3. a. State 0 is the start state, and both states 0 and 3 are final states with $\perp$ as the starting stack symbol:

$$(0, a, \perp, \text{push}(a), 1),$$
$$(1, a, a, \text{push}(a), 1),$$
$$(1, b, a, \text{pop}, 2),$$
$$(2, b, a, \text{pop}, 2),$$
$$(2, \Lambda, \perp, \text{nop}, 3).$$

4. a. Let the states be 0 and 1, where 0 is the start state and the final state with $\perp$ as the starting stack symbol. The PDA can be represented as follows:

$$(0, a, \perp, \langle \text{push}(X), \text{push}(a) \rangle, 1),$$
$$(1, a, a, \langle \text{pop}, \text{push}(X), \text{push}(a) \rangle, 1),$$
$$(1, b, a, \langle \text{pop}, \text{pop} \rangle, 1),$$
$$(1, b, X, \text{pop}, 1),$$
$$(1, \Lambda, \perp, \text{push}(X), 0).$$

5. Create a new start state s and a new empty stack state e. Then add the following instructions to the PDA of Example 12.1:

$$(s, \Lambda, Y, \text{push}(X), 0),$$
$$(2, \Lambda, X, \text{pop}, e),$$
$$(2, \Lambda, Y, \text{pop}, e),$$
$$(e, \Lambda, X, \text{pop}, e),$$
$$(e, \Lambda, Y, \text{pop}, e).$$

6. Create a new start state s and a new unique final state f. Then add the following instructions to the PDA of Example 12.2:

$$(s, \Lambda, Y, \text{push}(X), 0),$$
$$(0, \Lambda, Y, \text{nop}, f),$$
$$(1, \Lambda, Y, \text{nop}, f).$$

7. a. $(0, a, a, \text{pop}, 0),$
$(0, b, b, \text{pop}, 0),$
$(0, c, c, \text{pop}, 0),$
$(0, \Lambda, S, \langle \text{pop}, \text{push}(c) \rangle, 0),$
$(0, \Lambda, S, \langle \text{pop}, \text{push}(b), \text{push}(S), \text{push}(a) \rangle, 0).$

8. $S \to X_{01}, X_{01} \to a X_{01} X_{11}, X_{11} \to b, X_{01} \to \Lambda.$

9. a. The set of strings over $\{a, b\}$ containing an equal number of a's and b's such that for any letter in a string the number of b's to its left is less than or equal to the number of a's to its left.
b. $S \to X_{00}, X_{00} \to \Lambda \mid a A_{00} X_{00}, A_{00} \to b \mid a A_{00} A_{00}$, which can be simplified to $S \to \Lambda \mid a A S, A \to b \mid a A A.$ **c.** Yes.

Section 12.3

1. a. $S \to a \mid b A, A \to a \mid b a.$ **c.** $S \to a B, B \to a B c \mid b.$

2. An LL(3) grammar: $S \to a A \mid a a B, A \to a A \mid \Lambda, B \to b B \mid \Lambda.$ An LL(2) grammar: $S \to a C, C \to A \mid a B, A \to a A \mid \Lambda, B \to b B \mid \Lambda.$

3. a. $S \to a T, T \to b S \mid \Lambda.$ The grammar is LL(1).

4. a. $S \to c T, T \to a T \mid b T \mid \Lambda.$ The grammar is LL(1).

5. Let the letters E, R, T, V, and F denote procedures. Then define them as

E: call T; call R.
R: **if** lookahead $=$ "+" **then** match("+"); call T; call R **fi**.
T: call F; call V.
V: **if** lookahead $=$ "$*$" **then** match("$*$"); call F; call V **fi**.
F: **if** lookahead $= a$ **then** match(a)
        else if lookahead $=$ "(" **then** match("("); call E; match(")")
        else error
        fi.

6. a. The sentential forms that can occur in a rightmost derivation take the following forms, where $n \geq 0$: aA, ad, $ac^{n+1}A$, $ac^{n+1}d$, bB, bd, $bc^{n+1}B$, and $bc^{n+1}d$. In each case the handle is completely determined without scanning past it. So the grammar is LR(0).

8. The problem is with the handles defined by the productions $A \rightarrow ab$ and $B \rightarrow abb$. For example, the input strings ab and abb need one lookahead symbol beyond ab to determine which handle to use. But the strings $aabb$ and $aabbbb$ need two lookahead symbols beyond ab to determine the appropriate handle. Generalizing from these examples, we see that no fixed number of lookahead symbols can suffice to determine which handle to use.

Section 12.4

1. a. $S \rightarrow aA \mid aBb \mid a \mid ab$, $A \rightarrow aA \mid a$, $B \rightarrow aBb \mid ab$.

2. a. $S \rightarrow AC \mid RD \mid AB$, $T \rightarrow AT \mid b$, $R \rightarrow AE \mid AB$, $A \rightarrow a$, $B \rightarrow b$, $C \rightarrow BT$, $D \rightarrow AB$, $E \rightarrow RB$.
c. $S' \rightarrow AB \mid AC \mid BT \mid b \mid \Lambda$, $S \rightarrow AB \mid AC \mid BT \mid b$, $T \rightarrow BT \mid b$, $A \rightarrow a$, $B \rightarrow b$, $C \rightarrow SB$.

3. a. $S \rightarrow aTRB \mid aRB \mid cB \mid bA$, $A \rightarrow aTR \mid aR \mid c$, $B \rightarrow aT \mid a$, $T \rightarrow bT \mid b$, $R \rightarrow a$.

4. a. Let $z = a^m b^m a^m = uvwxy$. Show that the pumped variables v and x can't contain distinct letters. Then at least one of v and x must have the form a^i or b^i for some $i > 0$. Look at the different cases, and come up with contradictions showing that the pumped string uv^2wx^2y can't be in the language.
c. Let $L = \{a^p \mid p$ is a prime number$\}$, and assume that L is context-free. Let $z = a^p$, where p is a prime number larger than $m + 1$ from (12.13). Let $k = \mid u \mid + \mid w \mid + \mid y \mid$. Since $\mid vwx \mid \leq m$ and $\mid vx \mid \geq 1$, it follows that $k > 1$. For any $i \geq 0$ we have $\mid uv^iwx^iy \mid = k + i(p - k)$. Letting $i = k$, we get the equation $\mid uv^kwx^ky \mid = k + k(p - k) = k(1 + p - k)$, which can't be a prime number. This contradicts the requirement that each pumped string must be in L. Therefore, L is not context-free.

5. a. $\{xb^ny \mid x$ and y are strings of length n over $\{a, c\}^*\}$.
c. If $\{a^n b^n a^n \mid n \in \mathbb{N}\}$ is context-free, then by (12.17), $f^{-1}(\{a^n b^n a^n \mid n \in \mathbb{N}\})$ is context-free. So by (12.16) and Part (b) we must conclude that $\{a^n b^n c^n \mid n \in \mathbb{N}\}$ is context-free. But this is not the case. So $\{a^n b^n a^n \mid n \in \mathbb{N}\}$ can't be context-free.

Chapter 13

Section 13.1

1. Consider the general algorithm that repeatedly cancels the same letter from each end of the input string by replacing its occurrences by Λ. A Turing machine program to accomplish this follows, where the start state is 0.

$(0, a, \Lambda, R, 1)$ Replace a by Λ.

$(0, b, \Lambda, R, 4)$ Replace b by Λ.

$(0, \Lambda, \Lambda, S, \text{Halt})$ It's an even-length palindrome.

$(1, a, a, R, 1)$ Scan right.

$(1, b, b, R, 1)$ Scan right.

$(1, \Lambda, \Lambda, L, 2)$ Found the right end.

$(2, a, \Lambda, L, 3)$ Replace rightmost a by Λ.

$(2, \Lambda, \Lambda, S, \text{Halt})$ It's an odd-length palindrome.

$(3, a, a, L, 3)$ Scan left.

$(3, b, b, L, 3)$ Scan left.

$(3, \Lambda, \Lambda, R, 0)$ Found left end.

$(4, a, a, R, 4)$ Scan right.

$(4, b, b, R, 4)$ Scan right.

$(4, \Lambda, \Lambda, L, 5)$ Found the right end.

$(5, b, \Lambda, L, 3)$ Replace rightmost b by Λ.

$(5, \Lambda, \Lambda, S, \text{Halt})$ It's an odd-length palindrome.

3. This machine will remember the current cell, write Λ, and move right to the state that writes the remembered symbol. The start state is 0.

$(0, a, \Lambda, R, 1)$ Go write an a.

$(0, b, \Lambda, R, 2)$ Go write a b.

$(0, \Lambda, \Lambda, S, \text{Halt})$ Done.

$(1, a, a, R, 1)$ Write an a and go write an a.

$(1, b, a, R, 2)$ Write an a and go write a b.

$(1, \Lambda, a, S, \text{Halt})$ Done.

$(2, a, b, R, 1)$ Write a b and go write an a.

$(2, b, b, R, 2)$ Write a b and go write a b.

$(2, \Lambda, b, S, \text{Halt})$ Done.

4. a. The leftmost string is moved right one cell position, overwriting the # symbol. Then the machine scans left and halts with the tape head at the leftmost cell of the number. A Turing machine program with start state 0 follows.

$(0, 1, \Lambda, R, 1)$ Write Λ and go find #.

$(1, 1, 1, R, 1)$ Scan right.
$(1, \#, 1, L, 2)$ Overwrite # with 1.

$(2, 1, 1, L, 2)$ Scan left.
$(2, \Lambda, \Lambda, R, \text{Halt})$.

5. a. Complement the number while scanning it left to right. Then add 1. The start state is 0. The machine halts immediately when addition is complete.

$(0, 0, 1, R, 0)$ Complement while scanning left to right.
$(0, 1, 0, R, 0)$
$(0, \Lambda, \Lambda, L, 1)$ Go to add 1.

$(1, 0, 1, S, \text{Halt})$ No carry.
$(1, 1, 0, L, 1)$ Carry.
$(1, \Lambda, 1, S, \text{Halt})$ Carry.

c. Assume that the tape head is at the right end of the input string. The machine will overwrite the input string with the answer and halt with the tape head at the left end of the answer. The start state is 0.

Add the first bit:
$(0, 0, 1, L, 1)$ Add $1 + 0$, no carry.
$(0, 1, 0, L, 2)$ Add $1 + 1$, carry.

Add the second bit with no carry:
$(1, 0, 1, L, 4)$ Add $1 + 0$, done with add, move left.
$(1, 1, 0, L, 3)$ Add $1 + 1$, go to carry state.
$(1, \Lambda, 1, S, \text{Halt})$ Add 1, done with add.

Add the second bit with carry:
$(2, 0, 0, L, 3)$ Add $1 + 1 + 0$, go to carry state.
$(2, 1, 1, L, 3)$ Add $1 + 1 + 1$, go to carry state.
$(2, \Lambda, 0, L, 3)$ Add $1 + 1$, go to carry state.

Carry state:
$(3, 0, 1, L, 4)$ Done with add, move to left.
$(3, 1, 0, L, 3)$ Stay in carry state.
$(3, \Lambda, 1, S, \text{Halt})$ Done with add.

Move to left end of number:

$(4, 0, 0, L, 4)$ Move left.

$(4, 1, 1, L, 4)$ Move left.

$(4, \Lambda, \Lambda, R, \text{Halt})$ Done with add.

7. Let 0 be the start state and the "noncarry" state. State 1 will be the "carry" state. The tape head for the output tape will always be positioned at a blank cell. The first four instructions perform the normal add with no carry:

$$(0, (0, 0, \Lambda), (0, 0, 0), (L, L, L), 0)$$
$$(0, (0, 1, \Lambda), (0, 1, 1), (L, L, L), 0)$$
$$(0, (1, 0, \Lambda), (1, 0, 1), (L, L, L), 0)$$
$$(0, (1, 1, \Lambda), (1, 1, 0), (L, L, L), 1) \quad \text{Go to carry state.}$$

The next instructions copy the extra portion of the longer of the two numbers, if necessary:

$$(0, (0, \Lambda, \Lambda), (0, \Lambda, 0), (L, L, L), 0) \qquad \text{1st number longer.}$$
$$(0, (1, \Lambda, \Lambda), (1, \Lambda, 1), (L, L, L), 0)$$
$$(0, (\Lambda, 0, \Lambda), (\Lambda, 0, 0), (L, L, L), 0) \qquad \text{2nd number longer.}$$
$$(0, (\Lambda, 1, \Lambda), (\Lambda, 1, 1), (L, L, L), 0)$$
$$(0, (\Lambda, \Lambda, \Lambda), (\Lambda, \Lambda, \Lambda), (S, S, R), \text{Halt}) \quad \text{Done.}$$

The next four instructions perform the add with carry:

$$(1, (0, 0, \Lambda), (0, 0, 1), (L, L, L), 0) \quad \text{Go to noncarry state.}$$
$$(1, (0, 1, \Lambda), (0, 1, 0), (L, L, L), 1)$$
$$(1, (1, 0, \Lambda), (1, 0, 0), (L, L, L), 1)$$
$$(1, (1, 1, \Lambda), (1, 1, 1), (L, L, L), 1)$$

The next instructions add the carry to the extra portion of the longer of the two numbers, if necessary:

$$(1, (0, \Lambda, \Lambda), (0, \Lambda, 1), (L, L, L), 0) \qquad \text{1st number longer.}$$
$$(1, (1, \Lambda, \Lambda), (1, \Lambda, 0), (L, L, L), 1)$$
$$(1, (\Lambda, 0, \Lambda), (\Lambda, 0, 1), (L, L, L), 0) \qquad \text{2nd number longer.}$$
$$(1, (\Lambda, 1, \Lambda), (\Lambda, 1, 0), (L, L, L), 1)$$
$$(1, (\Lambda, \Lambda, \Lambda), (\Lambda, \Lambda, 1), (S, S, S), \text{Halt}) \quad \text{Done.}$$

9. A busy beaver with three states:

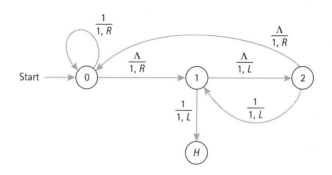

Section 13.2

1. a. $Z := X;\ Z := Z + Y.$
c. $\text{Temp} := X;$ **while** $\text{Temp} \neq 0$ **do** $S;\ \text{Temp} := 0$ **od**.
e. $A := X;\ B := Y;$
   while $B \neq 0$ **do**
      $A := \text{pred}(A);$
      $B := \text{pred}(B)$
   od;
   $Z := A.$
g. Use the fact that "$X \leq Y$" is equivalent to "$X < Y + 1$," which is equivalent to "$(Y + 1)$ monus $X \neq 0$."
i. Use the fact that "$X \neq Y$" is equivalent to "absoluteDiff$(X,\ Y) \neq 0$."
k. $Z := X + Y;$ **while** $Z \neq 0$ **do** $S;\ Z := X + Y$ **od**. A solution that does not use addition can be written as

$$
\begin{aligned}
&A := 1; \\
&\textbf{while } A \neq 0 \textbf{ do} \\
&\quad \textbf{if } X \neq 0 \textbf{ then} \\
&\quad\quad S \\
&\quad \textbf{else if } Y \neq 0 \textbf{ then } S \textbf{ else } A := 0 \textbf{ fi} \\
&\quad \textbf{fi} \\
&\textbf{od}.
\end{aligned}
$$

2. a. LE$(x,\ y)$ can be defined by any of the the following expressions.

$$
\begin{aligned}
\text{LE}(x,\ y) &= \text{monus}(1,\ \text{greater}(x,\ y)). \\
\text{LE}(x,\ y) &= \text{less}(x,\ \text{succ}(y)). \\
\text{LE}(x,\ y) &= \text{sign}(\text{monus}(\text{succ}(y),\ x)).
\end{aligned}
$$

3. a. f is the successor function.
c. $f(x, y) = $ if $x \leq y$ then $y - x$ else undefined.

5. a. a^i. **c.** a^i.

6. a. $a \to a$.
c. $* \to \Lambda$.
e. $Xa \to bX$, $Xb \to aX$, $X \to \Lambda$ (halt), $\Lambda \to X$.

7. a. $a \to a$.
c. $X*Y \to XY$.
e. $aX \to @b\#X$, $bX \to @a\#X$, $@X\#aY \to @Xb\#Y$, $@X\#bY \to @Xa\#Y$, $@X\# \to X$ (halt).

8. a. $X \to aXa$ and $X \to bXb$ with single axiom Λ.
c. $X*Y\#Z \to aX*bY\#cZ$ with axiom $*\#$.

Chapter 14

Section 14.1

1. Since f and g are computable, there are algorithms to compute f and g. An algorithm to compute $h(x)$ consists of running the algorithm for g on input x. If the algorithm halts with value $g(x)$, then run the algorithm for f on the input $g(x)$. If this algorithm halts, then output the value $f(g(x))$. If either algorithm fails to halt, then the algorithm for $h(x)$ fails to halt.

2. Since f_x is one of the computable functions, we can compute $h(x)$ by running the algorithm to compute $f_x(x)$. If the algorithm halts, then we'll return the value $h(x) = 1$. If the algorithm does not halt, then we're still OK, since we want $h(x)$ to run forever.

3. a. g is not computable because $f_n(n)$ may not be defined.
c. g is not computable because if $f_n(n)$ does not halt, there is no way to discover the fact and output the number 4.

5. Given two DFAs over the same alphabet, construct the minimum-state versions of each DFA. If they have a different number of states, then return 0. If they have the same number of states, then check to see whether the states of one table can be renamed to obtain the other table. If so, then return 1. Otherwise, return 0.

6. a. The sequence 1, 2, 2, 2, 2, 2 produces the equality *abbbbbbbbbb* = *abbbbbbbbbb*.
c. There is no solution.
e. The sequence 1, 5, 2, 3, 4, 4, 3, 4 will produce the equality *abababababbabbbabb* = *ababababbabbbabb*.

Section 14.2

1. Let both tapes be empty at the start. The machine starts by printing 1 on the first tape and $\#$ on the second tape. The 1 indicates $n = 1$, and $\#$ separates the empty string Λ from the next string to be printed. Then it executes the following loop forever: Scan 1's to the right printing a's; scan 1's to the left printing b's; scan 1's to the right printing c's; print 1 and # and scan 1's to the left. The start state is 0, and the tapes are initially blank.

$(0, (\Lambda, \Lambda), (1, \#), (S, R), 1)$ Initialize $n = 1$, print #.

$(1, (1, \Lambda), (1, a), (R, R), 1)$ Scan right printing a's.
$(1, (\Lambda, \Lambda), (\Lambda, \Lambda), (L, S), 2)$ Done with a's.

$(2, (1, \Lambda), (1, b), (L, R), 2)$ Scan left printing b's.
$(2, (\Lambda, \Lambda), (\Lambda, \Lambda), (R, S), 3)$ Done with b's.

$(3, (1, \Lambda), (1, c), (R, R), 3)$ Scan right printing c's.
$(3, (\Lambda, \Lambda), (1, \#), (L, R), 4)$ Done with string,
    increment n and print #.

$(4, (1, \Lambda), (1, \Lambda), (L, S), 4)$ Scan left.
$(4, (\Lambda, \Lambda), (\Lambda, \Lambda), (R, S), 1)$ Go print another string.

2. a. $S \rightarrow \Lambda \mid T, T \rightarrow TABC \mid ABC, AB \rightarrow BA, BA \rightarrow AB, BC \rightarrow CB, CB \rightarrow BC, AC \rightarrow CA, CA \rightarrow AC, A \rightarrow a, B \rightarrow b, C \rightarrow c$.

3. a. Modify the answer to Exercise 2a by replacing each of the six productions that permute two nonterminals. For example, replace $AB \rightarrow BA$ with the three productions $AB \rightarrow AX \rightarrow BX \rightarrow BA$, where X is a new nonterminal.

Section 14.3

1. $(u \vee v \vee x_1) \wedge (w \vee \neg x_1 \vee x_2) \wedge (x \vee \neg x_2 \vee x_3) \wedge (y \vee z \vee \neg x_3)$.

3. In the worst case there would be $2n^2 + 3n + 1$ clauses generated. If we numbered the clauses $C_1, C_2, \ldots, C_s$, where $s = 2n^2 + 3n + 1$, then in the worst case each clause C_k would be resolved with each of the clauses $C_1, \ldots, C_{k-1}$. So there would be at most $s(s + 1)/2 = O(n^4)$ resolution steps. Since there are at most $O(n^4)$ resolution steps and at most $O(n^2)$ clauses, it follows that there are at most $O(n^6)$ comparisons to see whether a resolvant is distinct from those clauses already listed.

4. a. $P \neq NP$. **c.** $P = NP$.

6. Any deterministic Turing machine can be thought of as a nondeterministic Turing machine (that always guesses and checks the right solution). So it follows that $DPSPACE \subseteq NPSPACE$. For the other direction, if $L \in NPSPACE$, then $L \in NSPACE(n^i)$ for some i. The given theorem then tells us that $L \in DSPACE(n^{2i})$, which implies that $L \in DPSPACE$. So $NPSPACE \subseteq DPSPACE$. So the two classes are equal.

References

Akra, M., and L. Bazzi, On the solution of linear recurrence equations. *Computational Optimization and Applications* 10 (1998), 195–210.

Appel, K., and W. Haken, Every planar map is four colorable. *Bulletin of the American Mathematical Society* 82 (1976), 711–712.

Appel, K., and W. Haken, The solution of the four-color-map problem. *Scientific American* 237 (1977), 108–121.

Apt, K. R., Ten years of Hoare's logic: A survey—Part 1. *ACM Transactions on Programming Languages and Systems* 3 (1981), 431–483.

Backus, J., Can programming be liberated from the von Neumann style? A functional style and its algebra of programs. *Communications of the ACM* 21 (1978), 613–641.

Bates, G. E., *Probability*. Addison-Wesley, Reading, MA, 1965.

Brady, A. H., The determination of the value of Rado's noncomputable function $\Sigma (k)$ for four-state Turing macines. *Mathematics of Computation 40* (1983), 647–665.

Chang, C., and R. C. Lee, *Symbolic Logic and Mechanical Theorem Proving*. Academic Press, New York, 1973.

Chomsky, N., Three models for the description of language. *IRE Transactions on Information Theory 2* (1956), 113–124.

Chomsky, N., On certain formal properties of grammar. *Information and Control 2* (1959), 137–167.

Church, A., An unsolvable problem of elementary number theory. *American Journal of Mathematics 58* (1936), 345–363.

Cichelli, R. J., Minimal perfect hash functions made simple. *Communications of the ACM 23* (1980), 17–19.

Cook, S. A., The complexity of theorem-proving procedures. *Proceedings of the 3rd Annual ACM Symposium on Theory of Computing* (1971), 151–158.

Coppersmith, D., and S. Winograd, Matrix multiplication via arithmetic progressions. *Proceedings of the 19th Annual ACM Symposium on the Theory of Computing* (1987), 1–6.

Cormen, T. H., C. E. Leiserson, and R. L. Rivest, *Introduction to Algorithms.* MIT Press, Cambridge, MA and McGraw-Hill, New York, NY, 1990. Second Edition with additional author, C. Stein, 2001.

Fischer, M. J., and M. O. Rabin, Super-exponential complexity of Presburger arithmetic. *Complexity of Computation,* ed. R. M. Karp. American Mathematical Society, Providence, RI, 1974, pp. 27–41.

Floyd, R. W., Algorithm 97: Shortest path. *Communications of the ACM 5* (1962), 345.

Floyd, R. W., Assigning meanings to programs. *Proceedings AMS Symposium Applied Mathematics, 19,* AMS, Providence, RI, 1967, pp. 19–31.

Frege, G., *Begriffsschrift, eine der arithmetischen nachgebildete Formelsprache des reinen Denkens.* Halle, 1879.

Galler, B. A., and M. J. Fischer, An improved equivalence algorithm. *Communications of the ACM 7* (1964), 301–303.

Garey, M. R., and D. S. Johnson, *Computers and Intractability: A Guide to the Theory of NP-Completeness.* W. H. Freeman, San Francisco, 1979.

Gentzen, G., Untersuchungen über das logische Schliessen. *Mathematische Zeitschrift 39* (1935), 176–210, 405–431; English translation: Investigation into logical deduction, *The Collected Papers of Gerhard Gentzen,* ed. M. E. Szabo. North-Holland, Amsterdam, 1969, pp. 68–131.

Gödel, K., Die Vollständigkeit der Axiome des logischen Funktionenkalküls. *Monatshefte für Mathematic und Physik 37* (1930), 349–360.

Gödel, K., Über formal unentscheidbare Sätze der Principia Mathematica und verwandter Systeme I. *Monatshefte für Mathematic und Physik 38* (1931), 173–198.

Graham, R. L., D. E. Knuth, and O. Patashnik, *Concrete Mathematics.* Addison-Wesley, Reading, MA, 1989.

Greibach, S. A., A new normal-form theorem for context-free phrase-structure grammars. *Journal of the ACM 12* (1965), 42–52.

Halmos, P. R., *Naive Set Theory.* Van Nostrand, New York, 1960.

Hilbert, D., and W. Ackermann, *Principles of Mathematical Logic.* (1938). Translated by Lewis M. Hammond, George G. Leckie, and F. Steinhardt. Edited by Robert E. Luce. Chelsea, New York, 1950.

Hoare, C.A.R., An axiomatic basis for computer programming. *Communications of the ACM 12* (1969), 576–583.

Kleene, S. C., General recursive functions of natural numbers. *Mathematische Annalen 112* (1936), 727–742.

Kleene, S. C., Representation of events by nerve nets. *Automata Studies*, ed. C. E. Shannon and J. McCarthy. Princeton University Press, Princeton, NJ, 1956, pp. 3–42.

Knuth, D. E., On the translation of languages from left to right. *Information and Control 8* (1965), 607–639.

Knuth, D. E., *The Art of Computer Programming. Volume 1: Fundamental Algorithms.* Addison-Wesley, Reading, MA, 1968; second edition, 1973.

Knuth, D. E., Two notes on notation. *The American Mathematical Monthly 99* (1992), 403–422.

Kruskal, J. B., Jr., On the shortest spanning subtree of a graph and the traveling salesman problem. *Proceedings of the American Mathematical Society 7* (1956), 48–50.

Kurki-Suonio, R., Notes on top-down languages. *BIT 9* (1969), 225–238.

Lewis, P. M., and R. E. Stearns, Syntax-directed transduction. *Journal of the ACM 15* (1968), 465–488.

Lin, S., and T. Rado, Computer studies of Turing machine problems. *Journal of the ACM 12* (1965), 196–212.

Lukasiewicz, J., *Elementary Logiki Matematycznej.* PWN (Polish Scientific Publishers), 1929; translated as *Elements of Mathematical Logic*, Pergamon, Elmsford, NY, 1963.

Mallows, C. L., Conway's challenge sequence. *The American Mathematical Monthly 98* (1991), 5–20.

Markov, A. A., The theory of algorithms. *Trudy Math. Inst. Steklov 42* (1954); English translation published in 1962.

Martelli, A., and U. Montanari, An efficient unification algorithm. *ACM Transactions on Programming Languages and Systems 4* (1982), 258–282.

Marxen, H., and J. Buntrock. Attacking the Busy Beaver 5. *Bulletin of the European Association for Theoretical Computer Science 40* (1990), 247–251.

Mealy, G. H., A method for synthesizing sequential circuits. *Bell System Technical Journal 34* (1955), 1045–1079.

Meyer, A. R., and L. J. Stockmeyer, The equivalence problem for regular expressions with squaring requires exponential time. *Proceedings of the 19th Annual Symposium on Switching and Automata Theory* (1972), 125–129.

Moore, E. F., Gedanken-experiments on sequential machines. *Automata Studies*, ed. C. E. Shannon and J. McCarthy. Princeton University Press, Princeton, NJ, 1956, pp. 129–153.

Morales-Bueno, R., Noncomputability is easy to understand. *Communications of the ACM 38* (1995), 116–117.

Myhill, J., Finite automata and the representation of events. WADD TR-57-624, Wright Patterson AFB, Ohio, 1957, pp. 112–137.

Nagel, E., and J. R. Newman, *Gödel's Proof.* New York University Press, New York, 1958.

Nerode, A., Linear automaton transformations. *Proceedings of the American Mathematical Society 9* (1958), 541–544.

Pan, V., Strassen's algorithm is not optimal. *Proceedings of the 19th Annual IEEE Symposium on the Foundations of Computer Science* (1978), 166–176.

Paterson, M. S., and M. N. Wegman, Linear Unification. *Journal of Computer and Systems Sciences 16* (1978), 158–167.

Post, E. L., Formal reductions of the general combinatorial decision problem. *American Journal of Mathematics 65* (1943), 197–215.

Post, E. L., A variant of a recursively unsolvable problem. *Bulletin of the American Mathematical Society 52* (1946), 246–268.

Prim, R. C., Shortest connection networks and some generalizations. *Bell System Technical Journal 36* (1957), 1389–1401.

Rabin, M. O., and D. Scott, Finite automata and their decision problems. *IBM Journal of Research and Development 3* (1959), 114–125.

Rado, T. On noncomputable functions. *Bell System Technical Journal 41* (1962), 877–884.

Rivest, R. L., A. Shamir, and L. Adleman, A method for obtaining digital signatures and public-key cryptosystems. *Communications of the ACM 21* (1978), 120–126.

Robinson, J. A., A machine-oriented logic based on the resolution principle. *Journal of the ACM 12* (1965), 23–41.

Rosenkrantz, D. J., and R. E. Stearns, Properties of deterministic top-down grammars. *Information and Control 17* (1970), 226–256.

Shepherdson, J. C., and H. E. Sturgis, Computability of recursive functions. *Journal of the ACM 10* (1963), 217–255.

Skolem, T., Über die mathematische logik. *Norsk Matematisk Tidsskrift 10* (1928), 125–142. Translated in ed. Jean van Heijenoort. *From Frege to Godel: A Source Book in Mathematical Logic 1879–1931*, Harvard University Press, Cambridge, MA, 1967, pp. 508–524.

Snyder, W., and J. Gallier, Higher-order unification revisited: Complete sets of transformations. *Journal of Symbolic Computation 8* (1989), 101–140.

Strassen, V., Gaussian elimination is not optimal. *Numerische Mathematik 13* (1969), 354–356.

Thompson, K., Regular expression search algorithms. *Communications of the ACM 11* (1968), 419–422.

Turing, A., On computable numbers, with an application to the Entscheidungsproblem. *Proc. London Math. Soc.*, series 2, *42* (1936), 230–265; correction in *43* (1937), 544–546.

Warren, D. S., Memoing for logic programs. *Communications of the ACM 35* (1992), 93–111.

Warshall, S., A theorem on Boolean matrices, *Journal of the ACM* 9 (1962), 11–12.

Whitehead, A. N., and B. Russell, *Principia Mathematica.* Cambridge University Press, New York, 1910.

Wos, L., R. Overbeek, E. Lusk, and J. Boyle, *Automated Reasoning: Introduction and Applications.* Prentice-Hall, Englewood Cliffs, NJ, 1984.

Greek Alphabet

A	α	alpha
B	β	beta
Γ	γ	gamma
Δ	δ	delta
E	ϵ	epsilon
Z	ζ	zeta
H	η	eta
Θ	θ	theta
I	ι	iota
K	κ	kappa
Λ	λ	lambda
M	μ	mu
N	ν	nu
Ξ	ξ	xi
O	o	omicron
Π	π	pi
P	ρ	rho
Σ	σ	sigma
T	τ	tau
Y	υ	upsilon
Φ	ϕ	phi
X	χ	chi
Ψ	ψ	psi
Ω	ω	omega

Symbol Glossary

Each symbol or expression is listed with a short definition and the page number where it first occurs. The list is ordered by page number.

$d \mid n$	d divides n with no remainder 6
$x \in S$	x is an element of S 14
$x \notin S$	x is not an element of S 14
$\ldots$	ellipsis 14
$\varnothing$	the empty set 14
$\mathbb{N}$	natural numbers 15
$\mathbb{Z}$	integers 15
$\mathbb{Q}$	rational numbers 16
$\mathbb{R}$	real numbers 16
$\{x \mid P\}$	set of all x satisfying property P 16
$A \subseteq B$	A is a subset of B 16
$A \subset B$	A is a proper subset of B 16
$A \cup B$	A union B 19
$A \cap B$	A intersection B 20
$A - B$	difference: elements in A but not B 22
$A \oplus B$	symmetric difference: $(A - B) \cup (B - A)$ 23
A'	complement of A 23

$\lvert A \rvert$	cardinality of A 26
$[a,\, b,\, b,\, a]$	bag, or multiset, of four elements 30
$(x,\, y,\, x)$	tuple of three elements 37
$(\,)$	empty tuple 37
$A \times B$	Cartesian product $\{(a,b) \mid a \in A \text{ and } b \in B\}$ 38
$\langle x,\, y,\, x \rangle$	list of three elements 40
$\langle\,\rangle$	empty list 40
$\mathrm{cons}(x,\, t)$	list with head x and tail t 41
$\mathrm{lists}(A)$	set of all lists over A 42
Λ	empty string 43
$\lvert s \rvert$	length of string s 43
A^{*}	set of all strings over alphabet A 43
LM	product of languages L and M 45
L^{n}	product of language L with itself n times 45
L^{*}	closure of language L 46
L^{+}	positive closure of language L 46
$R(a,\, b,\, c)$	$(a,\, b,\, c)$ is in the relation R 48
$x \; R \; y$	$R(x,\, y)$ or x is related by R to y 48
$f : A \to B$	function type: f has domain A and codomain B 77
$f(C)$	image of C under f 78
$f^{-1}(D)$	pre-image of D under f 78
$\lfloor x \rfloor$	floor of x: largest integer $\leq x$ 81
$\lceil x \rceil$	ceiling of x: smallest integer $\geq x$ 81
$\gcd(a,\, b)$	greatest common divisor of a and b 84
$a \bmod b$	remainder upon division of a by b 86
$\mathbb{N}_{n}$	the set $\{0,\, 1,\, \ldots,\, n-1\}$ 87
$\log_{b} x$	logarithm base b of x 89

$\ln x$	natural logarithm base e of x	91
χ_B	characteristic function for subset B	93
$f \circ g$	composition of functions f and g	97
f^{-1}	inverse of bijective function f	109
$x :: t$	list with head x and tail t	141
$\text{tree}(L, x, R)$	binary tree with root x and subtrees L and R	144
$A \to \alpha$	grammar production	180
$A \to \alpha \mid \beta$	grammar productions $A \to \alpha$ and $A \to \beta$	183
$A \Rightarrow \alpha$	A derives α in one step	183
$A \Rightarrow^+ \alpha$	A derives α in one or more steps	183
$A \Rightarrow^* \alpha$	A derives α in zero or more steps	183
$L(G)$	language of grammar G	184
$R \circ S$	composition of binary relations R and S	201
$r(R)$	reflexive closure of R	206
R^c	converse of relation R	206
$s(R)$	symmetric closure of R	206
$t(R)$	transitive closure of R	206
R^+	transitive closure of R	209
R^*	reflexive transitive closure of R	209
$[x]$	equivalence class of things equivalent to x	224
$\text{tsr}(R)$	smallest equivalence relation containing R	231
$\langle A, \prec \rangle$	irreflexive partially ordered set	241
$\langle A, \preceq \rangle$	reflexive partially ordered set	241
$x \preceq y$	$x \prec y$ or $x = y$	241
$x \prec y$	x is a predecessor of y	242
W_A	worst-case function for algorithm A	283
$\sum a_i$	sum of the numbers a_i	290

$n!$	n factorial: $n \cdot (n-1)\ldots 1$	315
$P(n, r)$	number of permutations of n things taken r at a time	315
$C(n, r)$	number of combinations of n things taken r at a time	319
$\binom{n}{r}$	binomial coefficient symbol	320
$P(A)$	probability of event A	327
$O(f)$	big oh: growth rate bounded above by that of f	370
$\Omega(f)$	big omega: growth rate bounded below by that of f	371
$\Theta(f)$	big theta: same growth rate as f	372
$o(f)$	little oh: lower growth rate than f	377
$\neg P$	logical negation of P	399
$P \wedge Q$	logical conjunction of P and Q	399
$P \vee Q$	logical disjunction of P and Q	399
$P \rightarrow Q$	logical conditional: P implies Q	399
$P \equiv Q$	logical equivalence of P and Q	403
$\vdash W$	turnstile to denote W is a theorem	454
$\exists x$	existential quantifier: there is an x	459
$\forall x$	universal quantifier: for all x	459
$W(x/t)$	wff obtained from W by replacing free x's by t	465
x/t	binding of the variable x to the term t	465
$W(x)$	W contains a free variable x	466
$\{P\}\, S\, \{Q\}$	S has precondition P and postcondition Q	527
$\square$	empty clause: a contradiction	566
$\{x/t,\, y/s\}$	substitution containing two bindings	574
ϵ	empty substitution	575
$E\theta$	instance of E: substitution θ applied to E	575
$\theta\sigma$	composition of substitutions θ and σ	575
$C\theta - N$	remove all occurrences of N from clause $C\theta$	582

$R(S)$	resolution of clauses in the set S 584
$C \leftarrow A, B$	logic program clause: C if A and B 596
$\leftarrow A$	logic program goal: does the program infer A? 597
$\langle A; s, a \rangle$	algebra with carrier A and operations s and a 623
$\overline{x}$	complement of Boolean algebra variable x 635
$\mathrm{lcm}(a, b)$	least common multiple of a and b 636
$R \bowtie S$	join of relations R and S 668
$x \equiv y \;(\mathrm{mod}\; n)$	congruence mod n: $x \bmod n = y \bmod n$ 676
$T(i, a) = j$	deterministic finite automaton transition 707
$T(i, a) = \{j, k\}$	nondeterministic finite automaton transition 710
a/x	Mealy machine: if a is input, then output x 717
i/x	Moore machine: if in state i, then output x 717
$\lambda(s)$	lambda closure of state s 732
$T_{\min}([s], a) = [T(s, a)]$	minimum-state DFA transition 740
$(i, b, C, \mathrm{pop}, j)$	pushdown automaton instruction 763
$(A \rightarrow y, p)$	handle of a sentential form 791
(i, a, b, L, j)	Turing machine instruction 812
$x \rightarrow y$	Markov string-processing production 835
$x \rightarrow y$	Post string-processing production 838
$s, t \rightarrow u$	Post system inference rule 839

Index